# Lecture Notes in Artificial Intelligence 7530

## Subseries of Lecture Notes in Computer Science

Jingsheng Lei   Fu Lee Wang   Hepu Deng
Duoqian Miao (Eds.)

# Artificial Intelligence and Computational Intelligence

4th International Conference, AICI 2012
Chengdu, China, October 26-28, 2012
Proceedings

Springer

Volume Editors

Jingsheng Lei
Shanghai University of Electric Power
School of Computer and Information Engineering
Shanghai 200090, China
E-mail: jshlei@shiep.edu.cn

Fu Lee Wang
Caritas Institute of Higher Education, Department of Business Administration
18 Chui Ling Road, Tseung Kwan O, Hong Kong, China
E-mail: pwang@cihe.edu.hk

Hepu Deng
RMIT University, School of Business Information Technology
City Campus, 124 La Trobe Street, Melbourne, VIC 3000, Australia
E-mail: hepu.deng@rmit.edu.au

Duoqian Miao
Tongji University, School of Electronics and Information
Shanghai 201804, China
E-mail: miaoduoqian@163.com

ISSN 0302-9743 e-ISSN 1611-3349
ISBN 978-3-642-33477-1 e-ISBN 978-3-642-33478-8
DOI 10.1007/978-3-642-33478-8
Springer Heidelberg Dordrecht London New York

Library of Congress Control Number: 2012946748

CR Subject Classification (1998): I.2.1, I.2.3, I.2.6, I.2.8-10, I.5.1-4, I.4.6-9, H.2.7-8, F.1, F.2.2, H.3.4, K.6.5, C.2.0, C.2.4, J.2

LNCS Sublibrary: SL 7 – Artificial Intelligence



*Typesetting:* Camera-ready by author, data conversion by Scientific Publishing Services, Chennai, India

Printed on acid-free paper

Springer is part of Springer Science+Business Media (www.springer.com)

# Preface

The 2012 International Conference on Artificial Intelligence and Computational Intelligence (AICI 2012) was held during October 26–28, 2012 in Chengdu, China. AICI 2012 received 724 submissions from 10 countries and regions. After rigorous reviews, 163 high-quality papers were selected for publication in the AICI 2012 proceedings. The acceptance rate was 23%.

The aim of AICI 2012 was to bring together researchers working in many different areas of artificial intelligence and computational intelligence to exchange new ideas and promote international collaboration. In addition to the large number of submitted papers and invited sessions, there were several internationally well-known keynote speakers.

On behalf of the Organizing Committee, we thank Xihua University and Leshan Normal University for its sponsorship and logistics support. We also thank the members of the Organizing Committee and the Program Committee for their hard work. We are very grateful to the keynote speakers, session chairs, reviewers, and student helpers. Last but not least, we thank all the authors and participants for their great contributions that made this conference possible.

October 2012

Jingsheng Lei
Fu Lee Wang
Hepu Deng
Duoqian Miao

# Organization

## Organizing Committee

### General Co-chairs

| | |
|---|---|
| WeiGuo Sun | Xihua University, China |
| Qing Li | City University of Hong Kong, Hong Kong, China |

### Program Committee Co-chairs

| | |
|---|---|
| Hepu Deng | RMIT University, Australia |
| Duoqian Miao | Tongji University, China |

### Steering Committee Chair

| | |
|---|---|
| Jingsheng Lei | Shanghai University of Electric Power, China |

### Local Arrangements Co-chairs

| | |
|---|---|
| Yajun Du | Xihua University, China |
| MingXing He | Xihua University, China |
| Jin Pei | LeShan Normal University, China |

### Proceedings Co-chairs

| | |
|---|---|
| Fu Lee Wang | Caritas Institute of Higher Education, Hong Kong, China |
| Ting Jin | Fudan University, China |

### Sponsorship Chair

| | |
|---|---|
| Zhiyu Zhou | Zhejiang Sci-Tech University, China |

## Program Committee

| | |
|---|---|
| Adi Prananto | Swinburne University of Technology, Australia |
| Adil Bagirov | University of Ballarat, Australia |
| Ahmad Abareshi | RMIT University, Australia |
| Alemayehu Molla | RMIT University, Australia |
| Andrew Stranier | University of Ballarat, Australia |
| Andy Song | RMIT University, Australia |
| An-Feng Liu | Central South University, China |
| Arthur Tatnall | Victoria University, Australia |

| | |
|---|---|
| Bae Hyeon | Pusan National University, South Korea |
| Baoding Liu | Tsinghua University, China |
| Carmine Sellitto | Victoria University, Australia |
| Caroline Chan | Deakin University, Australia |
| CheolPark Soon | Chonbuk National University, South Korea |
| Chowdhury Morshed | Deakin University, Australia |
| Chung-Hsing Yeh | Monash University, Australia |
| Chunqiao Tao | South China University, China |
| Costa Marly | Federal University of Amazonas, Brazil |
| Craig Parker | Deakin University, Australia |
| Daowen Qiu | Zhong Shan University, China |
| Dat Tran | University of Canberra, Australia |
| Dengsheng Zhang | Monash University, Australia |
| Edmonds Lau | Swinburne University of Technology, Australia |
| Elspeth McKay | RMIT University, Australia |
| Eng Chew | University of Technology Sydney, Australia |
| Feilong Cao | China Jiliang University, China |
| Ferry Jie | RMIT University, Australia |
| Furutani Hiroshi | University of Miyazaki, Japan |
| Gour Karmakar | Monash University, Australia |
| Guojun Lu | Monash University, Australia |
| Heping Pan | University of Ballarat, Australia |
| Hossein Zadeh | RMIT University, Australia |
| Ian Sadler | Victoria University, Australia |
| Irene Zhang | Victoria University, Australia |
| Jamie Mustard | Deakin University, Australia |
| Jeff Ang Charles | Darwin University, Australia |
| Jennie Carroll | RMIT University, Australia |
| Jenny Zhang | RMIT University, Australia |
| Jian Zhou | Tsinghua University, China |
| Jingqiang Wang | South China University, China |
| Jinjun Chen | Swinburne University of Technology, Australia |
| Joarder Kamruzzaman | Monash University, Australia |
| Kaile Su | Beijing University, China |
| Kankana Chakrabaty | University of New England, Australia |
| Konrad Peszynski | RMIT University, Australia |
| Kuoming Lin | Kainan University, Taiwan, China |
| Lemai Nguyen | Deakin University, Australia |
| Leslie Young | RMIT University, Australia |
| Liping Ma | University of Ballarat, Australia |
| Luba Torline | Deakin University, Australia |
| Maple Carsten | University of Bedfordshire, UK |
| Maria Indrawan | Monash University, Australia |

# Table of Contents

## Applications of Artificial Intelligence



## Applications of Computational Intelligence

## Data Mining and Knowledge Discovering

## Evolution Strategy



## Expert and Decision Support Systems

## Fuzzy Computation



## Information Security



## Intelligent Control

## Intelligent Image Processing

## Intelligent Information Fusion



## Intelligent Signal Processing



## Machine Learning

## Neural Computation



## Neural Networks

## Particle Swarm Optimization



## Pattern Recognition

# Expressive Secondo Performances of a Realtime Person-Computer Ensemble System

Tetsuya Mizutani[1], Yuki Shinagawa[2], Naoki Murakami[2], and Shigeru Igarashi[3]

[1] Faculty of Engineering, Information and Systems, University of Tsukuba, Japan
mizutani@cs.tsukuba.ac.jp
[2] Master's Program in Computer Science, University of Tsukuba, Japan
{ysinagaw,nmurakam}@cs.tsukuba.ac.jp
[3] Professor Emeritus, University of Tsukuba, Japan
igarashi@cs.tsukuba.ac.jp

**Abstract.** A person-computer ensemble system is one of time concerned cooperative systems, which performs the secondo in an ensemble played by a computer-controlled piano cooperating with the primo played by a person musician. For expressive performances, it is necessary to prepare very expressive secondo data before the actual cooperative performance, since the system is modified its expression in realtime using the agogic information calculated from the input stream of the primo. In this paper, a generating method of more expressive performance of the secondo, using the notions musical structures and structural functions, are introduced.

**Keywords:** Artificial Intelligence, Musical Informatics, Ensemble System, Expressive Performance.

## 1 Introduction

*A person-computer ensemble system* [7–11] is one of time concerned cooperative systems, which performs the *secondo* (the second part) in an ensemble played by a computer-controlled piano cooperating with the *primo* (the leading part) played by a person musician. This is a good example of intelligent realtime programs appropriate for formal verification and analysis. In [7, 8], the main part of this realtime system has been verified and analyzed formally by *$N\Sigma$-labeled calculus* [3], while in [10, 11], a formal representation and verification, by the calculus, of a matching algorithm between the realtime input stream of the primo performance from a MIDI keyboard and the sequence of notes of the primo on the score has been introduced.

Before the actual cooperating performance, the system learns the tendency of the expression that the person musician thinks and/or plans for performing the primo part by the person musician. The system calculates and prepares an expressive performance data of the secondo. At the actual performance, the system corrects and modifies this data with the actual input of the primo performance from the person musician. Thus, it is necessary for the expressive performance that the system prepares a suitable expression of the secondo, whereas there

J. Lei et al. (Eds.): AICI 2012, LNAI 7530, pp. 1–8, 2012.

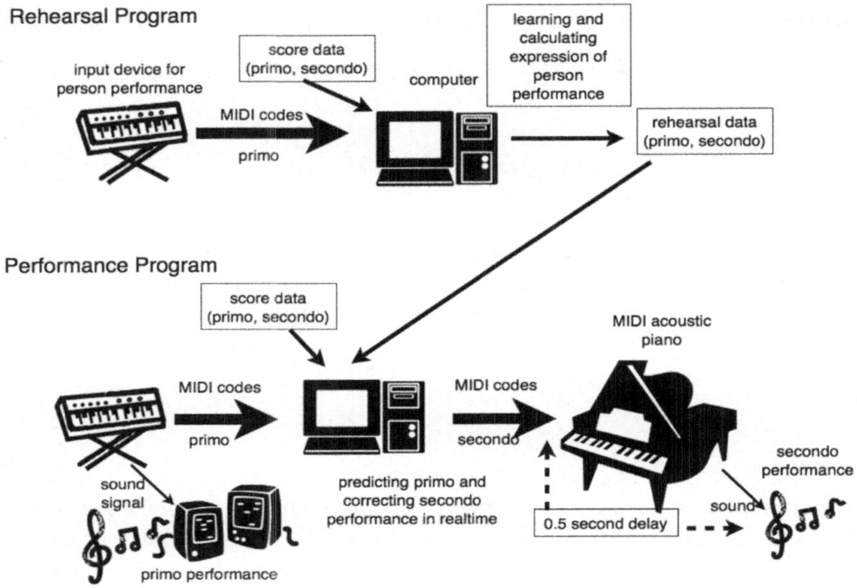


**Fig. 1.** The Ensemble System

frequently does not exist any sufficient clue from the primo data to calculate such expressive performance.

In this paper, a method for generating secondo performance data, with some experimental results of actual performances by a professional pianist, using the notions of *musical structures* [4, 13] and *structural functions* [4], the former of which are hierarchical structures of scores reflexing the phrasing analysis, and hence, which are generalizations of notion *phrase* or *motif* of music, while the latter of which represent passions of performers and listeners occurring in each element of each structure, especially in each motif or phrase.

## 2 A Realtime Person-Computer Ensemble System

The ensemble system consists of two programs; the *rehearsal* and the *performance* ones (see Figure 1).

The aim of the system is to perform the secondo in an ensemble to drive a MIDI acoustic piano by the performance program cooperating with performing the primo by a person pianist or other musician playing another MIDI instrument. Those two MIDI instruments are connected by a computer. MIDI (Musical Instrument Digital Interface) is a standard protocol for electronic musical instruments and computers to communicate, control, and synchronize with each other. A MIDI acoustic piano is an acoustic grand piano that can be played with MIDI codes. There is a 0.5[s] delay between the moment when each MIDI code as the input of the the piano and that when the corresponding key moves and actually sounds, since the actuator must adjust the timing of each group of notes which sound simultaneously. Hence, the program must output each MIDI code 0.5[s] before the moment when the actual performance is expected.

The program has the whole score data of ensemble. There are special notes called *reference notes* to measure the performance speed (or 'local tempo'), usually at the head of each or every other bar, depending on the structure and expected tempo of the score.

Before the cooperating performance, the rehearsal program records solo primo performed by the person musician. From this performance, the program calculates the *rehearsal data* which expresses the timing of performance of the whole reference notes, and *schedule* of performance of the secondo. During the cooperating performance, the program modifies and updates the schedule in realtime from the actual performance of the primo using the score data (including the reference notes), the rehearsal data and the schedule.

## 3 A Generating Method of the Secondo Expression

### 3.1 Problem

Let us consider the case that the ensemble system performs the Prelude in E minor, Op. 28, No. 4 by Chopin (Figure 2), in which the upper part is played as the primo by a person musician along with the lower one as the secondo by the program. As mentioned in Section 2, the system must prepares an expressive performance data of the secondo generated from the actual performance of the primo by the person musician before the actual cooperating performance.

Figure 3 is the actual performance data of the secondo performed by a famous pianist Martha Argerich, whose MIDI data is in Music Performance Expression Database of CrestMuse Project (CrestMusePEDB) [2], while Figure 4 is that generated by the rehearsal program. The measure of the horizontal axis of each figure indicates the position of each note in the score and the vertical one does the *length per beat* of each note, which means the performance length of each note per beat, calculated by $l \cdot b/n$, where $l$ is the length of the actual performance of each note, $n$ its note value, e.g. 1/4 for a quarter note, 1/8 for an eighth note, etc. and $b$ the note value of the beat of the piece, e.g. 1/4 if the piece is two-four or common time, or 1/2 if it is two-two time, etc.

Obviously, there are no clue in the primo to generate expression of the secondo performance. Thus the rehearsal program generates the expression as *in tempo* (in a uniform or constant tempo) between each two adjoint notes of the primo as Figure 4. But the expression of the actual person performance is very expressive as Figure 3. It is an important problem of the rehearsal program.

### 3.2 Musical Structures and Structural Functions

There are many studies to generate rich and expressive performance by computers. A typical and simple one to solve the problemcalled the *structural functions* based on the *musical structures* [4] is introduced.

**Fig. 2.** The Prelude in E minor, Op. 28, No. 4 by Chopin [6]

From the phrasing analysis [13], each score has its own *musical structure* which can be divided it into eight measures called *sentences*, four measures *phrase* and two measures *motif* in general (Figure 5).

In each motif or phrase, there exists at least 6 *structural functions*, which reflect passion of performers and listeners (Figure 6), as follows:

1. Inceptive: the beginning point of the structure,
2. Anacrusis: a range where music is getting tenser prior to the initiative,
3. Tiara: a climax range just before the initiative,
4. Initiative: the most tensional note,
5. Desinence: a range where tension is diminished following the initiative,
6. Conclusive: the ending point of the structure.

By analyzing music performances, there are the performance (rendition) rules for the agogic. Some essential rules are introduced below.

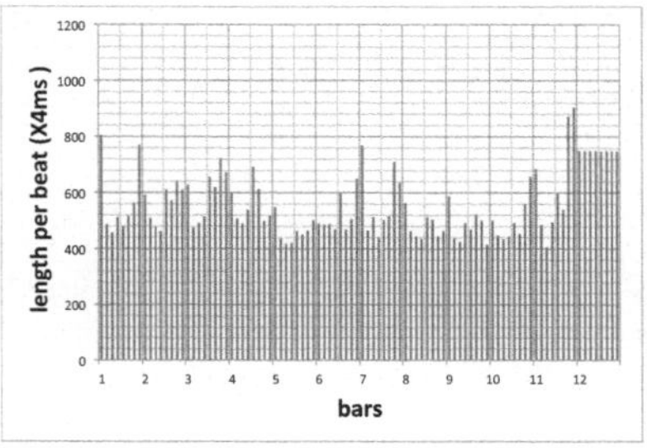


**Fig. 3.** The actual performance of the first 12 bars of the secondo of Prelude

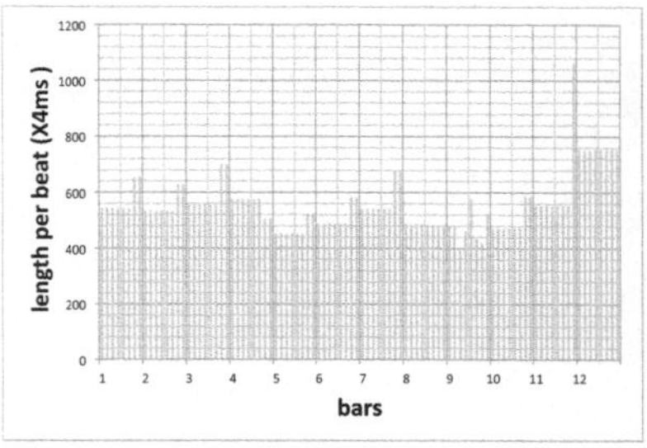


**Fig. 4.** The performance data generated by the rehearsal program

1. Inceptive and conclusive notes are performed longer than other ones.
2. In the anacrusis, the performance starts slower, and getting faster to the initiative.
3. The initiative note performs longer than the others. Tiara notes may be also done.
4. In the desinence, the performance is getting slower to the conclusive.

It must be noted that a musical expression depends on scores and performers, so that these rules have very many exceptions and they may not be applied to each individual score and performance.

From the rules and the actual performance (Figure 3), especially adopting and applying the structural functions with parabola approximation curves to each structure, i.e. motif or phrase, the expression as Figure 7 is obtained for the first 12 bars of the secondo of E minor Prelude. It seems "better" expression not only for ensemble but also for solo performances. Note that the curve cannot generate without the actual performance data since the rules only express the "quality" of expressions, while the "real" values of the performances are necessary to obtain the "quantity".

### 3.3 Experimental Results

Audition and performance experiments have been done by a professional musician who played the primo of Prelude with the three secondo performance data having the expressions mentioned above, one by Martha Argerich, one generated by the rehearsal program and one from the structural functions. After the experiment, the pianist remarked that the system should prepare more artistic expression and the last one is very artificial and unnatural, while it is relatively permissible for ensemble. Especially, when the system generates the expression, it should refer and approximate to that of the actual performance of the second bar by Argerich. And, needless to say, the second one, by the rehearsal program, is very unnatural and out of consideration.

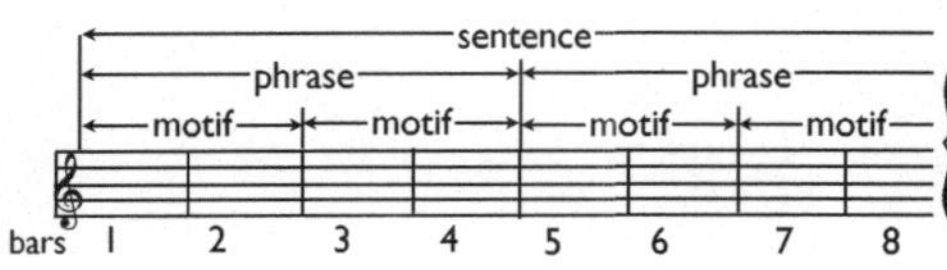


**Fig. 5.** The Musical Structure

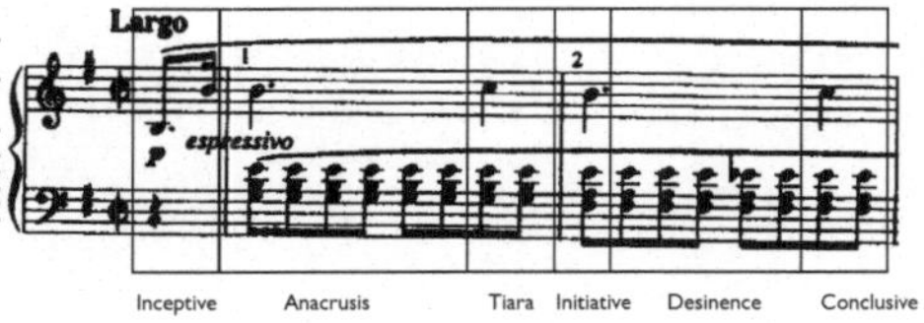


**Fig. 6.** The Structural Functions applied to the first motif of Prelude

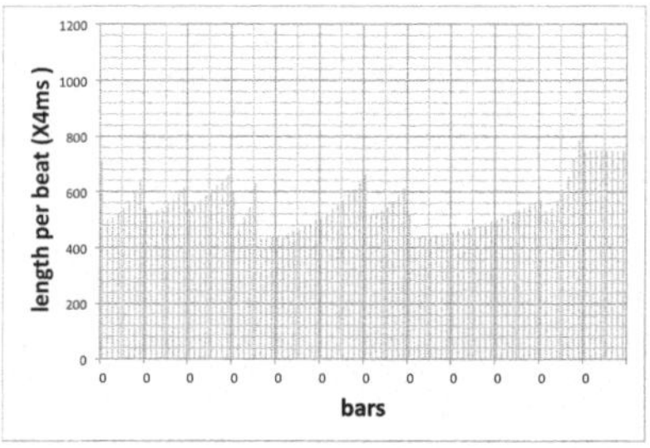


**Fig. 7.** The performance data by the expression rules from the structural functions

### 3.4 Tonal Tension Structures

The contemporary most famous theory of music in the musical informatics field is *GTTM* (*the generative theory of tonal music*) [5], which introduces the hierarchical musical structures. The notion *tonal tension and attraction* related with these structures is introduced by *TPS* (*the tonal pitch space*) [6], which is a succeeding study of GTTM. Each of them describes the musical emotion from the score for (the imaginary) *experienced listeners.* Figure 8 shows the result of the *reduction analysis* by GTTM of Prelude. The *tension values* introduced by TPS are attempted to be used for making the expression. Figure 9 is the values of Prelude. The authors of this article are attempt to introduce some generalized rules of expression from the tension values of scores for comparing and analyzing with the real expression data as Figure 3.

## 4 Related Works

There are many related works for automatic ensemble systems. Most of them, e.g. [12, 14, 15], etc. use HMM, Bayesian network models, or other statistical methods, or use the huge case-based database, to predict future rhythm and tempo of performance by person musician. A performance system that performs music to modify parameters of the performance rules is introduced in [1].

Compared with them, the system introduced in this paper uses *rehearsal* and the *rules of performance* to learn the expression pattern, and the realtime performance program only calculates the tempo of the primo performance by the person and modifies one of the secondo performance.

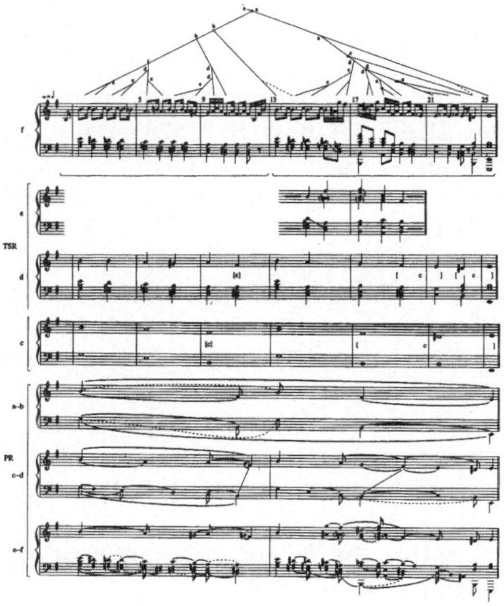


**Fig. 8.** Reductional Analysis of the E minor Prelude [6]

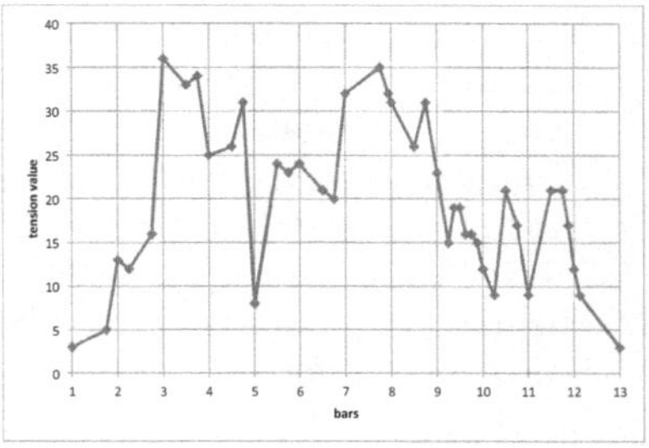


**Fig. 9.** Tension value of the first 12 bars of Prelude

## 5 Conclusion

Some examinations for preparing more expressive secondo performance from the score and the primo performance data in the person-computer ensemble system, especially the phasing analysis and tension analysis that is very essential for expressive ensemble performance, are introduced.

There remains some important future issues. The most important one is that the more good methods to obtain more artistic expression for such "difficult" pieces must be researched and developed. Additionally, the relation between the performance data and the tension values must be investigated.

## References


1. Bresin, R., Friberg, A., Sundberg, J.: Director Musices: The KTH Performance Rules System. In: Proceedings of SIGMUS, vol. 46, pp. 43–48 (2002)
2. CrestMuse Project: Music Performance Expression Database (CrestMusePEDB), `http://www.crestmuse.jp/pedb/index.htm`
3. Ikeda, Y., Mizutani, T., Shio, M.: Formal System and Semantics of $N\Sigma$-labeled Calculus. In: The 2009 International Conference on Artificial Intelligence and Computational Intelligence (AICI 2009), vol. III, pp. 270–274. IEEE (2009)
4. Igarashi, S.: Science of Music Expression. YAMAHA Music Media (2000) (in Japanese)
5. Lerdahl, F., Jackendoff, R.: A Generative Theory of Tonal Music. The MIT Press (1983)

6. Lerdahl, F.: Tonal Pitch Space. Oxford University Press (2001)
7. Mizutani, T., Igarashi, S., Shio, M., Ikeda, Y.: Labeled Calculi Applied to Verification and Analysis of Time-Concerned Programs I. TENSOR, N. S. 71, 172–186 (2009)
8. Mizutani, T., Igarashi, S., Shio, M., Ikeda, Y.: Labeled Calculi Applied Verification and Analysis of Time-Concerned Programs II. TENSOR, N. S. 71, 285–296 (2009)
9. Mizutani, T., Igarashi, S., Suzuki, T., Ikeda, Y., Shio, M.: A Realtime Human-Computer Ensemble System: Formal Representation and Experiments for Expressive Performance. In: Wang, F.L., Deng, H., Gao, Y., Lei, J. (eds.) AICI 2010, Part I. LNCS, vol. 6319, pp. 256–265. Springer, Heidelberg (2010)
10. Mizutani, T., Nishiyama, K., Igarashi, S.: A Matching Method between Music Scores and Performance Data in a Realtime Person-Computer Ensemble System. In: Deng, H., Miao, D., Lei, J., Wang, F.L. (eds.) AICI 2011, Part I. LNCS, vol. 7002, pp. 9–17. Springer, Heidelberg (2011)
11. Mizutani, T., Igarashi, S.: Formalization of a Realtime Person-Computer Ensemble System with an Effective Matching Algorithm. In: The 2nd International Conference on Computer Science and Automation Engineering (CSAE 2012), vol. 1, pp. 65–69 (2012)
12. Raphael, C.: Automated Rhythm Transcription. In: ISMIR 2001: International Symposium on Music Information Retrieval, pp. 99–107 (2001)
13. Riemann, H.: Musikalische Dynamik und Agogik – Lehrbuch der musikalischen Phrasirung. Breitkopf und Hartel (1884)
14. Takeda, H., Nishimoto, T., Sagayama, S.: Joint Estimation of Rhythm and Tempo of Polyphonic MIDI Performance Using Tempo Curve and Hidden Markov Models. Journal of Information Processing 48, 237–247 (2007) (in Japanese)
15. Widmer, G., Flossmann, S., Grachten, M.: Yqx plays chopin. AI Magazine 30, 35–48 (2009)

# Different-Level Schemes' Equivalence for Self-Motion Planning of Robot Manipulators

Yunong Zhang, Huarong Wu, Zhijun Zhang, Senbo Fu, and Yonghua Yin

School of Information Science and Technology
Sun Yat-sen University, Guangzhou 510006, China
zhynong@mail.sysu.edu.cn, {wuhuarong2010,983979597,550488516}@qq.com,
iloveyouzhijun@126.com
http://sist.sysu.edu.cn/~zhynong

**Abstract.** In this paper, an acceleration-level scheme is proposed for the self-motion planning (SMP) of redundant manipulators. Besides, the equivalence between the velocity-level and acceleration-level SMP schemes is discovered and proved by using Zhang *et al.*'s neural-dynamic method. Simply put, the equivalence exists when related parameters satisfy some conditions. For further verification, the two schemes are unified into one quadratic program (QP) formulation, which can be solved by the MATLAB routine "QUADPROG". Computer simulations based on a five-link planar robot arm performing the self-motion demonstrate both the efficacy of the proposed acceleration-level SMP scheme and the equivalence of the two different-level SMP schemes.

**Keywords:** Redundant robot manipulator, Self-motion planning (SMP), Control, Acceleration-level scheme, Equivalence.

## 1 Introduction

As one of the characteristics of redundant robot manipulators, self-motion is a special kind of motion planning and control that, with the end-effector motionless, the manipulator moves from one joint configuration (or to say, state) to another. That is, keeping the end-effector at certain position and/or orientation [i.e., with $r(t) \equiv r(0) \in R^m$, $\dot{r}(t) \equiv 0 \in R^m$ and $\ddot{r}(t) \equiv 0 \in R^m$], the manipulator can adjust its configuration in joint space from one state [e.g., an initial joint state, $\theta(0) \in R^n$] to another state [e.g., a desired joint state, $\theta_{\mathrm{d}} \in R^n$]. Many studies in recent years show that self-motion can be designed to achieve different tasks, such as, avoiding joint limits, singularities and obstacles. Thus, self-motion becomes an interesting topic in robotics in recent decades.

To implement such a self-motion, various self-motion planning (SMP) algorithms/schemes are developed by robot researchers [1,2]. It is worth pointing out that almost all of the existing SMP schemes are at the joint-velocity level. Evidently, such velocity-level SMP schemes lack the acceleration information, and thus do not apply to those robot manipulators controlled by the joint acceleration

J. Lei et al. (Eds.): AICI 2012, LNAI 7530, pp. 9–16, 2012.

or joint torque. In order to solve such a practical problem, a novel acceleration-level SMP scheme is proposed in this paper. For completeness of studies, we further investigate the relationship between our proposed acceleration-level SMP scheme and the existing velocity-level SMP scheme [1, 2]. It is discovered that there exists an equivalent relationship between a generalized velocity-level SMP scheme and the proposed acceleration-level SMP scheme, provided that related parameters satisfy some conditions. In addition, such an equivalence relationship is proved by using Zhang *et al.*'s neural-dynamic method [3, 4, 5]. Note that the equivalence between different-level SMP schemes is a very innovative and significant discovery in guiding the research of self-motion planning and control.

In order to verify the proposed acceleration-level SMP scheme and the equivalence between different-level SMP schemes based on computer simulations, we need to solve the two different-level SMP schemes at each time instant. It follows from the unification results [1, 2, 6] that such different-level SMP schemes can both be reformulated to be quadratic program (QP) problems, which can then be solved readily by MATLAB routine "QUADPROG" or many types of recurrent neural networks. For the sake of simplicity, in this paper, we adopt the MATLAB routine "QUADPROG". Via the solutions of the different-level SMP schemes, computer simulations based on a five-link planar robot arm are carried out to substantiate the efficacy of the proposed acceleration-level SMP scheme as well as the equivalence of the two different-level SMP schemes.

## 2 Schemes and Equivalence

For redundant robot manipulators with structure and parameters known, some well-known fundamental formulas are usually described as $f(\theta) = r$, $J(\theta)\dot{\theta} = \dot{r}$ and $J(\theta)\ddot{\theta} = \ddot{r}_{\mathrm{a}}$, with $\ddot{r}_{\mathrm{a}} = \ddot{r} - \dot{J}(\theta)\dot{\theta}$. Here, $\theta \in R^n$, $\dot{\theta} \in R^n$ and $\ddot{\theta} \in R^n$ denote the vectors of joint-angle, joint-velocity and joint-acceleration, respectively. Besides, $r \in R^m$, $\dot{r} \in R^m$ and $\ddot{r} \in R^m$ denote the vectors of end-effector pose (i.e., position and/or orientation), velocity and acceleration, respectively. In addition, $f(\cdot)$ is a differentiable nonlinear mapping from $R^n$ to $R^m$, $J(\theta) = \partial f(\theta)/\partial\theta \in R^{m\times n}$ is the Jacobian matrix, and $\dot{J}(\theta) \in R^{m\times n}$ is the time derivative of $J(\theta)$. According to existing studies [1,2], a velocity-level SMP scheme for redundant manipulators can be generalized and presented as follows.

**Scheme I** (Velocity-level SMP scheme). *A velocity-level SMP scheme with position-error feedback for redundant robot manipulators is formulated as*

$$\text{minimize} \quad (\dot{\theta} + c)^T(\dot{\theta} + c)/2, \tag{1}$$

$$\text{subject to} \quad J(\theta)\dot{\theta} = \dot{r} + k_1(r - f(\theta)), \tag{2}$$

$$\dot{r} = 0, \tag{3}$$

*where* $c = \mu_1(\theta - \theta_{\mathrm{d}})$ *with* $\mu_1$ *being a positive design parameter used to reflect the displacement magnitude between the current joint state* $\theta$ *and the desired joint state* $\theta_{\mathrm{d}}$*. As for Equation (2), it is improved from* $J(\theta)\dot{\theta} = \dot{r}$*: in consideration of the modeling and computational round-off errors, the feedback of Cartesian* *position error* $(r - f(\theta))$ *is introduced into the equality constraint* $J(\theta)\dot{\theta} = \dot{r}$*. Feedback-gain parameter* $k_1 > 0$ *are used to scale the magnitude of the manipulator response to the end-effector's Cartesian position error.*

Due to lack of the acceleration information, the velocity-level SMP scheme do not apply to those robot manipulators controlled by the joint acceleration or joint torque. To solve such a practically-existing problem, we propose a novel acceleration-level SMP scheme as follows.

**Scheme II** (Acceleration-level SMP scheme). *An acceleration-level SMP scheme with feedback for redundant robot manipulators is formulated as*

$$\text{minimize} \quad (\ddot{\theta} + q)^T(\ddot{\theta} + q)/2, \tag{4}$$

$$\text{subject to} \quad J(\theta)\ddot{\theta} = \ddot{r}_a + k_2(\dot{r} - J(\theta)\dot{\theta}) + k_3(r - f(\theta)), \tag{5}$$

$$\dot{r} = 0, \ \ddot{r} = 0, \tag{6}$$

*where* $q = \mu_2\dot{\theta} + \mu_3(\theta - \theta_d)$ *with positive design parameters* $\mu_2$ *and* $\mu_3$*. Equation (5) is improved from* $J(\theta)\ddot{\theta} = \ddot{r}_a$ *with two levels of feedback [i.e., Cartesian velocity error feedback* $(\dot{r} - J(\theta)\dot{\theta})$ *and Cartesian position error feedback* $(r - f(\theta))$*]. Similarly, feedback-gain parameters* $k_2 > 0$ *and* $k_3 > 0$*.*

For research completeness, we further investigate the connection and comparison between the velocity-level and acceleration-level SMP schemes. The analytical results show that there exists an equivalent relationship between the two different-level SMP schemes when the parameters satisfy some conditions. Such an equivalent relationship is summarized as the following theorem, and Zhang *et al.*'s neural-dynamic method is employed to prove such a theorem.

**Theorem** (Different-level SMP equivalence). *Velocity-level SMP scheme (1)-(3) is equivalent to acceleration-level SMP scheme (4)-(6), provided that*

*i) there exists a sufficiently large parameter* $\mu_4 > 0$ *satisfying* $\mu_2 = \mu_1 + \mu_4$ *and* $\mu_3 = \mu_1\mu_4$*; and,*
*ii) there exists a parameter* $k_4 > 0$ *satisfying* $k_2 = k_1 + k_4$ *and* $k_3 = k_1k_4$*.*

**Proof.** Comparing velocity-level and acceleration-level SMP schemes, we easily find that equation (3) in Scheme I is equivalent to equation (6) in Scheme II. Thus, we only need to prove that performance index (1) and equality constraint (2) in Scheme I are equivalent to performance index (4) and equality constraint (5) in Scheme II, respectively. The following proof is divided into two steps.

*Step 1.* The performance index of velocity-level SMP scheme depicted in (1) can be written as

$$(\dot{\theta} + c)^{\mathrm{T}}(\dot{\theta} + c)/2 = \|\dot{\theta} + \mu_1(\theta - \theta_{\mathrm{d}})\|_2^2/2, \tag{7}$$

where $\|\cdot\|_2$ denotes the two-norm of a vector. By following Zhang *et al.*'s neural-dynamic method [3,4,5], to minimize (7), we can define a vector-valued indefinite smooth error-function as follows:

$$e = \dot{\theta} + \mu_1(\theta - \theta_{\mathrm{d}}). \tag{8}$$

Then, to force the error function $e$ converge to zero, we can directly adopt Zhang *et al.*'s formula (i.e., $\dot{e} = -\mu_4 e$ with $\mu_4 > 0$ being a parameter) [3, 4, 5]. Substituting (8) into the right-hand side of such a formula yields

$$\dot{e} = -\mu_4(\dot{\theta} + \mu_1(\theta - \theta_{\mathrm{d}})). \tag{9}$$

Besides, since the time derivative of the error function $\dot{e} = \ddot{\theta} + \mu_1\dot{\theta}$, substituting it into the left-hand side of equation (9) and simple manipulations yield

$$\ddot{\theta} + (\mu_1 + \mu_4)\dot{\theta} + \mu_1\mu_4(\theta - \theta_{\mathrm{d}}) = 0. \tag{10}$$

By defining design parameters $\mu_2 = \mu_1 + \mu_4$ and $\mu_3 = \mu_1\mu_4$, equation (10) is

$$\ddot{\theta} + \mu_2\dot{\theta} + \mu_3(\theta - \theta_{\mathrm{d}}) = 0. \tag{11}$$

As the robot end-effector's path requirement and joint physical limits have to be considered into the SMP schemes, minimizing the performance index $\|\ddot{\theta} + \mu_2\dot{\theta} + \mu_3(\theta - \theta_{\mathrm{d}})\|_2^2/2$ is more consistent with the reality than forcing $\ddot{\theta} + \mu_2\dot{\theta} + \mu_3(\theta - \theta_{\mathrm{d}})$ to zero directly. Thus, the performance indices of velocity-level and acceleration-level SMP scheme are equivalent to each other. Note that such a conclusion requires that the design parameter $\mu_4$ is large enough. Specifically speaking, when $\mu_4$ is large enough, the design formula $\dot{e} = -\mu_4 e$ converges to zero rapidly, meaning that the velocity-level performance index (1) and the acceleration-level performance index (4) becomes equal rapidly. In this sense, the equivalence between them is better guaranteed in a practical manner.

*Step 2.* According to the velocity-level equality constraint (2), we can also define a vector-valued indefinite smooth error-function as follows:

$$\varepsilon = \dot{r} + k_1(r - f(\theta)) - J(\theta)\dot{\theta}. \tag{12}$$

To minimize $\varepsilon$, Zhang *et al.*'s neural-dynamic design formula, $\dot{\varepsilon} = -k_4\varepsilon$ (with $k_4$ being a positive design parameter), is applied to (12); i.e.,

$$\dot{\varepsilon} = -k_4(\dot{r} + k_1(r - f(\theta)) - J(\theta)\dot{\theta}). \tag{13}$$

Substituting $\dot{\varepsilon} = \ddot{r} + k_1(\dot{r} - J(\theta)\dot{\theta}) - \dot{J}(\theta)\dot{\theta} - J(\theta)\ddot{\theta}$ into the left-hand side of equation (13) yields the following:

$$\ddot{r} + k_1(\dot{r} - J(\theta)\dot{\theta}) - \dot{J}(\theta)\dot{\theta} - J(\theta)\ddot{\theta} = -k_4(\dot{r} + k_1(r - f(\theta)) - J(\theta)\dot{\theta}),$$

which can be rewritten as

$$J(\theta)\ddot{\theta} = \ddot{r} - \dot{J}(\theta)\dot{\theta} + (k_1 + k_4)(\dot{r} - J(\theta)\dot{\theta}) + k_1k_4(r - f(\theta)). \tag{14}$$

By defining feedback-gain parameters $k_2 = k_1 + k_4$ and $k_3 = k_1k_4$, equation (14) becomes the following simpler one:

$$J(\theta)\ddot{\theta} = \ddot{r} - \dot{J}(\theta)\dot{\theta} + k_2(\dot{r} - J(\theta)\dot{\theta}) + k_3(r - f(\theta)). \tag{15}$$

Evidently, equation (15) derived from velocity-level equality constraint (2) is exactly acceleration-level equality constraint (5), which proves that equality constraint (2) in Scheme I is equivalent to equality constraint (5) in Scheme II.

In summary, steps 1 and 2 collectively prove the equivalence between the velocity-level SMP scheme (1)-(3) and the acceleration-level SMP scheme (4)-(6); i.e., the velocity-level performance index (1) and equality constraint (2) are equivalent to the acceleration-level performance index (4) and equality constraint (5), respectively. In addition, the design parameters have to satisfy that $\mu_2 = \mu_1 + \mu_4$ and $\mu_3 = \mu_1\mu_4$ with $\mu_4 > 0$ sufficiently large; and that $k_2 = k_1 + k_4$ and $k_3 = k_1 k_4$ with $k_4 > 0$, which now completes the proof. □

It is worth pointing out that both of the velocity-level and the acceleration-level SMP schemes introduce the feedback to increase the robustness of the planning and control. In general simulations, the system modeling error and computational round-off error are small or even very tiny, thus the SMP is the main task with feedback negligible. When the redundant robot arm approaches its physical limits, singularities or environmental obstacles, the feedback mechanism begins to work and can reduce the end-effector position error effectively.

## 3 Unified QP Formulation

For the velocity-level SMP scheme (1)-(3), the minimization of its performance index $(\dot{\theta} + c)^{\mathrm{T}}(\dot{\theta} + c)/2$ is equivalent to minimizing $\dot{\theta}^{\mathrm{T}}\dot{\theta}/2 + c^{\mathrm{T}}\dot{\theta}$ given that $(\dot{\theta} + c)^{\mathrm{T}}(\dot{\theta} + c)/2 = \dot{\theta}^{\mathrm{T}}\dot{\theta}/2 + c^{\mathrm{T}}\dot{\theta} + c^{\mathrm{T}}c/2$. Thus, the velocity-level SMP scheme can be rewritten as the velocity-level QP formulation in terms of $\dot{\theta}$:

$$\text{minimize} \quad \dot{\theta}^{\mathrm{T}}\dot{\theta}/2 + c^{\mathrm{T}}\dot{\theta}, \tag{16}$$

$$\text{subject to} \quad J(\theta)\dot{\theta} = d_1, \tag{17}$$

where $d_1 = k_1(r - f(\theta))$. Note that equality constraint (17) results from the combination of equations (2) and (3) of the velocity-level SMP scheme.

Similarly, the acceleration-level SMP scheme (4)-(6) can be rewritten as the acceleration-level QP formulation in terms of $\ddot{\theta}$.

$$\text{minimize} \quad \ddot{\theta}^{\mathrm{T}}\ddot{\theta}/2 + q^{\mathrm{T}}\ddot{\theta}, \tag{18}$$

$$\text{subject to} \quad J(\theta)\ddot{\theta} = d_2, \tag{19}$$

where $d_2 = -\dot{J}(\theta)\dot{\theta} - k_2 J(\theta)\dot{\theta} + k_3(r - f(\theta))$. Equality constraint (19) results from equations (5) and (6) of the acceleration-level SMP scheme.

Following the above analysis procedures, we can see that the velocity-level QP formulation (16)-(17) and acceleration-level QP formulation (18)-(19) can both be unified into a QP formulation; i.e.,

$$\text{minimize} \quad x^{\mathrm{T}}Wx/2 + z^{\mathrm{T}}x, \tag{20}$$

$$\text{subject to} \quad Jx = d, \tag{21}$$

where $x = \dot{\theta}$, $W = I \in R^{n\times n}$, $z = c$, $J = J(\theta)$ and $d = d_1$ for the velocity-level SMP scheme, while $x = \ddot{\theta}$, $W = I \in R^{n\times n}$, $z = q$, $J = J(\theta)$ and $d = d_2$ for the acceleration-level SMP scheme.

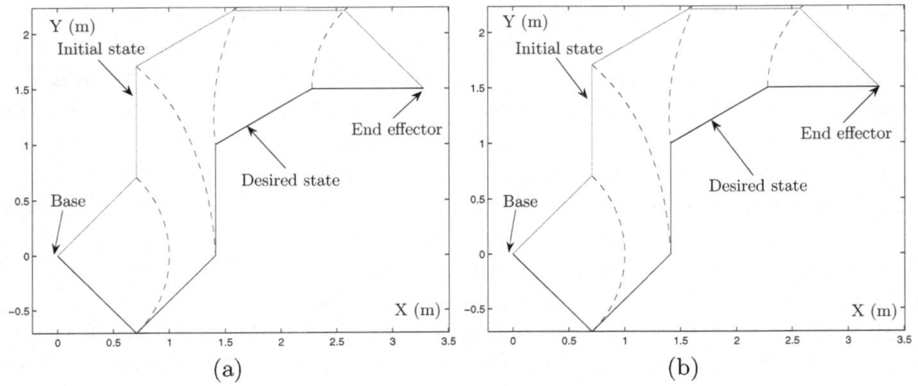


**Fig. 1.** Motion trajectories of a five-link planar robot arm performing self-motion, synthesized by (a) velocity-level SMP scheme (1)-(3) and (b) acceleration-level SMP scheme (4)-(6) (note that the two subplots appear to be the same)

As for the above unified QP formulation (20)-(21), we can readily solve it by MATLAB routine "QUADPROG" [7] or a suitable type of recurrent neural network [1,2,3,4]. For convenience, in the ensuing simulations, we adopt MATLAB routine "QUADPROG" to solve the unified QP formulation (20)-(21).

## 4 Simulative Verification

In this section, to verify the efficacy of the proposed acceleration-level SMP scheme and its equivalence to the velocity-level SMP scheme, the two different-level SMP schemes are applied to perform the self-motion of a five-link planar robot arm. Note that the five-link planar robot arm has 3 redundant degrees of freedom when it works in the 2-dimensional workspace (i.e., workplane). In simulations, we set the initial joint state $\theta(0) = [\pi/4, \pi/4, -\pi/3, -\pi/6, -\pi/4]^{\mathrm{T}}$ in radians (in short, rad), the desired joint state $\theta_{\mathrm{d}} = [-\pi/4, \pi/2, \pi/4, -\pi/3, -\pi/6]^{\mathrm{T}}$ rad, design parameters $\mu_1 = 2$, $\mu_2 = 202$ and $\mu_3 = 400$ (i.e., $\mu_4 = 200$) as well as $k_1 = 4$, $k_2 = 6$ and $k_3 = 8$ (i.e., $k_4 = 2$). Besides, the SMP-task duration $T = 5$ s. The computer-simulation results are shown in Figs. 1 and 2.

Fig. 1 illustrates the motion trajectories of the five-link planar robot arm performing the self-motion, which are synthesized by the velocity-level SMP scheme (1)-(3) and the acceleration-level SMP scheme (4)-(6). Both of the velocity-level and acceleration-level SMP schemes successfully realize the self-motion in the sense that the five-link manipulator moves from the initial joint state to the desired joint state as expected. In addition, the positioning errors of the robot end-effector, synthesized by the two SMP schemes, are both sufficiently tiny with the maximum error less than $5 \times 10^{-7}$ m. These substantiate the efficacy and accuracy of the velocity-level SMP scheme (1)-(3) and the proposed acceleration SMP scheme (4)-(6). Moreover, the motion trajectories corresponding to the velocity-level scheme (1)-(3) [i.e., in Fig. 1(a)] appear to be the same as those

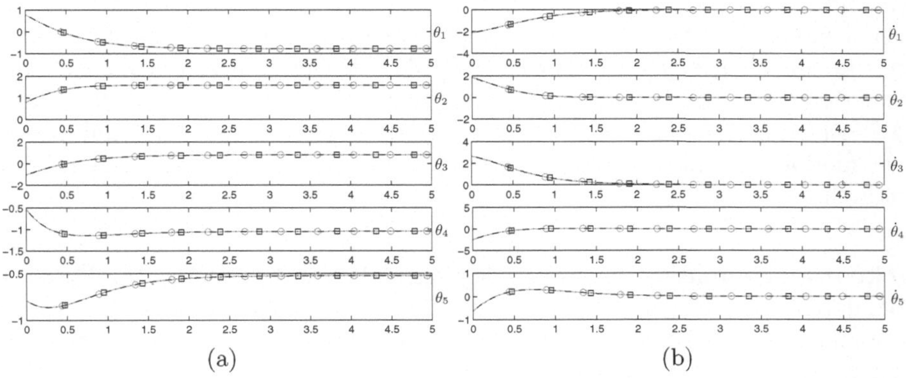


**Fig. 2.** Highly consistent joint-variable profiles of the five-link planar robot arm performing self-motion, synthesized by velocity-level scheme (1)-(3) (via solid curves with circle markers) and by acceleration-level scheme (4)-(6) (via dash-dotted curves with square markers): (a) Joint-angle profiles (rad) and (b) Joint-velocity profiles (rad/s)

corresponding to the acceleration-level scheme (4)-(6) [i.e., in Fig. 1(b)]. This demonstrates the equivalence between such two different-level SMP schemes.

Fig. 2 shows the joint-angle and joint-velocity profiles synthesized by velocity-level and acceleration-level SMP schemes when the five-link planar robot arm performs the self-motion. It can be observed from Fig. 2(a) that the joint angles of the two different-level SMP schemes not only converge to the desired joint values smoothly and quickly (less than 3 s), but also fit very well with each other. As for the joint-velocity profiles shown in Fig. 2(b), the same result can also be observed. These further demonstrate the efficacy of the two SMP schemes as well as their equivalence.

In summary, the above simulation results not only substantiate the efficacy and accuracy of the velocity-level SMP scheme (1)-(3) and the proposed acceleration-level SMP scheme (4)-(6), but also demonstrate the different-level SMP schemes' equivalence (i.e., the velocity-level SMP scheme is physically equivalent to the acceleration-level SMP scheme for redundant manipulators).

## 5 Conclusions

This paper has proposed a novel acceleration-level scheme for the self-motion planning of redundant robot manipulators. Compared to the velocity-level SMP scheme, such an acceleration-level SMP scheme generates and provides the joint-acceleration information, thus having wider applications. By further investigating the connection and comparison between the velocity-level SMP scheme and the proposed acceleration-level SMP scheme, we have shown the equivalence between the two different-level SMP schemes by using Zhang *et al.*'s neural-dynamic method, and then unified them into one QP formulation.

Computer-simulation results based on the five-link planar robot arm have further demonstrated the efficacy of the proposed acceleration-level SMP scheme and its equivalence to the velocity-level SMP scheme. Such a significant and novel discovery inspires us to investigate the velocity-level and acceleration-level SMP schemes as a whole, and motivates us to develop more meaningful mixed-level motion planning schemes for redundant manipulators in the future robotic research.

**Acknowledgements.** This work is supported by the National Natural Science Foundation of China under Grants 61075121 and 60935001, and also by the Fundamental Research Funds for the Central Universities of China. Besides, the corresponding author, Yunong, would like to thank and encourage the coauthors with the following thoughts: 1) "The steps of doing research are as follows: i) be a random walker (who tries to search papers and solutions globally); and ii) be a long thinker (who has and persists in clear directions)", 2) "The appearance of scientific research, in my opinion, is paper writing, book editing and theory proposing", and 3) "Science is the measurement on the degrees of philosophy".

## References

1. Zhang, Y.N., Zhu, H., Tan, Z.G., Cai, B.H., Yang, Z.: Self-Motion Planning of Redundant Robot Manipulators Based on Quadratic Program and Shown via PA10 Example. In: 2nd International Symposium on Systems and Control in Aerospace and Astronautics, Shenzhen, pp. 523–528 (2008)
2. Zhang, Y.N., Huang, Y., Guo, D.S.: Self-Motion Planning of Functionally Redundant PUMA560 Manipulator via Quadratic-Program Formulation and Solution. In: IEEE International Conference on Mechatronics and Automation, Changchun, pp. 2518–2523 (2009)
3. Zhang, Y.N., Jiang, D.C., Wang, J.: A Recurrent Neural Network for Solving Sylvester Equation with Time-Varying Coefficients. IEEE Trans. Neural Netw. 13(5), 1053–1063 (2002)
4. Zhang, Y.N.: Analysis and Design of Recurrent Neural Networks and their Applications to Control and Robotic Systems. Ph.D. Dissertation, Chinese University of Hong Kong (2002)
5. Zhang, Y.N., Wang, J., Xia, Y.S.: A Dual Neural Network for Redundancy Resolution of Kinematically Redundant Manipulators Subject to Joint Limits and Joint Velocity Limits. IEEE Trans. Neural Netw. 14(3), 658–667 (2003)
6. Zhang, Y.N., Tan, Z.G., Chen, K., Yang, Z., Lv, X.J.: Repetitive Motion of Redundant Robots Planned by Three Kinds of Recurrent Neural Networks and Illustrated with a Four-Link Planar Manipulator's Straight-Line Example. Robot. Auton. Syst. 57(6-7), 645–651 (2009)
7. Zhang, Y.N., Ma, S.G.: Minimum-energy redundancy resolution of robot manipulators unified by quadratic programming and its online solution. In: IEEE International Conference on Mechatronics and Automation, Harbin, pp. 3232–3237 (2007)

# Grey Incidence Optimization Model Based on Hybrid Differential Evolution Algorithm

Lifang Kong, Ying Zhao, and Xinbin Liu

Department of Basic Teaching, Air Force Logistic Academy,
Jiangsu Xuzhou 221002, China
klf030@163.com

**Abstract.** Because of complex relationships between of educational devotion and economic development, this thesis uses the sampling inspection result of Jiang Su province educational devotion data, presents nonlinear restoration method for establishing the basic reactions of education and economics, gains the minimal structural educational devotion through the proportion of citizen education investment and government education investment and economic development of relation model. Simulation results hybrid differential evolution algorithm and grey incidence based on entropy restoration algorithm is effective, efficient and robust for solving the optimization inverse proportion of educational devotion and gross domestic product problems.

**Keywords:** differential evolution algorithm, grey incidence, optimization model.

## 1 Introduction

There exists intricate relationship between education and economy. This paper offers a concrete presentation of this relationship by studying the data of Jiangsu province. This include establishing the fundamental relationship model between education and economy through the method of non-linear fitting, evaluating the boosting effect of family and government education investment on gross domestic product(GDP) through mutual spectrum analysis and getting the optimal family investment/government investment ratio. Then the optimal GDP value in the minimum education investment structure can be calculated if the results of family and government investment value are put into the relationship model which we built earlier. Also, the sensitivity of the various elements is tested by way of system dynamics and the results turns out to be satisfying.

## 2 Hybrid Differential Evolution Algorithm

Differential evolution algorithm is a method of self-organization and minimization. Very limited amount of input is needed. Its key idea is different from the traditional ES. The traditional method adopts the predetermined probability distribution function to decide the vector disturbance. The self-organization system in the differential

J. Lei et al. (Eds.): AICI 2012, LNAI 7530, pp. 17–24, 2012.

evolution algorithm takes two randomly selected vectors in the population to disturb the existed vector. Every vector in the population needs to be disturbed. If the cost of the new vector's function value is lower than its predecessor, the new vector will replace its predecessor. The differential evolution algorithm uses a vector population. The random disturbance of the vectors can take place independently. Therefore, it is natural and parallel.

### 2.1 Algorithm Description

Like other evolution algorithms, the differential evolution algorithm also operates on the population of the candidate solution. However, its multiplication method of the self-reference population is different from other evolution algorithms in that it adds the weighted difference vector of the two members in the population to the third member, and thus produces a new parameter vector. This step is called "variation". Then, in accordance with certain rules, blend the parameters of the variation vector with the parameters of the predetermined target vector to produce the experiment vector. This step can be named as "crossover".

### 2.2 Calculate Processes

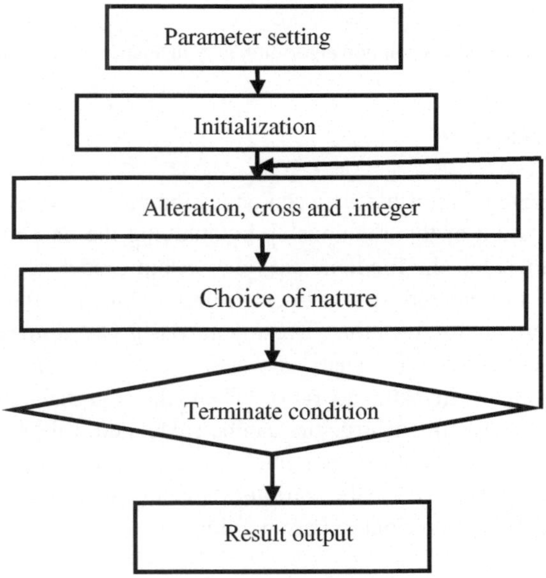


**Fig. 1.** The flow chart of the algorithm of differential evolution

### 2.3 Calculate Processes

The differential evolution algorithm takes the real-valued parameter vector whose dimension is D as the population of every generation. The individuals in the population can be presented as $x_{i,G}\,(i = 1,2,\cdots,NP\,)$ .In this formula $i$ is the sequence of the individual in the population; G represents the generation times and NP indicates the scale of the population, which remains stable in the minimum process.

To find the initial points of optimization search, the population has to be initialized. The traditional way of finding the initial population is to select randomly from the values within the boundary constraints. However, in the differential evolution algorithm, it is assumed that the initial population conforms to the uniform probability distribution. Suppose, the boundary of the parameter variations is $x_j^{(L)} < x_j < x_j^{(U)}$ .and $x_{ji,0} = rand\,[0,1]\,?(x_j^{(U)} x_j^{(L)}) + x_j^{(L)}$

$$(i = 1,2,\cdots,NP, j = 1,3,\cdots,D).$$

In the formula, rand[0,1] produces uniform random numbers in [0,1]. If the preliminary solution to the question can be got, the initial population can be obtained through adding random derivation of normal distribution to the preliminary solution.

For every target vector $i = 1,2,\cdots,NP$ , the derivation vector is produced like this $v_{i,G+1} = x_{r_1+G} + F\cdot(x_{r_2+G} - x_{r_3+G})$ ,in this formula, $r_1, r_2, r_3$ are different from each other and none of them is the same with $i$ which is the sequence of the target vector. Therefore, the condition of $NP \geq 4$ must be met.

To guarantee a wide diversity of disturbance parameter vector, this paper adopts the operation of crossover. After the crossover operation, the experiment vector is

$$\mu_{i,G+1} = \left(\mu_{1i,G+1}, \mu_{2i,G+1}, \cdots, \mu_{Di,G+1}\right)$$

$$\mu_{ji,G+1} = \begin{cases} v_{ji,G+1} & if\left(randb(j) \leq CR\right) or \quad j = rnbr(i) \\ x_{ji,G+1} & if\left(randb(j) > CR\right) or \quad j \neq rnbr(i) \end{cases}$$

$$(i = 1,2,\cdots,NP, j = 1,3,\cdots,D)$$

In this formula, the $j$ th estimated value of the random number generator, $randr(j) \in 1,2,\cdots,D$ , a randomly selected sequence, is to guarantee that $u_{i,G+1}$ can get at least one parameter from $v_{i,G+1}$ . The value range of CR, the crossover operator, is [0,1].

To test whether experiment vector $u_{i,G+1}$ will be a member of the next generation, the experiment vector is compared with the target vector $x_{i,G}$ ,in the population according to the greedy criterion. If the target function is to be minimized, the vectors with a smaller value will gain their position in the next generation. All the individuals of the next generation should be better than or at least the same with the current generation. Attention should be paid that in the differential evolution algorithm, the experiment vector can only be compared with the individuals one by one, but not every individual in the population.

It is important to guarantee that the new individuals should conform to the boundary condition. One simple solution to this question is to replace those individuals that do not conform to the boundary condition with the parameter vectors produced randomly in the boundary. Namely, if $\mu_{ji,G+1} < x_j^{(L)}$ or $\mu_{ji,G+1} > x_j^{(U)}$, then $x_{ji,G+1} = rand[0,1] \cdot (x_j^{(U)} - x_j^{(L)}) + x_j^{(L)} (i = 1,2,\cdots NP, j = 1,3,\cdots,D)$.

## 3 Grey Incidence Optimization Process

The Grey Correlation evaluation is mainly based on the model of R=Y×W. In this expression, R is the result vector of the objects (with a total number of M) being evaluated; W is the weight vector of the evaluation indexes (with a total number of N); E is the evaluation matrix of the indexes.

$\xi_i(k)$ is the correlation coefficient between the kth index and the kth optimal index of the ith object being evaluated. We will arrange the order in accordance with the result of R.

### 3.1 Decide the Optimal Index Set

In the expression of $F = [j_1^*, j_2^*, \cdots j_n^*]$, suppose $j_k^*$ is the optimal result of the kth index. In the optimal sequence, the result of every index can be the optimal result of the objects being evaluated and can also be the optimal result generally acknowledged by those who conduct the evaluation. After deciding the optimal index set, Matrix D can thus be constructed. Meanwhile, $j_k^i$ is the raw value of the kth index of the ith futures company.

### 3.2 The Standardization Processing of the Indexes

Due to the different dimensions and numbers of different indexes, direct comparison cannot be conducted. Therefore, standardization processing of the raw indexes is needed. Suppose the variation interval of the Kth index is $[j_{k1}, j_{k2}]$ and $j_{k1}$ is the minimum value of the Kth index among all the objects being evaluated while $j_{k2}$ is the maximum value. Then, the raw value in the above expression can be turned into dimensionless value $C_k^i \in (0,1)$ via the following expression.

$$C_k^i = \frac{j_k^i - j_{k1}}{j_{k2} - j_k^i}, \quad i = 1,2,\cdots m, \quad k = 1,2,\cdots,n$$

### 3.3 Calculate the Comprehensive Evaluation Results

According to the Grey System Theory, we set $\{C^*\}=[C_1^*, C_2^*, \cdots, C_n^*]$ as the reference sequence and $\{C\}=[C_1^i, C_2^i, \cdots, C_n^i]$ as the comparison sequence. The correlation coefficient between the Kth index of the Ith object being evaluated and the optimal index of the Kth index can thus be got.

$$\xi_i(k) = \frac{\min_i \min_k \left|C_k^* - C_k^i\right| + \rho \max_i \max_k \left|C_k^* - C_k^i\right|}{\left|C_k^* - C_k^i\right| + \rho \max_i \max_k \left|C_k^* - C_k^i\right|}$$

In this expression, $\rho \in (0,1)$, but the general practice is $\rho = 0.455$

In this way, the comprehensive evaluation result is $R = E * W$ .If $r_i$ is the maximum value, it suggests that $\{C\}$ is the closet to $\{C^*\}$. In other words, it suggests that the Ith object being evaluated is superior to other objects. We can thus sort out the objects in this way.

## 4 Data Restoration and Optimization

There exists the complex input-output relationship between education and economy. Education input refers to all the human resources, funds and materials in which funds are human resources and materials in the currency form. Education investment can be divided into family and government education investment. And economic growth refers to the increase in funds in the economic system. Gross domestic product predigests GDP, Family_edu and Gov_edu separately predigest citizen education investment and government education investment. Establish linear model: $LN(GDP) = a + b * family\_edu + c * gov\_edu$, $a$, $b$ and $c$ is estimation parameters. Using improved differential evolution and 1stopt software, result of Linear Model of gross domestic product is showed in figure2.

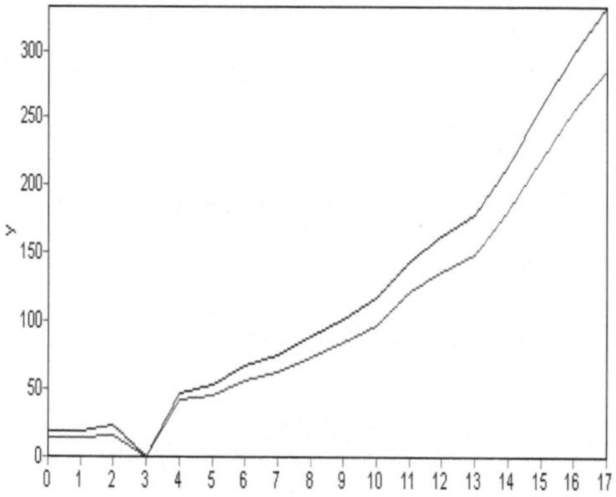


**Fig. 2.** Linear model of gross domestic product

| *Parameter* | *Best Estimate* |
|---|---|
| *a* | *7.78498726827953* |
| *b* | *0.00751409769678035* |
| *c* | *0.000233990079868227* |

The expression of the model is:

$$\ln(GDP) = 7.784987268 + 0.007514097696 * family\_edu + 0.0002339900798 * gov\_edu$$

After changing logarithm into exponent, the expression of the model is ：

$$GDP = e^{7.784987268+0.007514097696*family\_edu+0.0002339900798*gov\_edu}$$

However, when we restore logarithm model into exponent model, we find the errors become greater, which can be observed in the following:

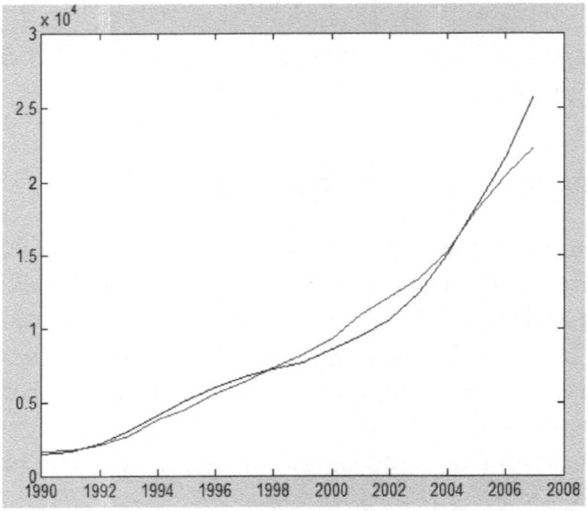


**Fig. 3.** Compare actual and model value of gross domestic product

This suggests that there exists some kind of distortion due to the truncation error of the data in the changing of function type. This paper moves away from the linear model built up for the convenience of calculation and establishes a non-linear model.

$$GDP = e^{a+b\cdot family\_edu+c\cdot gov\_edu}$$

Adopt the differential evolution algorithm again and we can get:

| *Parameters* | *Best Estimate* |
|---|---|
| *a* | *8.04948530247768* |
| *b* | *0.00955503446885053* |
| *x* | *-0.0033821266221795* |

It is found that the 9th value does not conform to the whole curve, so it is rejected. After this, the satisfying fitting curve is finally got.

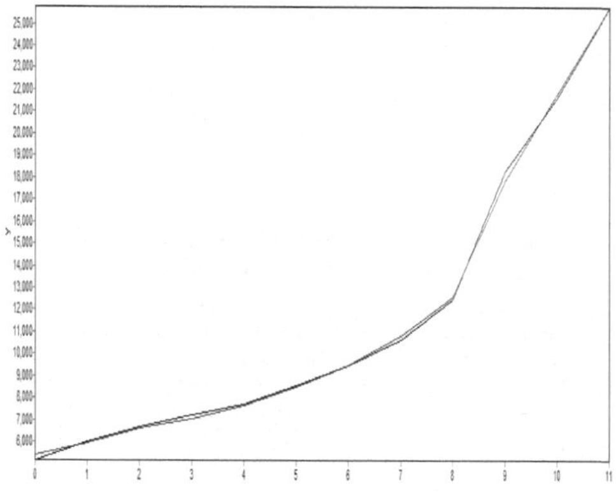

**Fig. 4.** Compare actual and model Value of gross domestic product

Correlation coefficient :

| *Parameters* | *Best Estimate* |
|---|---|
| *a* | *8.29723179769527* |
| *b* | *0.00396501901654512* |
| *x* | *0.00155402207728224* |

Model inspection:

Iterations: 2822
Elapsed Time (Hr:Min:Sec:Msec): 00:00:03:360
Algorithms: Hybrid Differential Evolution
Stop Reason: Convergence tolerance reached
Root of Mean Square Error (RMSE): 173.813447106416
Sum of Square Error (SSE): 302111.143950147
Correlation Coefficients (R): 0.999627007846776
R-Square (R^2): 0.999254154816699
Determination Coefficients (DC): 0.999254043877292
Chi-Square:9.8516783884214
F-Statistic: 10718.0865647643

From the evaluation, it can be observed that model I is very satisfying with the relevant coefficient reaching almost 1, which further proves that it is right to improve the model.

## 5 Grey Incidence Optimization Index

It can be observed that at this family education investment/government education investment ratio, GDP grows at the fastest rate. Considering the fact that family education investment is the total number of all the investments of all families and cannot be changed with human wills. Therefore, it changes with social development. Through entropy, gaining the weight of two indexes is showed in table1 and gaining final results through grey incidence are showed in table 2.

**Table 1.** The weight of citizen education investment and government education investment

| index | Family_edu | Gov_edu |
|---|---|---|
| weight | 0.5020 | 0.4980 |

Table 2 final results of grey incidence

| index | Synthesis educational index | family_edu | Gov_edu |
|---|---|---|---|
| Incidence of economics | 0.1700 | 0.0853 | 0.0846 |

## 6 Conclusion

In this paper, education is the input and economy is the output. The two indexes in evaluating education input are family and government education input and the index of economy is GDP. By fitting the tree indexes, the relationship between education and economy will be got. This paper predicts family education investment. Then, based on the minimum ration of education investment, the government education investment will be calculated. Then substitute the results in the education/economy model that is built. Then the GDP of Jiangsu Province in 2008 under the condition of minimum education investment ratio can be obtained.

## References


1. Liang, J.: Functional Procedure Neural Network. Dynamic of Continuous Discrete and Impulsive Systems-Series B-Applications & Algorithms 1(SI), 27–31 (2005)
2. Liang, J., Gao, J.-H.: Kernel Function Clustering Algorithm Optimized Parameters. In: The Forth International Conference on Machine Learning and Cybernetics, Guangzhou, China, vol. 7, pp. 4400–4404 (2005)
3. dos Leandro, S.C.: A quantum particle swarm optimizer with chaotic mutation operator. Chaos, Solitons and Fractals 37, 1409–1418 (2008)
4. Wang, X., Yang, J., Teng, X., Xia, W., et al.: Feature selection based on rough sets and particle swarm optimization. Pattern Recognition Letters 28(4), 459–471 (2007)
5. Niu, G., Lee, S.S., Yang, B.S., et al.: Decision fusion system for fault diagnosis of elevator traction machine. Journal of Mechanical Science and Technology 22(1), 85–95 (2008)
6. Gang, N., Tian, H., Bo-Suk, Y., et al.: Multi-agent Decision fusion for motor fault diagnosis. Mechanical Systems and Signal Processing (21), 1285–1299 (2007)
7. Chen, G., Liu, Y., Zhou, W.: Research on intelligent fault diagnosis based on time series analysis algorithm. The Journal of China Universities of Posts and Telecommunications 15(1), 68–74 (2008)
8. Rapuano, S., Harris, F.J.: An introduction to FFT and time domain windows. IEEE Instrumentation & Measurement Magazine 10(6), 32–44 (2007)
9. Tasgetiren, M.F., Suganthan, P.N., Chua, T.J., et al.: Differential evolution algorithms for the generalized assignment problem. In: IEEE Congress on Evolutionary Computation, CEC 2009, pp. 2606–2613 (2009)

# A Data Mining Based Approach to Research the Relationship between Kansei and Usability: A Case Study of Mobile Phones

Yongfeng Li and Liping Zhu

College of Mechanical and Electrical Engineering,
Jiangsu Normal University, Xuzhou, 221116, China
yfli_xznu@yahoo.com.cn, zhuliping@yahoo.cn

**Abstract.** Kansei and usability are two important aspects that should be considered in product development. In order to design the product that not only meets consumer's affective needs but also is easy to use, this paper presents a novel approach based on data mining to reveal the relationship between kansei and usability. Firstly, kansei image and usability evaluation indexes were determined. Secondly, within-subjects experimental design was applied to test kansei image, usability, and user's satisfaction. Finally, association rule and decision tree were utilized to mine the experimental data so as to discover the rules hidden in the data. A case study of mobile phones was conducted based on the proposed method. The results suggest that there is a significant relationship between kansei and usability which together influence user's satisfaction with product. This approach can provide designers with useful suggestions and solutions for product design.

**Keywords:** Data mining, Kansei, Usability, Mobile phones.

## 1 Introduction

With the advancement of technology and the improvement of quality of life, products become more and more complex. Consumers choose products not only considering whether the products meet their personality and social status, but also increasingly concerning about the ease of operation.

Kansei means consumer's psychological feeling and image. It refers to human sensitivity of a sensory organ where sensation or perception takes place in response to stimuli from external world [1]. ISO 9241-11 defines usability as the "Extent to which a product can be used by specified users to achieve specified goals with effectiveness, efficiency and satisfaction in a specified context of use [2]". Usability has become an important part of product design.

The relationship between kansei and usability has become one of the key issues in product design field. It has important theoretical significance and practical value for enterprises' product design.

J. Lei et al. (Eds.): AICI 2012, LNAI 7530, pp. 25–33, 2012.

## 2 Literature Review

Recently, some researches have been done on the relationship between kansei and usability. Sonderegger and Sauer [3] studied the influence of the design aesthetics in usability testing, and the results suggested that product appearance had a positive effect on performance, this means kansei impact on usability. But in their study, the subjects were asked to use a computer-simulated prototype of a mobile phone to complete the given tasks, and in this way it was very difficult to find problems in actual use. Seva, Gosiaco, Santos and Pangilinan [4] explored product affective quality and the apparent usability, they found that form elements were relevant to affective quality and apparent usability, and aesthetic perception of a product could enhance apparent usability. However, in their experiment subjects interacted with the product only through vision, not used the product to execute some task. Vergara, Mondragón, Sancho-Bru, Company and Agost [5] researched at different levels of user's interaction (seeing, touching, and using) with product, the relationship between quality, appearance, innovation, and so on, but the key indicators of usability, such as effectiveness and efficiency, were not discussed in detail.

Data mining was defined as the discovery of non-trivial, implicit, previously unknown, and potentially useful and understandable patterns from large data sets [6]. It has been widely used in the domain of product design study in recent years [7-9].

The purpose of this paper is to study the relationship between kansei and usability based on data mining technology. The experiment in this study has two stages of interaction: observing product stage and using product stage.

## 3 Methods

### 3.1 Data Mining

Association rule mining is a data mining method to find the interesting association or correlation among a large set of data items. Support, confidence and lift are main thresholds to measure an association rule. Apriori algorithm is one of the prevalent techniques used to find association rules. It operates in two phases. In the first phase, all large item sets are generated. The second phase generates rules from the set of all large item sets [10]. We adopted Apriori algorithm as a methodology for association rule.

A decision tree is a flow-chart-like tree structure that uses a tree-like graph or model of decisions and their possible consequences. The knowledge that was represented by decision tree can be extracted and expressed through the rules in the form of IF-THEN. These rules are easy to understand, and be widely used. For the current study, CHAID (Chi-Square Automatic Interaction Detector) was used because in comparison to other algorithms, CHAID does not restrict the number of branches from each node to a predetermined number. CHAID analysis uses chi-squared tests for each split to determine which variable best predicts the outcome variable for each subdivision and corrects for the multiple comparisons inherent in the procedure using Bonferroni adjusted $p$-values with an overall error rate of 0.05 [11].

### 3.2 Product Kansei Image

A total of 198 kansei words about mobile phones were collected from magazines, literature, manuals, experts, and experienced users. We extracted 5 representative kansei words by applying KJ method and factor analysis [12], which are gorgeous, friendly, practical, restrained, and simple.

### 3.3 Product Usability Evaluation

Product usability often takes task completion time, interaction efficiency, and number of errors as main evaluation indexes. Efficiency could be indirectly reflected by task completion time and the number of errors in executing tasks. Therefore, we took completion time and errors as usability evaluation indexes.

We based on two typical activities in mobile usage which were making a phone and short message service to set up usability test tasks. Taking into account the difference in signal strength and functional navigation interface of mobile phones, we only tested dialing phone numbers and writing messages, and did not include phone connecting, sending messages, and searching functional interface. The usability evaluation indexes included four aspects which were TDPN (Time of Dialing Phone Numbers), EDPN (Errors in Dialing Phone Numbers), TWM (Time of Writing Message), and EWM (Errors in Writing Message).

## 4 Experiment

### 4.1 Stimuli

Ninety-three mobile phones recently available on the current market in China were collected. After deleting the peculiar and similar, the remainders were classified according to shape. Eight typical mobile phones were determined as experimental samples. The brands of the 8 mobile phones were Nokia, LG, Desay, ZTE, Haier, and Meizu. To eliminate the influence of different colors and materials on the perception, these samples were similar in color and material, which are illustrated in Fig. 1.

Nokia E63 Nokia C6 Nokia 5310 LG GD300S Desay M800 ZTE U236 Haier V720 Meizu M8

**Fig. 1.** Experimental samples

### 4.2 Participants

Sixteen volunteer participants (8 male, 8 female; mean age 21.81; SD=0.83) were invited to participate in our experiment. All participants selected for the study had the experience using mobile phone at least 2 years. And they were college students receiving professional training in design, as it was felt that they would be more sensitive to kansei than average individuals and better able to concretely describe their thoughts about the product design. Eight of them had used the Nokia brand mobile phones, but had not used the types which were selected in this experiment. All subjects were not used the mobile phone brands which were LG, Desay, ZTE, Haier, and Meizu.

### 4.3 Procedures

A within-subjects design was used in the experiment. Every subject experimented with all products. The products and kansei words were presented in random order. The procedures involved the following five steps.

**Step1:** The participants were interviewed about the past use of mobile phones, including the use of time, quantity, and input methods, etc.

**Step2:** The mobile phones were placed on tables. The participants could look at the mobile phones but were not allowed to touch them. Then the participants rated the 5 kansei words for the 8 mobile phones.

**Step3:** The participants were required to use the mobile phones to write messages and to dial phone numbers. In order to overcome the learning effect, the sequence of the two tasks was counterbalanced. The participants dialed the given 8 phone numbers in accordance with the provided order, and used the same input method to write messages which contained 60 Chinese characters.

**Step4:** The participants assessed the satisfaction extent with the mobile phone.

**Step5:** The participants were asked the problems encountered when completing the tasks.

A 7-point Likert scale ranging from 1 (strongly disagree) to 7 (strongly agree) was used to evaluate the kansei and user's satisfaction. The entire experiment processes were videotaped and the time to complete the tasks were measured from the video tapes.

### 4.4 Data Preparation

As the experimental data were complex, in order to facilitate mining, they needed to be discrete. For the data of usability test, according to their distribution, we applied the methods shown in Table 1 to discretize them. For the evaluation value of kansei and satisfaction, we divided them into three categories. When the evaluations were 1, 2 and 3, the evaluation values were low, and we classified them into category 1; when the evaluations was 4, the evaluation value was medium, and we classified them into category 2; When the evaluations were 5, 6 and 7, the evaluation values were high, and we classified them into category 3. After discretizing the data of kansei, usability and satisfaction, frequency distribution of the categories are shown in Table 2, and the experimental data after discretization were listed in Table 3.

**Table 1.** Methods of discretizing data

| Items | Value range | Category |
|---|---|---|
| TDPN<br>(Time of Dialing Phone Numbers) | [0, 5.45)<br>[5.45, 7.20)<br>[7.20, +∞) | 1 (Short )<br>2 (Medium)<br>3 (Long) |
| EDPN<br>(Errors in Dialing Phone Numbers) | 0<br>(0, +∞) | 1 (Do not have)<br>2 (Have) |
| TWM<br>(Time of Writing Message) | [0, 200.00)<br>[200.00, 270.00)<br>[270.00, +∞) | 1 (Short )<br>2 (Medium)<br>3 (Long) |
| EWM<br>(Errors in Writing Message) | [0, 3)<br>[3, 7)<br>[7, +∞) | 1 (Few)<br>2 (Normal)<br>3 (Many) |

**Table 2.** Frequency distribution of discrete data

| Items | Category 1 | | Category 2 | | Category 3 | |
|---|---|---|---|---|---|---|
| | Frequency | % | Frequency | % | Frequency | % |
| Gorgeous | 38 | 29.69 | 60 | 46.88 | 30 | 23.44 |
| Friendly | 31 | 24.22 | 56 | 43.75 | 41 | 32.03 |
| Practical | 17 | 13.28 | 66 | 51.56 | 45 | 35.16 |
| Restrained | 34 | 26.56 | 42 | 32.81 | 52 | 40.62 |
| Simple | 25 | 19.53 | 60 | 46.88 | 43 | 33.59 |
| TDPN | 43 | 33.59 | 42 | 32.81 | 43 | 33.59 |
| EDPN | 67 | 52.34 | 61 | 47.66 | | |
| TWM | 42 | 32.81 | 42 | 32.81 | 44 | 34.38 |
| EWM | 43 | 33.59 | 42 | 32.81 | 43 | 33.59 |
| Satisfaction | 31 | 24.22 | 62 | 48.44 | 35 | 27.34 |

**Table 3.** Experimental data after discretization

| Items | ID 1 | ID 2 | ID 3 | … | ID 128 |
|---|---|---|---|---|---|
| Gorgeous | 3 | 3 | 2 | … | 3 |
| Friendly | 2 | 3 | 3 | … | 3 |
| Practical | 3 | 2 | 3 | … | 3 |
| Restrained | 2 | 3 | 2 | … | 1 |
| Simple | 2 | 3 | 2 | … | 3 |
| TDPN | 3 | 3 | 2 | … | 3 |
| EDPN | 2 | 2 | 1 | … | 2 |
| TWM | 2 | 2 | 1 | … | 3 |
| EWM | 3 | 3 | 2 | … | 2 |
| Satisfaction | 2 | 2 | 3 | … | 2 |

# 5 Results and Analysis

We applied Clementine 12.0 to analyze the data, and get association rules and decision tree rules. Using association rule and decision tree, we would analyze the relationship between kansei and usability, and extract useful information for product design.

Since there were lots of mining rules, we only selected the rules that are statistically very strong and contain substantial design information. For the association rules, minimum support, minimum confidence, and minimum lift were set as 0.05, 0.5, and 1.0, respectively. For the decision tree, we gleaned the rules that the number was greater than 8, and the percentage was greater than 50%.

## 5.1 The Relationship between Kansei and Usability

The relationship between kansei and usability can be illustrated by web graph in Clementine. The web graph represents a sort of graphical cross-tabulation in which thicker lines indicate relatively larger cell counts, and thinner lines indicate the opposite. For ease of observation, the gauge was set to 27, and the result is depicted in Fig. 2. It can be found that there are some strong relationships among "Practical – High value", "EDPN - Have", "Simple - High value", "Gorgeous - High value" and "Friendly – High value". Table 4 presents the detailed association rules.

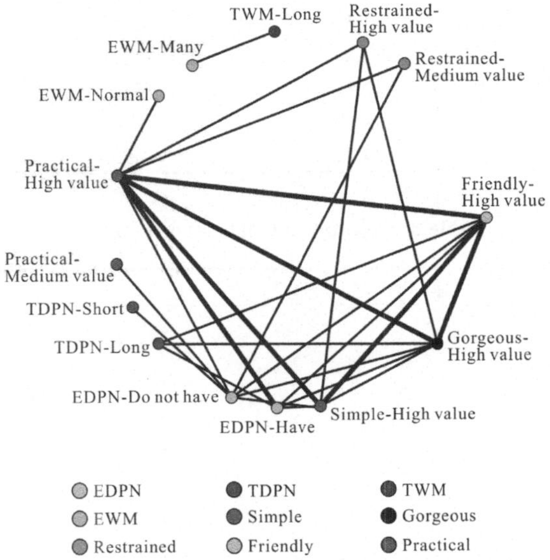


**Fig. 2.** Association graph for the relation between kansei and usability

In all extracted association rules, the rule that is representative and has the highest intensity is "Practical = High value ⇒ EDPN = Have". This rule conflicts with people's common sense that when a mobile phone's practical value is high, the user would not have errors dialing phone numbers. This indicates that there is a big gap between evaluation mobile phones only by vision and by the integration of vision, tactile sense, auditory sense, etc. These phones scores high in practicality, but scores lower in usability. Therefore, product designer should carefully analyze the action of people operating product, and study the interface of the product. Make users complete the task quickly and accurately with minimal attention load.

**Table 4.** Association rules between kansei and usability

| ID | Rules | Support % | Confidence % | Lift |
|---|---|---|---|---|
| 1 | Practical = High value ⇒ EDPN = Have | 30.47 | 59.09 | 1.24 |
| 2 | Practical = High value & Gorgeous = High value ⇒ EDPN = Have | 17.97 | 62.16 | 1.30 |
| 3 | Gorgeous = High value & Friendly = High value ⇒ TDPN = Long | 16.41 | 56.76 | 1.69 |
| 4 | Simple = Low value ⇒ TWM = Long | 10.16 | 52.00 | 1.58 |
| 5 | Friendly = Low value & Simple = Low value ⇒ TWM = Long | 6.25 | 61.54 | 1.88 |
| 6 | Gorgeous = High value & Restrained = Low value ⇒ TWM = Long | 6.25 | 80.00 | 2.44 |

## 5.2 Relationship of User Satisfaction with Product Kansei Image and Usability

Satisfaction evaluation was after kansei and usability test, so the evaluation reflected the integration of them. Table 5 shows the relationship of user satisfaction with kansei and usability.

**Table 5.** Relationship of user satisfaction with product kansei image and usability

| ID | Rules | (N, %) |
|---|---|---|
| 1 | IF (Practical = Low value) and (Gorgeous = Low value) THEN (Satisfaction = Low value) | (13, 84.62) |
| 2 | IF (Practical = Medium value) and (EWM = Normal or Few) and (Gorgeous = Medium value or High value) and (TWM = Medium or Short) THEN (Satisfaction = High value) | (16, 75.00) |
| 3 | IF (Practical = Medium value) and (EWM = Normal or Few) and (Gorgeous = Low value) THEN (Satisfaction = Medium value) | (11, 72.73) |
| 4 | IF (Practical = Medium value) and (EWM = Many) THEN (Satisfaction = Low value) | (16, 56.25) |
| 5 | IF (Practical = High value) and (Friendly = High value) and (Simple = Medium value) and (EWM = Normal or Few) THEN (Satisfaction = High value) | (9, 77.78) |
| 6 | IF (Practical = High value) and (Friendly = High value) and (Simple = High value) THEN (Satisfaction = High value) | (32, 93.75) |

From rule 1 and rule 6, we can infer that kansei has a direct impact on user's satisfaction. When "Practical" and "Gorgeous" evaluation are low, the user's satisfaction with the product is low, the percentage is 84.62%; When "Practical", "Friendly" and "Simple" evaluation are all high, user's satisfaction is high, the percentage is up to 93.75%. The main reason for these rules is product whose kansei evaluation is high make people feel good, which in turn makes them think more creatively, and makes it easier for people to find solutions to the problems they encounter. Thus when people use the product, the number of errors would be reduced, and the time would become shorter, and then user's satisfaction would increase.

Rules 2-5 suggest that usability has an important impact on user's satisfaction. When "Practical" evaluation is medium, and there are many errors in writing message, the user's satisfaction with the product would be low, and the percentage is 56.25%. When the errors in writing message are few or normal, the evaluation value of user's satisfaction with the product would mainly be high.

Therefore, designers must take both kansei and usability into account when designing product. Not only make product meet users' emotional needs, pleasing user in visceral level, but also make product be easy to operate, satisfying user in behavioral level[13]. In this way, the market competitiveness of product could be really enhanced.

## 6 Conclusions

Kansei and usability are two important factors of products. To increase the market competitiveness of products, designers must take them into account. In this paper, we applied data mining method to study the relationship between kansei and usability. Knowledge extract from the data mining result can provide designers with useful insights for product design.

To illustrate the approach, we conducted an experimental study on the mobile phones. The results suggest that there is a significant relationship between kansei and usability, such as "Practical – High value" and "EDPN - Have". Furthermore, kansei and usability together influence user's satisfaction with product. Although the mobile phones are chosen as a case study, this approach can be applied to other products.

Subjects used in the study were college students, and the number of subjects and specimen were limited as sufficient for the purpose of exploratory study. Thus the results may not produce generally acceptable features. Further work involving bigger size of populations and a variety of specimens should be considered.

**Acknowledgments.** This research is supported by the Natural Science Foundation of the Jiangsu Higher Education Institutions of China (Grant No. 10KJD460002). The authors would like to express their sincere thanks to the subjects for their participation and assistance in the experimental study.

## References

1. Yanagisawa, H., Murakami, T.: Factors Affecting Viewpoint Shifts When Evaluating Shape Aesthetics towards Extracting Customer's Latent Needs of Emotional Quality. In: 2008 ASME International Design Engineering Technical Conferences and Computers and Information in Engineering Conference, vol. 3, pp. 791–800. ASME, New York City (2008)
2. ISO 9241-11: Ergonomic Requirements for Office Work with Visual Display Terminals (VDTs). Part 11 - Guidelines for Specifying and Measuring Usability. International Standards Organization, Geneva (1998)
3. Sonderegger, A., Sauer, J.: The Influence of Design Aesthetics in Usability Testing: Effects on User Performance and Perceived Usability. Applied Ergonomics 41, 403–410 (2010)
4. Seva, R.R., Gosiaco, K.G.T., Santos, M.C.E.D., Pangilinan, D.M.L.: Product Design Enhancement Using Apparent Usability and Affective Quality. Applied Ergonomics 42, 511–517 (2011)
5. Vergara, M., Mondragón, S., Sancho-Bru, J.L., Company, P., Agost, M.-J.: Perception of Products by Progressive Multisensory Integration. A Study on Hammers. Applied Ergonomics 42, 652–664 (2011)
6. Anand, S.S., Büchner, A.G.: Decision Support Using Data Mining. Financial Times Pitman, London (1998)
7. Xia, S.S., Wang, L.Y.: Customer Requirements Mapping Method Based on Association Rule Mining for Mass Customization. Journal of Shanghai Jiaotong University (Science) 13, 291–296 (2008)
8. Bae, J.K., Kim, J.: Product Development with Data Mining Techniques: A Case on Design of Digital Camera. Expert Systems with Applications 38, 9274–9280 (2011)
9. Lin, P., Yang, C.: Impact of Product Pictures and Brand Names on Memory of Chinese Metaphorical Advertisements. International Journal of Design 4, 57–70 (2010)
10. Liao, S.-H., Chen, Y.-J., Deng, M.-Y.: Mining Customer Knowledge for Tourism New Product Development and Customer Relationship Management. Expert Systems with Applications 37, 4212–4223 (2010)
11. Horner, S.B., Fireman, G.D., Wang, E.W.: The Relation of Student Behavior, Peer Status, Race, and Gender to Decisions about School Discipline Using CHAID Decision Trees and Regression Modeling. Journal of School Psychology 48, 135–161 (2010)
12. Smith, S., Fu, S.-H.: Factor Analysis of Head-Up Display Presentation Images. In: 2009 ASME International Design Engineering Technical Conferences and Computers and Information in Engineering Conference, vol. 2, pp. 967–974. ASME, San Diego (2010)
13. Norman, D.A.: Emotional Design: Why We Love (or Hate) Everyday Things. Basic Books, New York (2004)

# Making Use of the Big Data: Next Generation of Algorithm Trading

Xiongpai Qin[1,2]

[1]MOE Key Lab. of Data Engineering and Knowledge Engineering (RUC), Beijing, 100872, P.R. China
[2] School of Information, Renmin University of China, Beijing, 100872, P.R. China
qxp1990@sina.com

**Abstract.** Algorithm trading is using computer programs to automate trading actions without much human intervention. Algorithm trading has been adopted by institutional investors and individual investors and made profit in practice. The soul of algorithm trading is the trading strategies, which are built upon technical analysis rules, statistical methods, and machine learning techniques. Big data era is coming, although making use of the big data in algorithm trading is a challenging task, when the treasures buried in the data is dug out and used, there is a huge potential that one can take the lead and make a great profit.

**Keywords:** Algorithm Trading, Technical Analysis, Statistical Methods, Machine Learning, Big Data.

## 1 Introduction to Algorithm Trading

Application of computer and communication techniques has stimulated the rise of algorithm trading. Algorithm trading is the use of computer programs for entering trading orders, in which computer programs decide on almost every aspect of the order, including the timing, price, and quantity of the order etc. In certain cases, computer can even start the execution of the order without any human intervention. Algorithm trading has been used in stock, futures and other financial markets by institutional and individual investors. [1]

There are many success stories of algorithm trading, the most noticeable one is James Simons and his company – Renaissance Technologies [2]. According to the Wiki page, Renaissance was started in 1982 by James Simons, it currently has more than $23 billion of assets under management. Since it's founding, Renaissance Technologies hedge fund has traded in financial markets around the world. The fund employed mathematical models to analyze data and execute trades. Most of trades are automated. The Medallion Fund of the company has averaged 35 percent returns annually since 1989 after paying the fees. Renaissance uses complex computer-based models to predict price movements of easy-to-trade financial assets. These models are based on analyzing as much data as can be collected, then searching for price moving patterns to make profitable predictions. By the end of 1999, cumulative returns of the fund were 2,478.6 percent [2], really high ones.

J. Lei et al. (Eds.): AICI 2012, LNAI 7530, pp. 34–41, 2012.

## 2 Architecture of an Algorithm Trading System

Figure 1 shows the data flow of a typical algorithm trading system. Firstly the trading system collects price data from the exchange (for cross market arbitrage, the system needs to collect price data from more than one exchange), news data from news companies such as Reuters, Bloomberg. Some algorithm trading systems may also collect data from the web for deep analysis such as sentiment analysis. While the data is being collected, the system performs some complicated analysis on the data to look for profitable chances with the expectation of making profit. Sometimes the trading system conducts a simulation to see what the actions may result in. Finally, the system decides on the buy/sell/hold actions, the quantity of order, and the time to trade, it then generates some trading signals. The signals can be directly transmitted to the exchanges using a predefined data format, and trading orders are executed immediately through an API exposed by the exchange without any human intervention. Some investors may like to take a look at what signals the algorithm trading system have generated, and he can initiate the trading action manually or simply ignore the signals. Human intervention is a double blade sword, on one hand it can screen away some unprofitable signals according to the experience of human, on the other hand human being is likely to make mistakes, they cannot trade in a consistent manner, because they will be tired, be over pessimistic or be over optimistic, one's mood will greatly affect the trading. In the author's opinion, if the algorithm trading is properly designed and thoroughly verified, it is better to let the system do the whole thing, from data analysis, to deciding on trading actions, and initiating the execution of trading orders.

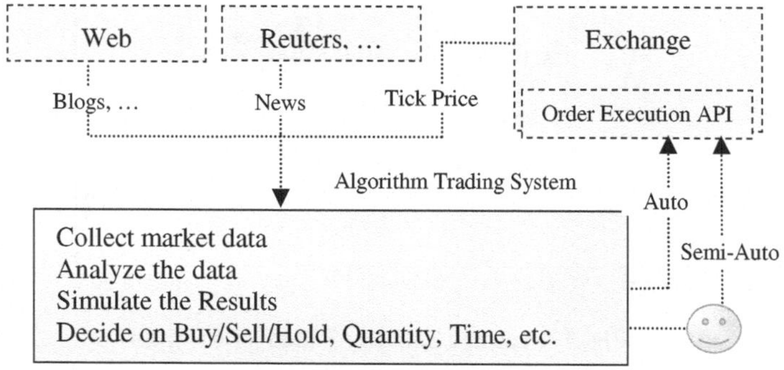


**Fig. 1.** Data Flows of an Algorithm Trading System

## 3 Trading Strategy – The Soul of Algorithm Trading

There are several standard modules in a proprietary algorithm trading system, including trading strategies, order execution, cash management and risk management. Trading strategies are the core of an automated trading system. Complex algorithms [3] are used to analyze data (price data and news data) to capture anomalies in market, to identify profitable patterns, or to detect the strategies of rivals and take advantages of the information. Various techniques are used in trading strategies to extract actionable information from the data, including rules, fuzzy rules, statistical methods, time series analysis, machine learning, as well as text mining. Some of the techniques and methods are presented as follows, their strengths and weaknesses are also discussed.

## 3.1 Technical Analysis and Rules

Some people believe that history will repeat itself, and they use technical analysis to trade. They look at the price chart of some financial asset and search for specific patterns, in the mean time they use some calculated indicators to confirm their judgments. A chart pattern is a unique shape of a chart that can create a trading signal or be a sign of future price movements. People use the patterns to identify price moving trends and trend reversals, and to generate buy and sell signals [4]. *"Head and Shoulders Pattern"* is a famous pattern that is extensile studied and widely used. Figure 2 show a typical Head and Shoulder Pattern: (a) Firstly the price rises to a peak and declines; (b) Then, the price rises above the former peak and declines again; (c) Finally, price rises again, but not to the second peak, and subsequently declines once more. The second peak forms the head, and the first and third peaks are shoulders. When some pattern is recognized by computer programs, some trading signals can be generated at the right moment.

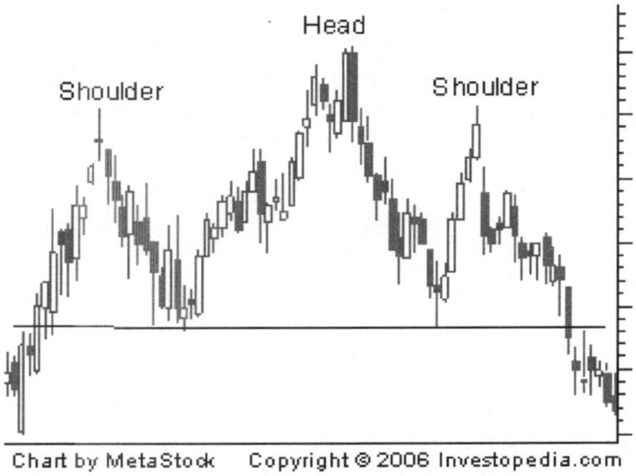


**Fig. 2.** Head and Shoulder Pattern of a Price[4]

With the time elapsing, some indicators can be calculated from price and trading volume. Some of the indicators are oscillation indicators such as RSI, which can be used to identify over bought and over sold situations. RSI takes the values between 0 and 100. People consider readings below 30 to be oversold, and readings above 70 to be overbought. Overbought conditions are usually companied by a pullback, while oversold conditions can be used to generate buying signals (refer to figure 3). This method might be a great way to trade in sideways market situations, but in trending markets one often misses great moves when using the RSI indicator only. Even worse is that traders might loss money because they are sometimes on the opposite side of the trend.

One problem is that the arbitrarily preset values of 30 and 70 as thresholds are not applicable to different financial assets (different stocks). A fixed value combination of 30 as low threshold and 70 as the high threshold can not always make profit, and may result in loss. Some researchers tune the parameters by combine fuzzy logic and technical analysis to find patterns and trends in financial indices [5], these trading strategies is more adaptable and the rules are still human understandable, which will make investor be confident of the system.

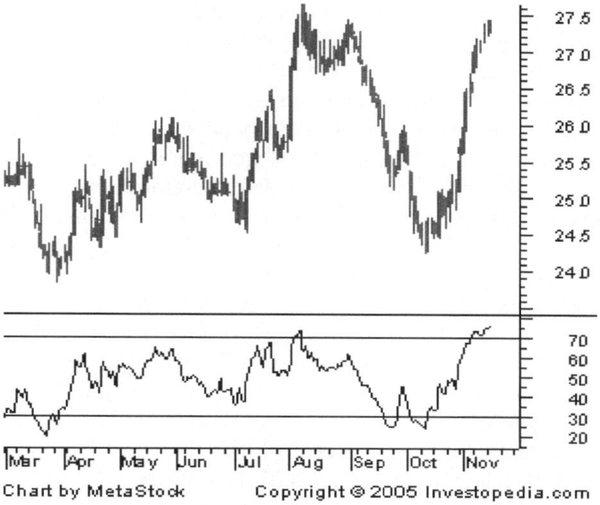


**Fig. 3.** The RSI Indicator and its Usage [6]

Many trading rules are studied, including the moving average cross trading strategy, when a shorter and a longer moving average (of a security's price) cross each other, a bullish or bearish signal is generated depending on the direction of the crossover. After the rules are made publicly accessible, more and more people master basic skills of technical analysis and using of indicators, then how can one win in financial market with simple trading strategies? People resort to more complicated techniques.

### 3.2 Using of Statistics

Statistics mathematics is widely used in time series prediction including algorithm trading. There are two types of time series forecasting including uni-variable and multi-variable. Arbitrage is some form of trading which involve more than one asset or more than one market, making profit by exploiting the mispricing. Statistics mathematics can be used in arbitrage. In statistical arbitrage, there is a statistical mispricing of one or more assets based on the expected value of these assets [7]. Statistical arbitrage conjectures the mispricings of price relationships that are true in expectation in the long run. According to some studies [8], statistical methods are inferior to machine learning methods.

### 3.3 Artificial Intelligence, Machine Learning Based Algorithm Trading

Artificial intelligence, machine learning and data mining techniques have long been used in algorithm trading. Various computational tools such as neural networks, genetic algorithms, support vector machines, case based reasoning and many others are used in prediction of stock markets as well as other financial markets.

Neural network techniques (NNs) [9] was claimed to be more effective in predicting of price of stocks than other techniques. The most valuable advantage of neural network techniques is that NNs can learn the non-linear relationship from the data itself rather than using some function to model the relationship. When enough data is given for model learning, the relationship between input variables and the output variables can be setup, and NNs are tolerant to incomplete and noisy data. However NNs also have some disadvantages, on one hand a neural network is a black box, it is opaque and can not be understood by human being, nor can it be interpreted, on the other hand when the neural network is fed with too much historical data, it may over fit to historical data and cannot yield an acceptable prediction results in the future data set.

Genetic algorithm (GA) is a natural evolution inspired technique. The technique is used to generate useful solutions to optimization and search problems. Genetic algorithm belongs to the larger class of evolutionary algorithm, which generates solutions to optimization problems by using natural evolution-like operations such as inheritance, selection, mutation, and crossover. In recent years, many researchers are interested in improving the performance of NNs by employing genetic algorithm. Genetic algorithms can be used to determine the optimal number of hidden layers, number of processing units, and selecting the most relevant feature sub set [10].

Support vector machine (SVM) is based on the statistical learning theory. Since it was developed in the late of 1970s, SVM has many successful applications in classification and regression task, including financial time series prediction [11]. Compared to NNs, SVM has shown to be very resistant to the over fitting problem, the final model can achieve a low variance in generalization using scenarios.

Many other machine learning techniques including case based reasoning, ensemble learning, reinforced learning etc. also have been put into use in financial applications. Some research works have tried to combine different machine learning techniques to create a more robust trading strategy [12] [13].

### 3.4 Text Mining for Algorithm Trading

In financial market, there are many factors that can affect the movement of a price. These factors can be obtained from news released by Reuters, Bloomberg and other news companies. Some news companies even generate the news in a machine readable format for easy consumption. Algorithm trading practitioners can buy the news data from them, when news data is fed into the trading system continuously, several techniques are employed to extract event information. Some researchers use prior knowledge such as how an event will affect the price movement, to make used of the events. Some other researchers proposed new statistical based news segmentation algorithms to identify the trends according to event information [14].

The internet contains up to date articles on political situation, social conditions, international events, government policies and investor's psychology as well as many other topics, which have something to do with overall macro economy situation and even some specific financial asset's price. Sentiment analysis, one of text mining techniques, can extract people sentiment buried in the text, happy or sad, optimistic or pessimistic. Sentiment of investors has impact on moving trend of some financial asset price, and can be use in algorithm trading for profit [15].

# 4 Making Use of the Big Data - Next Generation of Algorithm Trading

Big data is mentioned now and then in recent days. Big data is often characterized by four *Vs* [16]: (1) *Volume*, the volume of data that we collect for algorithm trading rises in a accelerated pace. Image that we are trying to capture arbitrage opportunities between an arbitrary pair of assets in a market, we should continuously collect price data of thousand of assets, if the data is in a finest granularity - *TICK* format (in high frequent trading, there may be hundreds of *TICKs* in one second), and then we are interested in cross market arbitrage, the data to be analyzed will be a big one. (2) *Velocity*: not only you are interested in algorithm trading, other people are interested in it too. When the trading are performed by machine, the speed of trading and the volume of trading will reach a level that is unseen before, which means that we should act on the newly generated data timely, otherwise we can not keep up with the market and lose money. (3) *Variety*: not only structured data (price data), but also semi-structured (machine readable news) and unstructured data (documents on the web) contain valuable information for success of algorithm trading. Unstructured data is estimated to account for 70% to 85% of the total data. Next generation of algorithm trading will be more data-driven.

In the big data era, algorithm trading practitioners face several challenges. (1) The first one is how to use large volume of *unstructured data*. Some researchers have combined technical analysis with sentiment analysis for stock price prediction. The together analysis of structured data (price, indicators) with unstructured data has shown great potential of making profit [17]. In front of a trading desk in London, Pal Hawtin monitors 340 million Twitter posts every day to assess the collective mood of the populace and generate a global sentiment score, which ranges from 1 to 50. The score is based on how pessimistic or how optimistic people seem to be according to their conversations. Then trading of millions of dollars of financial assets is based on the score - when every one is happy, buy; when anxiety rises, sell short. Hawtin is reported to achieve a gain of more than 7 percent in the first quarter of the year of 2012[18]. Large banks and funds in Wall Street also have been using this technique to make profit according the report of [19]. Analyzing hundreds of millions of posts is a big data problem. (2) Another challenge is that how we can process the *huge volume of data timely* which may span a long time period. It is likely that we will get new insights that we don't know before from the big data, which would be exciting. The new insights can then be used in trading strategies. When the algorithm trading system is overwhelmed with huge volume of data, we need a *parallel data processing platform* that can scale out easily to *process it timely*. Sampling is not enough, throwing away some data implies throwing away some valuable information which we should have exploited to make profit, and we need to perform analytics at a very granular level.

Traditional RDBMS alone can not deal with big data well because RDBMS are limited in scalability. MapReduce rises in recent years as a de-facto tool for big data processing [20]. Besides simple SQL summation, researches have migrated complex algorithms onto the MapReduce platform including OLAP, data mining, machine learning, information retrieval, multimedia data processing, science data processing, and many more algorithms can be run on the MapReduce platform. MapReduce and complex analytics algorithms running on it provide the infrastructure for data analysis task in algorithm trading.

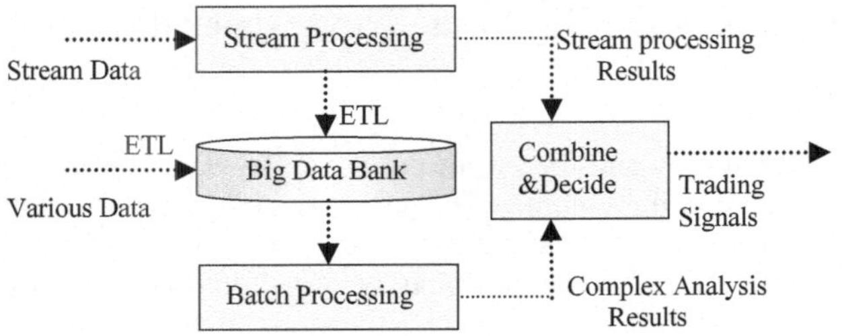


**Fig. 4.** Tackle the Big Data for Algorithm Trading

The next generation of algorithm trading system can be based on various data management techniques including RDBMS and MapReduce for high performance processing of large volume of various data (refer to figure 4). The stream data, including *TICK* price data and news feeding, should be processed timely; documents crawled from the web and extracted from others sources can be stored in the big data bank for bath processing. The big data bank is usually built upon HDFS (Hadoop Distributed File System). On-the-fly stream data processing results are combined with the results of batch processing of big data to decide on more profitable and less risky trading actions.

## 5 Conclusions

More and more complicated techniques are put into use in algorithm trading with an expectation that it can make profit and stop loss. Next generation of algorithm trading will be more data driven. Valuable information could be computed from the big data for the success of algorithm trading.

**Acknowledgements.** The work is supported by National Natural Science Foundation of China under Grant no.61170013.

## References

1. Algorithm Trading Wiki (2012), `http://en.wikipedia.org/wiki/Algorithmic_trading`
2. Renaissance Technologies Wiki (2012), `http://en.wikipedia.org/wiki/Renaissance_Technologies`
3. Nuti, G., Mirghaemi, M., Treleaven, P., Yingsaeree, C.: Algorithm Trading. Computer 44(11), 61–69 (2011)
4. Janssen, C., Langager, C., Murphy, C.: Technical Analysis: Chart Patterns (2012), `http://www.investopedia.com/university/technical/techanalysis8.asp#axzz1wJk7C3hk`
5. Cheung, W.M., Kaymak, U.: A fuzzy logic based trading system. In: Proceedings of the Third European Symposium on Nature-inspired Smart Information Systems (2007)

6. Relative Strength Index - RSI (2012), http://www.investopedia.com/terms/r/rsi.asp#axzz1wYkK714x
7. Statistical arbitrage Wiki (2012), http://en.wikipedia.org/wiki/Statistical_arbitrage
8. Yoo, P.D., Kim, M.H., Jan, T.: Financial Forecasting: Advanced Machine Learning Techniques in Stock Market Analysis. In: 9th Intl. Multi-topic Conference, pp. 1–7 (2005)
9. Wang, C.-P., Lin, S.-H., Huang, H.-H., Wu, P.-C.: Using neural network for forecasting TXO price under different volatility models. Expert Systems with Applications 39(5), 5025–5032 (2012)
10. Chan Lanz, C.W.J., Wing-Keung, W.: Automated Trading with Genetic-Algorithm Neural-Network Risk Cybernetics: An Application on FX Markets (2011), SSRN http://ssrn.com/abstract=1687763
11. Ullrich, C., Seese, D., Chalup, S.: Foreign Exchange Trading with Support Vector Machines. In: Advances in Data Analysis: Studies in Classification, Data Analysis, and Knowledge Organization, Part VII, pp. 539–546 (2007)
12. Kwon, Y.-K., Moon, B.-R.: A Hybrid Neuro genetic Approach for Stock Forecasting. IEEE Transactions on Neural Networks 18(3), 851–864 (2007)
13. Kwon, Y.-K., Sun, H.-D.: A Hybrid System Integrating a Piecewise Linear Representation and a Neural Network for Stocl Prediction. In: IFOST 2011, pp. 796–799 (2011)
14. Fung, G.P.C., Yu, J.X., Lam, W.: News Sensitive Stock Trend Prediction. In: Chen, M.-S., Yu, P.S., Liu, B. (eds.) PAKDD 2002. LNCS (LNAI), vol. 2336, pp. 481–493. Springer, Heidelberg (2002)
15. Byrne, C.: Sentiment analysis gives algorithmic trading an edge (2011), http://radar.oreilly.com/2011/05/sentiment-analysis-finance.html
16. Hopkins, B.: Blogging From the IBM Big Data Symposium - Big Is More Than Just Big (2011), http://blogs.forrester.com/brian_hopkins/11-05-13-blogging_from_the_ibm_big_data_symposium_big_is_more_than_just_big
17. Deng, S., Mitsubuchi, T., Shioda, K., Shimada, T., Sakurai, A.: Combining Technical Analysis with Sentiment Analysis for Stock Price Prediction. In: 9th International Conference on Dependable, Autonomic and Secure Computing, pp. 800–807 (2011)
18. Cha, A.E.: Big data' from social media, elsewhere online redefines trend-watching (2012), http://www.washingtonpost.com/business/economy/big-data-from-social-media-elsewhere-online-take-trend-watching-to-new-level/2012/06/06/gJQArWWpJV_story.html
19. Schmerken, I.: Wall Street Conquers Big Data on the Web (2011), http://www.wallstreetandtech.com/data-management/232200678
20. Lee, K.-H., Lee, Y.-J., Choi, H., Chung, Y.D., Moon, B.: Parallel data processing with MapReduce: a survey. SIGMOD Record 40(4), 11–20 (2011)

# Dynamic Causality Analysis on Default Mode Network

Rongrong Cao, Dongjuan Zhu, Qinqin Huang
Xunheng Wang, and Zongcai Ruan

Research Center for Learning Science, Southeast University,
Nanjing 210096, China
rwbdaisy@sina.com, 472762780@qq.com,
seu_04004414@gmail.com, {shwang,rzc.rcls}@seu.edu.cn

**Abstract.** In most studies of functional connectivity, the default mode network (DMN) is seen as a node and is analyzed with other nodes, or just examines the static interactions among the components of DMN. Few studies have analyzed the dynamic interactions among the components of DMN. In order to evaluate the dynamic connectivity within the components, Conditional-Granger causality analysis (CGCA) based on sliding window is applied to the components of DMN extracted by kernel independent component analysis (KICA). The results suggest that the connectivity of DMN changed significantly during the process of entering the resting-state. Especially, the left inferior parietal lobes (lIPL) are found significant activity and affect other regions at the beginning of entering the resting-state. Specifically, under the resting-state, lIPL's influence on other regions is suppressed by the medial prefrontal cortex (mPFC).

**Keywords:** default mode network (DMN), resting-state, kernel independent components analysis (KICA), Conditional-Granger causality analysis (CGCA).

## 1 Introduction

Functional magnetic resonance imaging (fMRI) based on blood oxygen level dependent (BOLD) signal is widely used in the area of imaging for its noninvasive, nonradioactive and high resolution[1]. Resting-state fMRI refers as no cognitive tasks during the fMRI scan, in other words, subjects should relax their bodies, keep their eyes closed and do not think of anything intentionally[2]. Default mode network (DMN) promoted in 2001[3] is the first resting-state network (RSN) detected, so DMN can better characterize the features of resting-state. For this reason, it has been a research focus of neuroscience since it was discovered. Previous studies found that DMN was associated with episodic memory processing [4]. Steven [5] made researches on vegetative patients. The results showed that the deeper the level of coma, the deeper the damage of DMN. In addition, some studies also found that the connectivity of DMN during the process of cognitive tasks was different between adults and adolescents and adults' connectivity is much weaker [6]. Based on those theories, we can predict the causal relationship among the regions of DMN must be special and important.

J. Lei et al. (Eds.): AICI 2012, LNAI 7530, pp. 42–49, 2012.

Independent component analysis (ICA) based on information theory is a classic tool to solve the blind source separation (BSS). It's popular in many fields, such as image processing and face recognition. However, common ICA has congenital deficiencies when it is applied to nonlinear data. Kernel ICA (KICA) [7] utilized the control function based on kernel and it's more flexible and robust in the field of nonlinear data. Here, KICA is employed to extract the regions of DMN. The results confirm that KICA is an effective method to extract the DMN.

To analyze the causality of fMRI data, granger causality analysis (GCA) [8] is brought to investigate the causal relationship between two regions or two voxels. GCA has been a hot topic in the field of neural network, for it is based on the measured signal and can be carried out without prior knowledge. While, GCA has a serious disadvantage that is it's restricted to two variables. In reality, the number of study targets is three or more, the causal relationship among them maybe directly or indirectly, but GCA can't resolve this situation. Fortunately, Conditional-Granger causality analysis (CGCA) [9] has been proposed to analyze multivariate causal relationships. Recently, RSNs are found to have interactions among themselves through CGCA, specially, self-referential and DMN play important roles in the causality among the RSNs [10] .

In this paper, in order to find the dynamic causality among the components of the DMN, KICA and CGCA based on sliding windows are applied to analyze the resting-state fMRI data. Our study found a particular dynamic process of the DMN, which has shown that CGCA based on sliding windows could be a potential tool to investigate dynamic causality among neural networks.

## 2 Materials and Methods

We will give an outline of two algorithms to analysis DMN. The first algorithm aims at reducing the time dimension of the fMRI data and extracting the DMN components; the second algorithm is aiming at evaluate the dynamic connectivity within the components.

### 2.1 PCA and KICA

For the purpose of lowering the computational complexity, principal component analysis (PCA) is utilized here to reduce the noise and temporal dimension before KICA performed. The significance of KICA is that it employs two control functions based on kernel, they are kernel canonical correlation analysis (KCCA) [7-11] and kernel generalized variance (KGV) respectively. The control functions above mentioned project signals into higher dimension space firstly and analyze the interdependence between the signals. Thus, nonlinear relationship can also be detected, just like the linear relationship. In this paper, KICA based on Gaussian kernel and KGV is employed. Fig.1 is the DMN extracted by KICA, its time course is presented in Fig.2 Through a large number of experiments, we verified that KICA can extract the DMN accurately and handle the resting-state fMRI data effectively.

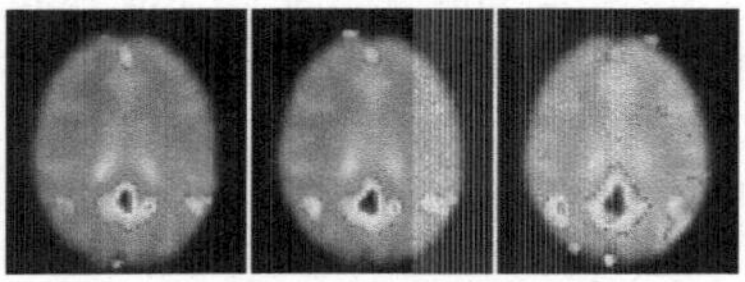

**Fig. 1.** DMN extracted by KICA

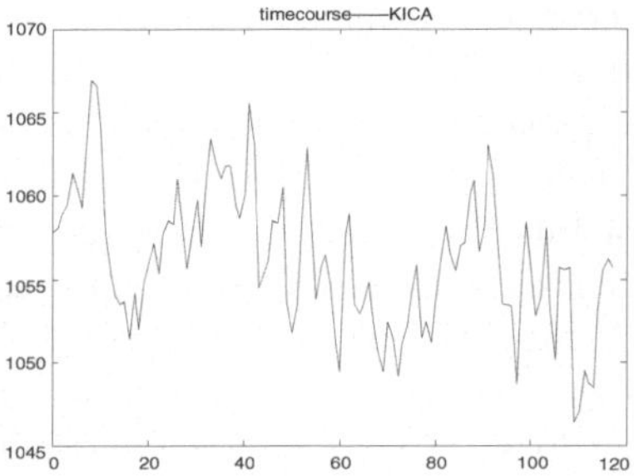


**Fig. 2.** Time course of DMN

## 2.2 Conditional-Granger Causality Analysis

After the DMN is obtained through KICA, CGCA is applied to detect the dynamic causality among the regions of the DMN. CGCA was proposed based on the theory of GCA, and it is suitable for multivariable causality analysis. Considering three temporal variables: $x(t)$, $y(t)$, $z(t)$. Now, we consider the causal relationship between $x(t)$ and $y(t)$, in the case that $z(t)$ is existing. Its binary regression model is presented as follows:

$$x(t) = \sum_{i=1}^{p} a_{1i} x(t-i) + \sum_{i=1}^{p} b_{1i} z(t-i) + \xi_{1t} \tag{1}$$

$$z(t) = \sum_{i=1}^{p} c_{1i} x(t-i) + \sum_{i=1}^{p} d_{1i} z(t-i) + \varepsilon_{1t} \tag{2}$$

$\xi_{1t}$ and $\varepsilon_{1t}$ are estimation errors, and the noise covariance matrix is defined as:

$$\mathbf{\Sigma}_1 = \begin{pmatrix} \mathrm{var}(\xi_{1t}) & \mathrm{cov}(\xi_{1t}, \varepsilon_{1t}) \\ \mathrm{cov}(\varepsilon_{1t}, \xi_{1t}) & \mathrm{var}(\varepsilon_{1t}) \end{pmatrix} \tag{3}$$

On the basis of binary regression model, we consider the multivariable regression model, $y(t)$ is involved in this model. The variables are described as:

$$x(t) = \sum_{i=1}^{p} a_{2i} x(t-i) + \sum_{i=1}^{p} b_{2i} y(t-i) + \sum_{i=1}^{p} c_{2i} z(t-i) + \xi_{2t} \tag{4}$$

$$y(t) = \sum_{i=1}^{p} d_{2i} x(t-i) + \sum_{i=1}^{p} e_{2i} y(t-i) + \sum_{i=1}^{p} f_{2i} z(t-i) + \eta_{2t} \tag{5}$$

$$z(t) = \sum_{i=1}^{p} g_{2i} x(t-i) + \sum_{i=1}^{p} h_{2i} y(t-i) + \sum_{i=1}^{p} m_{2i} z(t-i) + \varepsilon_{2t} \tag{6}$$

$\xi_{2t}$, $\eta_{2t}$ and $\varepsilon_{2t}$ are estimation errors. The new noise covariance matrix can be gained as:

$$\mathbf{\Sigma}_2 = \begin{pmatrix} \mathrm{var}(\xi_{2t}) & \mathrm{cov}(\xi_{2t},\eta_{2t}) & \mathrm{cov}(\xi_{2t},\varepsilon_{2t}) \\ \mathrm{cov}(\eta_{2t},\xi_{2t}) & \mathrm{var}(\eta_{2t}) & \mathrm{cov}(\eta_{2t},\varepsilon_{2t}) \\ \mathrm{cov}(\varepsilon_{2t},\xi_{2t}) & \mathrm{cov}(\varepsilon_{2t},\eta_{2t}) & \mathrm{var}(\varepsilon_{2t}) \end{pmatrix} \tag{7}$$

Where $p$ is the model order of the regression model. In our study, the value is 2 or 3 in general. At the same time, $y(t)$ is the reason of $x(t)$ which can be judged from the estimation errors from the two sets of equations. If $\xi_{1t}$ is larger than $\xi_{2t}$, it means that (4) can estimate $x(t)$ better than (1), so $y(t)$ is the granger reason of $x(t)$. On the contrary, If $\xi_{1t}$ is the same as $\xi_{2t}$, $y(t)$ has no direct influence on $x(t)$ or just indirect influence through $z(t)$.

The DMN has four regions: medial prefrontal cortex (mPFC), posterior cingulate cortex (PCC), left inferior parietal lobes (lIPL) and right inferior parietal lobes (rIPL). First, we extract the time courses of the region of interest (ROIs) and set six sliding windows, each window includes 100 time points, and the step size is 15 time points. It is shown in Fig.3.

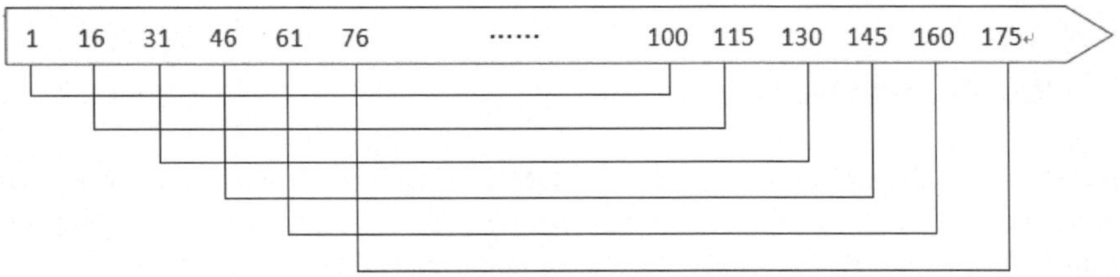


**Fig. 3.** Schematic diagram of sliding windows

## 2.3 Materials

Resting-state fMRI data was downloaded from the NYU CSC Test-Retest resource (http://www.nitrc.org/projects/nyu_trt/). Twenty-five (ten males, mean age 29.44+/-8.47) right-handed subjects took part in the experiment. All the participants are healthy and native English-speaking. During the scanning process, they were required to keep their eyes open and remain relaxed.

Images were acquired on a Siemens Allegra 3.0 Tesla scanner. Functional image is consisted of 197 continuous EPI functional volumes (TR=2000ms; TE=25ms; flip angle=90°, 39 slices, matrix=64×64; FOV=192mm; acquisition voxel size= 3×3×3 mm3). In our study, the first several volumes were discarded. In addition, a high-resolution T1-weighted volume using magnetization prepared gradient echo sequence (TR=2500ms; TE=4.35ms; T1=900ms; flip angle =8; 176 slices, FOV=256mm) was scanned for spatial normalization and localization.

## 2.4 Data Preprocessing

SPM8 (www.fil.ion.ucl.ac.uk/spm) was used to preprocess the data here. Main tasks of preprocess were included as follows: first, slice timing to correct difference in image acquisition time between slices; second, realign all EPI functional volumes to the mean volume of the same scan to reduce head motion correction; third, spatial normalization to transform all T1 weighted volumes to MNI152 standard brain space with spatial resolution of 3mm×3mm×4mm;forth, spatial registration to map all EPI functional volumes to corresponding normalized individual T1 weighted image; fifth, remove linear trends; sixth, remove high-frequency(>0.1 Hz) components; at last, spatial smoothing with a 8mm FWHM Gaussian kernel.

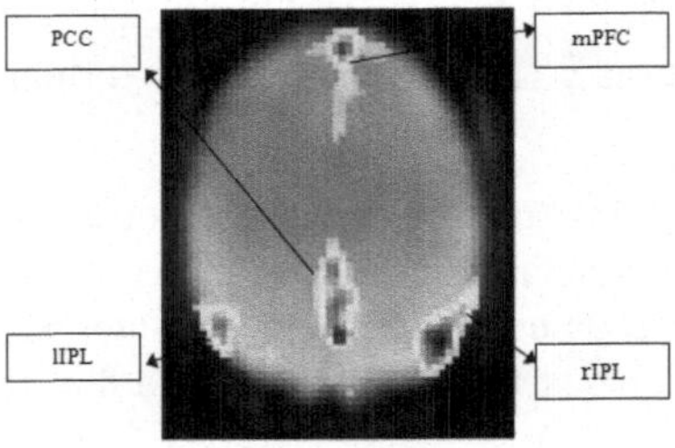


**Fig. 4.** ROIs of the DMN

## 3 Result and Disscussion

ROIs of the DMN are shown Fig.4. The DMN is associated with mind wandering, episodic memory and environmental monitoring [12-13]. These ROIs have important causality. In this study, we analyzed the dynamic causality among the ROIs for the first time, the results are presented in Fig.5~Fig.10 and detail information is described in Tabel.1~Tabel.6.

In the above figures, the red arrows indicate unidirectional causality, and the green arrows indicate bidirectional causality. From the data results, we found the ROIs have significant dynamic process. It can be concluded as follows: lIPL play an important role in the beginning of entering the resting-state. In window 1 and window 2, lIPL controlled the whole DMN and affected other ROIs. Generally speaking, lIPL and rIPL are similar in a way, while, results have shown great distinction between them. The lIPL is considered to have close relationship with language motion memory and have functional connectivity with left language memory system. On the other hand, rIPL would be activated when the brain is performing arithmetic operations [14]. So, the results of window 1 and window 2 indicated the subjects haven't entered resting-state completely in the beginning of the experiment. Maybe, he was considering what the doctor had said and how to complete the experiment.

The mPFC is the output of the network and PCC is the input of the network under the resting-state. Starting from the third window, as the subject slowly entering the resting-state, mPFC began to affect other ROIs, PCC and rIPL are the main inputs. This result is similar to the existing results [15]. It has shown that CGCA is an effective method to investigate the dynamic causality of the DMN.

In this study, another meaningful result has been gained. When the subject entered the resting-state completely, the coefficients of lIPL on PCC and rIPL are suppressed by mPFC. That is to say, mPFC's activity not only affected the activity of lIPL, also affected the signal transmission between lIPL and other two ROIs. It can be seen clearly in Fig.11. The special relationship confirmed that the core part of the DMN would inhibit the execution of cognitive tasks and make the brain stay in resting-state as possible [16].

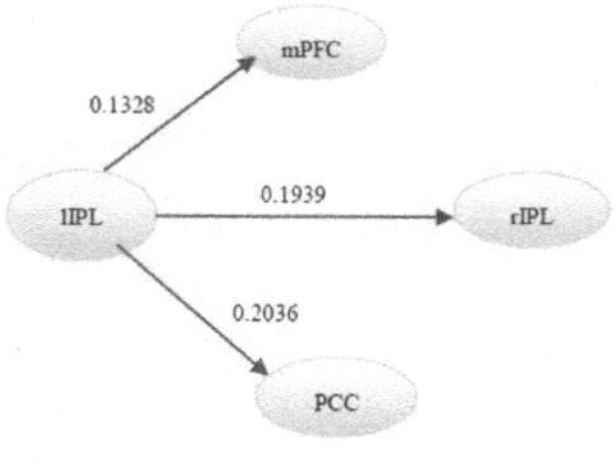


**Fig. 5.** Window 1

**Table 1.** Causal coefficients of window1

| | mPFC | PCC | lIPL | rIPL |
|---|---|---|---|---|
| mPFC | NAN | 0.0242 | 0.0011 | 0.0152 |
| PCC | 0.0329 | NAN | 0.0004 | 0.0370 |
| lIPL | 0.1328 | 0.2036 | NAN | 0.1939 |
| rIPL | 0.0043 | 0.0581 | 0.0222 | NAN |

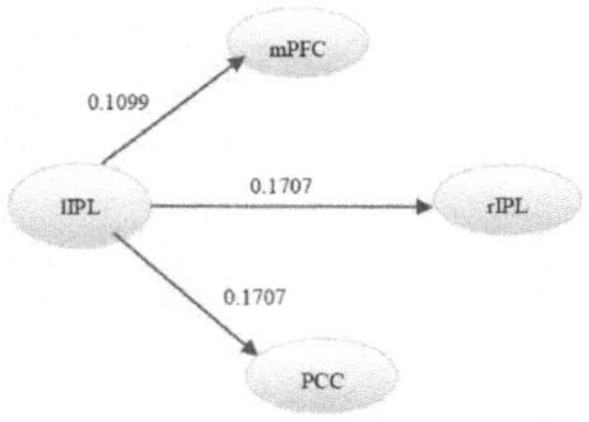


**Fig. 6.** Window 2

**Table 2.** Causal coefficients of window2

| | mPFC | PCC | lIPL | rIPL |
|---|---|---|---|---|
| mPFC | NAN | 0.0326 | 0.0037 | 0.0079 |
| PCC | 0.0327 | NAN | 0.0018 | 0.0542 |
| lIPL | 0.1099 | 0.1758 | NAN | 0.1707 |
| rIPL | 0.0128 | 0.0511 | 0.0141 | NAN |

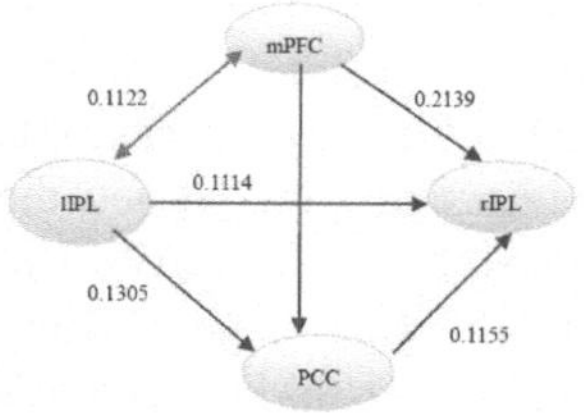


**Fig. 7.** Window 3

**Table 3.** Causal coefficients of window3

| | mPFC | PCC | lIPL | rIPL |
|---|---|---|---|---|
| mPFC | NAN | 0.1122 | 0.1967 | 0.2139 |
| PCC | 0.0370 | NAN | 0.0817 | 0.1155 |
| lIPL | 0.1091 | 0.1305 | NAN | 0.1114 |
| rIPL | 0.0122 | 0.0315 | 0.0005 | NAN |

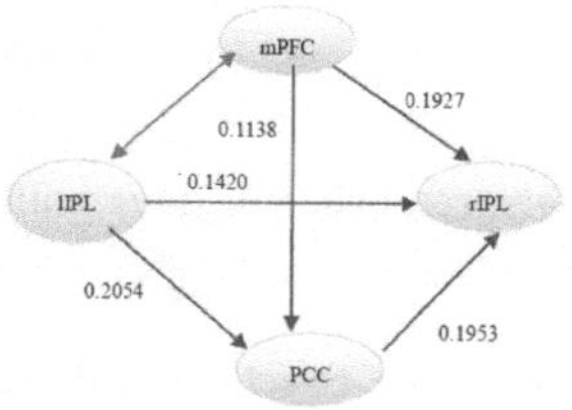


**Fig. 8.** Window4

**Table 4.** Causal coefficients of window4

| | mPFC | PCC | lIPL | rIPL |
|---|---|---|---|---|
| mPFC | NAN | 0.0326 | 0.0037 | 0.0079 |
| PCC | 0.0327 | NAN | 0.0018 | 0.0542 |
| lIPL | 0.1099 | 0.1758 | NAN | 0.1707 |
| rIPL | 0.0128 | 0.0511 | 0.0141 | NAN |

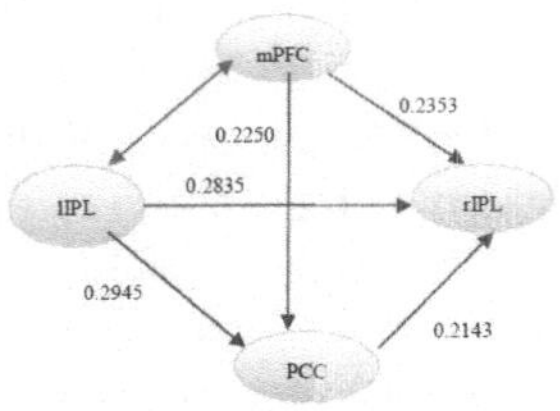


**Fig. 9.** Window 5

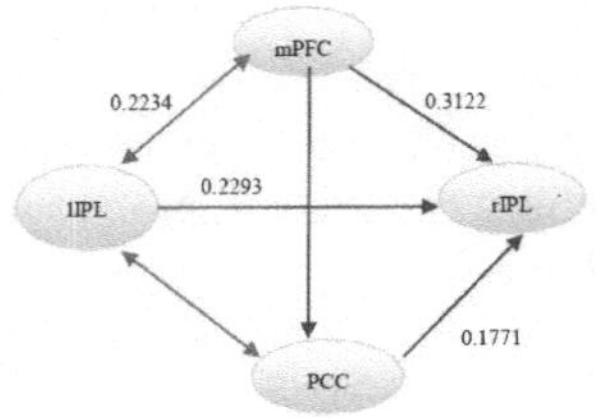


**Fig. 10.** Window6

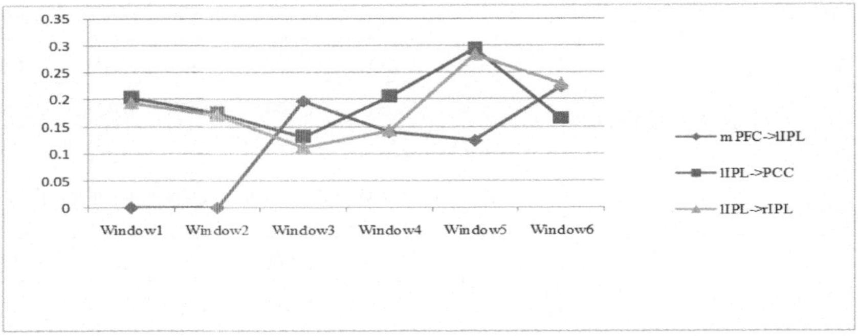


**Fig. 11.** The influence of lIPL

# 4 Conclusion

In this research, we focused on the dynamic causality within ROIs of the DMN. PCA and KICA were used to extract the DMN, and then set six sliding windows on the time course of the ROIs, CGCA was applied to detected the dynamic causality of the ROIs. The results confirmed mPFC and PCC were the signal output and the signal input respectively, which was found in previous studies. In addition, lIPL was found significant activity in the beginning period of entering the resting-state and its influence coefficients on PCC and rIPL were suppressed by mPFC. This discovery would provide important evidence to the study of revealing the dynamic causal relations among the ROIs of the DMN.

# References

1. Ogawa, S., Lee, T.M., Kay, A.R., et al.: Brain magnetic resonance imaging with contrast dependent on blood oxygenation. Proceedings of the National Academy of Sciences 87(24), 9868–9872 (1990)
2. Xunheng, W., Zhenyu, Z., Yun, J., et al.: Investigating the structure of default mode network with social network analysis. In: Human Brain Mapping, San Francisco, USA, vol. 101 (2009)

3. Raichle, M.E., Macleod, A.M., Snyder, A.Z., et al.: A default mode of brain function. Proceedings of the National Academy of Sciences 98(2), 676–682 (2001)
4. Greicius, M.D., Srivastava, G., Reiss, A.L., et al.: Default-mode network activity distinguishes Alzheimer's disease from healthy aging: Evidence from functional MRI. Proceedings of the National Academy of Sciences 101(13), 4637–4642 (2004)
5. Laureys, S.: The neural correlate of (un)awareness: lessons from the vegetative state. Trends in Cognitive Sciences 9(12), 556–559 (2005)
6. Fransson, P., Aden, U., Blennow, M., et al.: The Functional Architecture of the Infant Brain as Revealed by Resting-State fMRI. Cerebral Cortex 21(1), 145–154 (2011)
7. Bach, F.R., Jordan, M.I.: Kernel Independent Component Analysis. Journal of Machine Learning Research 3(3), 1–48 (2002)
8. Granger, C.W.J.: Investigating causal relations by econometric models and cross-spectral methods. Econometrica 37(3), 424–438 (1969)
9. Geweke, J.F.: Measures of conditional linear dependence and feedback between time series. Journal of the American Statistical Association 79(388), 907–915 (1984)
10. Liao, W., Mantini, D., Zhang, Z., et al.: Evaluating the effective connectivity of resting state networks using conditional Granger causality. Biological Cybernetics 102(1), 57–69 (2010)
11. Akaho, S.: A kernel method for canonical correlation analysis. International Meeting of Psychometric Society (IMPS 2001) (2001)
12. Gusnard, D.A., Akbudak, E., Shulman, G.L., et al.: Medial prefrontal cortex and self-referential mental activit: relation to a default mode of brain function. Proceedings of the Nattional Academy of Sciences 98(7), 4259–4264 (2001)
13. Buckner, R.L., Andrews-Hanna, J.R., Schacter, D.: The brain's default network: anatomy, function, and relevance to disease. Annals of the New York Academy of Sciences 1124, 1–38 (2008)
14. Chochon, F., Cohen, L., Moortele, P.F., et al.: Differential Contributions of the Left and Right Inferior Parietal Lobules to Number Processing. Journal of Cognitive Neuroscience 11(6), 617–630 (1999)
15. Uddin, L.Q., Clare Kelly, A.M., Biswal, B.B., et al.: Functional connectivity of Default Mode Network Components: Correlation, Anticorrelation, and Causality. Human Brain Mapping 30(2), 625–637 (2009)
16. Greicius, M.D., Menon, V.: Default-mode activity during a passive sensory task: Uncoupled from deactivation but impacting activation. Journal of Cognitive Neuroscience 16(9), 1484–1492 (2004)

# The Canny Edge Detection and Its Improvement

Xiaoju Ma, Bo Li* ,Ying Zhang, and Ming Yan

Kunming University of Science and Technology, Kunming, 650505, China
lbly9177@163.com

**Abstract.** To solve the problem of the traditional Canny edge detection operator has the weaknesses in excessive smoothing image and adaptability, and improved the parameter Sigma and the method to obtain high threshold. We did experiments with gray image of two cases with noise and without noise. The experimental results show that the improved Canny edge detection operators can balance eliminating noise from getting more edge information, which has the well continuity of the edge detection, and can detect the edge detail of the image. According to the image adaptive calculation, the improved algorithm has the advantage of low computational complexity, less calculation time.

**Keywords:** Canny edge detection, Edge gradient, Edge and texture.

## 1 Introduction

Edge information transfers the important information of the image, which is an important factor to affect the visual effect. Gray values adaptive to calculate low threshold edge detection. Edge detection is an important part of the image processing, by which we can acquire useful information of the target image and provide the basis for the target image processing. Therefore, the study of edge detection operators has always been very active, and also many practical edge detection methods [1] have been proposed.

Edge detection operators are mainly divided into the differential method, the optimal operator method, and the method based on mathematical morphology. First order differential method of Robert operator, Prewitt operator, Sobel operator, Kirsch operator, etc. Second order differential operator such as Laplace differential operator; the best operator method has the Laplacian of Gaussian, LoG and Canny operator.

The traditional Canny edge detection operator is still insufficient in excessive smoothing image and adaptability [2]. This paper consider and improve it from both local information and global information: (1) the original operator using a fixed parameter (standard deviation) of the Gauss filter, improved for Gaussian filter according to the local image variance and minimum variance adaptive parameter; (2) The edge detection which the original operator to set the gradient threshold, improved for the image based on the mean variance and average gray values adaptive to calculate low threshold edge detection.

---

* Corresponding author.

J. Lei et al. (Eds.): AICI 2012, LNAI 7530, pp. 50–58, 2012.

## 2 Traditional Canny Edge Detection Operator

The steps of the traditional Canny edge detection operator as follows:

Step 1: Using the Gauss function $G(x, y, \sigma)$ of smoothing filtering to the image f(x, y), Gauss function expression as formula (1):

$$G(x, y, \sigma) = \frac{1}{2\pi\sigma^2} \exp(-\frac{x^2 + y^2}{2\sigma^2}) \tag{1}$$

In practical application often use $\sigma = 1.4$ as Gauss template.

$$\frac{1}{115}\begin{bmatrix} 2 & 4 & 5 & 4 & 2 \\ 4 & 9 & 12 & 9 & 4 \\ 5 & 12 & 15 & 12 & 5 \\ 4 & 9 & 12 & 9 & 4 \\ 2 & 4 & 5 & 4 & 2 \end{bmatrix}$$

To compute the gradient of the point $(x, y)$ by formula (2) and (3):

$$\nabla g(x, y) = \nabla(G(x, y, \sigma) * f(x, y)) = \nabla(G(x, y, \sigma) * f(x, y) \tag{2}$$

$$E_x = \frac{\partial G(x, y, \sigma)}{\partial x} * f(x, y) \quad E_y = \frac{\partial G(x, y, \sigma)}{\partial y} * f(x, y) \tag{3}$$

Calculate the gradient amplitude $|\nabla g(x, y)|$ which reflects the size of the edge intensity .The calculation method of $|\nabla g(x, y)|$ which refers to gradient amplitude of Pixel $(x, y)$ are as formula (4) shown, the direction of the gradient is perpendicular to the edge direction.

$$|\nabla g(x, y)| = \sqrt{E_x^2 + E_y^2} \tag{4}$$

$$\theta(x, y) = \arctan(\frac{E_y}{E_x}) \tag{5}$$

Step 2: Doing gradient with "non-maxima suppression". In order to refine and enhance the gradient amplitude of roof, to all pixels, we reduce all that isn't the roof type of peak gradient amplitude in the gradient direction.

Gradient direction angle calculated from (5) show that each pixel of the image has four possible directions when connected with the adjacent vertex: 0 degrees (horizontal direction), 45 degrees (diagonal), 90 degrees (vertical direction), 135 degrees (negative diagonal). The edge direction must be the closest one in four kinds. Direction angle is classified to the four angles:

$0°$ : $0°$~$22.5°$, $157.5°$~$180°$ ; $45°$ : $22.5°$~$67.5°$ ;
$90°$ : $67.5°$~$112.5°$ ; $135°$ : $112.5°$~$157.5°$

For all the edge points in the image, if the gradient magnitude in the above direction angle direction is less than or equal to the amplitude between the two adjacent points along the line gradient, the value of $|\nabla g(x, y)|$ is assigned zero; If not, the value of $|\nabla g(x, y)|$ is not changed. After the non-maxima suppression, refined the previous wide ridge region, and made it become only one pixel wide. In the process of non-maxima suppression, did not change the roof height.

**Step 3:** Dual-threshold segmentation and edge stitching. Even though with non-maxima suppression, pseudo edge still exists. Namely, using a single threshold processing, it is difficult to choose the appropriate threshold. The solution is using dual-threshold, which combined the high threshold $T_h$ with the low threshold $T_l$ , usually the $T_l : T_h$ ratio is $2:3$.

## 3 The Improvement of Traditional Canny Edge Detection Operator

Traditional Canny edge detection operator has two shortcomings of over-smooth images and easy loss of edge details [9]. In recent years, three kinds of adaptive improvement method have been appeared to improve the insufficiency of the traditional Canny edge detection operator. The first one is making improvement for excessive smooth caused by single Gauss filter parameters; the second is making improvement for false edges or loss local edge caused by man-made setting level threshold; the third improvement is to integrate the first with the second.

This paper is based on the complexity and time consideration, on the basis of literature [1] and [2], is the third category to improve the traditional Canny edge detection operator. Specific improvements are: (1) on the basis of the iterative method to obtain parameter Sigma in literature [2], in order to achieve easy, we computed the value of Sigma utilizing the image of the local characteristics and global characteristic adaptive. (2) According to the image of the overall feature to set high and low threshold, in no need of extra hardware equipment than in the case of literature [1] of the lower computational complexity. The detailed description of these two aspects as follows.

**(1)The Improved Access Method about the Parameter Sigma of Gauss Filter**

When the Gauss filter start to smooth filter, the bigger the parameter of Sigma ( standard deviation) is , the higher frequency signal suppression effect, when avoid the false edge, this operator can make the edge blurred [3];The Sigma smaller, kept the edge detail information, but it will be diminished the ability of noise. The traditional Canny edge detection operator, selects the parameter Sigma man-made to the edge of the image region which does not have a consistent to fit, that may bring Gauss filter excessive image smoothing. Then, according to the local characteristics of image edge and the global image characteristic how to compute the value of Sigma, the method which the present paper will discuss the improvement of the adaptive selection of the parameter Sigma.

Firstly, we discuss the impact of different image region on the value of Sigma. Assuming that the original image is $I(x, y)$, the edge region is $I_e(x, y)$, the flat region is $I_s(x, y)$, we discuss this both two cases of noise-free and noisy.

①In no noise conditions. If the flat region $I_s(x, y)$ of a $N \times N$ window, every pixel gray approximately equal, variance is approximately zero, then do not need a large degree of smoothing, so we only need to take a very small value to the Gauss filter can be used; if the edge region $I_s(x, y)$ of a $N \times N$ window with the edge points, then the pixels within the window with the edge pixels mean the value of points will have bigger difference, variance will be larger, then the Sigma value only slightly larger than the flat region value.

②In noisy conditions. In flat and the edge of the region, the value of the pixels in the $N \times N$ window of pixels within the window mean spreads are larger, is also a great variance, so it is necessary to take a large value of Sigma.

From the above analysis, the value of Sigma is proportional to the window of the variance. According to the actual situation, not only to obtain the corresponding .The value of Sigma will be smoothed image but also preserve the image edge. Considering the minimum variance reflects the image of the overall characteristic, the $N \times N$ window variance reflects local characteristics, so the minimum variance and local variance as the parameter Sigma measurement standard. It can give attention to both the window within the local information and global information. In conclusion, the parameter of Sigma acquiring method as showed in formula (6):

$$\sigma = V/V_{\min} \tag{6}$$

The variance is

$$V = \sum_{n=1}^{N \times N} (I(x, y) - M)^2$$

The average gray value is

$$M = \frac{1}{N \times N} \sum_{n=1}^{N} I(x, y)$$

The minimum variance is $V_{\min} = \min(V)$ .There are two special cases to consider: one is a pixel value that equals to the minimum variance, namely $\sigma = 1$, but this value for slow changing edge is too big, and removed the noise and a lot of slow changing edge blurred out; another kind of circumstance is when the pixel variance is large, the Sigma is great, a lot of clear edges will be filtered out.

In order to solve the above problems, combined with the adaptive thought, add a weight to the parameter of Sigma ,the formula (1) revised as: $\sigma = kV/V_{\min}$ ,the value of $k$ is $1/M$ ( M in 0~ 1. ). Then $V$ is small, the value of Sigma should be smaller than by formula (6) the calculated value, it can prevent the slow changing edge blurring; when $V$ is large, the value of Sigma is also smaller than by formula

(6) the calculated value, so it does not overly smoothed image. Therefore, use $\sigma = kV/V_{\min}$ substitution of Gauss function type (6) is

$$G(x,y) = \frac{1}{2\pi(\frac{V}{M \times V_{\min}})^2} \exp(-\frac{x^2+y^2}{2(\frac{V}{M \times V_{\min}})^2}) \tag{7}$$

The value of Sigma is fully references each window gray information, the whole image of gray information, so that is conducive to effective at removing noise and retain more edge information.

**(2) The improvement of obtaining the high threshold**

In the Canny edge detection operator, the high and low threshold selection determines the detected edge information and its continuity. Edge detection threshold is higher starting point, the smaller high threshold is, the more retaining edge information, but it can also cause more false edges; the high threshold, reduced the false edges, but also ignored some edge information, causing edge discontinuities.

In order to detect all the details of the whole image edge, consider the overall image characteristics when choose the high threshold. Image mean variance and average gray value are reflected in its entirety parameter, so they can be used as high threshold selection standard. Above all, the value of $T_h$ can be

$$T_h = V_{ave}/M_{ave} \tag{8}$$

Among them, the image variance is

$$V_{ave} = \frac{1}{W \times H} \sum_{i=1}^{W \times H} V_i \; ,$$

The average gray is

$$M_{ave} = \frac{1}{W \times H} \sum_{i=1}^{W \times H} I_i(x,y) \; .$$

H and W are the height and width of the image. But this calculated high threshold will result in too many false edge , also join a weighting parameter $k$ , namely $T_h = kV_{ave}/M_{ave}$ , and through the experiment, the best value of $k$ between 0.2 and 0.5. In order to reduce the amount of computation, low threshold still using high and low threshold ratio is $2:3$ to calculate, namely the low threshold is $T_l = 2T_h/3$ .

## 4 The Description and Realization of the Three Improved Canny Edge Detection Operator

Above all, the operator calculates the value of Sigma automatically by the Gaussian function, using Sigma of the adaptive filter in this paper. When use adaptive filter to calculate the high and low threshold according to the gray level information of the image itself, to detect more edge information. The realization needs six steps as follows:

**Step 1:** According to the image $I$ , in accordance with the parameter of Sigma the calculation method to calculate the value of Sigma.

**Step 2:** Operator obtained after the smoothness of the image $I$ filtered image $I'$ .Image $I'$ get by convolution of image $I$ and the adaptive filter $G$ .

**Step 3:** Calculate the image gradient magnitude and direction.

**Step 4:** Return to step 3 in gradient non maxima suppression.

**Step 5:** According to the image $I'$ .In accordance with the high threshold value method to obtain a high threshold, and calculated the corresponding low threshold.

**Step 6:** Edge detection and connection. The process of realization as showed in figure 1.

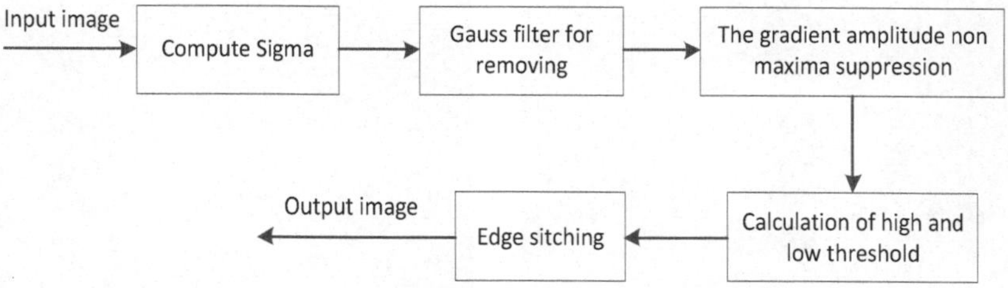


**Fig. 1.** The Flow Chart of the Improved Canny Edge Detection Operator

The core parts of the realization of ideas are as follows (MATLAB program fragments):

```
Input an image I
Find the size of I
%Gauss filter parameter Sigma automatic acquisition
For x=1 to width do
   For y=1 to Height do
        Calculate local variance.
Find the minimum variance.
        Calculate local mean.
        Calculate Sigma.
        Smooth I by Gaussian filter with Sigma.
    End for
End for
%High threshold improvement method for obtaining
For x=1 to width do
  For y=1 to Height do
Calculate the mean of the whole matrix' gray value.
        Calculate Mean variance.
        Calculate the high threshold.
    End for
End for
```

## 5 Experimental Results and Analysis

Select RGB color pictures of the car of tire as the experimental object, first carries on the color fusion, using the improved operator, Sobel operator and the traditional operator for experiments respectively, and finally compared the experimental results. We are first under noise-free conditions for image edge detection.

Experiment (1): At the case of no noise performance comparison of edge detection operator

The experimental results show in figure 2. As we can see from Figure 2, the Sobel operator can detect the horizontal direction and the vertical direction of the edge, the traditional Canny operator detects edge information is less, In this paper, our method have closed edge; the operator detects more edge information, and can detect the closed ring edge in the tire, the most important is detecting gray time-varying image edge efficiently.

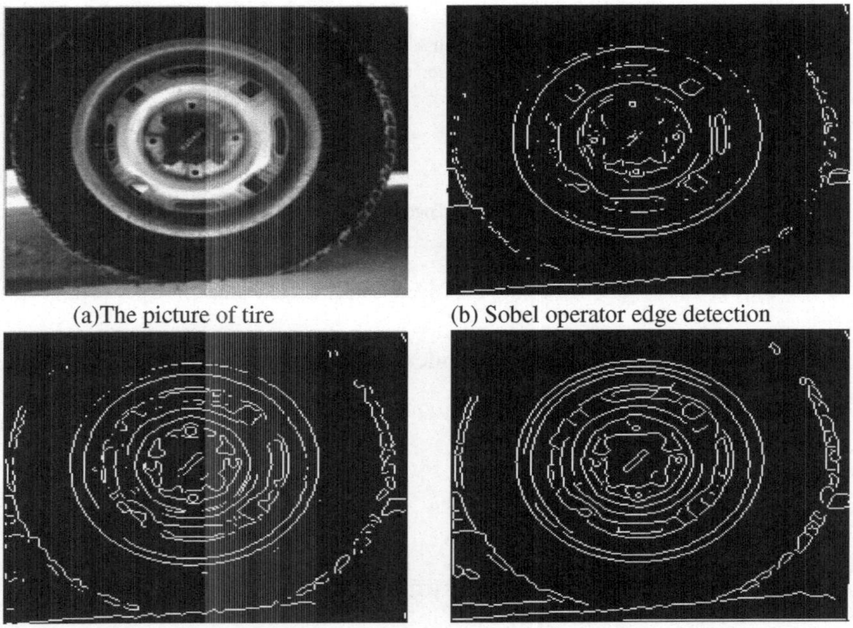

(a)The picture of tire (b) Sobel operator edge detection

(c) The traditional Canny operator edge detection (d) Improved Canny edge detection

**Fig. 2.** The Chart is without Noise Free Edge Detection Effect

Experiment (2): Adding noise under the edge detection operator performance comparison

Do experiments on the addition of noise in the image of tire .The experimental results show in Figure 3, the Sobel operator is very sensitive to noise, applied to image noise can seriously affect the quality of the traditional edge detection; Canny operators have the noise removal ability, but on the gray slow changing edge detects edge information is less, while the detected edge continuous closed is poor; and the operator can remove noise, gray slow changing edge detection to more edge information, the detected edge continuous closed property is better, and can detect the edge details.

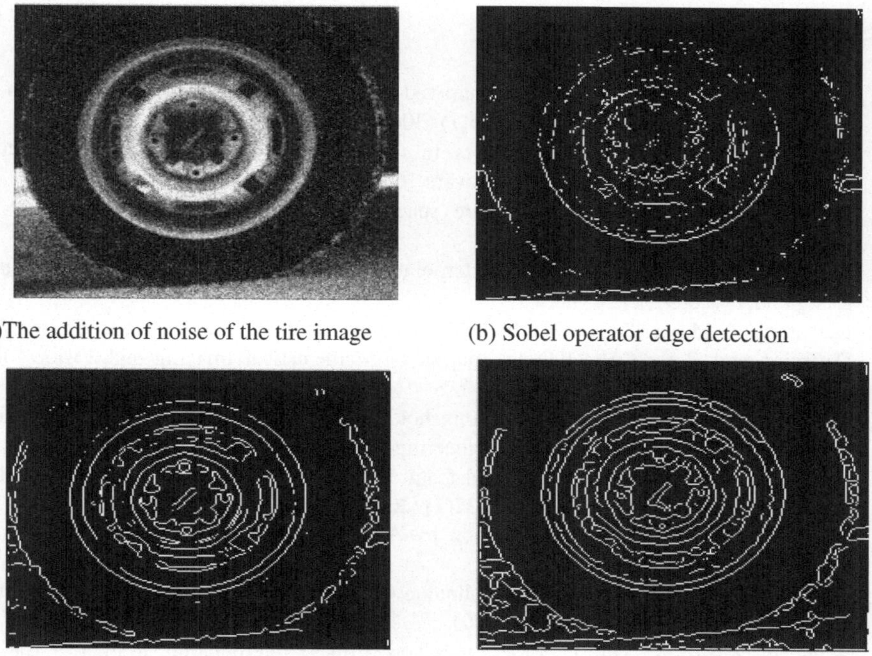

(a)The addition of noise of the tire image (b) Sobel operator edge detection

(c) The traditional Canny operator edge detection (d) Improved Canny edge detection

**Fig. 3.** The Chart is without Noise Free Edge Detection Effect

## 6 Conclusion

In this paper, according to the weaknesses of easy to lose the edge details with the traditional Canny edge detection operator ,we improved method to get the parameters and the high threshold, with noise and without noise , for tires of the car ,utilize Sobel operator, traditional Canny operator and improved Canny operator for edge detection, respectively.

Experimental results show that the improved Canny operator can eliminate noise and get more edge information. Detected edge continuity is better, and can detect the edge details of the whole image. According to the image adaptive calculation, the improved algorithm has the advantage of low computational complexity, less calculation time, etc.

**Acknowledgments.** This work is supported by the science and technology project about mobile police emergency disposal development in Yunnan province (project number is 2008CA012-4).

## References

1. Zhou, X., Ma, Q., Rong, X., et al.: Improved Canny operator edge detection operator of surveying and mapping. Engineering 17(1) (2008)
2. Chen, D., Liu, Z.: The edge features in the color image and their face detection performance evaluation. Journal of Software 16(5) (2005)
3. Lin, Z.: Edge detection in the feature space. Image and Vision Computing 29(2-3), 142–154 (2011)
4. Fu, X., Guo, B.: Overview of image interpolation technology. Computer Engineering and Design 30(1) (2009)
5. Demarcq, G., Mascarilla, L., Berthier, M., Courtellemont, P.: Application to Color Edge Detection and Color Optical Flow. Journal of Mathematical Imaging and Vision 40(3) (2011)
6. Chen, B., Li, J., Li, W.P.: Based on threshold and B spline interpolation of MR image enhancement algorithm. Computer Engineering and Applications 23(13) (2007)
7. Liu, X., Yang, X., Wang, J.: Statistical feature based with fast color image interpolation method. Chinese Journal of Electronic 32(1) (2004)
8. Llanas, B., Lantaon, S.: Edge detection by Adaptive splitting. Journal of Scientific is Computing 46(3) (2011)
9. Chen, Y., Wang, Y.: An Improved Technique for Watermarking Images and Video in the Wavelet Domain 6(5), 1661–1668 (2010)
10. Wang, L., Zhang, Y., Gu, Y.: Based-on adaptive image interpolation. Journal of Harbin Institute of Technology (1) (2005)
11. Xu, D., Zheng, Y., Gao, Y., Wang, D.: Parallel Computation for Discrete Orthogonal Moments of Images Using Graphic Processing Unit 9(3), 611–618 (2012)
12. Gelb, A., Hines, T.: Detection of edges from Nonuniform Fourier Data. Journal of Fourier Analysis and Applications 17(6) (2011)
13. Verma, O.P., Hanmandlu, M., Kumar, P., Chhabra, S., Jindal, A.: A novel bacterial foraging technique for edge detection. Pattern Recognition Letters 32(8) (2011)
14. Zhou, Z., Zheng, L., Xia, J., Yang, W., Lei, J.: Image Edge Detection Based on Improved Grey Prediction Model 6(5), 1501–1507 (2010)
15. Wang, X., Wang, Y., Tao, C., et al.: Image scaling algorithm based on edge detection. Bulletin of Science and Technology 9(5) (2005)
16. Li, Y., Gou, W., Li, B.: A New Digital Watermark Algorithm Based on the DWT and SVD. In: 2011 10th International Symposium on Distributed Computing and Application to Business, Engineering and Science, pp. 207–210. IEEE Computer Society (2011)

# Multi-modality Medical Image Registration Based on Improved I-alpha Information (SNI) with Gradient

Taizhe Tan, Hailing Liu, Yinwei Zhan, and Yuzhen Jin

Faculty of Computer, Guangdong University of Technology, Guangdong 510006, China
taizhetan@gdut.edu.cn, liuhailing0129@163.com,
1036423130@qq.com

**Abstract .** The paper introduces the development process of the mutual information for medical image registration, and analyzes the local maximum problems existence of information medical image registration based on the traditional mutual ,and using the SNI information get from the I-alpha information instead of the traditional mutual information to improve the speed and the accuracy of registration; Combining the improvement of spatial information to depict the gradient information with the SNI information. Use the new measure for the registration of different modes medical image from The Whole Brain Atlas database ( MRI, CT, SPECT). Results proved that the convergence and registration precision are greatly improved, and solution the robustness problems of the traditional mutual information for medical image registration commendably.

**Keywords:** Medical image registration, Mutual information, I-alpha information, SNI information, Gradient information.

## 1 Introduction

Nowadays, a variety of imaging equipment provides physicians with a different modality images from static to dynamic, qualitative to quantitative, flat to three-dimensional and structure to function[1]. However, each image can only get partial information about the patient's due to the different imaging principle of a variety of images. For example, CT images just to have high spatial resolution on bone tissue, and MRI images show soft tissue information while SPECT and PET images reflect organ function information[2 13]. It is necessary for image fusion to take full advantage of the information of each image, and the image registration is the premise of image fusion.

In recent years, a variety of image registration algorithms have emerge from two-dimensional to three-dimensional ,from single-mode images to multi-modal images, from external characteristics to internal characteristics, from the rigid registration to non-rigid registration[3 4 5 6]. The wavelet technology[7] is based on the characteristics and model require segmentation and feature extraction before registration, while you do not need to partition the image under gray-based approach to registration.

J. Lei et al. (Eds.): AICI 2012, LNAI 7530, pp. 59–66, 2012.

In 1993, Woods[8] uses the conditional entropy as a measure to registration, and achieves a software for PET-MR image registration. In 1997, Maes[9] and others give a normalized form of the mutual information entropy correlation coefficient. Studholme[10], etc. give another kind of normalized mutual information in 1999. In order to improve the problem of prone to local extremum in the mutual information registration process, Pluim[11], etc. propose that a new measure formed combing mutual information and the image gradient information. Rueckert[12], etc. propose the high order mutual information based on the gray symbiotic matrix but are too computationally intensive. In 2008, Yang Jinbao[1] used different f information and made compared of different measures in time, convergence and registration accuracy. Practice has proved time spent by the SNI short but its convergence performance of the worst.

So, we propose that a GSNI information for multi-modal medical image registration in this paper.

## 2 SNI Information for Image Registration

Mutual information algorithm uses the image gray-scale data directly for registration, and avoids the error caused by image segmentation. However, the mutual information calculated from the joint histogram of two images, is prone to local minima in the histogram of the evaluation process. Due to the insufficient of the mutual information, it proposed that the SNI information of I-alpha can be used for image registration instead of regular mutual information.

### 2.1 Mutual Information

It can be seen by the following diagram that the information of the data set Y contained by X is the same as that of data set X contained by Y, namely X ∩ Y-section. It means that this part of information is the public of the data set X and the data set Y, known as mutual information.

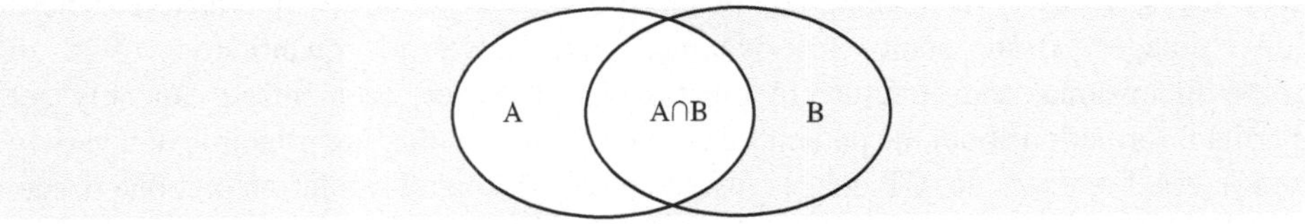


**Fig. 1.** Data sets X and Y

$p(x, y)$ is the joint probability distribution of the data set X, Y. $p(x)$ and $p(y)$ are the marginal probability distribution of X and Y, respectively. By the nature of the entropy we know:

$$H(X)-H(X|Y)=\sum_{x,y} p(x,y)\log\frac{p(x,y)}{p(x)p(y)}=H(Y)-H(Y|X) \tag{1.1}$$

Intuitively, the information of the data set X contained by Y is the same amount as that of data set Y contained by X， so we can use the entropy to describe the mutual information:

$$I(X;Y)=\sum_{x,y} p(x,y)\log\frac{p(x,y)}{p(x)p(y)} \tag{1.2}$$

## 2.2 I-alpha Information

For two images A and B, $p_{ij}$ is the probability of the joint gray level distribution, $p_i, p_j$ is the joint probability distribution which meets the independent condition respectively. the mutual information can be expressed as:

$$I(A,B)=\sum_{i,j} p_{ij}\ln\left(\frac{p_{ij}}{p_i p_j}\right) \tag{1.3}$$

I-alpha information can be expressed as:

$$I_\alpha(A,B)=\frac{1}{\alpha(\alpha-1)}\left(\left(\sum_{i,j}\frac{p_{ij}^{\alpha}}{(p_i p_j)^{\alpha-1}}\right)-1\right) \tag{1.4}$$

When a→1， it is defined as:

$$\lim_{\alpha\to 1} I_\alpha(A,B)=\lim_{\alpha\to 1}\frac{\sum_{i,j}\left\{(p_i p_j)^{\alpha-1} p_{ij}^{\alpha}\ln p_{ij}-p_{ij}^{\alpha}(p_i p_j)^{\alpha-1}\ln p_i p_j\right\}}{2\alpha-1}=\sum_{i,j} p_{ij}\ln\left(\frac{p_{ij}}{p_i p_j}\right) \tag{1.5}$$

From the above equation: the mutual information can be viewed as a special case of the I-alpha information.

## 2.3 Improved I-alpha Information-the SNI

Normally, Mutual information owns of such a nature:

$$I(A,A)=H(A)\; I(B,B)=H(B)$$

Also, the I-alpha information meets a similar nature, and the $I_\alpha(A,A)$ is defined as a entropy of A named the $H_\alpha$ entropy.

$$H_\alpha(A)=I_\alpha(A,A)=\frac{1}{\alpha(\alpha-1)}\left(\sum_i p_i^{2-\alpha}-1\right) \tag{1.6}$$

The joint entropy is defined as:

$$H_\alpha(A,B)=\frac{1}{\alpha(\alpha-1)}\left(\sum_{i,j}p_{ij}^{2-\alpha}-1\right) \tag{1.7}$$

Improved $I_\alpha$ Information-the SNI:

$$SNI_\alpha(A,B)=\frac{2I_\alpha(A,B)}{H_\alpha(A)+H_\alpha(B)} \tag{1.8}$$

Where D (0 << 1) values of 0.5 work best.

# 3 Combination of Gradient Information and SNI

In practice, mutual information makes the measure curve is not smooth enough[14] and the robustness of the image size. Therefore, the gradient information which is able to accurately depict spatial structure is added to the objective function .It can make fuller use of the image information.

## 3.1 Gradient Information

Due to the different modal images of different imaging principle, the gradient value of the same organization in a modal image may be greatly, but in another is not obvious. In this case, It will have to reduce its weight, while gradient values appear in the image of two at the same time will have to increase its weight. Thus, we set a threshold value of $\eta$ in the gradient information of Pluim, which guarantees the two images of the gradient can with subtle differences, and make the same part in the two images of the more prominent. The algorithm is explained as follows: the weight value $\omega$multiplied by $\min(|\nabla x(\sigma)|,|\nabla x'(\sigma)|)$ when the gradient difference of a corresponding pixel in different modes of images is greater than $\eta$. On the contrary, when the difference of the gradient is less than or equal to $\eta$ weights $\omega$ multiplied by $\max(|\nabla x(\sigma)|,|\nabla x'(\sigma)|)$. Finally, the overall weighting function written as:

$$\text{G (A, B)} = \sum \omega^* \tag{2.1}$$

$$\omega^*=\begin{cases}\omega\min(|\nabla x(\sigma)|,|\nabla x'(\sigma)|),\text{当}|\nabla x(\sigma)|-|\nabla x'(\sigma)|>\eta;\\ \omega\max(|\nabla x(\sigma)|,|\nabla x'(\sigma)|),\text{当}|\nabla x(\sigma)|-|\nabla x'(\sigma)|<\eta\end{cases} \tag{2.2}$$

$$\omega(\alpha)=\frac{\cos(2\alpha)+1}{2} \tag{2.3}$$

Gradient information includes the gradient range and direction. Calculate the gradient vector of the corresponding position of the image A and B at $\nabla A_{ij}(\sigma)$, $\nabla B_{ij}(\sigma)$ by the first derivative of the Gaussian function of variance for $\sigma$, the angle between them is:

$$\alpha_{A_{ij},B_{ij}}(\sigma)=\frac{\nabla A_{ij}(\sigma)\cdot\nabla B_{ij}(\sigma)}{|\nabla A_{ij}(\sigma)||\nabla B_{ij}(\sigma)|} \tag{2.7}$$

### 3.2 A New Measure of Joining Gradient Information into SNI Information

In this paper, it apply the improved I-alpha information which is SNI information instead of mutual information then to put the gradient information join to the objective function of the new measure:

$$I_{new}=GSNI=G(A,B)SNI(A,B) \tag{2.5}$$

Especially, the combination relations are multiplied rather than added because the multiplying makes the curve of the objective function more smooth; it can get the best effect when $\alpha$ (0< $\alpha$ <1) which is in SNI information is 0.5, then the curve of objective function gets more smooth.

## 4 Experiment

In order to verify the validity of the above methods, we select some head CT images, MRI images, SPECT images to experiment, and all images with a resolution of 255 × 255 pixels are from The Whole Brain Atlas. We compare the measure from traditional mutual information with that from SNI information and GSNI information. Part Images for experiment are as follows:

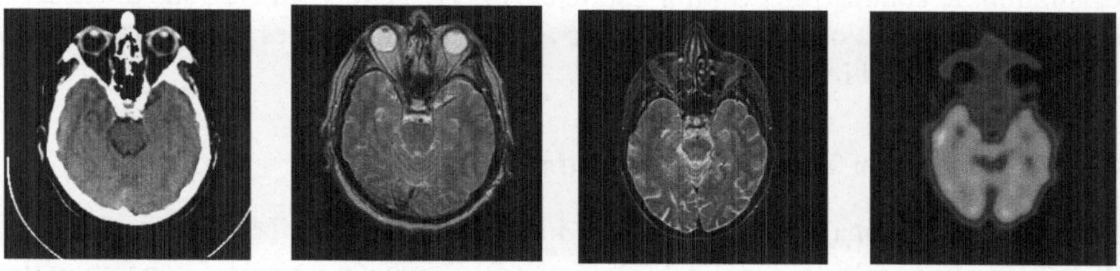

(a) CT image (b) MRI image (c) MRI image (d) SPECT image

**Fig. 2.** Used for registration images

Image registration steps are as follows:

1. Preprocessing on the images before registration: Convert the original images to grayscale and then the gray-scale compression (256 gray levels).
2. Apply rigid transformation[15] to transform the images then apply bilinear interpolation to interpolate the transformed image.

3. Apply the Powell optimization algorithm to find the optimal transformation parameters. According to the optimal parameters for transform we get registration image.

### 4.1 Different Measure of the Function Curve

Select figure 2(b) as an experimental reference image （Because of the gray balance of the image pixels， it can be a very good observation of the measurement curve）. And the rotation -20 degrees of the image as a floating image. Then rotate the floating image, For each Angle calculated the different measure value, get the measurement of function curve. The X-axis range [-20 20] .Calculated the value of the following measure: the traditional mutual information I, the SNI ($\alpha=0.5$) information in I-alpha, gradient information joined to the measure GSNI which threshold $\eta$ is 0.05. The measurement of the curve as shown below:

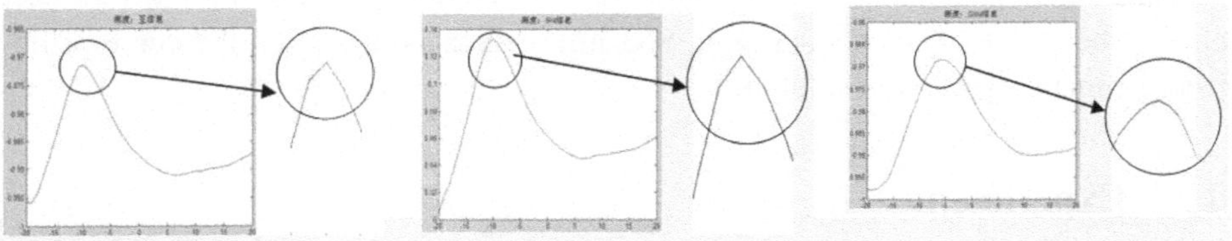

**Fig. 3.** Each measure function curve for floating image different rotation Angle (The right of the selected area enlarged image)

The curve of Mutual information and SNI is not smoothly, there will be a local extremum. SNI information rises faster than mutual information in the proximity of registration. In contrast, GSNI information curve is relatively smooth, and the abscissa of the maximum is closer to zero, so there are certain advantages in registration accuracy than the mutual information and SNI.

### 4.2 Computation Time and Registration Accuracy

Ⅰ: Select the CT images Fig 2 (a), MRI images Fig 2(b) ; Ⅱ: MRI images Fig 2(c), SPECT images Fig 2 (d); Ⅲ: Fig 2 (a), Fig 2 (d) to calculate the single operation time of each measure. And double counting to the five mean for accurate.

**Table 1.** Each measure computation time

Unit: seconds

| Measure | Mutual Information | SNI Information | GSNI Information |
|---|---|---|---|
| Ⅰ | 0.0498 | 0.0286 | 0.0365 |
| Ⅱ | 0.0408 | 0.0226 | 0.0326 |
| Ⅲ | 0.0476 | 0.0269 | 0.0347 |

In the above table, the fastest single computing time is SNI, GSNI added gradient information followed, the mutual information slowest; Mutual information contains a lot of arithmetic, resulting in the increase of time complexity; in addition to the SNI information GSNI information also contains a large number of multiplication which is a corresponding increase of computation time.

I: Select fig 2(a) as a reference image and rotated -20 degrees as a floating image Ⅱ :Do the same to fig2(c); Ⅲ: rotated fig 2(a) -20 degrees as the floating image and fig 2(b) as the reference, calculation 3 times of different measure for registration accuracy.

**Table 2.** The measurement accuracy of the registration（Unit: %）

| Measure | Mutual Information | SNI Information | GSNI Information |
|---|---|---|---|
| I | 80.76 | 78.75 | 94.29 |
| Ⅱ | 81.85 | 80.10 | 96.02 |
| Ⅲ | 79.65 | 79.06 | 94.03 |

It can be seen from the data in the table that SNI information and mutual information registration accuracy is about the same and mutual information is slightly better. GSNI information is significantly better than the mutual information and SNI.

### 4.3 Registration Results

Select the two MRI images in Figure 2 (b), (c) as a reference image, CT image (a) and SPECT images (d) as a floating image. Select the GSNI information as a similarity measure for image registration using bilinear interpolation, and apply Powell optimization algorithm which time efficiency has obvious advantages to image registration. Registration results as shown below:

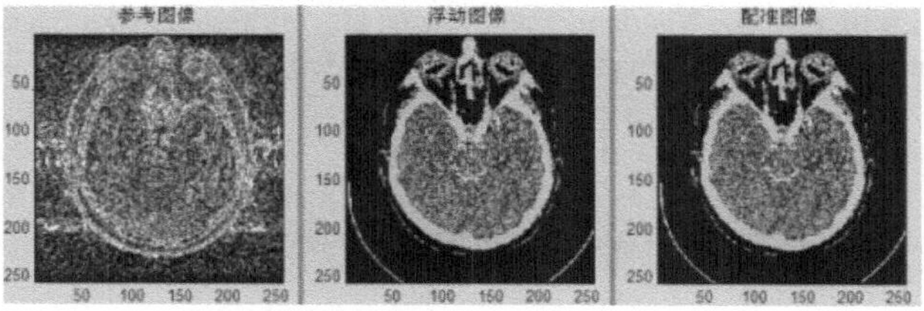


**Fig. 4.** CT images and MRI images registration

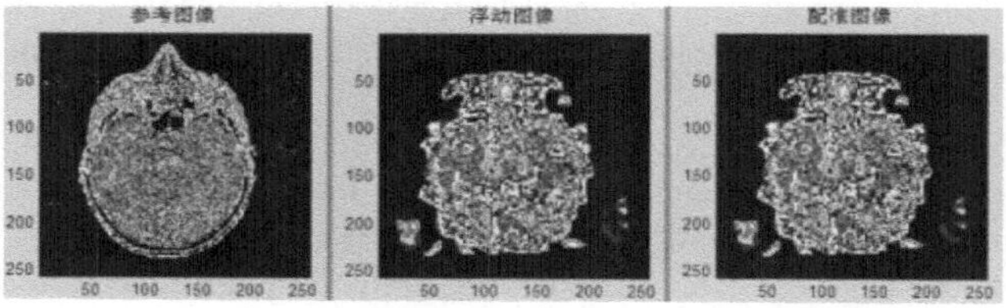


**Fig. 5.** SPECT images and MRI images registration

## 5 Conclusion

In this paper, the experiments proved that the SNI information to the dominant than the mutual information in terms of time efficiency while GSNI information can be a good solution to the problem of local minima of the mutual information and SNI, with good robustness. And its registration accuracy has greatly improved. Although time efficiency of GSNI is lower than the SNI , it is higher than that of the mutual information.

Overall, the GSNI information for image registration largely avoided the problem of local extremum, the convergence performance and registration accuracy are greatly improved. Registration is the premise of the medical image fusion, the future, we will study medical image fusion.

## References

1. Yang, J.: Based on gray level similarity measure of medical image registration technology research. Biomedical Engineering, Shandong University (2008)
2. Tian, J.: Medical image processing and analysis, pp. 54–59. The Electronic Industry Press, Beijing (2003)
3. Liao, X., Liang, L.: Medical image registration technology. Journal of Modern Electronic Technology (16), 107–112 (2009)
4. Huang, Y., Wang, S., Zhang, S.: Based on the regional characteristics of image registration method of automatic. Computer Engineering and Project 30(16), 3850–3855 (2009)
5. Peng, X., Chen, Q., Wei, B.: An efficient medical image registration method based on mutual information model. Fuzzy Systems and Knowledge Discovery 5, 2168–2172 (2010)
6. Tang, M.: Image Registration Based on Improved Mutual Information with Hybrid Optimizer. Chinese J. Biomed. Eng. 17(1) (March 2008)
7. Shi, H., Luo, S.: Image registration using the shift-insensitive discrete wavelet transform. Medical Image Analysis and Clinical Applications, 46–49 (2010)
8. Woods, R.P., Maziotta, J.C., Cherry, S.R.: MRI-PET registration with automated algorithm. Journal of Computer Assisted Tomography 17(4), 536–546 (1993)
9. Maes, E., Collignon, A., Vandermeulen, D., et al.: Multimodality image registration by maximization of mutual information. IEEE Trans. Med. Image 16(2), 187–198 (1997)
10. Studholme, C., Ct Hill, D.L., Hawkes, D.J.: Anovedap invariant entropy measure of 3D medical image alignment. Pattern Recognition 32, 71–86 (1999)
11. Pluim, J.P.W., Maintz, J.B.A., Viergever, M.A.: Image registration by maximization of combined mutual information and gradient information. IEEE Trans. Med. Image 19(8), 809–814 (2000)
12. Rueckert, D., Clarkson, M.J., Hill, D.L.G., et al.: Non-rigid registration using higher-order mutual information. In: SPIE Medical Imaging: Image Processing, vol. 3979, pp. 438–447 (2000)
13. Al-Azzawi, N.A., Sakim, H.A.M., Abdullah, W.A.K.W.: MR image monomodal registration based on the nonsubsampledcontourlet transform and mutual information. In: International Conference on Computer Applications and Industrial Electronics, ICCAIE, pp. 481–485 (2010)
14. Gorbunova, V., Durrleman, S., Lo, P., Pennec, X., de Bruijne, M.: Lung CT registration combining intensity, curves and surfaces. In: IEEE International Symposium on Biomedical Imaging: From Nano to Macro, pp. 340–343 (2010)
15. Chenoune, Y., Constantinides, C., El Berbari, R., Roullot, E., Frouin, F., Herment, A., Mousseaux, E.: Rigid registration of Delayed-Enhancement and Cine Cardiac MR images using 3D Normalized Mutual Information. Computing in Cardiology, 161–164 (2010)

# Design of Smart Home Control System Based on ZigBee and Embedded Web Technology

Lingling Li, Weicheng Xie, Ziyang He, Xin Xu, Changmin Chen, and Xiaorong Cui

School of Electrical and Information Engineering, Xihua University, 610039 Chengdu, Sichuan, China
lilinglingky@163.com, scxweicheng@yahoo.com.cn

**Abstract.** A design of Smart Home Remote Control System is proposed in this paper. It is based on ZigBee and Embedded Web Technology, to monitor and control the home environment remotely. The 32bits-ARM9 processor is the core of the system, and the Embedded Web Server Boa is established with Embedded Linux OS (Operating System). Combining with CGI (Common Gateway Interface) Technology, it is implemented to exchange data and information between the Server and remote users. And ZigBee wireless module CC2430 is to exchange data and information between information appliances and the Server. The experimental results reveal that Smart Home Control System can be achieved.

**Keywords:** Embedded Web Technology, Boa, ZigBee, CGI Technology, Embedded Linux OS, Smart Home, Remote Control.

## 1 Introduction

Science and technology are making progress with each passing day, in order to keep up with the pace of the times, the learning and working pressure of people is also growing, in order to ease the pressure, it can be considered to improve their living environment. In the past decades, with the rapid development of sensor technology, computer technology, network technology, communication technology, and control technology, we are entering the Information Age. It is the high time for people to enter the Age of Smart Home. Smart Home System generally includes several subsystems, namely video intercom, home security, network communications, interactive entertainment, intelligent lighting and appliance controlling. A whole Smart Home System is established with the subsystems by using some advanced technology, such as computer technology, network communication technology, wireless technology, integrated wiring technology and so on. There is no doubt that it will be propitious to our home life [1], [2]. Compared to traditional home, the biggest difference is the intelligence structure, which supports the remote home users to monitor and control home environment [2]. Wireless Sensor Networks occupy an important place of many new applications in the area of monitoring and controlling. It also involves the Smart Home.

J. Lei et al. (Eds.): AICI 2012, LNAI 7530, pp. 67–74, 2012.

## 2 General Design of Smart Home

The structure of Smart Home System can be roughly divided into three parts [3]. They are Intranet, Gateway and Extranet. Intranet is to connect a variety of household appliances; Gateway is in charge of the connection of family network and Extranet. Extranet is a residential LAN or the Internet. A plan of Smart Home is proposed in this paper. Architecture of the Smart Home Remote Control System, which is based on ZigBee technology and Embedded Web, is shown in figure 1. Intranet is established by ZigBee modules [4], [5], which assist Server in getting state of appliances and convey the commands from the Server to appliances via ZigBee network. Embedded ARM9 is the Server (Gateway), exchanging data between the remote users and the ZigBee coordinator.

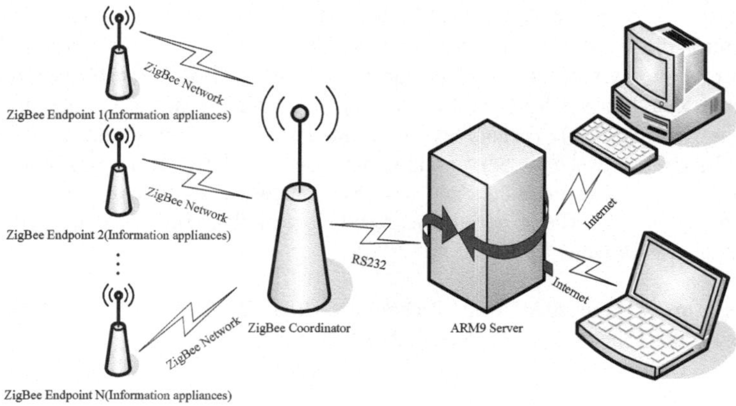


**Fig. 1.** Block diagram of Smart Home Remote Control System

## 3 Design of the Hardware System

### 3.1 ZigBee Wireless Home Network

The Intranet is composed of ZigBee star network. The main chip is CC2430-F128, which belongs to the CC2430 SOC family. The modules complete the wireless transmission of data in the 2.4GHZ ISM (Industrial Scientific and Medical) opening band [6]. Serial port of the CC2430 is configured in pins P0.2 and P0.3. It is the ZigBee2006 Protocol Stack released by TI to complete the Wireless Network. On the basis of the hardware connection, modify the Hardware Layer program to configure the Protocol Stack and write user program of Application Layer to achieve wireless communication. In the network, the ZigBee modules, connected to home appliances, are configured to RFD (Reduced Function Device) as terminals and the ZigBee module, connected to the ARM9, is configured to FFD (Full Function Device) as a coordinator.

### 3.2 Server System

Figure 2 shows hardware block diagram of Server in Smart Home System. The Server connects to ZigBee coordinator through the serial port and communicates with remote client through the Internet. Use DM9000 as a network card chip, which is adaptive 10/100M network. RJ45 wiring head contains coupling coil, so there is no necessary to use network transformer. Only with ordinary network cable can Server be connected to the router or switch [7].

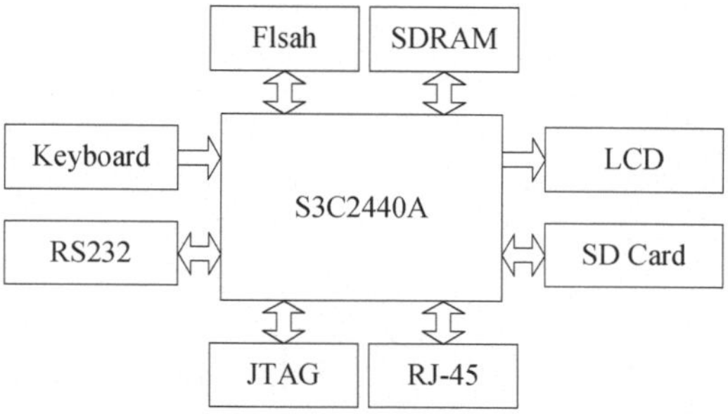


**Fig. 2.** Hardware block diagram of the Remote Control Server

## 4 Design of the Software System

Software development process includes the development of ZigBee Wireless Network nodes software, the setting up of Embedded Web Server Boa based on Embedded Linux OS and the development of application software on the Server.

### 4.1 ZigBee Network Nodes Software

ZigBee network is originally established by the coordinator node. Coordinator assigns ID to the terminal nodes, receives terminal nodes data, uploads data real-time to the ARM processor and delivers the order issued by the ARM processor to the terminal nodes. The main task of terminal node is to search the ZigBee network, gather real-time information of appliance sensors and upload to the coordinator node. The terminal node also receives the command from coordinator and implements it. Work flow chart of nodes is shown in figure 3 [8], [9].

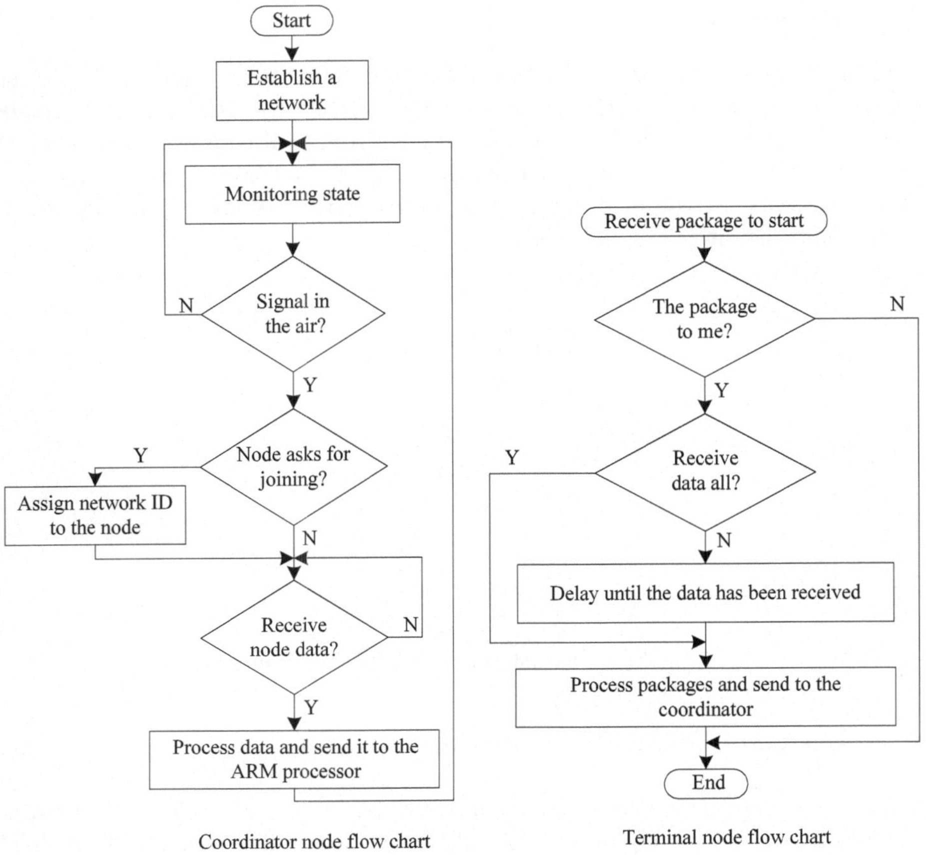


**Fig. 3.** Work flow chart of nodes

## 4.2 Set Up Web Server

The typical Embedded Web Server has three kinds, namely Httpd, Boa and Thttpd. As the simplest Web Server, Httpd has the following weak points: without supporting authentication and CGI Technology. It cannot provide dynamic page. While both Boa and Thttpd support these functions. Resource which is Thttpd required during operation is much larger than Boa [10]. Remote Control of home environment is related to the dynamic interaction of information, so Boa Server is appropriate. Boa is a very thin Web Server, and the executable code is only about 60KB. It is a single-tasking Web Server, which can respectively complete the users' request rather than fork a new process to handle concurrent connection requests.

For the safety of the Smart Home Remote Control, it is necessary to start HTTP authentication of Boa. The remote users of Smart Home Remote Control System must be certified. To start the Boa support for HTTP authentication, you must modify the configuration the file boa.conf. In boa.conf add:

```
Auth /www/smarthomecontrol.html

/etc/smarthomecontrol.passwd
```

It is to certify the user who asks for smarthomecontrol.html. Password file is in /etc/ directory, named smarthomecontrol.passwd. Boa does not provide multi-user authentication, there is no environment variable for HTTP authentication, so when a user logs on successfully, an environment variable must be provided to store the user name. First, add a variable named auth_name in the request structure used to store the user name. Second, in the function auth_authorize, copy the user name to the variable added in previous step. Then, in the appropriate location of the function complete_env, add the following statements:

```
# ifdef AUTH

if (req -> auth_name)

req -> cgi_env[req -> cgi_env_index+ + ] =

env_gen("AUTH NAME",req -> auth_name) ;

# endif
```

Where auth_name is string variable added in the request structure, AUTH_NAME is the name of the environment variable.

The Boa flow chart is shown in figure 4 [11]. Boa gets the HTTP requests from the socket, stores those in a request structure and saves them in the queue. First of all, the function get_request gets all the data from the socket and stores them in the request->header_line. Fill the basic environment variables in the remote address, port, host name, and local address to the request structure, and then process every request in the queue by the function process_requests. The queue here is the requests arrived at the same time. They will get different treatments according to the value of the status in the request structure. If the request is consistent with the HTTP protocol, the function process_option_line will be called to fill header information to the request structure to complete these environment variables, for example, document type, document length, the user authentication information. Password is encrypted with the encryption function crypt in the C language. This function uses the DES algorithm encrypted string or numeric values, so that the encrypted results can be saved in any form at any position. The Boa calls the function crypt to encrypt the input user information and compares with the information in the document to verify the legitimacy of the user. It is to acquire the user authentication information variables from the request structure. There is no distinction of the user name, so actually it is equivalent to a single user authentication model. If it does not pass the certification, the Boa would respond the client browser with the issue "you do not have the access to the information" and return. If it passes, the Boa would examine whether this request is a static page or a CGI program. If it is a static page, call the function init_get, and vice versa it is the CGI program, call the function init_cgi. For CGI programs, a series of function must be called to complete the CGI environment variables, the function create_common_env and complete_env is able to complete the registration of most of the CGI environment variables. Boa uses a PIPE (pipeline) mechanism for CGI processing, the actual CGI program is to redirect the output to the PIPE, and then Boa reads it from the PIPE and forwards to the client browser. After the entire process, return to the main function of an infinite loop waiting for the next socket connection.

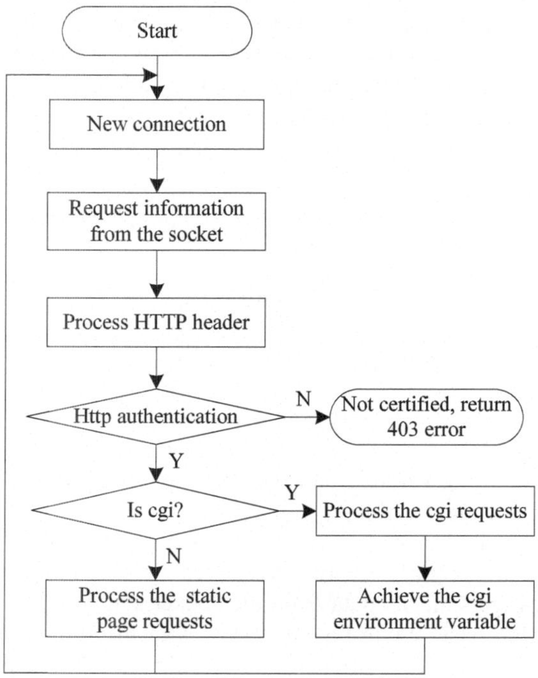


**Fig. 4.** Flow chart of the Boa Server

### 4.3 Implement Dynamic Web Pages

When the Embedded Web Server Boa has successfully completed, it is time to compile CGI programs, including the HTML form (the page the user sees) and the CGI script (running on the Server) [12]. CGI script can be written in any programming language, such as Shell scripting language, Perl, Fortran, Pascal, C language and so on. CGI script written in C language is at a higher execution speed and safer, for the C language program is compiled execution and cannot be modified.

## 5 Experiment Results

With the whole system accomplished, connect the Server to the Internet; enter the Server IP address in the browser to monitor the home environment. Take the control of air condition and lights as example. Figure 5 shows the login page for the remote monitoring. Every set of smart home system has a specific number of login accounts. Logs on it in the first step, and then gets into the Smart Home remote monitoring interface to monitor and control the home environment. Figure 6 shows the page of lights and air condition controlling and status displaying.

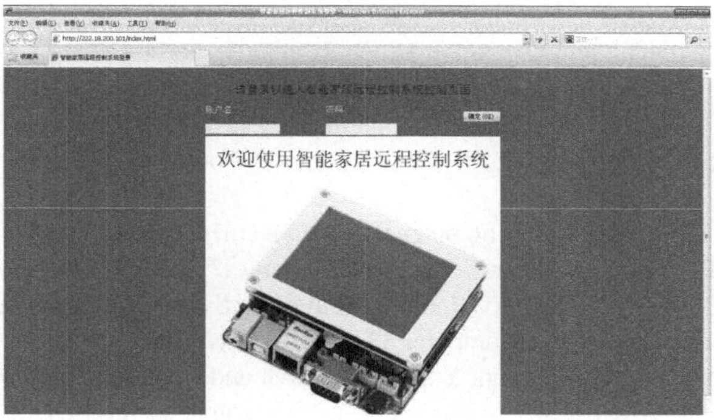


**Fig. 5.** Login page for the Smart Home System

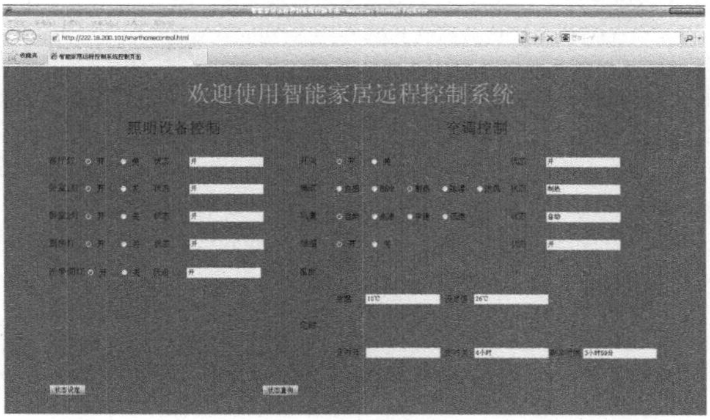


**Fig. 6.** Lights and air condition controlling

# 6 Conclusions

This paper presents a Smart Home Control System based on ZigBee wireless network technology and Embedded Web Technology to make home network more intelligent and automatic. It implements the proposed system and develops related hardware and software. The experimental results indicate that remote controlling home appliances and managing home appliances information is fairly good. And the expansion of the Smart Home System is very convenient. This study has certain reference significance for other remote monitoring system based on ARM development platform.

**Acknowledgments.** This work is supported by the Scientific Research Fund of Sichuan Provincial Education Department of China (No. 10ZA098), and the Innovation Fund of Postgraduate of XIHUA University of China (No. Ycjj201254).

## References


1. Jiang, Y., Fang, Z., Xu, D.: Design and Application of Wireless Sensor Network Web Server Based on S3C2410 and ZigBee Protocol, vol. 404 (2009), doi:10.1109/NSWCTC
2. Zhao, K.: Design of Smart Home Control Platform Based on ARM and ZigBee Wireless Technology. GuangXi University Master Degree Thesis (June 2009)
3. Huang, X.: Design of Wireless Smart Home System Based on CC2430. WuHan University of Technology Master Degree Thesis (March 2010)
4. Han, D.-M., Lim, J.-H.: Design and implementation of smart home energy management systems based on zigbee. J. IEEE Transactions on Consumer Electronics 56, 1417–1425 (2010)
5. Han, D.-M., Lim, J.-H.: Smart home energy management system using IEEE 802.15.4 and zigbee. J. IEEE Transactions on Consumer Electronics 56, 1403–1410 (2010)
6. Su, S., Lei, C., He, X.: Design of Remote Monitoring System Based on ARM Platform and ZigBee. J. Industrial Control Computer 23, 30–31+33 (2010)
7. Shi, F., Duan, C., Xiao, X., Qian, Y.: The Research and Implementation of Embedded Web Server Based on Wireless Sensor Networks. J. Microcomputer Information 27, 148–149+177 (2011)
8. Liu, L., Zhang, G.: Design of Intelligent Home Control System Based on ZigBee Wireless Technology. J. Computer Technology and Development 21, 250–253 (2011)
9. Jin, X.: Design of Wireless Measurement and Control System in Grain Storage Based on ZigBee. J. Measurement & Control Technology 30, 44–47 (2011)
10. Liu, Y., Cheng, X.: Design and Implementation of Embedded Web Server Based on Arm and Linux. In: ICIMA 2010 (2010)
11. Cao, L., Shi, J., Luan, J.: Boa Source Code Analysis and Its Application in Embedded System. J. Computer and Digital Engineering 33, 10–11+22 (2005)
12. Nong, S., Fan, Z.: Web Programming Based on Linux Platform(Linux 平台下的Web编程). Posts & Telecom Press, China (2000)

# Using Gaussian Potential Function for Underdetermined Blind Sources Separation Based on DUET*

Ye Zhang**, Kang Cao, Kangrui Wu, and Tenglong Yu

Department of Electronic and Information Engineering,
Nanchang University, Nanchang 330031, China
zhye901@126.com

**Abstract.** The DUET algorithm is a typical underdetermined blind source separation method, while the estimation of mixing parameters is an important part of DUET algorithm. In the presence of noise, a robust DUET algorithm is proposed to estimate the mixing parameters, i.e. the delay and attenuation between microphones and sources. The mixing parameters can be obtained by estimating the local maximum of the Gaussian potential function. Then the binary time–frequency mask can be constructed to recover the source signals by using the mixing parameters. From the experimental results on audio mixtures, the proposed algorithm is simple and highly effective, and the accuracy of the estimated source signals is higher than that of the original DUET algorithm.

**Keywords:** Blind Source Separation, DUET, Gaussian Potential Function.

## 1 Introduction

The problem of estimating unknown source signals from sensor measurements which are unknown mixtures of the source signals, commonly known as blind source separation(BSS), has been widely used in various fields such as biomedical signal analysis and processing, speech enhancement, image recognition and wireless communications. By assuming the sources are statistically independent and at most one source is Gaussian distributed, a powerful tool that has found increasing use is independent component analysis (ICA) [1],[2]. However, ICA cannot be used when there are more sources than mixtures, a case referred to as Underdetermined BSS (UBSS). Since the mixing system is not invertible; it is impossible to obtain the source signals by using simply inverting the mixing matrix. Even if we knew the mixing matrix exactly, it is difficult to recover the exact values of the source signals [3].

Recently, the Degenerate Un-mixing Estimation Technique (DUET)[4],[5],[6] has been proposed for UBSS problem by assuming that the sources can be represented

* Project supported by the National Nature Science Foundation of China (No. 61162014, 61141007).

** Corresponding author.

J. Lei et al. (Eds.): AICI 2012, LNAI 7530, pp. 75–81, 2012.

sparsely in a given basis and assumed that the source signals are disjoint in the time–frequency domain (W-disjoint orthogonal). DUET performs speech separation of any number of sources from only two mixtures. DUET exploits the fact that the ratio of the time–frequency representations of the mixtures can be used to partition the mixtures into the original sources. The key construct in DUET is the two-dimensional histogram which is used to estimate the mixing parameter i.e. the delay and attenuation between microphones and sources. The histogram is then smoothed, and peaks are located that correspond to distinct sources. Each peak is used to construct a binary time–frequency mask. The binary mask can then be used to recover an individual sound source from the mixture.

However, most of the DUET methods mentioned above assume in the absence of noise. In practice, the mixing signals usually could be intervened by the additive noise. In this paper, a potential function method was used to estimate delay and attenuation between microphones and sources, and then the time-frequency mask can be estimated to separate speech from speech-in-noise mixture. Computer simulation results show the validity of the algorithm.

## 2 The DUET Algorithm

For a two-channel microphone arrangement with $P$ sources, the incoming anechoic mixed signals $x_1$ and $x_2$ can be described as [4]

$$x_1(t) = \sum_{j=1}^{P} s_j(t) \tag{1}$$

$$x_2(t) = \sum_{j=1}^{P} \alpha_j s_j(t-\delta_j) \tag{2}$$

where $\delta_j$ is the arrival delay between sensors resulting from the angle of arrival, and $\alpha_j$ is a relative attenuation factor corresponding to the ratio of the attenuations of the paths between sources and sensors.

The DUET algorithm is based on the basic assumption that all of the sources have a sparse frequency spectrum for any given time. This implies that each time-frequency point in the spectrogram is associated with only one source. This property, which is essential for the DUET algorithm, is called the W-disjoint orthogonality property and can be described as

$$S_i(\omega)S_j(\omega) = 0, \quad \forall i,j,i \neq j \tag{3}$$

The assumptions of anechoic mixing and W-disjoint orthogonality allow us to rewrite the mixing equations (1) and (2) in the time–frequency domain as

$$\begin{bmatrix} X_1(\tau,\omega) \\ X_2(\tau,\omega) \end{bmatrix} = \begin{bmatrix} 1 \\ \alpha_j e^{-j\omega\delta_j} \end{bmatrix} S_j(\tau,\omega) \tag{4}$$

W-disjoint orthogonality is crucial to DUET because it allows for the separation of a mixture into its component sources using a binary mask. Consider the mask which is the indicator function for the support of $\hat{s}_j$

$$M_j(\tau,\omega) = \begin{cases} 1 & \hat{s}_j(\tau,\omega) \neq 0 \\ 0 & otherwise \end{cases} \tag{5}$$

To estimate the mixing parameters for each source, DUET computes for every time-frequency bin $(\tau,\omega)$. They are likely to come from the same source, if many time-frequency bins share similar values of $\alpha$ and $\delta$. The most delay and attenuation are found by building a 2d histogram $H(\alpha,\delta)$ from $\alpha(\tau,\omega)$ $\delta(\tau,\omega)$. Each peak in $H(\alpha,\delta)$ indicates a source with peak location corresponding to the estimated mixing parameters $(\alpha_j,\delta_j)$ for that source. Binary time-frequency masks are then built to partition the STFTs of the mixture by assigning each time-frequency bin to the estimated $(\alpha_j,\delta_j)$ which is closest to the local mixing parameters $(\alpha,\delta)$ extracted for that bin. Finally demix the sources by multiplying each mask with one of the mixtures use:

$$\hat{S}_j(\tau,\omega) = M_j(\tau,\omega) X_j(\tau,\omega) \tag{6}$$

Then convert each estimated source time–frequency representation back into the time domain.

This method can demix an arbitrary number of speech source signals given just two anechoic mixtures of the sources, providing that the time-frequency representations of the sources do not overlap. However, when too many time-frequency bins overlap between sources the source separation becomes unfeasible. This method has no good performance when separate the speech-in-noise mixture.

## 3 The Potential Function Method for DUET

Due to DUET algorithm haven't consider the noise in the speech. However noise is to be considered in real-world applications so a potential function algorithm for the detection of mixing parameters is proposed in this paper. The potential function algorithm following described in detail.

### 3.1 The Gaussian Potential Function

Similar to the way the STFT is built, we can obtain the parameters $(\alpha_j,\delta_j)$ for each time-frequency bins. To find the parameters we use a Gaussian potential function search

$$f(\alpha(t),\delta(t)) = \sum_{t=1}^{T}\sum_{i=1}^{P} \exp\left(-\frac{(\alpha(t)-\alpha_i)^2 + (\delta(t)-\delta_i)^2}{\sigma_i^2}\right) \tag{7}$$

where $T$ is the sampling points, $P$ is the source number.

We can find mixing parameters $(\alpha_i, \delta_i)$ by maximizing the equation (7). Take the partial derivative of $f(\alpha(t), \delta(t))$ with respect to $\alpha_i$, $\delta_i$

$$\frac{\partial f}{\partial \alpha_i} = \sum_{t=1}^{T} \sum_{i=1}^{P} \exp\left(-\frac{(\alpha(t)-\alpha_i)^2 + (\delta(t)-\delta_i)^2}{\sigma_i^2}\right)\left(\frac{2\alpha(t)-\alpha_i}{\sigma_i^2}\right) \tag{8}$$

$$\frac{\partial f}{\partial \delta_i} = \sum_{t=1}^{T} \sum_{i=1}^{P} \exp\left(-\frac{(\alpha(t)-\alpha_i)^2 + (\delta(t)-\delta_i)^2}{\sigma_i^2}\right)\left(\frac{2\delta(t)-\delta_i}{\sigma_i^2}\right) \tag{9}$$

The parameters $\alpha_i$ and $\delta_i$ are updated based on the previous estimate and the current gradient as,

$$\alpha_i^k = \alpha_i^{k-1} - \lambda_i \frac{\partial f}{\partial \alpha_i} \tag{10}$$

$$\delta_i^k = \delta_i^{k-1} - \lambda_i \frac{\partial f}{\partial \delta_i} \tag{11}$$

where $\lambda_i$ is step parameter. By maximizing the objective function (7) we can get the real mixing parameters, then use the binary time–frequency mask recover the source signals.

### 3.2 Steps of the Gaussian Potential Function Method

In summary, the essentials to the DUET algorithm based on the potential function method are:

1. Construct the time–frequency representation of both mixtures
2. Take the ratio of the two mixtures and extract local mixing parameter estimates by:

$$\begin{cases} \alpha(t) = \left| X_2(\tau,\omega) / X_1(\tau,\omega) \right| \\ \delta(t) = (-1/\omega)\angle(X_2(\tau,\omega) / X_1(\tau,\omega)) \end{cases} \tag{12}$$

3. Initialize $\alpha_i$ and $\delta_i$ randomly. Set iteration counter $k = 3000$ , $\varepsilon = 0.001$, $\lambda_i = 0.002$
4. Take the initial value into to the formula (7), make $\left| f_k(\alpha(t),\delta(t)) - f_{k-1}(\alpha(t),\delta(t)) \right|$ iterate three times, if each time that $\left| f_k(\alpha(t),\delta(t)) - f_{k-1}(\alpha(t),\delta(t)) \right| < \varepsilon$ exit, otherwise update $\alpha_i$ and $\delta_i$ by formula (10) and (11).
5. Demix the source by binary time–frequency mask like the original DUET in section 2.

# 4 Experiment

In this section, some simulations to blind source separation in the noise case are presented. All simulations were performed in MATLAB 2009b using a PC with Intel Pentium 4 CPU 2.93GHz under Microsoft Windows XP operating system. In these experiments, we compare the Gaussian Potential Function method with the classical DUET algorithm presented in [5]. In order to evaluate the estimation accuracy of the method we use of the signal-to-distortion ratio (SDR)[7] which is defined as:

$$SDR = 10\log_{10}\frac{\|s_i\|^2}{\|e_{noise}+e_{\text{int}\,erf}+e_{artif}\|^2} \tag{13}$$

And the Sources to Artifacts Ratio (SAR) which is defined as:

$$SAR = 10\log_{10}\frac{\|s_i+e_{\text{int}\,erf}+e_{noise}\|^2}{\|e_{artif}\|^2} \tag{14}$$

where $e_{\text{int}\,erf}$ is an allowed deformation of the sources which accounts for the interferences of the unwanted sources, $e_{noise}$ is an allowed deformation of the perturbating noise (but not the sources), and $e_{artif}$ is an "artifact" term that may correspond to artifacts of the separation algorithm such as musical noise. We use (13) and (14) as criteria to evaluate the performance of our Gaussian Potential Function method, the higher values indicate that the higher precision of the algorithm estimates. In order to verify the anti-noise performance of this new approach, we compared performance to the original DUET algorithm in [5]. Using $KmPs$ as the mixing form, where $P$ as the number of source signals and $K$ as the number of observed signals. e.g.'2m3s' represent the form of two observed signals three source signals.

**Experiment 1.** In this experiment, under the determined conditions (i.e.2m2s), we random initialized $\alpha_i = [1\ 1; 1.1\ 0.9]$ and $\delta_i = [0\ 0; -7\ 6]$ as the mixing matrix, then obtain the mixing signal. Initialize $\alpha_i$ and $\delta_i$ randomly, using the equation (7) obtain the parameter pairs, and then use time–frequency mask obtain the source signal. We compared the performance to the proposed method and the original method in the same conditions. We show the result in Table 1. We can see that both in no noise case and speech-in-noise mixture (with 50dB, 40dB, 30dB and 25dB noise), the performance of the proposed method obviously better than the original DUET method.

**Experiment 2.**In this experiment, under the underdetermined conditions (i.e.2m3s). We randomly select the $\alpha_i = [1\ 1\ 1; 1.1\ 1\ 0.9]$ $\delta_i = [0\ 0\ 0; -7\ 0\ 6]$ as the mixing matrix then do the same work as experiment one, We compared the performance of the two method in no noise case and speech-in-noise mixture (with 40dB, 30dB and 25dB noise ), We show the result in Table 2. We can see the performance of the proposed method obviously better than the original method.

**Table 1.** Separation results of simulation under determined conditions(2m2s)

| | | Original DUET algorithm | | Potential Function method | |
|---|---|---|---|---|---|
| | | SDR | SAR | SDR | SAR |
| In the no noise case | $s_1$ | 8.3549 | 8.6074 | 10.5208 | 10.6198 |
| | $s_2$ | 6.7937 | 7.1625 | 8.2588 | 8.2743 |
| 50dB noise case | $s_1$ | 7.5577 | 7.8021 | 10.4976 | 10.5946 |
| | $s_2$ | 5.0514 | 5.0852 | 8.2271 | 8.2420 |
| 40dB noise case | $s_1$ | 7.3687 | 7.6026 | 10.1731 | 10.2703 |
| | $s_2$ | 4.8438 | 4.8787 | 7.8799 | 7.8979 |
| 30dB noise case | $s_1$ | 5.1973 | 5.4011 | 7.9006 | 8.0134 |
| | $s_2$ | 3.2447 | 3.2728 | 5.3244 | 5.3431 |
| 25dB noise case | $s_1$ | 2.5990 | 2.7896 | 4.9884 | 5.0938 |
| | $s_2$ | 0.8715 | 0.9014 | 2.2055 | 2.2317 |

**Table 2.** Separation results of simulation under underdetermined conditions(2m3s)

| | | Original DUET algorithm | | Potential Function method | |
|---|---|---|---|---|---|
| | | SDR | SAR | SDR | SAR |
| In the no noise case | $s_1$ | 6.0184 | 6.1747 | 7.6495 | 7.7489 |
| | $s_2$ | 4.2895 | 4.5278 | 4.4074 | 4.7115 |
| | $s_3$ | 5.9682 | 6.0584 | 7.2495 | 7.3185 |
| 40dB noise case | $s_1$ | 5.8096 | 5.9819 | 7.3950 | 7.5021 |
| | $s_2$ | 4.0525 | 4.2798 | 4.2086 | 4.5150 |
| | $s_3$ | 5.6716 | 5.7694 | 6.9495 | 7.0276 |
| 30dB noise case | $s_1$ | 4.7129 | 4.9043 | 5.9937 | 6.1202 |
| | $s_2$ | 2.7772 | 2.9582 | 2.7801 | 3.0612 |
| | $s_3$ | 4.3693 | 4.5099 | 5.5093 | 5.6160 |
| 25dB noise case | $s_1$ | 2.8855 | 3.1406 | 4.1045 | 4.2697 |
| | $s_2$ | 0.7616 | 0.9454 | 0.9300 | 1.1973 |
| | $s_3$ | 2.4111 | 2.5246 | 3.6488 | 3.7403 |

## 5 Conclusion

In this paper, the improved DUET algorithm based on the potential function method has been proposed to solve the problem of underdetermined BBS when the sources are sparse signals and satisfied W-disjoint orthogonality. In this method, use the Gaussian Potential Function to estimate the mixing parameters then use time-frequency mask obtain the source signal. The experiment results show the validity and stability of the proposed algorithm. However in this method and the classical DUET algorithm in [5], we all assume speech signals satisfy W-disjoint orthogonality. Further research can be done in which the source signal dissatisfy the orthogonality.

## References


1. Common, P., Jutten, C.: Handbook of Blind Source Separation: Independent Component Analysis and Applications. Academic Press (2010)
2. Hyvarinen, A.: A unifying model for blind separation of independent sources. Signal Processing 85(7), 1419–1427 (2005)
3. Gavelin, R., Klomp, H., Priddle, C., Uddenfeldt, M.: Bind Source Separation. Signals and System group, Uppsala University, Sweden (June 2004)
4. Xie, L.: DUET Algorithm for Blind Source Separation. Harvard University Division of Engineering and Applied Sciences (May 18, 2005)
5. Rickard, S.: Blind Speech Separation, The DUET Blind Source Separation Algorithm, ch. 8, pp. 217–241. Springer (2007)
6. Rafii, Z., Pardo, B.: Degenerate unmixing estimation technique using the constant Q transform. Northwestern University EECS Department Evanston Plenum Press (2011)
7. Févotte, C., Gribonval, R., Vincent, E.: BSS EVAL toolbox user guide. IRISA, Rennes, France, Tech. Rep. 1706 (2005)
8. Wang, D.: Time-Frequency Masking for Speech Separation and Its Potential for Hearing Aid Design. Trends in Amplification 12(4), 332–353 (2008)
9. Frigui, H., Krishnapuram, R.: A Robust Competitive Clustering Algorithm with Applications in Computer Vision. IEEE Transactions on Pattern Analysis and Machine Intelligence 21, 450–465 (1990)
10. He, Z., Cichocki, A., Li, Y., Xie, S., Sanei, S.: Clustering Learning for Sparse Component Analysis. Signal Processing 89, 1011–1022 (2009)
11. Ferreol, A., Albera, L., Chevalier, P.: Fourth-order Blind Identification of Underdetermined Mixtures of Sources. IEEE Transactions on Signal Processing 53(5), 1640–1653 (2005)
12. Li, Y.Q., Amari, S.I., Cichocki, A., Ho, D.W.C., Xie, S.L.: Underdetermined Blind Source Separation Based on Sparse Representation. IEEE Transactions on Signal Processing 54(2), 423–437 (2006)
13. Fang, Y., Zhang, Y.: A Robust Clustering Algorithm for Underdetermined Blind Separation of Sparse Sources. Journal of Shanghai University (English Edition) 12(3), 228–234 (2008)
14. Washizawa, Y., Cichocki, A.: Sparse Blind Identification and Separation by Using Adaptive K-orthodrome Clustering. Neurocomputing 71(10-12), 2321–2329 (2008)

# A New Balancing Type of Wireless Sensor Network Routing Algorithm

Xiaochen Li, Xizhong Lou, Ting Peng, Jia Xu, Qian Zhou, and Daorong Wu

College of Information Engineering, China Jiliang University, Hangzhou, China
pandali89757@gmail.com

**Abstract.** In general, wireless sensor network works by a small battery-powered, or limited energy. Once the wireless sensor network is deployed, the energy of small sensor nodes can not be replaced. So, to improve energy efficiency and extend the survival time of the whole network is a crucial issue. This paper presents a new type of energy balancing algorithm for wireless sensor networks. The BCDCP-M algorithm draws on the main idea of the BCDCP routing protocol. When the base station divides the network, the new algorithm makes the number of cluster heads equal to the optimal number of cluster heads as far as possible. On cluster head election, not only the average energy of the sensor network, but also the remaining energy of the individual node must be taken into account. In data transmission, we use the multi-hop method to select the optimal path. The simulation results show that the survival time of the network in the new BCDCP-M algorithm is 19% longer than BCDCP.

**Keywords:** Wireless Sensor Networks, Energy Balance, Routing Protocols, Multi-hop, Residual Energy.

## 1 Introduction

Wireless sensor network is consisting of a large number of sensor nodes, which are small size, low cost, with wireless communication, sensing, data processing. The sensor nodes are generally composed by the sensing module, processing module, transceiver module, power module and other function modules. The power module is one of the most important modules, so to improve the energy efficiency of the whole network is a long-standing focus of the study. Wireless sensor network deployed in the area of detection with a large number of micro sensor nodes, composed of a multi-hop ad hoc network, which can perception, acquisition and processing of the information of the objects in the network through the wireless communication, and then sent to the observer [1-3].

The first major application of wireless sensor networks in the military field. With the development of science and technology, the sensor node cost is getting lower and lower, and the function has become increasingly powerful, make the sensor networks what are very expensive have been able to enter the civilian field. Such as

J. Lei et al. (Eds.): AICI 2012, LNAI 7530, pp. 82–89, 2012.

infrastructure security, logistics management, health care and so on [4-6]. With the progress and development of technology, more and more sensor networks will be applied to various aspects of social life.

The reference[7] proposed a hierarchical routing protocol called LEACH, in this protocol, the choice of the cluster head is randomly selected by the threshold, it is prone to make the cluster heads in the regional is too concentrated, or cause there is no cluster head around the nodes. Multi-hop is introduced in the reference [8], then BCDCP protocol is proposed, this has improved the performance of LEACH protocol, but his harsh condition is every node must know its precise location. This paper presents a new BCDCP-M routing algorithm, he fully utilized the advantages of LEACH and BCDCP protocol, the performance is better than BCDCP.

# 2 LEACH Protocol

## 2.1 Protocol Introduced

The LEACH protocol is short for "Low Energy Adaptive Clustering Hierarchy". The basic idea of the algorithm is based on select a cluster head randomly in the cycle way, make the energy load of the whole network evenly distributed to each sensor node. In this protocol, the reconstruction process of clusters is continuing cycling; each cluster reconstruction process can be described as round. Each round can be divided into two stages: the establishment of cluster and transfer data stability.

### 2.1.1 Cluster Establishment

At the beginning, each sensor node to choose a random number between 0 and 1, if the selected value is less than the threshold $T(n)$,this node will be selected as the cluster head, then the cluster head will broadcast the message to the non-cluster head nodes in the network, after the nodes received the message, they will join cluster with the strength of received message, and notify the cluster head. $T(n)$ is calculated as follows:

$$T(n)=\begin{cases}\dfrac{p}{1-p[r \bmod (1/p)]} & n\in G \\ 0 & else\end{cases} \quad (1)$$

Where $p$ denotes the cluster head accounted for the percentage of all nodes, $r$ denotes the current number of cycle round, $G$ denotes the collection of nodes which are not yet been elected as the cluster head in the last $1/p$ round.

### 2.1.2 Cluster Establishment

The cluster head node uses a TDMA time slot assigned to transmit data for the cluster members. Once cluster creation and TDMA schedule fixed, data transmission begins. Assume that the node has data to send, it will send data to the cluster head within its transmission time allocated. The cluster head must keep its receiver to receive all of the data of nodes in the cluster. When all data has been received, the cluster head will fuse the data.

### 2.2 Protocol Analysis

LEACH protocol is based on some assumptions, so there are some problems in the practical application: (1)In the LEACH protocol, the cluster head election is randomly generated, this randomness may lead to uneven distribution of cluster heads. Cluster head may be concentrated in a particular region, clustering would be meaningless. (2) In small-scale network, the performance of LEACH is more desirable, while in the large-scale network, nodes that far from the base station consume more energy than the nodes which close to the base station, this makes nodes death earlier, so that the whole network load is not balanced. (3) Cluster head selected did not consider the residual energy of nodes. If one node has little energy, and it happens to be selected as cluster head, because of the consumption of cluster head is relatively large, the data of the whole cluster will not be send to the sink node, thus will affect the lifetime of the network.

## 3 BCDCP Protocol

The BCDCP protocol is short for "Base station Controlled Dynamic Clustering Protocol". The key idea of BCDCP is to balance the distribution of cluster, not only the optimal number of cluster heads, but also the nodes of per cluster. Then the cluster head sent data to the base station through CH-to-CH with multi-hop.

The foundation of BCDCP lies in the realization that the base station is high-energy node with a large amount of energy supply. Thus, BCDCP utilizes the base station to control the coordinated sensing task performed by sensor nodes.

Cluster head selection in this protocol, using an iterative method, dividing the whole network into similar sub-clusters by the number of sensor nodes, and uses a CH-to-CH multi-hop routing scheme to transfer the data to the base station. The routing paths are selected by first connecting all the cluster head nodes using the minimum spanning tree approach that minimizes the energy consumption for each cluster head, and then randomly choosing one cluster head node to forward the data to the base station.

The disadvantage of this protocol is not suitable for the network that has a large number of sensor nodes. When the nodes of the sensor network is excessive, the data transmission within the cluster and data transmission between the cluster heads will consume more energy, which will lead to uneven energy consumption throughout the network, and reducing the lifetime of the sensor network.

## 4 BCDCP-M Algorithm

This paper proposed a new BCDCP-M improved routing algorithm based on BCDCP protocol. The BCDCP-M algorithm includes the establishment of cluster and transmission path, and data transfer.

### 4.1 Cluster Establish

In the BCDCP protocol, according to the energy of the sensor nodes, the base station select a set $S$ , the nodes in $S$ have a higher energy than average energy of the network, then uses an iterative manner, divides the network into Nch sub-clusters. This will cause the area of sub-clusters is uniform and energy consumption is not balance.

The BCDCP-M algorithm focuses on the equalization of the energy consumption, and is still let the base station divide network into sub-cluster. First, the base station calculates the optimal number of clusters, denoted by N, based on the number of sensor nodes in the whole network [7]. Then divides the network into N sub-clusters, make the sensor nodes in per cluster are approximately equal.

In the selection of cluster head, the sub-cluster will select the node has a higher energy than the average and is near the center of the cluster. This can reduce energy consumption within the cluster, and energy-intensive tasks to be borne by the high-energy sensor node can effectively balance the energy consumption and extend the lifetime of the network. The base station located far away form the network, so each sensor node can only be elected as cluster once.

### 4.2 Transmission Path

In the BCDCP-M algorithm, the general location of the cluster head is near the center of network, so the choice of the path is always single jump. If the cluster head does not meet the conditions of central location, the distance has beyond the threshold; we will select the multi-hop transmission path.

The transmission path between the cluster heads, still continued in reference [9], uses the minimum spanning tree method to link all cluster heads, this can minimize the energy consumption of each cluster head. Since the base station is far away from the network, so we select the path of single-hop method between cluster heads and base station.

### 4.3 Data Transmission

In data transmission, we mainly consider how to make the energy consumption of the transmission in the cluster minimized. The major energy consumption of a sensor node is in sensor module, data processing module and RF communication module. Since the energy consumption of sensor and data processing module is fixed, so we focus on the energy consumption in the RF communication module. We use the same RF communication module in the reference [7]. The energy consumption equation of receive and send K-bit data, and receive and send data between $r$ m are as follows:

$$E_T(k,r) = E_{Tx}k + E_{amp}(r)k \tag{2}$$

$$E_R(k) = E_{Rx}k \tag{3}$$

Where $E_T(k,r)$ denotes the total energy dissipated in the transmitter of the source node, and $E_R(k)$ represents the energy cost incurred in the receiver of the destination node. The parameters, $E_{Tx}$ and $E_{Rx}$ , are the per bit energy dissipations for transmission and reception, respectively. $E_{amp}(r)$ is the energy required by the transmit amplifier to maintain an acceptable signal-to-noise ratio in order to transfer data messages reliably. As is the case in reference [7], we use both the free-space propagation model and the two-ray ground propagation model to approximate the path loss sustained due to wireless channel transmission. Given a threshold transmission distance of $r_0$, the free-space model is employed when $r \le r_0$,and the two-ray model is applied for cases where $r > r_0$. Using these two models, the energy required by the transmit amplifier $E_{amp}(r)$ is given by:

$$E_{amp}(r) = \begin{cases} \varepsilon_{FS} r^4, r \le r_0 \\ \varepsilon_{TR} r^4, r > r_0 \end{cases} \tag{4}$$

Where $\varepsilon_{FS}$ and $\varepsilon_{TR}$ denote transmit amplifier parameters corresponding to the free-space and the two-ray models, respectively, and $r_0$ is the threshold distance given by:

$$r_0 = \sqrt{\varepsilon_{FS} / \varepsilon_{TR}} \tag{5}$$

We use the same simulation parameters in reference[7], the parameters are as follows: : $E_{Tx}$ = $E_{Rx}$ =50nJ/bit , $\varepsilon_{FS}$ =10pJ/b/m$^2$ , $\varepsilon_{TR}$ =0.0013pJ/b/m$^4$ , EDA=5 nJ/b/message.

When the data transmission path is established, the sensor nodes collect data and sent to their cluster head, the cluster head send the data to the highest-energy cluster head along the path established before, and then the cluster head with highest-energy sent the data of the whole network to base station. This can make the energy load evenly assigned to each cluster head.

## 5 Simulation Results

We carried out some simulations on BCDCP-M algorithm using MATLAB and compared its performance with that of LEACH and BCDCP protocol. Performance is measured by quantitative metrics of average energy dissipation, network lifetime. The simulation parameters are as follows:

We simulate $100m \times 100m$ network topologies with base station located at least 75m away from the nearest node. We set random network configurations with 500 nodes and each node is assigned an initial energy of 2J. Furthermore, the frame number of data transmitted each round is 40; the data size for each round is fixed at 500 bytes, of which 25 bytes is the length of the packet header. In the simulation process, we recorded the energy consumption and the survival time of per sensor node.

In figure 1, we compared the average node energy consumption of three different algorithms. It is concluded from figure 1, when the nodes of LEACH and BCDCP are all dead, the average node energy consumption in BCDCP-M algorithm is about 1.4J and 1.6J respectively. With the same energy consumption of 1J, the BCDCP-M algorithm can run about 85 rounds; LEACH and BCDCP can only run 40 and 60 rounds respectively.

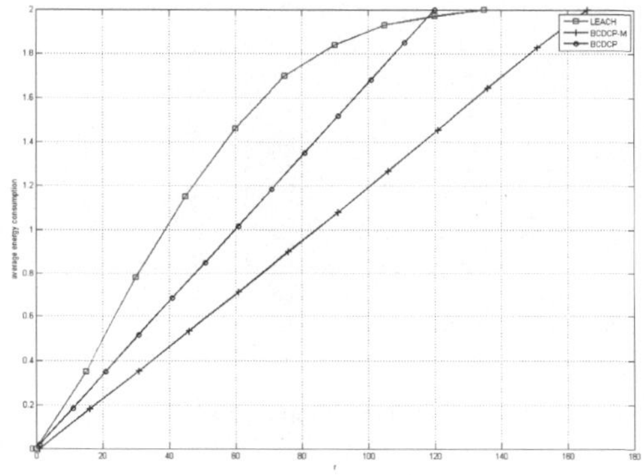


**Fig. 1.** Average Energy Consumption of Nodes

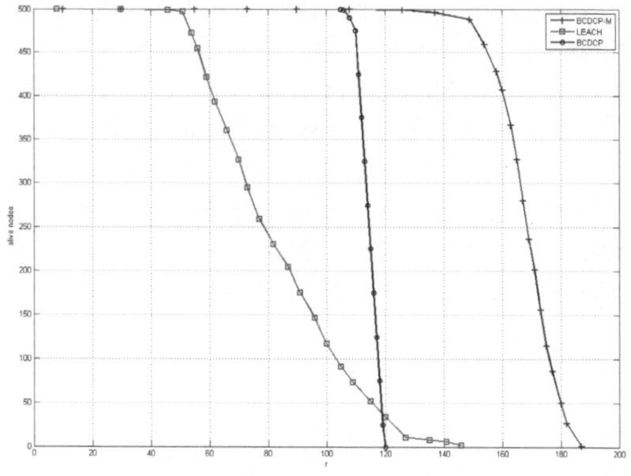


**Fig. 2.** Survival Node

We use the death time of the first node(FND) and death time of the half node(HND) this two sets of data to compare and analyze the BCDCP-M, BCDCP and LEACH protocol. From figures 2 and 3, we can see all the nods remain alive for 125 rounds, while the corresponding numbers for LEACH and BCDCP are 45 and 105. The BCDCP-M algorithm has extended 178% than LEACH on the complete running of the network, and compared with BCDCP, it has extended about 19%. In terms of the HND, the time of LEACH, BCDCP and BCDCP-M are 80, 115, 165 rounds. Compared with LEACH and BCDCP protocol, the BCDCP-M algorithm has extended 106% and 47%.

According to the above simulation, we can see the BCDCP-M algorithm has a lot of improvement than LEACH and BCDCP on FND and HND, it also reduces the energy consumption of nodes in each round. BCDCP-M algorithm proposed in this paper can effectively improve the lifetime of the network, and balance the energy consumption.

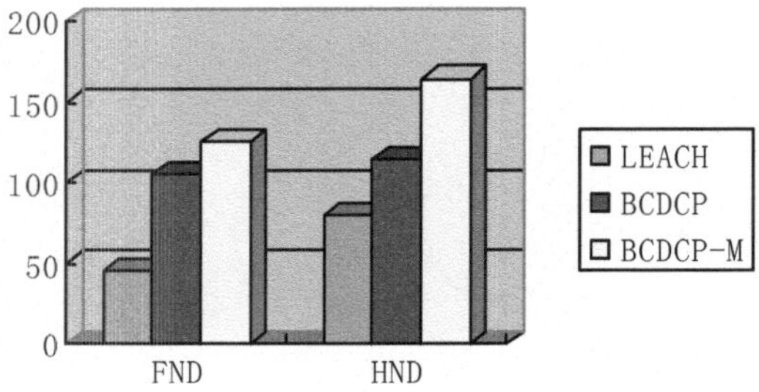


**Fig. 3.** FND and HND

## 6 Conclusions

In this paper, we make some improvements to balance the overall energy consumption in the wireless sensor networks, mainly in the following aspects: First, the base station calculates the optimal cluster head number N according to the number of the nodes in the whole network, and then divides the network into N sub-clusters. Second, the nodes which have higher energy and tend to be in the center of the cluster is selected as the cluster head. Third, the paths of multi-hop are optimized. The simulation results show that the BCDCP-M algorithm is more efficient on the balance of the energy consumption of the network than LEACH and BCDCP, and can extend the survival time of the networks.

## References

1. Akyildiz, I.F., Su, W., Sankarasubramaniam, Y., Cayirci, E.: A Survey on Sensor Networks. IEEE Communications Magazine 40(8), 102–114 (2002)
2. Younis, O., Fahmy, S.: Distributed clustering in ad-hoc sensor networks: a hybrid, energy-efficient approach. In: INFOCOM 2004. 23th Annual Joint Conference of the IEEE Computer and Communications Societies, West Lafayette (2004)
3. Kandris, D., Tsioumas, P., Tzes, A., Nikolakopulos, G., Vergados, D.D.: Power conservation through energy efficient routing in wireless sensor networks. Sensors 9, 7320–7342 (2009)
4. Saini, P., Sharma, A.K.: E-DEEC-enhanced distributed energy efficient clustering scheme for heterogeneous WSN. In: 2010 1st International Conference on Parallel, Distributed and Grid Computing (PDGC 2010), pp. 205–210 (2010)

5. Tang, L., Liu, S.: Improvement on LEACH Routing Algorithm for Wireless Sensor Networks. In: 2011 International Conference on Internet Computing and Information Services, pp. 199–202 (2011)
6. Lindsey, S., Raghavendra, C.S.: PEGASIS:Power-Efficient Gathering in Sensor Information Systems. In: Proceedings of the IEEE Aerospace Conference, pp. 1125–1130 (2002)
7. Heinzelman, W.B., Chandrakasan, A.P., Balakrishnan, H.: An application-Specific protocol architecture for wireless microsensor networks. IEEE Transaction on Wireless Communications 1(4), 660–670 (2002)
8. Muruganathan, S.D., Ma, D.C.F., Bhasin, R.I., Fapojuwo, A.O.: A centralized energy-efficient routing protocol for wireless sensor networks. IEEE Communications Magazine 43(3), S8–S13 (2005)
9. Shen, H.: Finding the k most vital edges with respect to minimum spanning tree. In: Proceeding of the IEEE 1997 National Aerospace and Electronics Conference, NAECON 1997, vol. 1(1), pp. 255–262 (1997)

# Development of Portable Gait Analysis System Based on a DSP

Juanjuan Xu and Jian Wu

Research Center of Biomedical Engineering, Graduate School at Shenzhen, Tsinghua University, Shenzhen, China
gnxujj@gmail.com, wuj@sz.tsinghua.edu.cn

**Abstract.** A real-time markers tracking system based on digital signal processor (DSP) was designed for portable gait analysis in this paper. In this system, the three-dimensional coordinates of markers and its relative angles were calculated and displayed on the screen in real time. Firstly, DSP detected the infrared markers pasted on the subject from two cameras in real time, and then calculated the markers' image mass centers by background subtraction. For each marker, we obtained its two mass centers with two cameras. Finally, DSP reconstructed each marker's 3D coordinates based on binocular vision principle, and displayed its position information on the screen at the same time. Experimental results showed that our system could obtain real-time three-dimensional position information of subject for his gait analysis.

**Keywords:** DSP, gait analysis, three-dimensional reconstruction, real-time display.

## 1 Introduction

Gait analysis, a special branch of the biomechanics, is the kinematic and dynamic analysis of limb and joint movement when a human is walking. It can offer a series of parameter values such as time, geometry, and mechanics and so on [1]. The postoperative curative effect of patient can be analyzed quantitatively by analyzing his gait index in gait analysis system. Now, all the mature gait analysis system, such as VICON, CODA Motion, process the gait analysis following the below: firstly save the video sequences collected by cameras, when the sequences are all collected, they are processed by computer to obtain gait parameters. However, a lot of space is needed to store these video sequences, and a specific data processor is allocated to deal with these data [2]. Such processor is not only expensive, but also complicated to operate and maintain. What's more, these devices occupy a large space, and are not easy to be moved.

To solve those problems mentioned above, we designed a portable gait analysis system based on DSP. Our system can track multiply markers pasted on the patient in real-time without saving video sequences as usual gait system done, and thus, it does not require a specific data processor to process the video data. Because our system is independent of the computer, it is easy to take for gait detection. In our system, the

J. Lei et al. (Eds.): AICI 2012, LNAI 7530, pp. 90–96, 2012.

human walking video sequences with markers collected from two cameras are directly loaded into DSP and the position of markers are calculated and saved. At the same time, the positions of the markers are displayed on the screen for observation.

# 2 Methods

## 2.1 Hardware Structure of System

The system consists of a DSP board, two cameras, several infrared markers and a display screen, as shown in Fig. 1. Two cameras collect the video images of subject's walk. DSP board is used to process the captured video images in real time, and the screen displays the positions of markers extracted by DSP. In our paper, we use SEED-VPM642 development board of SEED Company. In order to enhance the makers' contrast with the background, an optical filter is used in the camera.

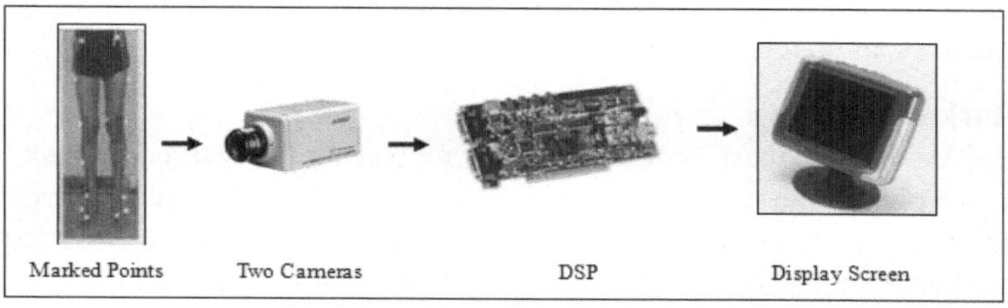


**Fig. 1.** The structure diagram of the system

## 2.2 Workflow of System

The work flowchart of the system is shown in Fig. 2

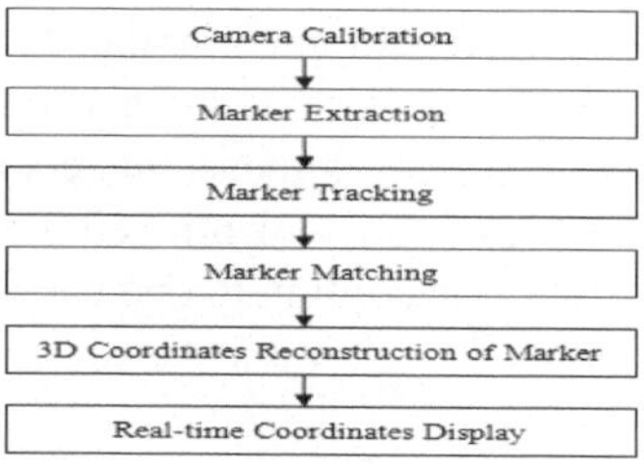


**Fig. 2.** The system's main flowchart

### 1) Camera Calibration

Direct Linear Transformation (DLT) method is used for camera calibration in this paper. We choose several specific points with knowing their 3D coordinates in the world coordinate system and their 2D image coordinates in each camera's image coordinate system, according to their relative coordinates, we build DLT equations of these points. And the DLT coefficients are obtained by solve DLT equation. Then we get the parameters of the camera indirectly.

**2) Marker Extraction**

In this paper, we adopt background subtraction method to extract five infrared markers when the experimental background hardly changes. Before subject walk, the background image is captured by cameras. After that, we subtract the gray-value of each pixel in the background image from the gray-value of the corresponding pixel in the walking image. If the difference is larger than the threshold, we record the coordinates of the pixel which composes the marked points; otherwise, the pixel will be dropped. In this way, we save lots of memory by only recording the markers image instead of the whole image. Then, the mass center of every marker can be calculated according to the recorded information. The method acts well in the constant background.

**3) Marker Tracking**

In fact, two cameras are unable to capture all the markers at all time because some markers may be out of the camera's field of view or be blocked. If system finds it happens, it will abandon this frame. DSP will not process image until the markers are all captured in an image.

**4) Marker Matching**

We consider that the mass center of marker represents the marker's center. For each marker, we obtain one mass center in one camera. Now there are ten centers of five markers in two cameras. We have to decide which two mass centers correspond to a same marker. Fortunately,, the markers pasted on the subject are arranged regularly in our system. They are arranged in two columns, and five markers in each column. Then, we just need to match the points by column, that is, the first point in one column matches the first one in the other column in the same direction, the second matching the second and so on. Therefore, every marker has two corresponding mass centers in two cameras.

**5) 3D Coordinates Reconstruction of Marker**

Now we know the image coordinates of each marker in two cameras, and previously obtained camera transform matrix of sizes 3*4. Then we will utilize binocular vision principle to calculate the markers' three-dimensional coordinates.

Here, we provide an overview of above mentioned method as follows.

For each point, we have Eq. (1) according to one camera,

$$Z_c \begin{pmatrix} u_i \\ v_i \\ 1 \end{pmatrix} = \mathbf{C} \begin{pmatrix} X \\ Y \\ Z \\ 1 \end{pmatrix} \tag{1}$$

where $\mathbf{C}$ is the 3*4 calibration matrix,

$$\mathbf{C} = \begin{pmatrix} a_{11} & a_{12} & a_{13} & a_{14} \\ a_{21} & a_{22} & a_{23} & a_{24} \\ a_{31} & a_{32} & a_{33} & a_{34} \end{pmatrix} \tag{2}$$

$(u_i \quad v_i)$ represents the image coordinates of a point and is already known; $(X \quad Y \quad Z)$ represents the world coordinates of a point and is to be solved; $(X_c \quad Y_c \quad Z_c)$ represents the camera coordinates of a point in the camera coordinates system.

The Eq. (1) is transform to Eq. (3) by dropping $Z_c$,

$$\begin{cases}(a_{11}-a_{31}u_i)X+(a_{12}-a_{32}u_i)Y+(a_{13}-a_{33}u_i)Z=a_{34}u_i-a_{14}\\(a_{21}-a_{31}v_i)X+(a_{22}-a_{32}v_i)Y+(a_{23}-a_{33}v_i)Z=a_{34}v_i-a_{24}\end{cases} \tag{3}$$

The Eq. (3) can be rewritten with two cameras, as Eq. (4) shows,

$$\begin{cases}(a_{11}-a_{31}u_i)X+(a_{12}-a_{32}u_i)Y+(a_{13}-a_{33}u_i)Z=a_{34}u_i-a_{14}\\(a_{21}-a_{31}v_i)X+(a_{22}-a_{32}v_i)Y+(a_{23}-a_{33}v_i)Z=a_{34}v_i-a_{24}\\(a'_{11}-a'_{31}u'_i)X+(a'_{12}-a'_{32}u'_i)Y+(a'_{13}-a'_{33}u'_i)Z=a'_{34}u'_i-a'_{14}\\(a'_{21}-a'_{31}v'_i)X+(a'_{22}-a'_{32}v'_i)Y+(a'_{23}-a'_{33}v'_i)Z=a'_{34}u'_i-a'_{24}\end{cases} \tag{4}$$

Finally we solve the Eq. (4) by standard least squares approach and get the 3D coordinates of a marker in the world coordinates system.

**6) Real-Time Coordinates Display**

3D coordinates of the markers are displayed on the screen in real-time, the precision of each data is kept two number after decimal point. We take the value -510.28 as an example to show how the value is displayed on the screen. Firstly, we built a character bank (such as number 0~9, negative sign (-) and decimal point (.)), then we design the character display size 30*30. After that, we allocate each character (5, 1, 0, 2, and 8), negative sign (-) and decimal point (.) to assigned memory. Finally, DSP outputs the corresponding image.

## 3 Results

Place two cameras at the end of the three-meters passage, the cameras facing the subject to collect subject walking image as far as possible. With the optical filters placed before the lens and the subject walks in clear background, the infrared markers can be extracted correctly. Set DSP in dual-channel input and single-channel output mode, then the screen can display the images collected by two cameras simultaneously. Fig. 3 shows the images displayed on the screen at the same time which are collected by two cameras without filters before the lens.

During the experiment, we substitute one leg skeleton model for the subject. Paste one marker in the waist, knee, ankle and the middle of these joints respectively. Place the model in the passage statically. Two cameras collect this image, and then DSP saves this static image and uses it as background image. After this, turn on the marker to glow the infrared light.

Move the model to simulate human walking, meanwhile, DSP extracts markers by background image subtraction. Then we can calculate the mass center of markers in order to reconstruct their three-dimensional coordinates. And DSP will display the results on the screen. Fig. 4 shows the image after background subtraction. Fig.5 shows the 3D coordinates of five markers were displayed on the screen.

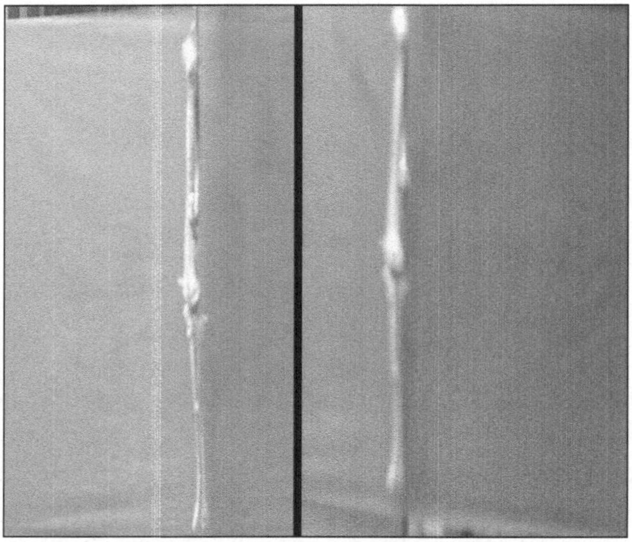

**Fig. 3.** Background image without filters, the left is collected by the first camera and the right is by the second camera

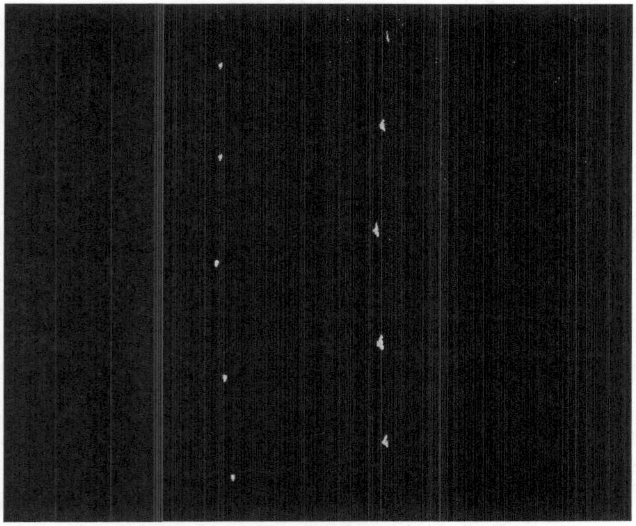

**Fig. 4.** The image obtained by background subtraction

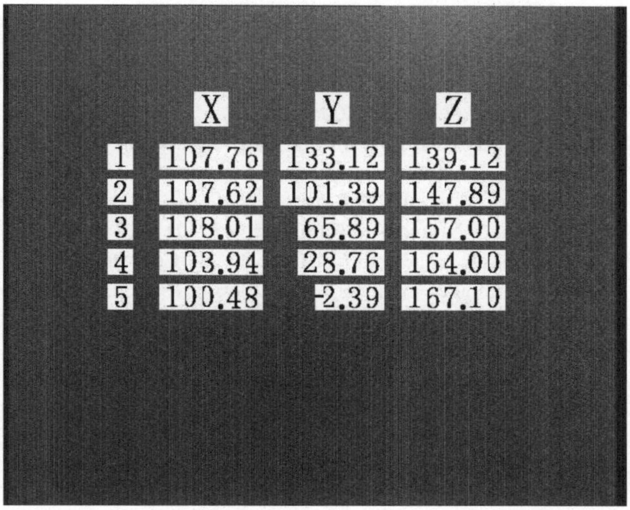


**Fig. 5.** The 3D reconstruction results displayed in real time

## 4 Discussions

A portable gait analysis system based on DSP is described in this paper. In this system, the three-dimensional coordinates of markers and its relative angles were calculated and displayed on the screen in real time. Because our system can track multiply markers pasted on the patient in real-time without saving video sequences as usual gait system done, and thus, it does not require a specific data processor to process the video data, and thus has significant savings in storage space. The method of three-dimensional reconstruction of markers by using DSP is proved to be effective according to the result of the experiment.

In this paper, we adopt the missing marker discarding method for motion tracking, and this method is so simple that some valid data is easy missing. It requires further improvement. In this paper, human gait is captured and the three-dimensional coordinates of markers are calculated in DSP, and then to calculate the gait parameters. All the gait parameter curves can be displayed on the screen in real time will be developed in our future work.

**Acknowledgments.** This work is partly supported by Shenzhen-Hongkong Innovative Circle Project of Study on Key Technologies of Computer-Assisted TKR Surgery and Postoperative Gait Analysis and Shenzhen key laboratory of non-destructive monitoring and minimally invasive medicine.

## References

1. Bai, Y., Zhou, J., Liang, J.: Application of Gait Analysis in Orthopaedic and Rehabilitation Medicine. J. Orthopedic Journal of China 10, 787–789 (2006)
2. Chen, D., Ding, H., Zhou, X.: A System of Measuring and Analyzing Gait of Human Walking. J. Chinese Journal of Biomedical Engineering 2, 133–141 (1997)

3. Zhao, L., Zhang, L., et al.: Review of the Research Technique in Human Gait Kinematics. J. Measurement & Control Technology 12, 1–3 (2007)
4. Hu, Q., Wang, W., Xia, S.: Human Gait Tracking Method Based on Multiple Cameras. J. Computer Engineering 22, 220–222 (2008)
5. Jia, Q., Xie, Q., Dong, B.: Real-time Motion Target Detecting and Tracking System Based on TMS320DM642. J. Computer Measurement & Control 3, 469–471 (2009)
6. Xue, Z., Jin, J., Ming, D., et al.: The Present State and Progress of Researches on Gait Recognition. J. Journal of Biomedical Engineering 5, 1217–1221 (2008)
7. Luo, S., Li, M.: Research on how to Get Object's 3D Coordinate on Two CCD Cameras Measure System. J. Computer Engineering and Design 19, 3622–3624 (2006)
8. Xia, N.: Camera Calibration of DLT Based on Method of Least Square. J. Information Technology 5, 98–100 (2006)
9. Qu, X., Zhang, L.: 3D Measurement Method Based on Binocular Vision Technique. J. Computer Simulation 2, 373–377 (2011)
10. Xue, C., Yang, G., Li, Z.: Study of Synchronous Movement of Dual CCD Based on DM642. J. Computing Technology and Automation 4, 24–27 (2006)

# Improved Neural Networks Based Method for Infrared Focal Plane Arrays Nonuniformity Correction

Dongjie Tan and An Zhang

School of Electronics and Information,
Northwestern Polytechnical University, Xi'an, 710072, China
tdj2011@sohu.com

**Abstract.** The non-uniform response in infrared focal plane array (IRFPA) detectors produces corrupted images with a fixed pattern noise. In this study, an improved neural networks based method for nonuniformity correction (NUC) is presented. In the improved method, the correction process of neural networks is decomposed from one step to two steps to fine the correction results. Besides, it uses a local median value of each neuron's output as the desired output for each neuron. Experimental results show that the improved algorithm can eliminate fixed stripe noise and stochastic noise in raw images and make infrared images more slippery.

**Keywords:** infrared focal plane array, nonuniformity correction, neural networks.

## 1 Introduction

Infrared detectors are widely used in a variety of applications such as defense, surveillance, remote sensing and astronomy. Usually, infrared imaging sensors are based on the infrared focal plane array (IRFPA) technology [1-3], because the IRFPA has advantages of simple configuration, high reliability and high sensitivity. An IRFPA can be considered as an array of independent detectors aligned at the focal plane of the imaging system. However, every detector on the IRFPA can have different responses under the same stimulus, because of detector material nonuniformity, material etching techniques, differences of signal processing circuits integrated into the IRFPA. This is called nonuniformity, leading then to the fixed-pattern-noise (FPN) superimposed on the true image. Even more, what makes matter worse is that the nonuniformity slowly varies over time. The nonuniformity greatly limits the application of the IRFPA, therefore, nonuniformity correction (NUC) is a necessary task to be performed in order to get higher quality infrared images. Generally, NUC algorithms can be divided into two categories: reference-based correction methods [4, 5] and scene-based correction methods [6-12]. NUC methods based on reference can be implemented real-timely. However, due to the temporal drift of the nonuniformity characteristic parameters, these algorithms must be often repeated to calibrate the parameters which must interrupt the system's operation.

J. Lei et al. (Eds.): AICI 2012, LNAI 7530, pp. 97–104, 2012.

NUC algorithms based on scene can update correction parameters got from the estimation of scene adaptively.

Neural networks based [6, 7] NUC method belongs to the latter which consists of an array of linear neurons that are connected to each detector's output, acting as the inverse model for each detector. Parameters of neurons (bias and weight) are updated through a steepest descent linear regression, using a local average value of each neuron's input as the desired output for each neuron. However, the correction precision of this algorithm is finite because there is only one step in the correction process of neural networks. And the correction effect is not perfect when the nonuniformity is serious, for example, there is fixed stripe noise.

In this study, an improved neural networks based NUC method is proposed. The main improvement is that the improved algorithm fines the correction process of neural networks which is changed from one step to two steps. Besides, it uses a local median value of each neuron's output as the desired output for each neuron because nonuniformity of the corrected image will be weakened gradually along with the increase of iteration and the result of median filer can prevent image from blurring. In this way, the improved algorithm will have higher correction precision and better correction effect.

## 2 Neural Networks Base NUC Method

Usually, an infrared detector is characterized by a linear model [8]. Then, for the $ij^{th}$ detector in the focal plane array, the readout signal $Y_{ij}$ at a given time $n$ can be expressed as:

$$Y_{ij}(n) = a_{ij}(n) \bullet X_{ij}(n) + b_{ij}(n) \tag{1}$$

where $a_{ij}(n)$ and $b_{ij}(n)$ are the gain and the offset of the $ij^{th}$ detector, and $X_{ij}(n)$ is the real infrared radiation. Equation (1) can be reordered as follows:

$$X_{ij}(n) = g_{ij}(n) \bullet Y_{ij}(n) + o_{ij}(n) \tag{2}$$

where the correction parameters $g_{ij}(n)$ and $o_{ij}(n)$ are related to the real gain and offset parameters of the detector, as expressed in the following expressions:

$$g_{ij}(n) = \frac{1}{a_{ij}(n)} \qquad o_{ij}(n) = -\frac{b_{ij}(n)}{a_{ij}(n)} \tag{3}$$

The main idea of the NUC algorithm based on neural networks relies in estimating the correction parameters of each detector using only the readout data $Y_{ij}(n)$.

In the algorithm, every neural cell is connected to an infrared detector. And a hidden layer is designed to compute the average value of local neighboring infrared detector output values. These average values are feedback to NUC layer to form the error function $E_{ij}(n)$ for each neuron defined as the difference between a desired target value $T_{ij}(n)$ and the corrected value $X_{ij}(n)$. At last, the method of steepest descent is used to update correction parameters.

The desired target value $T_{ij}(n)$ is calculated as the average value of the four nearest neighbor pixels:

$$T_{ij}(n) = \left[ Y_{i,j+1}(n) + Y_{i-1,j}(n) + Y_{i,j-1}(n) + Y_{i+1,j}(n) \right] / 4 \tag{4}$$

Then the error function $E_{ij}(n)$ can be defined in the following expression:

$$E_{ij}(n) = \frac{1}{2}\left[ X_{ij}(n) - T_{ij}(n) \right]^2 = \frac{1}{2}\left[ g_{ij}(n) \bullet Y_{ij}(n) + o_{ij}(n) - T_{ij}(n) \right]^2 \tag{5}$$

By using the method of steepest descent, correction parameters can be updated. The correction expressions are described as follows:

$$\begin{aligned} g_{ij}(n+1) &= g_{ij}(n) - \lambda \bullet \frac{\partial E_{ij}(n)}{\partial g_{ij}(n)} = g_{ij}(n) - \lambda \bullet \left[ X_{ij}(n) - T_{ij}(n) \right] \bullet Y_{ij}(n) \\ o_{ij}(n+1) &= o_{ij}(n) - \lambda \bullet \frac{\partial E_{ij}(n)}{\partial o_{ij}(n)} = o_{ij}(n) - \lambda \bullet \left[ X_{ij}(n) - T_{ij}(n) \right] \end{aligned} \tag{6}$$

where $n$ is frame number in an infrared image sequence and $\lambda$ is a learning rate which controls convergence speed. Then the traditional NUC algorithm based on neural networks can be summarized as follows:

**Step1:** Input the readout data $Y_{ij}(n)$.

**Step2:** Calculate the desired target value:

$$T_{ij}(n) = \left[ Y_{i,j+1}(n) + Y_{i-1,j}(n) + Y_{i,j-1}(n) + Y_{i+1,j}(n) \right] / 4 .$$

**Step3:** Correct the input data: $X_{ij}(n) = g_{ij}(n) \bullet Y_{ij}(n) + o_{ij}(n)$

**Step4:** Update correction parameters using the method of steepest descent:

$$\begin{aligned} g_{ij}(n+1) &= g_{ij}(n) - \lambda \bullet \left[ X_{ij}(n) - T_{ij}(n) \right] \bullet Y_{ij}(n) \\ o_{ij}(n+1) &= o_{ij}(n) - \lambda \bullet \left[ X_{ij}(n) - T_{ij}(n) \right] \end{aligned}$$

The principle of the traditional NUC algorithm based on neural networks is shown in Fig. 1.

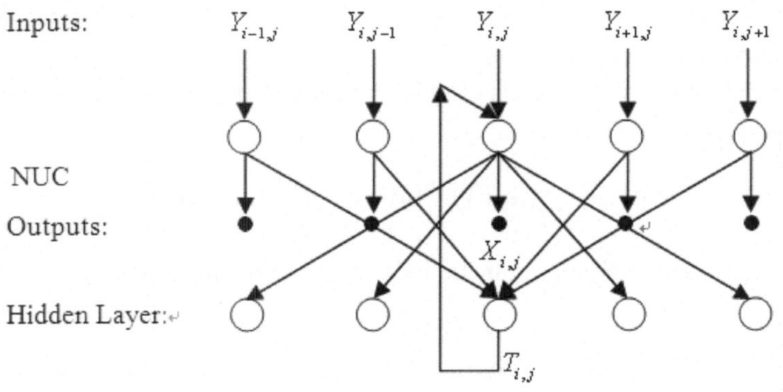


**Fig. 1.** The principle of the traditional NUC algorithm based on neural networks

## 3 Improved Neural Networks Based NUC Method

For traditional algorithm, there is only one step in NUC process:

$$X_{ij}(n) = g_{ij}(n) \bullet Y_{ij}(n) + o_{ij}(n)$$

We can fine the NUC process and change it form one step to two steps which are expressed as follows:

The first step:

$$\hat{X}_{ij}(n) = g_{ij}^{(1)}(n) Y_{ij}(n) + o_{ij}^{(1)}(n) \tag{7}$$

The second step:

$$X_{ij}(n) = g_{ij}^{(2)}(n) \hat{X}_{ij}(n) + o_{ij}^{(2)}(n) \tag{8}$$

The two pairs of parameters are like the inching button and speediness button of the experiment instrument. Parameters $g_{ij}^{(2)}(n)$ and $o_{ij}^{(2)}(n)$ find the rough range firstly, then Parameters $g_{ij}^{(1)}(n)$ and $o_{ij}^{(1)}(n)$ adjust values fine which can make the correction more precise.Then the error function $E_{ij}(n)$ can be defined in the following expression:

$$E_{ij}(n) = \frac{1}{2}[X_{ij}(n) - T_{ij}(n)]^2 = \frac{1}{2}[g_{ij}^{(2)}(n) \bullet (g_{ij}^{(1)}(n) \bullet Y_{ij}(n) + o_{ij}^{(1)}(n)) + o_{ij}^{(2)}(n) - T_{ij}(n)]^2 \tag{9}$$

Beside the main improvement, we also improve the estimation method of the desired output for each neuron. In the traditional algorithm, the desired output is calculated as the average value of each neuron's input. Actually, nonuniformity of the corrected image will be weakened gradually along with the increase of iteration and the result of median filer can prevent image from blurring. So the desired output $T_{ij}(n)$ can be calculated as the local median value of each neuron's output that is expressed as follows:

$$T_{ij}(n) = f_{median}\left[X_{i,j+1}(n), X_{i-1,j}(n), X_{i,j-1}(n), X_{i+1,j}(n)\right] \quad (10)$$

where $f_{median}(*)$ is the median filter function. Then NUC algorithm using improved neural networks can be summarized as follows:

**Step1:** Input the readout data $Y_{ij}(n)$.

**Step2:** Correct the input data:

The first step: $\hat{X}_{ij}(n) = g_{ij}^{(1)}(n)Y_{ij}(n) + o_{ij}^{(1)}(n)$

The second step: $X_{ij}(n) = g_{ij}^{(2)}(n)\hat{X}_{ij}(n) + o_{ij}^{(2)}(n)$

**Step3:** Calculate the desired target value:

$$T_{ij}(n) = f_{median}\left[X_{i,j+1}(n), X_{i-1,j}(n), X_{i,j-1}(n), X_{i+1,j}(n)\right].$$

**Step4:** Update correction parameters using the method of steepest descent:

$$\begin{aligned}
g_{ij}^{(1)}(n+1) &= g_{ij}^{(1)}(n) - \lambda_1 \bullet [X_{ij}(n) - T_{ij}(n)] \bullet g_{ij}^{(2)} \bullet Y_{ij}(n) \\
o_{ij}^{(1)}(n+1) &= o_{ij}^{(1)}(n) - \lambda_1 \bullet [X_{ij}(n) - T_{ij}(n)] \bullet g_{ij}^{(2)} \\
g_{ij}^{(2)}(n+1) &= g_{ij}^{(2)}(n) - \lambda_2 \bullet [X_{ij}(n) - T_{ij}(n)] \bullet \hat{X}_{ij}(n) \\
o_{ij}^{(2)}(n+1) &= o_{ij}^{(2)}(n) - \lambda_2 \bullet [X_{ij}(n) - T_{ij}(n)]
\end{aligned} \quad (11)$$

# 4 Experimental Results

## 4.1 The Quantitative Criterion for NUC

The sharpness [9] is used to evaluate the NUC results with different methods. The idea of this quantitative criterion is that successful NUC algorithm should attenuate high frequency energy due to fixed pattern noise. The sharpness criterion is given by

$$\rho = \frac{\|f * h\|_1}{\|f\|_1} \quad (12)$$

where $f$ respect a digital image, $h$ is a discrete Laplacian convolution kernel, $*$ refers to discrete convolution and $\| \ \|_1$ refers to an $L^1$ norm. The smaller an NUC algorithm's sharpness is, the better this algorithm performs.

## 4.2 Comparison of Experimental Results

Different NUC algorithms are tested with sequences of real infrared data including NUC algorithm based on neural networks, NUC algorithm based on high-pass filter [10], NUC algorithm using constant statistics method [9], NUC algorithm based on Kalman filter [11] and NUC algorithm using improved neural networks.

Because the size of the infrared picture is 128*128, the node number of input layer, hidden layer and output layer of neural networks are all 128*128. The traditional NUC algorithm based on neural networks uses a learning rate of $\lambda = 0.000024$. The NUC algorithm using improved neural networks uses learning rates of $\lambda_1 = 0.000016, \lambda_2 = 0.000008$.

We use the last frame corrected image as the correction result image which can be compared. Correction result images of different algorithms are shown in Fig. 2 and the sharpness value of different algorithms are shown in Table 1.

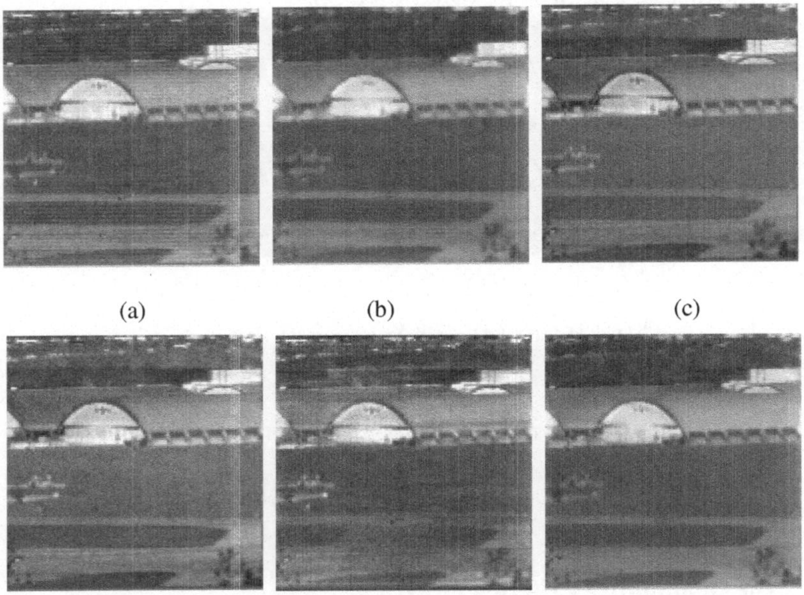

(a) (b) (c)

(d) (e) (f)

**Fig. 2.** Correction result images. (a) Raw image. (b) Corrected using traditional NUC algorithm based on neural networks. (c) Corrected using NUC algorithm based on high-pass filter. (d) Corrected using NUC algorithm using constant statistics method. (e) Corrected using NUC algorithm based on Kalman filter. (f) Corrected using NUC algorithm using improved neural networks.

**Table 1.** The sharpness value of different algorithms

| Algorithm | Sharpness |
|---|---|
| Unprocessed | 0.15320 |
| Traditional NUC algorithm based on neural networks | 0.09365 |
| NUC algorithm based on high-pass filter | 0.10920 |
| NUC algorithm using constant statistics method | 0.14283 |
| NUC algorithm based on Kalman filter | 0.14738 |
| NUC algorithm using improved neural networks | 0.08975 |

Fig. 2 shows that the improved algorithm can eliminate fixed stripe noise and stochastic noise in raw images and make infrared images more slippery. In these aspects, the improved algorithm performs better than the other four algorithms proving that the improved algorithm has higher correction precision and better correction effect. From the quantitative criterion for NUC shown in Table 1, the sharpness value of the improved algorithm is the least which prove that the improved algorithm is better than the other four algorithms which is consistent with above analysis.

# 5 Conclusion

In this work, an improved neural networks based NUC method is proposed. The improved algorithm fines the correction process of neural networks which is changed from one step to two steps. Besides, it uses a local median value of each neuron's output as the desired output for each neuron. From experimental results, the improved algorithm can eliminate fixed stripe noise and stochastic noise in raw images and make infrared images more slippery. By comparing the improved algorithm with other four algorithms, we can find that the improved algorithm has higher correction precision and better correction effect.

# References


1. Scribner, D., Kruer, M., Killiany, J.: Infrared Focal Plane Array Technology. Proceedings of the IEEE, 66–85 (1991)
2. Holst, G.: CCD Arrays, Cameras and Displays. JCD Publishing, Bellingham (1996)
3. Driggers, R., Cox, P., Edwards, T.: Introduction to Infrared and Electro-Optical Systems. ArtechHouse Publishers, Boston (1999)
4. Hormans, R., Hepfer Kenneth, C., Matthew, Z.: Uniformity Compensation for High Quantum Efficiency Focal Arrays. In: Proceedings of SPIE, pp. 154–164 (1996)
5. Yin, S., Liu, S.: The Multi-Point Non-Uniformity Correction Algorithms for IRFPA Based on Low Order Interpolation. Acta Photonic Asinica 31(6), 715–718 (2002)
6. Scribner, D.A., Sarkady, K.A., et al.: Adaptive Nonuniformity Correction for Infrared Focal-Plane Arrays Using Neural Networks. In: Proceedings of SPIE, pp. 100–109 (1991)
7. Scribner, D.A., Sarkady, K.A., et al.: Adaptive Retina-Like Preprocessing for Imaging Detector Arrays. Proceedings of the IEEE, 1955–1960 (1993)
8. Torres, S.N., Vera, E.M., et al.: Vera, et al: Adaptive Scene-Based Non-Uniformity Correction Method for Infrared-Focal Plane Arrays. In: Proceedings of SPIE, pp. 130–139 (2003)
9. Hardie, R.C., Baxley, F., et al.: Scene-Based Nonuniformity Correction with Reduced Ghosting Using a Gated LMS Algorithm. Optics Express 17(17), 14918–14933 (2009)
10. Chanlao, L., Linxun, T., et al.: The Improvement of Nonuniformity Correction Algorithm for Infrared Images Based on High-Pass Filter. Infrared Technology 28(8), 439–442 (2006) (in Chinese)

11. Torres, S.N., Hayat, M.M.: Kalman Filtering for Adaptive Nonuniformity Correction in Infrared Focal Plane Arrays. The Journal of the Optical Society of America A 20(3), 470–480 (2003)
12. Vera, E., Torres, S.: Fast Adaptive Nonuniformity Correction for Infrared Focal-Plane Array Detectors. EURASIP Journal on Applied Signal Processing 2005(13), 1994–2004 (2005)
13. Narayanan, B., Hardie, R.C., Muse, R.A.: Scene-Based Nonuniformity Correction Technique That Exploits Knowledge of the Focal-Plane Array Readout Architecture. Applied Optics 44(17), 3482–3491 (2005)

# Images Classifications Based on Color-Texture Feature

Kegang Wang[1,2], Liying Qi[3], and Guohua Geng[1]

[1] School of Information Science and Technology, Northwest University, China
Kg_wang@qq.com
[2] Department of Electronic and Information Engineering, AnKang University, China
akuwkg@aku.edu.cn
[3] Department of Mathematics, AnKang University, Ankang, China
648494871@qq.com

**Abstract.** This paper puts forwards a method of using MGabor filter-banks to extract Texture-Color features from digital images, and then to construct a group of support vector machines (SVM) classifiers to automatically and accurately classify color digital images. Successful experiments are conducted on the Simplicity and Brodatz image set and our own Ancient shards image sets. The experiments results show the proposed method can integrate the texture features and color information to further improve distinguishing ability of each category images.

**Keywords:** Image, Color, Texture, Classify, SVM.

## 1 Introduction

Texture is one of the important characteristics of image, which is a variation law about the pixel luminance or color of images on the vision, reflecting the relationship of the brightness in a local area and the whole color. In the study about image classification based on texture Yang jie[1] and others utilized wavelet transform, combining shape with texture, to extract the texture feature of image and realize classification. Wang run-sheng [2] applied Gabor transform and independent component analysis technique to extract texture feature. Xie shiming [3] have researched the description method about sparse texture. Shang yan [4] etc. presented the texture classification algorithm which is built on SVM method and double complex wavelet transform based on log-polar transform. In these studies, all extracted texture feature are gray images derived from color images. However, the color, which should have been regarded as the important feature of images, doesn't get enough attention.

## 2 The Model of Extracting Color Texture

Color is the important visual feature. It is generally acknowledged that color and texture of images are unrelated, so the two features are extracted and applied respectively. Actually, color variation of images can get rise to the strong visual

J. Lei et al. (Eds.): AICI 2012, LNAI 7530, pp. 105–112, 2012.

difference in image texture. When color image is transformed to gray mode, the images which should have been significant difference get the high similarity. The reason of the phenomenon is that the strongly different colors may possess the same gray level when color images are transformed to gray images[8], for example, gray level 120 may proceed from blue(0,158,250) and also from green(92,154,0), which may lead to the situation that after one image which's major color components are blue(0,158,250) and green(92,154,0) is transformed to it's gray mode, the obvious color difference will be cleaned away because of the same gray level. Therefore, because of the loss of color information, the visual diversity among color images is decreased and the visual feature of images is weakened in the process of gray transformation for image. So, the integration of color and texture information of images and computing the color-texture feature value to prevent from the loss of color information in gray transformation process before texture extraction may be a better practice. According to the above idea, this paper presents a color image texture extraction model based on color-texture feature, such as Figure 1.

In the model showed as Figure1, color space of the original color image is changed into the more practical color space according to the application requirements without gray transformation, for instance, RGB color space may be transformed to HSV color space which is more accord with human vision, and in the new color space image is layered extraction, thus, a series of pseudo-gray images can be attained. The textures of each pseudo-gray image reflect the most relevant texture characteristic between the original and the pseudo-gray color tones. Finally, the texture feature of the group of pseudo-gray images make up the color-texture feature of original image.

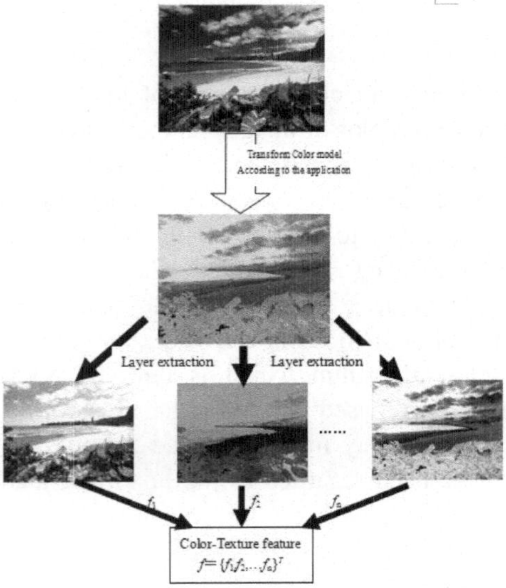


**Fig. 1.** The Model of extracting Color Texture

In the model, one image is changed into a group of pseudo-gray images through image layering, and the texture feature extraction for one image becomes extraction for a group of images, which will increase the time of extraction. Therefore, in practical application, choosing the rational extraction method is very important to improve the efficiency of extraction.

## 3 Color-Texture Feature Extraction Based on MGabor Transformation

### 3.1 Modified Gabor Transformation

The general form of two-dimensional Gabor function is as follows,

$$g_{uv}(x,y)=\frac{k^2}{\sigma^2}\exp(-\frac{k^2(x^2+y^2)}{2\sigma^2})[\exp(ik\cdot\begin{bmatrix}x\\y\end{bmatrix})-\exp(-\frac{\sigma^2}{2})] \tag{1}$$

The process of extracting texture feature of image f through Gabor wavelet transform is following,

$$F_{u,v}(x,y)=f*g_{u,v}(x,y)=\int_{-\infty}^{+\infty}f(\xi,\eta)\overline{g}(x-\xi,y-\eta)d\xi d\eta \tag{2}$$

$$M_{u,v}=\int_{-\infty}^{+\infty}\int_{-\infty}^{+\infty}|F_{u,v}(x,y)|dxdy \tag{3}$$

$$\sigma_{u,v}=\int_{-\infty}^{+\infty}\int_{-\infty}^{+\infty}|F_{u,v}(x,y)-M_{u,v}|^2dxdy \tag{4}$$

$\overline{g}$ represents the complex conjugation of $g$. The texture feature of image comprise a group of energy distribution means $M_{u,v}$ and variances $\sigma_{u,v}$ which are worked out according to the above process with a group of different direction and center frequency Gabor filter.

The normal Gabor transform is interrelated with direction parameters, which's selection range is $[0,2\pi)$ . The variation in direction responds to the direction information in image. Generally a group of direction parameters should be selected from the range. Further, MGabor transform [6] eliminate the dependency of Gabor transform to direction. The definition of MGabor transform is as follows.

$$\begin{aligned}G_v(x,y)&=\frac{1}{2\pi}\int_0^{2\pi}g_{u,v}(x,y)du\\&=\frac{1}{2\pi\sigma^2}\int_0^{2\pi}k^2\exp(-\frac{k^2(x^2+y^2)}{2\sigma^2})[\exp(ik\cdot\begin{bmatrix}x\\y\end{bmatrix})-\exp(-\frac{\sigma^2}{2})]du\\&\approx\frac{1}{2\pi K\sigma^2}\sum_{n=0}^{K-1}e^{-\frac{1}{2}(\frac{(x\cos(\frac{2\pi n}{K})+y\sin(\frac{2\pi n}{K}))^2}{\sigma^2}+\frac{(-x\sin(\frac{2\pi n}{K})+y\cos(\frac{2\pi n}{K}))^2}{\sigma^2})}\\&\quad\times e^{2\pi i k_v(x\cos(\frac{2\pi n}{K})+y\sin(\frac{2\pi n}{K}))}\\&=\frac{e^{-\frac{x^2+y2}{2\sigma^2}}}{2\pi K\sigma^2}\sum_{n=0}^{K-1}e^{2\pi i k_v(x\cos(\frac{2\pi n}{K})+y\sin(\frac{2\pi n}{K}))},k_v=2^{-\frac{v+2}{2}}\pi\end{aligned} \tag{5}$$

Where the direction domain is discredited, $K$ represents the total scattering number. Referring to the texture extraction method of normal Gabor filter, the texture feature of image $f$ is defined as follows.

$$M_v = \int_{-\infty}^{+\infty}\int_{-\infty}^{+\infty} | f * G_v(x,y) | \, dxdy \tag{6}$$

$$\sigma_v = \int_{-\infty}^{+\infty}\int_{-\infty}^{+\infty} | f * G_v(x,y) - M_v |^2 dxdy \tag{7}$$

**Fig. 2.** M-Gabor filter-banks

The above M-Gabor is unrelated with direction parameter. In the below example, five center frequencies $k_{0,1,2,3,4} = \frac{\pi}{2}, \frac{\sqrt{2}\pi}{4}, \frac{\pi}{4}, \frac{\sqrt{2}\pi}{8}, \frac{\pi}{8}$ are selected, and the total scattering number is set at 64. Using the above parameters, M-Gabor filters-bank is constructed. In Figure2, these filters are ranged according the center frequencies $\frac{\pi}{2}, \frac{\sqrt{2}\pi}{4}, \frac{\pi}{4}, \frac{\sqrt{2}\pi}{8}, \frac{\pi}{8}$ from left to right. The upper part of center frequency is the real part of filter and the under part is imaginary part of filter.

### 3.2 Color-Texture Extraction Based on M-Gabor Transform

In the light of the above color-texture extraction model, let $I$ be color image, and RGB color space model is changed into HSV color model[7], the Hue, Saturation and Value component image which be let as $I_H$, $I_S$, $I_V$ are extracted and used as a group of gray images. Utilizing the constructed MGabor filter-bank containing five filters, the texture features are extracted from the three layer component images respectively, and then the color-texture feature is built for the original color image.

$$\begin{aligned}
M_{H,v} &= \int_{-\infty}^{+\infty}\int_{-\infty}^{+\infty} | I_H * G_v(x,y) | \, dxdy \qquad & \sigma_{H,v} &= \int_{-\infty}^{+\infty}\int_{-\infty}^{+\infty} | I_H * G_v(x,y) - M_{H,v} |^2 \, dxdy \\
M_{S,v} &= \int_{-\infty}^{+\infty}\int_{-\infty}^{+\infty} | I_S * G_v(x,y) | \, dxdy \qquad & \sigma_{S,v} &= \int_{-\infty}^{+\infty}\int_{-\infty}^{+\infty} | I_S * G_v(x,y) - M_{S,v} |^2 \, dxdy \\
M_{V,v} &= \int_{-\infty}^{+\infty}\int_{-\infty}^{+\infty} | I_V * G_v(x,y) | \, dxdy \qquad & \sigma_{V,v} &= \int_{-\infty}^{+\infty}\int_{-\infty}^{+\infty} | I_V * G_v(x,y) - M_{H,v} |^2 \, dxdy
\end{aligned} \tag{8}$$

In the process, every value in $k_{0,1,2,3,4} = \frac{\pi}{2}, \frac{\pi}{4}, \frac{\sqrt{2}\pi}{8}, \frac{\pi}{8}$ is used for *V*.

According to the above method, the color-texture feature vector of image ***I*** $(M_{H,0},\sigma_{H,0}, M_{H,1},\sigma_{H,1}, \ldots, M_{H,4},\sigma_{H,4}, M_{S,0},\sigma_{S,0}, M_{S,1},\sigma_{S,1}, \ldots, M_{S,4},\sigma_{S,4}, M_{V,0},\sigma_{V,0}, M_{V,1},\sigma_{V,1}, \ldots, M_{V,4},\sigma_{V,4})^{\mathrm{T}}$ can be attained.

## 4 Color-Texture Classifications Based on SVM

The description of Support Vector Machines is as follows, for the training sample set: $T = \{(x_1, y_1), \cdots, (x_l, y_l)\} \in (x \times y)^l, x_i \in X = R^n, y_i \in Y = \{1, -1\}, i = 1, 2, \cdots, l$
Constructing and solving the optimization problem:

$$\min_{\alpha} \frac{1}{2} \sum_{i=1}^{l} \sum_{j=1}^{l} y_i y_j \alpha_i \alpha_j K(x_i, x_j) - \sum_{j=1}^{l} \alpha_j$$

$$s.t. \sum_{i=1}^{l} y_i \alpha_i = 0, 0 \le \alpha_i \le C, i = 1, \cdots, l$$

In the above expression, $K(x, x')$ is kernel function, $C$ is penalty parameter. The optimal solution of the problem is

$$\alpha^* = (\alpha_1^*, \alpha_2^*, \cdots, \alpha_l^*)^T$$

Computing $w^* = \sum_{i=1}^{l} y_i \alpha_i^* K(x_i, x)$ and selecting $0 < \alpha_j^* \in \alpha^*$ , then computing $b^* = y_j - \sum_{i=1}^{l} y_i \alpha_i^* K(x_i, x_j)$, constructing separating hyper plane, finally, the two classification decision function $f(x) = \mathrm{sgn}((w^* \cdot x) + b^*)$ can be attained according to the hyper plane.

Using pairing classification, there are $M(M-1)/2$ decision functions $f_{i,j}(x), 1 \le i, j \le l, i \ne j$ for *M* categories shards samples. The category of samples is determined by voting mechanism.

The penalty factor *C* and kernel function parameters have a greater impact for performance of the SVM classifiers. Up till the present moment, there is no unified and effective way to SVM parameter optimization, the most common method is to use M-fold cross validation (M-fold cross validation), we obtain parameter (*C*, *g*) verify the classification accuracy on the training set under Rate, thus provide the parameters that we can choice the kernel function parameter g and the penalty factor *C*. In our experiments, the original data is divided into *M* group average, respectively, for each subset of data to do a validation set, the other subset of M-1 group as training sets of data, thus we can get hold of M models, compute average classification accuracy for these M models validation set as the classifier performance finally. If more than one set (*g*, *C*), select the minimum penalty factor *C* which can achieve the highest validation classification accuracy as the best parameters of the group, thus to avoid the problem of "over fitting" and "less fitting" [7].

## 5 Results Analysis

We tested the Brodatz standard image bank, Simplicity image bank and ourselves ancient shards image set by means of the above method.

In Brodatz image bank(Fig 3), 10 categories of texture images including brick, wood, building and cloud are selected. In each category, 20 images form a subset of Brodatz. For the subsets, with K-multiple cross-validation method, 10 images of each category are chosen as training samples and the other 10 are used as testing samples each time.

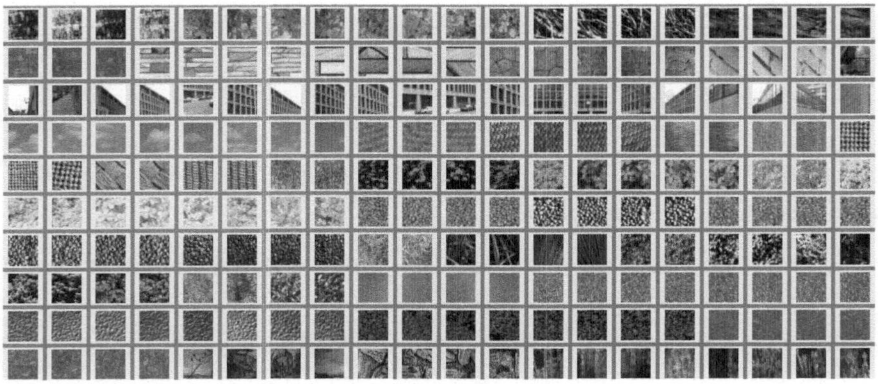

**Fig. 3.** Subset of Brodatz

**Fig. 4.** Subset of Simplicity image bank

In Simplicity image bank(Fig 4), there are 10 categories of images, which include bus, horse, flower, dinosaur, building, elephant, people, beach, scenery and dish, 100 images in each category. With K-multiple cross-validation method, 50 images of each category are chosen as training samples and the other 50 are used ad testing samples.

In the ancient shards image set(Fig 5), there are ten categories,136 images, and the size of each image is 256×256. These categories include 9 images at least and 14 images at most. When shard images are tested, half of images in a category are regarded as training samples and the other half are used as testing samples.

In addition, we adopt the classification accuracy as the performance indicator. The results are showed in the Table1. In the test, Gabor transform includes 12 standard Gabor filters [5] with 4 directions and 3 center frequencies.

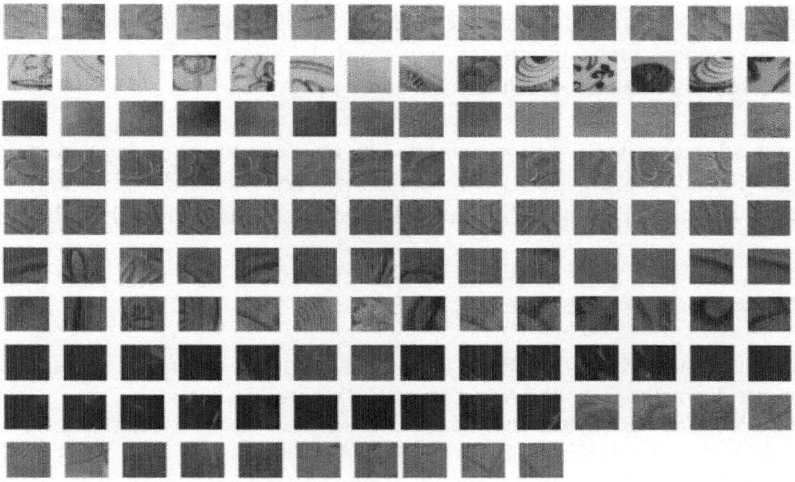

**Fig. 5.** Ancient shards image set

**Table 1.** Comparison of accurate-rate between Gray-Texture and Color-Texture

| Image Set and feature | | Gabor | M-Gabor |
|---|---|---|---|
| Brodatz | Gray-texture | 85.5% | 89.64% |
| | Color-texture | 87.79% | 95.43% |
| Simplicity | Gray-texture | 81.5% | 86.35% |
| | Color-texture | 86.62% | 90.67% |
| Ancient shards | Gray-texture | 60.41% | 85.32% |
| | Color-texture | 82.35% | 92.03% |

Analyzing the data of Table1, for Brodatz image set, comparing with gray-texture classification method, the Gabor transform improved the classification accuracy of 2.29% . While using MGabor transform, the classification accuracy for color-texture feature reaches 95.43% which is 5.79% higher than gray-texture method. For the ancient shard images, the normal Gabor filter make the classification accuracy of image color-texture 21.94% higher than the gray-texture method. When using MGabor transform, the classification accuracy is 6.71% higher as compared to adopting the gray-texture method. for Simplicity image set, comparing with gray-texture classification method, the Gabor transform improved the classification accuracy of 5.12% . While using MGabor transform, the classification accuracy for color-texture feature reaches 91.67% which is 4.32% higher than gray-texture method.

## 6 Conclusions

Color-texture feature provides stronger the recognizing ability than gray-texture feature. Despite that image layering may be take longer time to extract texture feature, the smaller number of filter used by MGabor transform does not increase the extraction time of texture feature about whole image too much. Integration of color and texture information can enhance category recognizing ability of texture feature, so in the practical application for ancient shard classification, the importance of color information should attract much attention.

**Acknowledgment.** This work is supported by the key project of the National Natural Science Foundation of China (No.61172170) and Scientific Research Program Funded by Shaanxi Provincial Education Department (Program No.11JK1053).

## References

1. Yang, J., Chen, X.-Y., Xu, R.-C.: Shape and texture-based image classification using wavelet. Journal of Computer Applications 27(2), 373–375 (2007)
2. Chen, Y., Wang, R.-S.: A Method for Texture Classification by Integrating Gabor Filters and ICA. Acta Electronica Sinica 35(2), 299–303 (2006)
3. Xie, S.-P., Hu, M.-L.: Parse Texture Representation and Classification. Application Research of Computers 24(3), 306–308 (2007)
4. Shang, Y., Lian, Q.-S.: Rotation invariant texture classification algorithm based on Log-Polar and DT-CWT. Computer Engineering and Applications 43(11), 48–50 (2007)
5. Wei, N., Geng, G., Zhou, M.: Content-based Image Retrieval Using Gabor Filters. Computer Engineering 31(8), 10–11 (2005)
6. Sastry, C.S., Ravindranath, M., Pujari, A.K., Deekshatulu, B.L.: A modified Gabor function for content based image retrieval. Pattern Recognition Letters 28, 293–300 (2007)
7. Shi, F., Wang, X., Yu, L.: 30 neural networks Case analyses Using Matlab. Beihang University Press, Beijing (2010)
8. Gonzalez, R.C., Woods, R.E.: Digital Image Processing. Publishing House of Electronics Industry, Beijing (2005)

# A Synchronization Strengthen RFID Authentication Protocol Based on Key Array

Yu Bai and Yanlong Liu

School of Electronic and Information Engineering
Tianjin University
Tianjin 300072, P.R. China
baiyu1978@tju.edu.cn

**Abstract.** If traditional RFID privacy authentication protocols are applied to multi-server environment, such as RFID credit card system and supply chain, the tags will be vulnerable to insider attack, an attacks from an internal legal entity who impersonates as another legal entity to do authority-exceeding violation. In 2011, H.Ning, et.al's proposed a RFID authentication protocol based on key array to prevent authority-exceeding violation. But their scheme adopts static key mechanism, which may be a security risk. In this paper, we proposed an improved scheme that use key array to avoid insider attack. Furthermore, counter are also employed to resolve synchronization issue. The security analysis demonstrates that our protocol is reliable in advanced RFID systems. Additionally, the performance is evaluated and compared with other related protocols to show that our protocol can improve the reliability and efficiency of RFID systems with insignificantly increased complexity.

**Keywords:** RFID, authentication, key array, synchronization.

## 1 Introduction

Radio Frequency IDentification (RFID) makes it possible to remote identification of objects through radio frequency. They are emerging in a variety of applications such as access control, avert counterfeit trade, tracking goods and assets, transport payments [1,2]. However, RFID system also results in significant user privacy and security issues [3-5]. Without protection, the private information related to a tag, such as an item's ID or user's location can be exposed to attackers, because most tags do not have any built-in feature to prevent attackers [6]. In order to solve this issue, many crypto-based RFID authentication protocols have been proposed [7-12]. However, all these schemes focus on the threats from the external illegal attackers, but ignore the attacks from the internal legal entities. In many mainstream protocols, legal readers are allowed to access entire data stored in tags. But the access control mechanism can result in insider attack in multi-server environment, such as RFID credit card system and supply chain, because internal legal entities have no trust in each other. Therefore, it is essential for authenticated entities to access the specified field areas in a tag (TID)[13].

J. Lei et al. (Eds.): AICI 2012, LNAI 7530, pp. 113–119, 2012.

In 2011, H.Ning,et.al [13]proposed a RFID authentication protocol can prevent authority-exceeding violation by key array. Meanwhile, pseudonyms can conserve memory and improve scalability efficiently. But, their protocol adopts static key mechanism, and key is always unchanged, which may be a security risk. In the paper, we refined the protocol in [13], get an improvement RFID authentication protocol suitable for multi-server system, which can resist various attacks and ensure privacy. In view of security, the dynamic key mechanism was employed, instead of static one. As each authentication finished, authentication key will be update. The dynamic key mechanism strengthens the security of system, but which can also result in synchronization issue. If it is fail for keys stored in server and tag to update synchronously, further RFID communication would be impossible. In our scheme, this problem was solved primly.

The remainder of this paper is organized as follows: we first introduce the application conditions for our new protocol in section 2. Then, we present the details of the improved protocol in section 3, followed by security analysis and performance evaluation of the protocol in section 4. Finally, we end with conclusion in section 5.

## 2 Backgrounds

In multi-server environment, a RFID system is composed of tags, readers and servers which keep databases of all the tags' identity information. The communication channel between a reader and the server is assumed to be secure, employing high speed local Ethernet, while the wireless communication channel between a reader and a tag is vulnerable.

According to the function of tag, RFID can be divided into four classes. (1) Class 1: the most basic RFID tags can only transmit an identification number; (2) Class 2: the more complex RFID tags compliant to the ISO 14443 standards; (3) Class 3: contactless smart cards, which can even be programmed to some degree. Besides, Class 3 tag additionally contains a battery, which can achieve a larger communication range. (4) Class 4: Their advantages compared to passive RFID tags are the higher transmission range, permanently powered clocks, the use of environment sensors and a possible multi-hop communication. The major drawback of RFID tags is their limited amount of available energy, which is obtained wirelessly from the reader (class 1 and 2) or from a battery (class 3). According to the requirements of hardware and software, the class 4 RFID tag was presumed in the scheme.

## 3 The New Protocol

In order to resolve such security and privacy problems from both externals and internal attacks, the server assigns a key for each of readers and tags. Moreover, random numbers are generated by the tag and reader for each session.

### 3.1 Prior Conditions

In the new protocol, it is assumed that the system consists of $N$ readers and $M$ tags. Initialization, two sets of ($PIDR_i$, $k^R_{ij}$, $R\_cout_{ij}$, $i$ =1…$N$, $j$ = 1…$M$) and ($PIDT_j$, $k^T_{ij}$,

$T\_cout_{ij}$) are assigned to the reader $R_i$ and $T_j$, respectively. Where, $PIDR_i$, $k^R_{ij}$ and $R\_cout_{ij}$ are the pseudonym, key and counter of reader $R_i$; while, the meanings of $PIDT_j$, $k^T_{ij}$ and $T\_cout_{ij}$ stored in tag $T_j$ are similar. $R\_cout_{ij} = T\_cout_{ij} = 0$, $k^R_{ij} = k^T_{ij}$, initially. We use one-way hash function $H$ to update key $k^T_{ij}$ and $k^R_{ij}$. $R\_cout_{ij}$ and $T\_cout_{ij}$, denoted respectively the update times of key $k^R_{ij}$ in $R_i$ and $k^T_{ij}$ in $T_j$, which are helpful for key update synchronously.

When a reader queries a tag, the tag first checks whether the reader is in its records. In a RFID system, a tag can be read by all readers. We establish key array $S_i$ in server as follows:

$$S_i = [s_{i1},\dots, s_{ij},\dots,s_{iM}]$$

Where, $s_{ij} = \{PIDR_i, k^R_{ij}, R\_cout_{ij}\}$.Moreover, to achieve authentication to readers, we build key array $Q_j$ in tag $j$ as follows:

$$Q_j = [q_{1j},\dots, q_{ij},\dots,q_{Nj}]$$

where $q_{ij} = \{PIDT_j, k^T_{ij}, T\_cout_{ij}\}$. Besides, all of readers, servers and tag in RFID system share a key $k_u$ to prevent external attacks.

### 3.2 Description of the Protocol

There are three phases in our improved scheme - the reader identification phase, the tag authentication phase and the server/reader authentication phase. The scheme is depicted in Figure 1. The communication details are described as follows.

**Reader Identification.** In this phase, symmetrical encryption is used to protect reader information [3].

(1) Reader $R_i$ generates a random number $r_{Ri}$, encrypts $PIDR_i \| r_{Ri}$ using $k_u$, and then sends $\{PIDR_i \| r_{Ri}\}_{ku}$ to tag $T_j$. Where □ denotes concatenate operator.
(2) As tag $T_j$ receives $\{PIDR_i \| r_{Ri}\}_{ku}$, it decrypts it with $k_u$ to obtain $PIDR_i$. Then the tag checks if there is $PIDR_i$ in its record $Q_j$. If no matching record is found, the reader is not authorized and the session is aborted. Otherwise, the tag can find the corresponding key $k^T_{ij}$ shared by itself and reader $R_i$ though $PIDR_i$. Thus, $R_i$ is successfully authenticated by $T_j$. Then $T_j$ generates a random number $r_{Tj}$; Thereafter, $T_j$ encrypts $r_{Tj} \| r_{Ri}$ by $k^T_{ij}$ to obtain $\{r_{T_j} \| r_{R_i}\}_{k^T_{ij}}$, and continues to encrypt its pseudorandom identifier $PIDT_j$, $\{r_{T_j} \| r_{R_i}\}_{k^T_{ij}}$ and $T\_cout_{ij}$ by $k_u$ to obtain $\{PIDT_j \| \{r_{T_j} \| r_{R_i}\}_{k^T_{ij}} \| T\_cout_{ij}\}_{k_u}$. The new ciphertext will be responded to $R_i$. Or else, $T_j$ will stop the authentication process. And then, $T_j$ update key by hash operation, $k^{T\prime}_{ij} = H(k^T_{ij})$, meanwhile $T\_cout_{ij}' = T\_cout_{ij} + 1$.
(3) Upon receiving $T_j$'s response, reader $R_i$ extracts $PIDT_j$ and $T\_cout_{ij}$ from $\{PIDT_j \| \{r_{T_j} \| r_{R_i}\}_{k^T_{ij}} \| T\_cout_{ij}\}_{k_u}$ with $k_u$. Then, $R_i$ forwards $PIDT_j$, $R\_cout_{ij}$ and $T\_cout_{ij}$ to its server for further authentication.

**Tag Authentication.** As receives the information from $R_i$, the server finds the corresponding key $k^R_{ij}$ with the help of $PIDT_j$, $R\_cout_{ij}$ and $T\_cout_{ij}$ in key array $S_i$. In view of synchronization issue, server compares $R\_cout_{ij}$ with $T\_cout_{ij}$, firstly. If $R\_cout_{ij} = T\_cout_{ij}$, then server delivers $k^R_{ij}$ to $R_i$, or else hash $k^R_{ij}$ $c$ times and $R\_cout_{ij}' = R\_cout_{ij} + c$ , where $c = T\_cout_{ij} - R\_cout_{ij}$, then server delivers $k'^R_{ij} = \underbrace{H(H(\cdots H}_{c}(k^R_{ij})))$ to $R_i$. $R_i$ decrypts $\{r_{T_j} \| r_{R_i}\}_{k^T_{ij}}$ to obtain $r_{Ri}$ and $r_{Tj}$ with $k'^R_{ij}$. $R_i$ checks whether the current computed $r_{Ri}$ equals the previous generated $r_{Ri}$ in process (1). If the two values are identical, $T_j$ will be successfully authenticated by $R_i$, then $R_i$ message server to $R\_cout_{ij}$ +1. Then $R_i$ continues to encrypt $r_{Tj}$ to obtain $\{r_{T_j}\}_{k^R_{ij}}$ , and forwards the ciphertext to $T_j$. Otherwise, $T_j$ is considered as an imitative entity, and $R_i$ will stop the authentication process with an error code.

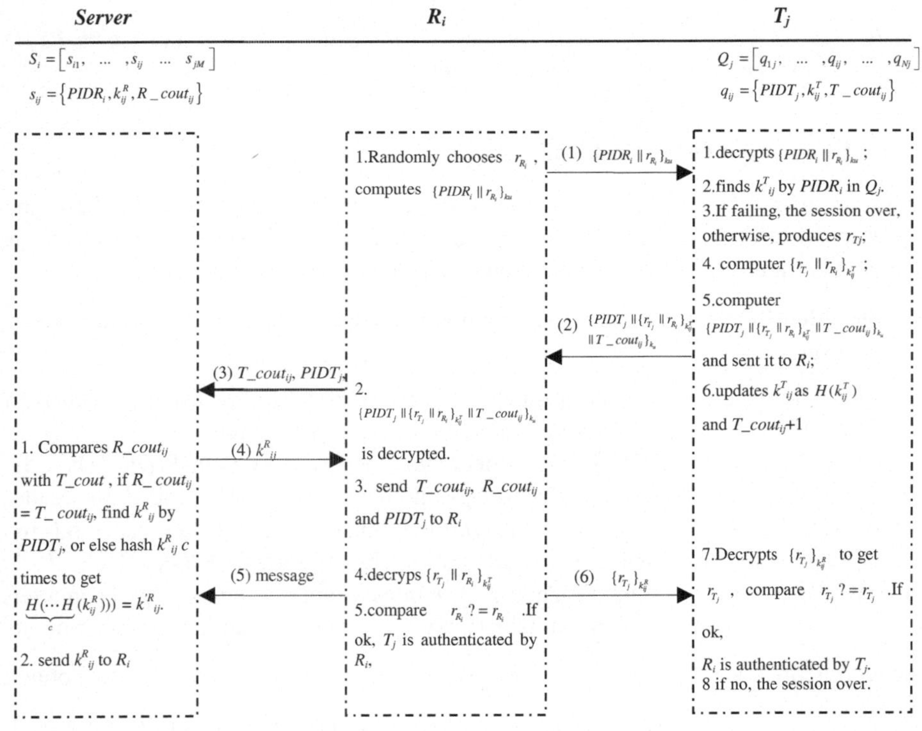


**Fig. 1.** The new protocol

**Server/Reader Authentication**

While receiving $\{r_{T_j}\}_{k^R_{ij}}$ , $T_j$ extracts $r_{Tj}$ by decryption with $k^T_{ij}$. $T_j$ checks whether the current computed $r_{Tj}$ equals the previous generate $r_{Tj}$ in process (2). If the two values are identical, $T_j$ sent the data specified to $R_i$. Otherwise, $R_i$ will be considered as an imitative entity, and $T_j$ will stop the authentication with an error code. Thus, the entire authentication is accomplished.

# 4 Evaluation

In this section, we evaluate the security of the new protocol and compare its performance with other schemes in terms of security attributes.

As stated in Section 3, tags and readers can successfully authenticate each other. This section will analyze the security of the new protocol by the required properties and the possible attacks as follows:

- Mutual authentication: our scheme provides mutual authentication between tags and readers by checking whether the computed value equals to the previously generated value. In the protocol, the reader is firstly identified by the tag. The tag checks whether it has the record $PIDR_i$ of reader $R_i$ by decrypting $\{PIDR_i \| r_{R_i}\}_{ku}$ from reader $R_i$. Only the recorded reader can make the tag find its pseudonym $PIDR_i$ and their shared secret $key^T_{ij}$, and generate random number $r_{T_j}$. Only the genuine tag has right pseudonym $PIDT_j$ and $key^T_{ij}$ to compute $\{PIDT_j \| \{r_{T_j} \| r_{R_i}\}_{k^T_{ij}} \| T\_cout_{ij}\}_{k_u}$. Servers can authenticate $T_j$ by $PIDT_j$ and $key^T_{ij}$. If servers fail to find the record matching with $PIDT_j$, or $key^R_{ij}$ fail to make $R_i$ get positive answer in its process (6), the tag $T_j$ would be considered as counterfeit and the authentication over. Or else, $T_j$ can be authenticated by the server. Four-round authentication is executed to block an unauthorized access, including the double searches in the access lists and the reader-tag mutual verifications.
- Resistance to eavesdropping: Throughout the protocol, the messages that adversary can acquire via eavesdropping are $\{PIDR_i \| r_{Ri}\}_{ku}$ and $\{PIDT_j \| \{r_{T_j} \| r_{R_i}\}_{k^T_{ij}} \| T\_cout_{ij}\}_{k_u}$. The information of the reader and tag $<PIDR_i, PIDT_j, key^T_{ij}>$ can not be retrieved as these messages are protected by one-way hash function.
- Forward secrecy: The data transmitted between the server and the tag is well protected by random numbers ($r_{T_j}$, $r_{R_i}$) and dynamic keys. Even if an attacker can compromise tag $T_j$, the attacker still can not infer the communication in previous sessions since the tag and the reader communicate by unpredictable variables. For each reading, random numbers $r_{Ri}$ and $r_{Tj}$ are generated by the reader and tag. Moreover, the shared keys are updated after each successful authentication. The applying of hash function keeps the keys unpredictable. Thus compromising a tag can not lead to compromising the previous communications of the same tag.
- Resistance to tracking attack: The computation of $\{PIDR_i \| r_{Ri}\}_{ku}$ involves random number $r_{Tj}$ generated by the tag, so tag's response $\{PIDT_j \| \{r_{T_j} \| r_{R_i}\}_{k^T_{ij}} \| T\_cout_{ij}\}_{k_u}$ is different in next sessions. The reader identification information $\{PIDR_i \| r_{Ri}\}_{ku}$ is also various in different sessions because of the random number $r_{Ri}$.
- Resistance to Dos attack: If a tag updates its key but the server not, de-synchronization will arises. In next sessions, the legal tag can not pass the server's authentication, leading to Dos attack. In our scheme, an attacker can not gain the tag's authentication, so it is difficult for attacker to badger the tag update its key. Moreover, synchronization can be resumed with the help of $R\_cout_{ij}$ and $T\_cout_{ij}$.

Because tag authenticates reader firstly, and do $T\_cout_{ij}$ +1 before server do this, $T\_cout_{ij}$ must be not less than $R\_cout_{ij}$, namely $T\_cout_{ij} \geq R\_cout_{ij}$. Because $k^T{}_{ij}$ is equal to $k^R{}_{ij}$ initially, $\underbrace{H(H\cdots H}_{R\_count_{ij}}(k_{ij}^R)))$ must equal to $\underbrace{H(H\cdots H}_{T\_count_{ij}}(k_{ij}^R)))$ , only if $T\_cout_{ij} = R\_cout_{ij}$. In this way, when a server find $T\_cout_{ij} \neq R\_cout_{ij}$, it can know that synchronization issue occur, $k^T{}_{ij} \neq k^R{}_{ij}$. The server will compute $c = T\_cout_{ij} - R\_cout_{ij}$, then do $k'^R{}_{ij} = \underbrace{H(H\cdots H}_{c}(k_{ij}^R)))$ ,to make $k^T{}_{ij} = k'^R{}_{ij}$. Thus, the Dos attack caused by de-synchronization will be solved.

- Resistance to replay attack: In the protocol, the reader and the tag generate random numbers $r_{Ri}$ and $r_{Tj}$, respectively. Even if the attacker has acquired the message of the previous session, as $<\{r_{T_j} \| r_{R_i}\}_{k_{ij}^T}$ 〈〉 $\{PIDT_j \| \{r_{T_j} \| r_{R_i}\}_{k_{ij}^T} \| T\_cout_{ij}\}_{k_u}$ and $\{r_{T_j}\}_{k_{ij}^R}>$ change every session, it is impossible for an attacker to replay intercepted massage in next sessions and pass the authentication.
- Resistance to insider attack: reader $R_i$ and tag $T_j$ share the key $(k^T{}_{ij}, k^R{}_{ij})$, exclusively. Even if an insider attacker acquires other reader's identifier, it can not acquire the corresponding keys shared by the reader and tag. So our scheme can withstand the insider attack.

Table 1 shows the comparison of existing protocols. It can be seen that the improved protocols can meet the required properties and resist all possible attacks, especially can prevent insider attack that most of other protocols can not.

**Table 1.** Comparisons of various attributes

| Scheme | M | P1 | P2 | A1 | A2 | A3 | A4 |
|---|---|---|---|---|---|---|---|
| Lim et al. [7] | Yes | Yes | Yes | Yes | Yes | Yes | No |
| Yang et al. [8] | Yes | Yes | No | No | Yes | Yes | No |
| Peris-Lopez.[12] | Yes | Yes | No | Yes | Yes | Yes | No |
| Chen et al.[9] | No | Yes | Yes | No | No | No | No |
| Yeh et al.[11] | Yes | Yes | Yes | Yes | Yes | Yes | No |
| H. Ning et al.[13] | Yes | Yes | Yes | Yes | No | Yes | Yes |
| Our protocol | Yes | Yes | Yes | Yes | Yes | Yes | Yes |

M: mutual authentication, P1: eavesdropping resistance, P2: forward secrecy, A1: tracking attack resistance, A2: Dos attack resistance, A3: replay attack resistance, A4: insider attack resistance.

# 5 Conclusions

In this paper, we proposed an improvement of H. Ning et al.' protocol. The improved scheme can resist all the possible attacks that break the security of the previous protocol, especially Dos attack. The suggested protocol can be applied to RFID systems with more than one reader/server. One possible future application is RFID-enabled credit card.

With more features incorporated into tags, RFID devices will be widely used in general purpose identifications. Research benefiting RFID security today will contribute to the development of security in ubiquitous computing systems in the future.

**Acknowledgment.** This work is partially supported by Tianjin Natural Science Foundation (09JCYBJC00700) of China. The authors also gratefully acknowledge the helpful comments and suggestions of the reviewers, which have improved the presentation.

## References

1. Miles, S.B., Sarma, S.E., Williams, J.R.: RFID Technology and Applications. Cambridge University Press, New York (2008)
2. Yao, X., Dai, H., Zhang, Z., Ge, D., Chen, Y.: RFID-enhanced integrated manufacturing for job-shop floor problems. ICIC Express Letters, Part B: Applications 2(2), 319–324 (2011)
3. Lu, M.C., Ku, C.Y., Hwang, L.C., Chao, H.M.: Using smart card in RFID infrastructure to protect consumer privacy. International Journal of Innovative Computing, Information and Control 7(4), 1777–1788 (2011)
4. Weis, S.A., Sarma, S.E., Rivest, R.L., Engels, D.W.: Security and Privacy Aspects of Low-Cost Radio Frequency Identification Systems. In: Hutter, D., Müller, G., Stephan, W., Ullmann, M. (eds.) Security in Pervasive Computing. LNCS, vol. 2802, pp. 201–212. Springer, Heidelberg (2004)
5. Juels, A.: RFID security and privacy: A research survey. IEEE Journal on Selected Areas in Communication 24(2), 381–394 (2006)
6. Lin, I.C., Wang, C.W., Luo, R.K., You, H.C.: An efficient mutual authentication protocol for RFID systems. International Journal of Innovative Computing, Information and Control 7(6), 3097–3106 (2011)
7. Lim, J., Oh, H., Kim, S.: A New Hash-Based RFID Mutual Authentication Protocol Providing Enhanced User Privacy Protection. In: Chen, L., Mu, Y., Susilo, W. (eds.) ISPEC 2008. LNCS, vol. 4991, pp. 278–289. Springer, Heidelberg (2008)
8. Yang, J., Park, J., Lee, H., Ren, K., Kim, K.: Mutual authentication protocol for low cost RFID. In: Proc. of the Workshop on RFID and Lightweight Cryptography, pp. 17–24 (2005)
9. Chen, Y., Chou, J.S., Sun, H.M.: A novel mutual-authentication scheme based on quadratic residues for RFID systems. Computer Networks 52(12), 2373–2380 (2008)
10. Lin, I.C., Yang, C.W., Tsaur, S.C.: Nonidentifiable rfid privacy protection with ownership transfer. International Journal of Innovative Computing, Information and Control 6(5), 2341–2351 (2010)
11. Yeh, T.C., Wu, C.H., Tseng, Y.M.: Improvement of the RFID authentication scheme based on quadratic residues. Computer Communications 34(3), 337–341 (2011)
12. Peris-Lopez, P., Hernandez-Castro, J.C., Estevez-Tapiador, J.M., Ribagorda, A.: An Efficient Authentication Protocol for RFID Systems Resistant to Active Attacks. In: Denko, M.K., Shih, C.-S., Li, K.-C., Tsao, S.-L., Zeng, Q.-A., Park, S.H., Ko, Y.-B., Hung, S.-H., Park, J.-H. (eds.) EUC-WS 2007. LNCS, vol. 4809, pp. 781–794. Springer, Heidelberg (2007)
13. Ning, H., Liu, H., Mao, J., et al.: Scalable and distributed key array authentication protocol in radio frequency identification-based sensor systems. IET Communications 5(12), 1755–1768 (2011)

# Study of a Fuzzy Comprehensive Evaluation Problem in Cloud Computing

Wenchuan Yang[1,a,*], Bei Jia[1,b], and Bowei Cao[1,c]

[1] Beijing University of Posts and Telecom., Beijing 100876, China
[2] Xi'an Communication Institute, Shaanxi, Xi'an, 710106, P.R. China
yangwenchuan@bupt.edu.cn

**Abstract.** Under the cloud computing environment, the privacy-preserving fuzzy comprehensive evaluation problem is proposed and solved, which has not been studied in the area of secure multi-party computation. Besides, through the analysis of the protocol's security, an improved protocol whose security is higher is given based on blind-and-permute protocol and privacy-preserving protocol of finding the maximum component's position, whose security is stated using the simulation paradigm at last. The experimental results map out the correct possibilities for our algorithm.

**Keywords:** Cloud Computing, Fuzzy, Comprehensive Evaluation, Protocol.

## 1 Introduction

A secure multi-party computation (SMC) problem deals with computing a function on any input in a distributed network where each participant holds one of the inputs, and that no more information is revealed to a participant in the computation than can be inferred from that participant's input and output[1]. SMC problem was initially suggested by Andrew C. Yao and then extended by Goldreich, Micali and Wigderson in paper [2], where a general SMC protocol of calculating any functionality was given using circuit evaluation protocols. But Goldreich points out that using the solutions derived by these general results for special cases of multiparty computation can be impractical; special solutions should be developed for special cases for efficiency reasons. So special cases of SMC studied extensively [4], including Yao's millionaires' problem, Secure multi-party computational geometry problems, privacy-preserving data mining and so on. The author has ever studied secure multi-party ranking problem. Nowadays, the research of secure multi-party computation becomes a subject of great importance in the international cryptographic circle.

Fuzzy comprehensive evaluation is to give some object a comprehensive evaluation using methods of fuzzy mathematics, which has already been studied extensively. But the problem of privacy-preserving has not been proposed and solved until now.

* Corresponding author.

J. Lei et al. (Eds.): AICI 2012, LNAI 7530, pp. 120–125, 2012.

## 2 Preliminary Definitions

Let us consider such a scenario: Alice is a server, and Bob is a client. Alice launches a network connection to Bob, then Bob detect whether the connection is an abnormity or not. Assume that the factor set is $U=\{u_1,\ldots,u_n\}$. And there are $m$ orders of evaluation, which is denoted $V=\{v_1,\ldots,v_m\}$. They can complete the detection by fuzzy comprehensive evaluation. But Alice does not like to tell Bob about his network connection features so as not to be used by malicious intruders. Then Bob has a weight vector to the factor set, and he also does not like to tell Alice, or Alice may invade successfully by reduplicate calculation. And how to carry out fuzzy comprehensive evaluation at this scenario is a problem[5].

So we have the following Preliminary definitions.

***Definition 2.1 Semi-honest model***
A semi-honest model consists of semi-honest parties. Such a semi-honest party follows the protocol properly with the exception that it keeps a record of all its intermediate computations which may be used to learn more information than it is allowed.

Let $f:\{0,1\}^*\times\{0,1\}^*\mapsto\{0,1\}^*\times\{0,1\}^*$ be a probabilistic polynomial-time functionality, where $f=(f_1,f_2)$. And let $\Pi$ be a two-party protocol for computing *f*. During the execution of $\Pi$ on (*x,y*), the views of the *i*th (*i*=1,2) party are denoted $VIEW_1^{\pi}(x,y)=(x,r^1,m_1^1,m_2^1,\ldots,m_t^1)$ and $VIEW_2^{\pi}(x,y)=(y,r^2,m_1^2,m_2^2,\ldots,m_t^2)$.

***Definition2.2 Security under Semi-Honest Model:*** The model whose security we consider here is that of two-party computation of any functionality under semi-honest model. Now we present in brief the security definition as follows. The definition presented here are according to Goldreich in paper [4].

Here $r^i$ represents the outcome of the *i*th party's internal coin tosses, and $m_j^i$ represents the *j*th message received by the *i*th party. $j=1,2,\ldots,t$ The output of the *i*th party during an execution of $\Pi$ on (*x,y*) is denoted $OUTPUT_i^{\pi}(x,y)$ that is implicit in the *i*th party's view of the execution. $i\in\{1,2\}$.

When there exist probabilistic polynomial-time algorithms $S_1$ and $S_2$, $f$ is securely computed with $\Pi$ under semi-honest model, such as

$$\{(S_1(x,f_1(x,y)),f_2(x,y))\}_{x,y\in\{0,1\}^*}\equiv\{(VIEW_1^{\pi}(x,y),OUTPUT_2^{\pi}(x,y))\}_{x,y\in\{0,1\}^*} \quad (1)$$

$$\{(f_1(x,y),S_2(y,f_2(x,y)))\}_{x,y\in\{0,1\}^*}\equiv\{(OUTPUT_1^{\pi}(x,y),VIEW_2^{\pi}(x,y))\}_{x,y\in\{0,1\}^*} \quad (2)$$

Where I $|x|=|y|$, and $\equiv$ denotes computational indistinguishable [4].

The formal Definition above uses the simulation paradigm. Anyway, it suffices to simulate the view of each semi-honest party, since anything which can be obtained after participating in the protocol is obtainable from the view [4].

## 3 Comprehensive Evaluation Model

In this section, we add privacy-preserving requirements into the existing model and structure a model of privacy-preserving fuzzy comprehensive evaluation in a distributed network as follows[6].

***Definition 3.1 Fuzzy Comprehensive Evaluation:*** Assume that the three basic elements of fuzzy comprehensive evaluation are: factor set $U=\{u_1,...,u_n\}$, evaluation set $V=\{v_1,...,v_m\}$ and weight vector to the factor set $A=(a_1,...,a_n)$. Alice holds the fuzzy evaluation matrix $R=(r_{ij})_{n\times m}$, Bob holds the weight vector *A*, and they both keep their values confidential to each other. They want to get the element's subscript $j_0$ through a secure protocol, such that $b_{j_0}=\max(b_1,...,b_m)=\max(B)=\max(A\circ R)$. Furthermore, $v_{j_0}$ is Bob's final evaluation to Alice. The model can be described with Figure 1 as follows.

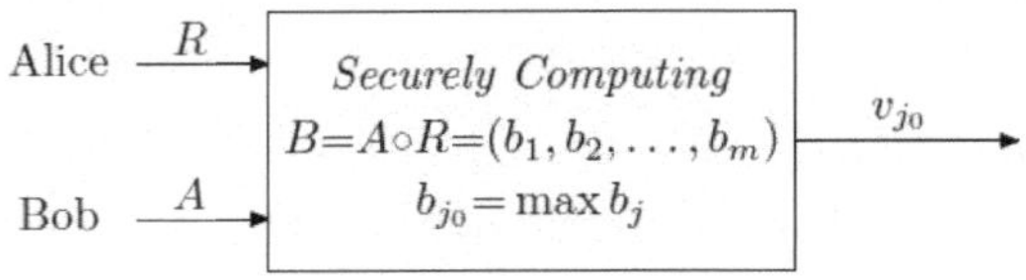


**Fig. 1.** Model of fuzzy comprehensive evaluation

As is shown in Figure 1, in the existing model, both participants know the fuzzy evaluation matrix Rand weight vector *A*. While in the privacy-preserving one, Alice keeps secret *R* and Bob keeps secret *A*. After the computation, they only know the final evaluation result, but Alice does not know *A*, Bob does not know *R*. As a result, the above problem can be formulated as a secure two-party computation problem. Furthermore, it requires that Alice and Bob calculate vector $B=A\circ R$ secretly without leaking their secret information to each other.

## 4 Fuzzy Comprehensive Evaluation

### 4.1 Simple Protocol and Its Algorithm

In order to calculate $B=A\circ R$, $b_j=\overset{n}{\underset{i=1}{\vee}}(a_i\wedge r_{ij})$, $j=1,2,...,m$, a direct thought is: we only need to compare each pair of their secret components by making use of millionaires' protocol repeatedly, and taking corresponding operation of taking the larger value or the smaller one. If we lower the security to only caring about not to leak specific values of *R* and *A*, then it has to carry out such simple operations above [7].

***Protocol 4.1 Fuzzy comprehensive evaluation***

***Input:*** Alice has a fuzzy evaluation matrix $R=\left(r_{ij}\right)_{n\times m}$. Bob has a weigh vector $A=(a_1,...,a_n)$.

***Output:*** Bob's final evaluation to Alice, denoted $v_{j_0}$. For each $j=1,2,...,m$, Alice and Bob conduct the following steps:

***Step1.*** Alice chooses a random permutation function $\pi_{1j}$ and a random vector $\eta_{1j}=(\eta_{11}^{(j)},\eta_{12}^{(j)}..,\eta_{1n}^{(j)})$. And then Alice upon input ( $\pi_{1j}$, $\eta_{1j}$ ) and Bob upon input $A=(a_1,a_2,...,a_n)$ execute blind-and-permute protocol so that Bob (but not Alice) obtains a new vector $\pi_{1j}(A+\eta_{1j})$, denoted $\pi_{1j}(A+\eta_{1j})=(a'^{(j)}_{\pi_{1j}(1)},a'^{(j)}_{\pi_{1j}(2)},...,a'^{(j)}_{\pi_{1j}(n)})$, where, subscript $\pi_{1j}(i)$ represents the element in the position $\pi_{1j}$ in the new vector being replaced by the element in the position $i$ in the $a'^{(j)}_{\pi_{1j}(i)}=\pi_{1j}(a_i+\eta_{1i}^{(j)})$ original one through permutation function $\pi_{1j}(i)$ ; $a'^{(j)}_{\pi_{1j}(i)}=\pi_{1j}(a_i+\eta_{1i}^{(j)})$ equals where $\pi_{1j}(a_i+\eta_{1i}^{(j)})$ represents changing $\pi_{1j}(i)$ from position i to position $i=1,2,...,n$

***Step2.*** Alice calculates $\pi_{1j}(R_j^T+\eta_{1j})$, denoted $\pi_{1j}(R_j^T+\eta_{1j})=(r'^{(j)}_{\pi_{1j}(1)j},r'^{(j)}_{\pi_{1j}(2)j},...,r'^{(j)}_{\pi_{1j}(n)j})$, where $R_j=\left(r_{1j},r_{2j},...,r_{nj}\right)^T$ .And then Alice and Bob compare $r'^{(j)}_{\pi_{1j}(i)j}$ and $a'^{(j)}_{\pi_{1j}(i)}$ using millionaires' protocol for each $i=1,2,...,n$ ,and the results are only known by Bob. If $r'^{(j)}_{\pi_{1j}(i)j}\geq a'^{(j)}_{\pi_{1j}(i)}$, Bob records $b^{(2)}_{\pi_{1j}(i),j}=a'^{(j)}_{\pi_{1j}(i)}$.

***Step3.*** Suppose all the data recorded by Bob in step 2) are denoted $b^{(2)}_{\pi_{1j}(i_1),j},b^{(2)}_{\pi_{1j}(i_2),j},...,b^{(2)}_{\pi_{1j}(i_{k_j}),j}$, $k_j\in\{1,2,...,n\}$. Firstly, Bob chooses a random permutation function $\pi_{2j}$ and a random vector $\eta_{2j}=(\eta_{21}^{(j)},\eta_{22}^{(j)},...,\eta_{2n}^{(j)})$. Next, Bob constructs a vector *Bj* with length *n*, and he puts $b^{(2)}_{\pi_{1j}(i_1),j},b^{(2)}_{\pi_{1j}(i_2),j},...,b^{(2)}_{\pi_{1j}(i_{k_j}),j}$ the positions $\pi_{1j}(i_1),\pi_{1j}(i_2),...,\pi_{1j}(i_{k_j})$ of *Bj*, other positions $k\in\{1,2,...,n\}$ are put values $-\eta_{2k}^{(j)}$. And then Bob calculates $\pi_{2j}(B_j+\eta_{2j})$. At this time $b^{(2)}_{\pi_{1j}(i_1),j},b^{(2)}_{\pi_{1j}(i_2),j},...,b^{(2)}_{\pi_{1j}(i_{k_j}),j}$ are changed into the position of the new vector $\pi_{2j}(B_j+\eta_{2j})$ denote $b'^{(2)}_{\pi_{2j}(\pi_{1j}(i_1)),j}$ , $b'^{(2)}_{\pi_{2j}(\pi_{1j}(i_2)),j}$ ,..., $b'^{(2)}_{\pi_{2j}(\pi_{1j}(i_{k_j})),j}$ ,where, $b'^{(2)}_{\pi_{2j}(\pi_{1j}(i_l)),j}=\pi_{2j}(b^{(2)}_{\pi_{1j}(i_l),j}+\eta^{(j)}_{2\pi_{1j}(i_l)})$ , $l=1,...,k_j$.

***Step4.*** Alice upon input $\pi_{1j}(\eta_{1j})$ and Bob upon input ( $\pi_{2j}$, $\eta_{2j}$ ) use blind-and-permute protocol so that Alice (but not Bob) gets $\pi_{2j}(\pi_{1j}(\eta_{1j})+\eta_{2j})$. And then Bob tells Alice to take out values (but not tell Bob) in positions $\pi_{2j}(\pi_{1j}(i_1)),\pi_{2j}(\pi_{1j}(i_2)),...,\pi_{2j}(\pi_{1j}(i_{k_j}))$ of vector $\pi_{2j}(\pi_{1j}(\eta_{1j})+\eta_{2j})$ and named them as $\xi^{(1)}_{\pi_{2j}(\pi_{1j}(i_1)),j},...,\xi^{(1)}_{\pi_{2j}(\pi_{1j}(i_{k_j})),j}$ where , $\xi^{(1)}_{\pi_{2j}(\pi_{1j}(i_l)),j}=\pi_{2j}(\pi_{1j}(\eta_{1i_l}^{(j)})+\eta^{(j)}_{2\pi_{1j}(i_l)})$ , $l=1,...,k_j$.

### 4.2 Security Analysis

***Security***

In the process of calculating $b_j = \vee_{i=1}^{n}(a_i \wedge r_{ij})$ （$j = 1,2,...,m$） Alice and Bob use millionaires' protocol to compare each pair of $r_{ij}$ 与 $a_i$, the smaller one can be determined without leaking the specific values of the secret inputs except the information that who keeps the smaller one. And then, the largest one among the *n* values $a_i \wedge r_{ij}, i = 1,...,n$ need to be determined. In general case, the above *n* values are separated into two parts, some of which are owned secretly by Alice, the rest are owned secretly by Bob. They only need to compare the largest value on their hands by making use of millionaires' protocol, and the only information leaked in this process is who owns the largest value. Even if the above *n* values are all kept by one of them. Let's assume that Alice keeps all the values. She determines the largest values among the *n* values alone, so Bob will not get any valuable information from this step[8].

In the process of calculating the $\max_{j=1,...,m} b_j$, if both of the parties recorded some data, they only need to compare the largest one among the data they hold using millionaires' protocol, what is leaked here is only the subscript of the largest component in the fuzzy comprehensive evaluation vector *B*. Even if there is a party did not record any data, the owner who recorded data only need to output the subscript of the largest component, and it obviously won't leak any additional information except this subscript.

Anyway, Protocol 4.1 is secure under semi-honest model if we only care about protecting parties' secret inputs. But if we require higher security of not revealing each pair's relative size when using millionaires' protocol, we need to improve the simple version to reach this requirement.

***Correctness***

Protocol 4.1 is obviously correct based on the correctness to the existing method of fuzzy comprehensive evaluation.

Therefore, Protocol 4.1 is secure under semi-honest model.

## 5 Conclusions

The growth of internet brings tremendous opportunities to cooperative computation, so privacy-preserving fuzzy comprehensive evaluation has a wide application prospect with the growth of people's demands to security. A simple protocol and an improved protocol of privacy-preserving fuzzy comprehensive evaluation with higher security under semi-honest model are proposed, whose correctness and security analysis are given in this paper. Privacy-preserving fuzzy comprehensive evaluation protocols in this paper could meet security requirements in most area when using fuzzy comprehensive evaluation model, even if these parties are mutually untrust, they also can stop the process through our protocol.

## References


1. Lindell, Y., Pinkas, B.: Privacy Preserving Data Mining. In: Bellare, M. (ed.) CRYPTO 2000. LNCS, vol. 1880, pp. 36–54. Springer, Heidelberg (2000)
2. Luo, Y.L., Huang, L.S., et al.: An algorithm for privacy-preserving boolean association rule mining. Acta Electronica Sinica 33(5), 900–903 (2005) (in Chinese with English abstract)
3. Huang, Y.Q., Lu, Z.D., Hu, H.P.: A method of security improvement for privacy preserving association rule mining over vertically partitioned data. In: Proceedings of the 9th International Database Engineering & Application Symposium, Washington, pp. 339–343 (2005)
4. Xiao, Q., Luo, S.S., Chen, P., et al.: Research on the problem of secure multi-party ranking under semi-honest model. Acta Electronica Sinica 36(4), 709–714 (2008) (in Chinese with English abstract)
5. Chen, J., Liao, J.X., Chen, J.L.: A novel replacement algorithm based on multi-layer fuzzy decision policy with multi-factor. Journal on Communications 21(4), 25–29 (2005) (in Chinese with English abstract)
6. Mu, C.P., Huang, H.K., Tian, S.F., et al.: Intrusion-detection alerts processing based on fuzzy comprehensive evaluation. Journal of Computer Research and Development 42(10), 679–1685 (2005) (in Chinese with English abstract)
7. Goldreich, O.: Foundations of Cryptography: Basic Applications. Cambridge University Press, Cambridge (2001)
8. Atallah, M.J., Du, W.: Secure Multi-party Computational Geometry. In: Dehne, F., Sack, J.-R., Tamassia, R. (eds.) WADS 2001. LNCS, vol. 2125, pp. 165–179. Springer, Heidelberg (2001)

# Knowledge-Guided Clustering of Large-Scale Time Series under Wavelet Transformation

Xiao Wang[1,*], Fusheng Yu[1], Huixin Zhang[2], and Yuming Liu[1]

[1] School of Mathematical Sciences, Beijing Normal University, Laboratory of Mathematics and Complex Systems, Ministry of Education Beijing 100875, The People's Republic of China
[2] School of Statistics, Capital University of Economics and Business, Beijing 100070
wangxiao19871125@163.com

**Abstract.** The clustering of a group of large-scale time series with same lengths is a challenging problem. Facing with this problem, the existing clustering algorithms usually show high computation cost and low efficiency. In this paper, a knowledge-guided clustering approach is proposed for this problem. In this new approach, the given group of large-scale time series is first changed into some new groups of subsequences by segmentation. All the new groups have same sizes to that of the given group of large-scale time series. In each new group, all the subsequences have same lengths. Different subsequences are obtained from different large-scale time series and start from the same index. Thus, the clustering of the given group of large-scale time series is changed into the clustering of the last new group of subsequences which is implemented by the guidance of the cluster knowledge obtained from the previous new groups. In order to obtain better performance, we perform Haar wavelet transformation on the given group of large-scale time series, and the clustering is carried on the transformed time series. The simulation experiments are given and the results show that the new clustering approach exhibits with high efficiency in revealing the cluster property of the original group of large-scale time series.

**Keywords:** large-scale time series, knowledge-guidance, wavelet transformation, collaborative clustering.

## 1 Introduction

Large-scale time series exist in many domains such as medicine, economy, atmosphere and experimental sciences etc. Many research tasks (prediction, clustering and association rule mining) can be carried on such kind of data [1]. In this paper we focus on the clustering of large-scale time series.

Large-scale time series is the sequential data with high-dimension. This characteristic makes the clustering of large-scale time series distinct from most methods in literature, and leads to many new questions. For a given group of large-scale time series, we usually regard each large-scale time series as a point in a multidimensional feature space and use the existing methods such as those appeared in

* Corresponding author.

J. Lei et al. (Eds.): AICI 2012, LNAI 7530, pp. 126–133, 2012.

[2, 3]. But, these methods usually present high time complexity. Besides that, the existing algorithms [4, 5] usually have no mechanism of taking means of the acquired experience and knowledge. If we can provide such a mechanism for utilizing prior knowledge, we can avoid using large-scale data by replacing some data with the corresponding knowledge in them.

Motivated by this idea, by taking into account Horizontal Collaborative Fuzzy C-Means (HC-FCM) [6], in this paper we present a knowledge-guided clustering approach to dealing with the clustering problem of large-scale time series.

In this new approach, the given group of large-scale time series is first changed into some new groups of subsequences by segmentation. All the new groups have same sizes to that of the given group of large-scale time series. In each new group, all the subsequences have same lengths, different subsequences are obtained from different large-scale time series and start from same index. Thus, the original clustering problem of the given group of large scale time series is changed into the clustering problem of the last new group of subsequences. Furthermore, the new clustering is implemented by the guidance of the cluster knowledge obtained from the previous new groups of subsequences obtained by Fuzzy C-Means (FCM) [7].

In order to obtain better performance, in our new approach, before clustering we apply Haar wavelet transformation to each new group of large-scale time series for feature extraction [8, 9]. This manner accelerates the speed of our algorithm by better compressing the original data.

The rest of this paper is organized as follows: Section 2 outlines all necessary prerequisites; we present the knowledge-guided clustering algorithm of group of large-scale time series under Haar Wavelet Transformation in Section 3; we then show the performance of our new clustering algorithm by two experiments on real stock data in Section 4; Section 5 summarizes and concludes this paper.

## 2 Preliminaries

In this section, we briefly review the algorithms of FCM [10], HC-FCM [6] and the relevant aspects of Haar wavelet transformation [8, 9].

### 2.1 FCM

Suppose there are $N$ patterns to be clustered into $c$ clusters, each pattern is described by an $n$-dimension vector, and all the $N$ vectors comprise dataset $X$. Let $U$ be the corresponding partition matrix and $V$ be the vector of the prototypes. FCM implements the clustering by solving the following optimization problem:

$$Min\ \ Q=\sum_{k=1}^{N}\sum_{i=1}^{c}u_{ik}^{2}d_{ik}^{2},\quad s.t.\ \sum_{i=1}^{c}u_{ik}=1\ (k=1,2,\cdots,N)$$

where $d_{ik}$ is the distance between the $k$th pattern $x_k$ and the $i$th cluster's prototype $v_i$.

The optimal partition matrix $U$ and prototype vector $V$ can be obtained through the following cyclic iteration formulas $(s=1,2,\cdots,c;t=1,2,\cdots,N)$:

$$u_{st} = 1 \Big/ \sum_{j=1}^{c} \left(\frac{d_{st}}{d_{jt}}\right)^2, \quad v_s = \sum_{j=1}^{N} u_{sj}^2 x_j \Big/ \sum_{j=1}^{N} u_{sj}^2 \tag{1}$$

## 2.2 HC-FCM

If we have a collection of datasets with interactive relationship existing at different organizations, we should consider that the other sets can impact the result of the set which will be clustered. In order to deal with this problem, W. Pedrycz proposed the horizontal collaborative fuzzy clustering algorithm[6] which implements the clustering on a dataset of some patterns with the collaboration of some knowledge obtained from other dataset(s) about the same patterns described in different feature space(s).

Here, we give a brief review of HC-FCM.

As for the dataset $X$ with $N$ patterns to be clustered, given that the datasets $X[jj]$ are of the same $N$ patterns but described in different feature spaces whose dimensions are $n_{[jj]}$ respectively. Let $U[jj]$, $V[jj]$ be the partition matrix, the prototype of dataset $X[jj]$ respectively, and $\alpha_{[jj]}$ is the interactive coefficient between $X$ and $X[jj]$ $(jj = 1, 2, \cdots, p)$. The objective function of HC-FCM is:

$$Min\ Q = \sum_{k=1}^{N} \sum_{i=1}^{c} u_{ik}^2 d_{ik}^2 + \sum_{jj=1}^{p} \alpha_{[jj]} \sum_{k=1}^{N} \sum_{i=1}^{c} (u_{ik} - u_{ik}[jj])^2 d_{ik}^2$$

$$s.t. \ \sum_{i=1}^{c} u_{ik} = 1 \ (k = 1, 2, \cdots, N)$$

The optimal partition matrix $U$ and prototype vector $V$ can be obtained through the following cyclic iteration formulas:

$$u_{st} = \varphi_{st} / (2 + 2\psi) + \left(1 - \sum_{j=1}^{c} \varphi_{jt} \Big/ (2 + 2\psi)\right) \Big/ \sum_{j=1}^{c} (d_{st}^2 / d_{jt}^2)$$
$$(s = 1, 2, \cdots, c; t = 1, 2, \cdots, N) \tag{2}$$
$$v_{sj} = (A_{sj} + C_{sj}) / (B_s + D_s) \quad (s = 1, 2, \cdots, c;\ j = 1, 2, \cdots, n_{[p+1]})$$

where $\varphi_{st} = \sum_{jj=1}^{p} \alpha_{[jj]} u_{st}[jj]$, $\psi = \sum_{jj=1}^{p} \alpha_{[jj]}$, $A_{sj} = \sum_{k=1}^{N} u_{sk}^2 x_{kj}$, $B_s = \sum_{k=1}^{N} u_{sk}^2$,

$$C_{sj} = \sum_{jj=1}^{p} \alpha_{[jj]} \sum_{k=1}^{N} (u_{sk} - u_{sk}[jj])^2 x_{kj}, \ D_s = \sum_{jj=1}^{p} \alpha_{[jj]} \sum_{k=1}^{N} (u_{sk} - u_{sk}[jj])^2.$$

## 2.3 Haar Wavelet Transformations

Haar wavelet transformation is one of the simplest and basic transformations from the space domain to a local frequency domain [8]. We use it to the time series transformation in our proposed method. For a given time series $X = (x_1, x_2, \cdots, x_n)$

satisfying $n = 2^j$ ( $j$ is an integer), we outline the procedure of Haar wavelet transformation on $X$ as follows:

**Step 1:** Averaging: calculate the average of the adjacent time series values; result in a lower resolution time series $X' = (\frac{x_1+x_2}{2}, \frac{x_3+x_4}{2}, \cdots, \frac{x_{2^j-1}+x_{2^j}}{2})$;

**Step 2:** Differencing: let $c_i = x_{2^i-1} - \frac{x_{2^i-1}+x_{2^i}}{2}$ $(i = 1, 2, \cdots, j)$ record the missing information of the original time series, and call $c_i$ the corresponding detail coefficient; the original time series $X$ can be represented as $(\frac{x_1+x_2}{2}, \frac{x_3+x_4}{2}, \cdots, \frac{x_{2^j-1}+x_{2^j}}{2}, c_1, c_2, \cdots, c_j)$;

**Step 3:** Recursive stage: let $X = X'$, turn to step 1 to do the repeat until the dimension of time series $X$ reduces to 1.

More details about Haar wavelet transformation or wavelet transformation are referred to papers [8, 9].

# 3 Knowledge-Guided Clustering of Large-Scale Time Series under Wavelet Transformation

In this section, we will probe into the clustering of large-scale time series and present the knowledge-guided wavelet-based clustering algorithm for that.

Assume that we are given a dataset $X$, a group of $N$ large-scale time series with equal length $n$. This means that we are facing with a clustering problem of $N$ patterns described in a feature space with $n$ -dimension.

According to the chronological order, we segment $X$ into $p$+1new groups $X[jj]$ satisfying $X[ii] \underset{ii \neq jj}{\cap} X[jj] = \varnothing$. Each group $X[jj]$ is consisted of $N$ subsequences with same length $n_{[jj]}$, which are cut out from $N$ given large-scale time series. Thus, we can say the $N$ patterns are described in $p$+1different feature spaces $X[jj]$ with dimension $n_{[jj]}$ satisfying $\sum_{jj=1}^{p+1} n_{[jj]} = n$ $(jj = 1, 2, \cdots, p+1)$.

After changing the form of the original clustering problem, we can link the new clustering problem with HC-FCM.

In order to extract features and reduce dimensions, we first carry out Haar wavelet transformation on each new group $X[jj]$ resulting in new group $X'[jj]$. Then, HC-FCM is applied to $X'[jj]$ s. $X'[p+1]$ (called collaborated dataset) will be guided by the clustering knowledge obtained from $X'[jj]$ s (called collaborating dataset). Let $\alpha_{[jj]}$ be the interactive coefficient between $X'[p+1]$ and $X'[jj]$, then the knowledge-guided wavelet-based clustering algorithm can be outlined by the following five steps $(jj = 1, 2, \cdots, p)$:

**Step 1:** Problem description: dataset $X$ consisting of $N$ large-scale time series with equal length $n$, common cluster number $c$ for all clustering problems;

**Step 2:** Partition: according to the chronological order, divide $X$ into $p$+1groups of subsequences $X[jj]$, each of which describes the same $N$ patterns in a feature space with dimension $n_{[jj]}$ $(jj = 1, 2, \cdots, p+1)$;

**Step 3:** Wavelet transformation: for each $X[jj]$, apply Haar wavelet transformation to all subsequences in $X[jj]$ resulting in $X'[jj]$ $(jj = 1, 2, \cdots, p+1)$;

**Step 4:** Initialization: select interactive coefficient $\alpha_{[jj]}$, distance function, and max number of the iteration (itermax); use formula (1) to initialize partition matrix $U[jj]$, prototype vector $V[jj]$ of transformed group of subsequences $X'[jj]$, and partition matrix $U$, prototype $V$of the last group $X'[p+1]$; and then let iterative number be equal to 1 $(jj = 1, 2, \cdots, p)$;

**Step 5:** Iteration stage: compute new prototypes and new partition matrix by formula (2), then let iterative number= iterative number+1; if iterative number reaches itermax, the iteration will stop; otherwise, go on to the next iteration.

## 4 Experimental Examples

In order to show the performance of the knowledge-guided wavelet-based clustering algorithm, we design two experiments in this section. Experiment I is used to show the feasibility of the new clustering approach, and Experiment II is carried out to show the performance of the new clustering algorithm in time efficiency.

In the following two experiments, the group of large-scale time series $X$ is the close price data of Shanghai Stock Exchange. It has $N$=180 patterns described in $n$=1152 dimension space and will be clustered into $c$=9clusters. According to the chronological order, $X$ are segmented into two new groups $X[1]$ and $X[2]$ with dimensions $n_{[1]}$, $n_{[2]}$satisfying $X[1] \cap X[2] = \varnothing$. $X'[i]$ is resulted from carrying out Haar wavelet transformation on $X[i]$ ($i$=1,2). $\alpha$ is used to denote the interactive level between $X'[2]$ and $X'[1]$.

### 4.1 Experiment I

To illustrate the feasibility of our new clustering algorithm, in this experiment, we do the comparison between the knowledge-guided wavelet-based clustering algorithm and FCM by the difference of the partition matrixes ($\|U_2 - U_1\|$), where $U_1$, $U_2$ denote the final optimized partition matrix of FCM, the knowledge-guided wavelet-based clustering algorithm respectively. It is obvious that less difference of partition matrixes ($\|U_2 - U_1\|$) will indicate better performance of our new clustering algorithm.

Similar to HC-FCM, the interactive coefficient ($\alpha$) plays an important role in the knowledge-guided wavelet-based clustering algorithm. That is to say, we will get different collaborative clustering results under different interactive coefficients. So, it is necessary to examine the feasibility of the knowledge-guided wavelet-based clustering algorithm under different interactive coefficients. In this experiment, the collaborated dataset and the collaborating dataset resulted from segmentation are of same dimensions, and six different interactive coefficients are chosen to compare the clustering results of the knowledge-guided wavelet-based clustering algorithm with that of FCM algorithm. The results are listed in Table 1.

**Table 1.** Comparison of knowledge-guided wavelet-based clustering algorithm and FCM under different collaborative levels ( $\alpha$ )

| Collaborative levels ( $\alpha$ ) | Difference of partition matrices $\|U_2 - U_1\|$ |
|---|---|
| 10 | 3.6653 |
| 5 | 3.3355 |
| 1 | ***1.3183*** |
| 0.5 | 1.9097 |
| 0.1 | 1.8407 |
| 0 | 1.6712 |

Table 1 show that the result of the knowledge-guided wavelet-based clustering algorithm is close to that of the FCM algorithm for any collaborative level coefficients. This means that our new proposed algorithm has good feasibility.

We can find that the difference of partition matrices does not present monotonicity with respect to the collaborative level coefficient, especially, that the difference of partition matrices ($\|U_2 - U_1\|$) reaches its minimum 1.3183 at $\alpha$ =1. In this experiment, an optimal interactive coefficient leads to the best collaborative clustering among the given six interactive coefficients can be found. How to select a best interactive coefficient in HC-FCM is an interesting problem. Paper [11] gave a discussion about this topic.

### 4.2 Experiment II

For the clustering methods of large-scale time series, time efficiency is an important testing indicator. From our clustering algorithm in section 3, one can see that the clustering problem of large-scale time series group is solved by segmentation. Different segmentations usually affect the result of our proposed clustering algorithm. So, it is necessary to examine the performance by comparing the knowledge-guided wavelet-based clustering algorithm with FCM algorithm in the respect of time cost under different segmentations.

In this experiment, we will give three different segmentations of the given group of large-scale time series $X$: the first is that the dimension of collaborating dataset $X[1]$is less than that of collaborated dataset $X[2]$; the second is that the dimension of collaborating dataset $X[1]$ is equal to that of collaborated dataset $X[2]$; the last is that the dimension of collaborating dataset $X[1]$ is more than that of collaborated dataset $X[2]$. Furthermore, in experiment I we get the conclusion that the new clustering algorithm reaches the optimal result when the collaborative level $\alpha = 1$. So in this experiment, assume $\alpha = 1$, and more detailed information of comparison is listed in Table 2.

**Table 2.** Comparison of knowledge-guided wavelet-based clustering algorithm and FCM under different segmentations

| Segmentations | Time cost | |
|---|---|---|
| | Knowledge-guided Wavelet-based Clustering Algorithm | FCM |
| $n_{[1]}$=288, $n_{[2]}$= 864 | 1.1113 s | 7.3697 s |
| $n_{[1]}$=576, $n_{[2]}$= 576 | 1.5881 s | |
| $n_{[1]}$= 864, $n_{[2]}$= 288 | 2.0869 s | |

From the above Table 2, we can see that for the collaborating datasets with lower dimensions, the knowledge-guided wavelet-based clustering algorithm exhibits higher efficiency. Furthermore for any of the three segmentations given in this experiment, the knowledge-guided wavelet-based clustering algorithm proposed in this paper spends less time than FCM algorithm, especially in the first segmentation the time consumed by the new clustering algorithm is about one seventh of that of FCM algorithm. This means that our new clustering algorithm is of high efficiency in clustering of group of large-scale time series.

## 5 Conclusions

This paper aims at probing into the clustering implementation of the group of large-scale time series. First, considering the disaster of dimensions, we perform Haar wavelet transformation on the original time series. Thus, the original clustering problem is changed into the clustering of the group of transformed time series. Then, by segmenting all the transformed time series in same manner, we get some new groups of subsequences. All the new groups have same size to that of the transformed time series. In each new group, all the subsequences have same lengths. Different subsequences are obtained from different transformed time series and start from same index. Thus, the clustering of the group of transformed time series is changed into the clustering of one new group of subsequences. But, the clustering of the right new group of subsequences is guided by the cluster knowledge obtained in the other groups of subsequences. Based on the above idea, we formulate the knowledge-guided wavelet-based clustering algorithm. Experiments carried on the close price data of Shanghai Stock Exchange show that the knowledge-guided wavelet-based clustering of large-scale time series is of high efficiency and applicable in real problems.

In this paper, we assume that all the segmented groups of large-scale time series have no intersection and each of them was clustered into same number of clusters. In future, we will study the clustering of group of large-scale time series with knowledge guidance under dynamic clustering number.

**Acknowledgments.** This work is supported by Natural Science Foundation of China (Project 60775032 and Project 10971243), and Beijing Natural Science Foundation (Project 4112031: Clustering and forecasting of large scale temporal data based on knowledge-guidance and optimal granulation of information). It is also sponsored by the priority discipline of Beijing Normal University.

## References

1. Liao, T.: Clustering of time series data-a survey. Pattern Recognition 38, 1857–1874 (2005)
2. Dennis, Shasha: Time Series in Finance: The Array Database Approach (April 2000), `http://www.cs.nyu.edu/cs/faculty/shasha/papers/jagtalk.html`
3. Last, M., Klein, Y., Kandel, A.: Knowledge discovery in time series databases. IEEE Transactions on Systems, Man, and Cybernetics 31, 160–169 (2000)
4. Bradley, P., Bennett, K., Demiriz, A.: Constrained K-Means Clustering. Microsoft Research Technical Report (2006)
5. Ng, R.T., Han, J.: CLARANS: a Method for Clustering Objects for Spatial Data Mining. IEEE Transactions on Knowledge and Data Engineering 14(5), 1003–1016 (2002)
6. Pedrycz, W.: Knowledge-based Clustering: from Data to Information Granules. A John Wiley & Sons, Inc. Publication (2005)
7. Bezdek, J.C.: Pattern Recognition with Fuzzy Objective Function Algorithms. Plenum Press, New York (1981)
8. Daubechies, I.: Ten Lectures on Wavelets. SIAM, Philadelphia (1992)
9. Chan, Y.T.: Wavelet Basis. Kluwer Academic Publishers (1995)
10. Luo, C.: Introduction to Fuzzy Set. Beijing Normal University Press (2005)
11. Wang, X., Yu, F., Zhang, H.: Implement the Horizontal Collaborative Fuzzy Clustering by Optimizing the Weight. In: 2012 the 24th Chinese Control and Decision Conference (2012)

# A Significance-Driven Framework for Characterizing and Finding Evolving Patterns of News Networks

Leiming Yan, Jinwei Wang, Jin Han, and Yuxiang Wang

School of Computer & Software, Jiangsu Engineering Center of Network Monitoring
Nanjing University of Information Science & Technology
Nanjing 210044, China
yanleiming@gmail.com, wjwei_2004@163.com,
hjhaohj@126.com, wyx71wyx71@gmail.com

**Abstract.** Social network analysis has become extremely popular in recent years. What are the most significant evolving behaviors in a social network? It is very difficult to find significant evolving behaviors from a large network in a long evolving time interval. Besides, verifying and evaluating enormous dynamic patterns extracted from a large social network by experts are also too hard to generalize well. In this work, a significance-driven framework is proposed to characterize the evolution of local topology and find dynamic patterns with evidently statistical significance for temporally varying news report networks. Two significance indices—potential index and evolving score are introduced for evaluating evolving patterns. Finally, we present a systematic analysis of one real news network, which demonstrates that the method we proposed can find the evolving characteristic and extract significant dynamic patterns from news networks.

**Keywords:** Social network, Statistical significance, Evolving pattern, Network motif.

## 1 Introduction

With an increasing emphasis towards studying complex network systems, social networks such as social interaction networks (e.g. Facebook), online multimedia networks (e.g. Flickr, Youtube), e-mail networks, co-authorship networks, WWW networks and news networks (e.g. Google news, Digg) have gained much attention in recent years. This has led to the development of computational and mathematical techniques allowing modeling of social networks. These networks evolve over time and their evolution is one of the major areas of research today. The study of these complex networks can provide insight into their structure, properties and behaviors.

Analyzing and understanding the evolving profiles of news networks, a kind of special networks, is very useful and significant for public services. When an event occurs, it attracts attention of information sources to publish related documents along its lifespan. According to the life-cycle perspective, the number of daily news stories about a disaster will increase until it reaches a peak, and then decrease. Studies show

J. Lei et al. (Eds.): AICI 2012, LNAI 7530, pp. 134–141, 2012.

that the number of news stories on a disaster (e.g. plane crashes, industrial accidents, terrorist attacks) directly affects audience response, and leads to negative effects even [1]. The news networks can be used to analyze the relationship between disaster response, relief and aid, and analyze the interactions between the entities in news stories, analyze public sentiment even, and so on.

An essential task of analyzing a media network or news network is to characterize the evolving of the popular events, or find topics that evolve over time in the network. Analyzing the evolving patterns of news stories network have attracted the interest of several groups [1,2,3]. Recently, some models and algorithms are proposed to capture the evolution of network structure, the diffusion of information on the network and the burst of news stories [4,5].

However, two huge challenges arise from the difficulty in mining dynamic patterns. The first is where the significant evolutions are since there are enormous nodes, edges and when a notable dynamic behavior occurs in a network with a large number of timestamps. The mining algorithms have to search the evolving behaviors of nodes and edges in different times. Therefore, the mining process is expensive in CPU time and memory consumption. A rational solution is to characterize efficiently the evolution of network and extract the dynamic behaviors with the help of evolving features.

The second difficulty is how to verify the evolving patterns extracted efficiently. A pragmatic method is to conform to prior knowledge so as to demonstrate the evolving patterns found are realistic. As the [6] indicates, this approach is reasonable but does not generalize well because an algorithm can find more patterns in a large news network than an expert can inspect. Maybe measuring the statistical significance is an alternative solution.

In this paper, we propose a significance-driven framework for characterizing and extracting evolving patterns in news report networks. We begin by extracting network motifs and calculating their evolving significance. Network motifs[7] are specific small subgraphs, which occur in networks at numbers significantly higher than those in randomized networks. Two significance indices are defined to characterize the evolution and discover the changes of topology with evident significance. The whole process of finding dynamic behaviors is directed by significance score.

## 2 Related Work

Recently, much work has focused on understanding the dynamic behaviors of news stories networks. Jiuchang Wei et al. propose a method to identify three disaster news growth models. This method is proven to be valid using the 112 disasters occurring between 2003 and 2008. The factors that influence the likelihood of the growth models include disaster types, newsworthy material, disaster severity, and economic development of the affected area [1]. Chen et al. propose a concept called life profile, modeled by a hidden Markov model, to model the activeness trends of events. In addition, a general event detection framework, LIPED, is introduced, which utilizes the learned life profiles and the burst-and-diverse characteristic to adjust the event detection thresholds adaptively, can detect the activeness trends of events [2]. Kristina Lerman et al. propose a social dynamic model to predict the popularity of news [3].

To gain a deep understanding of the evolution of network structures and functions, there have been some novel studies. Sitaram Asur et al. [8] presented an event-based characterization of critical behavioral patterns for temporally varying interaction graphs. Benevenuto et al. [9 ] presented an in-depth analysis of user workloads in online social networks. Lancichinetti et al. [10 ] presented a systematic empirical analysis of the statistical properties of communities in five categories of real large networks.

## 3 Methodology

We transform an evolving news network *DG* to a series of static snapshots at different timestamps, then the social network *DG* can be defined as $DG = < G_1, G_2, ..., G_{|T|} >$, where $G_t$ is a snapshot at timestamp *t*, and *T* is a set of timestamps. The evolving pattern is defined as a two-tuples pattern <*P, T*>, where *P* is the evolving behaviors in the time interval *T*.

The framework consists of three stages. In the first stage, motifs are counted and extracted from these snapshots respectively. We construct the evolving vectors of motifs to characterize the evolving behaviors of motifs, and cluster motifs with correlated evolving vectors. These motifs dominate the evolving behaviors together. In the second stage, using significance indices, the evolving features are depicted and significant evolution and time interval are captured. In the last stage, the dynamic behaviors are extracted by clustering the substructures being isomorphic with co-evolving motifs.

**Intuition**

The motivation of this work is that the complex network systems are not isolated, and therefore the evolutionary processes of nodes and edges must be affected by environmental conditions, such as friendships, environmental stability and community effect itself. For instance, Lars Backstrom et al. [11] find that the tendency of an individual to join a community is influenced not just by the number of friends he or she has within the community, but also crucially by how those friends are connected to one another. This result suggests the evolution of networks can be seen as the evolution of motifs since they construct the whole network.

**Motifs Detection**

Network motifs are specific frequent subgraphs with potential functional properties, which occur in networks at numbers significantly higher than those in randomized networks, and can be seen as the basic building blocks of complex networks.

The statistical significance of a motif can be measured through comparing the network local structures with the randomized networks. Z-Score[7] is a significance index for network motifs, which is defined as

$$Z_i=(Nreal_i-Avg_i)/std_i \quad (1)$$

where $Nreal_i$ is the number of the motifs observed in the network, and $Avg_i$ and $std_i$ are the mean and standard deviation of the counts of motif $i$ in an ensemble of random networks with the same degree sequence. Generally, if a motif's Z-Score is less than a threshold, it can be omitted for the lower significant.

The SP [7] is the value of Z-scores normalized to length 1:

$$SP_i=Z_i/(\Sigma Z_j^2)^{1/2} \quad (2)$$

The normalization emphasizes the relative significance of subgraphs, rather than the absolute significance. This is important for comparison of networks of different sizes, because motifs in large networks tend to display higher Z-scores than motifs in small networks.

### 3.1 Clustering Significant Evolving Motifs

The evolution of networks can be seen as the evolution of motifs since they construct the whole network. As a consequence of nodes and edges being created or deleted, different types of motifs are added or deleted in the network also. So, we first detect the motifs in network.

In this stage, the first step, for each network snapshot, all the possible motifs with 3 or 4 nodes are enumerated and compared to the average count of random networks. There have been many tools for detecting and counting network motifs from static networks, such as MFinder[7], FANMOD [12], etc.

We now define *evolving vector* to describe the evolving tendency of a motif.

**Evolving Vector**

The evolving vector of a motif, composed of a list of Z-Score in time order, such as $V=< z_1, z_2, \ldots, z_t >$, can be used to analyze the evolving trends and cluster correlated motifs.

The second step is to cluster those motifs with correlated evolving vectors in appropriate time interval. Here, the correlation between motifs means these motifs co-evolve in the networks. The motifs with correlated evolving vectors will hold correlated evolving actions.

Different networks have different distributions of evolving network motifs, and clustering correlated evolving vectors in a time interval can characterize the evolving features of motifs in social networks.

### 3.2 Characterizing Significant Evolutions

The characteristic of behaviors can be depicted by the co-evolving motifs. We define two evolution measures to evaluate the statistical significance of evolution tendencies comparing with the changes in randomized networks in a certain time interval.

**Potential Index**

The potential index measures the tendency of the current motifs evolving to the next timestamp. A motif cluster is highly conservational if its potential does not change much over time. Let $M$ represent one motif cluster at a time interval $T$, motif $x_i \in M$, The potential index ($PI$) of a motif cluster at timestamp $t$ evolving to a state at $t+1$ time can be defined as:

$$PI(M,t) = \frac{1}{|M|}\sum_{i=1}^{|M|}\left|SP_t(x_i) - SP_{t+1}(x_i)\right| \tag{3}$$

where $|M|$ is the number of motifs in the cluster $M$. Specially, when there is only a single motif in the cluster, the $PI(M, t)$ is also used to examine the evolution significance of a single motif.

The value waving of $PI(M,t)$ in a certain time interval will reveal some evolving signals: when and which motifs dominate an evolution with evident significance.

**Evolving Significance Score**

The evolving significance score counts the cumulated potential index of motif cluster evolving over a period of time, which is a measure for evaluating a continuous pattern at a time interval. The evolving significance score ( *E-Score*) is defined as

$$E-Score(M,T) = \sum_{i=0}^{|T|-1} \gamma^i \cdot PI(M,t+i) \tag{4}$$

where $0 \leqslant \gamma < 1$ is a attenuation factor, which means forgetting very old affects and taking new ones into account at a discount $\gamma^i$.

When $\gamma$ is set as 1, the *E-Score( M, T)* with the same motif cluster $M$ will increase with the time interval growth. While $\gamma$ is set as less than 1, it is crucial to select an appropriate time interval for maximizing the *E-Score( M, T)*, which means the evolving patterns with higher significance can be obtained by adjusting the time interval. For example, when $\gamma$=0.6, an evolution process at time interval [4, 6] will deduce a higher significant evolving pattern than the one at interval [3, 7]. Naturally, the particular $\gamma$ value can be decided by the features and other prior knowledge of real networks.

## 4 Experimental Analysis

In this section, we show the effectiveness of our framework with experiments on one real network dataset.

For each network snapshot, all the possible motifs with 3 or 4 nodes are enumerated and compared to the average count over 1000 random networks. We use a tool named FANMOD [12] to obtain the motifs at different size from network snapshots, and a Biclustering algorithm MI-TSB [13] to cluster the co-evolving motifs by evolving vectors.

### 4.1 DaysALL Datasets

DaysALL (The Reuters terror news network) is based on all stories released during 66 consecutive days by the news agency Reuters concerning the September 11 attack on the U.S., beginning at 9:00 AM EST 9/11/2001. The vertices of a network are words (terms); there is an edge between two words iff they appear in the same text unit (sentence). The weight of an edge is its frequency. The network has 13308 vertices (different words in the news) and 148035 edges.

The dataset can be downloaded from Pajek websit (http://pajek.imfm.si/ ).

## 4.2 Analyzing DaysAll Network

**Stage 1:** Clustering Motifs

DaysAll network is an undirected temporal network. There are only two types of motifs with 3 nodes, i.e. motif △ and V .We obtain only one 3-motif △ (id 238) and two 4-motifs with Z-Score larger than 2.0 and p-value less than 0.05 from DaysAll network in 15 days.

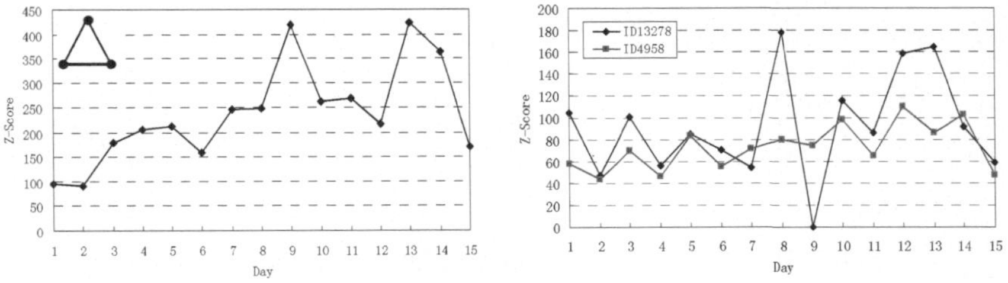


**Fig. 1.** The Z-Score vectors of motif 238, 13278 and 4598 in DaysAll over 15 days

**Table 1.** The potential index of motif 238, 13278 and 4598

| ID | Motif | Days | | | |
|---|---|---|---|---|---|
| | | 7→8 | 8→9 | 9→10 | 12→13 |
| 238 |  | 0.002184 | 0.173108 | 0.157769 | 0.206572 |
| 13278 | 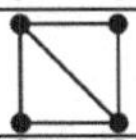 | 0.308701 | 0.446953 | 0.290409 | 0.014986 |
| 4598 | 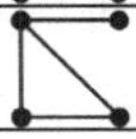 | 0.028955 | 0.020317 | 0.081611 | 0.082072 |

**Stage 2:** Finding Significant Evolutions

The potential index will help find notable evolutions. Table 1 list the potential index of 3 motifs. The *PI(238, 8)* and *PI(13278, 8)* in table 1 evidently indicate that the evolution from day 8 to 9 is statistically significant , which means the evolution is dominated by motif 238 and 13278. Fig 2 demonstrates the evolving feature clearly. As the figure shows, the scale of network experienced a sharp decrease from day 8 to 9. Then, what changes are worth being noticed? The *PI(238, 8)* and *PI(13278, 8)* suggest that distributions of motif 238 and 13278 contribute to the topology evolution.

**Stage 3:** Identifying Evolving Patterns

Given $\gamma$=0.8, time interval T=[7, 9], then we compute the three motifs' E-Scores respectively, E-Score((238), [7,9])=0.241462, and E-Score((13278), [7,9])=0.852124, while E-Score((4598), [7,9])=0.09744. Clearly, these E-Scores suggest the evolutions introduced by motif 238 and 13278 are more statistically evident than the one of motif 4598, and more worthy of mining and exploring. After clustering the substructures which are isomorphic with motif 238 and 13278, we can obtain some patterns in a certain day.

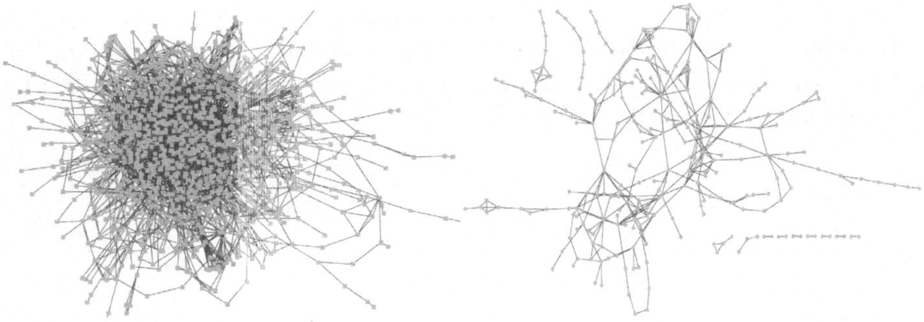

day 8 day 9

**Fig. 2.** The structural evolution of Daysall network form day 8 to 9

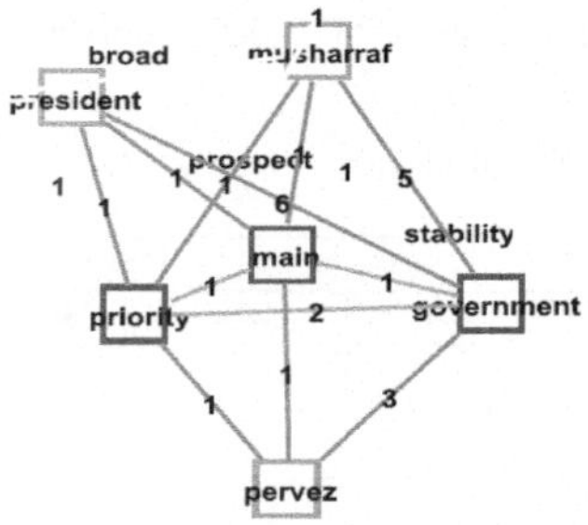


**Fig. 3.** One cluster of motif 238 and 13278 in Day 9

The Fig. 3 shows the clustering results of motifs in Day 9. We get a small community in Fig. 3, which includes *Pervez Musharraf*, the president of Pakistan, and A triangle motif (ID 238) with the word *main*, *priority* and *government*. As we know, at that time, the government of Pakistan announced that join the counter-terrorism alliance led by USA. This pattern shows *Pervez Musharraf* and the Pakistani government became the hot topics at the 9$^{th}$ day after 911 Event.

## 5 Conclusion

In this paper, we present a significance-driven method for exploring and characterizing the evolution of news networks. Two significance measures for evolving patterns are defined, which can offer new and interesting insights for the characteristic of dynamic behaviors in networks. We demonstrate the validity of our method in analyzing one different real news networks—DaysAll network. Our study shows that the evolving features of networks can be characterized and found by the indices we introduce; consequently, the evolving patterns with apparent significance can be extracted efficiently.

**Acknowledgement.** This work was supported by Research Foundation of Nanjing University of Information Science & Technology (20100391), A Project Funded by the Priority Academic Program Development of Jiangsu Higher Education Institutions (PAPD).

## References

1. Wei, J., Zhao, D., Liang, L.: Estimating the growth models of news stories on disasters. Journal of the American Society for Information Science and Technology 60(9), 1741–1755 (2009)
2. Chen, C.C., Chen, M.C., Chen, M.-S.: An adaptive threshold framework for event detection using HMM-based life profiles. ACM Transactions on Information Systems 27(2), 1–35 (2009)
3. Lerman, K., Hogg, T.: Using a model of social dynamics to predict popularity of news. In: WWW 2010, pp. 621–630 (2010)
4. Lin, C.X., Zhao, B., Mei, Q., Han, J.: PET: A Statistical Model for Popular Events Tracking in Social Communities. In: Proceedings of the 16th ACM SIGKDD International Conference on Knowledge Discovery and Data Mining (KDD 2010), pp. 929–938 (2010)
5. Kotov, A., Zhai, C., Sproat, R.: Mining named entities with temporally correlated bursts from multilingual web news streams. In: Proceedings of the Fourth ACM International Conference on Web Search and Data Mining (WSDM 2011), pp. 237–246 (2011)
6. Spiliopoulou, M.: Evolution in social networks: a survey. In: Social Network Data Analytics, pp. 149–175. Springer (2011)
7. Milo, R., et al.: Network motifs: simple building blocks of complex networks. Science 298, 824–827 (2002)
8. Asur, S., Parthasarathy, S., Ucar, D.: An event-based framework for characterizing the evolutionary behavior of interaction graphs. ACM Trans. Knowl. Discov. Data 3(4), 1–36 (2009)
9. Benevenuto, F., Rodrigues, T., Cha, M., Almeida, V.: Characterizing user navigation and interactions in online social networks. Information Sciences 195, 1–24 (2012)
10. Lancichinetti, A., Kivela, M., Saramaki, J.: Characterizing the community structure of complex networks. Plos One 5(8) (August 12, 2010)
11. Lars, B., Dan, H., Jon, K., Lan, X.: Group formation in large social networks: Membership, growth, and evolution. In: Proceedings of the Twelfth ACM SIGKDD International Conference on Knowledge Discovery and Data Mining (KDD), pp. 44–54 (2006)
12. Wernicke, S., Rasche, F.: FANMOD: a tool for fast network motif detection. Bioinformatics 22(9), 1152–1153 (2006)
13. Yan, L., Sun, Z., Wu, Y., Zhang, B.: Biclustering nonlinearly correlated time series gene expression data. Journal of Computer Research and Development 45(11), 1865–1873 (2008)

# Analysis of Association Rule Mining on Quantitative Concept Lattice

Dexing Wang, Qian Xie, Dongmei Huang, and Hongchun Yuan

School of Information Technology, Shanghai Ocean University,
Shanghai 201306, China
{dxwang,dmhuang,hcyuan}@shou.edu.cn, Akaich2000@163.com

**Abstract.** In the process of association rule mining on rough set, it is always needed to deleting the reduplicative rows or columns, so supports and confidences of association rules cannot be obtained accurately. While the Hasse diagram of quantitative concept lattice contains all the objects and attributes information, supports of nodes can be obtained visually from the lattice, and the vivid association rule mining can be realized. Association rule mining algorithm on quantitative concept lattice effectively avoids the combinatorial explosion problem existing in rough set. Confidences of rules can be obtained accurately via the supports of relative concept nodes, and it can also effectively avoid the problem of information loss existing in rough set reduction, thus the efficiency of association rule mining can be improved.

**Keywords:** Association Rule Mining, Quantitative Concept Lattice, Rough Set.

## 1 Introduction

In 1993, Agrawal et al. [1] firstly proposed the problem and solution of association rule mining in customer transactions database. Since then, many typical methods [2, 3, 4] of association rule mining have been put forward.

Rough set [5] is an extensive tool of association rule mining. It mines rules chiefly by seeking the minimum reduction, which includes the attribute reduction and the value reduction. Skowron discernibility matrix [6], the heuristic algorithm of Pawlak [5, 7] based on attribute significance, and the reduction algorithm based on mutual information [8] are typical methods of attribute reduction. And many feasible algorithms [9, 10, 11] of attribute reduction have been proposed to reduce the time complexity. Value reduction methods [12, 13] seek a minimum range to keep the same capacity with original values, and decision rules can be obtained finally.

As another effective tool of association rule mining, concept lattice [14] sets up its hierarchical structure based on the binary relation between attributes and objects. Each node in the Hasse diagram of concept lattice corresponds to a formal concept, which includes the extent and intent. The extraction of association rules is realized through the relationship between intents, which also reflects the direct relationship of including and included between extents. In [15, 16], effective methods of association rule mining were proposed, as well as the improved efficiency.

J. Lei et al. (Eds.): AICI 2012, LNAI 7530, pp. 142–149, 2012.

However, finding out all attribute reductions or the minimum reduction has been testified to be the NP-hard problem [17], in the meantime, providing that all attribute reductions or the minimum reduction have been found, association rule mining from attribute reductions is the very difficult work because how to generate and express association rules is not very an easy task. In contrast, quantitative concept lattice is built via quantifying the extent of concept lattice, so it contains all the extent information. The association rule mining algorithm on quantitative concept lattice searches the concept nodes in a breath-first traversal way, and the efficiency of rules mining can be improved.

The paper is organized as follows: In section 2, we review basic notions of quantitative concept lattice, then discuss the association rule mining algorithm and its good performance compared with rough set in section 3, in addition, association rule mining on quantitative concept lattice and rough set are illustrated and comparison between the two models is made in section 4, finally, we reach the conclusion in section 5.

## 2 Basic Notions of Quantitative Concept Lattice

**Definition 1.** *Suppose* $I = \{i_1, i_2, i_3 \dots i_n\}$ *is an item set,* $A \subset I$, $B \subset I$, $A \cap B = \varnothing$. *An association rule is a logical implication as* $A \Rightarrow B$. *If this rule is true in the item set D, we can get Support* $(A \Rightarrow B) = P(A \cup B) = Support(A \cup B)$, and *Confidence* $(A \Rightarrow B) = P(B|A) = Support(A \cup B) / Support(A)$.

**Definition 2.** *A formal context (U, A, I) consists of two sets U and A, and a relation I between U and A. U is called the objects set and A is called the attributes set. I is the binary relationship between U and A. Two operations are defined on* $N \subseteq U$ *and* $B \subseteq A$ *respectively in (U, A, I):* $N^* = \{a | a \in A, \forall n \in N, nIa\}$ *and* $B^* = \{n | n \in U, \forall a \in B, nIa\}$, *which represent objects in B and attributes in N. If* $N = B^*$, $N = B^*$, *we call C= (N, B) is a concept, of which N is called the extent of C, and B is called the intent of C.*

**Definition 3.** *The partial ordered set of all the concepts in (U, A, I) is recorded as L (U, A, I), which is called concept lattice (GCL). We call* $C' = (|N|, B)$ *is the quantitative concept of C= (N, B), and |N| is the cardinality of N. The lattice constituted by quantitative concepts is defined as Quantitative Concept Lattice (QCL).*

**Definition 4.** $(N_1, B_1)$ *and* $(N_2, B_2)$ *are two concepts in QCL. If* $N_1 \subseteq N_2$ *(namely* $B_2 \subseteq B_1$*), we call* $(N_1, B_1)$ *the sub-concept of* $(N_2, B_2)$, *and* $(N_2, B_2)$ *the sup-concept of* $(N_1, B_1)$, *recorded as* $(N_1, B_1) \leq (N_2, B_2)$. *If* $N_1 \subseteq N_2$, *there exists no a concept (N, B), such that satisfies* $N_1 \subseteq N \subseteq N_2$, $N_1 \subseteq N_2$ *is called a direct-sub-concept-direct-sup-concept-relation between* $N_1$ *and* $N_2$, *the Hasse diagram of concept lattice is generated according to the partial order relation: If* $N_1 \subseteq N_2$ *is a direct-sub-concept-direct-sup-concept, there exists an edge from* $N_2$ *to* $N_1$.

**Theorem 1.** *In the quantitative concept lattice, if the node $C = (N, B)$ exists with m sup- concept nodes, namely $C_1 = (N_1, B_1)$, $C_2 = (N_2, B_2)$... $C_m = (N_m, B_m)$, then the association rule of $t \Rightarrow B\text{-}t$ can be deduced when it satisfied $t \in \{B\text{-}B_1 \bigcup B_2 ... B_m\}$.*

**Theorem 2.** *If the node $C = (N, B)$ exists with two sup- concept nodes, namely $C_1 = (N_1, B_1)$, $C_2 = (N_2, B_2)$, then the association rule of $t_1t_2 \Rightarrow B\text{-}t_1t_2$ can be deduced when it satisfied $\exists t_1 \in \{B_1\text{-}B_1 \bigcap B_2\}$ and $\exists t_2 \in \{B_2\text{-}B_1 \bigcap B_2\}$.*

(1) *If concepts $C_1$and $C_2$ satisfy $C_2 \in sup\ (C_1)$, we can get the rule: $B_2 \Rightarrow B_1\text{-}B_2$, its confidence is $N_1/N_2$. Otherwise, $B_1\text{-}B_2 \Rightarrow B_2$, and its confidence is 100%.*

(2) *If concepts $C_1$ and $C_2$ do not satisfy $C_2 \in sup\ (C_1)$, and there are nonempty common maximum-sub-concept $C = (N, B)$, namely $C \in$ (max-sub-concept $(C_1) \wedge$ max-sub-concept $(C_2)$) $\wedge\, C \neq \varnothing$. Then the association rules between $B_1$ and $B_2$ can be obtained, namely $B_1 \Rightarrow B\text{-}B_1$, $B_2 \Rightarrow B\text{-}B_2$, and their confidences are $N/N_1$, $N/N_2$ respectively.*

From the definitions and theorems, the hierarchical relations between the concepts are shown to us, which can be used as an efficient tool for association rule mining and knowledge acquisition.

# 3 Association Rule Mining Algorithm on Quantitative Concept Lattice

## 3.1 The Association Rule Mining Algorithm

Beginning with the top concept node, we make a breath-first search downward to search for concept nodes. Users can set the minimum support threshold and minimum confidence threshold for association rules. And if the confidence of an association rule is greater than the minimum confidence threshold, output the corresponding association rule. The algorithm is described as follows:

Input: quantitative concept lattice-QCL; confidence_threshold

Output: association rules

```
Void Association-Rule-Ming-QCL (QCL, C,
confidence_threshold)
{C0 :=( all,Ø);
 Set null (Q);
 FOR each Ci of C0 direct-sub-concept;
   Enqueue (Q,Ci);
   WHILE not empty (Q) DO
   {L'_node:=Outqueue(Q);
  C_node:=First-direct-sub-concept (L, L'_node);
  WHILE (C_node < > null) OR (C_node < > (Ø, all)) DO
  {IF (C_node not in Q) and (Mark [C_node] ==1) THEN
     IF Extent (C_node) / Extent (Ci) >=
Confidence_threshold
```

```
        THEN output the association rules according to
Theorems 1 and 2;
    Enqueue (Q, C_node) ;}
      C_node:=Next-direct-sub-concept (L, L'_node) ;}}
```

### 3.2 Analysis of the Algorithm Performance

The time complexity of association rule mining based on quantitative concept lattice is $O(n + e)$, in which $n$ is the number of concept nodes, and $e$ is the number of edges between nodes. Let the attributes set be $A$, and let the tuples (objects) set be $U$. The structure of quantitative concept lattice is an incremental construction algorithm [18], and its time complexity is $O(|A||U|^2)$. It is based on the attribute, inserting into the initial quantitative concept one by one, and adjusting to structure longitudinally.

In rough set, the time complexity of the algorithm based on discernibility matrix [19] is $O(2^{|x|}|A||U|lg|U|)$, in which $x$ is the subset of condition attribute set. In the reduction algorithm of Pawlak, the number of cycle is $|A|$ at most, and the worst complexity is $O(|A|^2|U|lg|U|)$. The time complexity of reduction algorithm based on mutual information is $O(|A|^2|U|^2)$. In the worst case, the time complexity of value reduction is $O(|A|^2|U|^2)$. So it is a large workload for big scale databases.

The association rule mining algorithm of quantitative concept lattice is not NP-hard problem, and it significantly improves the mining efficiency. Of course, time spent on the association rule mining is mainly the construction time. But for users, objective and efficient rules mining on quantitative concept lattice are the advantages.

## 4 Analysis of Association Rule Mining on Quantitative Concept Lattice

### 4.1 An Example and Association Rule Mining on Rough Set

In the decision table below, {A, B, C, D, E} is condition attributes set, and {F} is decision attributes set.

**Table 1.** Decision Table

| | A | B | C | D | E | F |
|---|---|---|---|---|---|---|
| 1 | $a_1$ | $b_2$ | $c_3$ | $d_2$ | $e_1$ | $f_2$ |
| 2 | $a_2$ | $b_1$ | $c_1$ | $d_2$ | $e_2$ | $f_2$ |
| 3 | $a_1$ | $b_2$ | $c_3$ | $d_2$ | $e_1$ | $f_2$ |
| 4 | $a_2$ | $b_1$ | $c_2$ | $d_1$ | $e_2$ | $f_1$ |
| 5 | $a_2$ | $b_1$ | $c_2$ | $d_2$ | $e_3$ | $f_1$ |
| 6 | $a_1$ | $b_2$ | $c_2$ | $d_1$ | $e_2$ | $f_1$ |
| 7 | $a_2$ | $b_1$ | $c_1$ | $d_1$ | $e_2$ | $f_1$ |
| 8 | $a_2$ | $b_1$ | $c_2$ | $d_2$ | $e_3$ | $f_1$ |

Association rule mining based on rough set generally includes the following steps: (a) Reduction of reduplicative rows or columns. (b) Attribute reduction, namely to remove relative unnecessary attributes. (c) Value reduction, is to get the minimum reduction; (d) Extraction of the minimum rule set according to the minimum reduction. It is the first step to delete reduplicative tuples, which is represented by the equivalence class, namely $[1]_A=[3]_A$ and $[5]_A=[8]_A$.

According to the discernibility matrix algorithm, and the heuristic algorithms of Pawlak and reduction algorithm based on mutual information, we can get two reduction sets: {C, D} and {D, E}. Through the reduction of values, we will get a minimum reduction that has the consistent classification ability with original decision table. Merging the repeat rows in the value reduction table, then we get the radically simplify forms as shown in Table 2 and 3. The corresponding rules refer to the minimum reduction are: rule 1: $c_3 \Rightarrow f_2$; rule 2: $c_1 d_2 \Rightarrow f_2$; rule 3: $c_2 \Rightarrow f_1$; rule 4: $d_2 \Rightarrow f_1$; rule 5: $e_1 \Rightarrow f_2$; rule 6: $d_2 e_2 \Rightarrow f_2$; rule 7: $d_1 \Rightarrow f_1$; rule 8: $e_3 \Rightarrow f_1$. And their confidences are 1.

**Table 2.** Value Reduction of {C, D}

| | C | D | F |
|---|---|---|---|
| 1 | $c_3$ | * | $f_2$ |
| 2 | $c_1$ | $d_2$ | $f_2$ |
| 3 | $c_2$ | * | $f_1$ |
| 4 | * | $d_2$ | $f_1$ |

**Table 3.** Value Reduction of {D, E}

| | D | E | F |
|---|---|---|---|
| 1 | * | $e_1$ | $f_2$ |
| 2 | $d_2$ | $e_2$ | $f_2$ |
| 3 | $d_1$ | * | $f_1$ |
| 4 | * | $e_3$ | $f_1$ |

In particular, the heuristic methods of reduction tend to delete the attributes with small significance, which may produce an important role in correct classification of the entire decision table. Data information corresponding to different rows and columns may be exactly the same, but if we delete the reduplicative rows, it may have an impact on the accuracy of rule mining.

## 4.2 Association Rule Mining on Quantitative Concept Lattice

The Hasse diagram of quantitative concept lattice is constructed by Table 1 directly, as shown in Fig.1. With the Hasse diagram, it is easy to find the support of each attribute set and realize the visualization of knowledge. For example, the support of $\{d_2\}$ is 5/8 in the Hasse diagram, and the support of $\{a_2 b_1\}$ is 5/8. But they become

3/6 and 4/6 after deleting reduplicative rows in the rough set. So the support and confidence of rules from rough set is not accurate.

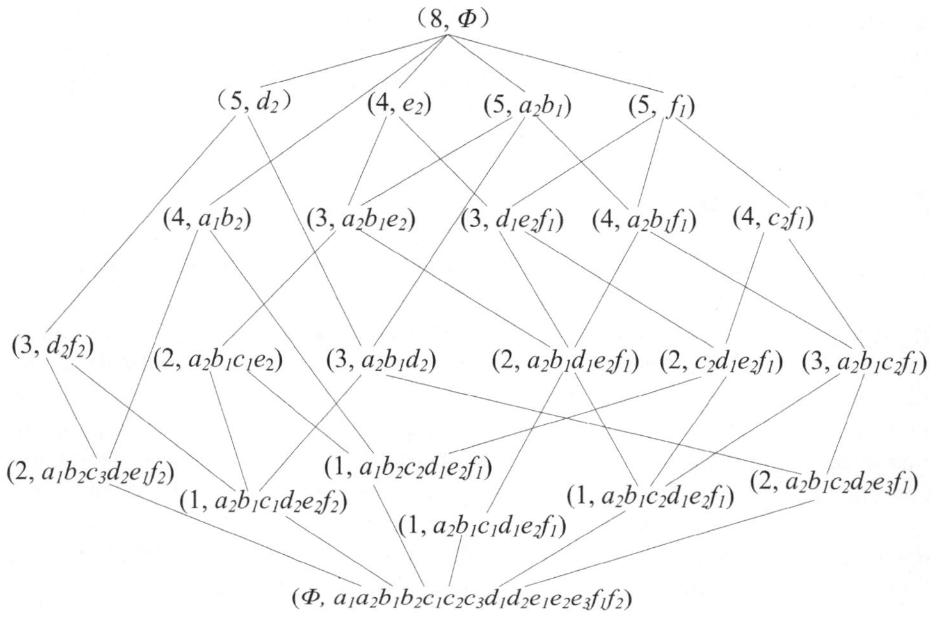


**Fig. 1.** The Hasse diagram of quantitative concept lattice

In the Hasse diagram, the supports of $\{d_2\}$ and $\{d_2f_2\}$ are 5/8 and 3/8, so we can directly get the rule 4: $d_2 \Rightarrow f_2$, which is extracted from the nodes (5, $d_2$) and (3, $d_2f_2$), and confidence of the rule is 3/5. This is virtually the implication rule which comes from rule 2 and rule 6 of rough set. Rule 2: $c_1d_2 \Rightarrow f_2$ and rule 6: $d_2e_2 \Rightarrow f_2$, can also be extracted from the concept nodes (2, $a_2b_1c_1e_2$), (3, $d_2f_2$) and (2, $a_2b_1c_1d_2e_2f_2$). The confidence of these two rules is 1.

The association rule 3 $c_2 \Rightarrow f_1$, according to the algorithm in section 3, can be extracted from the concept nodes (5, $f_1$) and (4, $c_2f_1$), and the confidence of this rule is 1. According to concept nodes (4, $e_2$), (5, $f_1$) and (3, $d_1e_2f_1$), we can get the implication rule $d_1 \Rightarrow f_1$, which is the same with rule 7, and its confidence is 1. The concept node (2, $a_2b_1c_2d_2e_3f_1$) is the largest common sub-concept of node (3, $a_2b_1d_2$) and node (3, $a_2b_1c_2f_1$). So we can get the rule 8: $e_3 \Rightarrow f_1$.

Similarly, the concept node (2, $a_1b_2c_3d_2e_1f_2$) is also the largest common sub-concept of the nodes (3, $a_1b_2$) and (3, $d_2f_2$). So the association rule $c_3e_1 \Rightarrow f_2$ can be extracted, and its confidence is 1 too. Actually, this result includes the rule 1: $c_3 \Rightarrow f_2$ and the rule 5: $e_1 \Rightarrow f_2$. The two relative reduction sets, according to rough set theory, are just to ensure the classification ability, but can't make sure that the final rules are objective.

Base on the analysis above, we find that each rule extracted by rough set, can be expressed concisely and visually by quantitative concept lattice. Association rules based on rough set are deterministic and each tuple in value reduction table corresponds to a non-redundant rule. So its confidence is only to explain whether the rule exist or not but failing to show the supports and confidences accurately. Correspondingly, in the model of quantitative concept lattice, we can easily find the support of each concept node by its extent information, and it contains all the information for being constructed directly via the original decision table, so objective mining results can be easy to obtain.

## 5 Conclusion

Compared with the process of association rule mining on rough set, this paper concluded the advantage of association rule mining on quantitative concept lattice: (1) By means of Hasse diagram, the relationship between objects and attributes can be showed vividly. (2) The mining algorithm based on quantitative concept lattice avoids the NP-hard problem which is difficult to overcome in searching for the minimum reduction in rough set theory, so the efficiency of association rule mining can be improved. (3) Confidences can be obtained accurately by the supports of relative concept nodes, and it can also effectively avoid the problem of information loss existing in rough set reduction, thus it is an objective process of mining. So the quantitative concept lattice can play a significant role in association rule mining. But how to improve its construction efficiency is still needs more work.

**Acknowledgement.** This paper is supported by the national 973 program of China under grant No.2012CB316200 and the innovation program of shanghai municipal education commission under grant No. 12ZZ162.

## References

1. Agrawal, R., Imielinski, T., Swami, A.: Mining association rules between sets of Items in large databases. In: Proceedings of ACM SIGMOD Conference on Management of Data, Washington, DC, pp. 207–216 (1993)
2. Hsu, P.Y., Chen, Y.L., Ling, C.C.: Algorithms for mining association rules in big databases. Information Sciences 1(166), 31–47 (2004)
3. Chen, G.Q., Liu, H.Y., Yu, L., Wei, Q., Zhang, X.: A new approach to classification based on association rule mining. Decision Support Systems 42(2), 674–689 (2006)
4. Yang, G.F., Mabu, S., Shimada, K., Hirasawa, K.: A novel evolutionary method to search interesting association rules by keyword. Expert Systems with Applications 38(10), 13378–13385 (2011)
5. Pawlak, Z.: Rough sets. Communication of the ACM 38, 89–95 (1995)
6. Skowron, A.: The rough sets and evidence theory. Fundaments Informatae 13, 245–262 (1990)
7. Pawlak, Z.: Rough sets: theoretical aspects of reasoning about data. Kluwer Academic Publishers, Dordrecht (1991)
8. Wang, G.Y., Hu, H., Yang, D.: Decision table reduction based on conditional information entropy. Chinese J. Computer 25(7), 1–8 (2002) (in Chinese)

9. Li, J., Wang, X., Fan, X.W.: Improved binary discernibility matrix attribute reduction algorithm in customer relationship management. Procedia Engineering 7, 473–476 (2010)
10. Li, F., Yin, Y.Q.: Approaches to knowledge reduction of covering decision systems based on information theory. Information Sciences 179(11), 1694–1704 (2009)
11. Liang, J.Y., Wang, F., Dang, C.Y., Qian, Y.H.: An efficient rough feature selection algorithm with a multi-granulation view. International Journal of Approximate Reasoning 53(6), 912–926 (2012)
12. Ziarko, W.: Variable precision rough set model. Journal of Computer and System Sciences (46), 39–59 (1993)
13. Ziarko, W.: Introduction to the special issue on rough sets an knowledge discovery. Computer Intelligence (11), 223–226 (1995)
14. Wille, R.: Restructuring lattice theory: An approach based on hierarchies of concepts. In: Rival, I. (ed.) Ordered Sets, pp. 445–470. Reidel, Dordrecht (1982)
15. Du, Y.J., Li, H.M.: Strategy for mining association rules for web pages based on formal concept analysis. Applied Soft Computing 10(3), 772–783 (2010)
16. Nguyen, L.T.T., Vo, B., Hong, T.P., Thanh, H.C.: Classification based on association rules: A lattice-based approach. Expert Systems with Applications 39(13), 11357–11366 (2012)
17. Wang, S.K.M., Ziarko, W.: On optimal decision rules in decision tables. Bulletin of Polish Academy of Sciences 33, 693–696 (1985)
18. Godin, R.: Incremental concept formation algorithm based on Galois (concept) lattice. Computational Intelligence 11(2), 246–267 (1995)
19. Hu, K.Y., Lu, Y.C., Shi, C.Y.: Advances in rough set theory and its applications. Journal of Tsinghua University 41(1), 64–68 (2001) (in Chinese)

# Composite Rough Sets*

Junbo Zhang[1,2], Tianrui Li[1], and Hongmei Chen[1]

[1]School of Information Science and Technology, Southwest Jiaotong University, Chengdu 610031, China
[2]Department of Computer Science, Georgia State University, Atlanta, GA 30303, USA
JunboZhang86@163.com, jbzhang@cs.gsu.edu, {trli,hmchen}@swjtu.edu.cn

**Abstract.** There are multiple kinds of data in information systems, e.g., categorical data, numerical data, set-valued data, interval-valued data and missing data. Such information systems are called as composite information systems in this paper. To process such data, composite rough sets are introduced, composite relation is defined and composite classes are used to drive approximations from composite information systems. Lower and upper approximations of a concept are the basis for rule acquisition and attribute reduction in rough set theory. To intuitively compute the approximations, positive, boundary and negative regions, matrix-based method is presented in composite rough sets. A case study validates the feasibility of the proposed method.

**Keywords:** Composite Rough Sets, Information Systems, Matrix.

## 1 Introduction

Rough set theory, proposed by Z. Pawlak, is a powerful mathematical tool to describe the dependencies among attributes, evaluate the significance of attributes, and derive decision rules [13,14], and plays an important role in the fields of data mining, machine learning and pattern recognition [6,9,16,19,20].

Since the classical rough set model can only be used to deal with categorical data, many extended rough set models were developed for different kinds of data [2,3,6,8]. For example, Hu et al. generalized classical rough set model with neighborhood relations to deal with numerical data [6]. Guan et al. defined tolerance relation and used the maximal tolerance classes to drive optimal decision rules from set-valued information systems [3]. Leung et al. defined $\alpha-$tolerance relations and employed the $\alpha-$misclassification rate for rule acquisition from interval-valued information systems[8]. Grzymała-Busse [2] proposed characteristic relations for missing data in incomplete information systems.

* This work is supported by the National Science Foundation of China (Nos. 60873108, 61175047, 61100117), the Fundamental Research Funds for the Central Universities (No. SWJTU11ZT08), the Doctoral Innovation Foundation of Southwest Jiaotong University (No. 2012ZJB), and the Young Software Innovation Foundation of Sichuan Province (No. 2011-017), China.

J. Lei et al. (Eds.): AICI 2012, LNAI 7530, pp. 150–159, 2012.

In real-applications, there are multiple kinds of data in information systems, e.g., categorical data, numerical data, set-valued data, interval-valued data and missing data. Such information systems are called as composite information systems. Most of the rough set based methods are failed to deal with more than two kinds of data. To solve this problem, Abu-Donia proposed multi knowledge based rough approximations using a family of finite number of relations [4]. Furthermore, we here present the composite rough set model, define a composite relation and use composite classes to drive approximations from composite information systems. To intuitively compute the approximations, positive, boundary and negative regions, matrix-based methods are proposed in composite information systems.

The remainder of the paper is organized as follows. Section 2 provides the basic concepts of rough sets. Section 3 defines composite information system. Section 4 gives matrix-based rough sets approaches in composite information systems. The paper ends with conclusions and further research topics in Section 5.

## 2 Rough Set Model

In this section, we first briefly review the concepts of rough set model as well as their extensions. [3,6,13,14].

### 2.1 Classical Rough Set Model

Given a pair $K = (U, R)$, where $U$ is a finite and non-empty set called the universe, and $R \subseteq U \times U$ is an an indiscernibility relation on $U$. The pair $K = (U, R)$ is called an approximation space. $K = (U, R)$ is characterized by an information system $IS = (U, A, V, f)$, where $U$ is a non-empty finite set of objects; $A$ is a non-empty finite set of attributes; $V = \bigcup_{a \in A} V_a$ and $V_a$ is a domain of attribute $a$; $f : U \times A \to V$ is an information function such that $f(x, a) \in V_a$ for every $x \in U$, $a \in A$. In the classical rough set model, $R$ is the equivalence relation. Let $B \subseteq A$ and $[x]_{R_B}$ denote an equivalence class of an element $x \in U$ under the indiscernibility relation $R_B$, where $[x]_{R_B} = \{y \in U | x R_B y\}$.

Classical rough set model is based on the equivalence relation, the elements in an equivalence class satisfy reflexive, symmetric and transitive. It also does not allow the non-categorical data (*e.g.*, numerical data, set-valued data, interval-valued data and missing data) and requires the information table should be complete. However, non-categorical data appears frequently in real applications [2,5,7,15]. Therefore, it is necessary to investigate the situation of non-categorical data in information systems. In what follows, we just introduce two rough set models [3,6], which will be used in our examples. More rough set models for dealing with non-categorical data are available in the literatures [1,2,5,7,8,12,15,18].

### 2.2 Neighborhood Rough Set Model

**Definition 1.** *Let $B \subseteq C$ be a subset of attributes, $x \in U$. The neighborhood $\delta_B(x)$ of $x$ in the feature space $B$ is defined as*

$$\delta_B(x) = \{y \in U | \Delta_B(x, y) \leq \delta\} \tag{1}$$

*where $\Delta$ is a distance function. For $\forall x, y, z \in U$, it satisfies:*
(I). $\Delta(x, y) \geq 0, \Delta(x, y) = 0$ *if and only if* $x = y$;
(II). $\Delta(x, y) = \Delta(y, x)$;
(III). $\Delta(x, z) \leq \Delta(x, y) + \Delta(y, z)$.

There are three metric functions widely used in pattern recognition. Considered that $x$ and $y$ are two objects in an $m$-dimensional space $A = \{a_1, a_2, \cdots, a_m\}$. $f(x, a_j)$ denotes the value of sample $x$ in the $j$th attribute $a_j$. Then a general metric, named Minkowsky distance, is defined as

$$\Delta_P(x, y) = \left(\sum_{j=1}^{m} |f(x, a_j) - f(y, a_j)|^P\right)^{1/P} \tag{2}$$

where (2) is called: (a) Manhattan distance $\Delta_1$ if $P = 1$; (b) Euclidean distance $\Delta_2$ if $P = 2$; (c). Chebychev distance if $P = \infty$ [17].

The detailed theorems and properties on neighborhood rough sets can be found in [6].

### 2.3 Set-Valued Rough Set Model

**Definition 2.** [3] *In the set-valued information system $(U, A, V, f)$, for $b \in A$, the tolerance relation $T_b$ is defined as:*

$$T_b = \{(x, y) | f(x, b) \cap f(y, b) \neq \emptyset\}. \tag{3}$$

*and for $B \subseteq A$, the tolerance relation $T_B$ is defined as follows:*

$$T_B = \{(x, y) | \forall b \in B, f(x, b) \cap f(y, b) \neq \emptyset\} = \bigcap_{b \in B} T_b. \tag{4}$$

When $(x, y) \in T_B$, we call $x$ and $y$ are indiscernible or $x$ tolerant with $y$ w.r.t. $B$. Let $T_B(x) = \{y | y \in U, yT_Bx\}$, we call $T_B(x)$ the tolerance class for $x$ w.r.t. $T_B$.

The detailed theorems and properties on set-valued rough sets can be found in [3].

## 3 Composite Rough Set Model

In many practical issues, there are multiple kinds of data in information systems, called composite information systems. A composite information system can be written as $CIS = (U, A, V, f)$, where

$$\begin{cases} U, & \text{a non-empty finite set of objects;} \\ A = \bigcup A_k, & \text{a union of attribute sets,} \\ & \text{where } A_k \text{ is an attribute set with the same data type;} \\ V = \bigcup_{A_k \in A} V_{A_k}, & V_{A_k} = \bigcup_{a \in A_k} V_a, V_a \text{is a domain of attribute } a; \\ f : U \times A \to V, & \text{namely, } U \times \bigcup A_k \to \bigcup V_{A_k}, \\ & \text{where } U \times A_k \to V_{A_k} \text{is an information function,} \\ & f(x, a) \text{ denotes the value of object } x \text{ on attribute } a. \end{cases}$$

Table 1 illustrates a composite information system, where $U = \{x_1, x_2, x_3, x_4, x_5, x_6\}$, $A = \{a_1, a_2, a_3, a_4\}$. According to the data type, the attribute set $A$ can be partitioned into $A_1, A_2, A_3$, where $A_1 = \{a_1\}$, $A_2 = \{a_2\}$, $A_3 = \{a_3, a_4\}$ are categorical, set-valued and numerical attribute sets, respectively.

**Table 1.** A composite information system

| $U$ | $a_1$ | $a_2$ | $a_3$ | $a_4$ |
|---|---|---|---|---|
| $x_1$ | $y$ | $\{1,2\}$ | 0.2 | 0.1 |
| $x_2$ | $y$ | $\{1\}$ | 0.2 | 0.3 |
| $x_3$ | $y$ | $\{0\}$ | 0.1 | 0.1 |
| $x_4$ | $y$ | $\{0,1,2\}$ | 0.1 | 0.2 |
| $x_5$ | $n$ | $\{1\}$ | 0.1 | 0.3 |
| $x_6$ | $n$ | $\{0,2\}$ | 0.2 | 0.2 |

**Definition 3.** *Given $x, y \in U$ and $B = \bigcup B_k \subseteq A$, $B_k \subseteq A_k$, the composite relation $CR_B$ is defined as*

$$CR_B = \{(x, y) | (x, y) \in \bigcap_{B_k \in B} R_{B_k}\} \tag{5}$$

*where $R_{B_k} \subseteq U \times U$ is an indiscernibility relation defined by an attribute set $B_k$ on $U$* [14].

When $(x, y) \in CR_B$, we call $x$ and $y$ are indiscernible w.r.t. $B$. Let $CR_B(x) = \{y | y \in U, \forall B_k \in B, y R_{B_k} x\}$, we call $CR_B(x)$ the composite class for $x$ w.r.t. $CR_B$.

*Example 1.* A composite information system $CIS = (U, A, V, f)$ is presented in Table 1. Let $B = \bigcup_{k=1,2,3} B_k$, where $B_1 = \{a_1\}$, $B_2 = \{a_2\}$, $B_3 = \{a_3, a_4\}$. We set the neighborhood $\delta = 0.15$. According to the introduction in Section 2, it is easy to know $R_{B_1}$, $R_{B_2}$ and $R_{B_3}$ are equivalence relation, neighborhood relation and tolerance relation by the attribute sets $B_1$, $B_2$ and $B_3$, respectively. By Definition 3, $\forall x_i \in U$, $CR_B(x_i) = \bigcap_{B_k \in B} R_{B_k}(x_i)$. The results are listed in Table 2.

**Table 2.** Results of composite classes

| $x_i$ | $R_{B_1}(x_i) = [x_i]_{R_{B_1}}$ | $R_{B_2}(x_i) = T_{B_2}(x_i)$ | $R_{B_3}(x_i) = \delta_{B_3}(x_i)$ | $CR_B(x_i)$ |
|---|---|---|---|---|
| $x_1$ | $\{x_1, x_2, x_3, x_4\}$ | $\{x_1, x_2, x_4, x_5, x_6\}$ | $\{x_1, x_3, x_4, x_6\}$ | $\{x_1, x_4\}$ |
| $x_2$ | $\{x_1, x_2, x_3, x_4\}$ | $\{x_1, x_2, x_4, x_5\}$ | $\{x_2, x_4, x_5, x_6\}$ | $\{x_2, x_4\}$ |
| $x_3$ | $\{x_1, x_2, x_3, x_4\}$ | $\{x_3, x_4, x_6\}$ | $\{x_1, x_3, x_4, x_6\}$ | $\{x_3, x_4\}$ |
| $x_4$ | $\{x_1, x_2, x_3, x_4\}$ | $\{x_1, x_2, x_3, x_4, x_5, x_6\}$ | $\{x_1, x_2, x_3, x_4, x_5, x_6\}$ | $\{x_1, x_2, x_3, x_4\}$ |
| $x_5$ | $\{x_5, x_6\}$ | $\{x_1, x_2, x_4, x_5\}$ | $\{x_2, x_4, x_5, x_6\}$ | $\{x_5\}$ |
| $x_6$ | $\{x_5, x_6\}$ | $\{x_1, x_3, x_4, x_6\}$ | $\{x_1, x_2, x_3, x_4, x_5, x_6\}$ | $\{x_6\}$ |

**Definition 4.** *Given a composite information system $CIS = (U, A, V, f)$, $\forall X \subseteq U$, $B \subseteq A$, the lower and upper approximations of $X$ in terms of composite relation $CR_B$ are defined as*

$$\underline{CR_B}(X) = \{x \in U | CR_B(x) \subseteq X\} \tag{6}$$

$$\overline{CR_B}(X) = \{x \in U | CR_B(x) \cap X \neq \emptyset\} \tag{7}$$

Here, these two approximations divide the universe $U$ into three disjoint regions: the positive region $POS_{CR_B}(X)$, the boundary region $BND_{CR_B}(X)$ and the negative region $NEG_{CR_B}(X)$, respectively.

$$\begin{cases} POS_{CR_B}(X) = \underline{CR_B}(X) \\ BND_{CR_B}(X) = \overline{CR_B}(X) - \underline{CR_B}(X) \\ NEG_{CR_B}(X) = U - \overline{CR_B}(X) \end{cases} \tag{8}$$

*Example 2.* We continue Example 1. Let $X = \{x_1, x_3, x_4, x_5\}$. Since $CR_B(x_1) \subseteq X, CR_B(x_2) \nsubseteq X, CR_B(x_3) \subseteq X, CR_B(x_4) \nsubseteq X, CR_B(x_5) \subseteq X$ and $CR_B(x_6) \nsubseteq X$, then $\underline{CR_B}(X) = \{x_1, x_3, x_5\}$. Since $\forall i \in \{1, 2, 3, 4, 5\}, CR_B(x_i) \cap X \neq \emptyset$ and $CR_B(x_6) \cap X = \emptyset$, then $\overline{CR_B}(X) = \{x_1, x_2, x_3, x_4, x_5\}$. Hence, $POS_{CR_B}(X) = \underline{CR_B}(X) = \{x_1, x_3, x_5\}$, $BND_{CR_B}(X) = \overline{CR_B}(X) - \underline{CR_B}(X) = \{x_2, x_4\}$ and $NEG_{CR_B}(X) = U - \overline{CR_B}(X) = \{x_6\}$.

## 4 Matrix-Based Rough Sets Approaches in Composite Information Systems

In this section, we present the matrix representation of the lower and upper approximations in the composite information system.

Before this, we review the matrix-based approaches in rough sets. In 2006, a set of axioms were constructed to characterize classical rough set upper approximation from the matrix point of view by Liu [10]. Followed by Liu's work, Zhang et al. defined a basic vector $H(X)$, which was induced from the relation matrix. And four cut matrices of $H(X)$, denoted by $H^{[\mu,\nu]}(X)$, $H^{(\mu,\nu]}(X)$, $H^{[\mu,\nu)}(X)$ and $H^{(\mu,\nu)}(X)$, were derived for the approximations, positive, boundary and negative regions intuitively in set-valued information systems [21]. In this paper, we follow their work and use matrix-based approaches to deal with composite data. The most difference is to construct the relation matrix with a composite relation.

**Definition 5.** [21] *Given two $\mu \times \nu$ matrices $Y = (y_{ij})_{\mu\times\nu}$ and $Z = (z_{ij})_{\mu\times\nu}$, Minimum of two matrices is defined as*

$$\min(Y, Z) = (\min(y_{ij}, z_{ij}))_{\mu\times\nu} \tag{9}$$

**Definition 6.** [21] *Let $U = \{x_1, x_2, \ldots, x_n\}$, and $X$ be a subset of $U$. The characteristic function $G(X) = (g_1, g_2, \ldots, g_n)^T$ ($T$ denotes the transpose operation) is defined as*

$$g_i = \begin{cases} 1, & x_i \in X \\ 0, & x_i \notin X \end{cases} \tag{10}$$

*where $G(X)$ assigns 1 to an element that belongs to $X$ and 0 to an element that does not belong to $X$.*

*Example 3.* Let $U = \{x_1, x_2, x_3, x_4, x_5, x_6\}$ and $X = \{x_1, x_3, x_4, x_5\}$. Then $G(X) = (1, 0, 1, 1, 1, 0)^T$.

**Definition 7.** [11] *Given an information system $IS = (U, A, V, f)$. Let $B \subseteq A$ and $R_B$ be an indiscernibility relation on $U$, $M_{n\times n}^{R_B} = (\zeta_{ij})_{n\times n}$ be an $n\times n$ matrix representing $R_B$, called the relation matrix w.r.t. $B$. Then*

$$\zeta_{ij} = \begin{cases} 1, & (x_i, x_j) \in R_B \\ 0, & (x_i, x_j) \notin R_B \end{cases} \tag{11}$$

**Lemma 1.** *Given a composite information system $CIS = (U, A, V, f)$, where $A = \bigcup A_k$. Let $B = \bigcup B_k \subseteq A$, $B_k \subseteq A_k$ and $CR_B$ be a composite relation on $U$ by the attribute set $B$, $M_{n\times n}^{CR_B} = (m_{ij})_{n\times n}$ be an $n \times n$ matrix representing $CR_B$, called the relation matrix w.r.t. $B$. Then*

$$M_{n\times n}^{CR_B} = \min_{B_k \in B} M_{n\times n}^{R_{B_k}}. \tag{12}$$

**Corollary 1.** *Let $M_{n\times n}^{CR_B} = (m_{ij})_{n\times n}$ and $CR_B$ be a composite relation on $U$. Then $m_{ii} = 1$, $1 \leq i, j \leq n$.*

*Example 4.* Continuation of Example 1. Table 1 shows a composite information system $CIS = (U, A, V, f)$. Let $B = \bigcup_{k=1,2,3} B_k$, where $B_1 = \{a_1\}$, $B_2 = \{a_2\}$, $B_3 = \{a_3, a_4\}$. By Definition 7, we have

$$M_{6\times 6}^{R_{B_1}} = \begin{bmatrix} 1 & 1 & 1 & 1 & 0 & 0 \\ 1 & 1 & 1 & 1 & 0 & 0 \\ 1 & 1 & 1 & 1 & 0 & 0 \\ 1 & 1 & 1 & 1 & 0 & 0 \\ 0 & 0 & 0 & 0 & 1 & 1 \\ 0 & 0 & 0 & 0 & 1 & 1 \end{bmatrix}, M_{6\times 6}^{R_{B_2}} = \begin{bmatrix} 1 & 1 & 0 & 1 & 1 & 1 \\ 1 & 1 & 0 & 1 & 1 & 0 \\ 0 & 0 & 1 & 1 & 0 & 1 \\ 1 & 1 & 1 & 1 & 1 & 1 \\ 1 & 1 & 0 & 1 & 1 & 0 \\ 1 & 0 & 1 & 1 & 0 & 1 \end{bmatrix}, \text{ and}$$

$$M_{6\times 6}^{R_{B_3}} = \begin{bmatrix} 1 & 0 & 1 & 1 & 0 & 1 \\ 0 & 1 & 0 & 1 & 1 & 1 \\ 1 & 0 & 1 & 1 & 0 & 1 \\ 1 & 1 & 1 & 1 & 1 & 1 \\ 0 & 1 & 0 & 1 & 1 & 1 \\ 1 & 1 & 1 & 1 & 1 & 1 \end{bmatrix}.$$

According to Lemma 1, $M_{6\times 6}^{CR_B} = \min_{k=1,2,3} M_{n\times n}^{R_{B_k}} = \begin{bmatrix} 1 & 0 & 0 & 1 & 0 & 0 \\ 0 & 1 & 0 & 1 & 0 & 0 \\ 0 & 0 & 1 & 1 & 0 & 0 \\ 1 & 1 & 1 & 1 & 0 & 0 \\ 0 & 0 & 0 & 0 & 1 & 0 \\ 0 & 0 & 0 & 0 & 0 & 1 \end{bmatrix}$.

**Definition 8.** [21] *Let $B \subseteq A$ and $CR_B$ be a composite relation on $U$, $\Lambda_{n\times n}^{CR_B}$ be an induced diagonal matrix of $M_{n\times n}^{CR_B} = (m_{ij})_{n\times n}$. Then*

$$\Lambda_{n\times n}^{CR_B} = \text{diag}(\frac{1}{\lambda_1}, \frac{1}{\lambda_2}, \cdots, \frac{1}{\lambda_n}) = \text{diag}(\frac{1}{\sum_{j=1}^{n} m_{1j}}, \frac{1}{\sum_{j=1}^{n} m_{2j}}, \cdots, \frac{1}{\sum_{j=1}^{n} m_{nj}}) \tag{13}$$

*where $\lambda_i = \sum_{j=1}^{n} m_{ij}$, $1 \leq i \leq n$.*

**Corollary 2.** $\Lambda_{n\times n}^{CR_B} = \text{diag}(\frac{1}{|CR_B(x_1)|}, \frac{1}{|CR_B(x_2)|}, \cdots, \frac{1}{|CR_B(x_n)|})$ *and* $1 \leq |CR_B(x_i)| = |\bigcap_{B_k \in B} R_{B_k}(x_i)| \leq n, 1 \leq i \leq n$.

*Example 5.* Continuation of Example 4. Here we compute the induced diagonal matrix of $M_{6\times 6}^{CR_B}$ by Definition 8. $\Lambda_{6\times 6}^{CR_B} = \text{diag}(1/2, 1/2, 1/2, 1/4, 1, 1)$.

**Definition 9.** [21] *The $n$-column vector called basic vector, denoted by $H(X)$, is defined as:*

$$H(X) = \Lambda_{n\times n}^{CR_B} \bullet (M_{n\times n}^{CR_B} \bullet G(X)) = \Lambda_{n\times n}^{CR_B} \bullet \Omega_{n\times 1}^{CR_B} \tag{14}$$

*where $\bullet$ is dot product of matrices and $\Omega_{n\times 1}^{CR_B} = M_{n\times n}^{CR_B} \bullet G(X)$.*

*Example 6.* Continuation of Examples 3 and 5. Since $\Lambda_{6\times 6}^{CR_B} = \text{diag}(1/2, 1/2, 1/2, 1/4, 1, 1)$ and $G(X) = (1, 0, 1, 1, 1, 0)^T$, then

$$\Omega_{6\times1}^{CR_B} = M_{6\times6}^{CR_B} \bullet G(X) = \begin{bmatrix} 1 & 0 & 0 & 1 & 0 & 0 \\ 0 & 1 & 0 & 1 & 0 & 0 \\ 0 & 0 & 1 & 1 & 0 & 0 \\ 1 & 1 & 1 & 1 & 0 & 0 \\ 0 & 0 & 0 & 0 & 1 & 0 \\ 0 & 0 & 0 & 0 & 0 & 1 \end{bmatrix} \bullet \begin{bmatrix} 1 \\ 0 \\ 1 \\ 1 \\ 1 \\ 0 \end{bmatrix} = \begin{bmatrix} 2 \\ 1 \\ 2 \\ 3 \\ 1 \\ 0 \end{bmatrix}, \text{ and}$$

$$\begin{aligned} H(X) &= \Lambda_{6\times6}^{CR_B} \bullet \Omega_{6\times1}^{CR_B} \\ &= \mathrm{diag}(1/2, 1/2, 1/2, 1/4, 1, 1) \bullet (2, 1, 2, 3, 1, 0)^T \\ &= (1, 1/2, 1, 3/4, 1, 0)^T. \end{aligned}$$

**Definition 10.** [21] *Let $0 \leq \mu \leq \nu \leq 1$. Four cut matrices of $H(X)$, denoted by $H^{[\mu,\nu]}(X)$, $H^{(\mu,\nu]}(X)$, $H^{[\mu,\nu)}(X)$ and $H^{(\mu,\nu)}(X)$, are defined as follows.*

(1) $H^{[\mu,\nu]}(X) = (h'_i)_{n\times1}$

$$h'_i = \begin{cases} 1, & \mu \leq h_i \leq \nu \\ 0, & else \end{cases} \tag{15}$$

(2) $H^{(\mu,\nu]}(X) = (h'_i)_{n\times1}$

$$h'_i = \begin{cases} 1, & \mu < h_i \leq \nu \\ 0, & else \end{cases} \tag{16}$$

(3) $H^{[\mu,\nu)}(X) = (h'_i)_{n\times1}$

$$h'_i = \begin{cases} 1, & \mu \leq h_i < \nu \\ 0, & else \end{cases} \tag{17}$$

(4) $H^{(\mu,\nu)}(X) = (h'_i)_{n\times1}$

$$h'_i = \begin{cases} 1, & \mu < h_i < \nu \\ 0, & else \end{cases} \tag{18}$$

*Remark 1. These four cut matrices are Boolean matrices.*

**Lemma 2.** *Given any subset $X \subseteq U$ in a composite information system $IS = (U, A, V, f)$, where $U = \{x_1, x_2, \ldots, x_n\}$. $B \subseteq A$ and $CR_B$ is a composite relation on $U$. $H(X) = (h_1, h_2, \cdots, h_n)^T$ is the basic vector. Then the lower and upper approximations of $X$ in the composite information system can be computed from the cut matrix of $H(X)$ as follows.*

*(1) The n-column boolean vector $G(\underline{CR_B}(X))$ of the lower approximation $\underline{CR_B}(X)$:*

$$G(\underline{CR_B}(X)) = H^{[1,1]}(X) \tag{19}$$

*(2) The n-column boolean vector $G(\overline{CR_B}(X))$ of the upper approximation $\overline{CR_B}(X)$:*

$$G(\overline{CR_B}(X)) = H^{(0,1]}(X) \tag{20}$$

**Corollary 3.** *The positive region $POS_{CR_B}(X)$, the boundary region $BND_{CR_B}(X)$, and the negative region $NEG_{CR_B}(X)$ can also be generated from the cut matrix of $H(X)$, respectively as follows.*

(1) *The n-column boolean vector $G(POS_{CR_B}(X))$ of the positive region:*

$$G(POS_{CR_B}(X)) = H^{[1,1]}(X) \tag{21}$$

(2) *The n-column boolean vector $G(BND_{CR_B}(X))$ of the boundary region:*

$$G(BND_{CR_B}(X)) = H^{(0,1)}(X) \tag{22}$$

(3) *The n-column boolean vector $G(NEG_{CR_B}(X))$ of the negative region:*

$$G(NEG_{CR_B}(X)) = H^{[0,0]}(X) \tag{23}$$

*Example 7.* Continuation of Example 6. $H(X) = (1, 1/2, 1, 3/4, 1, 0)^T$. The approximations, positive, boundary and negative regions from $H(X)$ are computed as follows.

$$\begin{cases} G(\underline{CR_B}(X)) & = H^{[1,1]}(X) = (1,0,1,0,1,0)^T \\ G(\overline{CR_B}(X)) & = H^{(0,1]}(X) = (1,1,1,1,1,0)^T \\ G(POS_{CR_B}(X)) & = H^{[1,1]}(X) = (1,0,1,0,1,0)^T \\ G(BND_{CR_B}(X)) & = H^{(0,1)}(X) = (0,1,0,1,0,0)^T \\ G(NEG_{CR_B}(X)) & = H^{[0,0]}(X) = (0,0,0,0,0,1)^T \end{cases} \Rightarrow \begin{cases} \underline{CR_B}(X) & = \{x_1, x_3, x_5\} \\ \overline{CR_B}(X) & = \{x_1, x_2, x_3, x_4, x_5\} \\ POS_{CR_B}(X) & = \{x_1, x_3, x_5\} \\ BND_{CR_B}(X) & = \{x_2, x_4\} \\ NEG_{CR_B}(X) & = \{x_6\} \end{cases}$$

Compared with Example 2, it is easy to verify that the matrix-based method for computing approximations has the same result with the traditional method.

## 5 Conclusions

In this paper, we defined composite information systems that contained multiple kinds of data. Composite relation was defined and composite classes were used to compute approximations from such information systems. Followed by work of Zhang et al. [21], we also use $H(X)$ to drive the approximations, positive, boundary and negative regions intuitively. Be different from constructing relation matrix in terms of tolerance relations [21], this paper presented the method for constructing relation matrix in terms of multiple relations. One of our future work is to study on rule acquisition and attribute reduction in composite information systems. Another is to design a parallel method to mine knowledge from composite and massive data.

## References

1. Dubois, D., Prade, H.: Putting fuzzy sets and rough sets together. In: Slowiniski, R. (ed.) Intelligent Decision Support, pp. 203–232. Kluwer Academic, Dordrecht (1992)
2. Grzymała-Busse, J.W.: Characteristic Relations for Incomplete Data: A Generalization of the Indiscernibility Relation. In: Tsumoto, S., Słowiński, R., Komorowski, J., Grzymała-Busse, J.W. (eds.) RSCTC 2004. LNCS (LNAI), vol. 3066, pp. 244–253. Springer, Heidelberg (2004)

3. Guan, Y., Wang, H.: Set-valued information systems. Information Sciences 176(17), 2507–2525 (2006)
4. Abu-Donia, H.M.: Multi knowledge based rough approximations and applications. Knowledge-Based Systems 26, 20–29 (2012)
5. Hu, Q., Xie, Z., Yu, D.: Hybrid attribute reduction based on a novel fuzzy-rough model and information granulation. Pattern Recognition 40, 3509–3521 (2007)
6. Hu, Q., Yu, D., Liu, J., Wu, C.: Neighborhood rough set based heterogeneous feature subset selection. Information Sciences 178(18), 3577–3594 (2008)
7. Kryszkiewicz, M.: Rough set approach to incomplete information systems. Information Sciences 112(1-4), 39–49 (1998)
8. Leung, Y., Fischer, M.M., Wu, W.Z., Mi, J.S.: A rough set approach for the discovery of classification rules in interval-valued information systems. International Journal of Approximate Reasoning 47(2), 233–246 (2008)
9. Li, T., Ruan, D., Wets, G., Song, J., Xu, Y.: A rough sets based characteristic relation approach for dynamic attribute generalization in data mining. Knowledge-Based Systems 20(5), 485–494 (2007)
10. Liu, G.: The axiomatization of the rough set upper approximation operations. Fundamenta Informaticae 69(3), 331–342 (2006)
11. Liu, G.: Axiomatic systems for rough sets and fuzzy rough sets. International Journal of Approximate Reasoning 48(3), 857–867 (2008)
12. Mi, J.S., Zhang, W.X.: An axiomatic characterization of a fuzzy generalization of rough sets. Information Sciences 160(1-4), 235–249 (2004)
13. Pawlak, Z.: Rough Sets: Theoretical Aspects of Reasoning about Data, System Theory, Knowledge Engineering and Problem Solving, vol, vol. 9. Kluwer Academic Publishers, Dordrecht (1991)
14. Pawlak, Z., Skowron, A.: Rough sets: Some extensions. Information Sciences 177(1), 28–40 (2007)
15. Qian, Y., Dang, C., Liang, J., Tang, D.: Set-valued ordered information systems. Information Sciences 179(16), 2809–2832 (2009)
16. Qian, Y., Liang, J., Pedrycz, W., Dang, C.: Positive approximation: An accelerator for attribute reduction in rough set theory. Artificial Intelligence 174(9-10), 597–618 (2010)
17. Wilson, D.R., Martinez, T.R.: Improved heterogeneous distance functions. Journal of Artificial Intelligence Research 6, 1–34 (1997)
18. Yao, Y.Y.: Relational interpretations of neighborhood operators and rough set approximation operators. Information Sciences 111(1-4), 239–259 (1998)
19. Zhang, J., Li, T., Ruan, D., Gao, Z., Zhao, C.: A parallel method for computing rough set approximations. Information Sciences 194(1), 209–223 (2012)
20. Zhang, J., Li, T., Ruan, D., Liu, D.: Neighborhood Rough Sets for Dynamic Data Mining. International Journal of Intelligent Systems 27(4), 317–342 (2012)
21. Zhang, J., Li, T., Ruan, D., Liu, D.: Rough sets based matrix approaches with dynamic attribute variation in set-valued information systems. International Journal of Approximate Reasoning 53(4), 620–635 (2012)

# Analysis of Requirement and Constrained Model of Inter-Satellite-Link TT&C Scheduling Problem on Navigation Constellation

Jing Li[1], Hong-jun Hu[1,2], Pei-jun Yu[1], and Jun Zhu[1]

[1] State Key Laboratory of Astronautic Dynamics, Xi'an Satellite Control Center, Xi'an China
[2] School of Astronautics, Northwestern Polytechnical University, Xi'an China
Carol_lee_0727@sina.com

**Abstract.** The TT&C (Telemetry, Tracking and Command) requirement and task description method are analyzed based on the background of navigation constellation with the inter-satellite-link. Aiming at the requirement, the corresponding math model of constrainted condition is studied. The constrainted model is verified by application instance and proved that it can be applied to schedule TT&C link on the navigation constellation.

**Keywords:** Navigation constellation, Optimization scheduling, TT&C requirement, Constrainted model, Inter-Satellite-Link.

## 1 Introduction

The traditional techniques of navigation and position have been replaced by satellite navigation and position for the merit of all places, all time, high precision and speediness. It has been applied widely in the fields of military and civilian [1].

Satellite navigation system is composed of navigation constellation, ground TT&C network and user position facility. Navigation constellation is composed of some navigation satellites, included GEO (Geostationary Earth Orbit) satellite, IGSO (Inclining Geostationary Synchronized Orbit) satellite and MEO (Medium Earth Orbit) satellite. The ground TT&C network is composed of master control station, constellation monitoring station and date injection station. The user position facility is composed of receiver, timing clock and date processor [2].

By far, the representative satellite navigation system is GPS. GPS has three operation modes. They are: the ground independent operation mode, the constellation autonomous navigation mode and the distribution of crosslink telemetry and command messages mode [3][4][5].

The "contact one, contact all" paradigm improves system responsiveness and flexibility in navigation constellation system [6]. Owing to TT&C task can be operated by ISL (Inter-Satellite-Link), it is difference between ISL TT&C scheduling requirement and traditional TT&C task.

In order to support navigation constellation TT&C, it is a basic work to analyze the requirement and constraint model. It can guarantee the TT&C resources be used efficiently.

J. Lei et al. (Eds.): AICI 2012, LNAI 7530, pp. 160–169, 2012.

It is indicated by means of literature index, there are no open literature about navigation constellation TT&C requirement analysis, at present. The reference [7] shows the research result about the TT&C requirement of America deep space TT&C network user. The priority, flexibility, specification and preference are described in the literature. The reference [8] shows the research result about the periodic task model of computation and communication. The information of constraint, priority and time can be described by the model. The reference [9] shows the task description method with HTN (Hierarchy Task Network), which can describe the basic task by means of decomposing complicated task. The reference [10] shows the further research result of task description problem for deep space TT&C network. A kind of user description language is designed according to the requirement of TT&C service.

In the paper, the TT&C requirement of navigation constellation is analyzed. The constraint model of TT&C requirement is established. The constraint model is proved suitable for scheduling ISL TT&C by an application instance.

## 2 Analysis of TT&C Requirement

The research will be developed basic on the background of the navigation constellation composed of three GEO satellites, three IGSO satellites and twenty four MEO satellites.

### 2.1 TT&C Task of ISL

There are three kinds of ISL in the navigation constellation, viz. MEO-MEO link, MEO-GEO link and MEO-IGSO link.

During the on orbit management, the TT&C task of ISL can be included as follow: (1) Constellation monitoring. The healthy state of constellation can be monitored by telemetry parameters transmitted from visible satellites. (2) Constellation operation and control. The commands and injecting data can be sent to operate and control satellites by ISL. (3) Combined orbit determination. The satellite to satellite ranging data from ISL is combined with the ground measuring data to determine the orbit of constellation.

### 2.2 Characteristic Analysis of ISL

There two types ISL according to the mode of duration, they are: permanent ISL and impermanent ISL. Permanent ISL means the ISL will be connected during whole system period, the three conditions will be meet: (1) Geometry visible permanently. (2) Signal reachable. (3) The ISL can be established in the range of elevation and azimuth. Impermanent ISL means the ISL will be connected sometimes and disconnected sometimes. Impermanent ISL is formed mainly between differ orbit plane. The constellation topology will be resulted in change frequently because of impermanent ISL, so that the ISL will be switched frequently.

The space parameters of ISL can be described by satellite to satellite range, satellite to satellite elevation and satellite to satellite azimuth. The definition is similar with ground station to satellite.

A Walker constellation can be described by a group of parameters, marked as: *i:N/P/F*, expressed separately as: orbit inclination *i*, satellite number *N*, orbit plane number *P* and phase parameter *F*. Assume, the configuration code of a walker constellation is *N/P/F*, the right ascension of ascending node and ascending node angular range of the any satellite m in the constellation will be expressed as:

$$\Omega_m = \frac{360}{P}(P_m - 1) \tag{1}$$

$$u_m = \frac{360}{S}(N_m - 1) + \frac{360}{N}F(P_m - 1) \quad (P_m = 1,2\cdots,P),(N_j = 1,2,\ldots S-1) \tag{2}$$

Where, $S$ is the satellite number of each orbit plane. $P_m$ is the orbit plane number in which the satellite is located. $N_m$ is the satellite number inside the orbit plane $P_m$.

$$S = \frac{N}{P} \qquad P_m = \frac{m}{S} + \mathbf{1} \qquad N_m = m - (P_m - \mathbf{1})S$$

Inside the Walker constellation, all satellites are symmetrical in geometry. The geometrical characteristic of all satellites can be expressed by the geometrical characteristic of any satellite. Therefore, we will discuss the characteristic of the first satellite which is located in the first orbit plane, namely satellite 1.

Inside the constellation, the two orbit plane near by the orbit plane in which satellite 1 is located are plane 2 and plane *P* respectively. Assume, there is a satellite located orbit plane 2, namely satellite *j*. The difference phase angular between satellite 1 and satellite *j* is:

$$\Delta u_{1j} = \frac{360}{S}(N_j - 1) + \frac{360}{N}F \tag{3}$$

The difference phase angular between satellite 1 and satellite *k* locate in orbit plane *P* is:

$$\Delta u_{1k} = \frac{\mathbf{360}}{S}(N_k - \mathbf{1}) + \frac{\mathbf{360}}{N}F(P - \mathbf{1}) \tag{4}$$

$$\Delta U = \Delta u_{1j} + \Delta u_{1k} \tag{5}$$

Lets $\Gamma = \frac{\Delta U}{360}$. If $\Gamma$ is integer, the geometrical characteristic of ISL between satellite *j* to satellite 1 and satellite *k* to satellite 1 are consistent. And satellites will appear in couples. To combining formula (3) and (4) to, $\Gamma = \frac{\Delta U}{360}$ we have:

$$\Gamma = \frac{(N_j + N_k - 2)N + SFP}{SN} = \frac{N_j + N_k + F - 2}{S} \tag{6}$$

We can know from formula (6) that the suitable value of $N_j$ and $N_k$ can be found to make $\Gamma$ become integer.

Therefore, the ISL of each satellite will be established in couple inside the Walker constellation. The geometrical characteristic of ISL is consistent. The changes of geometric parameters of each pair of ISL are same except phase, because of periodicity movement.

## 2.3 Topology Requirement of Satellite to Ground Combined Orbit Determination

In order to simplify the problem, the estimated parameter vector $X_0$ is determined as position and velocity of satellite which is located in inertial reference frame.

$$\mathbf{X} = \begin{bmatrix} \mathbf{r} \\ \dot{\mathbf{r}} \end{bmatrix} \tag{7}$$

Here, $\boldsymbol{r}$ and $\dot{\boldsymbol{r}}$ are position vector and velocity vector of satellite, $\boldsymbol{r} = [x \quad y \quad z]^T$ .Assume, the initial value of estimated vector is named as $\boldsymbol{X}^*$. The corrected estimated vector is $\Delta X$, namely, $X = X^* + \Delta X$. So, the satellite dynamics equation can be expressed as follow:

$$\Delta \mathbf{X}_k = \boldsymbol{\Phi}(\mathbf{t}_k, \mathbf{t}_{k-1})\Delta \mathbf{X}_{k-1} + \mathbf{W}_k \tag{8}$$

Hence, $W_k$ is the error vector, $t_k$ is the observing time, and $\Phi(t_k, t_{k-1})$ is the state transition matrix, the approximation is

$$\boldsymbol{\Phi}(\mathbf{t}_k, \mathbf{t}_{k-1}) = \mathbf{I} + \mathbf{A}_{t_k}(\mathbf{t}_k, \mathbf{t}_{k-1}) = \mathbf{I} + \mathbf{A}_{t_k}\Delta \mathbf{t} \tag{9}$$

If the range measuring value is $\rho_k$ at the moment of $t_k$, we have

$$\rho_k = \mathbf{R}_k + \varepsilon_k \tag{10}$$

Hence, $\boldsymbol{R}_k$ is the actual distance from satellite to TT&C node. We have

$$\mathbf{R}_k = \left[(\mathbf{x} - \mathbf{X}_i)^2 + (\mathbf{y} - \mathbf{Y}_i)^2 + (\mathbf{z} - \mathbf{Z}_i)^2\right]^{1/2} \tag{11}$$

$\boldsymbol{\varepsilon}_k$ denotes the measuring error. The observation equation can be Taylor expanded at point of $\boldsymbol{X}^*$.

$$\Delta \rho_k = \rho_k - R_k\big|_{X^*} = H_k \Delta X + \varepsilon_k \tag{12}$$

Where, $[X_i, Y_i, Z_i]$ denotes the position vector of each TT&C nodes, $[X_k \quad Y_k \quad Z_k]_k^T = r$ denotes the position vector of satellites, $\boldsymbol{H}_k$ denotes the designing matrix.

$$\mathbf{H}_k = \begin{bmatrix} \dfrac{\mathbf{x}_k - \mathbf{X}_k}{\mathbf{R}_k} & \dfrac{\mathbf{y}_k - \mathbf{Y}_k}{\mathbf{R}_k} & \dfrac{\mathbf{z}_k - \mathbf{Z}_k}{\mathbf{R}_k} \end{bmatrix} = \frac{1}{\mathbf{R}_k}[\Delta \mathbf{x}_k \quad \Delta \mathbf{y}_k \quad \Delta \mathbf{z}_k] \tag{13}$$

If the observation data at different ephemeris time is normalized initial ephemeris time by formula (8), we have

$$y = \begin{bmatrix} \Delta\rho_0 \\ \Delta\rho_1 \\ \vdots \\ \Delta\rho_M \end{bmatrix} \quad \mathbf{H} = \begin{bmatrix} \mathbf{H}_0\boldsymbol{\Phi}(\mathbf{t}_0,\mathbf{t}_0) \\ \mathbf{H}_1\boldsymbol{\Phi}(\mathbf{t}_1,\mathbf{t}_0) \\ \vdots \\ \mathbf{H}_M\boldsymbol{\Phi}(\mathbf{t}_M,\mathbf{t}_0) \end{bmatrix} \quad \varepsilon = \begin{bmatrix} \varepsilon_0 \\ \varepsilon_1 \\ \vdots \\ \varepsilon_M \end{bmatrix} \tag{14}$$

Where, $M$ denotes the number of TT&C nodes. The batched orbit determination model can be expressed as

$$\Delta X_0 = \left(H^T H\right)^{-1} H^T y \tag{15}$$

$$\hat{\sigma}^2 = \varepsilon^T / \varepsilon / (m-n) \tag{16}$$

$$P_0 = \left(H^T H\right)^{-1} \hat{\sigma}^2 \tag{17}$$

Here, $m$ denotes the number of observation data, $n$ denotes the number of estimable parameters, $\hat{\sigma}$ denotes the estimate of mean of variance, $P_0$ denotes the estimate of covariance matrix.

Assume, $\sigma_x^2$, $\sigma_y^2$ and $\sigma_z^2$ are expressed as the variance of satellite position sub-vector. The estimate of position error for orbit determination can be obtained by formula (17).

$$\sigma_r = \left(\sigma_x^2 + \sigma_y^2 + \sigma_z^2\right)^{1/2} = \left[tr\left(H^T H\right)^{-1}\right]^{1/2} \cdot \hat{\sigma}^2 \tag{18}$$

Where, denominator is the $|\mathbf{H}|$ of matrix $\mathbf{H}$, the elements of matrix $\mathbf{H}$ denotes the direction cosine of satellite related to the line of sight of TT&C nodes. So, the relationship between the $|\mathbf{H}|$ of matrix $\mathbf{H}$ and the space distribution of TT&C station is tightness, especially for GEO satellite. The direction cosine of GEO satellite TT&C node is not change basically, the right coefficient is more important to matrix $\mathbf{H}$. The matrix $\mathbf{H}$ can be expanded as follow

$$H = \begin{bmatrix} \frac{x-X_1}{R_1} & \frac{y-Y_1}{R_1} & \frac{z-Z_1}{R_1} \\ \frac{x-X_2}{R_2} & \frac{y-Y_2}{R_2} & \frac{z-Z_2}{R_2} \\ \vdots & \vdots & \vdots \\ \frac{x-X_M}{R_M} & \frac{y-Y_M}{R_M} & \frac{z-Z_M}{R_M} \end{bmatrix} \tag{19}$$

If the $|\mathbf{H}|$ will be considered only, we have

$$|H| = \frac{1}{R_1 R_2 \cdots R_M} \begin{bmatrix} x & y & z & 1 \\ X_1 & Y_1 & Z_1 & 1 \\ X_2 & Y_2 & Z_2 & 1 \\ \vdots & \vdots & \vdots & \vdots \\ X_M & Y_M & Z_M & 1 \end{bmatrix} \tag{20}$$

Here, the larger the $|\mathbf{H}|$ is, the smaller the orbit determination precision is affected by ranging error.

Therefore, the bigger the flare angle from TT&C nodes to ISL is, the more beneficial to orbit determination, under the condition of satellite visible related to TT&C nodes.

### 2.4 Requirement of Real Time Operation and Control

The ability of ISL data transmission and the level of ISL management and optimization scheduling are determined by ability of real time operation and control inside the navigation constellation. Therefore, after the data transmission rate of ISL has been determined, the key problem will be management and optimization scheduling of ISL.

The satellite network is expressed with figure function $G(V, E, W)$. The set of nodes network can be expressed by $V=\{v_1, v_2, \cdots, v_n\}$. The set may denote satellites, ground stations or other facilities. The set of ISL can be expressed with $E=\{e_1, e_2, \cdots, e_n\}$. The set may denote ISL or UDL (User Data Link). The number of nodes in the network is n. The QoS (Quality of Service) router is made with k kinds of additivity constraint, $k \geq 2$. There are k kinds of independent measure are corresponded with each link $e(u, v)$, $w_j(e, t)$, $1<j<k$. The $w$ denotes the measure parameter set of network link. The $w_i(e, t_j)=\{w_1(e, t_j), w_2(e, t_j), \cdots, w_k(e, t_j)\}$ denotes the measure set of link e at the moment of ti. For the convenience to calculate and to discuss the problem, the time will be discretized. The ti denotes the $i$th segment of time, $t_i=(t_i, t_{i+1})$. Here, $\Delta t$ denotes time interval. The condition of discrete model of dynamic topology network must be met, when the $\Delta t=t_i - t_{i+1}$ is selected, viz. $LT<\Delta t<t_{max}$. There are k kinds of measure parameters for router $P$ from source node $S$ to terminal node $D$, $w_j(p,t_i)=\sum_{e\in p} w_j(e,t_i)$, $1<j<k$, the requirement of QoS measure parameter will be met by parameter $P$, $w_j(p,t_i<M_j)$, $1<j<k$, Mj denotes one of the measure parameter of QoS, that is QoS constraint. There is also a cost function $c(e,t_i)$ for each link. The cost function of router $P$ can be expressed as $c(p,t_i)=\sum_{e\in p} c(e,t_i)$. The purpose of above analysis is to search router $p_o$, to make

$$\begin{cases} w_j(p,t_i)<M_j, & 1<j<k \\ c(p_o,t_i)=min(c(p,t_i)\backslash w_j(p,t_i)<M_j) \end{cases} \tag{21}$$

The math expression of the problem will be $min(c(p,t))$

$$\text{s.t.} \quad \begin{matrix} w_1(p,t)\leq M_1 \\ w_2(p,t)\leq M_{21} \\ \vdots \\ w_k(p,t)\leq M_k \end{matrix} \tag{22}$$

To the ISL of navigation constellation, the measure parameters will be selected as the selection of nodes of telemeter and command data transmission, the selection of number of mono-satellite link, the selection of topology of mono-satellite link.

## 3 Constraint Model of TT&C Scheduling

### 3.1 Description of ISL TT&C Scheduling Problem

To the problem of TT&C scheduling of navigation constellation ISL, we will make the following description [11]:

Assumption: A set of navigation constellation $U$ is given. And it includes finite amount ( $|U|$ =$N_U$) satellite $U_i \in U$ (i=1, ..., $N_U$); A set of ground TT&C equipment $V$ is given. And it includes finite amount ( $|V|$ =$M_V$) equipment $V_i \in V$ (i=1, ..., $M_V$); A set of ISL $L$ established for mono-satellite inside the navigation constellation is given. And it includes finite amount ( $|L|$ =$H_L$) ISL $L_i \in L$ (i=1, ..., $H_L$). At any time, the satellite position inside the constellation can be expressed as ($x_i^U(t)$, $y_i^U(t)$, $z_i^U(t)$).

At the same time, the finite amount waiting carried out tasks are included in the task area. So that, the task set $T$( $|T|$ =NT) is composed. Every task $T_i \in T$ (i=1, ..., NT), in the set $T$ has a unique ID ((identification), which possess of space position ($x_i^T$,$y_i^T$,$z_i^T$).

The result of ISL TT&C resource scheduling is to assign 1~j satellites to the equipment during the visible window. Here, j is satellite number which can be tracked simultaneously by the TT&C equipment or the ISL. The task routing Pi can be expressed as

$$P_i = \{(x_{i1}^T, y_{i1}^T, z_{i1}^T), \cdots, (x_{in}^T, y_{in}^T, z_{in}^T)\} \tag{23}$$

Formula (23) shows the space position when the tasks are carried out in turn. That means, a sequential task set $\Theta_i=\{T_{i1},T_{i2},\ldots,T_{in}\}$ should be specified to a equipment $V_i$ or ISL $L_i$.

### 3.2 Analysis of Constraint Condition of ISL TT&C Scheduling

The ISL TT&C resource scheduling is a constraint optimization problem. The constraint condition of ground TT&C facility can refer to the literature [11] and [12]. The following constraint conditions should be considered on ISL mainly.

- **Maximal Executing Capacity**

That is the maximal targets number for the ground TT&C facility. The data transmission nodes of telemetry and command will be selected according to the number. If the task routing is $P_k$ for the equipment $V_k$ and $P_k$ consumes capacity $Q(P_k)$. The maximal executing capacity of $V_k$ is $q_{Vk}$. The constraint is expressed as below

$$Q(P_k) \le q_{Vk}(\forall V_k \in V) \tag{24}$$

- **Maximum Operating Range of ISL**

It depends on the specification of the equipment on the satellite. If the task routing is $P_k$ for the ISL $L_k$ and the maximum TT&C distance related $P_k$ is $D(P_k)$. The maximum operating range of $L_k$ is $D_{Tk}$. The constraint is expressed as below

$$D(P_k) \le D_{Tki}(\forall L_k \in L) \tag{25}$$

- **Maximum Number of ISL for Mono-satellite**

It depends on the number of ISL antenna on satellite, operating system and visible situation between satellites. If the task routing is $P_k$ for the ISL $L_k$ and the maximum number of ISL is $N(P_k)$. If the number of ISL $L_k$ is $N_{Uk}$, the constraint can be expressed as

$$N(P_k) \le N_{Uki}(\forall L_k \in L) \tag{26}$$

- **Tracking Window**

The TT&C task can only be carried out during the satellite visible window. For the satellite $U_i \in U$ and ISL $L_j \in L$, if the visible window for task $T_k$□ $\Theta_k$ is ($BL_{ij}$, $EL_{ij}$), and the position of satellite $U_i$ is ($x^T{}_k$, $y^T{}_k$, $z^T{}_k$) at the moment $S_{ik}$. The constraint is expressed as

$$BL_{ij} \le S_{ik} \le EL_{ij}(\forall i = 1,2\cdots,N_T; \forall k = 1,2,\cdots,N_U) \tag{27}$$

- **ISL Maintenance Time**

To make sure no assign any task to the ISL $L_k$ when it is abnormal or maintenance. The start maintenance time is $SDT_k$, the end maintenance time is $EDT_k$, and the minimum maintenance time is $\Delta T_k$. The constraint is expressed as

$$SDT_k \ge EDT_k + \Delta T_k \tag{28}$$

- **Priority Relationship**

In the preparing task, the two specify tasks $T_1$ and $T_2$ have the priority relationship. Assume: Task $T_1$ will be executed at $S_{t1}$, and task $T_2$ will be executed at $S_{t2}$.The interval between two tasks is $\Delta_k$. The constraint is expressed as

$$S_{t2} \ge S_{t1} + \Delta_k \tag{29}$$

- **Resource Scheduling**

We should insure every task will be scheduled. Suppose the task number which is not assigned is no_service. The constraint is expressed as below.

$$\cup_{i=1}^{N_U} \Theta_i = T \quad or \quad no\_service = 0 \tag{30}$$

# 4 Description of Application Instance

The constraint model proposed above is used to optimize ISL TT&C resource scheduling, so as to verify its rationality.

The simulation scenery is composed of several parts.

- The navigation constellation is composed of thirty satellites, they are: three GEO satellites, three IGSO satellites and twenty four MEO satellites.
- The ground TT&C network is composed of three ground stations located inner of China.
- Four sets of antenna are installed on the MEO satellite, so as to establish four pairs of ISL between MEO satellite and MEO satellite, MEO satellite and GEO satellite, MEO satellite and IGSO satellite.
- The maximal operating range of ISL is 50000km. According to the constraint condition above, we will simulate as follow:

- SY station is used to manage three IGSO satellites, and the three IGSO satellites are data transmission nodes. KS and JMS station are used to manage three GEO satellites, and the three GEO satellites are data transmission nodes.
- The permanent ISLs are established as MEO to MEO ISL.

Figure 1 shows the TT&C strategy by which the permanent ISL is established between MEO-s11 and other MEO satellites. Here, SY station is used as ground TT&C node, IGSO satellite is used as space TT&C node.

Figure 2 shows the simulation result of TT&C task scheduling strategy based on Figure 1. The algorithm of neural network is used for TT&C scheduling simulation. The three IGSO satellites are monitored and managed for twenty four hours by SY station. The MEO-s11 is managed one hour continuously. The IGSO satellite is used as space TT&C node. The other four MEO satellites are managed in turn through permanent ISL by MEO-s11 satellite.

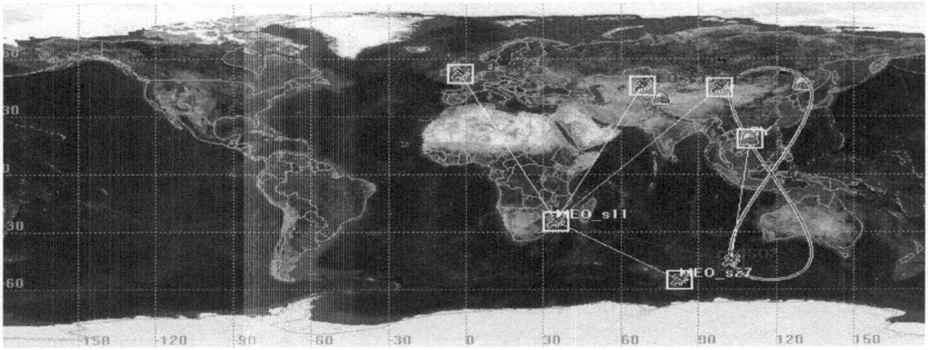


**Fig. 1.** Sketch map of TT&C strategy

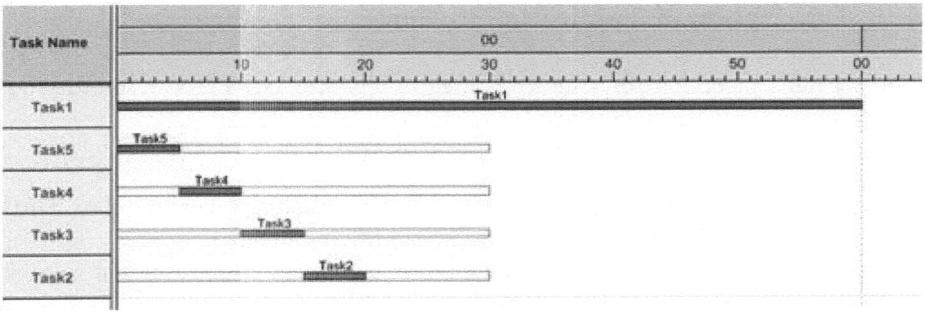


**Fig. 2.** Simulation result of TT&C task scheduling

# 5 Conclusion

It is a foundation that analysis of requirement and constraint model of ISL TT&C scheduling problem on the navigation constellation. It possesses of very strong applying background. The GEO+IGSO+MEO navigation constellation is configurated for researching scenery in the paper. The TT&C requirement and task description method are analyzed. Aiming at the requirement, the corresponding math model of constraint condition is studied. The simulation result shows that the requirements and constraint model are feasible and reasonable. It will provide technique support for the research of TT&C strategy determining and task scheduling further.

## References

1. Zhang, G.-X.: Application of modern small satellite. Posts and Telecom Press, Beijing (2009)
2. Wen, Y.-L.: Analysis and simulation of satellite navigation system. Chinese Astronautics Press, Beijing (2009)
3. Raian, J.A.: Highlights of GPS II-R Autonomous Navigation. In: ION 58th Annual Meeting and CIGTF 21st Guidance Test Symposium, CA, June 24-26 (2002)
4. Maine, K.P., Anderson, P., Langer, J.: Crosslinks for the Next-Generation GPS. In: IEEE Aerospace Conference Proceedings, Los Angeles, CA, March 8-15, pp. 1589–1596 (2003)
5. Maine, K.P., Anderson, P., Bayuk, F.: Communication Architecture for GPS III. In: IEEE Aerospace Conference Proceedings, Los Angeles, CA, March 6-13, pp. 1532–1539 (2004)
6. Luba, O., Boyd, L., Gower, A.: GPS III System Operations Concepts. IEEE Aerospace Systems Magazine 20(6), 10–18 (2005)
7. Cesta, A.: Heuristic scheduling of the DRS communication system. Engineering Application of Artificial Intelligence 8(2), 147–156 (1995)
8. Peng, D.T., Abdelzahe, T.F.: Assignment and scheduling communication periodic tasks in distributed real-time system. IEEE Transactions on Software Engineering 23(12), 745–758 (1997)
9. Chad, H.: From task definitions and plan traces to HTN methods. Report (2006)
10. Clement, B.J., Johnston, M.D.: The deep space network scheduling problem. In: 20th National Conference on Artificial Intelligence (AAAI 2005), and the 17th Innovative Applications of Artificial Intelligence Conference (IAAI 2005), Pittsburgh, PA, vol. 3 (2005)
11. Li, J., Yu, P.-J., Liu, J.-P.: The TT&C Resource Scheduling Method Based on Multi-satellite. In: IEEE ICIS 2010, China, pp. 431–435 (2010)
12. Ling, X.-D., Wu, X.-Y.: Analysis of requirement and task description method of multi-satellite TT&C scheduling problem. Computer Engineering and Applications 45(20), 195–198 (2009)

# NHFCNNC: A New Fuzzy Neural Network Algorithm Based on Matrix and Its Application

Zhiyong Liu[1,2], Xinqing Geng[2], and Guishi Deng[1]

[1] Institute of Systems Engineering, Dalian University of Technology, Dalian Liaoning 116024, China

[2] Department of Mathematics, Anshan Normal University, Anshan Liaoning 114005, China

brove67@yahoo.com.cn

**Abstract.** A new fuzzy neural network algorithm (NHFCNNC) based on matrix is presented, which is applied to the text clustering. The main defect of traditional methods of fuzzy neural network is to know the number of clustering in advance. A new cohesion formula is put forward in NHFCNNC algorithm. The lines (or columns) of the membership matrix of NHFCNNC algorithm are sorted and transformed; the membership matrix is blocked to realize hierarchical clustering. The experiment shows the precision and the efficiency of clustering NHFCNNC are higher than traditional fuzzy neural network.

**Keywords:** fuzzy neural network, text mining, matrix transformation.

## 1 Introduction

Clustering is an algorithm procedure used to partition a set of observations into clusters (i.e. groups, or classes) so that patterns within a class are more similar to each other (according to a predefined measure of similarity) than patterns in the rest of the classes. Text mining is about looking for patterns in text. It may be defined as the process of analyzing text to extract information that is useful for particular purposes.

Treatments of many classical approaches to this problem include the texts by Kohonen[1], Bezdek[2], Duda and Hart[3], Tou and Gonzalez[4], Hartigan[5], and Dubes and Jain[6]. Kohonen's work has become particularly timely in recent years because of the widespread resurgence of interest in the theory and applications of neural networkstructures. However, Kohonen clustering networks (KCNs) suffer from several major problems.

The kohonen clustering network (KCN) clustering is closely related to the fuzzy c-Means(FCM) algorithms. Since fuzzy c-Means algorithms are optimization procedure because the objective function is approximately minimized. The integration of FCM and KCN is one way to address several problems of KCN.KCN is not suitable for fuzzy data.

FKCN is self-organizing algorithm, since the "size" of the updated neighborhood is automatically adjusted during learning, and FKCN usually terminates in such a way of minimized objective function of FCM. However, the structure of FKCN is fixed. The number of clustering is unknown. The defect of FCL algorithm is complicated.

J. Lei et al. (Eds.): AICI 2012, LNAI 7530, pp. 170–176, 2012.

NHFCNNC algorithm is based on matrix, which makes the lines (or columns) of membership matrix sorted or transformed. The matrix is blocked as submatrix to clustering. Cohesion formula is used as clustering criterion to overcome the defect of the tradition fuzzy neural network algorithm. NHCNNC possesses the advantage of high speed and accuracy.

## 2 Character Representation of Texts

Vector Space Model (VSM) is wide used in character representation of texts. In the model, the space of texts is considered as vectors space of orthogonal vectors. Each text means one of feature vector

$V(d)=(t_1,w_1(d);\cdots;t_i,w_i(d);\cdots;t_n,w_n(d))$ Where $t_i$ is term, $w_i(d)$ is the weight in d. $w_i(d)$ is function of appearing frequency $tf_i(d)$ of $t_i$ in d., namely, $w_i(d)=\psi(tf_i(d))$.

TF·IDF is a $\psi(tf_i(d))$ method of determining the weights of the terms. The formula is defined as

$$\psi=tf_i(d)\times\log(\frac{N}{n_i}) \quad . \tag{1}$$

Where N is the number of texts, $n_i$ is the number of $t_i$ in texts.

## 3 NHFCNNC Model

The membership matrix is sorted and transformed in the neural network structure. the membership matrix is blocked to realize hierarchical clustering

### 3.1 Determining the Membership

According to FCC algorithm [7], Suppose input vectors $X=(X_1,X_2...,X_p)^T$, $x_j=(x_{j1},x_{j2},...,x_{js})\in R^s$, $p$ is the number of dataset. The input vectors are partitioned c classes, $2\le c<p$. The p input vectors belong to c classes, whose membership can be represented as follow:

$$u=\begin{pmatrix} u_{11} & u_{12} & \dots & u_{1p} \\ \dots & \dots & \dots & \dots \\ u_{c1} & u_{c2} & \dots & u_{cp} \end{pmatrix} \quad . \tag{2}$$

Each column means the membership of the j vector $X_j$ belonging to c classes. $u_{ij}$ means the membership of the i competitive neuron belonging to the j input vector. u satisfies three constraint.

$$\begin{cases} (1) \quad \sum_{i=1}^{c} u_{ij} = 1 \quad j = 1,2,\dots,p \\ (2) \quad 1 \ge u_{ij} \ge 0 \quad i = 1,2,\dots,c; \quad j = 1,2,\dots p \\ (3) \quad p > \sum_{j=1}^{p} u_{ij} \quad i = 1,2,\dots,c \end{cases} . \tag{3}$$

Fuzzy clustering centers $v_i$ (i=1,2,…,c) mean the center vectors of every class in input space, namely

$v_i=(v_{i1},v_{i2},\dots,v_{is})\in R^s$, fuzzy clustering center vectors are defined as[8]

$$v_i = \frac{\sum_{j=1}^{p} (u_{ij})^m X_j}{\sum_{j=1}^{p} (u_{ij})^m} . \tag{4}$$

Where m is exponential constant, $v_i$ is the weight of the i neuron.

Input vector $X_j$ belongs to the i competitive neuron, whose membership is defined as

$$u_{ij} = \frac{1}{\sum_{k=1}^{c} \left(\frac{d_{ij}^{(w)}}{d_{kj}^{(w)}}\right)^{\frac{2}{m-1}}} . \tag{5}$$

## 3.2 Matrix Partitioning

The lines (or columns) of the membership matrix A are sorted and transformed, as matrix B.

$$A = \begin{bmatrix} u_{11}\ u_{12} \dots u_{1n} \\ u_{21}\ u_{22} \dots u_{2n} \\ \dots\ \dots\ \dots\ \dots \\ u_{n1}\ u_{n2}\ \dots\ u_{nn} \end{bmatrix}, \quad B = \begin{bmatrix} A_{11} & . & A_{12} \\ \dots & . & \dots \\ & d & \\ A_{21} & . & A_{22} \end{bmatrix}$$

To find out partition point d, the definition of $M(A_{pq})$ as followed:

$$M(A_{pq}) = \sum_{i=(p-1)*d+1}^{d+(m-d)*(p-1)} \left( \sum_{j=(q-1)*d+1}^{d+(m-d)*(q-1)} u_{ij} \right), 1 \le p \le 2, 1 \le q \le 2 . \tag{6}$$

d is the line(or column) pertained to the partition point, meets formula (7)

$$f_d = M^{(d)}(A_{11}) * M^{(d)}(A_{22}) - M^{(d)}(A_{12}) * M^{(d)}(A_{21}) . \tag{7}$$

When the partition point is acquired, the membership is blocked.

FK algorithm is as followed.

Input: A

1) Output: submatrix $A_{11}$,$A_{12}$

For i=1:n-1

Begin

Find out the column the maximum pertained to from i+1 column to n column of the i line.

Swap the i+1 line and column

Swap the i+1column and line

End

2) For d=1: n-1

Compute $M^{(d)}(A_{11})$,$M^{(d)}(A_{12})$,$M^{(d)}(A_{21})$,$M^{(d)}(A_{22})$

Compute $M^{(d)}(A_{11}) * M^{(d)}(A_{22}) - M^{(d)}(A_{12}) * M^{(d)}(A_{21})$

3)Find out the d value of $F_d$

4)output $A_{11}$,$A_{22}$

### 3.3 Definition of Cohesion

$$T(C_k) = \frac{1}{M_k} \times \sum_{1 \le i_k < j_k \le n_k} u_{i_k j_k} . \tag{8}$$

Where $C_k$ is the k cluster, $u_{i_k j_k}$ is interpreted as the membership of $x_{j_k}$ in the $i_k$-th partitioning cluster of $x$, $n_k$ is the text number of $C_k$, $M_k = n_k * (n_k - 1)/2$ .

### 3.4 NHFCNNC Algorithm

(i)Suppose initial number of clustering k=2 and T, the membership matrix is initialized according to formula (3), iterations p=1.

(ii)Random select vector $X_j$ from input vectors.

(iii) According to formula (4), the initial cluster vectors are acquired, which are used as the weights of neurons.

(iv)According to formula (5), the membership is computed. The neuron with the largest membership is considered as the winner; the neuron with the second largest membership is considered as the second winner.

(v) According to formula (4) (5), fuzzy cluster center vectors (weights) and membership of the winner and the second winner are modulated, and at the same time, fuzzy cluster vectors (weights) are updated by formula (9).

$$\Delta v_i = \begin{cases} \alpha_c (X_j - v_i) & the\ l\arg est\ u_{ij} \\ \alpha_r (X_j - v_i) & the\ \sec ond\ l\arg est\ u_{ij} \end{cases} . \quad (9)$$

Where $0 \le \alpha_r < \alpha_c \le 1$

(vi) The learning rate is updated, the formula is defined as

$$\begin{cases} \alpha_c = \alpha_c \times \alpha \\ \alpha_r = \alpha_r \times \alpha \end{cases} . \quad (20)$$

(vii) Repeat step ii)-viii), until the training of all samples are over.

(viii) FK algorithm is carried out with the membership, then $A_{11}$,$A_{12}$ and $F_d$ are acquired.

(w) Compute T($A_{11}$),if (T($A_{11}$)&&T($A_{22}$))>T, then clustering is over, $k_{p0}$ is the number of clustering;

else if T($A_{11}$)<T, then A=$A_{11}$,goto step (ii),p=p+1,enter the next period of training; when iteration is over, $k_{p1}$ is the number of clustering.

else if T($A_{22}$)<T, then A=$A_{22}$,goto step(ii),p=p+1,enter the next period of training; when iteration is over, $k_{p2}$ is the number of clustering.

(m)Compute $\sum_{a=1,i=0}^{a=p,i=2} k_{ai}$ , which is the total number of clustering.

## 4 Realization of Text Clustering with NHFCNNC Algorithm

The texts are segmented in the preprocess phase, and VSM model is denoted. The text vectors are used as the input of fuzzy neuron network. NHFCNNC algorithm is applied to clustering. The winner neuron and the second winner neuron are acquired by comparing membership. The membership has nothing with the dimension. So NHFCNNC algorithm doesn't need dimensionality reduction, which avoids the defect that text mining of high dimension must dimensionality reduction.

## 5 Experiments

We collect 1000 documents that are obtained from www.nlp.org.cn, Chinese natural language open platform. The text vectors are 2356 dimensionality after segmentation. Agriculture, economy, education, computer, spaceflight are selected. Each class is 200 pieces of texts.

When $\alpha_c$=0.9, $\alpha$=0.9, $\alpha_r$=0.5$\alpha$, $\varepsilon$=0.01, T=0.95, the initial number of clustering is 2,m is 2, the right number of clustering is 5.

When $\alpha_c$=0.9, $\alpha$=0.5, $\alpha_r$=0.5$\alpha$, $\varepsilon$=0.01, T=0.95,the initial number of clustering is 2,m is 2, the right number of clustering is 5.The convergent rate is faster than the network in which $\alpha$=0.9 is used, and the precision is lower than $\alpha$=0.9's network.

To demonstrate the superiority of NHFCNNC algorithm, NHFCNNC algorithm makes a contrast experiment with FKCN algorithm. In the same experimental condition, NHFCNNC need 3 training cycles and execution time is 5minutes; FKCN need 5 training cycles and execution time is 20 minutes. So the efficiency of NHFCNNC is higher than RPCL algorithm. The total number of segmented texts in table 1 and table 2 means the number of texts belonging to a class after executing the algorithm. The sum of the right statistical number and wrong statistical number is the fact number of texts of a class.

Clustering result is as follow.

**Table 1.** Clustering result of NHFCNNC

| | Agricultu-re | Econo-my | Educ-ation | Com-puter | Space flight |
|---|---|---|---|---|---|
| Wrong statistical number | 20 | 14 | 10 | 18 | 25 |
| Right statistical number | 180 | 186 | 190 | 182 | 175 |
| Total number of segmented texts | 209 | 210 | 195 | 191 | 195 |
| Precision% | 86% | 86% | 97% | 95% | 90% |
| Call% | 90% | 93% | 95% | 91% | 88% |

**Table 2.** Clustering result of FKCN

| | Agricu-lture | Econ-omy | Educa-tion | Comp-uter | Space flight |
|---|---|---|---|---|---|
| Wrong statistical number | 27 | 35 | 45 | 42 | 57 |
| Right statistical number | 173 | 165 | 155 | 158 | 143 |
| Total number of segmented texts | 204 | 206 | 196 | 191 | 190 |
| Precision% | 85% | 80% | 79% | 82% | 75% |
| call% | 87% | 83% | 78% | 79% | 72% |

Compare table 1 with table 2, the precision of NHFCNNC is higher than that of FKCN in documentation. According to membership; the winner neuron and the second winner neuron are acquired in NWFNN model. The winner neuron is not according to distance. The membership has nothing with the dimension of text vectors. The two neurons update the weights and accelerate convergent rate. The lines (or columns) of the membership matrix of NHFCNNC algorithm are sorted and transformed; the membership matrix is blocked to realize hierarchical clustering. In view of the diversity and complexity of Chinese semantic, NWFNN model applies the formula of the membership and fuzzy center vectors of FCC algorithm to represent the texts belonging to the classes.

Compared with FKCN, NWFNN improves the precision and convergent rate.

## 6 Conclusion

NHFCNNC algorithm overcomes the defect that FCM is suitable for spheroid space and partitional fuzzy clustering algorithm and fuzzy neuron network is to know in advance.

NHFCNNC algorithm is more suitable for text mining than FCC algorithm and FKCN algorithm.

## References

1. Kohonen, T.: Self-Organization and Associative Memory, 3rd edn. Springer, Berlin (1989)
2. Bezdek, J.: Pattern Recognition with Fuzzy Objective Function Algorithms. Plenum, New York (1981)
3. Duda, R., Hart, P.: Pattern Classijication and Scene Analysis. Wiley, New York (1973)
4. Tou, J., Gonzalez, R.: Pattern Recognition Principles. Addison-Wesley, Reading (1974)
5. Hartigan, J.: Clustering Algorithms. Wiley, New York (1975)
6. Dubes, R., Jain, A.: Algorithms that Cluster Data. Prentice Hall, Englewood Cliffs (1988)
7. Zeng, H., Liu, X.: Improvement and comparison with learning methods of Fuzzy central clustering algorithms. Sichuan Institute of Light Industry and Chemical Technology 17(1), 1–8 (2004)
8. Geng, X., Wang, Z.: NFCNNC: A new fuzzy competitive neuron network clustering model with its application in text clustering. Journal of the China Society for Scientific and Technical Information 25(3), 296–300 (2006)
9. Liu, Z., Deng, G.: A hierarchical clustering algorithm based on matrix. Journal of Zhengzhou University 42(2), 39–42 (2010)

# Research of Emergency Logistics Distribution Routing Optimization Based on Simulated Annealing Ant Colony Optimization

Teng Fei[1], Li-yi Zhang[1,*], Jin Zhang[2], Yanqin Li[3], Xiaopei Liu[1], and Cheng Zhu[1]

[1]Information Engineering, College Tianjin University of Commerce, Tianjin, China
[2]Office of Equipment, First Hospital of Shanxi University of Medicine, Taiyuan, China
[3]Automation, Institute of Disaster Prevention, Beijing, China
{feiteng,zhangliyi,liuxiaopei,zhucheng}@tjcu.edu.cn

**Abstract.** With the analysis about the factor influencing emergency logistics distribution path optimization, this article builds up the mathematical model based on which aims to take the shortest distribution time special traffic situation, and uses ant colony algorithm based on simulated annealing to solve the model, obtaining a optimization algorithm about emergency logistics. The research about emergency logistics distribution path optimization in critical situation has important theoretical significance and practical value.

**Keywords:** path optimization, emergency logistics, simulated annealing, ant colony algorithm.

## 1 Introduction

Emergency logistics means taking the shortest time to transport relief supplies and people to disaster area to carry out relief after natural disasters. Its outstanding feature is time urgency which calls for the decision maker completing the scheduling solution in the shortest time with the fastest speed to transport relief supplies to disaster area with less cost and time. In this situation, the choice about the best path in emergency logistics is particularly important, which directly relates to people's life and property safety Thus, the research about emergency logistics distribution path optimization in critical situation has important theoretical significance and practical value.

## 2 Emergency Logistics Distribution Path Optimization Model Based on Special Traffic Conditions

### 2.1 Basic thought about Building Up Model

When considering the number of vehicles $l$ taking part in delivery at material storage or distribution center is less than the number of affected points $n$, the distribution paths

* Corresponding author.

J. Lei et al. (Eds.): AICI 2012, LNAI 7530, pp. 177–183, 2012.

are circuit paths beginning from material storage or distribution center, passing several affected points, and finally arriving at distribution center. The number of circuit paths is determined by the number of vehicles taking part in delivery, and the number of affected points which vehicles pass by is based on the load of vehicles. The problem have to be solved is how to make the $l$ vehicles choose suitable affected points based on special traffic conditions and satisfy the hard time window requirements of these affected points to complete delivery with shortest time, which actually is a vehicle routing problem with hard time window.

## 2.2 Suppose of Model

In order to stand out the emergency, take these supposes based on the thoughts of building model, there are several vehicles start from arterial storage or distribution center, there exist link path between each two affected points, the load of vehicles can satisfy the demand of each affected point, the time constraints and unloading time of each affected point are known, the traffic condition between each affected point is determined by traffic coefficient which is bigger means traffic condition is worse. With the considering that affected degree is different from each affected point, which means the urgent degree of the demand of relief supplies is different, it gains shortage coefficient. The bigger the shortage coefficient is, the worse the disaster affects, and the shorter delivery time takes with the influence of people's subjective factors.

## 2.3 Build Up Model

Suppose there are $n$ affected points whose coordinates $(x_i, y_i)$ are known. The distance between affected point $i$ and affected point $j$ is $d_{ij}$, and the travel time is $t_{ij}$. The supplies of affected point $i$ must be delivered before $T_i$. The unloading time of affected point $i$ is $\tau_i$. $v$ is the average speed in the situation without road barriers, which is the average speed in ideal traffic condition. $t_i$、$t_j$ are the time the vehicles arriving at affected point $i$ and affected point $j$. $\omega_{ij}$ is the judgment of effectiveness of the annexation between affected point $i$ and affected point $j$. $\pi_i$ is the judgment whether the unloading time of affected point $i$ is included in the total time. $\varphi_{ij}$ is the traffic condition between affected point $i$ and affected point $j$. $\mu_i$ the relief supplies shortage coefficient of affected point $i$. Thus, the model is:

$$\min t = \sum_{i=1}^{n}\sum_{j=1}^{n}\sum_{k=1}^{l} t_{ij} w_{ijk} + \sum_{i=1}^{n}\sum_{k=1}^{l} \tau_i \pi_{ij} \tag{1}$$

$$t_{ij} = \frac{d_{ij}}{v} \varphi_{ij} \mu_j \tag{2}$$

$$\sum_{i=1}^{n} q_i \varphi_{ik} \le Q \tag{3}$$

$$\sum_{k=1}^{l} \varphi_{ik} = 1 \tag{4}$$

$$\sum_{i=1}^{n} w_{ijk} = \varphi_{jk} \tag{5}$$

$$\sum_{j=1}^{n} w_{ijk} = \varphi_{ik} \tag{6}$$

$$t_j = t_i + t_{ij} w_{ijk} + \tau_i \pi_{ik} \tag{7}$$

$$t_i \le T_i \tag{8}$$

$$w_{ijk} = \begin{cases} 1 & car\,k \quad from \quad i \quad to \quad j \\ 0 & other \end{cases} \tag{9}$$

$$\varphi_{ik} = \begin{cases} 1 & distribution \quad of \quad i \quad by \quad car\,k \\ 0 & other \end{cases} \tag{10}$$

$$\pi_{ik} = \begin{cases} 1 & i \quad is \quad the \quad last \quad affect\,piont \quad of \quad car \quad k \\ 0 & other \end{cases} \tag{11}$$

Where, Formula (2) gives the definition of the vehicles' travel time between each two affected points, and $v$ is the average speed in the situation without road barriers. Formula (3) guarantees that load of each vehicle are no more than the load capacity. Formula (4) formulates each affected point must be delivered by only one vehicle. Formula (5) and Formula (6) is the relationship between variable $w_{ijk}$ and variable $\varphi_{ik}$. Formula (7) is the arrival time of affected point $j$, $t_i$ is the arrival time of Affected point $i$, $\tau_i$ is the unloading time. Formula (8) is the time constraints of vehicles arriving at affected point $i$ which can not lag time $T_i$. Formula (9) and Formula (10) are the definition of variable $w_{ijk}$ and $\varphi_{ik}$. Formula (11) is the judgment whether the unloading time of affected point $i$ will be included in the total time.

## 3 Solved the Model by the Simulated Annealing Ant Colony Optimization

Based on type (12), calculate the transition probability of ant $k$.

$$P_{ij}^{k} = \begin{cases} \dfrac{[\tau_{ij}(t)]^{\alpha} \cdot [\eta_{ik}(t)]^{\beta}}{\sum\limits_{s \in allowed_k} [\tau_{is}(t)]^{\alpha} \cdot [\eta_{is}(t)]^{\beta}} & 若 j \in allowed_k \\ 0 & other \end{cases} \tag{12}$$

Based on type (13), calculate the increment $\Delta t$.

$$\Delta t = T_k(S') - T_k(S) \tag{13}$$

$$\tau_{ij}(t+1) = (1-\rho)\tau_{ij}(t) + \Delta\tau_{ij}(t) \tag{14}$$

$$\Delta\tau_{ij}(t) = \sum_{k=1}^{m} \Delta\tau_{ij}^{k}(t) \tag{15}$$

$$\Delta\tau_{ij}^{k}(t) = \begin{cases} Q/L_k & \text{section k of ants pass edge ij in the course of this tour} \\ 0 & else \end{cases} \tag{16}$$

Where, $\rho$ is the volatile factor of global pheromone, of which the value is $[0,1)$ usually, determine the volatile speed of pheromone $\Delta\tau_{ij}(t)$ is the pheromone's increment of this circling in the pathway $ij$ . $\Delta\tau_{ij}^{k}(t)$ indicates the pheromone that ant $k$ releasing in the pathway $ij$ during the circling. $Q$ is constant, $L_k$ is the loop length formed by ant $k$ traveling around.

As shown in Fig.4. is the flow chart that simulated annealing ant colony optimization solves the mode.

# 4 The Simulation Calculation

The coordinate of the Emergency Distribution Center is known as（14.5, 13.0）. There are 3 vehicles at Emergency Distribution Center, which will deliver supplies to 20 affected points. The load of each vehicle is 10t. The average travel speed of vehicles is $v = 45km/h$.Diagram 1 indicates the coordinate data of each affected point, quantity demanded, terminate time, and discharge time and shortage of each affected point. The traffic condition of each affected point is provided by Document [1]. The begin temperature is T_Max=30, and the end temperature is T_Min=1. Maximum Iterations is 100 with the fixed state temperature. The max unchanged coefficient is 2. Document [2] provides the basis of the values of the parameters $\alpha$、$\beta$、$\rho$. According to the results of many times experiments, when the values of the parameters are $\alpha$、$\beta$、$\rho$, the operation result is the best. The distances between each affected point and between the Emergency Distribution Center and each point can be calculated by the distance formula (17).

$$d_{ij} = \sqrt{(x_i - x_j)^2 + (y_i - y_j)^2} \tag{17}$$

**Table 1.** Experimental Date

| NO. | coordinate | Require ments | End time | unload time | Shortage degree |
|---|---|---|---|---|---|
| 1 | (12.8,8.5) | 0.1 | 4.5 | 0.1 | 0.99 |
| 2 | (18.4,3.4) | 0.4 | 8.5 | 0.2 | 0.98 |
| 3 | (15.4,16.6) | 1.2 | 2.5 | 0.5 | 0.65 |
| 4 | (18.9,15.2) | 1.5 | 4.0 | 0.4 | 0.76 |
| 5 | (15.5,11.6) | 0.8 | 3.5 | 0.3 | 0.97 |
| 6 | (3.9,10.6) | 1.3 | 4.5 | 0.5 | 0.85 |
| 7 | (10.6,7.6) | 1.7 | 4.5 | 0.2 | 0.67 |
| 8 | (8.6,8.4) | 0.6 | 5.0 | 0.4 | 0.83 |
| 9 | (12.50,2.10) | 1.2 | 8.5 | 0.1 | 0.94 |
| 10 | (13.80,5.20) | 0.4 | 7.0 | 0.3 | 0.92 |
| 11 | (6.7,16.9) | 0.9 | 3.5 | 0.5 | 0.86 |
| 12 | (14.8,2.6) | 1.3 | 6.5 | 0.4 | 0.87 |
| 13 | (1.8,8.7) | 1.3 | 7.5 | 0.1 | 0.90 |
| 14 | (17.1,11.0) | 1.9 | 5.5 | 0.3 | 0.73 |
| 15 | (7.4,1.0) | 1.7 | 7.5 | 0.2 | 0.69 |
| 16 | (0.2,2.8) | 1.1 | 6.5 | 0.2 | 0.88 |
| 17 | (11.9,19.8) | 1.5 | 5.5 | 0.3 | 0.95 |
| 18 | (13.2,15.1) | 1.6 | 5.5 | 0.1 | 0.78 |
| 19 | (6.4,5.6) | 1.7 | 5.0 | 0.4 | 0.87 |
| 20 | (9.6,14.8) | 1.5 | 6.0 | 0.2 | 0.93 |

Fig.1 indicates an optimal solution optimal curve under the same situation between the Basic Ant Colony Optimization and Simulated Annealing Ant Colony Optimization. The simulation indicates that, combine simulated annealing Optimization with basic ant colony Optimization to form a double search structure including basic ant colony search and simulated annealing search. Simulated annealing ant colony Optimization can avoid the disadvantage that basic ant colony Optimization is easy to convergence in the non-global optimal solution when solving optimization problem, which improves search ability and optimal quality.

Fig. 2 is the path of the Ant Colony Optimization. Fig. 3 is the path of the Simulated Annealing Ant Colony Optimization.

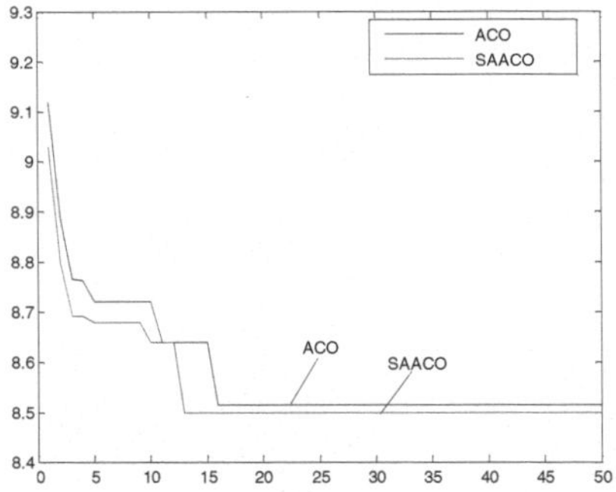


**Fig. 1.** An optimal solution optimal curve

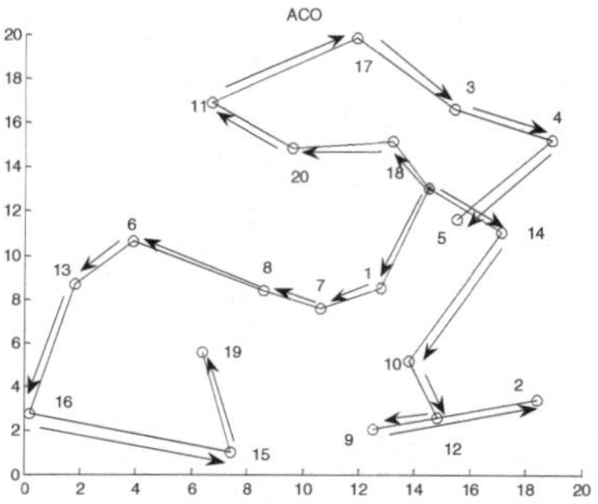


**Fig. 2.** The path of the Ant Colony Optimization

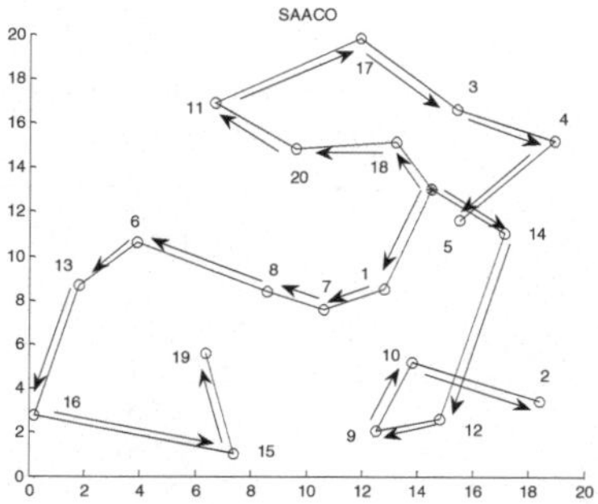


**Fig. 3.** The path of the simulated annealing ant colony Optimization

**Acknowledgment.** This work was financially supported by Funding Issues of Soft Science Research Projects in Shanxi Province (2010041077-3).

## References

1. Fei, T.: Research of ACO in the medical devices logistics distribution routing optimization. Taiyuan University of Technology (2010) (in Chinese)
2. Ye, Z., Zheng, Z.: Study on the parameters $\alpha,\beta,\rho$ in Ant colony algorithm——An example to TSP. J. Wuhan University (Information Science) 29(7), 597–601 (2004)
3. Jiang, X.-Z., Gao, S., Chen, J.-Z.: Hybrid algorithm combining ant colony optimization algorithm with simulated annealing algorithm optimization. Computer Engineering and Design 29(6), 1491–1493 (2008)
4. Zhu, J., Rui, T., Jiang, X.: Simulated annealing ant colony algorithm for QAP. Computer Engineering and Applications 47(14), 34–36 (2011)
5. Liu, B., Meng, P.: Simulated annealing-based ant colony algorithm for traveling salesman problems. Journal of Huazhong University of Science and Technology (Nature Science Edition) 37(11), 26–30 (2009)

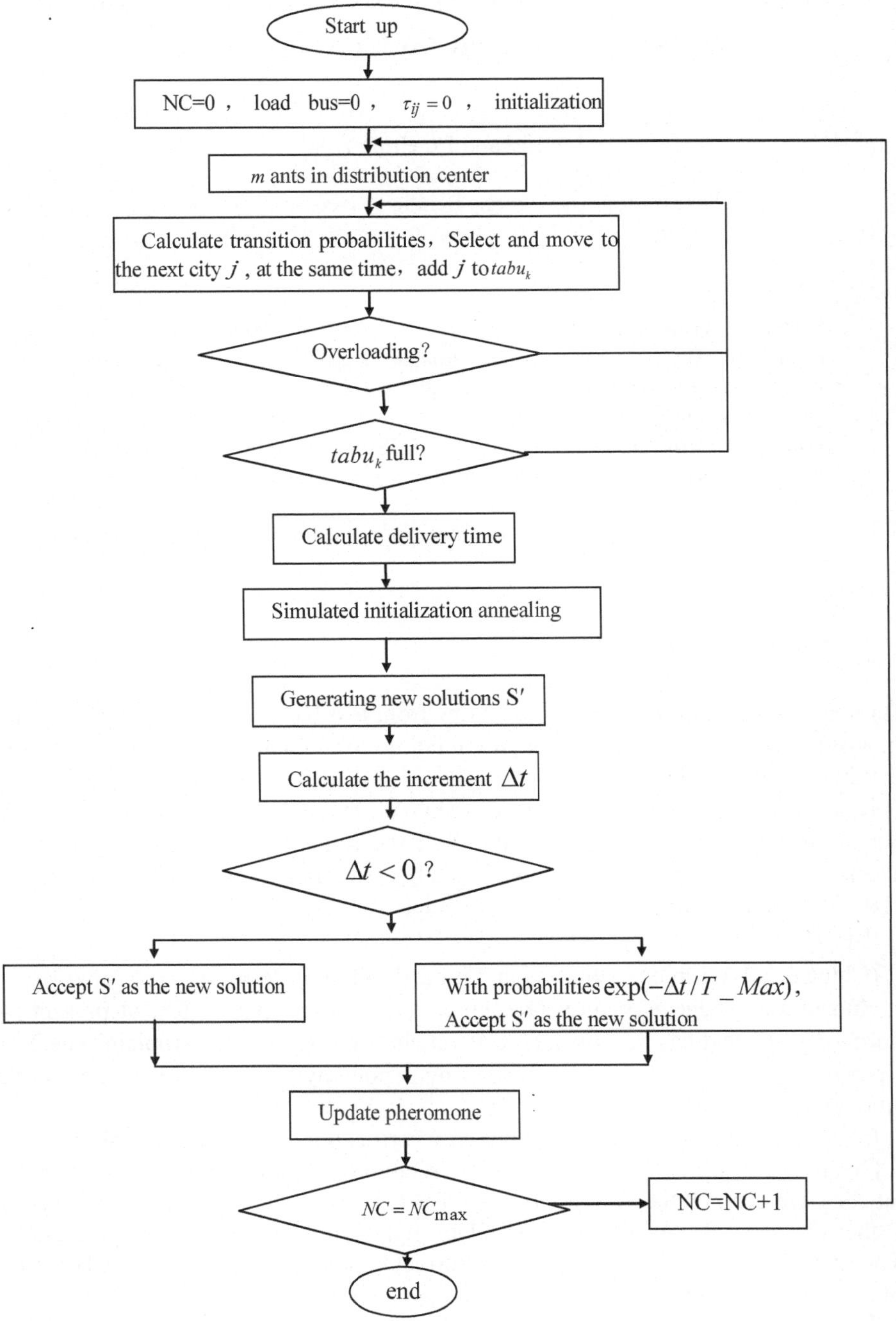
Start up
NC=0 , load bus=0 , $\tau_{ij}=0$ , initialization
$m$ ants in distribution center
Calculate transition probabilities, Select and move to the next city $j$ , at the same time, add $j$ to $tabu_k$
Overloading?
$tabu_k$ full?
Calculate delivery time
Simulated initialization annealing
Generating new solutions S′
Calculate the increment $\Delta t$
$\Delta t < 0$ ?
Accept S′ as the new solution
With probabilities $\exp(-\Delta t / T\_Max)$, Accept S′ as the new solution
Update pheromone
$NC = NC_{\max}$
NC=NC+1
end

# A Spatial Structure Analysis of Candidate Chinese Hyponymy Based on Concept Space

Lei Liu and Lu Hong Diao

College of Applied Sciences, Beijing University of Technology
{liuliu_leilei,diaoluhong}@bjut.edu.cn

**Abstract.** Verification of hyponymy relations is a basic problem in knowledge acquisition. We present a spatial structure analysis of candidate Chinese hyponymy based on concept space. Firstly, we give the definition of concept space about a group of candidate hyponymy relations. Secondly we analyze the basic spatial structure of concept space. Finally experimental results show that the relation between the precision of hyponymy and spatial structure.

**Keywords:** hyponymy, relation acquisition, concept space, hyponymy verification.

## 1 Introduction

Automatic acquisition of semantic relations from text has received much attention in the last ten years. Especially, hyponymy relations are important in accuracy ontology learning, knowledge bases and lexicons [1].

Hyponymy is a semantic relation between concepts. Given two concepts X and Y, there is the hyponymy between X and Y if the sentence "X is a (kind of) Y" is acceptable. X is a hyponym of Y, and Y is a hypernym of X. We denote a hyponymy relation as hr(X, Y), as in the following example:

中国是一个发展中国家 ---hr(中国,发展中国家)

(China is a developing country ---hr(China, developing country) )

Human knowledge is mainly presented in the format of free text at present, so processing free text has become a crucial yet challenging research problem. The error hyponymy relations in the phase of acquiring hyponymy from free text will affect the building of hyponymy lexicon.

In our research, the problem of hyponymy verification is described as follows:

Given a set of candidate hyponymy relations acquired based on pattern or statistics, we denoted these relations as CHR= $\{(c_1, c_2), (c_3, c_4), (c_5, c_6), \ldots\}$, where $c_i$ is the concept of constituting candidate hyponymy relation. The problem of hyponymy verification is how to identify correct hyponymy relations from CHR using some specific verify methods.

The rest of the paper is organized as follows. Section 2 describes related work in the area of hyponymy acquisition, section 3 gives the definition of concept space and the structure of concept space, section 4 analyzes the relation between the precision of hyponymy and spatial structure and finally section 5 concludes the paper.

J. Lei et al. (Eds.): AICI 2012, LNAI 7530, pp. 184–191, 2012.

## 2 Related Work

There are two main approaches for automatic/ semi-automatic hyponymy acquisition. One is pattern-based, and the other is statistics-based. The former uses the linguistics and natural language processing techniques to obtain hyponymy patterns, and then makes use of pattern matching to acquire hyponymy, and the latter is based on corpus and statistical language model, and uses clustering algorithm to acquire hyponymy.

At present the pattern-based approach is dominant, and its main idea is the hyponymy can be extracted from text as they occur in detectable syntactic patterns. The so-called patterns include special idiomatic expressions, lexical features, phrasing features, and semantic features of sentences. Patterns are acquired by using the linguistics and natural language processing techniques [6-8].

There have been many attempts to develop automatic methods to acquire hyponymy from text corpora. One of the first studies was done by Hearst [2]. Hearst proposed a method for retrieving concept relations from unannotated text (Grolier's Encyclopedia) by using predefined lexico-syntactic patterns, such as

| | |
|---|---|
| ...$NP_1$ is a $NP_2$... | ---hr($NP_1$, $NP_2$) |
| ...$NP_1$ such as $NP_2$... | ---hr($NP_2$, $NP_1$) |
| ...$NP_1$ {, $NP_2$}*{,} or other $NP_3$ ... | ---hr ($NP_1$, $NP_3$), hr ($NP_2$, $NP_3$) |

Other researchers also developed other ways to obtain hyponymy. Most of these techniques are based on particular linguistic patterns.

Morin and Jacquemin produced partial hyponymy hierarchies guided by transitivity in the relation, but the method works on a domain-specific corpus [3].

Llorens and Astudillo presented a technique based on linguistic algorithms, to construct hierarchical taxonomies from free text. These hierarchies, as well as other relationships, are extracted from free text by identifying verbal structures with semantic meaning [4].

Sánchez presented a novel approach that adapted to the Web environment, for composing taxonomies in an automatic and unsupervised way. It uses a combination of different types of linguistic patterns for hyponymy extraction and carefully designed statistical measures to infer information relevance [5].

## 3 Concept Space

### 3.1 Building Concept Space

As we know, Chinese is a language different from any western language. A Chinese sentence consists of a string of characters which do not have any space or delimiter in between. So there are still many error relations in the acquired hyponymy relations from free text. They must be verified further for the building of hyponymy lexicon.

Firstly we initially acquire a set of candidate hyponymy relation from large Chinese free text based on Chinese lexico-syntactic patterns [9]. Then we build concept space using those candidate hyponymy relations.

**Definition 1:** The concept space is a directed graph G = (V, E, W) where nodes in V represent concepts of the hyponymy and edges in E represent relationships between concepts. A directed edge ($c_1$, $c_2$) from $c_1$ to $c_2$ corresponds to a hyponymy from concept $c_1$ to concept $c_2$. Weights in W are used to represent the degree of certainty.

**Definition 2:** For each node c in a graph G, (c, c') $\in$E, c' is a direct hypernym concept of c, and c is a direct hyponym concept of c', the set of direct hypernym concept of c is denoted by $\mu^h(c)$, the set of direct hyponym concept of c is denoted by $\mu_h(c)$。 The number of direct hypernym concept of c is denoted by $|\mu^h(c)|$, and the number of direct hyponym concept of c is denoted by $|\mu_h(c)|$.

**Definition 3:** For each edge ($c_1$, $c_2$) in a graph G= (V, E, W), ($c_1$, $c_2$) $\in$E, if $|\mu^h(c_1)| =1$, $|\mu_h(c_1)| =0$, $|\mu^h(c_2)| =0$, $|\mu_h(c_2)| =1$, then ($c_1$, $c_2$) is an isolated edge. If ($c_1$, $c_2$) $\in$E is not an isolated edge, ($c_1$, $c_2$) is denoted by adjacent edge.

### 3.2 The Structure of Concept Space

With the concept space scale increases, its structure becomes more complex, but this complex spatial structure can be split into a number of simple structures.

Let ($c_1$, $c_2$) is an adjacent edge in the concept space G,.

According to the definition of the adjacent edge, ($c_1$, $c_2$) is an adjacent edge , at least one of the following conditions are met: (i) $|\mu^h(c_1)|>1$; (ii) $|\mu_h(c_1)| >0$; (iii) $|\mu^h(c_2)| >0$; (iv) $|\mu_h(c_2)| >1$. The adjacent structure of ($c_1$, $c_2$) is divided into: 2-adjacency, 3-adjacency, and 4- adjacency.

**(1) 2-adjacency:** $|\mu_h(c_1)| =0$, and $|\mu^h(c_2)| =0$. There are three basic 2-adjacency structures, as shown in **Fig. 1**. The dashed arrow is used to represent ($c_1$, $c_2$).

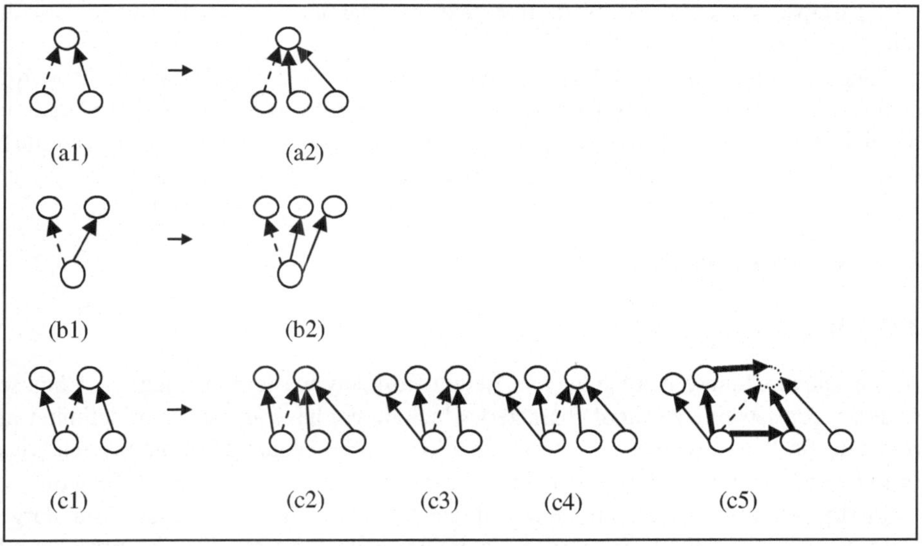


**Fig. 1.** The analysis of 2-adjacency

**Fig. 1(a)**: $|\mu^h(c_1)| =1$, $|\mu_h(c_2)| =2$, in **Fig. 1(a1)**. This structure can be extended to $|\mu^h(c_1)| =1$, $|\mu_h(c_2)| >2$, as shown in **Fig.1(a2)**;

**Fig. 1(b)**: $|\mu^h(c_1)| =2$, $|\mu_h(c_2)| =1$, in **Fig. 1(b1)**. This structure can be extended to $|\mu^h(c_1)| >2$, $|\mu_h(c_2)| =1$, as shown in **Fig. 1(b2)**;

**Fig. 1(c)**: $|\mu^h(c_1)| =2$, $|\mu_h(c_2)| =2$, in **Fig. 1(c1)**. This structure can be extended to:

$|\mu^h(c_1)| >2$, $|\mu_h(c_2)| =2$, as shown in **Fig. 1(c2)**;

$|\mu^h(c_1)| =2$, $|\mu_h(c_2)| >2$, as shown in **Fig. 1(c3)**;

$|\mu^h(c_1)| >2$, $|\mu_h(c_2)| >2$, as shown in **Fig. 1(c4)**;

In particular, if $|\mu^h(c_1) \cap \mu_h(c_2)| >1$, then there is redundant edges, as shown in **Fig. 1(c5)**, where the thick line arrows indicate the edge;

(2) **3-adjacency:** There are two basic 3-adjacency structures, as shown in **Fig. 2**. The dashed arrow is used to represent $(c_1, c_2)$.

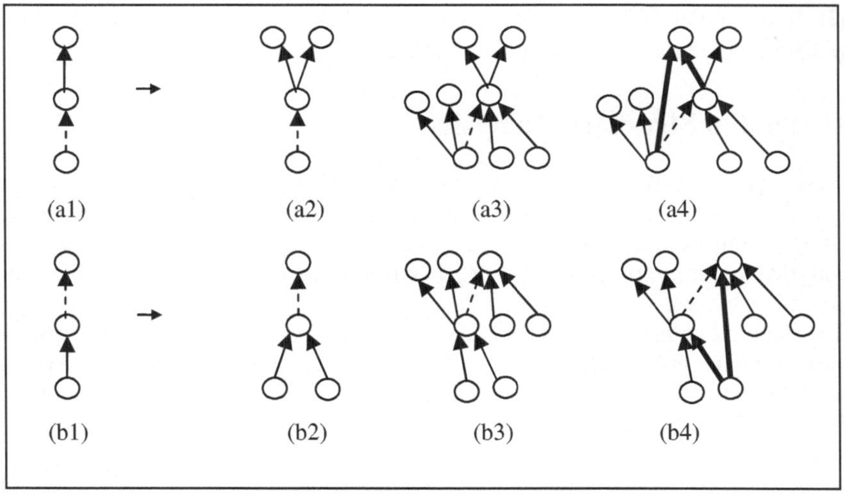


**Fig. 2.** The analysis of 3-adjacency

**Fig. 2(a):** $|\mu_h(c_1)|=0$, $|\mu^h(c_1)| =1$, $|\mu_h(c_2)|=1$, $|\mu^h(c_2)|=1$, in **Fig. 2(a1)**, This structure can be extended to $|\mu_h(c_1)|=0$, $|\mu^h(c_1)| =1$, $|\mu_h(c_2)|=1$, $|\mu^h(c_2)|>1$, as shown in **Fig. 2(a2)**; If it exists a 2-adjacency structure, this structure can be extended to **Fig. 2(a3)**. In particular, if $\mu^h(c_1) \cap \mu_h(c_2)| >1$, then there is redundant edges, as shown in **Fig. 2(a4)**, where the thick line arrows indicate the edge;

**Fig. 2(b):** $|\mu^h(c_2)|=0$, $|\mu^h(c_1)| =1$, $|\mu_h(c_2)| = 1$, $|\mu_h(c_1)| = 1$, in **Fig. 2(b1)**, This structure can be extended to $|\mu_h(c_1)|=0$, $|\mu^h(c_1)| =1$, $|\mu_h(c_2)|=1$, $|\mu_h(c_1)|>1$, as shown in **Fig. 2(b2)**; If it exists a 2-adjacency structure, this structure can be extended to **Fig. 2(b3)**. In particular, if $|\mu_h(c_1) \cap \mu_h(c_2)| >1$, then there is redundant edges, as shown in **Fig. 2(b4)**, where the thick line arrows indicate the edge;

(3) **4-adjacency:** As shown in **Fig. 3**. The dashed arrow is used to represent $(c_1, c_2)$.

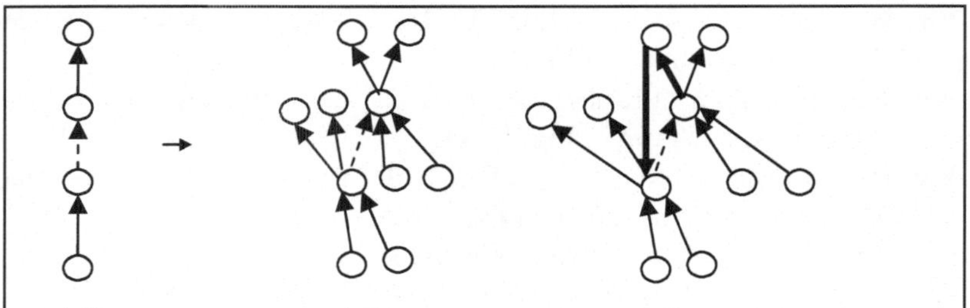


**Fig. 3.** The analysis of 4-adjacency

**Fig. 3(a):** $|\mu_h(c_1)| = 1$, $|\mu^h(c_1)| = 1$, $|\mu_h(c_2)| = 1$, $|\mu^h(c_2)| = 1$, in **Fig. 3(a1)**, If it exists 2-adjacency structure and 3-adjacency structure, this structure can be extended to **Fig. 3(a2)**. In particular, if $|\mu_h(c_1) \cap \mu^h(c_2)| > 1$, then there is redundant edges, as shown in **Fig. 2(b4)**, where the thick line arrows indicate the edge.

## 4 Spatial Structure Analysis

We used about 8GB of raw corpus from the Chinese Web pages. Raw corpus is preprocessed in a few steps, including word segmentation, part of speech tagging, and splitting sentences according to periods. Then we acquired more than 60,000 candidate hyponymy relations (precision 73.3%) from processed corpus by matching Chinese hyponymy patterns.

The detailed result is shown in Table 1. The percentage of edge is the ratio of the number special edges and the number of all edges in concept space. The percentage of node is the ratio of the number of node meeting special edges and the number of all nodes in concept space.

**Table 1.** The result of edge analysis

| Edge structure category | The number of edges | The percentage of edge | Precision | The number of node | The percentage of node |
|---|---|---|---|---|---|
| all the edges | 62,201 | 100% | 73.3% | 40,390 | 100% |
| isolated edge | 2,677 | 4.3% | 65% | 5,331 | 13.2% |
| adjacent edge | 59,524 | 95.7% | 73.5% | 35,059 | 86.8% |
| **2-adjacency** | 16,064 | 25.8% | 76% | 17,045 | 42.2% |
| **(a)** | 7,066 | 11.4% | 75% | 10,178 | 25.2% |
| **(b)** | 1,733 | 2.8% | 71% | 2,867 | 7.1% |
| **(c)** | 7,265 | 11.7% | 78% | 6,381 | 15.8% |
| **3-adjacency** | 32,440 | 52.1% | 77% | 23,506 | 58.2% |
| **(a)** | 22,165 | 35.6% | 79% | 18,135 | 44.9% |
| **(b)** | 10,273 | 16.5% | 73% | 6,825 | 16.9% |
| **4-adjacency** | 11,020 | 17.7% | 57% | 2,221 | 5.5% |

It can be seen from Table 1, the number of isolated edges is very low (4.3%) and the number of adjacent edges is very high (95.7%). It indicates that the correlation of knowledge.

**(1) Isolated edge**

The precision of isolated edges is 65%.The correct rate is lower than the average rate (8.3%). Analysis from the isolated side ($c_1$, $c_2$), $c_1$ is generally the instance concept, and $c_2$ is generally the class concept. Both the level of the difference is small. For example:

(上海现代人剧社，民间职业剧团) ((Shanghai modern theaters, private professional troupes) )

(大泉湾煤矿, 村办股份制煤矿)(( Big Spring Bay coal mine, village-run joint-stock coal mine))

**(2) 2-adjacency**

**(a)** The percentage of edges satisfying this structure is 11.4%, Analysis from the ($c_1$, $c_2$), $c_1$ is generally the instance concept, and $c_2$ is generally the class concept. Both the level of the difference is small. For example:

$c_2$::灌木树种 $c_1$:百里香杜鹃、大卫蔷薇、高山绿线菊、灰梅子

($c_2$: Shrub species $c_1$: thyme cuckoo, David Rose, Alpine Green Line Ju, gray plum)

**(b)** The percentage of edges satisfying this structure is only 2.8%, Analysis from the ($c_1$, $c_2$), this structure does not meet the usual structure of semantic relations.

(甲骨文, 考古资料) (甲骨文, 软件制造商)

((Oracle, The archaeological data) (Oracle, the software manufacturer))

Here "Oracle" is a concept of polysemous words.

**(c)** The percentage of edge satisfying this structure is only 11.7%. Analysis from the ($c_1$, $c_2$), this structure satisfies the structure (a) and (b) at the same time.

$c_2$:: 动物药材 $c_1$: 豹骨, 鹿茸, 熊胆, 麝香

($c_2$: Animal ingredients $c_1$: leopard bone, antler, bear bile, musk)

$c_2$:: 名贵药材, 国家禁止出口货物, 芳香药 $c_1$:麝香

($c_2$: Rare medicinal herbs, the state banned the export of goods, aromatic medicine c1: musk)

**(3) 3-adjacency**

The percentage of edges satisfying this structure is 52.1%(3-adjacency(a) 35.6%, 3-adjacency(b) 16.5%). Analysis from the 3-adjacency(a), $c_1$ is generally the instance concept, and $c_2$ is generally the class concept. Analysis from the 3-adjacency(b), $c_1$ and $c_2$ are both generally the class concept. This case is reasonable, indicating that more hyponymy ($c_1$, $c_2$), $c_1$ is generally the instance concept, and $c_2$ is generally the class concept.

**(4) 4-adjacency**

The percentage of edges satisfying this structure is 17.7% and the percentage of nodes meeting this structure is 5.5%. This is caused by two reasons. Firstly 4-adjacency structure is the most complex in all structures, and therefore consistent with a relatively small number of edges of this structure. Secondly nodes are reused, so it makes the number of nodes have a low proportion.

Analysis from the 4-adjacency, $c_1$ and $c_2$ are both generally the class concept. This case is reasonable, indicating that more hyponymy ($c_1$, $c_2$), $c_1$ is generally the instance concept, and $c_2$ is generally the class concept.

Analysis from the 4-adjacency, the precision is relatively low(57%), the pass is worse, and it is easy to form the side of the loop structure,. For example:

(灾难, 历史) (历史, 回忆) (回忆, 美丽) (美丽, 灾难)

((disaster, history) (history, memories) (memories, beauty) (beauty, disaster))

## 5 Conclusion

In this paper, we present a spatial structure analysis of candidate Chinese hyponymy based on concept space. Firstly, we give the definition of concept space about a group of candidate hyponymy relations. Secondly we analyze the basic structure of concept space. Experimental results show that the relation between the precision of hyponymy and spatial structure.

We can define a set of hyponymy features based on the spatial structure, and verify candidate hyponymy relations. There are still some inaccurate relations in the result. In future, we will combine some methods (such as web page tag etc.) to the further verification of hyponymy.

**Acknowledgments.** This work is supported by the National Natural Science Foundation of China under Grant No. 61105040; the Beijing University of Technology Science Foundation (grant nos. X4006012201101, 006000514311002).

## References

1. Beeferman, D.: Lexical discovery with an enriched semantic network. In: Proceedings of the Workshop on Applications of WordNet in Natural Language Processing Systems, ACL/COLING, pp. 358–364 (1998)
2. Hearst, M.A.: Automated Discovery of WordNet Relations. In: Fellbaum, C. (ed.) To Appear in WordNet: An Electronic Lexical Database and Some of its Applications, pp. 131–153. MIT Press (1998)
3. Morin, E., Jacquemin, C.: Projecting corpus-based semantic links on a thesaurus. In: Proceedings of the 37th Annual Meeting of the Association for Computational Linguistics, pp. 389–396 (1999)
4. Lloréns, J., Astudillo, H.: Automatic Generation of Hierarchical Taxonomies from Free Text Using Linguistic Algorithms. In: Bruel, J.-M., Bellahsène, Z. (eds.) OOIS 2002. LNCS, vol. 2426, pp. 74–83. Springer, Heidelberg (2002)
5. Sánchez, D., Moreno, A.: Pattern-ed automatic taxonomy learning from the Web. AI Communications 21(3), 27–48 (2008)
6. Mititelu, V.B.: Hyponymy Patterns Semi-automatic Extraction, Evaluation and Inter-lingual Comparison. In: Sojka, P., Horák, A., Kopeček, I., Pala, K. (eds.) TSD 2008. LNCS (LNAI), vol. 5246, pp. 37–44. Springer, Heidelberg (2008)

7. Costa, R.P., Seco, N.: Hyponymy Extraction and Web Search Behavior Analysis Based on Query Reformulation. In: Proceedings of the 11th Ibero-American Conference on AI: Advances in Artificial Intelligence, pp. 1–10 (2008)
8. Hattori, S., Ohshima, H., Oyama, S., Tanaka, K.: Mining the Web for Hyponymy Relations Based on Property Inheritance. In: Zhang, Y., Yu, G., Bertino, E., Xu, G. (eds.) APWeb 2008. LNCS, vol. 4976, pp. 99–110. Springer, Heidelberg (2008)
9. Liu, L., Zhang, S., Diao, L., Cao, C.: An Iterative Method of Extracting Chinese ISA Relations for ontology learning. Journal of Computers 5(6), 870–877 (2010)

# Prediction of Energy Consumption Based on Grey Model - GM (1,1)

Xiuli Yu and Zhen Lu

School of Automation Science and Electrical Engineering
Beijing University of Aeronautics and Astronautics
Beijing, China
alicexiuli@163.com, zhenluh@buaa.edu.cn

**Abstract.** The prediction of energy consumption plays an important role in energy management system of enterprise. This paper presents an algorithm of grey model-GM(1,1) to forecast the energy consumption of enterprise. In this article, the principle of grey prediction is analyzed and grey model- GM(1,1) is established, at the same time, the validation of method is verified by making use of the sampled data of compressed air consumption from steel workshop. The average relative error of grey model-GM(1,1) is no more than 1%. The result shows that grey model-GM(1,1) has higher prediction precision and the trend of energy consumption can be reflected accurately in actual energy consumption forecasting.

**Keywords:** Grey system, grey prediction, grey model-GM(1,1), energy consumption of enterprise, Matlab.

## 1 Introduction

With the increasing of energy consumption, energy prices continue to rise. In large manufacturing enterprises, the energy management system that is taken as the part of management information system plays more and more important role, at the same time, the ability to forecast a trend is very significant in energy management system.

In 1982, Professor Deng Julong, a Chinese scholar, founded the theory of grey system[1]. Grey system theory is characterized by small samples, uncertainty and poor information. In fact, some researches have great difficulty in accurate data acquisition, which is the biggest bottleneck restricting research work. For example, the moving average and the exponential smoothing need a large amount of data, if the underlying data is too little, so the reliability of prediction result is greatly reduced. Similarly, if the regression analysis and the regression equation are short of a large amount of data, equation parameters is incorrect. In Markov prediction, state transfer probability is the basis of Markov model, in general, we cannot determine the typical distribution of random variables, and can only use the frequency instead of the probability. Probability theory points out that this replacement is meaningful on the premise of a large sampled values.

In all, the traditional prediction method needs a large number of basic data, and the application of these methods is greatly restricted because of the shortage of data, and

J. Lei et al. (Eds.): AICI 2012, LNAI 7530, pp. 192–199, 2012.

grey system can be applied in modeling by a small amount of data, and then a satisfactory conclusion could be obtained.

Grey system theory considers random processes as grey processes varying in a definite range and time-related, and it regards all sorts of random variables as grey variables changing within a certain limits, so it introduces the method of Grey Data Generation to trim disorder original data. By generating a random sequence can transform into a gray model to meet the conditions of a regulatory sequence [2].

Energy consumption is a dynamic process that changes constantly, which can be regarded as a Grey system, it happens under special condition and it is random and uncertainty. But its latency, development and occurrence is continuous, comparability and relativity, so its occurrence is relative with its past and actuality, this is why we can forecast its occurrence with the analysis of past and actuality. Energy consumption has the characters of grey system, so it can be analyzed by grey system theory [3].

In this paper, we first introduce basic concept of Grey system and the algorithm of Grey Model-GM(1,1), and then discuss its application of energy consumption forecasting model, use several(no less than 4) known data to forecast the energy consumption. With the comparison with actual data, we can find that the error is small enough that Grey Model-GM(1,1) can be actually used.

## 2 The Principle and Calculation Method of Grey Model-GM(1,1)

### 2.1 Principle of Grey Model- GM (1,1)

The grey system prediction only use little data to predict the next data. In such a system, unknown system's information can be determined by using known information. Experimental data have some important characteristics, for example, the sampled data are very few and incomplete.

In this article, grey sequences (generated from original data) are used in forecasting model. We can analyze the historical data, then find the relation and rules between these data, and make use of Grey model-GM(1,1) to forecast the energy consumption of enterprise, include the trend of report form about daily data, weekly data, monthly data, yearly data , and so on. With these forecasting data, enterprise energy department can make some energy plans, and try to improve the economic efficiency and reduce the cost.

### 2.2 Calculation Method of Grey Model- GM (1,1)

Accumulated Generating Operator-AGO is an important method to turn grey system into white system, and it plays an important role in grey system theory. With accumulating, we can find out the development trend of grey data under this method and the rules within the messy original data can be discovered[4-6].

Assume $X^{(0)}$ be the original sequence, means $X^{(0)} = \{X^{(0)}(i), i = 1,2,...,n\}$ , consists of a non-negative data. The first step is to generate a cumulative sequence (1-AGO, Accumulated Generating Operator ) for the establishment of grey forecasting model. The basic expression in AGO:

$X^{(1)} = \{X^{(1)}(k), k = 1,2,...,n\}$

Defined as the 1-AGO of $X^{(0)}$

$$X^{(1)}(k) = \sum_{i=1}^{k} X^{(0)}(i) = X^{(1)}(k-1) + X^{(0)}(k) \quad (1)$$

Suppose $X^{(1)}$ meets the following equation:

$$\frac{dX^{(1)}}{dt} + aX^{(1)} = u \quad (2)$$

Formula (2) is GM(1,1), variable a and variable u are coefficient. Define variable $\hat{a}$ as: $\hat{a} = [a,u]^T$.

Thus we can figure out parameter a and u of GM(1,1). The roots of ordinary differential equations can be calculated on the basis of the formula. The particular solution is:

$$\hat{X}^{(1)}(k+1) = (X^{(0)}(1) - \frac{u}{a})e^{-ak} + \frac{u}{a} \quad (3)$$

$$\text{Or} \quad \hat{X}^{(1)}(k) = (X^{(0)}(1) - \frac{u}{a})e^{-a(k-1)} + \frac{u}{a} \quad (4)$$

Where $k$ is a time series.

### 2.3 Identification Algorithms

Here is parameter sequence $\hat{a}$ , $\hat{a} = [a,u]^T$ , $\hat{a}$ can be calculated by the following formula:

$$\hat{a} = (B^T B)^{-1} B^T Y_n \quad (5)$$

Where : $B$ - data array; $Y_n$ - data column

$$B = \begin{bmatrix} -\frac{1}{2}(X^{(1)}(1) + X^{(1)}(2)) & 1 \\ -\frac{1}{2}(X^{(1)}(2) + X^{(1)}(3)) & 1 \\ \cdots & \\ -\frac{1}{2}(X^{(1)}(n-1) + X^{(1)}(n)) & 1 \end{bmatrix} \quad (6)$$

$$Y_n = (X^{(0)}(2), X^{(0)}(3),..., X^{(0)}(n))^T \quad (7)$$

## 2.4 Recovery of Predictive Value

Because the result that is calculated by GM(1,1) model is an accumulation value, we must have revert the data sequence $\hat{X}^{(1)}(k+1)$ ( or $\hat{X}^{(1)}(k)$ ). Define $\hat{X}^{(0)}(k)$ :

$$\hat{X}^{(1)}(k)=\sum_{i=1}^{k}\hat{X}^{(0)}(i)=\sum_{i=1}^{k-1}\hat{X}^{(0)}(i)+\hat{X}^{(0)}(k)$$
$$\hat{X}^{(0)}(k)=\hat{X}^{(1)}(k)-\sum_{i=1}^{k-1}\hat{X}^{(0)}(i) \quad (8)$$

Because

$$\hat{X}^{(1)}(k-1)=\sum_{i=1}^{k-1}\hat{X}^{(0)}(I),\ \text{So}\ \ \hat{X}^{(0)}(k)=\hat{X}^{(1)}(k)-\hat{X}^{(1)}(k-1) \quad (9)$$

## 2.5 Accuracy Test Method of Grey System Model

Usually, a method that can be used to prediction must be validated through the model accuracy test. A model can be checked by various accuracy tests to determine whether it is reasonable or not. Here are the residual error, relative error, average relative error, mean square deviation.

**i. Residual Error Test**

The original series:

$$X^{(0)}=\{x^{(0)}(1),x^{(0)}(2),\cdots,x^{(0)}(n)\}$$

The corresponding model simulation time series:

$$\hat{X}^{(0)}=\{\hat{x}^{(0)}(1),\hat{x}^{(0)}(2),\cdots,\hat{x}^{(0)}(n)\}$$

Residual Error Series

$$\varepsilon^{(0)}=\{\varepsilon(1),\varepsilon(2),\cdots\varepsilon(n)\}=\{x^{(0)}(1)-\hat{x}^{(0)}(1),x^{(0)}(2)-\hat{x}^{(0)}(2),\cdots,x^{(0)}(n)-\hat{x}^{(0)}(n)\} \quad (10)$$

**ii.Relative Error Sequence**

$$\Delta=\left\{\left|\frac{\varepsilon(1)}{x^{(0)}(1)}\right|,\left|\frac{\varepsilon(2)}{x^{(0)}(2)}\right|,\cdots,\left|\frac{\varepsilon(n)}{x^{(0)}(n)}\right|\right\}=\{\Delta_k\}_1^n \quad (11)$$

(1) For $k<n$, known $\Delta_k=\left|\frac{\varepsilon(k)}{x^{(0)}(k)}\right|$ as relative error of K, known $\Delta_n=\left|\frac{\varepsilon(n)}{x^{(0)}(n)}\right|$ as the filter relative error, $\overline{\Delta}=\frac{1}{n}\sum_{k=1}^{n}\Delta_k$ average relative error of simulation;

(2) Where: $1-\overline{\Delta}$ is average relative accuracy, $1-\Delta_n$ is the filtering precision;

(3) Given $\alpha$ , when $\overline{\Delta} < \alpha$ and $\Delta_n < \alpha$ are established, called model as residual lattice model.

**iii. Residual Sum of Squares**

$$s = \varepsilon^{\mathrm{T}} \varepsilon = [\varepsilon(2) \quad \varepsilon(3) \quad \ldots\ldots \quad \varepsilon(n)] \bullet \begin{bmatrix} \varepsilon(2) \\ \varepsilon(3) \\ \ldots. \\ \varepsilon(n) \end{bmatrix} \tag{12}$$

**iv. Definition 2**

Assume $X^{(0)}$ as the original series, $\varepsilon^{(0)}$ as the residual error sequence.
Defined $X^{(0)}$ mean:

$$\overline{x} = \frac{1}{n} \sum_{k=1}^{n} x^{(0)}(k) \tag{13}$$

Defined $x^{(0)}$ variance :

$$s_1^2 = \frac{1}{n} \sum_{k=1}^{n} (x^{(0)}(k) - \overline{x})^2 \tag{14}$$

Defined residual error mean:

$$\overline{\varepsilon} = \frac{1}{n} \sum_{k=1}^{n} \varepsilon(k) \tag{15}$$

Defined residual variance:

$$s_2^2 = \frac{1}{n} \sum_{k=1}^{n} (\varepsilon(k) - \overline{\varepsilon})^2 \tag{16}$$

## 3 Application of Grey Model-GM(1,1) in Energy Consumption

Using historical statistical data, we can establish grey forecasting model with every target that evaluate the trend of energy consumption. Let's use the original data of compressed air consumption and those of power consumption to establish forecasting model-GM(1,1) for each kind of energy consumption, and compare the statistical data with forecasting data, analyze error between forecasting data and the reality. Using the GM(1,1) method, we can establish the forecasting model of the energy consumption.

Data source: Statistical data are a few sampled data of Compressed air from energy management system of steel workshop. we can forecast the next daily data of energy consumption. The same, we can use the weekly or monthly data to forecast the data of weekly or monthly report form, and so on.

**Table 1.** The daily data of compressed air consumption in second steel smelting workshop and forecasting error

| Serial number | Actual data $x^{(0)}(k)$ | Predictive data $\hat{x}^{(0)}(k)$ | Residual error $\varepsilon(k)=x^{(0)}(k)-\hat{x}^{(0)}(k)$ | Relative error $\Delta_k=\left\lvert\frac{\varepsilon(k)}{x^{(0)}(k)}\right\rvert$ |
|---|---|---|---|---|
| 1 | 1802818.25 | 1802818.25 | 0 | 0 |
| 2 | 1812935.88 | 1811400 | 1523.1 | 0.0840% |
| 3 | 1822165.63 | 1819600 | 2531.6 | 0.1389% |
| 4 | 1830537.00 | 1827900 | 2644.4 | 0.1445% |
| 5 | 1830553.13 | 1836200 | -5635.6 | -0.3079% |
| 6 | 1840396.25 | 1844500 | -4126.2 | -0.2242% |
| 7 | 1851399.38 | 1852900 | -1494.7 | -0.0807% |
| 8 | 1861947.13 | 1861300 | 643.5 | 0.0346% |
| 9 | 1872421.88 | 1869800 | 2670.5 | 0.1426% |
| 10 | 1879511.88 | 1878200 | 1274.4 | 0.0678% |

average relative error is 0.1225%.

**Table 2.** The weekly data of compressed air consumption in Blacksmith - Heavy Forging workshop and forecasting error

| Serial number | Actual data $x^{(0)}(k)$ | Predictive data $\hat{x}^{(0)}(k)$ | Residual error $\varepsilon(k)=x^{(0)}(k)-\hat{x}^{(0)}(k)$ | Relative error $\Delta_k=\left\lvert\frac{\varepsilon(k)}{x^{(0)}(k)}\right\rvert$ |
|---|---|---|---|---|
| 1 | 2912722.50 | 2912722.50 | 0 | 0 |
| 2 | 2971516.75 | 2979100 | -7546.4 | -0.2540% |
| 3 | 3032858.25 | 3025700 | 7135.5 | 0.2353% |
| 4 | 3081365.50 | 3073100 | 8252.4 | 0.2678% |
| 5 | 3113650.00 | 3121200 | -7595.8 | -0.2440% |

average relative error is 0.2002 %.

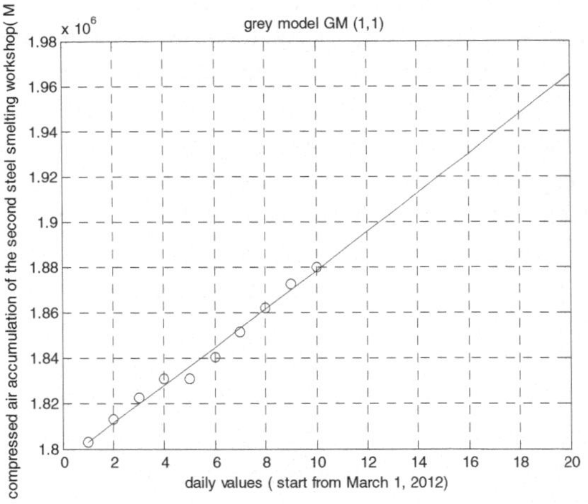


**Fig. 1.** Prediction result of daily compressed air consumption in second steel smelting workshop

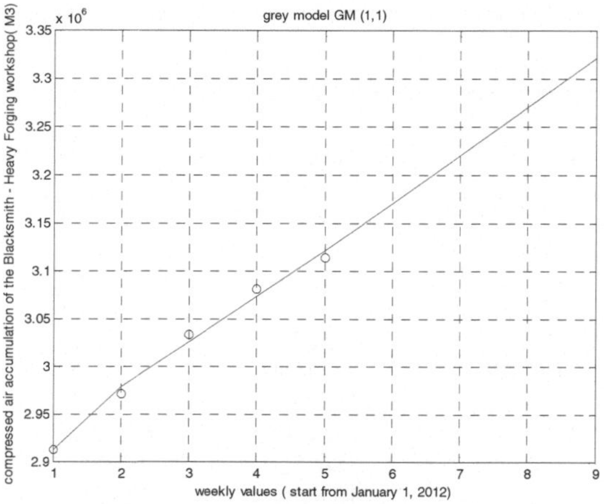


**Fig. 2.** Prediction result of weekly compressed air consumption in Blacksmith - Heavy Forging workshop

The grey forecasting curves are shown in Fig.1 and Fig2. The data of power and other energy consumption can be predicted by using the same method.

Above experimental results are only a part of verified result, we can draw the conclusion that the average relative error between forecasting result and original data is no more than 1%, and basically satisfy our requirement.

## 4 Conclusions

In this paper, the principle of grey model-GM(1,1) is studied, at the same time, an algorithm of grey model- GM (1.1) is realized on the platform of MATLAB. On the basis of the analysis for experimental result, draw the following conclusions:

Not large amount data is required in grey model-GM(1.1), just a few data is used to generate the midterm data, which can weaken the interference of random factors, then the forecasting data can be obtained. Meanwhile, when the data have the same change trend, the error is small, and the error will get larger when the original data have oppositely changes, especially within two back-to-back data.

Experimental results show that grey model-GM(1,1) can be applied for the forecasting of energy consumption, and it is worthy of expanding and applying in the energy consuming prediction and energy programming because of its simple calculation and higher accuracy.

**Acknowledgment.** This article was supported by Scientific and technological cooperation project between the Ministry of Science and Government. (Project number: 2008DFA81280), and the author acknowledges Professor Lu Zhen gratefully, for his guide the work in this paper.

## References

1. Deng, J.: Foundation of grey theory. Huazhong University of Science and Technology Press, Wuhan (2002)
2. Liu, S.-F., Dang, Y.-G., Fang, Z.-G., et al.: Grey system theory and its application. Science Press, Beijing (2010) (in Chinese)
3. Shi, Y.: Prediction of energy consumption based on grey MarkoV model. Journal of Shandong Lnstltute of Light Industry 23(2), 63–65 (2009)
4. Yuan, J.-B., Li, X.: Study on highway passenger volume forecast by the method of the GM(1,3)–Markov chain model. Journal of Transport Science and Engineering 27(4), 68–72 (2011)
5. Zheng, S.-Q., Ma, J.-Z., Guan, J.: Application of a multi-variable grey model in prediction. Journal of Hebei University (Natural Science Edition) 26(4), 350–353 (2006) (in Chinese)
6. Liu, A., Zhao, S., Zhang, Y.-P.: Yield Forecast Based on Grey-Markov Model. Computer Technology and Development 17(6), 191–196 (2007)

# Fuzzy Decision Making Based on Fuzzy Logic with Contradictory Negation, Opposite Negation and Medium Negation

Shanshan Wang, Zhenghua Pan, and Lei Yang

School of Science, Jiangnan University, Wuxi, China
Wangss_89@163.com

**Abstract.** FLCOM is a fuzzy propositional logic with contradictory negation, opposite negation and medium negation which corresponds to fuzzy set FSCOM. Based on theories of FLCOM, this paper took a financial decision making for example, adopted Distance Ratio Function to give an evaluation to fuzzy propositional, gave a method to ascertain value of λ in the semantic model of FLCOM, gave significance of fuzzy production rule based on FLCOM and discussed the application of FLCOM in fuzzy decision making.

**Keywords:** Fuzzy logic, FLCOM, contradictory negation, opposite negation, medium negation, fuzzy decision making.

## 1 Introduction

In representation and processing of uncertain knowledge and their negations, logic with one kind of negation can't satisfy needs of knowledge processing development. Many scholars suggested some different research ideas. Wagner and some others put forward to distinguish strong negation and weak negation in computing information processing system[3][4][5]; Ferré considered distinguishing the extension and intension of negation in logical concept analysis[6]; Kaneiwa put forward an expanding description logic $ALC_{\sim}$ with classical logic and strong negation[7].

Zhenghua Pan indicated that there were three kinds of negative relations in fuzzy knowledge[8]. Then he put forward a fuzzy set with the three kinds of negations which is called FSCOM[2]. To describe the three kinds of negations from view of logic, presently, Pan put forward a new fuzzy propositional logic which is called FLCOM.

In order to indicate applicability of FLCOM when dealing with fuzzy information in practice, this paper researches the fuzzy decision making based on FLCOM.

## 2 Fuzzy Propositional Logic

Since Aristotle, the inconsistent relation of concepts had been distinguished into contradictory relation and opposite relation. The contradictory relation of concepts means, the inconsistent relation of two concepts in the same genus concept, their

J. Lei et al. (Eds.): AICI 2012, LNAI 7530, pp. 200–208, 2012.

extension repulse mutually, and the sum of their extension are equal to extension of the genus concept. The opposite relation of concepts means, the inconsistent relation of two concepts in the same genus concept, their extension repulse mutually, but the sum of their extension are less than extension of the genus concept. The relation between a concept and its negation is inconsistent relation. All kinds of knowledge in the real world, there are ″medium″ concepts between two opposite concepts. It means concepts which partly possess objects' essential attribute reflected by both sides of concepts. We consider the relation between the medium concept and opposite fuzzy concepts is a negative relation, which called medium negative relation.

So the relation between a fuzzy concept and its negation, there are three situations: Contradictory negative relation in Fuzzy Concept (CFC); Opposite negative relation in Fuzzy Concept (OFC); Medium negative relation in Fuzzy Concept (MFC).

In order to expound the three kinds of negations integrally, from the view of logic, we define a new fuzzy propositional logic FLCOM.

## 2.1 Fuzzy Propositional Logic FLCOM with Three Kinds of Negations

**Definition 1[1].** (I) Let $S$ is a nonempty set, its elements are called atomic propositions or atomic formulas, ″$\neg$″ , ″$╕$″ , ″$\sim$″ , ″$\rightarrow$″ , ″$\wedge$″ and ″$\vee$″ are connectives, ″(″ and ″)″ are brackets. Formulate:

(a) For each $A \in S$, A is a fuzzy proposition;

(b) If A and B are propositions, then $\neg A$, $╕A$, $\sim A$, $(A \rightarrow B)$, $(A \vee B)$ and $(A \wedge B)$ are propositions.

(c) The set of entire fuzzy propositions generated by (a) and (b) is called $\Im(S)$ (or$\Im$). The element in $\Im$ is called fuzzy formula (or formula).

(II) Following formulas are called axioms:

(A1) : $A \rightarrow (B \rightarrow A)$, (A2) : $(A \rightarrow (A \rightarrow B)) \rightarrow (A \rightarrow B)$, (A3) : $(A \rightarrow B) \rightarrow ((B \rightarrow C) \rightarrow (A \rightarrow C))$;

($M_1$) : $(A \rightarrow \neg B) \rightarrow (B \rightarrow \neg A)$, ($M_2$) : $(A \rightarrow ╕B) \rightarrow (B \rightarrow ╕A)$, (H) : $\neg A \rightarrow (A \rightarrow B)$;

(C) : $((A \rightarrow \neg A) \rightarrow B) \rightarrow ((A \rightarrow B) \rightarrow B)$, ($\vee_1$) : $A \rightarrow A \vee B$, ($\vee_2$) : $B \rightarrow A \vee B$, ($\wedge_1$) : $A \wedge B \rightarrow A$;

($\wedge_2$) : $A \wedge B \rightarrow B$, (Y$╕$) : $╕A \rightarrow \neg A \wedge \neg \sim A$, (Y$\sim$) : $\sim A \rightarrow \neg A \wedge \neg ╕A$.

(III) Inference Rule MP (*Modus Ponens*): $A \rightarrow B$ and A can infer B.

Formal system consist of (I), (II) and (III) is called Fuzzy propositional Logic with Contradictory negation, Opposite negation and Medium negation (FLCOM).

**Definition 2[1].** In FLCOM, relation of fuzzy propositions $\neg A$, $╕A$ and $\sim A$ is:

$$\neg A = ╕A \vee \sim A$$

## 2.2 The Semantic Interpretation of FLCOM

**Definition 3[1].** Let $\lambda \in (0, 1)$. Mapping$\partial$**:** $\Im \rightarrow [0, 1]$ is called a $\lambda$–evaluation of$\Im$, if

(a) $\partial(A) + \partial_\lambda(\neg A) = 1,$ (1)

(b) $\partial(\sim A) =$

$$\begin{cases} \frac{2\lambda-1}{1-\lambda}(\partial(A)-\lambda)+1-\lambda, & \text{when } \lambda\in[½, 1) \text{ and } \partial(A)\in(\lambda, 1] \quad (2) \\ \frac{2\lambda-1}{1-\lambda}\partial(A)+1-\lambda, & \text{when} \lambda\in[½, 1) \text{ and } \partial(A)\in[0, 1\text{-}\lambda) \quad (3) \\ \frac{1-2\lambda}{\lambda}(\partial(A)+\lambda-1)+\lambda, & \text{when } \lambda\in(0, ½] \text{ and } \partial(A)\in(1\text{-}\lambda, 1] \quad (4) \\ \frac{1-2\lambda}{\lambda}\partial(A)+\lambda, & \text{when } \lambda\in(0, ½] \text{ and } \partial(A)\in[0, \lambda) \quad (5) \\ \partial(A), & \text{others} \quad (6) \end{cases}$$

(c) $\partial(A\vee B) = \max(\partial(A), \partial(B)), \partial(A\wedge B) = \min(\partial(A), \partial(B)),$ (7)

(d) $\partial(A\rightarrow B) = \Re(\partial(A), \partial(B))$, there $\Re$: $[0, 1]^2 \rightarrow [0, 1]$ is a binary function. (8)

**Definition 4 (λ–tautology) [1].** Let $\Im$ is a λ–evaluation set of $\Sigma$, $\forall A\in\Sigma$.If for each $\xi\in\Im$, always existent $\xi(A) = 1$, then A is called $\Im$–tautology. If $\lambda > ½$, and for each $\xi\in\Im$, always existent $\xi(A)\geq \lambda$, then A is called λ–tautology.

## 3 The Application of FLCOM in Fuzzy Decision Making

We take a fuzzy decision making for example to discuss how to use FLCOM to handle the fuzziness and its different kinds of negation in practical problem.

For example: In the real world, whether investor uses the superfluous money every month to deposit in the bank or invest in stock depends on his monthly income, savings and decision rules as follows:

1) If the investor has little savings, then no matter how much monthly income he has, he should deposit the superfluous money in the bank.

2) If the investor has much savings, and he has much monthly income, then he should invest the superfluous money in stock.

3) If the investor has much savings and he has moderate monthly income, then he should use the most superfluous money to invest in stock and few to deposit in the bank.

4) If the investor has moderate savings and he has moderate monthly income, then he should use the most superfluous money to deposit in the bank and few to invest in stock.

Suppose an investor who has savings of 170 thousand Yuan, and monthly income of 8000yuan, then, how to determine the investor's investment strategy?

### 3.1 The Distinction and Formal Denotation of Fuzzy Proposition and Its Different Negations in the Decision Rules

Clearly, propositions in above are all fuzzy propositions. We use FLCOM to represent.

We abstract above fuzzy propositions to formulas: MUCHsavings($x$): represent "Investor X has much savings"; MUCHincome($x$): represent "Investor X has much monthly income"; ╕MUCHsavings($x$): represent "Investor X has little savings"; ╕MUCHincome($x$): represent "Investor X has low monthly income"; ~MUCHsavings($x$): represent "Investor X has moderate savings"; ~MUCHincome($x$): represent "Investor X has moderate monthly income".

The formal denotations of decision making behavior are as follows:

INVESTMENT (stocks): "The investor invest the superfluous money in stock"; INVESTMENT (savings): "The investor deposit the superfluous money in the bank"; MORE (stocks, savings): "The investor use the most superfluous money to invest in stock and few to deposit in the bank"; MORE (savings, stocks): "The investor use the most superfluous money to deposit in the bank and few to invest in stock".

So, we can use logical expression to express the decision rules:

1) ╕MUCHsavings($x$)→ INVESTMENT (savings)
2) MUCHsavings($x$)∧MUCHincome($x$)→ INVESTMENT (stocks)
3) MUCHsavings($x$)∧ ~MUCHincome($x$)→
   (INVESTMENT (stocks)∧INVESTMENT(savings)∧MORE(stocks, savings))
4) ~MUCHsavings($x$)∧ ~MUCHincome($x$) →
   (INVESTMENT (stocks)∧INVESTMENT(savings)∧MORE(savings, stocks))

## 3.2 The Truth-Value Measurement of Fuzzy Proposition in the Decision Rules

We conduct a random searching about cognitions of above fuzzy propositions to people living in Yangtze River delta and the vicinage. And then we calculate the average value of the data in the same area. The results are as Table 1:

**Table 1.** People's viewpoints of " The investor has much monthly income"and so on

| area | The investor has much monthly income (yuan) | The investor has low monthly income (yuan) | The investor has much savings (thousand yuan) | The investor has little savings (thousand yuan) |
|---|---|---|---|---|
| Jiangsu | ≥11000 | ≤1340 | ≥160 | ≤82 |
| Shanghai | ≥14400 | ≤2000 | ≥210 | ≤100 |
| Anhui | ≥5000 | ≤920 | ≥100 | ≤56 |
| Shandong | ≥7000 | ≤1100 | ≥124 | ≤68 |

In order to improve the accuracy of the data, we take an elasticity value to each type of the data. (Data of investor has much monthly income: ±500yuan; data of investor has low monthly income: ±100yuan; data of investor has much savings: ±20 thousand yuan; data of investor has little savings: ±10 thousand yuan.)

By definition 3, we defined ∂(MUCHincome($x$)) as the truth-value function of data $x$ to MUCHincome($x$), and ∂(MUCHsavings($x$)) as the truth-value function of data $x$ to MUCHsavings($x$).

If investor X lives in Shanghai, and $\partial$(MUCHincome($x$)) $\geq$ t (t$\in$[0, 1]), then in other areas, always $\partial$(MUCHincome($x$)) $\geq$ t, if investor X lives in Anhui, and $\partial$(MUCHincome($x$)) $\leq$ t (t$\in$[0, 1]), then in other areas, always $\partial$(MUCHincome($x$)) $\leq$ t. So it is the same to savings. According to these features of the data, the Euclidean distance of one-dimension $d(x, y) = |x-y|$ and Distance Ratio Function[9], we can get:

$$\partial(\text{MUCHincome}(x)) = \begin{cases} 0, & \text{when } x \leq \alpha_F+\varepsilon_F \\ \dfrac{d(x,\alpha_F+\varepsilon_F)}{d(\alpha_F+\varepsilon_F,\alpha_T-\varepsilon_T)}, & \text{when } \alpha_F+\varepsilon_F < x < \alpha_T-\varepsilon_T \\ 1, & \text{when } x \geq \alpha_T-\varepsilon_T \end{cases} \quad (9)$$

(or $\partial$(MUCHsavings($x$)))

In it, $\alpha_T$ is maximal data of "The investor has much monthly income (or much savings)", $\varepsilon_T$ is elasticity value of this data, $\alpha_F$ is minimum data of "The investor has low monthly income (or little savings)", $\varepsilon_F$ is elasticity value of this data.

If $x$ is the monthly income data, then $\alpha_T$=14400, $\varepsilon_T$ = 500, $\alpha_F$ = 920, $\varepsilon_F$ = 100, so

$$\partial(\text{MUCHincome}(x)) = \begin{cases} 0, & \text{when } x \leq 1020 \\ \dfrac{d(x,1020)}{d(1020,13900)}, & \text{when } 1020 < x < 13900 \\ 1, & \text{when } x \geq 13900 \end{cases} \quad (10)$$

If $x$ is the saving data, then $\alpha_T$=210, $\varepsilon_T$=20, $\alpha_F$=56 and $\varepsilon_F$=10,so

$$\partial(\text{MUCHsavings}(x)) = \begin{cases} 0 & \text{when } x \leq 66 \\ \dfrac{d(x,66)}{d(66,190)}, & \text{when } 66 < x < 190 \\ 1, & \text{when } x \geq 190 \end{cases} \quad (11)$$

From these, for any data $x$ of investor's monthly income or savings, we can get the truth-value by corresponding truth-value function. And $\partial$(╕MUCHincome($x$))(or $\partial$(╕MUCHsavings($x$))) can be got by 1−$\partial$(MUCHincome($x$))(or $\partial$(MUCHsavings($x$))). $\partial$(~MUCHincome($x$)) and $\partial$(~MUCHsavings($x$)) can be got by formula (2) to (6) in Definition 3. But it needs to ascertain the value of parameter$\lambda$.

### 3.3 Ascertain Value of $\lambda$ in the Semantic Model of FLCOM

Now we give a method to ascertain the value of $\lambda$ in actual application of FLCOM.

By Table 1, the minimum value of "The investor in Jiangsu has much monthly income" is 11000, the maximal value of "The investor in Jiangsu has low monthly income" is 1340. By corresponding truth-value function:

$$\partial(\text{MUCHincome}(11000)) = \frac{d(11000,1020)}{d(1020,13900)} = 0.775,$$

$$\partial(╕\text{MUCHincome}(11000)) = 1 - \partial(\text{MUCHincome}(11000)) = 0.225,$$

$$\partial(\text{MUCHincome }(1340))=\frac{d(1340,1020)}{d(1020,1390)}=0.025,$$

$$\partial(\neg\,\text{MUCHincome }(1340))=1-\partial(\text{ MUCHincome }(1340))=0.975.$$

Because of the regional economic disparities and the fuzziness of fuzzy proposition, both of ∂(MUCHincome(11000)) and ∂(╕ MUCHincome (1340)) are not equal to one. So, we take $\lambda=\frac{1}{2}(\partial(\text{MUCHincome}(11000))+\partial(\neg\,\text{MUCHincome}(1340)))=0.875$. For an investor in Jiangsu, if $\partial(\text{MUCHincome}(x))\geq\lambda$, then he has much monthly income.

Similarly, we can get the value of λ of "The investor has much monthly income" and "The investor has much savings" in each areas as Table 2:

**Table 2.** Value of λ of "The investor has much monthly income" and "The investor has much savings" in each area

| | Jiangsu | Shanghai | Anhui | Shandong |
|---|---|---|---|---|
| The investor has much monthly income | 0.875 | 0.962 | 0.655 | 0.729 |
| The investor has much savings | 0.815 | 0.863 | 0.637 | 0.726 |

We average the value of λ of the same proposition as the value of λ of data to the corresponding propositions, so we get the results as Table 3:

**Table 3.** Value of λ of "The investor has much monthly income (or much savings)"

| Fuzzy proposition | The investor has much monthly income | The investor has much savings |
|---|---|---|
| Value of λ | 0.805 | 0.760 |

The significance of value of λ to fuzzy propositions in the example are as follows:

(1) To any investor X, if $\partial(\text{MUCHincome}(x))$(or $\partial(\text{MUCHsavings}(x)))\geq\lambda$, then the investor has much monthly income (or much savings).
(2) To any investor X, if $(1-\lambda)\leq\partial(\text{MUCHincome}(x))$(or $\partial(\text{MUCHsavings}(x)))\leq\lambda$, then the investor has moderate monthly income (or moderate savings ).
(3) To any investor X, if $\partial(\text{MUCHincome}(x))$(or $\partial(\text{MUCHsavings}(x)))\leq 1-\lambda$, then the investor has low monthly income (or little savings).

### 3.4 Fuzzy Production Rule and Its Achievement Based on FLCOM

The general form of fuzzy production rule is: $P_1, P_2, \ldots P_n \to Q$, CF, $(\tau_1, \tau_2, \ldots \tau_n)$.

$P_i$ ($i = 1, 2,\ldots, n$) is fuzzy proposition, which indicate inference premise; Q indicate conclusions or actions. CF($0<CF<1$) is rule confidence, $\tau_i$ ($0 \leq \tau_i \leq 1$, $i = 1, 2, \ldots, n$) indicates corresponding threshold value of $P_i$. Based on FLCOM, significance of this fuzzy production rule is: To inference rule with confidence of CF, when fuzzy proposition $P_i$ satisfies the corresponding threshold value, the rule can be used and the conclusions Q can be reasoned.

Apparently, above fuzzy production rule can be used for the decision rules in this example. Rule confidence can be ascertained by random searching or expert determination. We may as well suppose CF=0.9 in this example.

According to 3.3(1), it's easy to see that significance of $\lambda$ of "The investor has much monthly income (or much savings)" is similar to the significance of the corresponding threshold value. And according to 3.3(3), when $\partial(\text{MUCHincome}(x))$ (or$\partial(\text{MUCHsavings}(x)))\leq 1-\lambda$, which can be translate into

$$\partial(\neg\,\text{MUCHincome}(x))(\text{or}\partial(\neg\,\text{MUCHsavings}(x)))$$
$$=1-\partial(\text{MUCHincome}(x))(\text{or}\partial(\text{MUCHsavings}(x)))\geq\lambda,$$

the investor has low monthly income (or little savings), then we get the corresponding threshold value as Table 4:

**Table 4.** Threshold value of "The investor has much (or low) monthly income " and "The investor has much (or little) savings"

| Fuzzy propositions | The investor has much (or low) monthly income | The investor has much (or little) savings |
|---|---|---|
| Threshold value $\lambda$ | 0.805 | 0.760 |

The significance of the threshold value are: For investor X's monthly income data (or savings data) $x$, if $\partial(\text{MUCHincome}(x))(\text{or }\partial(\text{MUCHsavings}(x)))\geq\lambda$, then investor X has much monthly income (or mqiuuch savings); if $\partial(\neg\,\text{MUCHincome}(x))$ (or $\partial(\neg\,\text{MUCHsavings}(x)))\geq\lambda$, then investor X has low monthly income (or little savings).

According to 3.3(2), when $1-\lambda\leq\partial(\text{MUCHincome}(x))(\text{or}\partial(\text{MUCHsavings}(x)))\leq\lambda$, the investor has moderate monthly income (or savings), we can get results as Table 5:

**Table 5.** Threshold value of "The investor has moderate monthly income(or savings)"

| Fuzzy propositions | The investor has moderate monthly income | The investor has moderate savings |
|---|---|---|
| Threshold value $\tau_1$ | 0.195 | 0.240 |
| Threshold value $\tau_2$ | 0.805 | 0.760 |

There the significance of threshold value is: For investor X's monthly income data (or savings data) $x$, if $\tau_1\leq\partial(\text{MUCHincome}(x)(\text{or }\partial(\text{MUCHsavings}(x)))\leq\tau_2$, then investor X has moderate monthly income (or moderate savings).

Next, we discuss how to determine the investor's investment strategy. We can get:

$$\partial(\text{MUCHincome}(8000))=\frac{d(8000,1020)}{d(1020,13900)}=0.545,$$

$$\partial(\text{MUCHsavings}(170))=\frac{d(170,66)}{d(66,190)}=0.839,$$

$$\partial(\neg\,\text{MUCHincome}(8000))=1-\partial(\text{MUCHincome }(8000))=0.455,$$

$$\partial(\neg\,\text{MUCHsavings}(170))=1-\partial(\text{MUCHsavings }(170))=0.161.$$

By Definition 3, the value of λ of "The investor has much monthly income" is 0.805, and 0.805>1/2, ∂(MUCHincome (8000))=0.545∈(1−λ, λ), so

$$\partial(\sim\text{MUCHincome }(8000))=\partial(\text{MUCHincome}(8000))= 0.545.$$

The value of λ of "The investor has much savings" is 0.760, and 0.760>1/2, ∂(MUCHsavings (170))=0.839∈(λ, 1], so

$$\partial(\sim\text{MUCHsavings}(170))=\frac{2\lambda-1}{1-\lambda}(\partial(\text{MUCHsavings}(170))-\lambda)+1-\lambda=0.411.$$

For decision rule 1): ╕ MUCHsavings($x$)→INVESTMENT (savings), 0.9, 0.805.

Because ∂(╕ MUCHsavings(170)))=0.161<0.805, decision rule 1) cannot be used.

For decision rule 2):

MUCHsavings($x$)∧MUCHincome($x$)→INVESTMENT (stocks), 0.9, (0.760, 0.805).

Because ∂(MUCHsavings (170)) = 0.839 > 0.760, ∂(MUCHincome(8000))=0.545< 0.805, according to fuzzy production rule, decision rule 2) cannot be used.

For decision rule 3): MUCHsavings($x$)∧ ~MUCHincome($x$)→ (INVESTMENT (stocks) ∧INVESTMENT (savings)∧MORE(stocks, savings)), 0.9, (0.760, (0.195,0.805)).

Because ∂(MUCHsavings(170))=0.839>760, 0.195<∂(MUCHincome(8000)) =0.545<0.805, according to fuzzy production rule, decision rule 3) can be used.

For decision rule 4): ~MUCHsavings($x$) ∧~MUCHincome($x$)→ (INVESTMENT (stocks) ∧INVESTMENT (savings)∧MORE(savings, stocks)), 0.9, ((0.240,0.760), (0.195,0.805)).

0.195<∂(MUCHincome (8000))=0.545<0.805, but, ∂(MUCHsavings(170))= 0.839 is not between 0.240 and 0.760, so, decision rule 4) cannot be used.

So, investor in the example can choose decision rule 3) as his investment strategy.

## 4 Conclusions

We discuss fuzzy propositional logic FLCOM with three kinds of negations. Specifically, we discuss the distinction and formal denotation of fuzzy propositions and their negations, give methods to calculate the truth-value of fuzzy formulas and ascertain the value of λ in the semantic model of FLCOM. At last, we discuss fuzzy reasoning and fuzzy decision making by fuzzy production rule.

**Acknowledgments.** This work was supported by the National Natural Science Foundation of China (60973156) and the Program for Innovative Research Team of Jiangnan University.

## References

1. Pan, Z.: Fuzzy Propositional Logic with Contradictory Negation, Opposite Negation and Medium Negation. In: Proceedings of the 13th European Conference on Logics in Artificial Intelligence (2012) (in Press)

2. Pan, Z., Yang, L., Xu, J.: Fuzzy Set with Three Kinds of Negations and Its Applications in Fuzzy Decision Making. In: Deng, H., Miao, D., Lei, J., Wang, F.L. (eds.) AICI 2011, Part I. LNCS, vol. 7002, pp. 533–542. Springer, Heidelberg (2011)
3. Wagner, G.: Web Rules Need Two Kinds of Negation. In: Bry, F., Henze, N., Małuszyński, J. (eds.) PPSWR 2003. LNCS, vol. 2901, pp. 33–50. Springer, Heidelberg (2003)
4. Analyti, A., Antoniou, G., et al.: Negation and negative information in the W3C resource description framework. Annals of Mathematics, Computing & Teleinformatics 1(2), 25–34 (2004)
5. Dung, P.M., Mancarella, P.: Production systems need negation as failure. IEEE Transactions on Knowledge and Data Engineering 14(2), 336–353 (2002)
6. Ferré, S.: Negation, Opposition, and Possibility in Logical Concept Analysis. In: Missaoui, R., Schmidt, J. (eds.) Formal Concept Analysis. LNCS (LNAI), vol. 3874, pp. 130–145. Springer, Heidelberg (2006)
7. Kaneiwa, K.: Description logic with contraries, contradictories, and subcontraries. New Generation Computing 25(4), 443–468 (2007)
8. Pan, Z., Zhang, S.: Differentiation and processing on contradictory relation and opposite relation in knowledge. In: Proceedings of IEEE-The 3rd International Conference on Natural Computation (ICNC 2007), Shanghai, vol. 4, pp. 334–338 (2007)
9. Hong, L., Xiao, X., Zhu, W.: Measure of Medium Truth Scale and Its Application (I). Chinese Journal of Computers 29(12), 2186–2193 (2006) (in Chinese)

# An Approach of Affection Thinking Based on Ant Colony Strategy

Xiangbing Zhou[1,2], Hongjiang Ma[1], and Fang Miao[1]

[1] Department of Computer Science, Aba Teachers College,
Sichuan Wenchuan, 623002, China
[2] School of information Science and Technology,
Chengdu University of Technology Chengdu, 610059, China
{3dsmaxmaya,mhj69,mfang}@163.com

**Abstract.** Affective computing has got some well results in some fields such as artificial intelligence, virtual reality, human-computer interaction, pattern recognition. But those affective models are presented from a one-dimensional angle, and autonomous regulation is in charge of the affective running which prevents further improvement for the affective process efficiency. In the paper, Employing travelers' affection as a case this study proposes an affective thinking model based on ant colony strategy to help affection running. In this model, we introduce an affection space to satisfy traveler's affection features by considering two different aspects: tourism products and tourism consumption. A satisfied thinking running model is proposed in the affective space which are known as affection thinking and we defined some different features in affection thinking. Meanwhile, in order to better intelligentize the model, we employ ant colony to do the affection process which results in a new approach: affection ant colony. We adopt the VTA case under the WSMO to test, and the results show that this approach is not only efficient, but also is better than others approaches.

**Keywords:** Ant Colony Algorithm, Affection Thinking, Tourism consumption.

## 1 Introduction

Application of affective computing is developing with Web 2.0 and outstretching to other fields. It is significant to improve intelligent information processing. However, how to ensure application level of affective computing and how to obtain travelers' psychology dynamic by considering sentiment analysis have now become hot topics nowadays[1-2].

The affective computing application mainly focuses on sentiment analysis; it is a multidisciplinary integrated research field containing many study tasks such as subjective and objective recognition and sentiment classification[3-4], etc. At present, sentiment analysis research tasks have been focused on affective characteristics, in this scenario, affection space model is set up simplex with less expansibility and diversity, so it is hard to express deeper affective intelligence [5-6]. Therefore, this paper raises an

J. Lei et al. (Eds.): AICI 2012, LNAI 7530, pp. 209–218, 2012.

approach of affection thinking based on ant colony strategy[15][16]. This approach uses a binary-variables and *N*-dimensional affective space and affective thinking model so that we can cover both perspectives, and introduce ant colony strategy into the model to realize the optimized affective activities. To speed up the application of the proposed affection thinking in a context of realistic scale, we employed VTA (Virtual Travel Agency) services which fits WSMO(Web Service Modeling Ontology). With a number of simulation experiments, we validate the approach and compare its effectiveness against an approach based on non-affective ant colony.

## 2 Affection Thinking Model

Sentiment plays an important role in human activity, it has crucial influence on one's memory, thinking, character, cognition and judgment [7][8]. When emotion and cognition interact, overlap and interdepend, cognition will affect feelings, and feelings will decide the perception of things and action, and so on. For now, through the researches, emotion characteristics have been introduced to computers, and the interaction of the two has been realized so as to form a new research area: affective computing. This study analyze travelers' affective competence on tourism consumption products based on affective computing in the paper. But emotions will gradually decay as time changes, it can be constantly activated in different environment and feelings. This paper assume the emotional angle that travelers have on tourism products and tourism consumption to analyze travelers' attitude, character (personality), motive, acceptability, likes and tropism on tourism consumption. Consequently, we set up an affection space from two angles: emotional cognition and emotional expression[9][10][11][12][14].

**Definition 1.** Suppose travelers' affective space is a *N*-dimensional vector space: $AS=(S_1[X_{11}, X_{12}], S_2[X_{21}, X_{22}],\ldots, S_n[X_{n1}, X_{n2}])$,where, $S_n(X_{n1}, X_{n2})=S_n[X_{n1}(\alpha_{attitude}, \alpha_{personality}, \alpha_{motive}, \alpha_{acceptability}, \alpha_{likes}, \alpha_{tropism}), X_{n2}(\beta_{joy}, \beta_{sadness}, \beta_{anger}, \beta_{surprise}, \beta_{fear}, \beta_{disgust})]$

Vector $X_{n1}$ ($\alpha_{attitude}, \alpha_{personality}, \alpha_{motive}, \alpha_{acceptability}, \alpha_{likes}, \alpha_{tropism}$)respectively show travelers' attitude, personality (character), motive, acceptability, likes and tropism on tourism products. Vector $X_{n2}$ ($\beta_{joy}, \beta_{sadness}, \beta_{anger}, \beta_{surprise}, \beta_{fear}, \beta_{disgust}$) respectively show travelers' own state of joy, sadness, anger, surprise, fear and disgust.

If $\forall \mathbf{X}_{\mathbf{n}1}\alpha[\cdot] \in [0,1], \forall \mathbf{X}_{n2}\beta[\cdot] \in [0,1]$ ,so $\mathbf{X_{1n}X_{2n}}^T \in [0,1]$. When ($t_1<t_2$) during $t_1$ to $t_2$, travelers' emotion change space **AS** of tourism consumption products will generate more affection space which have several $X_{n1}$ and $X_{n2}$ vectors which can form a limited, controllable affection space. Accordingly, $X_{n1}$ and $X_{n2}$ together will decide the change of **AS**, the result of their joint action is shown in the following matrix **R**:

$$\mathbf{R}_{n\times n} = \begin{bmatrix} \theta(x_{1,1}) & \theta(x_{2,1}) & \cdots & \theta(x_{n,1}) \\ \theta(x_{2,1}) & \theta(x_{2,2}) & \cdots & \theta(x_{n,2}) \\ \vdots & \vdots & \vdots & \vdots \\ \theta(x_{n,1}) & \theta(x_{2,n}) & \cdots & \theta(x_{n,n}) \end{bmatrix} \tag{2.1}$$

$\theta(x_{i,j}) = \mathbf{X_{1n}}\mathbf{X_{2n}}^T$ ,and $n$ *is* related to the emotional change of time ($t_1$, $t_2$), when frequency rises the **R** space rises, vice-versa. Then if we divide ($t_1$, $t_2$) into M emotional moments, we can produce a M-dimensional emotion activity space to **R**, and then we get affection space $\mathbf{AS}'_{m\times m} = (\mathbf{R}_{n\times n})_{m\times m}$.

But because the emotional change is random, the change of vector $X_{n1}$ and $X_{n2}$ is also random, so the action result **R** turned out random. And because the emotional changes are related to the change of time and environment, and the switch of feelings and scenes, so we employ the following probability formula to calculate the interaction of vector $X_{n1}$ and $X_{n2}$ :

$$\phi(x_{i1}, x_{j2}) = \frac{\sum_{i,j=1}^{n} \kappa(\omega_{i1} \mid \omega_{j2}) \Big/ m}{n} \tag{2.2}$$

In the (2.2), $\kappa(\omega_{i1} \mid \omega_{j2})$ represents the emotional change probability (using Bayes formula) when vector $X_{n1}$ and $X_{n2}$ interact, $n$ is the $N$-dimensional space, $m$ is a number of time $t_1$,$t_2$ ($t_1$<$t_2$) . That makes each element $\theta(x_i,x_j)$ in **R** become function of $\varphi$, so we say $R(\varphi)=(\theta(x_i,x_j)(\varphi))_{m\times n}$. If $R(\varphi)$ is integrable, we can define the emotional intensity $\delta$ in the affective space as the following formula:

$$\delta(x_i, x_j) = \max\left( \int_{t_2}^{t_1} \mathbf{R}(\varphi) d\varphi = \left( \int_{t_2}^{t_1} \theta(x_i, x_j)(\varphi) d\varphi \right)_{m\times n} \right) \tag{2.3}$$

Vector $X_{n1}$ and $X_{n2}$ in **AS** can only generate a interaction matrix **R** when the emotions in the established affection space **AS** are activated, so we need to set up a thinking model in the affection space.

*Remark 1.* $S_n(X_{n1}, X_{n2})$ describes the demand feeling and cognitive feeling the tourists have on the tourism products.

*Remark 2.* The rise of $\delta(x_i,x_j)$ means tourists' consumption demand for tourism products increases in a time.

**Definition 2.** Suppose affection thinking is a 7-dimensional vector $AT=(T_{\text{activate}}, T_{\text{shift}}, T_{\text{move}}, T_{\text{map}}, T_{\text{attenuate}}, T_{\text{update}}, T_{\text{aware}})$, and each vector respectively represents activate (invoke), shift, move, map, attenuate, update and aware, the relationship among them are shown in Fig.1.

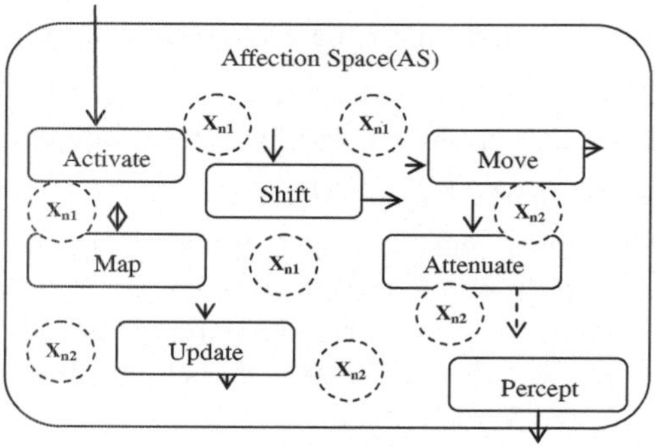


**Fig. 1.** Affection thinking model

(1) $\boldsymbol{T}_{activate}$ means that one emotion in the affection space **AS** has been activated, shown as $\boldsymbol{T}_{activate}$: $(\varphi(x_{i1}, x_{j2}))$•**AS,** namely it is related to the probability of the interaction between vector $\boldsymbol{X}_{n1}$ and $\boldsymbol{X}_{n2}$.

(2) $\boldsymbol{T}_{shift}$ means that one emotion in the affection space **AS** has mutually switched to another, shown as $\boldsymbol{T}_{shift}$: $(\varphi(x_{i1}, x_{j2}))$ •$\boldsymbol{T}_{shift}'$.

(3)$\boldsymbol{T}_{move}$ means that one emotion in the affection space **AS** shall replace another emotion when need. If the current emotion is $a$, the emotion needs to be replaced is $b$, so we can say $\boldsymbol{T}_{move}$: $(a, b)$ •$(b, a)$.

(4) $\boldsymbol{T}_{map}$ means the mapping of $\boldsymbol{X}_{n1}$ and $\boldsymbol{X}_{n2}$ in affection space **AS**, shown as $\boldsymbol{T}_{map}$: $\boldsymbol{X}_{n1} \leftrightarrow \boldsymbol{X}_{n2}$, namely the mapping of tourists' cognition of tourism consumer products and their own emotional state.

(5) $\boldsymbol{T}_{attenuate}$ means the emotional attenuation in affection space **AS**. It is pointed out in [5] that the emotion attenuation in psychology is close to the index function $y=e^{-x}$, so the emotional attenuation function of $\boldsymbol{T}_{attenuate}$ in time $(t_1, t_2)$ is defined as:

$$E(x_i, x_j) = \delta e^{-a_i(t_2 - t_1)} \tag{2.4}$$

In the (2.4), $E$ means the emotional attenuation function in the time $(t_1,t_2)$, $a_i \in [0, 1)$ means the emotional attenuation coefficient that control the rate of decay, and we have different $a_i$ values according to different times. For example, we have $m$ $a_i$ values in $m$ moments in the time $(t_1, t_2)$, we also have $m$ $X = e^{-a_i(t_2 - t_1)}$ values to form the vector $\boldsymbol{AE}=(X_m)$. Now, we put (2.3)into(2.4), and we get:

$$\boldsymbol{E}(x_i, x_j)=\delta(x_i, x_j)\boldsymbol{AE}^{-1} \tag{2.5}$$

In the (2.5), when the line or column is impossible to calculate, we fill in 0.99 when emotional intensity is the strongest (immediately determine a tourism consumption), and we fill in 0 when emotional intensity is the lowest (do not want to consume at all).

(6) $\boldsymbol{T}_{update}$ represents the emotional update. It means the emotional update in activated state, boosting, strengthening or weakening of the emotional intensity. If an random emotional state vector $I$ update to the emotional state vector $j$ in **AS**, we compute it using the following formula:

$$\tau_{(i,j)} = (1-a_i)\cdot\tau_{(i,j)}(t) + \Delta\tau_{(i,j)}(t)$$

$$\Delta\tau_{(i,j)}(t) = \sum_{k=1}^{m} \Delta\tau_{(i,j)}^{k}(t) \quad (2.6)$$

In the (2.6), (1-$a_i$) represents the residual factors of emotional attenuation, $\tau_{(i,j)}(t)$ represents the emotional intensity of the moment $t\in(t_1, t_2)$, and $\Delta\tau_{(i,j)}(t)$ represents the emotional intensity when the emotion updates. At the initial moment $\Delta\tau_{(i,j)}(t) = 0$, $\Delta\tau_{(i,j)}^{k}(t)$ means the combined action emotion intensity of $\boldsymbol{X}_{n1}$ and $\boldsymbol{X}_{n2}$ at the k*th* times. We calculate $\Delta\tau_{(i,j)}(t)$ according to the following formula:

$$\Delta\tau_{(i,j)}(t) = \begin{cases} E(x_i, x_j)/t \\ 0 \end{cases} \quad (2.7)$$

In the (2.7), $\boldsymbol{E}(x_i, x_j)/t$ represents the emotional intension rate at the time $t\in(t_1, t_2)$, 0 represents the case to the exclusion $\boldsymbol{E}(x_i, x_j)/t$ . We use $\delta(x_i,x_j)$ to calculate the emotional intensity of different times.

(7) $\boldsymbol{T}_{aware}$ means the consciousness of objects in **AS**, to eventually obtain tourists' emotional consciousness through thinking, and can be expressed as $\boldsymbol{T}_{aware}$: $\tau_{(i,j)}$•**AS.**

**Definition 3.** Affection thinking be defined as: *ATM*=(***AS***, ***AT***, $\boldsymbol{A}_{requirement}$), as for ***AS*** and ***AT*** refer to **Definition 1-2**, $\boldsymbol{A}_{requirement}$ means tourists' requirements.

*Remark 3.* ***AT*** is performed in ***AS*** on condition that *Base*:($\boldsymbol{T}_{activate}$, $\boldsymbol{T}_{shift}$, $\boldsymbol{T}_{move}$, $\boldsymbol{T}_{map}$)→***AS***

*Remark 4.* The attenuation of $\boldsymbol{T}_{attenuate}$ updates the emotion by increasing emotional intensity.

## 3 Affection Thinking Mechanism Based on Ant Colony Strategy

The previous section established an affective thinking expression method from the angle of affection thinking model, but how to express tourists' emotion in emotional thought model to have better tourism consumption and promptly satisfy the needs of their own is another problem needs to be solved in this paper. Therefore, this paper adopts ant colony strategy [10][15][16] to help affection thinking running. Ants carry these feeling information and automatically choose the best emotion to make the tourism products consumption decision through the ant colony.

**Definition 4.** The affection thinking information that ant carries is four-tuple:

$$ANT=(id,\ \delta(x_i, x_j),\ E(x_i, x_j),\ \tau_{(i,j)})$$

The items in the *ANT* respectively represents the ant number, emotional intensity, emotional attenuation and emotional update, see the expression of each item in

**Definition 1-2.** When meet the following expression, ants start to optimize.

$$\boldsymbol{A}_{\text{requirement}}\ (S_n\text{: } \boldsymbol{T}_{\text{activate}}) \rightarrow ANT$$

At this time, if the ant $k$ is activated from emotion $S_i$ to $S_j$ in the affection space, the calculation formula is shown as following:

$$S_\delta = \begin{cases} \arg \max\limits_{x_j \in allowed_{x_i}} \{[E(x_i, x_j)]^\alpha [\tau_{(i,j)}]^\beta\} & \text{if } q \le q_0 \\ (3.2) & \text{else} \end{cases} \tag{3.1}$$

$$p_{ij}^k(x_i, x_v) = \begin{cases} \dfrac{[E(x_i, x_v)]^\alpha \cdot [\tau_{(i,v)}]^\beta}{\sum\limits_{u \in J_k(x_i)} [E(x_i, x_u)]^\alpha \cdot [\tau_{(i,u)}]^\beta} & \text{if } x_v \in J_k(x_i) \\ 0 & \text{else} \end{cases} \tag{3.2}$$

In the (3.2), $q_0$ is a random number in [0, 1] ,$J_k(x_i)$ represents the emotion set to be activated in affection space, $allowed_{x_i}$ represents the emotion set that has been activated in affection space. Then, according to the decay degree of $E(x_i, x_j)$ , we can infer the following formula:

$$E(x_i, x_v) \leftarrow (1-\gamma) \cdot E(x_i, x_v) + \gamma\left( \max_{x_j \in allowed_{x_i}} \tau_{(i,j)} + \Delta E(x_i, x_v) \right) \tag{3.3}$$

At this moment, the increment of attenuation degrees can be compute using the following formula:

$$\Delta E(x_i, x_v) = \begin{cases} \omega_1 \dfrac{W}{\chi_{k_{gb}}} + \omega_2 \dfrac{W}{\chi_{k_{ib}}} \\ 0 & \text{esle} \end{cases} \tag{3.4}$$

In the (3.4), if $(x_i, x_v)$ respectively represents the ants number of the corresponding global optimal and iterative optimal, so, $\chi_{k_{gb}}$ and $\chi_{k_{ib}}$ respectively represents the emotional intensity of the ants, namely we acquire the best emotional thinking model. Among them, i$\omega_1$ and $\omega_2$ are the respectively weight, and $\omega_1+\omega_2=1$, usually we say $W$=10.

In order to balance the emotional attenuation, we can apply vaccination and immune selection method [9] to choose the strongest emotional intensity to maintain the balance. When the emotional intensity perfectly fix the emotional attenuation, the corresponding bits meet the demands; if not, do random repair through 1 and integers between $a$ and $b$. Then we can reach the bits acceptable probability after repair according to simulated annealing:

$$P_{ij} = e^{-\delta(x_i, x_j)/T_{ab}} \Big/ \sum_{i=1}^{a} \sum_{j=1}^{b} e^{-\delta(x_i, x_j)/T_{ab}} \tag{3.5}$$

In the (3.5) ,$P_{ij}$ can iterate between the given [a, b], the annealing temperature difference $T_{ab}$=$b$-$a$.

## 4 Experimental Results

The ultimate goal of $\boldsymbol{A}_{\text{requirement}}$(showed in **Definition 3**)is to obtain the most satisfactory tourism consumption products according to their own schedule. If a traveller has 3 days for tourism, he will need to schedule reasonably to satisfy his own travel demands. His basic needs can be described as follow:

*Traveler demand = {tourism product selection, tourism route selection, vehicle selection, schedule choice, catering accommodation selection, ticket booking selection}*

The six basic choices are all associated with emotion, and each choice would affect the final decision of emotional thinking because of different characteristics. The different selection characteristics are listed List 1:

List 1 Choice characteristics of tourists' demand consumption

*tourism products={geographic and neighboring environment , scenic spot above 4A , humanistic atmosphere, supporting facility };tourist routes={the shortest path, the widest tourist routes, security , roads unblocked };vehicle={convenience , speedy , less cost , security};schedule={ rationality, effectiveness , availability , quality}; accommodation={inexpensive, comfortable lodging, delicious food , reasonable collocation};ticket booking={booking method, vehicle selection, the cheapest, method of payment}*

The existence of consumption characteristics have direct influence on their emotion. Now we are going to analyze the suggested emotional thinking based on the VTA case under WSMO. It is the example in WSMO development which can provide end users services of booking and buying line tickets in Austria and Germany, etc. That is to complete the emotional thinking mechanism based on ant colony strategy according to the demands and consumption characteristics, and combined with the VTA application model. Now we get tourism consumption choice model facing the emotional thinking as shown in Fig.2.

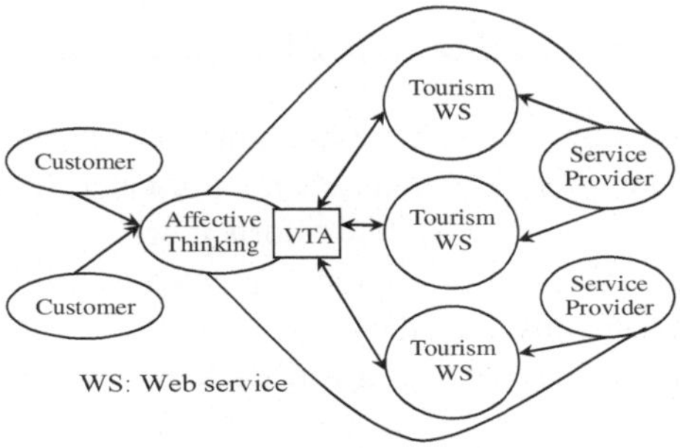


**Fig. 2.** The tourism consumption preference pattern of affection thinking facing the ant colony strategy

In Fig.2, when a tourist requests a service in VTA, we analyze the request emotion which are obtained according to the characteristics shown in List 1. Now according to the services provided by VTA, we do simulation test by respectively design services by stages with 10 services, 30 services, 50 services, 70 services, 90 services and 100 services. The configuration of the test computer is: 2.4 Intel dual-core CUP, 3 GB internal memory, 160 GB hard drive. We set initial ants according to travelers' request $k$=24($k$≥6), $\gamma$=0.5, $\alpha$=1, $\beta$=5, $\omega_1$=0.3, $\omega_2$=0.7, $W$=10, $a$=5, $b$=15.

We design service counters and timer are designed in simulation experiment analysis, its meaning include: record service request number ($r$), service response number($s$), successful service composition number($q$), the time they take, and analyze using the ATACA (Affection Thinking Ant Colony Algorithm), the common ACA (Ant Colony Algorithm) and the ATNGACA (Affection Thinking Annealing Strategy Ant Colony Algorithm). The three algorithms would test and reach the service composition number of the optimal tourism consumption selection and the time it cost as is shown in Table 1-2. Fig.3-4 is the service composition number of the optimal tourism consumption selection scheme and its relationship with ant colony iteration, and the relationship of ants number and the optimal combination scheme.

**Table 1.** The service composition number of the best tourism consumption selection scheme $q$(Unit: Number)

| | 10($r$=5, $s$=5) | 30($r$=10, $s$=20) | 50($r$=30,$s$=20) | 70($r$=40,$s$=30) | 90($r$=50,$s$=40) | 100($r$=50,$s$=50) |
|---|---|---|---|---|---|---|
| ATACA | 7 | 22 | 38 | 54 | 68 | 70 |
| ACA | 7 | 18 | 33 | 42 | 55 | 63 |
| ATNGACA | 8 | 25 | 41 | 59 | 75 | 80 |

**Table 2.** The time of the best tourism consumption selection scheme(Unit: ms)

| | 10($r$=5, $s$=5) | 30($r$=10,$s$=20) | 50($r$=30, $s$=20) | 70($r$=40,$s$=30) | 90($r$=50, $s$=40) | 100($r$=50,$s$=50) |
|---|---|---|---|---|---|---|
| ATACA | 11.25 | 15.01 | 20.55 | 34.56 | 36.28 | 39.45 |
| ACA | 11.95 | 16.21 | 21.78 | 35.69 | 38.65 | 40.09 |
| ATNGACA | 12.06 | 16.76 | 22.11 | 36.28 | 36.78 | 42.03 |

From Table 1-2 we know that the service composition number of the best and available tourism consumption selection scheme presented by ATNGACA is much bigger than that of ATACA and ACA, but it costs more time than the other two. The cost of time is worth of more optimization composition results. But the maximum time difference is less than 3 ms after all.

In Fig. 3, the three algorithms have been verified to optimize the service composition numbers (namely the adaptive value) of users' requirements. And we know that the affection ant colony algorithm with the annealing strategy brought the best effect. The solution adaptive values are basically equilibrium after iterated for 400 times. The relationship between the ants number and the three algorithms is verified in Fig. 4. Seeing from the needs of verification requirements in this article, 26 ants would get the best adaptive value. That means the bigger ants number would not bring the better results, and the overmuch ants number would descend the system performance.

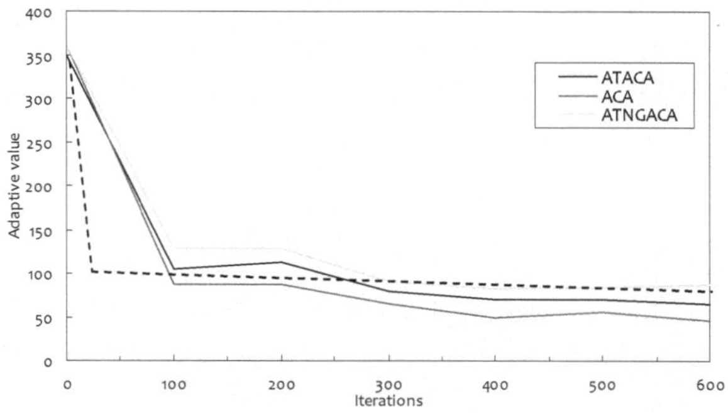


**Fig. 3.** between adaptive value (Service compositions) and iterations for option of tourism consumption

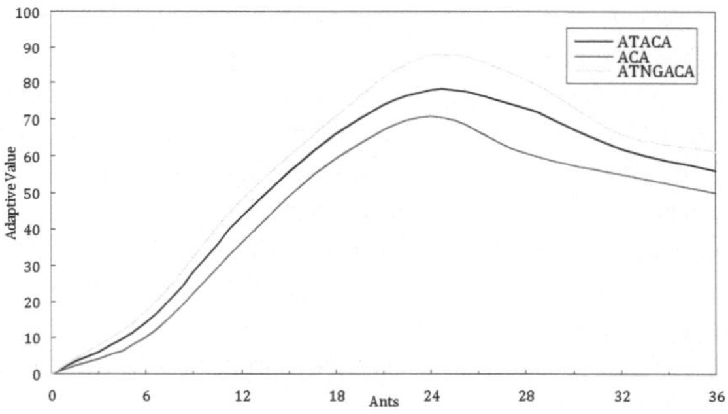


**Fig. 4.** between adaptive value (Service compositions) and ants for option of tourism consumption

# 5 Conclusions

From the two aspects of emotional tendency and emotional expression, this paper proposed an affection space based on emotional intensity, and intercommunication of the two can be realized in the space. On this basis, we establish a thinking model which meet the emotional operation and allow thinking to run emotionally so as to form an emotional thinking model. Ant colony strategy is introduced into the model to facilitate thinking to process information more intellectively. Use the change of emotional thought to be the transfer and pheromone update of ant colony, so that the ant colony optimization ability is enhanced, thus an emotional ant colony algorithm is formed. Finally, through verifying the VTA case under WSMO, the method in this study is proved effective, feasible and can provide better service demand results. Nevertheless, the thinking operational mode in this thesis is simplified to some degree, performance degrades with the increase of user number. Affection space aspects are waiting to be improved in the following further researches.

**Acknowledgments.** This work is supported by the Natural Sciences of Education and Science Office Bureau of Sichuan Province of China under Grant No. 2010JY0J41,11ZB152.

## References

1. Chen, M.-C., Chiu, A.-L., Chang, H.-H.: Mining changes in customer behavior in retail marketing. Expert Systems with Applications 28(4), 773–781 (2005)
2. Forgas, J.P., East, R., Chan, N.Y.M.: The use of computer-mediated interaction in exploring affective influences on strategic interpersonal behaviours. Computers in Human Behavior 23(2), 901–919 (2007)
3. Riloff, E., Wiebe, J.: Learning extraction patterns for subjective expressions. In: Collins, M., Steedman, M. (eds.) Proc. of the EMNLP, pp. 105–112. ACL, Morristown (2003)
4. Kim, S.M., Hovy, E.: Automatic detection of opinion bearing words and sentences, http://acl.ldc.upenn.edu/I/I05/I05-2011.pdf
5. Li, H., He, H., Chen, J.: A Multi-layer Affective Model Based on Personality, Mood and Emotion. Journal of Computer-Aided Design & Computer Graphic 23(4), 725–730 (2011)
6. Yoon, Y., Uysal, M.: An examination of the effects of motivation and satisfaction on destination loyalty: A Structural Model. Tourism Management 26(1), 45–56 (2005)
7. Xue, W., Wang, Z., Meng, Z.: A New Method for Simulating Human Emotions. Journal of University of Science and Technology Bejing 10(2), 72–74 (2003)
8. Ishihara, H., Fukuda, T.: Individuality of agent with emotional algorithm. In: Proceedings of the IEEE International Conference on Intelligent Robots and Systems, Piscataway, pp. 1195–1200. IEEE Press, NJ (2001)
9. Velasquez, J.D.: Modeling emotions and other motivations in synthetic agents. In: Pro. of the AAAI Conf. 1997, Rhode Island (1997)
10. Ortony, A., Clore, G.L., Collins, A.: The cognitive structure of enotions. Cambridge University Press, Cambridge (1990)
11. Picard, W.: Affective computing. MIT Press, Cambridge (1997)
12. Kshirsagar, S., Magnenat-Thalmann, N.: A multilayer personality model. In: Proceedings of the 2nd International Symposium on Smart Graphics, pp. 107–115. ACM Press, New York (2002)
13. Roseman, I.J., Jose, P.E., Spindel, M.S.: Appraisals of emotioneliciting events: Testing a theory of discrete emotions. Journal of Personality and Social Psychology 59, 899–915 (1990)
14. Yang, Y., Kamel, M.S.: An aggregated clustering approach using multi-ant colonies algorithms. Pattern Recognition 39(7), 1278–1289 (2006)
15. Dorigo, M.: Optimization, learning and natural algorithms. Ph.D. Thesis, Department of Electronics, Plitecnico diMilano, Italy (1992)
16. Dorigo, M., Maniezzo, V., Colorni, A.: Ant system: optimization by a colony of cooperating agents. IEEE Transaction on Systems, Man, and Cybernetics-Part B 26, 29–41 (1996)

# Studies on the Covering Rough Set and Its Matrix Description

Xiaogai Wang, Yingcang Ma, Lu Wang, and Juanjuan Zhang

School of Science, Xi'an Polytechnic University, Xi'an 710048, China
{gaizi.521,wanglu_841210}@163.com,
{mayingcang,zhang79jj}@126.com

**Abstract.** This paper mainly studies on the covering rough set based on the close friend element. Firstly, the upper and lower approximations of the covering rough set based on the close friend element are defined, while the properties are discussed. Secondly, we define the binary relation is induced by a covering called the close friend relation and its properties are studied. Finally, we give the matrix description of the covering rough set based on the close friend element, and prove that the upper and lower approximations obtained from the matrix are same to from the definition of covering rough set based on the close friend element, which give a new way to describe the covering rough set.

**Keywords:** Covering rough set, Minimum description, Close friend element, Lower and upper approximations, Relationship matrix.

## 1 Introduction

Rough set theory (RST), proposed by Pawlak[1] in 1982, is one of the effective mathematical tools for processing fuzzy and uncertainty knowledge. Nowadays, RST has been applied to a variety of fields such as artificial intelligence, data mining, pattern recognition and intelligent information processing [2-7]. In theory, a variety of extension of the definition of upper and lower approximation operators of rough set was studied [8-10].

The classical rough set theory is based on the equivalence relation on the domain. In many practical problems, it is difficult to construct equivalence relation between objects or between objects is essentially no equivalence relation, so this equivalence relation as the basis of the classic rough set model cannot fully meet the actual needs. For this people do a lot of significance promotion of Pawlak rough set, such as the equivalence relation relax to the general binary relation, covering etc. The combination of rough and fuzzy theory is extended to fuzzy rough set theory. Covering rough set model, proposed by Zakowski[11] in 1983, and the further studies can be found in [12-14]. In addition, the matrix forms of lower and upper approximation also an important content. In [15-17], the matrix descriptions of the upper and lower approximations of rough set are given, which can use the matrix to depict the rough set.

The paper is organized as follows. After this introduction, in section 2, upper and lower approximations of covering rough set based on the close friend element are

J. Lei et al. (Eds.): AICI 2012, LNAI 7530, pp. 219–226, 2012.

given, and its properties are stuided. In section 3, the definition of a close friend element relation are given and its properties are introduced. In section 4, the matrix description of upper and lower approximation operators of covering rough set based on a close friend element are given, and finally gives a conclusion.

## 2 The Covering Rough Set Based on the Close Friend Element

**Definition 1.**[13] (Covering and Covering approximation space) Let *U* be a domain of discourse and *C* a family of nonempty subsets of *U*. If $\bigcup C = U$ , *C* is called a covering of *U*. We call the ordered pair a covering approximation space.

**Definition 2.**[13] (Minimal description) Let <*U*, *C*>be a covering approximation space,

$$Md(x) = \{K \in C \mid x \in K \wedge (\forall S \in C \wedge x \in S \wedge S \subseteq K \Rightarrow K = S)\}$$

is called the minimal description of *x*.

**Definition 3.** (Close friend element). Let <*U*, *C*> be a covering approximation space, for any $x \in U$ , $\bigcup\{K \mid x \in K \wedge K \in Md(x)\}$ is called the close friends of *x* and denoted as *CF*(*x*).

**Definition 4.** (Lower and upper approximations of covering) Let *C* be a covering of *U*, for any $X \subseteq U$ , we define:

$$L(X) = \{x \mid (CF(x)) \subseteq X\}, H(X) = \{x \mid (CF(x)) \cap X \neq \varnothing\}$$

are the lower and the upper approximation of *X* respectively.

**Example 1.** Let $U = \{a,b,c,d,e\}$ and $C = \{\{a,b\},\{a,d\},\{b,c\},\{c,d,e\},\{e\}\}$ be a covering of *U*, Let $X = \{b,c,d\}$ , then $L(X) = \{c\}$ , $H(X) = \{a,b,c,d\}$ .

**Theorem 1.** The lower and upper approximations of covering in Definition 4 have the following properties:

(1) $L(X) \subseteq X \subseteq H(X)$ ;

(2) $L(\varnothing) = \varnothing$ , $H(\varnothing) = \varnothing$ ;

(3) $L(U) = U$ , $H(U) = U$ ;

(4) $L(X \cap Y) = L(X) \cap L(Y)$ , $H(X \cup Y) = H(X) \cup H(Y)$ ;

(5) $\sim L(\sim X) = H(X)$ , $\sim H(\sim X) = L(X)$ ;

(6) $X \subseteq Y \Rightarrow L(X) \subseteq L(Y)$ , $H(X) \subseteq H(Y)$ .

**Proof**

(1) Let $\forall x \in L(X) \Rightarrow CF(x) \subseteq X$ , because $x \in CF(x)$ , so $x \in X$ .Then $L(X) \subseteq X$ . Re-established $\forall x \in X$ , then $CF(x) \cap X \neq \varnothing$ , so $x \in H(X)$ . Then $X \subseteq H(X)$ .

(2) We can see from (1) that $L(\varnothing) \subseteq \varnothing$ , while $\varnothing \subseteq L(\varnothing)$ , so $L(\varnothing) = \varnothing$ . Suppose

$H(\varnothing) \neq \varnothing$ , it certainly existence $x$ make $x \in H(\varnothing)$ , that is $CF(x) \cap \varnothing \neq \varnothing$ , but $CF(x) \cap \varnothing = \varnothing$ , contradicts the assumption. Therefore, $H(\varnothing) = \varnothing$ .

(3) By (1), that $L(U) \subseteq U$ , and because when $x \in U$ , there is $CF(x) \subseteq U$ , so $x \in L(U)$ , that is $U \subseteq L(U)$ .Therefore $L(U) = U$ . and from (1) we known $U \subseteq H(U)$ , but $H(U) \subseteq U$ . Then $H(U) = U$ .

(4) Let $\forall x \in L(X \cap Y) \Leftrightarrow CF(x) \subseteq (X \cap Y)$

$$\Leftrightarrow CF(x) \subseteq X \wedge CF(x) \subseteq Y$$

$$\Leftrightarrow x \in L(X) \cap x \in L(Y)$$

So $L(X \cap Y) = L(X) \cap L(Y)$ .

Let $\forall x \in H(X \cup Y) \Leftrightarrow CF(x) \cap (X \cup Y) \neq \varnothing$

$$\Leftrightarrow (CF(x) \cap X) \cup (CF(x) \cap Y) \neq \varnothing$$

$$\Leftrightarrow (CF(x) \cap X) \neq \varnothing \wedge (CF(x) \cap Y) \neq \varnothing$$

$$\Leftrightarrow x \in H(X) \vee x \in H(Y)$$

$$\Leftrightarrow x \in H(X) \cup x \in H(Y) .$$

So $H(X \cup Y) = H(X) \cup H(Y)$ .

(5) Because $\forall x \in L(X) \Leftrightarrow CF(x) \subseteq X \Leftrightarrow CF(x) \cap X^C = \varnothing$ $\Leftrightarrow x \notin H(\sim X) \Leftrightarrow x \in \sim H(\sim X)$ . With $\sim X$ instead X, that was $H(\sim X) = \sim L(X)$ .

(6) Let $X \subseteq Y$ , then $X \cap Y = X$ , so $L(X \cap Y) = L(X)$ . We known from (4) that $L(X) \cap L(Y) = L(X)$ . Therefore $L(X) \subseteq L(Y)$ .We can have $X \subseteq Y \Rightarrow H(X) \subseteq H(Y)$ by the same reason.

**Proposition 1.** The following properties do not hold:

(7) $L(L(X)) = L(X)$ , $H(H(X)) = H(X)$ ;

(8) $L(\sim L(X)) = \sim L(X)$ , $H(\sim H(X)) = \sim H(X)$ .

**Proof**

(7) Let $U = \{a,b,c,d,e\}$ , $C = \{\{a,b\},\{a,d\},\{b,c\},\{c,d,e\},\{e\}\}$ , $X = \{b,c,d\}$ , then $L(X) = \{c\}$ but $L(L(X)) = \varnothing$ , so $L(L(X)) = L(X)$ does not hold.

Let $C = \{\{a,b\},\{a,c\},\{c,d\}\}$ , $X = \{a,b\}$ , then $H(X) = \{a,b,c\}$ , but $H(H(X)) = \{a,b,c,d\}$ , so $H(H(X)) = H(X)$ does not hold.

(8) Let $U = \{a,b,c,d,e\}$ , $C = \{\{a,b,c,d\},\{a,b\},\{e\}\}$ is a covering of $U$, $X = \{a,b\}$ , then $L(X) = \{a,b\}$ , $\sim L(X) = \{c,d,e\}$ , $L(\sim L(X)) = L(\{c,d,e\}) = \{e\}$ , So $L(\sim L(X)) = \sim L(X)$ does not hold.

Let $U = \{a,b,c\}$ , $C = \{\{a,b\},\{b,c\}\}$ is a covering of $U$, $X = \{a,b\}$ , $H(X) = \{a,b\}$ , $\sim H(X) = \{c\}$ , $H(\sim H(X)) = \{b,c\}$ , so $H(\sim H(X)) = \sim H(X)$ ,does not hold.

## 3 Induced Binary Relations by the Covering

**Definition 5.** Let $C$ be a covering of $U$, we define: $xRy \Leftrightarrow y \in CF(x)$, called a close friend element relation induced by covering $C$. $R$ corresponds to the relation matrix is denoted by $M_R$, that is , if $x_i R x_j$, $R(x_i, x_j) = 1$, else is 0.

**Example 4.** Let $U = \{a,b,c\}$, $C = \{\{a,b\},\{b,c\}\}$ is a covering of $U$, then

$$M_R = \begin{pmatrix} 1 & 1 & 0 \\ 1 & 1 & 1 \\ 0 & 1 & 1 \end{pmatrix}$$ is the relationship matrix of $R$.

**Definition 6.**[18] $R$ is binary relation:

(1) $R$ is reflexive, if $\forall a \in U, aRa$;

(2) $R$ is irreflextive, if $\forall a \in U, aR^c a$, $R^c$ is the complement relation of $R$;

(3) $R$ is symmetric, if $aRb \Rightarrow bRa$;

(4) $R$ is antisymmetric, if $aRb$ and $bRa \Rightarrow a = b$;

(5) $R$ is asymmetric, if $aRb \Rightarrow bR^c a$;

(6) $R$ is complete, if $aRb$ or $bRa$ for $a \neq b$;

(7) $R$ is strongly complete, if $aRb$ or $bRa$;

(8) $R$ is transitive, if $aRb$ and $bRc \Rightarrow aRc$;

(9) $R$ is negatively transitive, if $aR^c b$ and $bR^c a \Rightarrow aR^c c$;

(10) $R$ is Ferrers relation, if $aRb$ and $bRc \Rightarrow aRd$ or $dRc$;

(11) $R$ is semitransitive, if $aRb$ and $cRd \Rightarrow aRd$ or $cRb$;

(12) $R$ is serial, if $\forall a \exists b, s.t. aRb$;

(13) $R$ is Euclidean, if $aRb$ and $aRc \Rightarrow bRc$.

**Theorem 2.** $R$ is the close friend relationship induced by the covering $C$. $R$ is reflexive and serial.

**Proof**

(1) For $\forall a \in U$, we have $a \in CF(a)$, so reflexive established.

(12) Because for $\forall a \in U$, there is $a \in U$, such that, therefore serial holds.

**Proposition 2.** For Definition 6, a close friend of relation $R$ induced by the cover $C$ does not satisfy the properties except (1) and (12).

**Proof**

(2) We known from (1), (2) does not hold.

Let, $U = \{a,b,c,d\}$ $C = \{\{a,b\},\{b,d\},\{c,d\},\{d\}\}$ is a covering of $U$, then $R = \{\{a,b\},\{a,b,d\},\{c,d\},\{d\}\}$, $R^C = \{\{c,d\},\{c\},\{a,b\},\{a,b,c\}\}$.

(3) Here $d \in CF(c)$, but $c \notin CF(d)$, therefore (3) does not hold..

(4) $a \in CF(b)$ and $b \in CF(a)$, but $a \neq b$, therefore (4) does not hold.

(5) $b \in CF(a)$ ,but $a \notin \sim CF(b)$, therefore (5) does not hold.

(6)On $a \in U$ , there is $aRa$, and therefore does not meet(6).

(7)For $a,d \in U$, $a \notin CF(d)$ and $d \notin CF(a)$, therefore (7) does not hold.

(8) $d \in CF(b)$, $b \in CF(a)$, but $d \notin CF(a)$, so (8) does not hold.

(9) $d \notin \sim CF(a)$, $a \notin \sim CF(c)$, but $d \notin \sim CF(c)$, so(9) does not hold.

(10)When $c,d$ satisfies $c\overline{R}a$, $c\overline{R}b$, $d\overline{R}a$, $d\overline{R}b$, (10) does not hold. For example Let, $U=\{a,b,c,d\}$ $C=\{\{a,b\},\{c,d\}\}$, satisfies $aRb$ and $cRd$, but $a\overline{R}d$ and $c\overline{R}b$.

(11)Let $U=\{a,b,c,d\}$, $C=\{\{a\},\{a,b\},\{b,c\},\{d\}\}$ is a covering of *U*, satisfies $aRb$ and $bRc$, but $a\overline{R}d$ and $d\overline{R}c$.

(13) $d \in CF(b)$ and $d \in CF(c)$, but $b \notin CF(c)$, so(13)does not hold.

## 4 The Upper and Lower Approximation of the Matrix Description

For $\forall X \subseteq U$, the characteristic function of *X*, denoted *X*(*x*), namely:

$$X(x)=\begin{cases}1, & x \in X\\ 0, & x \notin X\end{cases}.$$

**Definition 7.** Let *C* is a covering of *U*, $M_R = R(x_i, x_j)_{nn}$ is the relation matrix of *R*, a close friend element induced by *C*, we call $\sim M_R = (1-R(x_i,x_j))_{nn}$ is the complement relationship of $M_R$, then for $\forall X \subseteq U$, the upper and lower approximation of *X* can be expressed as:

$$H_R(x) = \vee_{y\in u} (M_R(x,y) \wedge X(y)),\ L_R(x) = \wedge_{y\in u} (1-M_R(x,y) \vee X(y)).$$

Then define:

$$H_R(X)=\{x \mid H_R(x)=1\},\ L_R(X)=\{x \mid L_R(x)=1\}.$$

**Theorem 3.** Let<*U*, *C*>be a covering approximation space, $M_R$ is the relation matrix of *R*, the binary relation induced by *C*, then for $\forall X \subseteq U$, there is:

(1) $H_R(X)=H(X)$;

(2) $L_R(X)=L(X)$.

**Proof:**

(1) First, we prove that $H_R(X) \subseteq H(X)$. If $x \in H_R(X)$, that is $H_R(x_i)=1$.

Then $H_R(x_i) = \vee_{y\in u} (R(x_i, y) \wedge X(y))$

$$= \vee_{x_i R y} (R(x_i, y) \wedge X(y)) \vee (\vee_{x_i \bar{R} y} (R(x_i, y) \wedge X(y)))$$

$$= \vee_{x_i R y} X(y) = 1$$

That there is $X(y) = 1$, that is $y \in X$, and because $x_i R y$, so $y \in CF(x_i)$, and thus $CF(x_i) \cap X \neq \varnothing$. Therefore $x_i \in H(X)$, therefore the $H_R(X) \subseteq H(X)$.

Then we prove that $H(X) \subseteq H_R(X)$. If $x_i \in H(X)$,

$$H_R(x_i) = \vee_{y \in u} (R(x_i, y) \wedge X(y))$$

$$= \vee_{x_i R y} (R(x_i, y) \wedge X(y)) \vee (\vee_{x_i \bar{R} y} (R(x_i, y) \wedge X(y)))$$

$$= \vee_{x_i R y} X(y)$$

By known we have $y \in CF(x_i)$ and $y \in X$, that is $R(x_i, y) = 1$ and $X(y) = 1$, so $H_R(x_i) = 1$, and so $H(X) \subseteq H_R(X)$.

(1) is proved.

(2) First, we prove that $L_R(X) \subseteq L(X)$. If $x \in L_R(X)$, that is $L_R(x_i) = 1$,

Then $L_R(x_i) = \wedge_{y \in u} (1 - R(x_i, y) \vee X(y))$

$$= \wedge_{x_i R y} ((1 - R(x_i, y)) \vee X(y)) \wedge (\wedge_{x_i \bar{R} y} (1 - (R(x_i, y)) \vee X(y)))$$

$$= \wedge_{x_i R y} X(y) = 1$$

The above formula shows that there must be $X(y) = 1$, $x_i R y$, that is $y \in CF(x_i)$, there must be $y \in X$, that is $CF(x_i) \subseteq X$. Therefore $x_i \in H(X)$, so $L_R(X) \subseteq L(X)$.

Then we prove that $L(X) \subseteq L_R(X)$。 If $x_i \in L(X)$,

$$L_R(x_i) = \wedge_{y \in u} (1 - R(x_i, y) \vee X(y))$$

$$= \wedge_{x_i R y} ((1 - R(x_i, y)) \vee X(y)) \wedge (\wedge_{x_i \bar{R} y} (1 - (R(x_i, y)) \vee X(y)))$$

$$= \wedge_{x_i R y} X(y)$$

So we have $y \in CF(x_i) \subseteq X$, that is $R(x_i, y) = 1$ and $X(y) = 1$, so $L_R(x_i) = 1$, so $H(X) \subseteq H_R(X)$, therefore $L(X) \subseteq L_R(X)$. (2) is proved.

From Definition 7 and Theorem 3, If the collection written in the n-dimensional column, the value of its characteristic function 0,1 constitutes a column vector, then ,

$$H_R(X) = \vee(M_R \wedge X), \quad L_R(X) = \wedge(\sim M_R \vee X).$$

Thus the upper and lower approximation of covering rough set can be portrayed by matrix。 See the following cases.

**Example 5.** Let $U=\{a,b,c,d,e\}$, $C=\{\{a,b\},\{a,d\},\{b,c\},\{c,d,e\},\{e\}\}$, we have $X=\{b,c,d\}$, then $L(X)=\{c\}$, $H(X)=\{a,b,c,d\}$.

$L_R(X)=\wedge(\sim M_R \vee X)$

$$=\wedge\begin{pmatrix}0&0&1&0&1\\0&0&0&1&1\\1&0&0&1&1\\0&1&1&0&1\\1&1&1&1&0\end{pmatrix}\vee\begin{pmatrix}0\\1\\1\\1\\0\end{pmatrix}=\begin{pmatrix}0\\0\\1\\0\\0\end{pmatrix},$$

Thus $L(X)=L_R(X)=\{c\}$.

$H_R(X)=\vee(M_R \wedge X)$

$$=\vee\begin{pmatrix}1&1&0&1&0\\1&1&1&0&0\\0&1&1&0&0\\1&0&0&1&0\\0&0&0&0&1\end{pmatrix}\wedge\begin{pmatrix}0\\1\\1\\1\\0\end{pmatrix}=\begin{pmatrix}1\\1\\1\\1\\0\end{pmatrix},$$

Thus $H(X)=H_R(X)=\{a,b,c,d\}$.

## 5 Conclusion

This paper gives a new upper and lower approximations of covering rough set and their properties are discussed. Moreover, in order to facilitate characterization and calculation of the relationship matrix to portray on the upper and lower approximations. We also explored the properties of a close friend element induced by the covering, rich study covering rough sets. The next we will study on the algebra structure of the covering rough set and its applications.

**Acknowledgments.** This work is partially supported by Scientific Research Program Funded by Shaanxi Provincial Education Department (Program No.12JK0878) and Doctor Scientific Research Foundation Program of Xi'an Polytechnic University.

## References

1. Pawlak, Z.: Rough sets. International Journal of Computer and Information Science 11(5), 314–356 (1982)
2. Zhang, W.X., Wu, W.Z., Liang, J.Y., et al.: Rough Set Theory and Approaches. Science Press (2001)
3. Wang, G.Y.: Rough Set Theory and Knowledge Acquisition. Xi'an Jiaotong University Press (2001)

4. Pawlak, Z., Skowrongs, A.: Rudiments of rough sets. Information Sciences 177(1), 3–27 (2007)
5. Pawlak, Z., Skowrons, A.: Rough sets: Some extensions. Information Sciences 177(1), 28–40 (2007)
6. Wong, S.K.M., Ziarko, W.: On optimal decision rules in decision tables. Bulletin of Polish Academy of Sciences 32(11/12), 693–696 (1985)
7. Skowrons, A., Rauszer, C.: The discernibility matrices and functions in information system. In: Slowingski, R. (ed.) Intelligent Decision Support Handbook of Applications and Advances of the Rough Sets Theory. Kluwer Academic Publishers (1992)
8. Banerjee, M., Pal, S.K.: Roughness of fuzzy set. Information Sciences 93, 235–246 (1996)
9. Yao, Y.Y.: Two views of the theory of rough sets in finite universes. International Journal of Approximate Reasoning 15, 291–317 (1996)
10. Nanda, S., Majumdar, S.: Fuzzy rough sets. Fuzzy Sets and Systems 45, 157–160 (1992)
11. Zakowski, W.: Approximation in the space(U,∏). Demonstration Mathematica 16, 761–769 (1983)
12. Zhu, W., Wang, F.Y.: Reduction and axiomization of covering generalized rough set. Information Sciences 152, 217–230 (2003)
13. Zhu, W., Wang, F.Y.: Reduction and axiomization of covering generalized rough sets. Information Sciences 152, 217–230 (2003)
14. Zhu, W., Wang, F.Y.: On Three Types of Covering Rough Sets. IEEE Transactions on Knowledge and Data Engineering 19(8), 1131–1144 (2007)
15. Liu, G.L.: Fuzzy approximation space on the rough fuzzy set. Fuzzy Sets Systems and Mathematics 16, 75–78 (2002)
16. Lei, X.W.: Matrix method of rough set theory. Computer Engineering and Applications 42(17), 73–75 (2006)
17. Yang, Y.: Rough set definition of the matrix. Computer Engineering and Applications 43(14), 1–6 (2007)
18. Fodor, J., Roubens, M.: Fuzzy preference modelling and multicriteria decision support. Kluwer Academic Publishers (1994)

# Yard Allocation for Outbound Containers Based on the Unified Neutral Theory of Biodiversity and Biogeography

Zhenyu Min, Wenbin Hu*, Long Xu, Chang Xia, and Kaikai Wang

School of Computer Science, Wuhan University, Wuhan 430070, China
hwb@whu.edu.cn

**Abstract.** Yard allocation is the key point of container terminal management. This paper solves the problem by minimizing the distance between the berth and the yard, offering multiple work ways, avoiding loading or collecting at the same time. The solution of yard allocation is similar with the model which all individual organisms have the same possibility to be killed in Unified Neutral Theory of Biodiversity and Biogeography. Based on the improvement of the neutral theory, this paper proposed a multiple group model to reduce the solution set. To guarantee the adjacent container group not to be put in the same yard and assure a consecutive and parallel loading process, a strategy which kills anyone of the species exist prey relationship is put forward. And a greedy strategy is proposed to select an island to accelerate the ecological balance among islands. All these strategies make ecological selection more instructive and faster to find an optimal solution. The experiments show that the model proposed in this paper owns a good performance.

**Keywords:** yard allocation, container terminal, the neutral theory, multiple group model, greedy strategy, GA.

## 1 Introduction

Container terminal is a place for loading and unloading inbound and outbound containers. Services based on the port are of vital importance. Besides the characteristics of huge investment in port construction, expensive loading and unloading equipment, relatively high handling costs, container distribution system also has such features as strong stochastic, poor flexibility and uncertain equipment operating time. Therefore, the study of optimizing the scheduling of the container yard causes great concern at home and abroad.

However, allocating for outbound containers is the kernel part of the management of the yard, and a stacking plan for outbound containers is also necessary. A proper block plan will lay a good foundation for the scheduling of the containers and improve the efficiency of the yard. A well block plan is aimed at allocating blocks for the vessels, locating container groups to the specified blocks, and minimizing the total distance between the berths and blocks, avoiding two or more vessels loading at the

* Corresponding author.

J. Lei et al. (Eds.): AICI 2012, LNAI 7530, pp. 227–238, 2012.

same time, avoiding collection while loading and providing multiple work ways while shipment. In this way, the utilization rate and the efficiency of the yard will own a good improvement.

Oversea and native scholars have done a lot about the yard allocation for outbound containers problem. Zhang[1] developed a strategy for yard allocation using block plan. And several rules for block allocation were proposed. Outbound containers should be allocated near the berths. Two or more vessels should avoid loading at the same time of period in a block. That shipment when collecting should be avoided. The amount of containers in a block has its limits. Mi[2] proposed a rolling-horizon strategy via objective programming, solving the yard allocation problem by heuristic rules and distribute genetic algorithm. The paper optimizes the objective function via minimizing the total distance between the blocks and the berths, balancing the workload among the blocks and reducing the parking time of vessels. Bazzazi[3] solved the block allocation problem using genetic algorithm. The variables like size or type of a container were considered. The paper optimizes the objective function via balancing the workloads among the blocks and minimizing the parking time of vessels. Kim[4] discussed a yard planning decision system, including storage space, yard cranes and traffic areas. It aimed at balancing the workloads among the blocks and providing the ability to modify current yard plans by detecting blocks and periods with overloaded workloads. Zhao[5] proposed a best roadstead efficiency model for the yard allocation. The main objective is minimizing the distance between the block and the berth and making the loading operations more efficient. Yan [6] developed a model which was combined by a heuristic rule and a parallel genetic algorithm, using the way of objective programming under the rolling-horizon approach. The objective function is to minimize the distance between the yard crane and the berth. Gao[7] focused on the storage space allocation operation on the gate side of container yards.

Based on the research above, this paper studies the Unified Neutral Theory of Biodiversity and Biogeography proposed by Hubbell and develops and applies the modified model to the block plan for outbound containers. The neutral theory holds the opinion that every individual survives and dies in a totally random way. Therefore, the process of achieving the ecological balance via ecological selection has the global optimization capability and can be apply to the block allocation model and find the optimal solution. This paper focuses on the block allocation plan for outbound containers in a container terminal. To guarantee the utilization of the yard and the efficiency of loading at the same time, this paper solves the problem by minimizing the total distance between the berth and the yard, offering multiple work ways, avoiding loading or collecting at the same time. The experiments show that the model proposed in this paper owns a good performance.

## 2 Formulation of the Block Plan Based on the Unified Neutral Theory of Biodiversity and Biogeography

According to this paper, the assumptions are as follows.

(1) This paper mainly focuses on the outbound containers.

(2) The berths are assigned to the vessel before its arrival.

(3) The historical collecting or loading information of vessels are known and can be the guidance of the block plan.

(4) The trend of the amount outbound containers in a block is known and can be got from the historical information of the port.

(5) Containers belong to the same type, size, vessel, unloading port, weight level are called a container group.

(6) Containers in a container group should be allocated in a block to assure the continuity of loading operation.

This paper studies the Unified Neutral Theory of Biodiversity and Biogeography proposed by Hubbell and develops and applies the modified model to the block plan for outbound containers. The modified model takes a block as an island and takes a container group as a species. Therefore, the process of allocating container groups to blocks is like the process of ecological selection which species select the best islands to achieve ecological balance.

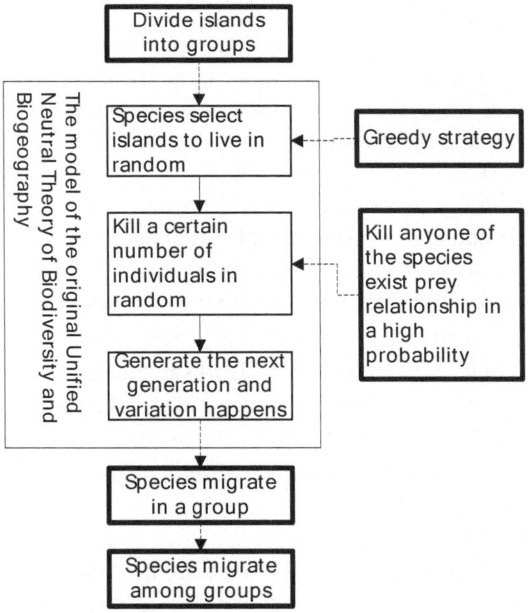


**Fig. 1.** Description of the modified model in this paper

When the original unified neutral model is used to solve the block allocation problem, it will fall into local optimal solution if the amount of blocks is great. Therefore, grouping strategy is used. As is shown in Fig.1, blocks are divided into several equivalent groups and the conception of migration in a group and migration among groups is taken in this paper. Migration in a group represents that species migrate from island A to another island B in the same group with A, assuring that species have the chance to move to other islands in the same group to avoid dying out when killed in an island after the initial selection of islands. Migration among groups represents all species in island A migrate to island B in another group to assure that all species have the equal chance to live in all islands. Greedy strategy is used to select the optimal living island for species. To guarantee the adjacent container group not to be put in the same yard and assure a consecutive and parallel loading process, a strategy which kills anyone of the species exist prey relationship is put forward.

The unified neutral theory points out that every individual in a colony has the same probability of being born, dying and variance and has nothing to do with the species the individual belongs to. The difference among species only relates to the amount of individuals the species belongs to. In this paper, each species has only one individual, after a certain times of iteration, all islands will result into a balanced state, which is the optimal solution this paper will get.

# 3 The Block Plan of Outbound Containers

## 3.1 Definition and Symbols

S: Species, which is equivalent to a container group in this paper. The same species has the same attributes while prey relationship exists between different species.
Ss: Let Ss denote the amount of species.
$s_i$: Let $s_i$ denote the size of container group i.
$p_i$: Let $p_i$ denote the unloading port of container group i.
$n_i$: Let $p_i$ denote the number of containers of container group i.
$w_i$: $w_i$ represents the weight level of container group i.
$m_i$: The history trend of the cumulative coming containers.$m_{i_1} = 0.1$, $m_{i_2} = 0.5$, $m_{i_3} = 1$, $m_{i_4} = 0.1$, $m_{i_5} = 0$. $m_{i_2} = 0.5$ represents the cumulative coming containers in the second stage is 50%. That is to say that there are 40% of containers coming to the yard during the first and second stage. Concentration period is from stage $m_{i_1}$to$m_{i_3}$, while loading period is from $m_{i_3}$to $m_{i_5}$.
L: Islands, which is equivalent to the blocks of outbound container in this paper. The capacity of an island is limited.
Ll: The amount of islands.
$B_n$: The number of pre-allocated blocks for the coming vessels during the block plan.
$lm_i$: The prediction number of containers in the $i_{th}$ block. $lm_{ij}$ is the prediction number of containers in the $i_{th}$ block during the $j_{th}$ period.
$c_i$: The free space of the $i_{th}$ island, which can be used to hold some individuals.
$d_i$: The distance between the $i_{th}$ island and the berth.
$cll_i$: In the $i_{th}$ island, if two vessels are going to load at the same time, then $cll_i$ is used to represent the value of collision.
$clc_i$: In the $i_{th}$ island, if a vessel are going to load while another are going to collect at the same time, then $clc_i$ is used to represent the value of collision.

$dis_{i,j}$ : The prey relationship between species i and j is expressed as $dis_{i,j} = \begin{cases} 1/|w_i - w_j|, & p_i = p_j, s_i = s_j \\ 0, & else \end{cases}$,

G: The number of islands in a group. The situation of '$Ll\%G = 0$' is considered in this paper.

$M_i$: It stands for the ith group, which is the congregation of a certain number of islands. All islands are divided into several groups.

N: The congregation of all islands. And the formula $N = \bigcup_{i=1}^{L_l/G} M_i$ is satisfied.

m_kr: The probability of killing an individual.

m_vr: The probability of generating a new individual via variance while generate the next generation.

m_sr: The probability of migrating within a group, which means that let all species in an island migrate to another island in the same group.

m_ir: The probability of migrating among groups, which means that let some species in an island of group A migrate to another island in group B.

$\delta_{i,j}$: $\delta_{i,j} = \begin{cases} 1, & \text{species i chooses island j} \\ 0, & else \end{cases}$, the decision variable.

### 3.2 The Objective Function

(1) To satisfy multiple work ways while load containers, containers groups should be located evenly in blocks and distributed in different blocks.

Firstly, judge if the amount of assigned containers in each block satisfies the uniform distribution.

$$u_1 = \sum_{i=1}^{L_l} \sum_{j=1}^{S_s} (\delta_{j,i} n_j) / L_l \tag{1}$$

$$\sigma_1 = \sqrt{\sum_{i=1}^{L_l} \left(\sum_{j=1}^{S_s} \delta_{j,i} n_j - u_1\right)^2 \Big/ L_l - 1} \tag{2}$$

Let $f1 = \max\{u_1/\sigma_1\}$. To ensure that the adjacent container groups are assigned to different blocks, the prey relationship '$dis_{i,j}$' is taken in this paper.

$$f2 = \max\{1 \Big/ \sum_{i=1}^{L_l} \sum_{j=1}^{S_s} \sum_{k=1}^{S_s} (dis_{j,k} \delta_{j,i} \delta_{k,i})\} \tag{3}$$

(2) Minimize the total distance between a block and the berth.

$$f3 = \max\{1 \Big/ \sum_{i=1}^{L_l} \sum_{j=1}^{S_s} (n_j d_i)\} \tag{4}$$

(3) Avoid several vessels loading in the same block at the same time of period.

$$f4 = \max\{1 / \sum_{i=1}^{L_l}((\prod_{j=1}^{S_s}\delta_{j,i})cll_i)\} \tag{5}$$

(4) Avoid other vessels collecting containers while a vessel loads containers in the same block.

$$f5 = \max\{1 / \sum_{i=1}^{L_l}((\prod_{j=1}^{S_s}\delta_{j,i})clc_i)\} \tag{6}$$

(5) Avoid the number of containers to be allocated in a block exceeding the capacity of it.

$$f6 = \max\{1 / \sum_{i=1}^{L_l}\sum_t(lm_{i_t} + \sum_{j=1}^{S_s}\delta_{j,i}m_{j_t}n_j - c_i)\} \tag{7}$$

(6) The fitness function.

This article takes each objective function projected onto the normal distribution function to normalize each target. 1000 feasible solutions are generated in random at first and evaluated via functions f1-f6, finally expectation $uf_i(1 \le i \le 6)$ and variance $\sigma f_i(1 \le i \le 6)$ are got.

$$fitness = \sum_{i=1}^{6}\Phi((f_i - uf_i)/\sigma f_i), 0 \le fitness \le 6 \tag{8}$$

### 3.3 Constraints

(1) The constraints of blocks. The number of selected blocks can't exceed the pre-selected blocks of the specified vessel.

$$\sum_{i=1}^{Ll}(\prod_{j=1}^{Ss}\delta_{j,i}) = Bn \tag{9}$$

(2) The constraints of the amount of containers. The amount of containers can't exceed the capacity of a block in any time of period.

$$lm_{i_k} + \sum_{j=1}^{Ss}(\delta_{j,i}m_{j_k}n_j) \le c_i \tag{10}$$

### 3.4 Steps of Algorithm

Step1: Divide all outbound container blocks into n groups by using optimal sub-population genetic algorithm.

Step2: Begin to find an optimal solution by using the modified unified neutral theory iteratively. If the iterative process is not end, then turn to Step 2.1, or turn to Step 3.

```
   Step2.1: Follow these operations for every dividing
group.
      Step2.1.1: Regard blocks in a group as a number of
   islands. Regard container groups as a number of
   species. Calculate the pre-allocated blocks for a
   vessel by its length (The number is determined by the
   relationship between the berth and the bridge crane
   and the relationship between the bridge crane and the
   block). The number of pre-allocated blocks is also
   regarded as the number of islands.
      Step2.1.2: Assign islands for all species.
      Step2.1.3: Use the modified unified neutral theory
   within group iteratively. If the iteration is not end,
   then turn to Step2.1.4, or turn to Step4.
      Step2.1.4: Kill species in random.
       Step2.1.5: Generate the next generation.
      Step2.1.6: Revise the results of step2.1.4 and
   step2.1.5.
      Step2.1.7: Species migrate within groups.
   Step2.2: Save the optimal solutions until now.
   Step2.3: Species migrate among groups.
   Step2.4: Save the optimal solutions until now.
  Step3: Generate the optimal solution.
  Step4: Finish the algorithm.
```

# 4 Experiments

Experiments are conducted via the real data of a port in Shanghai in this paper. This port owns 50 blocks, with 50 bays in each block. The historical information of the containers is the port is known. Experiments are carried out under the configuration of Windows Vista$^{TM}$ Home Premium, with AMD Turion(tm) 64 X2 Mobile Technology TL-62 2.10GHz, 2GB RAM.

The grouping strategy proposed in this paper is finished by the optimal sub-population genetic algorithm. After a certain number of experiments, the parameters in this algorithm are set as follows: the size of sub-population is 10%, the size of population is 60%, the probability of crossover is 0.6, the probability of variance is 0.01 and the group size is 10 (then 50 blocks can be divided into 5 groups). Using these parameters, the grouping strategy owns a good performance.

## 4.1 Analysis of Parameters

In order to make sure the optimal value of the fitness function, experiments are conducted to test the proper value of m_kr, m_vr, m_sr, m_ir.

AF(average fitness): $(\sum_{i=1}^{n} fitness)/n$, n=10 is used in this paper.

AR(average runtime): $(\sum_{i=1}^{n} runtime)/n$, n=10 is used in this paper.

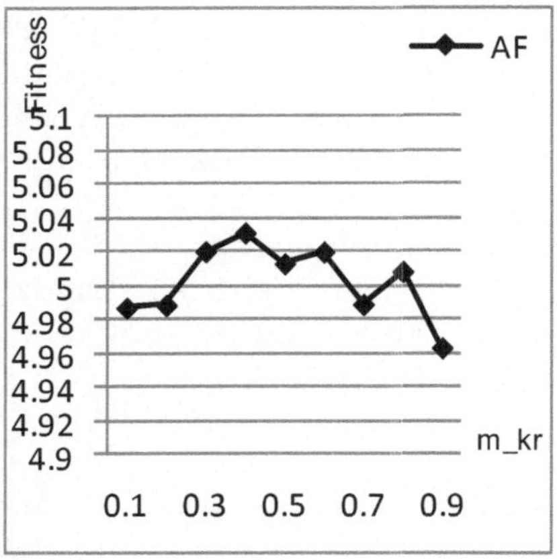


**Fig. 2.** m_kr-fitness analysis

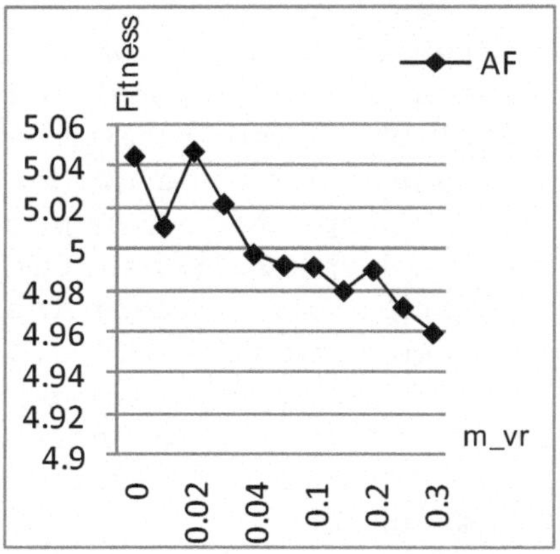


**Fig. 3.** m_vr-fitness analysis

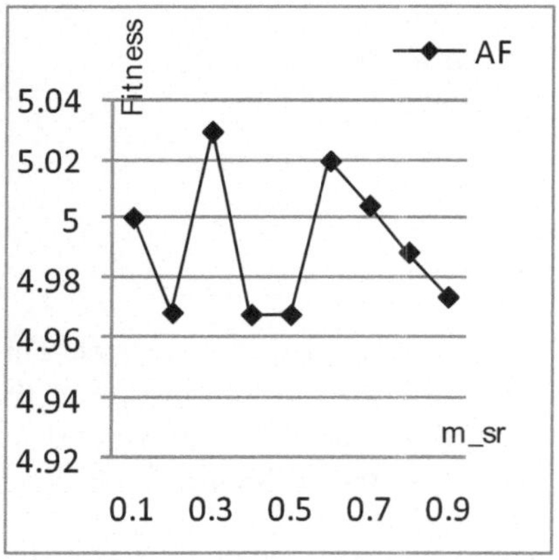


**Fig. 4.** m_sr-fitness analysis

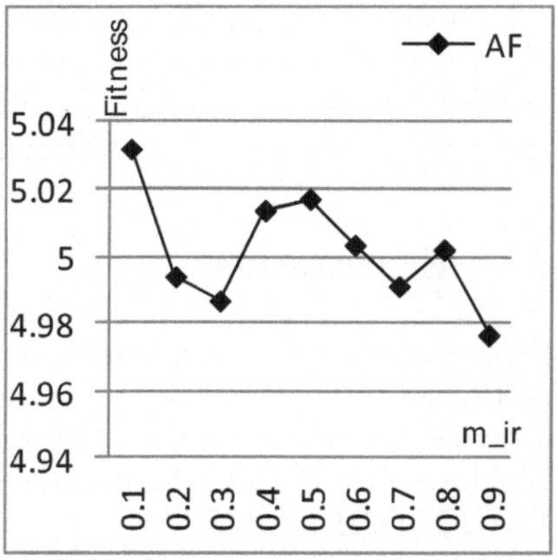


**Fig. 5.** m_ir-fitness analysis

As is shown in Fig2-5, the algorithm owns a good performance when m_kr=0.4, m_vr=0.02, m_sr=0.3 and m_ir=0.1. So these parameters are used to do block plan based on the modified unified neutral theory.

### 4.2 Iteration and Comparison Experiments

Iteration is used to solve the optimal solution in this paper. The best solution comes out at the end of iteration.

(1) The iteration experiments of the optimal sub-population genetic algorithm.

Set the number of times of iteration to be 1 when execute the grouping strategy. Set the number of times of iteration to range from 10 to 1000 when proceed the optimal sub-population algorithm, with 10 to be the step value. The results of experiments are as follows.

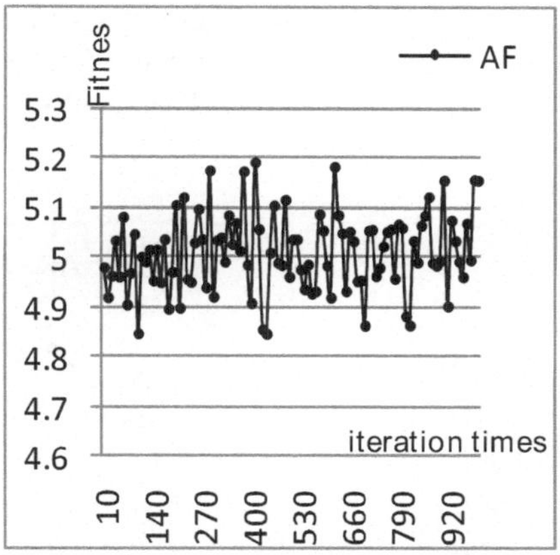


**Fig. 6.** Sub-population GA iteration times VS fitness

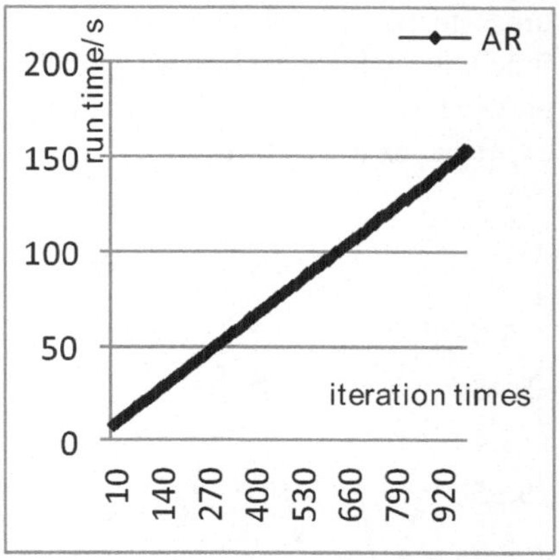


**Fig. 7.** Sub-population GA iteration times VS AR

Fig.6 shows that grouping strategy using the sub-population GA owes an optimal fitness value when the iteration time is 410.

(2) The iteration experiments of the within-group strategy in the modified unified neutral theory.

Set iteration times to be 410 when execute the sub-population GA. Set the iteration times to be 1 when execute the among-groups strategy. Set the number of times of iteration to range from 10 to 1000 when proceed the within-group strategy, with 10 to be the step value. The results of experiments are as follows.

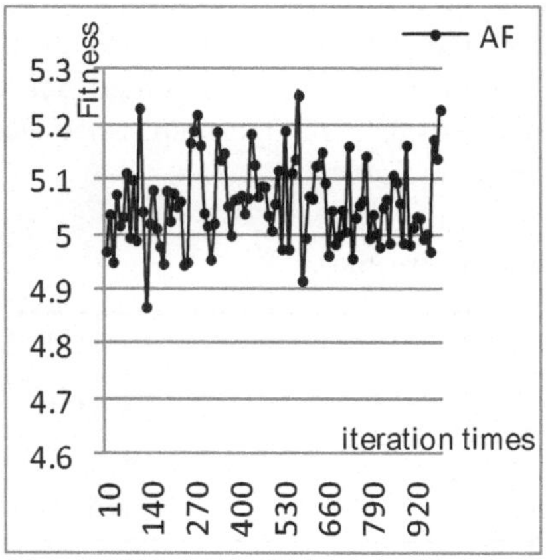


**Fig. 8.** Iteration times within group VS fitness

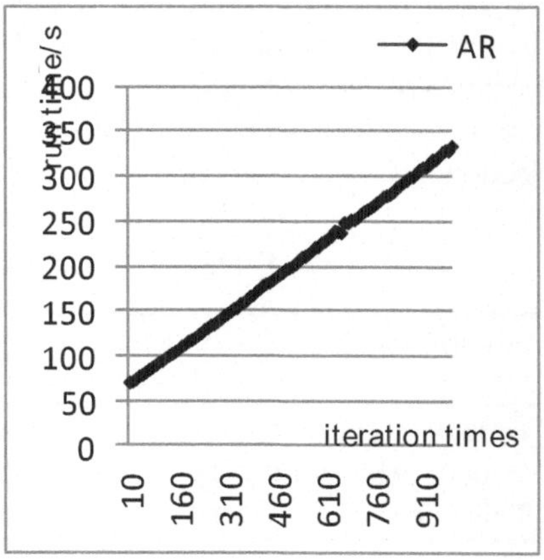


**Fig. 9.** Iteration times within group VS AR

Fig.8 shows that the optimal fitness value is ideal, with 410 to be the sub-population GA iteration times and 580 to be the within-group iteration times.

(3) The iteration experiments of the among-group strategy in the modified unified neutral theory.

Set iteration times to be 410 when execute the sub-population GA. Set the iteration times to be 580 when execute the within-group strategy. Owing to the calculation time will increase in a linear growth, so set iteration times to range from 5 to 95 when proceed the among-groups strategy, with 10 to be the step value. The results of experiments are as follows.

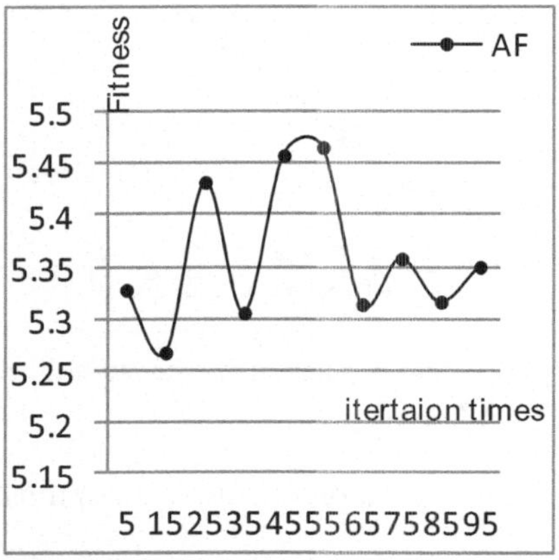


**Fig. 10.** Iteration times among groups VS fitness

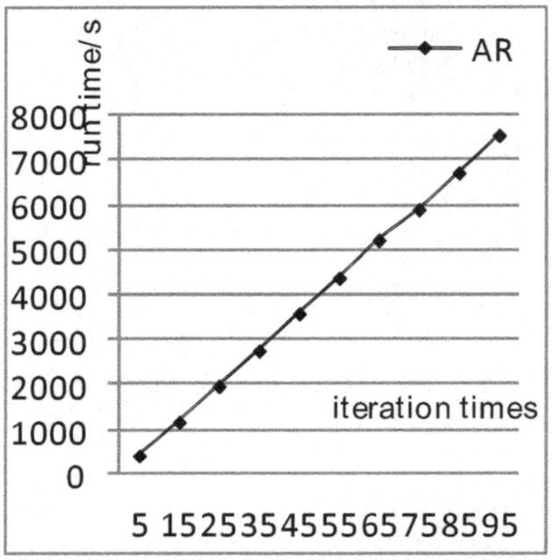


**Fig. 11.** Iteration times among groups VS AR

Fig.10 shows that the optimal fitness value is got when the sub-population GA iteration times are 410 and the within-group iteration times are 580 and the among-groups iteration times are 55.

(4) Comparison experiments with other algorithms

To verify the model proposed in this paper, the genetic algorithm and greedy algorithm are used to compare with the algorithm based on the modified unified neutral theory. The same objective function is taken as the two algorithms'. See more in Section 3.2. The results of experiments are shown in Table 1.

**Table 1.** The results of comparison experiments

| Algorithm | Experiment Times | Value of Objective Function | Time(s) |
|---|---|---|---|
| Greedy Algorithm | 10 | 4.868008 | 10 |
| Genetic Algorithm | 10 | 5.345722 | 277 |
| This paper | 10 | 5.465745 | 4380 |

The value of objective function in this table is the average value of 10 times' experiments. As is shown in this table, greedy algorithm has the smallest fitness value because of falling into the local optimal solution too early. However, the block plan based on GA or the algorithm in this paper owns a good performance. In general, on the condition of consuming a certain amount of computing time, the algorithm proposed in this paper owns a good performance.

## 5 Conclusion

The classic ecological unified neutral theory model is a completely random process. It considers that the individual has an equal right to life and groups achieve ecological balance through a completely random birth and death process, which has in common with the block plan model in this paper. Mathematical model and an algorithm are established to solve the block allocation problem based on the result of studying the original unified neutral theory, modifying some parts according to the characteristic of block plan and applying the modified model to the block allocation problem. In order to search for the solutions in a global scope, grouping strategy, a strategy which kill anyone of the species exist prey relationship and a greedy strategy which select an island to accelerate the ecological balance among islands are proposed based on the original one. All these strategies make ecological selection more instructive and faster to find an optimal solution. The experiments show that the model proposed in this paper owns a good performance on the condition of consuming a certain computing time. However, the computing time will grow in a linear growth with the increase of the problem scope.

**Acknowledgments.** This work is partially supported by National Natural Science Foundation, China(No.70901060); Hubei Province Natural Science Foundation (No. 2011CDB461); State Key Lab of Software Engineering Open Foundation (No.SKLSE2010-08-15) ;Youth Plan Found of Wuhan City(No.201150431101)and the Fundamental Research Funds for the Central Universities. The authors also gratefully acknowledge the helpful comments and suggestions of the reviewers, which have improved the presentation.

## References

1. Zhang, Y., Zhou, Q., Zhu, Z., Hu, W.: Storage planning for outbound container on maritime container terminals. In: Proceedings of the 2009 IEEE International Conference on Automation and Logistics, ICAL 2009, pp. 320–325 (2009)
2. Mi, W., Yan, W., He, J., Chang, D.: An investigation into yard allocation for outbound containers. The International Journal for Computation and Mathematics in Electrical and Electronic Engineering 28(6), 1442–1457 (2009)
3. Bazzazi, M., Safaei, N., Javadian, N.: A genetic algorithm to solve the storage space allocation problem in a container terminal. Computers and Industrial Engineering 56(1), 44–52 (2009)

4. Kim, K.H., Zhang, X.-H.: Distributed framework for yard planning in container terminals. Journal of Zhejiang University: Science A 11(12), 992–997 (2010), ISSN: 1673565X, E-ISSN: 18621775, doi:10.1631/jzus.A1001527
5. Su, Z.: Export Containers Block Allocation Based on Best homestead Efficiency. Tsinghai University (2007)
6. Yan, W., Xie, C., Chang, D.: Stockpile allocation strategy for container terminals based on parallel genetic algorithm. Journal of Shanghai Maritime University, 2 (2009)
7. Gao, P.: Research on Optimization Problem on Container Yard Operation Scheduling. Dalian University of Technology (2005)

# An Auto-Adapted Method to Generate Pairwise Test Data Set

Penghui Fan, Shuyan Wang, and Jiaze Sun

School of Computer Science and Technology,
Xi 'an University of Posts and Telecommunications Xi 'an, 710121, China
fanpenghui999@163.com, {wsylxj,sunjiaze}@126.com

**Abstract.** The pairwise test data set generation is one of key issues of combinatorial testing. This paper presents a novel auto-adapted method to generate a pairwise test data set. In this method, all test cases are made at a time, which is called "all-tests-at-a-time". Firstly, generate a certain number of test data sets; these test data sets have the same number of test cases. Secondly, chose the best data set and check whether it satisfy the requirements, if not ,go to next step, else the best is selected and the algorithm is end. Thirdly, update every data set: calculate the "repeat number" of each test case in a data set, chose two or three test cases according to the "repeat number"; update the selected test cases relies on "main factors" of each data set. Moreover, the classic examples are used to illustrate the performance of the proposed method. Compared with the existing algorithms, this paper provides an effective pairwise test suite generation method which updates test cases depend on the data set's coverage not any one independent case; it takes into the relationship of every test case consideration not like the traditional methods which also find the current best case. It can help the data set improve its coverage quickly.

**Keywords:** Combinational testing, Pairwise Test, Auto-Adapted Method, Test Data Set Generation, All-Tests-at-a-Time.

## 1 Introduction

Software testing is a key part of the software development process, it takes a lot of time in a software development. Therefore, it is necessary to improve the testing efficiency and performance. Interaction testing is widely used in software test process, and the most important problem of the interaction testing is to find the appropriate interaction test suites, that can be described by the set covering problem which is well known to be NP-complete(Nondeterministic Polynomial time) problem. For example, it supposed a system with 13 parameters, every parameter has 10 different discrete values, it has 10^13=10,000,000,000,000 different Combinatorial sets, so, exhaustive testing using all possible combinations of input values for a system is not feasible, in consider of the time and cost. Fortunately, according to Kuhn's report, many software mistakes are caused by a few parameters, 70% of them by two, and 90% by three [1]. So pairwise testing is able to find most of software mistakes and improve the stability of a system.

J. Lei et al. (Eds.): AICI 2012, LNAI 7530, pp. 239–246, 2012.

Pairwise testing is a combinatorial technique which selects a subset of all possible test case input combinations. It requires all possible pairs of input parameters to be covered by at least once in a perfect data set. An example can explain pairwise covering question. Suppose a system has 4 parameters and every parameter has 3 different values. The first group are a1,a2 and a3,the second are b1,b2,and b3,the third are c1,c2 and c3,and the fourth are d1,d2 and d3, each two different values from different group consist a pair, a pair is not relevant to its order,<a1,b1> and <b1,a1> are the same pair. There are all 54 different pairs:

<a1,b1>,<a1,b2>,<a1,b3>,<a1,c1>,<a1,c2>,<a1,c3>,<a1,d1>,<a1,d2>,<a1,d3>
<a2,b1>,<a2,b2>,<a2,b3>,<a2,c1>,<a2,c2>,<a2,c3>,<a2,d1>,<a2,d2>,<a2,d3>
<a3,b1>,<a3,b2>,<a3,b3>,<a3,c1>,<a3,c2>,<a3,c3>,<a3,d1>,<a3,d2>,<a3,d3>
<b1,c1>,<b1,c2>,<b1,c3>,<b1,d1>,<b2,d2>,<b1,d3>,<b2,c1>,<b2,c2>,<b2,c3>
<b2,d1>,<b2,d2>,<b2,d3>,<b3,c1>,<b3,c2>,<b3,c3>,<b3,d1>,<b3,d2>,<b3,d3>
<c1,d1>,<c1,d2>,<c1,d3>,<c2,d1>,<c2,d2>,<c2,d3>,<c3,d1>,<c3,d2>,<c3,d3>

A pairwise test case are consisted by 4 different values from each group, for example <a1,b2,c3,d2>,and it covers 6 pairs <a1,b2>,<a1,c3>,<a1,d2>,<b2, c3>,<b2,d2> and <c3,d2>.we can generate a pairwise test data set, which is consisted by a certain amount of test cases covering all 54 pairs . For example:

<a2,b1,c2,d3>,<a1,b3,c3,d3>,<a2,b2,c3,d2>
<a1,b2,c2,d1>,<a3,b3,c2,d2>,<a1,b1,c1,d2>
<a2,b3,c2,d1>,<a3,b2,c1,d3>,<a3,b1,c3,d1>

However, it is difficult to generate a data set consisted of only 9 cases to cover all pairs without optimization Algorithm. For more complex system, the scale expands rapidly and the solution becomes more difficult.

The generation of pairwise test sets with a minimal size is a combinational optimization problem and can be described by the set covering problem which is well known to be NP-complete. And several deterministic algorithms have been published. While NP-complete problems do not admit efficient deterministic solutions in practice, generally speaking, the NP-complete problems can be solved approximately by heuristic approximate algorithm. Such as, Ant Colony Optimization (ACO)[2], solution space tree (SST)[3] and Simulated Bee Colony Algorithm (SBCA)[4].

In this paper, a novel auto-adapted method, which concentrates on our new colony update strategies from one generation to the next, is proposed to solve the problem of pairwise test case generation. From testing results of the classic example of pairwise test case generation problem, the novel auto-adapted method is obviously feasible.

In the next section, we outline and describe basic conception and related work. In Section 3, the auto-adapted method is described concretely, and the typical examples are employed to evaluate the performance of the novel method. Finally, conclusion is given in the 4th section.

## 2 Related Work

Many researchers have studied kinds of methods to generate a test data set. David M. Cohen and Siddhartha R. Dalal propose the automatic efficient test generator

(AETG)[5] framework based on a random greedy algorithm. SHI L and NIE C H invented a new method SST [3], in which all of the available test data sets were represented as a tree solution space and each data set is a path from root to leaf. The method creates a smallest path which covered all of the pairwise pairs with backtracking algorithm. Yu L and Raghu Kacker proposed In-Parameter-Order-General (IPOG) method[6], on which a testing tool called FireEye is built for t-way testing. Our team also has done lots of related work in generating test data set. We have made an intensive study of relative algorithms such as genetic algorithm (GA), polyethylene terephthalate glycol (PETG), and Simulated Annealing (SA). Especially in Discrete Particle Swarm Optimization (DPSO)[7][8], we have a lot of experiments and theory research.

One-test-at-a-time theory has been applied in above research activities. One-test-at-a-time is a method which generates one test case to cover as much pairs as possible in one calculation, which is also combined with some intelligent strategies. The method is similar to greedy algorithm which finds the best at once. The main steps are described as follow:

Step 1: Generate a certain number of test cases (generally less than 100, depending on the scale of specific problem).

Step 2: Find the best case which covers most uncovered pairs.

Step 3: Judge the current generating cases can cover all pairs. If they covers all pairs, the algorithm will be end, else go to step 2.

The main advantage of One-test-at-a-time method is to save software and hardware resources, and it can get a data set in finite time. At the beginning, it is an effective way to be able to generate good test cases in little time.

# 3 Our Work

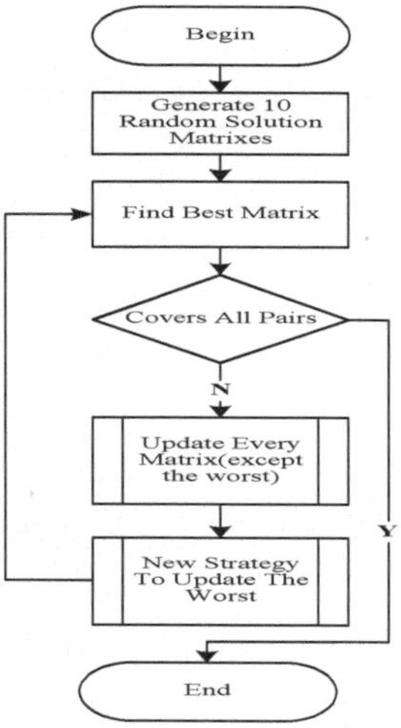


**Chart 1.** Main Steps

The impacts of One-test-at-a-time model are likely to generate optimal test data set slowly. Because the method can only ensure the case could cover the most pairs in current time, but it ignores the relationship of a test case with another which can affect the efficiency in most of situations. For example, if a test case can covers 20 pairs generated current time, and the test case generated next time can covers 15, the total is 35. While, if they are generated at a time. Maybe, the 2 test cases cover 36 in total. Especially, when most of pairs are covered, the effective of a new one is far less than the first or second one. We have done several experiments by this means, the result demonstrated that, for a system which a test case can cover 76 pairs at most, each test case of first 3 can covers more than 70 pairs, but the last 2 or 3 test cases can only covers 4 or 5 pairs.

This paper outlines and describes a new method which can generate a test case data set at a time. We called it "All-Tests-at-a-Time". It is described as chart 1.

Step 1: Generates a certain number of test case data sets which includes equal amount of test cases, the number depends on the specific problem which is usually greater than 3.

Step 2: calculate covering pairs of every data set and select the best data set which covers most pairs.

Step 3: if the best data set can't cover enough pairs to satisfy the requirements, update every data set. The strategies of worst data set differ to the other data sets. Else, the algorithm will be end.

The following section will explain the details and focus on the updating strategies (step 3) – core of our algorithm.

### 3.1 Details of Algorithm

In order to describe the problem clearly, it is supposed a system with n parameters and every parameter has m different values, the total pairs (tp) is:

$$tp = \frac{1}{2}n(n-1)m^2 \tag{1}$$

**Step 1:** Create a specific number of test case data sets. It is supposed there are p test cases in every test data set. The test data set is a matrix with n*p, it is called

"Solution" matrix, and every column is a test case, the possible value of the i-th row is: m(i-1)+1…mi.

Create a matrix with 3*pairs called "ObjectM", the first and second rows are the pairs, and the third row is a flag which represents coverage of every pair. "1" represents the pair is covered, and "0" not covered, it is initialized as "0".

**Step 2:** calculate "covering pairs" of every data set and chose the best one.

1) Visit every test case of current data set and calculate the value of ObjectM's third column.

2) Count the sum number of covering pairs(cp).

$$cp = \sum_{i=1}^{\frac{1}{2}n(n-1)m^2} \mathbf{ObjectM}(3,i) \tag{2}$$

3) Repeat the first two steps, calculates every *cp*.

**Step 3:** Judge the best data set whether meet the conditions. If $\max(cp) < \frac{1}{2}n(n-1)m^2$, update every data set except the worst.

1) Create a vector with 1*$p$ called RN (repeat number), which records the quantity of pairs the current test case and at least one other test case cover commonly. The $i$-th element represents the $i$-th test case's repeat number and its scope is: $0 \le \mathbf{RN(1,i)} \le \frac{1}{2}n(n-1)$.

2) Find the relative columns with max 3 "repeat number" in RN, the relative columns make up the new matrix SC (Selected-Cases) ordered by RN.

3) To update SC (1:n,1) and SC (1:n,2). Before updating, 2 vectors *Frequency* and *FrequencyCover* should be calculated. *Frequency* is a 1**mn* vector, it indicates the frequency of each possible value $(1,2,3...mn)$ emerging in the test case data set. *FrequencyCover* is also a 1**mn* vector, it indicates the frequency of each possible value $(1,2,3...mn)$ emerging in the covered pairs. Because every value of SC(1:n,1) has *m* possible. The next, calculate possibility to turn to each possible value. The possibilities are consisted of 3 factors - fre, frecov and bal (balance factor). For SC (i,1), the new value may be $m(i-1)+1, m(i-1)+2, ...or\ mi$. The value with max probability would be selected. The details are described in table 1.

**Table 1.** Update follow 2 columns of SC

```
af=p/m;
for i=1:m
   afc= FrequencyCover(1,m(i-1)+1:mi)/m;
   oldvalue= SC (i,1);
   for  j=m(i-1)+1:mi
        if  oldvalue==j
            fre=af/Frequency(1,oldvalue)*rand(0,1);
            frecov=
afc/FrequencyCover(1,oldvalue)*rand(0,1);
            bal=1/3*rand(0,1);
        else
            fre=(af-
Frequency(1,j))/Frequency(oldvalue,1)*rand(0,1);
            frecov=(afc-FrequencyCover(1,j)/
FrequencyCover(oldvalue)*rand(0,1);
            bal=1/3*rand(0,1);
        end
        new_vector(1,j-m(i-1))=fre+frecov+bal;
   end
   SC (i,1)=MaxIndexOf(new_vector)+m(i-1);
End
```

4) To update SC(1: *n*,3),it is the test case with max "repeat number ", it is key column for a data set. Each value will be updated:

SC(*i*,3)=MinIndexOf(*FrequencyCover*(*m*(*i*-1)+1:*mi*)) (3)

where *i* is 1,2,3…*n*;

5) Make the new SC replace the old relative columns in "solution" matrix.

Repeat the first 5 steps and update other "solution" matrixes except the worst one.

The worst data set is updated in another method.

1) Make the worst data set means the best one.

2) Calculate the vector RN and select the column with max "repeat number".

3) Update the selected column as follow; the details are described in table 2.

**Table 2.** Update column of worst solution

$startNum = rand(1, \frac{1}{2}n(n-1)m^2);$

```
flag_matrix=zeros(1,n);
```

$for \quad start = 1 : \frac{1}{2}n(n-1)m^2$

```
      if(ObjectM(3,startNum)==0)
         first_value=ObjectM(1,startNum);
         first_index=IndexOf(first_value);
         second_value=ObjectM(2, startNum);
         second_index=IndexOf(second_value);
          if(flag_matrix(1,first_index)==0)
             MaxRepeat(first_index,1)=first_value;
             flag_matrix(1,first_index)=1;
          end
         if(flag_matrix(1,second_index==0))
             MaxRepeat(second_index,1)=second_value;
             flag_matrix(1,second_index)=1;
         end
      end
      startNum=startNum+1;
```

$if\,(startNum > \frac{1}{2}n(n-1)m^2)$

```
          startNum=1;
      end
end
```

If the updated worst data set also can't cover all pairs, go to step 2 and begin a new cycle.

If the data set covering $\frac{1}{2}n(n-1)m^2$ are created or the loop times are more than 300, the algorithm runs the end.

### 3.2 Examples

The following 2 examples illustrate a practical application of the method. For a system that $n = 4, m = 3$. There are total 4 parameters and 12 different possible values; from 1 to 12. The sum pairs are 54. Select $p = 10$ because of the small scale. The best test data set are composed by 9 test cases by referring to some publications such as [9]. So the "Solution Matrix" is 4*9.By our method, the algorithm has updated 8 times and get the perfect solution in table 3 which can cover all 54 pairs.

**Table 3.** The final solution matrix (n=4, m=3)

| | | | | | | | | |
|---|---|---|---|---|---|---|---|---|
| 2 | 2 | 1 | 2 | 1 | 3 | 3 | 1 | 3 |
| 5 | 4 | 5 | 6 | 6 | 6 | 5. | 4 | 4 |
| 8 | 7 | 7 | 9 | 8 | 7 | 9 | 9 | 8 |
| 11 | 12 | 10 | 10 | 12 | 11 | 12 | 11 | 10 |

Another experiment is also applied by us. It is supposed that a software system with 13 parameters and every parameter has 3 different values. There are 39 different possible values and the *tp* =702.Referring some publications such as [9], the minimum quantity of test case data set is 15. 30 test case data sets are selected. It is executed by 300 times, Fig. 1 display the result.

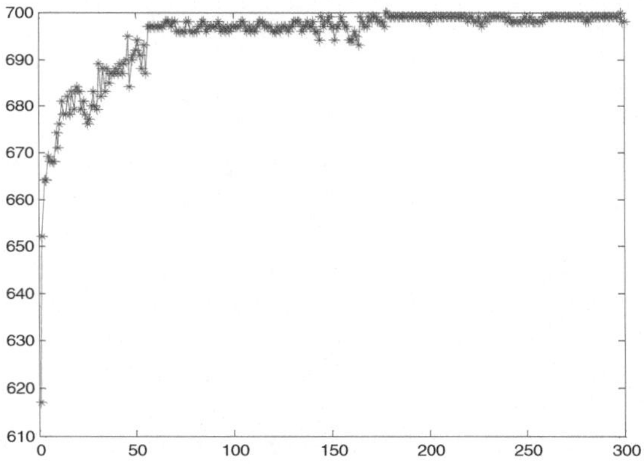


**Fig. 1.** Covering pairs (n=13, m=4)

From Fig.1, the best covering pairs is 700, 2 pairs are not covered. When the algorithm are executing 60 times, the covering number rose slowly.

# 4 Conclusions

The experiments show that it is an effective method for pairwise test case generation problem. It adopts "All-Tests-at-a-Time" strategy which takes the relationship of test cases as an important factor. Some parameters are produced to describe the relationship. RN influences the importance of every test case directly and can help to find the worst terrible test case. Frequency and FrequencyCover reflect the coverage of every possible value in different angles, which indicate every test case's evolution. Moreover, its time complexity is limited, for a system with 4 parameters and every parameter has 3 different values.it needs about 0.1 second in a normal computer (Windows XP professional sp3, Pentium(R) Dual-Core E5800, Matlab2008a). The experiment with 13 parameters also completed in about 40 seconds. The experiments show that our method is a feasible method to generate a perfect pairwise data set.

According to some mature algorithms, our method is not better, but we point to a new direction to generate pairwise data set, which differs to most traditional methods.

However, our works are limited; maybe more factors could be considered to direct the test case. We will try to find a way to combine our new method with other algorithms such as PSO, Ant Colony Optimization (ACO) and SA. Furthermore, our method will be expanded to solve *t*-way test case generation problem.

**Acknowledgments.** This work was supported in part by project "Research on Key Problem of Combinatorial Software Testing optimization Based on Swarm Intelligence" (61050003) from National Natural Science Foundation of China, by project "Smart Combinatorial Embedded Soft Testing Platform" (2009K08-26) from project Key Technologies R&D Programmed Foundation of Shaanxi Province, and by the Scientific Research Program of Shaanxi Provincial Education Department (12JK0732).

## References

1. Kuhn, D., Reilly, M.: An investigation of the applicability of design of experiments to software testing. In: 27th Annual NASA Goddard/IEEE Software Engineering Workshop, pp. 91–95 (2002)
2. Xiang, C., Qing, G.: Building Prioritized Pairwise Interaction Test Suites with Ant Colony Optimization. In: 2009 Ninth International Conference on Quality Software, pp. 347–352 (2009)
3. Shi, L., Xu, B., Nie, C.: Pairwise Test Data Generation Based on Solution Space Tree. Chinese Journal of Computers, 849–857 (2006)
4. McCaffrey, J.D.: Generation of Pairwise Test Sets using a Simulated Bee Colony Algorithm. In: IEEE IRI 2009, Las Vegas, Nevada, USA, July 10-12, pp. 115–119 (2009)
5. Cohen, D.M., Dalal, S.R., Fredman, M.L., Patton, G.C.: The AETG System: An Approach to Testing Based on Combinatorial Design. IEEE Transactions on Software Engineering, 437–444 (1997)
6. Lei, Y., Kacker, R., Richard Kuhn, D., Okun, V., Lawrence, J.: IPOG: A General Strategy for T-Way Software Testing. In: Proceedings of the 14th Annual IEEE International Conference and Workshops on the Engineering of Computer-Based Systems (2007)
7. Sun, J., Wang, S.: A Novel Chaos Discrete Particle Swarm Optimization Algorithm for Test Reduction. In: IWSE 2010, pp. 623–632 (2010)
8. Sun, J., Wang, S.: Generation of Pairwise Test Sets Using a Novel DPSO. In: GCN 2011, pp. 479–487 (2011)
9. Li, X., Zhang, W.: Test Case Generation Based on Annealing Immune Genetic Algorithm. Computer Simulation, 171–174 (2008)

# The Application of Evolutionary Algorithm in B-Spline Curved Surface Fitting

Jixin Yang, Fang Liu, Xueheng Tao, Xuejun Wang, and Jinshi Cheng

Dalian Polytechnic University, School of Mechanical Engineering and Automation, 116034 Dalian, China

**Abstract.** This paper proposes combining B-spline curved surface fitting with evolutionary algorithm to improve the fitting efficiency and precision. In the process of selecting curved surface control points, taking the minimum error sum of squares as the fitness standard, the optimal basis curved surface is obtained by optimizing control points constantly, which is partitioned into small blocks according to its precision. Control points of each piece are fast and precisely reversed, then x, y and z values of control point are used as gene chromosomes. Genetic operation is continuous reiteration until original curve surface reconstruction is achieved.

**Keywords:** evolutionary algorithm, B-spline curve, curved surface fitting.

## 1 Introduction

In the engineering practice, there are many surface complex objects which are expressed precisely by the measurement extracting a series of data points. There are a lot of curved surface fitting methods widely applied to CAGD (computer aided geometric design), computer graphics and computer vision, and so on. The methods are mainly divided into four kinds: quadrilateral region surface reconstruction, trilateral region surface reconstruction, subdivision surface reconstruction and hierarchical model surface reconstruction [1-2], while represented by B-spline surface and NURBS surface, is the common quadrilateral region surface reconstruction method.

B-spline curved surface is constructed with approaching a set of control points. B-spline curved surface has two obvious advantages: permitting that surface is controlled by local, and the degree of multinomial is independents of the number of control points [3]. Geometry reconstruction based on quadrilateral region surface is the common method in B-spline curved surface fitting reconstruction. After the object surface triangular mesh model is rebuilt from point cloud, how to convert it to the surface used by B-splin is the most important. Usually for scattered data points, quadrilateral region surface fitting process is structuring a base surface, then the data point triangular mesh is parameterization. The sum of error squares from metrical data to target surface is minimized to judge the precision of obtained surface. But the computation of common solving process is usually excessive and no purpose. This paper proposes improving the algorithm computation speed and efficiency using evolutionary algorithm robust to simplify solving process, and proves this method feasibility and advantage.

J. Lei et al. (Eds.): AICI 2012, LNAI 7530, pp. 247–254, 2012.

## 2 B-Spline Curved Surface Fitting

In the computation process, reverse control points are obtained through the curved surface in two stages. According to data points of complex surface seem as rectangular array $p_{i,j}$ (i=0, 1,.., r; j=0, 1,…, s), control points $d_{i,j}$ (i=0, 1,…, m; j=0, 1,…, s) are reversed. Knot vectors are obtained by average standard accumulation chord length parameterization from vertical and transverse, so that a B-spline fitted curved surface which approaches to known complex surface is constructed. However, because the quantity and location of selected data points may exist in noise and vibration, the fitting surface will excessively deviate from the original surface at some points. According to error judgment condition, the corresponding control points are adjusted in real time to modify the local shape to achieve better fitting effect. In engineering practice, dual cubic B-spline surface is usually used, because its good local features. Thus this paper adopts dual cubic B-spline curved surface as example and using quadrilateral region surface fitting.

The specific steps of quadrilateral region surface fitting as follows:

1. Boundary preprocessing;
2. Parameterization of space scattered points;
3. Constructing triangle mesh;
4. Constructing mathematics model
5. Solving mathematics model

In order to obtain the approaching surface model of scattered points, the number of scattered points must be more than the unknown control vertex (namely, l>m×n) so that unknowns are less than or equal to equations. Objective function is:

$$I^r = \alpha \sum^{J} [S^r(u_k^{r-1}, v_k^{r-1}) - P_k] + \beta[S_{uu}^r(u_k^{r-1}, v_k^{r-1})^2 + S_{vv}^r(u_k^{r-1}, v_k^{r-1})^2 + 2S_{uv}^r(u_k^{r-1}, v_k^{r-1})^2] \quad r \tag{1}$$

$$\begin{aligned} S^r(u,0) &= p(u,0); & S^r(u,1) &= p(u,1) \\ S^r(0,v) &= p(0,v); & S^r(1,v) &= p(1,v) \\ S_v^r(u,0) &= p_v(u,0); & S_v^r(u,1) &= p_v(u,1) \\ S_u^r(0,v) &= p_v(0,v); & S_u^r(1,v) &= p_v(1,v) \end{aligned} \tag{2}$$

Where, r stands for the iteration times of surface, $S^r(u,v)$ stands for the surface after r times iteration, and $S^r(u,v) = \sum_{i=0}^{m}\sum_{j=0}^{n} N_{i,3}(u)N_{j,3}(v)V_{i,j}^r$; $S_{uu}^r(u,v)$, $S_{vv}^r(u,v)$, $S_{uv}^r(u,v)$ are respectively second order partial derivative and mixed partial derivative of surface $S^r(u,v)$; $p_k(k = 0,1,\dots,l)$ is scattered point; $(u_k^{r-1}, v_k^{r-1})$ is parameter value of $p_k$ in k-1 times iterative surface; $\alpha$ stands for approximation weight; $\beta$ stands for faring weight. In addition, $\alpha + \beta = 1$, the larger $\alpha$ is, the closer surface is to data points, and surface is less fairing. The larger $\beta$ is, the faster surface is from data points, and surface is more fairing. So $\alpha$ and $\beta$ are related to the balance between fairing and precision.

As has been mentioned above, the mathematical model of objective function is the quadratic optimal problem with complex constraint conditions. The ultimate purpose is that objective function $I^r$ meets constraint conditions, and is minimum using the optimizing method. Thus control points $V_{i,j}^r(i = 0, 1, .., m; j = 0, 1, ..., n)$ of objective curved surface after r iterations are obtained.

The optimization of objective function $I^r$ was divided into two steps. And constraint conditions only were related to boundary curve and cross-boundary tangent vector which could be determined by control points. According to (1), dealing with the control points of objective curved surface boundary and sub-boundary, the surface was obtained.

## 3 B-Spline Curved Surface Fitting Based on Evolutionary Algorithm

Currently common method in evolutionary algorithm theory is new Darwin model that life reproduction history is explained by a series of physical process. New Darwin model theory is summarized in the following respects: individual is the main object of selection; hereditary mutation is occasional phenomenon to a great extent, and random process plays an important part evolution; the major genetic change is the result of gene recombination, mutation is in a small degree; deformation mutation may take place in gradual evolution process; not all deformation is the result of natural selection. Evolution is the adaptive and multifarious process, not only the change of gene coding; the selection is probabilistic, nor definitive [5-8]. Above viewpoints of evolution compose the theoretical foundation of simulated evolutionary study.

The surface fitting based on evolutionary algorithm consists of two sections: optimizing fitting curve boundary and optimizing fitting basis surface. Optimizing fitting curve boundary could be found in reference [4], so I didn't repeat them here.

For a start, basis surface should be set up. The first step of inner surface reconstruction was scattered points parameterization, which was designed as triangle mesh. Then the first surface (base surface) was initially fitted as Fig.1. At present common surface mesh partition method was classified as mapping method and direct method. Advantages of the former were simple implement, efficient, strong

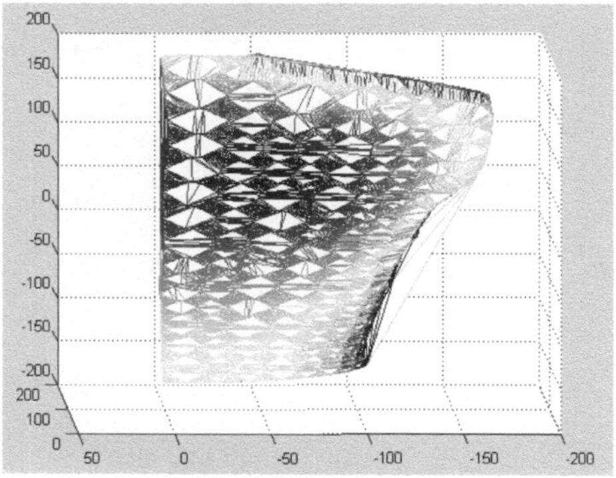


**Fig. 1.** Triangle mesh

controllability of mesh density, local mesh refinement and re-meshing. So this paper adopted Delaunay triangulation method to fit scattered points.

This paper took the surface of manikin as example. Because manikin was complex model which couldn't be expressed by single surface, surface was refined. The mesh controlled by B-spline was constructed from rough to precious. Finally B-spline surface gradually approached to given scattered points. Specific operational flow chart as Fig. 2 :

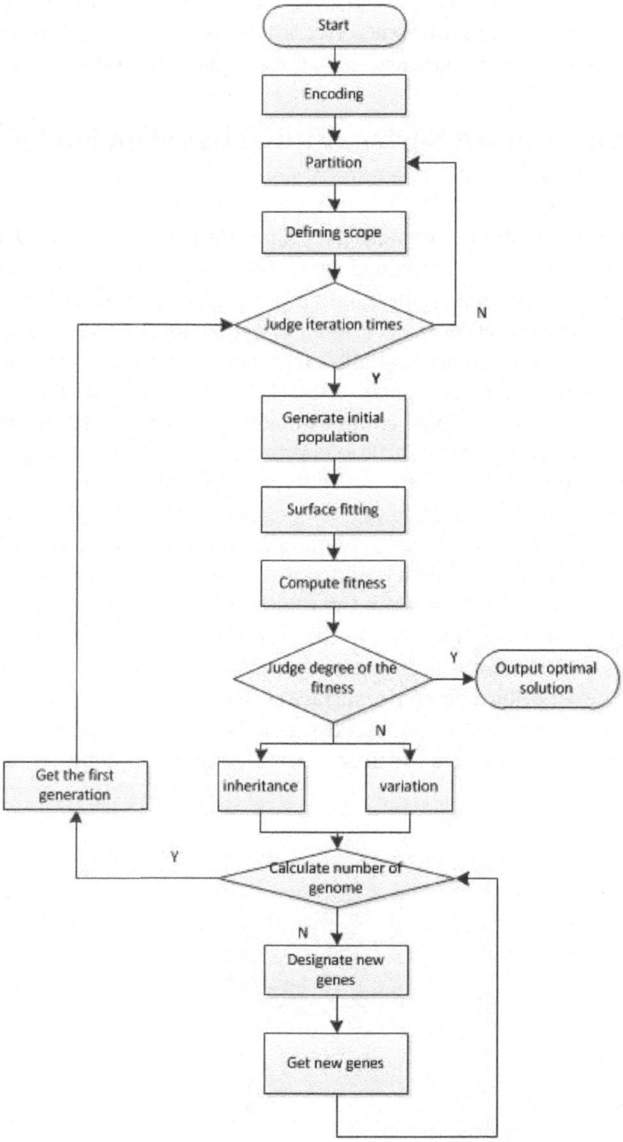


**Fig. 2.** Specific operational chart

Surface control points were reversed by applying least square method to (1) and (2), and initial population was produced with encoding control points. Then fitness function $f = {}^{1}/_{|(F(x,y,z))^2 + (y(x,y) - y)^2 + (x - x_i)^2 - \varepsilon|}$ took the maximum, namely, the final control points were obtained to fit the base surface.

At first encoding the reversed control points, conterminal rectangular areas were obtained by partitioning control points. In order to keep the surface continuous, the numerical values of top left corner and bottom right corner of each piece were recorded to confirm the refining boundary. The data points in sub-block composed the initial population. Then iteration times were judged whether beyond the given limit (the initial iteration time was zero). If beyond the limit, the procedure went back to last computer repeating partition operation. If not, zero generation population (including n genes) was randomly selected from the initial population (including N genes). And then fitness was obtained. The repeat stopped until the fitness reached the given standard so that the fitting surface was obtained. If the fitness didn't meet the requirement, the genetic and mutation operation would be performed on the data based on degree of fitness. Then the number of new generation genes was judged. If it was less than the number of initial population, a random selection was made from the remaining data to ensure n genes, and then the first generation could be got. Next, the procedures re-entered the circles until the stop requirement was met. Through the above steps, the optimal fitting surface could be obtained. The partition mode was shown in Fig. 3. Four interconnected boundaries were all expressed by B-spline curves. The inner scattered points were fitted by curved surface fitting, and the continuity between neighboring surfaces should be guaranteed.

Fig. 3 was an example of scattered point segmentation. The whole surface was divided into four adjacent zones, and it was supposed that the boundary of every zone had been computed. So the interpolation boundary curves and the curved surface approaching the inner scattered points could be fitted. The fitting precision of the curved surface was computed as fitness. The fitted curved surface meeting the demand was obtained after several iterations.

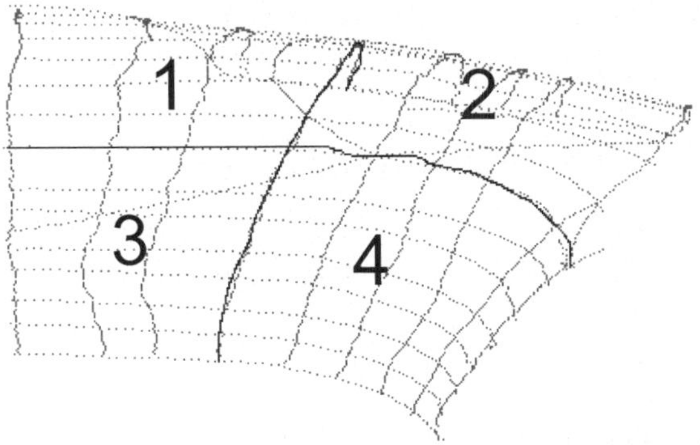


**Fig. 3.** Partition method

The paper took double cubic B-spline curved surface as example, the equation was:

$$p(u,v) = \sum_{i=1}^{m} \sum_{j=1}^{n} d_{i,j} N_{i,3}(u) N_{j,3}(v) \tag{3}$$

Where, knot vector U and V, and the control points of boundary $\{d_{i,1}\}_{i=1}^{m}$, $\{d_{i,n}\}_{i=1}^{m}$, $\{d_{1,j}\}_{j=1}^{n}$, $\{d_{m,j}\}_{j=1}^{n}$ were computed. Curved surface data points $\{Q_i\}_{i=1}^{L}$ were parameterized as $\{w_{i,} = (u_i, v_i)\}_{i=1}^{L}$. In order to solve the equation, let control points $d_{i,j}(i = 2,3,\ldots,m-1; j = 2,3,\ldots,n-1)$ approach to $\{Q_i\}_{i=1}^{L}$ in the way of least square method as follow:

$$f = \sum_{k}^{L}[Q_k - p(u_k, v_k)]^2 = \sum_{k}^{L}[Q_k - \sum_{i=1}^{m} \sum_{j=1}^{n} d_{i,j} N_{i,3}(u_k) N_{j,3}(v_k)]^2 = \min \tag{4}$$

$$\begin{aligned} f = \sum_{k}^{L}[Q_k - \sum_{i=1}^{m} d_{i,1} N_{i,3}(u_k) N_{1,3}(v_k) - \sum_{i=1}^{m} d_{i,n} N_{i,3}(u_k) N_{n,3}(v_k) \\ - \sum_{j=2}^{n-1} d_{i,n} N_{i,3}(u_k) N_{n,3}(v_k) - \sum_{j=2}^{n-1} d_{i,n} N_{i,3}(u_k) N_{j,3}(v_k) \\ - \sum_{j=2}^{n-1} d_{i,n} N_{m,3}(u_k) N_{j,3}(v_k) - \sum_{j=2}^{n-1} d_{i,n} N_{m,3}(u_k) N_{j,3}(v_k)]^2 \end{aligned} \tag{5}$$

$$\begin{aligned} \text{Let: } r_k = Q_k - \sum_{i=1}^{m} d_{i,1} N_{i,3}(u_k) N_{1,3}(v_k) - \sum_{i=1}^{m} d_{i,n} N_{i,3}(u_k) N_{n,3}(v_k) \\ - \sum_{j=2}^{n-1} d_{i,n} N_{1,3}(u_k) N_{j,3}(v_k) - \sum_{j=2}^{n-1} d_{i,n} N_{1,3}(u_k) N_{j,3}(v_k) \end{aligned} \tag{6}$$

$$\begin{aligned} & d_{(i-2)\times(n-2)+j} = d_{i,j}, t_{(i-2)\times(n-2)+j}(w_k) = N_{i,3}(u_k) N_{j,3}(v_k) \\ & (i = 2,3,\ldots,m-1; j = 2,3,\ldots,n-1) \end{aligned} \tag{7}$$

Thus, following equations could be obtained:

$$\begin{aligned} f &= \sum_{k=1}^{L}[r_k - \sum_{i=2}^{m-1} \sum_{j=2}^{n-1} d_{i,j} N_{i,3}(u_k) N_{j,3}(v_k)]^2 \\ &= \sum_{k=1}^{L}[r_k - \sum_{s=1}^{(m-2)\times(n-2)} d_s t_s(w_s)]^2 \end{aligned} \tag{8}$$

Where,

$$F = \begin{bmatrix} t_1(w_1) & \cdots & t_{(m-2)\times(n-2)}(w_1) \\ \vdots & \ddots & \vdots \\ t_1(w_L) & \cdots & t_{(m-2)\times(n-2)}(w_L) \end{bmatrix}_{L\times(m-2)\times(n-2)}$$

$$D = \begin{bmatrix} d_1 \\ \vdots \\ d_{(m-2)\times(n-2)} \end{bmatrix}_{(m-2)\times(n-2)\times 3} \qquad R = \begin{bmatrix} r_1 \\ \vdots \\ r_L \end{bmatrix}_{L\times 3}$$

(4) could be rewritten as $f = (R - FD)^T(R - FD) = \min$, to make the object function f minimum, its derivative on the control vertex D must be zero. So, the equation could be computed as follow:

$$D = F^{-}R \tag{9}$$

Supposed that the value of f is a, every surface was re-fitted as f1, f2, f3, f4 which were compared with a, and $f_n - a = \min, (n = 1,2,3,4)$ was viewed as the fitness to make the judgment. The inheritance and variation operation was applied to the control points in the curved surface. Thus, the optimal control points were selected to fit the n-th surface. If the time of iteration was more than 30 or $f_n - a > 0, (n = 1,2,3,4)$, the n-th surface was divided to four equal pieces. When dividing the surface, the top left corner and lower right corner coordinates of the curved surface should be recorded as the boundary points. Thus, the curved surface was easy to splice after fitting and the continuity of the whole surface could be ensured. The procedure above was repeated until the optimal curved surface was obtained. Taking the manikin in this paper as example, the fitting result was shown in Fig. 4:

The B-Spline curve fitting based on evolutionary algorithm was one kind of adaptive algorithm. It increased the fitting precision. It was objective that if the given precision had been reached, the procedure stopped outputting the surface result at once to reduce the computation cost.

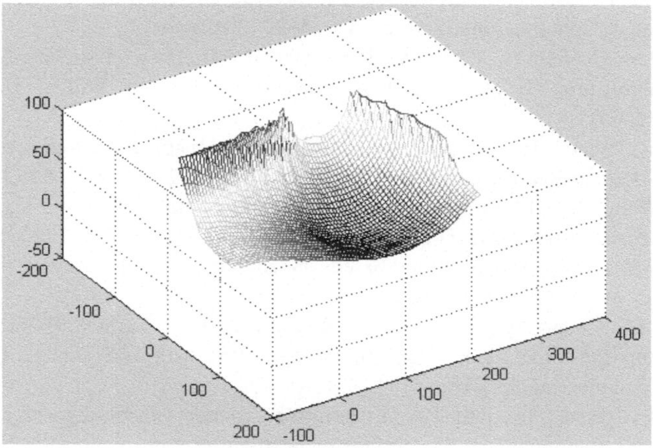


**Fig. 4.** Experimental result

# 4 Conclusion

This paper introduced the four boundaries fitting in B-Spline curved surface fitting. Firstly, boundaries of the curved surface were fitted using the theory of optimal curve. Then the control points in the curved surface were optimized and judged using evolutionary algorithm. The continuity of the curved surface should be guaranteed when improving the precision of the three dimensional surface, so the curved surface was partitioned based on the value of the precision. The optimal procedure was repeated until the fitted curved surface meeting requirements was obtained. The paper took data points which gathered from manikin as example, and the method above was applied to realizing the fitting process, with the result that the method was proved to be superior and feasible.

**Acknowledgement.** Supported by Science and Technology department of Liaoning province (201102009) Supported by Educational Commission of Liaoning province (2009A086, 05L071) Supported by Science and Technology Bureau of Dalian (2011A17GX075, 2005A10GX104, 2003A1GX172, 2010A16GX087).

## References

1. Peng, F., Zhou, Y., Zhou, J.: Algorithm of sculptured surface fitting based on interpolation and approximation. Journal of Engineering Graphics (4), 87–96 (2002)
2. Zhao, L., Ji, L., Wang, L., Huang, F.: Point cloud data surface reconstruction method in reverse engineering. Electronic Test (2), 19–22 (2010)
3. Zhou, J., Zhou, T.: The application of B-spline in design of automobile headlamp. China Illuminating Engineering Journal 11(4), 31–35 (2000)
4. Xin, B., Chen, J.: A survey and taxonomy on hybrid algorithms based on particle swarm optimization and differential evolution. Journal of Systems Science and Mathematical Sciences 31(9), 1131–1150 (2011)
5. Hu, P., Liu, M., Li, B.: Adaptive shape-preserving curve fitting on surfaces. Journal of Information & Computational Science (1), 15–23 (2012)
6. Gu, C., Pan, G., Shi, G., Chen, X.: Parameter identification of surface fitting based on genetic algorithm. Geomatics and Information Science of Wuhan University 34(8), 983–991 (2009)
7. Zhang, X., Wang, R., Song, L.: A novel evolutionary algorithm-seed optimization algorithm. Pattern Recognition and Artificial Intelligence 21(5), 667–682 (2008)
8. Ma, Z., Wang, L.: The influences of parameters changing in generic algorithms on the results of curve fitting. Ningxia Engineering Technology 1(8), 52–54 (2009)
9. Park, H.: B-spline surface fitting based on adaptive knot placement using dominant columns. Computer-Aided Design 43, 258–264 (2011)
10. Santos, J.C., Cheng, Y., Dias, M.M., Rodrigues, A.E.: Surface B-splines fitting for speeding up the simulation of adsorption processes with IAS model. Computers and Chemical Engineering 35, 1186–1191 (2011)
11. Gálvez, A., Iglesias, A., Puig-Pey, J.: Iterative two-step genetic-algorithm-based method for efficient polynomial B-spline surface reconstruction. Information Sciences 182, 56–76 (2012)

# Self-adaptive Optimization for Traffic Flow Model Based on Evolvable Hardware

Peng Ke[1,2], Yuanxiang Li[1], and Xin Nie[1]

[1]State Key Laboratory of Software Engineering, Wuhan University, Wuhan, China
[2] College of Computer Science and Technology,
Wuhan University of Science and Technology, Wuhan, China
Ke_peng_cn@yahoo.com.cn
{yuanxiangli,niexinzhulin}@qq.com

**Abstract.** BML model is a kind of cellular automata model, which is used to simulate and analyze the traffic system in the road network structure. The simulation and evolutionary optimization of the model implemented by software are optimized slowly and very low efficiently, so that it limited the ability enormously of traffic flow model to be used in some high real-time and high-speed occasion. In view of this question, we present the architecture of an EHW-based cellular automata model, a cellular automata model implemented in Evolvable Hardware platform and intended for the on-line evolution of the traffic flow model. And then it can adjust the rule of the traffic light signal according to the real-time state of traffic flow. After a careful analysis of the comparison result, the self-adaptive optimization for traffic flow model based on Evolvable Hardware is proved to be very useful and can meet the needs in the research and design of intelligent traffic system.

**Keywords:** Cellular Automata, BML, Evolvable Hardware, Software and Hardware Co-evolution, Intelligent Traffic System.

## 1 Introduction

At present, the cellular automata (Cellular Automata,CA) model [1], is widely used in the study of the theories of road traffic flow, is one of the theoretical models of traffic flow. Among these models, proposed by Biham,Middleton and Levine, the two-dimensional traffic flow of ideal models (BML model) [2] has become an important foundation for the study on characteristics of traffic flow in the urban road network structure. Based on BML model, Nagatani[3] studies on the asymmetric distribution of vehicles, while Fukui and others [4] study on the condition of vehicles' non-uniform travelling. In addition, experts have also made several other improvements, to simulate more sensible traffic behavior.

In recent years, the evolutionary algorithms (Evolutionary Algorithms,EA) [5] has become a hot spot in the study of engineering optimization and computer mind. Literature [6,7] apply the evolutionary algorithms to the optimization of the traffic flow

J. Lei et al. (Eds.): AICI 2012, LNAI 7530, pp. 255–262, 2012.

model and the analysis and forecast of urban traffic flow control. However, software-based traffic-flow simulation and evolutionary optimization are generally inefficient and thus greatly limit the real-time practical ability of traffic flow models. This is mainly because the evolution of cell populations is dealt wiht a serial approach, which consumes a lot of time during the evaluation phase of adaptive values of the evolutionary algorithm.

This article embarks from the angle of intrinsic parallelism of cellular automata model to conduct a study on the adaptive optimization of traffic flow model with the basis of evolvable hardware (Evolvable Hardware,EHW). Utilizing the field-programmable gate array (Field Programmable Gate Array,FPGA) greatly improves the efficiency of the evolution of cellular automata, and on the evolutionary hardware platform, expresses the BML model as a self-adaptive, reconfigurable hardware circuit module, and implements the on-line evolution of BML model through hardware-software co-evolution of configuration strings of target circuit . Meanwhile the models have been improved, through the self-adaptive control of traffic light signals based on real-time traffic conditions, which indicates the feasibility of being used for intelligent traffic control field .

## 2 BML Model of Parallel Evolution

### 2.1 BML Model

Urban traffic network is generally interconnected, with a lot of intersections, which forms a certain road network structure. Therefore, the two-dimensional model, the CA model applied to road network, comes into being. In 1992, under the guide of cellular automata theory, Biham, Middleton and Levine proposed a two dimensional model of urban traffic flow (BML model). In BML model, the urban transport network is represented in a two-dimensional square lattice matrix of NxN: n is the side length of the square lattice, each grid point has 3 states, represented by $S_{i,j}=\{0,1,2\}$. And 2 indicates that there is a vehicle travelling from North to West in the lattice point of Row I Column j; 1 indicates that a vehicle travelling from West to East in the point; 0 indicates that no vehicles in the point. Requirements are that in every odd time step the West-Eastern vehicle moves forward with a lattice point, while in every even time step the North-Southern vehicle move forward with a point. If the point front of a vehicle is occupied by other vehicles, even if the vehicle is obstructed to move forward at the time and thus make room for others, it can only remain motionless, not following.

Usually, the software simulation of the evolution of cellular automata is, on the whole matrix, to scan the cells one by one and then to calculate the state value for the next generation of cell populations. This method of serial execution cannot take full advantage of the inside characteristics of parallel evolution of cellular automata, restricts the evolutionary rate of cell populations instead. In some real-time, high-speed applications, the cellular automata should be evolved with the method of parallel computing.

## 2.2 FPGA-Based Implementation of BML Model

BML model stipulates that cells in the lattice points in accordance with the parity of the time step, respectively have the horizontal and vertical evolution, and each cell takes the three states {0,1,2}, with the neighbor radius of 1. Therefore, the evolutionary rules of the two-dimensional cellular automata can be defined as follows:

(1) East-west traffic Rule :

If $S_i^t$ =2, then $S_i^{t+1}$ =2

If $S_i^t$ =1, then $S_i^{t+1} = \begin{cases} 0 & S_{i+1}^t = 0 \\ 1 & S_{i+1}^t \neq 0 \end{cases}$

If $S_i^t$ =0, then $S_i^{t+1} = \begin{cases} 0 & S_{i-1}^t \neq 1 \\ 1 & S_{i-1}^t = 1 \end{cases}$

(2) North-south traffic Rule_ns:

If $S_i^t$ =1, then $S_i^{t+1}$ =1

If $S_i^t$ =0, then $S_i^{t+1} = \begin{cases} 0 & S_{i-1}^t \neq 2 \\ 2 & S_{i-1}^t = 2 \end{cases}$

if $S_i^t$ =2, then $S_i^{t+1} = \begin{cases} 0 & S_{i+1}^t = 0 \\ 2 & S_{i+1}^t \neq 0 \end{cases}$

According to the evolutionary rules defined above, a cell unit can be designed a three-input-output hardware module, with the enter value of the current state of the cell and its left and right neighbors, the output value of the next time status of the cell. Due to the evolutionary rules in a single cell is not the same in the odd and even time step, which should meet the the formula(1)and formula(2) respectively, so these two rules should be written in the two modules.When the cellular automata begin to evolve, these two modules will be executed in turn in a clock cycle of equidistant assignment.

As during the process of the evolution of cellular automata cell populations adopt the same evolutionary rules at the same time, the row-processing module can be designed to deal with each cell of the parallel processing of each row (when using Rule_ns, that is, each column of cells). Similarly, from row to row (row to column) is also parallel evolution, so we re-package the upper row (column) processing module. In this way, the cell populations in a two-dimensional cellular automata can have the parallel evolution of the state of each cell.

## 2.2 Experimental Results and Analysis

The development board of Xilinx's Virtex-II Pro series, the FPGA chip including model XC2VP30-6ff896 and the integrated development environment ISE9.1i are adopted to achieve the parallel evolution of the BML model. And the language of Verilog HDL is used to write the hardware functional module. The initial state of the cell populations is input by an external signal. Logically, the cell populations are arranged in a two-dimensional matrix. However, in the process of implementation, this lattice matrix is represented by row with a one-dimensional vector, each lattice point with three states. In this way, to set a 20 × 20 cell matrix evolution, a 800bit register variable is used to represent the cell populations. Once the initial state of the cell is input, the top-level module will, in accordance with the odd and even time step, decomposite the cell vector and extract all the single cells and their neighbors, then pass them to the underlying module to achieve the parallel evolution of all cells.

The following is a 20 × 20 cross road network simulation. Through the code input, composite and layout, the usage of system resources is shown in Table 1.

**Table 1.** FPGA system resource usage

| Resources | Number of occupied | Number of available | The proportion |
|---|---|---|---|
| Slice Flip Flops | 776 | 27,392 | 2% |
| 4 input LUTs | 2288 | 27,392 | 8% |
| occupied Slices | 1183 | 13,696 | 8% |
| bonded IOBs | 364 | 556 | 65% |
| GCLKs | 1 | 16 | 6% |

We use the Java language to perform the software simulation of cellular automata model under the same conditions. In terms of 20 × 20 cell lattice matrix, the statistic results of performance differences of cellular automata implemented in software and hardware platform can be shown in Table 2:

**Table 2.** 20x20 cell population evolution takes

| Resources | Elapsed time | Evolution of algebra |
|---|---|---|
| software | 94ms | 1000 |
| hardware | 0.062ms | 1000 |

From the above results, the evolutionary process of cellular automata by parallel hardware implementation is much faster than that by software implementations,which is a practical method of improving the evolutionary efficiency of cellular automata. In the process of optimizing the evolutionary rules of cellular automata by using evolutionary algorithms,the time evaluating the adaptive values of individuals will be greatly reduced. Therefore, in terms of the acceleration of models' self-adaptive optimization and the enhancement of real-time practical capability, it is an effective method to build the traffic flow model the evolution of hardware platform.

# 3 Evolution of Hardware-Based Adaptive Optimization

## 3.1 Improvement of the BML Model

The advent of the BML model has aroused widespread attention. How to improve and expand the BML model and take the various transport factors into consideration with the purpose of making the cellular automata traffic model closer to the actual situation of urban traffic,is the current direction we should strive for.

In the classical BML model, all the lattice points ,from its transport function, in fact are equivalent to intersections. Each intersection has traffic lights, which all turn to green lights at some point to the North-Southern vehicles, and then to the same rhythm transform the traffic lights. The equitime transform of traffic light signals is difficult to meet real needs. As example, in some main roads, driving traffic is significantly more than that in the small streets of the vertical direction, or the vertical direction can be regarded as the crosswalk. In this way, this kind of equitime traffic signal seetings may cause the slow moving traffic, even congestion in one direction, while in the other direction the sparse traffic, or very few pedestrians, will not take full advantage of the urban traffic resources and even increase the amount of crossings.

Therefore, during the improving process of BML model, the following adjustments are made:

① The boundary conditions of the cell is set to a fixed boundary value of 0;
② Through a configuration string, the calling sequence of the evolution of the rules Rule_we and Rule_ns is dynamically regulated.

This dynamic adjustment is achieved through a reconfigurable circuit configuration string. When the configuration string is downloaded to the FPGA, it can decide the calling sequence of the two hardware functional modules Rule_we and Rule_ns in order to achieve the distribution of traffic light signals under the traffic flow model. On the evolutionary hardware platforms, through evolving an optimal configuration string, and then setting the optimal time configuration of traffic light signals, to make all vehicles in a current area and time leave in a short time is functional as to balance the interests of the traffic in all directions, and even to ease traffic congestion.

## 3.2 Hardware and Software co Evolution Strategies

We use a genetic algorithm to optimize traffic flow model.The genetic algorithm is a computational model to simulate the Darwinian genetic selection and the biological process of evolution of natural elimination, and is also the optimal solution by simulating the search in natural evolutionary process. The algorithm embarks from an arbitrary initialization groups, through selection, crossover and mutation, then enters into the better and better region in the search space from generation to generation, until reaches the optimal solution point, which in essence is a efficient parallel search method for problem-solving.

The algorithm steps are as follows:

① Configuration string populations are randomly generated.
② Configuration strings are downloaded to the FPGA, together with the already well-generated evolution rule modules Rule_we and Rule_ns, generates the circuit.
③ Enter the initial state of the cell population.
④ Calculate the adaptive value of this configuration string, and returns the result. Continue to evaluate the next individual in the population of this generation.
⑤ When the individuals in the population of the generation are all evaluated, after a series of genetic manipulation, a new generation of configuration string population is generated.And then perform the step □, until meet the hereditary algebra to find the optimal solution..
⑥ The purpose of the genetic algorithm is to find the optimal timing configuration of the traffic light signals, under the current condition all the vehicle leave as soon as possible. Here, we do not consider the length of time of a car leaving the section, but all the vehicles instead.In other words, the time that the states of cell lattice matrix are all evolved into 0 is the shorter the better, so the evaluation function F of the adaptive value is defined as:

$$S_i^{t+1} = f\left(S_i^t, S_N^t\right) = 0 \tag{3}$$

In formula(3), f is the evolution rule of cellular automata, and S tN indicates the state of neighboring cells of cell i at time to find the t qualified in type(3)for all $1 \leq i \leq N \times N$, that is, the adaptive value of the configuration string. And the minimum individual of the t in the population is taken as the current optimal configuration string.

# 4 Experimental Results and Analysis

## 4.1 Model of Online Evolution

We still use the development board of the above experiments to achieve the online evolution of the traffic flow model. In this FPGA chip, the processor of IBM PowerPC405 model is integreted. Therefore, based on the co-evolution strategies of hardware and software, the improved BML model is optimized. Firstly, the traffic rules of traffic flows, namely the evolution rules of cellular automata Rule_we and Rule_ns, are represented in a hardware circuit module, and together with the registers of configuration string are packaged in the upper module and finally written to the FPGA chip. Subsequently, on the PowerPC processor the initialization of the system is completed to identify the optimal configuration string, according to the aforementioned system optimization algorithm.

The genetic manipulations of optimization algorithm set the parameters as follows:

**Table 3.** Set algorithm parameters

| Population size | Hybrid probabilistic | Mutation probability | Genetic algebra | Gene length |
|---|---|---|---|---|
| 20 | 0.3 | 0.1 | 100 | 4×32 |

Taking the simulation of 20 × 20 road transport network into account, we will set the size of configuration string as four 32-bit integer variables with the length of 128. That is to say, we evaluate every individual of a configuration string and make judgments based only on the cellular automata evolution of the result of generations of 128. If it still can not meet formula (3), then take t = 128. As a result, the individual's adaptive value is the worst.

### 4.2 Results and Analysis

We did the numerical experiments on condition after setting the traffic density in models as 0.3 and 0.4. The initial distribution of the traffic is generated randomly by the system. And to simulate the difference between the main road and secondary roads, the east-west direction and north-south direction of traffic flow ratio is set to 2:1 and 3:1 seperatively. For each case, there are 30 experiments. The optimization results of traffic flow model are as follows (Figure 1, Figure 2) as shown.

In the following figure, the imaginary lines indicate the time which all the vehicles in a region take to leave under the conditions of the equitime-assigned traffic light signals in the classical BML model, while the solid lines indicate the time which all the vehicle in a region take to leave after the self-adaptive optimization. The results obtained from the experiment show that ,when traffic flow of all directions in the sections is irregularly distributed or a direction for the sidewalk, through the dynamic adjustment of the traffic signals, the traffic of the region can achieve a faster flow.

This shows that, under such initial distribution of a car, any timing strategies of traffic signals is not able to significantly improve the efficiency of traffic flow in the region. For this kind vehicle distribution that can easily cause the slow traffic flow, even congestion, the further study of its mechanism is neccessary.

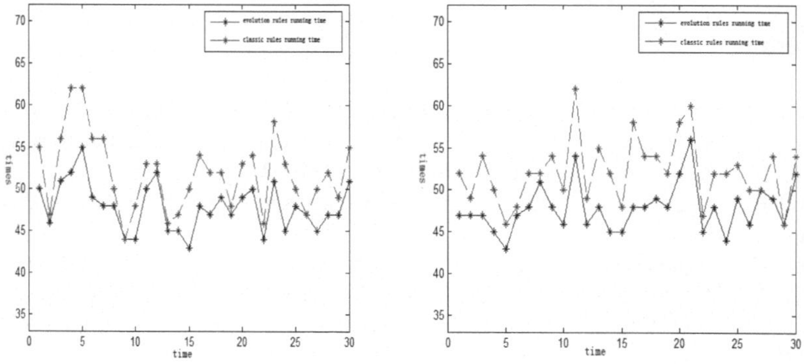


(a) vehicles 2:1 (b) vehicles 3:1

**Fig. 1.** Vehicle density is 0.3

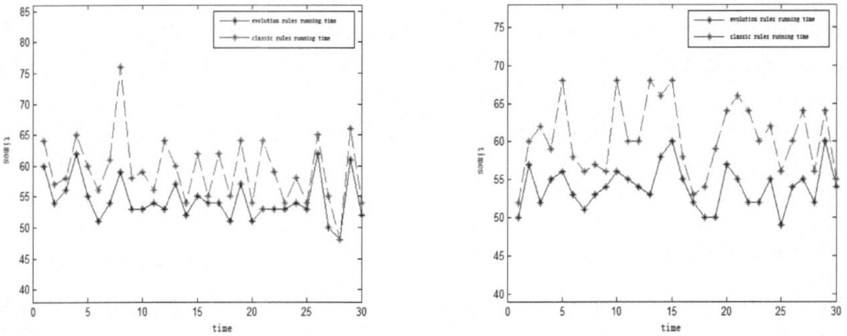


(a) vehicles 2:1 (b) vehicles3:1

**Fig. 2.** Vehicle density is 0.4

# 5 Conclusion

Based on the basic principles of the BML model, this article studies on the optimization and adaptive evolution of hardware-based traffic flow model applied to real-time intelligent transportation systems. Through the parallel implement of evolution rules of cellular automata on the FPGA, it greatly improves the speed of evaluating the adaptive value of traffic flow model during the evolutionary optimization. Subsequently, a kind of improvement strategies for the model has been proposed for the study of traffic lights on the impact of urban traffic flow. Experimental results show that the self-adaptive optimization method of the evolution of hardware-based traffic flow model has its broad application prospects in the development of the field of intelligent transportation systems.

# References

1. Wolfram, S.: A New Kind of Science. Wolfram Media, Champaign Illinois (2002)
2. Biham, O., Middleton, A.A., Levine, D.A.: Self-organization and a dynamical transition in traffic-flow models. Phys. Rev. A 46, R6124–R6127 (1992)
3. Nagatani, T.: Anisotropic Effect on Jamming Transition in Traffic-Flow Model. Phys. Soc. Japan 62, 2656–2662 (1993)
4. Fukui, H.O., Ishibashi, Y.: Flow of Cars Crossing with Unequal Velocities in a Two-Dimensional Cellular Automaton Model. Phys. Soc. Japan 65, 2514–2517 (1996)
5. Pan, Z., Kang, L., Chen, Y.: Evolutionary computation. Tsinghua University Press, Beijing (1998)
6. Guan, H., et al.: Traffic flow prediction based on hierarchical genetic optimized algorithm. In: 3rd International Conference on Innovative Computing Information and Control, ICICIC 2008 (2008)
7. Song, Y., Hu, W., et al.: Combined prediction research of city traffic flow based on genetic algorithm. In: 8th International Conference on Electronic Measurement and Instruments, ICEMI, pp. 3862–3865 (2007)

# An Improved Decision Tree Algorithm Using Rough Set Theory in Clinical Decision Support System

Qingshan Li, Jian'guo Zhang, and Hua Chu

Software Engineering Institute, Xidian University, 710071 Xi'an, China
qshli@mail.xidian.edu.cn

**Abstract.** In the Clinical Decision Support System (CDSS), over-fitting phenomenon may appear when decision tree algorithm was used. For this problem, this paper will make use of the Rough Set theory to the training set for attribute reduction, the decision tree built by using the decision tree algorithm was used to predict the test data. In this paper, 46 copies of coronary heart disease clinical data were used to test the improved algorithm. Comparing the accuracy of the algorithm and the improved algorithm, we can know that, the improved algorithm has a better recognition rate for the diagnosis of coronary heart disease, and effectively solves the over-fitting phenomenon in the Decision Tree Algorithm.

**Keywords:** Clinical Decision Support System (CDSS), Rough Set (RS), Decision Tree Algorithm.

## 1 Introduction

Compared with the neural network classification, Decision Tree Algorithm is easy to extract an explicit rules, the amount of calculation is relatively small, does not require any background knowledge, important decision attributes can be displayed and has a higher classification accuracy rate, etc. However, data with noise will affect the classification efficiency and the accuracy of the results. The most commonly used decision tree construction algorithm doesn't consider the impact of redundant information in the training data set, which will result in over-fitting and the classification accuracy rate will reduce.

To prevent over-fitting, usually two ways can be used to solve the problem. One is to prune the resulting decision tree, and its essence is to eliminate the anomalies, noise and redundant information in the training set. Another is to use the rough set theory to reduce training set attributes, remove the redundant information. As building the decision tree needs a process, the advantages of the second method would be obvious reflected when the training set is relatively large. Because the second method reduced the attributes of the training set, and got rid of duplicate samples, which can greatly accelerate the process of constructing a decision tree.

This paper mainly uses the attribute reduction algorithm of the rough set theory to preprocess the clinical data, and solve the over-fitting problem. This paper is organized as follows: Section II describes the related theory. Section III details of

J. Lei et al. (Eds.): AICI 2012, LNAI 7530, pp. 263–270, 2012.

attribute reduction and decision tree construction issues. Section IV presents case studies. Section V describes the conclusion.

## 2 Related Theory

### 2.1 Rough Set Theory

Rough Set Theory was a theory of data analysis and processing, founded by Polish scientists Z. Pawlakin 1982[1]. The first International Conference about the Rough Set Theory held in Poland in 1992, the rough set theory was listed as a new computer science research by ACM in 1995.

At present, the research of rough set theory mainly in three aspects:

In theory, the use of abstract algebra to study rough set algebra space, the particular algebraic structure [2].The use of topological to describe the rough space [3]. There is a study of the combination of rough set theory and other methods of soft computing or artificial intelligence methods, for example, fuzzy theory, neural networks, support vector machines, genetic algorithms[4]. For the limitations of the framework in the classical rough set theory, broadens the framework of rough set theory, extends the classical rough set theory built on the equivalence to the similar relationship even the general relations of rough set theory [5].

In application, rough set theory has been applied in many fields, Clinical diagnosis[6];Fault diagnosis in power system and other industrial processes[7];Prediction and control[8];Pattern recognition and classification[9]; Learning and data mining[10]; Image Processing;other.

Definition 1 A information system S = (U, A, V, f), U is the non-empty finite set of the sample; A is the non-empty finite set of attribute; V is Range set of the attribute; f is information function. If A=C∪D, C∩D=Φ, C is condition attribute set, D is decision attribute set, then call S the decision table. Expressed with S=(U,C∪D) or S=(U,C∪{d}), d is the single decision attribute.

Definition 2 In the decision table S, if B∈C∪D, the Indistinguishable relation of B on the set U is: IND(B)={(x, y)∈U×U | a ∈B:f(x, a)=f(y, a)}. IND(B) represent that object x and y are indistinguishable on the sub-set B of attribute set C∪D.U/IND(B) represent all the equivalence classes families composed by equivalence classes, according to equivalence relation IND(B).

Definition 3 For X⊆U, B⊆C, the lower approximation of X under B was defined: B(X)={Y | (Y∈U/IND(B))∩(Y⊆X)}; the upper approximation of X under B was defined: B(X)={Y | (Y∈U/IND(B))∩(Y∩X =Φ)}. the upper approximation of X under B was called B positive domain of X, POSB(X)=B(X), represents the set of objects in set U which can be included in set X, according to knowledge B.

Definition 4 The dependence between condition attribute set C and decision attribute set D was defined: kc(D)=Card(POSc(D))/Card(U),Card() is the cardinality, POSc(D)=U{C(X): X∈IND(D)}.

Definition 5 The importance of condition attribute c was defined: Sgf(c, C, D)= kc(D)-kC-{c}(D). Sgf(c, C, D)∈{0,1}, if Sgf(c, C, D)=0,attribute c is redundant, can be omitted for the decision attribute set D. Otherwise, can't be omitted.

## 2.2 Decision Tree Algorithm

Decision tree algorithm was originally proposed by Quinlan, is a widely used data mining algorithm and machine learning algorithm. Decision tree use a "divide and conquer" strategy. It will change a complex problem into simpler problems and re-use this tactic to solve sub-problems, and then complex issues will be collapsed.

Here is the basic algorithm to generate the decision tree by the training samples:

```
Create node N;
If samples belong to category C then
Return N as leaf node, marked with category C;
If attribute_list is none, then
Return N as leaf node, marked the most common
category in samples;
Select the attribute with the highest
information gain among attribute_list;
Mark node N with test_attribute;
For each known value a_i among test_attribute;
Group out a branch with the condition of
test_attribute=a_i by node N;
S_i is the sample set of test_attribute=a_i in
samples;
If s_i is none , then
Increase a leaf, marked the most common category
in samples;
Else increase a node returned by
Generate_decision_tree(s_i,
attribute_list-test_attribute)
```

In the formation of the decision tree, the most important part is the choice of splitting attributes. One more common method is to use the information gain to select the test attribute.

According to the definition of information theory, suppose S to be the set of s data samples. Assume that class label attribute has m distinct values, define m different categories $C_i$ (i=1, 2…m).Suppose $s_i$ is the sample number of category $C_i$. The desired information of classification is given by the follow formula:

$$I(S_1, S_2, ..., S_m) = -\sum_{i=1}^{m} p_i \log_2(P_i) \quad \textbf{(1)}$$

$p_i$ is the probability of any sample belongs to category $C_i$, and estimate with $s_i/s$. Note that the base of the logarithmic function is 2, because the information is binary.

Let attribute A have V different values{$a_1$,$a_2$…$a_v$}.and can divide set S into V sub-sets by attribute A {$s_1$,$s_2$…$s_v$ }; $s_j$ includes such samples in set S, their values are $a_i$ on attribute A. Suppose select attribute A as the properties of this classification. These sub-sets correspond to the branches grow out from the nodes included in set S.

Suppose $s_{ij}$ is the number of category $C_i$ in sub-set $s_j$ . The entropy of dividing the sub-set according to attribute A is given by the follow formula:

$$E(A)=\sum_{i=1}^{v}\frac{S_{1j}+...+S_{mj}}{S}(S_{1j},...,S_{mj}) \quad (2)$$

Item $\frac{S_{1j}+...+S_{mj}}{S}$ is the weight of the j-th sub-set, and equals to (The value of A is $a_j$) sample number of sub-set divided by the total number of set S. The smaller the entropy value, the higher the purity of the subset divided. The given sub-set $S_j$,

$$I(S_{1j},...,S_{mj})=-\sum_{i=1}^{m}p_{ij}\log_2(P_{ij}) \quad (3)$$

Among these, $p_{ij}=\frac{S_{ij}}{|s_j|}$ is the probability that the sample belongs to category $C_i$.

The information of encoding obtained on the branch A, the information gain of node is:

$$Gain(A)=I(S_{1j},...,S_{mj})-E(A) \quad (4)$$

## 3 Improved Decision Tree Algorithm

The quality level of the data after pretreatment has a significant influence on the accuracy and performance, and data validity of the model. The original indicators of coronary heart disease in Table 1:

**Table 1.** Eight indicators of coronary heart disease

| No | Attribute Name | Data Type | Attribute Value |
|---|---|---|---|
| $X_1$ | Age | Value | |
| $X_2$ | History of hypertension | Classification | 1 (Has), 0 (No) |
| $X_3$ | History of hypertension | Classification | 1 (Has), 0 (No) |
| $X_4$ | Smoke | Classification | 1 (Has), 0 (No) |
| $X_5$ | History of high cholesterol | Classification | 1 (Has), 0 (No) |
| $X_6$ | Animal fat intake | Classification | 1 (Has), 0 (No) |
| $X_7$ | Body Mass Index (BMI) | Value | |
| $X_8$ | Type A personality | Classification | 1 (Has), 0 (No) |
| Y | Vascular disease | Classification | 1 (Healthy), 0 (Pathological) |

According to Table 1 , We can put all the records as the domain A={K1,K2,…,K40}, Any column indicates a property ,constitutes a partition of elements on the domain, In the division of each class have the same properties. and the attributes can be divided into 2 categories, one is condition attribute, like X1, X2,…, X8, the other is decision attribute, like the final column Y: is coronary heart disease or not. We will consider all the basic knowledge: X1, X2,..., X8 are necessary or not? If we remove X1,the basic knowledge, in the knowledge system, and the knowledge system will be A/(R-R1)={K1, {K2,K10,K17}, {K7,K13}, {K14,K25},{K19,K31},{K28,K30}...,K40} and the union of these subset. If we use the new knowledge system to express"1", the upper approximation is:{K1, K3, K4, K5, K6, K8,K9}, the lower approximation is:{K1-K10, K13, K14, K17, K19, K28 ,K30, K31}, the upper approximation and the lower approximation are not the same with the original in the knowledge system, and the upper approximation of "0" is {K11,K12,K15, K16,K18,K20,K21,K22,K23,K24, K25,K26, K27,K29,K32, K40}, the approximations of "0" are changed, So removing the attribute "X1" will affect the knowledge expression, and can't be removed. Similarly, "X2" attribute can't be removed too. If we remove the X3, the basic knowledge in the knowledge system, and the knowledge system will be A/(R-R3) = {{K1, K19}, K2, K3,..., K40} and the union of these subset. If we use the new knowledge system to express"1", the upper and lower approximations are: {K21, K22,...,K40},they are the same with the originals. The upper and lower approximations of "0" are {K1,K2,...,K20},they are the same with the originals, This shows that removing property "X3" will not change the knowledge expression of coronary heart disease, so that attribute "X3" is redundant and can be deleted. For attributes "X4","X5","X6","X7","X8", "X7" is redundant, and can be removed. Finally, we get the simplified knowledge base X1, X2, X4, X5, X6, X8.

C4.5 algorithm is an extension of ID3 algorithm. This method uses the gain ratio instead of gain. Gain ratio Gain Ratio was defined:

$$Gain\_ratio(A) = Gain(A) / Split\_Infox(A) \tag{5}$$

$$Split\_Infox(A) = SUM\left(\frac{|T|}{|T_i|}\right)LOG\left(\frac{|T_i|}{|T|}\right) \tag{6}$$

Choose the attribute which has the greatest gain ratio as a new tree node, and create branches for each attribute value, divide the training set into subsets, each branch contains a subset, and repeat this process for each branch until all the training data under a node belong to the same class or there is no remaining property to be divided.

# 4 Experimental Analysis

## 4.1 Experimental Data

Experimental data from 54 patients with coronary heart original sample. Diagnosis of coronary heart disease in eight indicators shown in Table 1. Eliminate inconsistencies and duplication of the original sample concentration of the sample, got 46 samples, take one of the 40 sample data as training set(see Table 2,Y=0:non-coronary heart disease, Y=1: coronary heart disease);The remaining six samples as test data set (see Table 3).

**Table 2.** Training set samples of coronary heart disease

| 1 | 2 | 3 | 4 | 5 | 6 | 7 | 8 | Y | 1 | 2 | 3 | 4 | 5 | 6 | 7 | 8 | Y |
|---|---|---|---|---|---|---|---|---|---|---|---|---|---|---|---|---|---|
| 2 | 0 | 0 | 0 | 0 | 0 | 1 | 0 | 0 | 2 | 1 | 1 | 1 | 0 | 1 | 2 | 1 | 1 |
| 2 | 1 | 0 | 1 | 0 | 0 | 1 | 0 | 0 | 3 | 0 | 0 | 1 | 1 | 1 | 2 | 1 | 1 |
| 3 | 0 | 0 | 1 | 0 | 1 | 1 | 1 | 0 | 2 | 0 | 0 | 1 | 1 | 1 | 1 | 0 | 1 |
| 3 | 0 | 1 | 1 | 0 | 0 | 2 | 1 | 0 | 3 | 1 | 1 | 1 | 1 | 1 | 3 | 1 | 1 |
| 3 | 0 | 1 | 1 | 1 | 0 | 1 | 0 | 0 | 2 | 0 | 0 | 1 | 0 | 0 | 1 | 1 | 1 |
| 2 | 0 | 0 | 0 | 0 | 0 | 1 | 1 | 0 | 2 | 0 | 1 | 0 | 1 | 1 | 1 | 1 | 1 |
| 1 | 0 | 0 | 1 | 0 | 0 | 1 | 0 | 0 | 2 | 0 | 0 | 1 | 0 | 1 | 1 | 0 | 1 |
| 1 | 0 | 1 | 0 | 0 | 0 | 1 | 1 | 0 | 2 | 1 | 1 | 1 | 1 | 0 | 1 | 1 | 1 |
| 1 | 0 | 0 | 0 | 0 | 0 | 2 | 1 | 0 | 3 | 1 | 1 | 1 | 0 | 1 | 1 | 1 | 1 |
| 4 | 1 | 0 | 1 | 0 | 0 | 1 | 0 | 0 | 3 | 1 | 1 | 1 | 1 | 0 | 1 | 1 | 1 |
| 3 | 0 | 1 | 1 | 0 | 0 | 1 | 1 | 0 | 3 | 0 | 1 | 0 | 0 | 0 | 1 | 0 | 1 |
| 1 | 0 | 0 | 1 | 0 | 0 | 3 | 1 | 0 | 2 | 1 | 1 | 1 | 1 | 0 | 2 | 1 | 1 |
| 2 | 0 | 0 | 1 | 0 | 0 | 1 | 0 | 0 | 3 | 1 | 0 | 1 | 0 | 1 | 2 | 1 | 1 |
| 1 | 0 | 0 | 1 | 0 | 0 | 1 | 1 | 0 | 3 | 1 | 1 | 1 | 1 | 1 | 2 | 0 | 1 |
| 3 | 1 | 1 | 1 | 1 | 0 | 1 | 0 | 0 | 4 | 0 | 0 | 1 | 1 | 0 | 3 | 1 | 1 |
| 2 | 1 | 1 | 1 | 1 | 0 | 2 | 0 | 0 | 3 | 1 | 1 | 1 | 1 | 0 | 3 | 1 | 1 |
| 3 | 1 | 0 | 1 | 0 | 0 | 1 | 0 | 0 | 4 | 1 | 1 | 1 | 1 | 0 | 3 | 0 | 1 |
| 2 | 1 | 1 | 0 | 1 | 0 | 3 | 1 | 0 | 3 | 0 | 1 | 1 | 1 | 0 | 1 | 1 | 1 |
| 2 | 0 | 1 | 0 | 0 | 0 | 1 | 0 | 0 | 4 | 0 | 0 | 1 | 0 | 0 | 2 | 1 | 1 |
| 2 | 0 | 0 | 1 | 1 | 0 | 1 | 1 | 0 | 1 | 0 | 1 | 1 | 1 | 0 | 2 | 1 | 1 |

**Table 3.** Comparison of diagnostic test set samples

| NO | $X_1$ | $X_2$ | $X_3$ | $X_4$ | $X_5$ | $X_6$ | $X_7$ | $X_8$ | Y | Decision results | | | |
|---|---|---|---|---|---|---|---|---|---|---|---|---|---|
| | | | | | | | | | | ID3 | C4.5 | RS+ID3 | RS+C4.5 |
| 1 | 2 | 0 | 1 | 1 | 0 | 1 | 2 | 1 | 1 | 1 | 1 | 1 | 1 |
| 2 | 2 | 1 | 1 | 1 | 0 | 0 | 2 | 1 | 1 | 1 | 0 | 1 | 1 |
| 3 | 2 | 1 | 0 | 1 | 0 | 0 | 1 | 1 | 1 | 1 | 0 | 1 | 1 |
| 4 | 3 | 1 | 1 | 0 | 1 | 0 | 3 | 1 | 1 | 1 | 0 | 1 | 1 |
| 5 | 2 | 0 | 0 | 0 | 0 | 0 | 1 | 0 | 0 | 0 | 0 | 0 | 0 |
| 6 | 2 | 0 | 0 | 0 | 0 | 0 | 2 | 0 | 0 | 0 | 0 | 0 | 0 |

## 4.2 Analysis of Test Results

First, the 40 training sets were constructed using ID3 and C4.5 decision tree algorithm, the results are shown in Fig 1 and Fig 2 Secondly, the 40 training sets using rough set attribute reduction method of reduction, elimination of redundant features of the $X_3$ and $X_7$ two properties, then delete the duplicate samples, leaving only 33 samples. Of the reduction after the 33 training sets were constructed using ID3 and C4.5 decision tree algorithm, the results are shown in Fig 3 and Fig 4 The four rules of decision tree generated for the six test sets, respectively, to predict, making the results as shown in Table 3.

Use the training set after eliminating redundant attributes by attribute reduction to conduct the decision tree can avoid over-fitting phenomenon, more importantly, to help experts in the diagnosis of coronary heart disease may be considered excluded from family history of hypertension and body mass index two redundant targets . From the resulting decision tree can clearly see that animal fat intake ($X_6$), high cholesterol history ($X_5$), A-type personality ($X_8$) and age ($X_1$) factors for coronary heart disease and other indicators of the importance of the law, this is basically in keeping with the medical knowledge.

Note: ⊘on-coronary heart disease ◎ronary heart disease

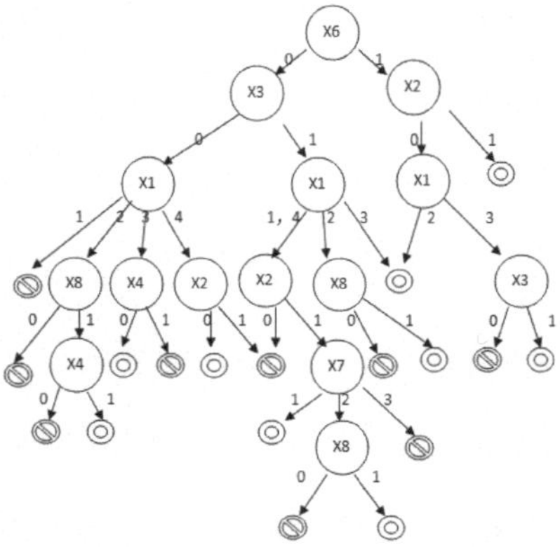


**Fig. 1.** ID3 Generated decision tree

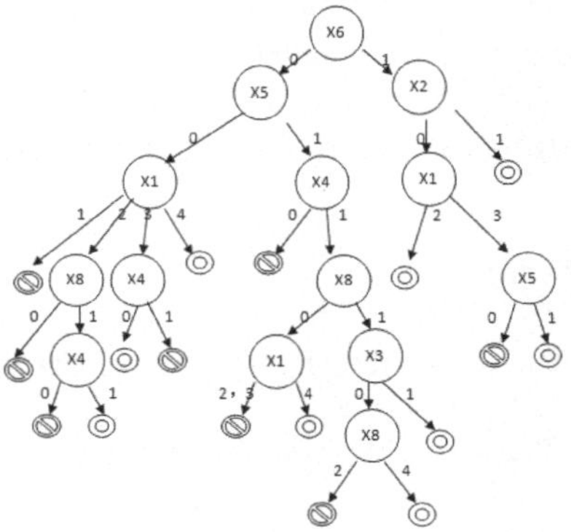


**Fig. 2.** C4.5 Generated decision tree

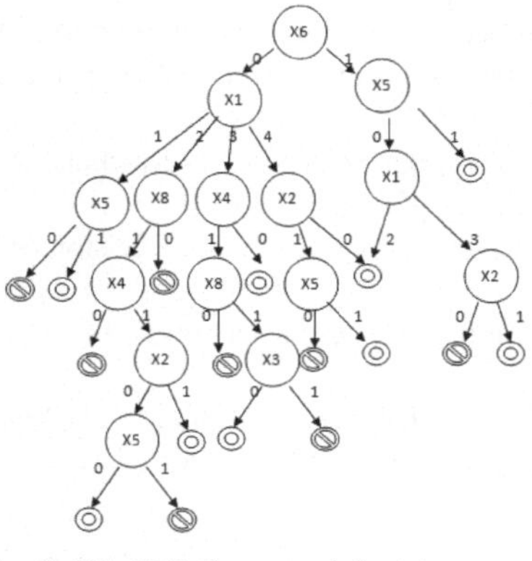


**Fig. 3.** RS+ID3 Generated decision tree

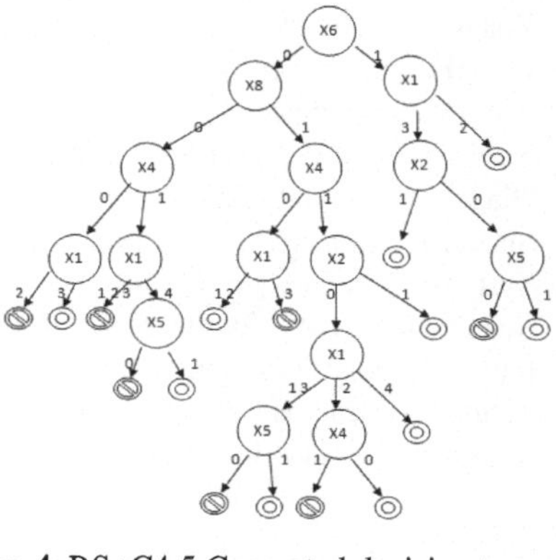


**Fig. 4.** RS+C4.5 Generated decision tree

# 5 Conclusion

In practice, the diagnosis of a disease indicators (attributes) are many, and some indicators are redundant or secondary, directly use of the training set contains redundant information to construct a decision tree in addition to increased storage and computing burden, but also may affect the accuracy of decision tree classification.

The rough set theory can ensure information integrity in the context of the sample set attribute reduction and data filtering to remove the interference of irrelevant data and decision-making, and to construct a decision tree decision tree algorithm is more compact.

In this paper, there are cases of coronary heart disease, the test introduces rough set theory and decision tree algorithm to the diagnosis of coronary heart disease, has established a number of effective rules for the interpretation of these risk factors in coronary heart disease.

**Acknowledgments.** This work is supported by the National High Technology Research and Development 863 Program of China (2012AA02A603)、 the National Natural Science Foundation of China (61173026),the Defense Pre-Research Project of the 'Twelfth Five-Year-Plan'of China(513***301), and the Fundamental Research Funds for the Central Universities of China.

## References

1. Pawlak, Z.: Rough sets. International Journal of Information and Computer Science 11(5), 314–356 (1982)
2. Zhu, F., He, H.: The Axion of Rough Set. Chinese Journal of Computers 23(3), 330–333 (2000)
3. Chen, D., Zhang, W.: Rough Set and Topological space. Journal of Xi'an Jiao Tong University 35(12), 1313–1315 (2001)
4. Chena, W.C., Changb, N.-B., Chen, J.: Rough Setbased hybrid fuzzy-neural controller design for industrial wastewater treatment. Water Research (37), 95–107 (2003)
5. Zhang, W., Wu, W.: Theory and method of RougSet. Science Publication, Beijing (2000)
6. Liu, H., Qiu, T.: Knowledge discovery and its application in clinical medicine. Journal of Biomedical Engineering 21(4), 677–680 (2004)
7. Yuan, X., Zhao, Z., Qu, L.: On the Application of Rough Set Theory in Mechanical Fault Diagnosis. Journal of Xi'an Jiaotong University 35(9), 954–957 (2001)
8. Shena, L., Loh, H.T.: Applying rough sets to market timing decisions. Decision Support Systems (37), 583–597 (2004)
9. Swiniarskia, R.W., Skowron, A.: Rough set methods in feature selection and recognition. Pattern Recognition Letters (24), 833–849 (2003)
10. Zhang, W., Xue, H., Zhang, H., Peng, W.-X.: A New Algorithm for Data Mining Based on Rough Set Theory. Journal of Northwestern Polytechnical University 20(3) (2002)

# An AHP-Based Assessment Model for Clinical Diagnosis and Decision

Qingshan Li, Lihang Zhang, and Hua Chu

Software Engineering Institute, Xidian University, 710071 Xi'an, China
qshli@mail.xidian.edu.cn

**Abstract.** There are many scholars trying to diagnose the disease using intelligent computers and computer networks, in order to meet the growing needs of patients. Most of them diagnose disease based on the patients' symptoms, the collected medical data and some algorithms, but they can't figure out the probability of every possible disease. This paper presents an assessment model by using AHP method which can evaluate these possible diseases. The assessment model can calculate the correlation between each disease and every possible symptom by the collected medical data. According to the correlation and the actual severity of symptoms, we can evaluate the exact probability of every possible disease.

**Keywords:** Clinical Decision Support System (CDSS), Analytic Hierarchy Process, Assessment Model.

## 1 Introduction

With economic development, the current medical standards can't meet everyone's needs. The researchers begin to use the computers and computer networks to treat patients. Now there are two major methods, one is for a certain disease, doctors make tables into the computer according to their own knowledge and then the patients use the tables to treat themselves; another is for some symptoms, computer use some algorithms to give several possible diseases according to the symptoms. Both the two methods have their own limitations. The former does not take into account all the patient's current symptoms such as the symptoms caused by the psychological; the latter does not take into account the severity of every symptoms and can't give the most likely disease.

This paper presents an evaluation model based on AHP. The model collect the relevant information of diseases and symptoms firstly, then calculate the correlation which is reflect by weight between diseases and symptoms by the AHP method, finally, the model according to these weights and current situations evaluates the treatment decisions given by the computers.

The main contributions are: this paper presents an assessment model which can evaluate the treatment decisions given by the computers and use AHP method in medical fields. Rest of this article is organized as follows. Section 2 describes the involved theories. Section 3 introduces detail issues about the AHP-based assessment

J. Lei et al. (Eds.): AICI 2012, LNAI 7530, pp. 271–278, 2012.

model. Section 4 is testing and analyzing. Section 5 concludes the paper and look to the future work.

# 2 Relevant Theories

## 2.1 Delphi

Delphi method is an intuitive forecasting technique and is a development of expert meetings prediction method; its core is carried out several rounds of an anonymous letter of inquiry to seek the views of experts. The comments which are collected by the prediction team and evaluation team will be sent to every expert as a reference and the experts will analyze and judge them, then out forward a new view. The conclusions and solutions will be more consistent and reliable after repeated so many times. In order to ensure the quality of expert advice, in most cases it need to set up a group which goal is to assess and predict according to the size of the project. The group's main tasks are: developing project evaluation; forecasting topics; establishing the communication and prediction table; selecting the experts and completing the inquiry form; collecting and analyzing the experts' advice and results and so on.

## 2.2 Analytic Hierarchy Process

AHP (Analytic Hierarchy Process), was formally proposed by the Saaty in the mid-1970s, is a combination of qualitative and quantitative, systematic, hierarchical methods of analysis. Its application has over economic, management, distribution of energy policy, behavioral science, military, transportation, agriculture, education, human resources, health, environment and other fields.

AHP is a flexible and practical multi-criteria decision making methods when quantitative analyzing qualitative issues. Its characteristics are that it can make varieties of factors of the complex problems orderly, hierarchy and principled; it can effective combine the experts' opinion and the analysts' objective results based on some objective reality of subjective judgments structure (mainly pairwise comparison); it can make a quantitative description about the importance between the two elements in one level. Then, using mathematical calculates the weights to reflect the relative importance of each level's elements. Lastly, calculate the relative weights of all the elements and sort them through the total order of all levels.

# 3 Assessment Model

The proposed AHP-based evaluation model includes three parts: the first part screens the composition of disease by using Delphi method and establish pairwise comparison matrix of symptoms; the second part adjust the matrix through AHP method and calculates the weight of each case; the third part obtains the results of the assessment based on the weight and the actual situation.

### 3.1 Gather Information and Build Pairwise Comparison Matrix

Assessment model presented in this paper need to collect the information consists of two parts, one is the number of symptoms of each disease and the name of symptoms; another is the degree of importance between every two symptoms.

Collect information using Delphi method. Firstly, put the data of disease and symptoms which are collected by us into tables. Secondly, send these tables to 25 experts and receive these after they delete part of symptoms based on their medical knowledge and clinical experience. Thirdly, collate data follow the consensus of experts and then redistributes them. After 3-4 rounds of surveys, experts' opinion will be close to agreement and the rest symptoms can be used to build the matrix.

Each disease has a comparison matrix and table 1 shows a simple comparison matrix.

**Table 1.** A simple comparison matrix

| | A | B | C |
|---|---|---|---|
| A | 1 | 3 | 1/2 |
| B | 1/3 | 1 | 1 |
| C | 2 | 1 | 1 |

A, B, C representative symptoms and the value of matrix is made by the current assessment scale. The assessment scales of AHP have five levels include equally important, slightly important, fairly important, very important and absolutely important. The levels can be quantified into 1, 3, 5, 7, 9. Table 2 shows the 1-9 scale meaning.

**Table 2.** The scale of AHP method

| Scale | Define |
|---|---|
| 1 | equally important |
| 3 | slightly important |
| 5 | fairly important |
| 7 | very important |
| 9 | absolutely important |
| 2, 4, 6, 8 | the median value |

At home and abroad there are 1-9scale、1-5scale、1-15scale(1, 5, 8, 11, 15)、$x^2$scale(1, 9, 25, 49, 81)、9／9—9／1scale、10／10-18／2scale、the index scale and so on. From a psychological point of view, too much grading would be beyond people's ability to judge, not only increasing the difficulty of making judgments, and easily providing false data. Among the many experimental psychology research shows that ordinary people can identify the properties of the level and the number of things generally between 5 and 9(7$\pm$2 rule) when comparing a certain property of a group at the same time. Saaty and others compare correctness of the results from different scales by experimental methods and show that the most appropriate scale is 1-9 scale after 26 experimental methods.

After choosing an assessment scale, the pairwise comparison matrix can be built by Delphi and then these matrixes can be used to calculate the weights of symptoms.

## 3.2 Parameter Calculation and Adjust the Consistency Matrix

This section includes two parts, one is to calculate the weights of symptom, the value of C.R. and so on, and another is to adjust the matrix when the matrix doesn't meet the consistency.

### 3.2.1 Weights and Parameters of the Calculation

At present the calculation method of AHP used mainly in the following categories [7]: and method (arithmetic average), root method (geometric average), Eigenvalue method (EM), Least-squares method (LSM) and Logarithmic Least square (LLSM). In this model, the and method, the root method and the Eigenvalue method will be used to calculate the weights considering that the result of the LSM and LLSM are same and LSM has not yet entered the application phase as nonlinear optimization problems. Because and method and root method are relatively simple, they can be widely used in some of the less demanding of precision items. Eigenvalue method can be widely used in large-scale projects as it has high accuracy and the use of the computer calculation is not too large.

Comparison matrix is generally impossible to have exactly the same with user thought due to the complexity of objective things, the limitations and diversity of subject knowledge. Saaty sum up the steps of test the consistency through theoretical research and social practice:

1) Calculate the consistency index (C.I).

$$C.I. = \frac{\lambda_{\max} - n}{n - 1} \tag{1}$$

In the formula: n representative the order of the comparison matrix and $\lambda_{\max}$ representative the largest eigenvalue of matrix.

2) Find the average random consistency index R.I.

According to Wharton School and Dark Ridge National Laboratory studies, the reciprocal matrix which produced by assessment scales have different C.I. value under different order and the value increase when the order of matrix increase. So we usually use R.I. not by calculate their own, but by the table which has been calculated before.

**Table 3.** R.I. values

| Order n | 1 | 2 | 3 | 4 | 5 | 6 | 7 | 8 | 9 | 10 | 11 |
|---|---|---|---|---|---|---|---|---|---|---|---|
| R.I. | 0 | 0 | 0.52 | 0.89 | 1.12 | 1.26 | 1.36 | 1.41 | 1.46 | 1.49 | 1.52 |

The computer choose 17 value by random from 1-9 scale and make a reciprocal matrix by fill with a triangular matrix of n-th order of n(n-1) elements. The R.I.value is average of many C.I. values which calculated by the formula 3-7. Saaty gives the values of R.I. of matrix in order 1 to 15 in the size 100 to 500. Table 3.3 shows the R.I. values and the calculation results in the size of 1000.

3) Calculate the consistency radio (C.R).

$$C.R. = \frac{C.I.}{R.I.} \tag{2}$$

When C.R. < 0.1, which meaning the consistency of decision-makers is less 10% than the consistency of randomly matrix, the consistency of randomly matrix can be accepted. On the contrary, the comparison matrix needs some appropriate correction in order to maintain a certain degree of consistency when C.R. $\geq 0.1$. But for 1,2-order matrix, C.R.=0.

### 3.2.2 Adjust the Consistency Matrix

The consistency of the matrix which is built by AHP method is often difficult to meet compliance requirements. In recent years some scholars have proposed some methods of adjustments which can be grouped into two categories: one is called mechanical method and another is called subjective method.

This article use mechanical method. There are many mechanical adjustment algorithms have been developed by many scholars in recent years and weighted arithmetic average method which is proposed by Ma WY[8] and weighted geometric average method which is proposed by XuZi, Wei C P[9] are the most widely used. This article uses the two method and another method named vector angle approach which can simplify the two method.

Suppose k is the number of iterations, C.R. is the initial volume, λ is adjust parameter of vector and assume that k=0, $C.R.^*=0.1$, $\lambda \in (0,1)$. The steps of adjust matrix as follows:

1. Suppose a comparison matrix A,
2. Calculate eigenvalue λ of A and eigenvectors W.
3. Calculate the consistency index and the consistency ratio by using formula C.I., C.R.
4. If C.R. < $C.R.^*$, turn to the last step, or turn to the next step.
5. Get $A_0$ after normalizes the matrix A. $A_0$ is the column vector of matrix $A_0$. Then calculate the cosine between W and $a_i$.

$$\cos\theta_i = \frac{\langle W, a_i \rangle}{|W||a_i|} \tag{3}$$

W is the eigenvectors of matrix A.

Calculate the variable r when $\cos\theta_r = \min\{\cos\theta_i\}$. Suppose matrix A can be calculated by the follow two methods:

$$b_{ij} = \begin{cases} \left( a_{ir}^{\lambda} \left[ \dfrac{\omega_i}{\omega_r} \right] \right)^{1-\lambda} \text{..........} j = r \\ \left( a_{rj}^{\lambda} \left[ \dfrac{\omega_r}{\omega_j} \right] \right)^{1-\lambda} \text{..........} i = r \\ a_{ij} \text{..............................} else \end{cases} \tag{4}$$

$$b_{ij} = \begin{cases} \lambda a_{ir} + (1-\lambda)\dfrac{\omega_i}{\omega_r} \text{..........} j = r \\ \dfrac{1}{\lambda a_{jr} + (1-\lambda)\dfrac{\omega_j}{\omega_r}} \text{.........} i = r \\ a_{ij} \text{.....................................} else \end{cases} \tag{5}$$

1) Weighted geometric average method, as formula (4) show.
2) Weighted arithmetic average method, as formula (5)show.
6. Suppose k = k+1 and turn to the step of 2.
7. Output the matrix $A_l$ which meet the consistency.

Parameter λ values from 0 to 1, when the value of λ is larger, the more the number of iterations required, but the adjusted matrix is more closely to the original matrix. 0.9 is the appropriate value through the practice. The convergence speed of weighted geometric average method is faster than weighted arithmetic average method when use the same λ according to some experiments. The weighted geometric average method can maintain the matrix of the reciprocal nature without transform, which mean that weighted geometric average method is better than weighted arithmetic average method.

### 3.3 Assessment Results

The assessment results are given by the weights and the actual situations. Suppose one patient has symptoms of A, B, C and he may get the disease of 1, 2, and 3 after treated by computer. Through the evaluation process based on AHP, we can get each symptom's weight which representative the degree of importance of all the symptoms. Suppose the weights are $\omega_{ij}$ , i=A, B, C, D, E , j=1, 2, 3and A>B>C>D>E. Then we can calculate the possibility $\rho_j$ of every disease.

$$\rho_1 = \omega_{A1} * A + \omega_{B1} * B + \omega_{C1} * C + \omega_{D1} * D + \omega_{E1} * E$$
$$\rho_2 = \omega_{A2} * A + \omega_{B2} * B + \omega_{C2} * C + \omega_{D2} * D + \omega_{E2} * E$$
$$\rho_3 = \omega_{A3} * A + \omega_{B3} * B + \omega_{C3} * C + \omega_{D3} * D + \omega_{E3} * E$$

According to the value of $\rho_1$, $\rho_2$, $\rho_3$, we can give the assessment results. For example, if $\rho_1 > \rho_2 > \rho_3$, we can know that the possibility of getting disease 1 is the biggest.

# 4 Experiment

This article presents a simple program which is created by JAVA to test the AHP-based evaluation model. Suppose the disease is blood disorders and the collection of symptoms are case[]={headache, dizziness, tinnitus, chest pain, chest tightness, tongue skew}.

First, the calculating of the assessment method should be selected including the method of establishing the matrix, the method of calculating weight and the method of adjusting the matrix, as shown in Figure 1. Then, the comparison matrix can be built by the calculation, as shown in Figure 2.

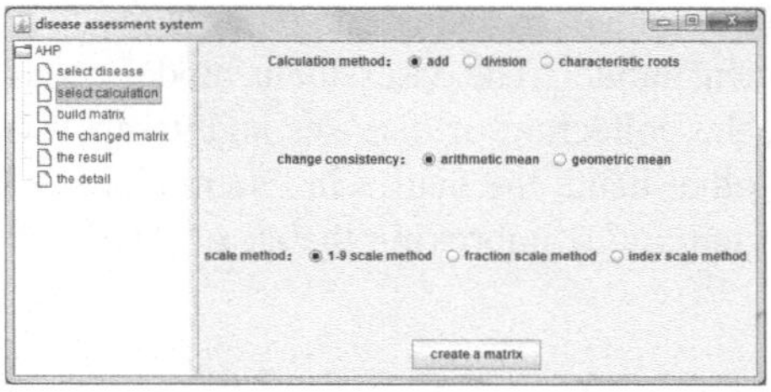


Fig. 1. Select the calculation

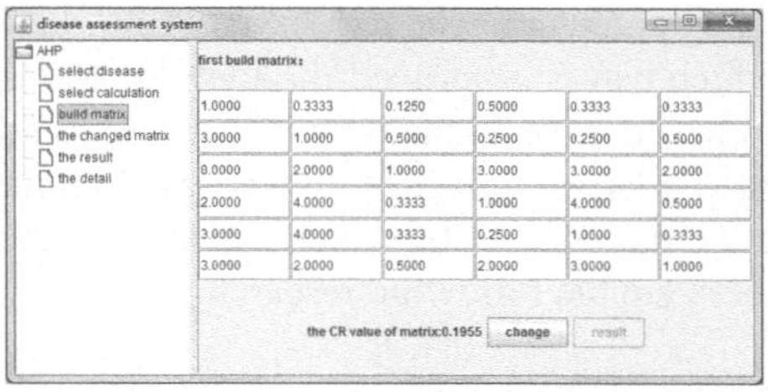


Fig. 2. The comparison matrix

The value of the C.R. in figure 2 does not meet the consistency, which means the matrix needs to be adjusted. The changed matrix in figure 3 meets the consistency, which C.R. < 0.1.

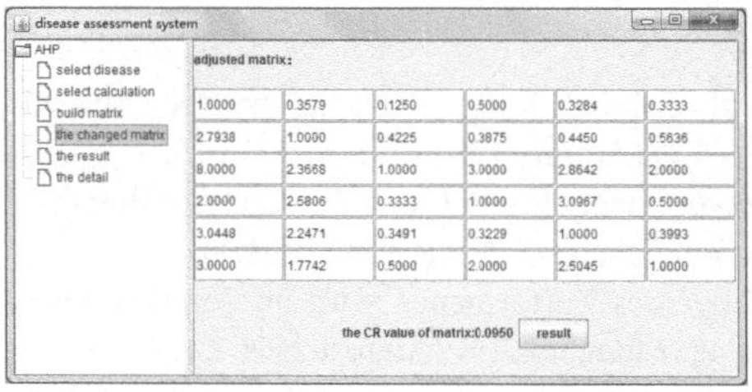


Fig. 3. Select the calculation

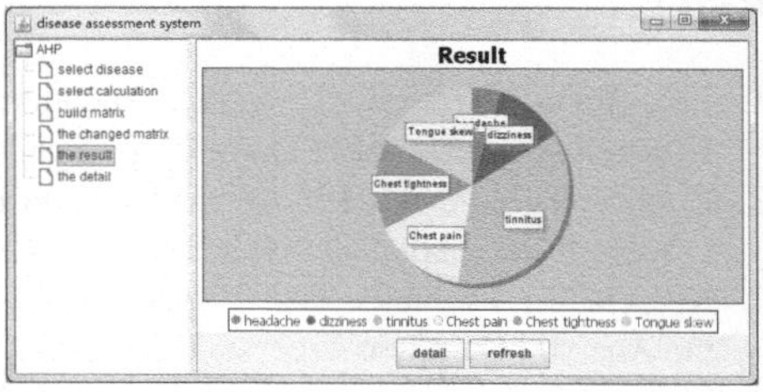


Fig. 4. The comparison matrix

Weights can be calculated by the adjusted matrix, as shown in figure 4.

The possibility of getting blood disorders can be calculated by the assessments results and the actual situations.

# 5 Related Work

AHP method has been used in many research and applications. For example: Mohammad used AHP for natural disasters, physical risk, economic risk and political risk and other risk assessments; Young-Woo Lee used the AHP framework for the evaluation of military weapon systems[2]; Christos used the AHP method to analysis the quality of service of different telecommunication companies [4]; Jukka Korpela used the AHP to establish a framework to analysis the logistics management [5]; Elliot used the AHP to establish the medical technology assessment[6]. These have fully showed that the AHP method is widely used and has high research value.

## 6 Conclusion

This article presents an assessment model based on the AHP method, having three parts including collecting information by the Delphi method, calculate the weights by the AHP method, giving assessment results according to the weights and the actual situations, which can be used to the doctors or the general users when treating by using computer. The reliability of the current treatment by using computer is too low, which can be improved through the assessment model. The assessment model needs to be improve in many ways, on one hand, the collecting of diseases and symptoms need a large number of clinical data, on the other hand, the matrix in this model need to be more scientific which can refer to the continuous improving theory of AHP. All these are the follow-up research.

**Acknowledgments.** This work is supported by the National High Technology Research and Development 863 Program of China (2012AA02A603), the National Natural Science Foundation of China (61173026), the Defense Pre-Research Project of the 'Twelfth Five-Year-Plan' of China(513***301), and the Fundamental Research Funds for the Central Universities of China.

## References

1. Lee, Y.-W., Ann, B.-H.: Static Valuation of Combat Force Potential by the Analytic Hierarchy Process. IEEE Trans. Eng. Manga. 38(3) (August 1991)
2. Douligeris, C., Pereira, I.J.: A Telecommunications Quality Study Using the Analytic Hierarchy Process. IEEE Journal on Selected Areas in Communications 12(2) (February 1994)
3. Korpela, J., Tuominen, M.: Benchmarking Logistics Performance with an Application of the Analytic Hierarchs Process. IEEE Trans. Eng. Manga. 43(3) (August 1996)
4. Sloane, E.B.: Using a Decision support system Tool for Healthcare Technology Assessments—Applying the Analytic Hierarchy Process to Improve the Quality of Capital Equipment Procurement Decisions. IEEE Engineering In Medicine and Biology Magazine (May/June 2004)
5. Xu, J., Wu, W.: Multiple Attribute Decision. Qing Hua Publication, Beijing (2006)
6. Ma, W.Y.: A practical approach to modifying pair wise comparison matrices and two criteria of modificatory effectiveness. Journal of Systems Science & Systems Engineering 3(4), 334–338 (1994)
7. Xu, Z.S., Wei, C.P.: A consistency improving method in the analytic hierarchy process. European Journal of Operational Research 116, 443–449 (1999)

# The Solution of Smart Home Indoor Positioning Based on WiFi

Songjuan Zhang* and Lilei Qi

Nanyang Institute of Technology, Nanyang, Henan, 473004, China
zhangsongjuannylg@163.com

**Abstract.** LAN gradually to the wireless technology in the direction of multi-play development in the multi-play in the process of rapid development, driven by a wide range of applications for a variety of wireless technologies, WIFI is one of them. WiFi the most important advantage is that do not require wiring can be exempted from the constraints of wiring. The extensive application of the Internet will make home control has become more automated, intelligent and humane. This paper proposes the solution of smart home indoor positioning based on WiFi. Experimental data prove that the proposed methods are effective and reasonable.

**Keywords:** smart home, indoor positioning, WiFi.

## 1 Introduction

With the growing demand for mobile communications, allowing access to specific location information demonstrated the importance of a wide range of pervasive computing and applications. In the outdoor environment, the Global Navigation Satellite System (GNSS)-based Global Positioning System (GPS) or the Big Dipper positioning system is in order to satisfy outdoor positioning requirements. However, these techniques can not be a good use of the indoor positioning system; you must use some alternative technologies. To achieve this goal, based on the IEEE 802.11 wireless LAN (Wi-Fi) technology is in order to provide highly cost-effective solutions. Practical application has been proposed many algorithms programs, including measurement of the receive signal-to-noise ratio (SNR), and the use of a wider range of the received signal strength indication (RSSI).

With the continuous development of science and technology, local area network is gradually to the wireless, multi-play direction, widely used in the multi-play in the process of rapid development, led to a variety of wireless technologies, WIFI is one of the one [1]. At present, the smart home industry is in full swing, can be predicted that the future development of smart home will no longer be confined to the remote control of appliances, lighting, etc., embedded intelligent terminals, wireless WIFI technology and Internet-wide application will the home control has become more

* Author Introduce: Songjuan Zhang, Female, Lecturer, Master, Nanyang Institute of Technology, Research area: Computer network, data mining.

J. Lei et al. (Eds.): AICI 2012, LNAI 7530, pp. 279–284, 2012.

automated, intelligent and humane, will change the mode of traditional smart home, smart home onto a stage of rapid development.

Indoor positioning must be considered indoor environment shown by the transmission channel characteristics, due to the effects of walls and obstacles will bring multipath weak absorption and shadowing a series of questions. Therefore, based on the geometric angle measurement techniques, such as angle of arrival (AOA), to reach phase (POA), time of arrival (TOA) or time of arrival difference (TDOA) are not well used in the indoor positioning system. In recent years, not with the radio frequency identification technology (Radio Frequently Identification RFID) to improve and promote, and by virtue of its light weight, low power consumption and strong ability to identify the unique advantages of gradually used in a variety of occasions, such as identity engineering controls and location tracking, and other fields.

The focus of this research is to study the Wi-Fi wireless transmission technology and RFID identification technology combining RFID tags to achieve the solution of the Wi-Fi positioning system. According to the initial search, the domestic Wi-Fi-based RFID location technology research is still rare, the present work will explore WiFi RFID positioning system-based positioning scheme. At present, the smart home industry is in full swing, can be predicted that the future development of smart home will no longer be confined to the remote control of appliances, lighting, etc., embedded intelligent terminals, wireless WIFI technology and Internet-wide application will the home control has become more automated, intelligent and humane, will change the mode of traditional smart home, smart home onto a stage of rapid development. This paper proposes the solution of smart home indoor positioning based on WiFi.

## 2 The Principle and Application of WIFI Technical

WiFi Wireless Fidelity full name, its biggest advantage is that a higher transmission speed can reach 54Mbps. Addition, its effective distance is very long, its main characteristics: speed, reliability, communication distance up to 305 meters in open areas,; in a closed area, the communication distance of 76 meters to 122 meters, convenient and existing wired Ethernet integration, lower cost of networking, highlight the advantages of Wi-Fi technology is the following.

First, the wide coverage of the radio waves, WiFi radius of up to 100 meters office since Needless to say, can also be used in the whole building [2]. Second, the transmission speed is very fast, you can reach 11mbps, in line with the needs of personal and social information. In contrast with the wired network have many advantages: no cabling, WiFi main advantage is that does not require wiring from wiring constraints, making it ideal for mobile office users need, and has broad market prospects. It has special industry from the traditional health care, inventory control and management services to more industries to expand away, and even began to enter the field of family and educational institutions, is shown by equation 1, as is shown by equation1.

$$P\{X_1 = x_1, \cdots, X_n = x_n\} = \prod_{i=1}^{n} p(x_i; \theta_1, \cdots, \theta_k) \tag{1}$$

Health and safety, use of wireless network is not as direct contact with the human body should be absolutely safe. Simple set up, the general set up wireless network with wireless LAN and an AP, so will be able to wireless mode, with both the wired architecture to share network resources, erection cost and complexity of procedures is far below the traditional wired network. If only a few computers, peer-to-peer network, can also be not to the AP, only require each computer is equipped with a wireless card. AP the Access Point referred to, generally translated as "wireless access point", or "bridge". It is mainly played in the media access control layer MAC wireless stations and wired Local Area Network Bridge. With the AP, just like the wired network Hub in general, wireless stations can be quickly and easily connected to the network. In particular, for the use of broadband WiFi even more advantages, wired broadband networks (ADSL, residential LAN, etc.) to the home, connecting to an AP, then a wireless network card can be installed on the computer. Ordinary families have an AP is enough, even the neighborhood of the user's authorization, you do not need to add ports to a shared Internet, is shown by equation 2.

$$L(x_1, x_2, \cdots, x_n; \hat{\theta}) = \max_{\theta \in \Theta} L(x_1, x_2, \cdots, x_n; \theta) \tag{2}$$

Radio Frequency Identification (RFID) technology is emerging in the 1990s, an automatic identification technology, it is mainly through the label corresponding to the unique ID number to identify markers. Similar to the sensor technology, RFID technology is considered to be the Internet of Things (The Internet of Things) a supporting technology. Some people believe that the former is only recognition, processing capacity, while the latter can be perceived items to be processed. And compared to traditional magnetic card, IC card, RF card of the biggest advantages is that the non-contact, Therefore, the identification of work without human intervention can work in all kinds of environments. RFID technology can identify the high-speed moving objects can also identify multiple tags, quick and easy operation.

RFID is a wireless system, only two basic components, the system used to control, detect and track objects. The system consists of a reader and tag. RFID technology uses radio frequency non-contact two-way transmission of data between the reader and tag, has reached the target recognition and data exchange purposes. The most basic RFID system consists of three parts: electronic tags (Tag), reader (Reader), and a miniature antenna of the radio frequency signal passed between tag and reader, as are shown by equation3.

$$\overline{\bigcup_{j=-\infty}^{\infty} V_j} = L^2(R), \quad \bigcap_{j=-\infty}^{\infty} V_j = \{0\} \tag{3}$$

Long-distance work, do not look at the working distance of the wireless WiFi 802.11b actual working distance can reach more than 100 meters in the network construction is complete, and solve the problem of high-speed mobile data error correction, error problems WiFiI equipment and devices, equipment and base station switching and safety certification have been well resolved [3]. All in all, the home and small office network users' demand for mobile connectivity is the engine of growth of the wireless LAN market, as is shown by equation 4.

The first stage, it is the contribution stage. Select the part of the training data to establish the decision tree, decision tree until each leaf node, including breadth-first class mark so far. The second phase is adjustment phase. Test the decision tree using the remaining data, if established by the decision tree can not correctly answer the research, we want to adjust a pruning decision tree and add nodes until a correct decision tree, so that in the decision tree for each internal the node comparison of property values, the conclusion in the leaf nodes. A path from the root to the leaf node corresponds to a rule, whole grain decision tree corresponds to a set of disjunctive expression rules. In general, for a given sample, extracted from the samples of random sampling in the customer decision tree model analysis, the remaining customers to test the model accuracy, and lists the calculation results and rules. Indicators to determine the accuracy of regression is in order to assess the coefficient. Generally speaking, the return to the assessment factor can be greater than that regression is more accurate, as is shown by equation4 [3].

$$
\begin{aligned}
V_j &= V_{j-1} \oplus W_{j-1} \\
&= V_{j-2} \oplus W_{j-2} \oplus W_{j-1} \\
&\cdots \\
&= V_0 \oplus W_0 \oplus W_1 \oplus \cdots \oplus W_{j-2} \oplus W_{j-1}
\end{aligned} \tag{4}
$$

Voice and WLAN, to meet the large traffic, the number of users of 3G technology, WLAN network based on IP technology is more suitable to carry out the broadcast voice services (PTT, multi-party conferencing, long distance calls, advertising, etc.).Wide area coverage and regional coverage data services under the coverage relative to the 3G technology, fast-moving maintains the characteristics of the data rate of 144kbit WIFI technology to meet user demand for high-speed data transmission has an absolute advantage in a given area.

## 3 The Solution of Smart Home Indoor Positioning Based on WiFi

WiFi is a wireless network by the AP (the Access Point) and Wireless LAN, the AP is commonly referred to as a network bridge or access point is a bridge between traditional wired local area networks and wireless local area network so any computer with wireless PC can share the resources of a wired local area network or wide area network via the AP to its working principle is equivalent to a built-in wireless transmitter HUB or routing, wireless network card is responsible for receiving the transmitted signal by the AP client side devices.

WIFI in the smart home, the application includes a home gateway, and a number of wireless communication sub-node, a wireless transmitter module on the home gateway, each child node contains a wireless network receiver module, through which The wireless transceiver module, the data transmission between the gateway and the child node. The smart home system solutions, including WIFI intelligent gateway is the indoor unit. WIFI smart gateway is the family of an intelligent hub, after a smart gateway, wireless RF modules with the collection of all child nodes to communicate, to achieve control of home appliances; intelligent gateway control through a Web network, enabling remote control of appliances.

Compared to traditional smart home system using the cable distribution network, WIFI technology to reduce wiring trouble has better scalability, mobility [4]. Therefore, the wireless intelligent control mode is the inevitable choice of the smart home development is the nature of equation 5.

$$PV(x) = \frac{1}{y_2 - y_1} \sum_{y=y_1}^{y_2} I(x, y) , \quad x \in (x_1, x_2) \tag{5}$$

The WIFI intelligent gateways include the following applications: digital video intercom, the introduction of WIFI purpose, WIFI intelligent gateway as a mobile terminal device, so you can easily in the living room, bedroom and home anywhere in the intercom control. B, security alarm, once alarm occurs, alarm information will be uploaded in a timely manner to the management center, but also through SMS, telephone owners, and can automatically capture and other functions. Information, district management software, property managers can edit all kinds of information such as text, images, video, weather, real-time information is sent to home owners in the intelligent gateway, owners can browse through the intelligent gateway prompts all kinds of information on the central server, as is shown by equation6.

$$W_f(a,b) = \frac{1}{\sqrt{|a|}} \int_{-\infty}^{\infty} f(x) \overline{\psi\left(\frac{x-b}{a}\right)} dx \tag{6}$$

Positioning of indoor Wi-Fi Terminal in accordance with the actual situation of the indoor planning the interior of the terminal maps, stored in the database of information processing; AP access point and then the indoor distribution, we need to set a fixed location reference tags as a measurement reference point to help with alignment, set the number of access points according to specific indoor. When the Wi-Fi terminal into the interior, the RFID reader inside the AP access point to be awakened, and a wireless wake-up signal is sent by the reader, wake-up inside the terminal of the Wi-Fi RFID tags. The label received a wake-up signal wake-up from sleep mode, and then compare their own ID number and ID number in the received signal are the same ID number does not match the label to re-enter the Sleep mode, as is shown by figure1.

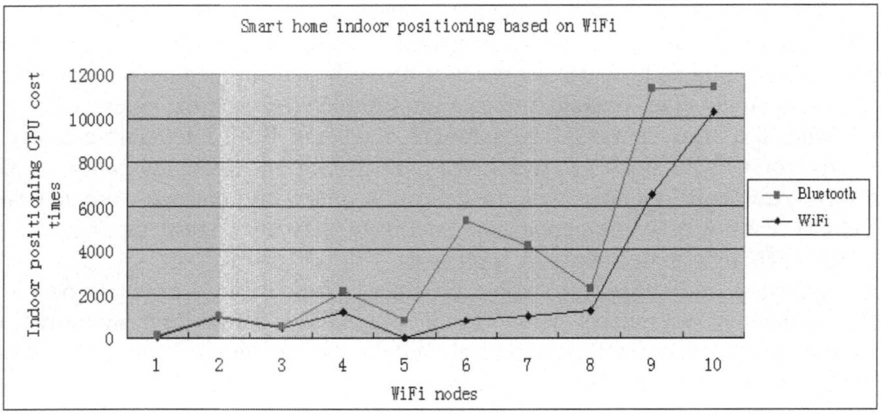


**Fig. 1.** The solution of smart home indoor positioning based on WiFi

# 4 Summary

In short, through the use of WIFI technology, the smart home equipment and building intercom link up to provide than traditional smart home more comfortable, safe and convenient smart home living space, optimizing the way people live, thus gives users a new, comfortable home life.

# References


1. Attaullah, H., Javed, M.Y.: QoS based Vertical handover between UMTS, WiFi and WiMAX Networks. JCIT 4(3), 59–64 (2009)
2. Lai, C., Li, H., Zhang, Y., Cao, J.: A Fast Seamless Handoff Scheme between IEEE 802.11 and EPS Networks Using Optimized Proxy Mobile IPv6. JDCTA 5(8), 81–91 (2011)
3. Choi, K.-H., Jang, K.-S., Shin, H.-J.: RFID-ACP: RFID-based Digital Content Identification and Authentication Mechanism in Smart Home Environments. JDCTA 5(6), 129–141 (2011)
4. Liu, C., Liu, D., Wang, S.: Dealing with Uncertainty in Situation-aware Computing System. JCIT 5(9), 175–189 (2010)

# Modified Cosine Similarity Measure between Intuitionistic Fuzzy Sets

Chao-Ming Hwang[1] and Miin-Shen Yang[2,*]

[1] Department of Applied Mathematics, Chinese Culture University, Taipei, Taiwan
[2] Department of Applied Mathematics, Chung Yuan Christian University, Chung-Li, Taiwan
msyang@math.cycu.edu.tw

**Abstract.** Similarity of intuitionistic fuzzy sets (IFSs) is an important measure to indicate the similarity degree between IFSs. Recently, Ye (2011) proposed a similarity measure between IFSs based on the cosine concept. Although this cosine similarity measure has good concept and merit, the measure is not satisfied the definition of a similarity between IFSs and not presented well for analyzing IFS data. In this paper, we modify the cosine similarity measure between IFSs. This modified similarity measure between IFSs is not only to satisfy the definition of a similarity between IFSs, but also to improve the efficiency of the Ye's measure. An example is used to demonstrate this phenomenon.

**Keywords:** Fuzzy set, Intuitionistic fuzzy set, Similarity measure, Cosine similarity measure.

## 1 Introduction

Fuzzy sets, first introduced by Zadeh [1], give an approach for treating fuzziness. In fuzzy sets, the degree of nonbelongingness is just the complement to 1 of the membership degree. However, humans who expresses the degree of membership of a given element in a fuzzy set very often does not express a corresponding degree of nonmembership as the complement to 1. Thus, Atanassov [2] introduced the concept of an intuitionistic fuzzy set (IFS) which is a generalization of a fuzzy set. Since an IFS can present the degrees of membership and nonmembership with a degree of hesitancy, the knowledge and semantic representation become more meaningful and applicable [3-4]. These IFSs have been widely studied and applied in various areas such as decision making problems [5], medical diagnosis [6], pattern recognition [7].

Similarity measures are an important tool for determining the degree of similarity between two objects. Different similarity measures between fuzzy sets have been proposed and similarity measures between IFSs are also widely studied in the literature. Dengfeng and Chuntian [8] proposed some similarity measures between IFSs used in pattern recognition problem. Liang and Shi [9] proposed similarity measures between IFSs in which they used numerical comparisons to show that Liang and Shi's similarity measures are more reasonable than those of Dengfeng and Chuntian. Mitchell [10] interpreted IFSs as ensembles of ordered fuzzy sets from the

* Corresponding author.

J. Lei et al. (Eds.): AICI 2012, LNAI 7530, pp. 285–293, 2012.

statistical point. Hung and Yang [11] proposed several similarity measures between IFSs based on Hausdorff distance which are well used with linguistic variables. Xu and Chen [12] gave a comprehensive overview of distance and similarity measures between IFSs. Recently, Ye [13] proposed a cosine similarity measure between IFSs. Although this cosine similarity measure has good concept and merit, the measure cannot satisfy the definition of a similarity between IFSs. In this paper, we modify this cosine similarity measure such that the modified measure can satisfy the definition of a similarity between IFSs. Some examples are used to demonstrate the efficiency of the modified cosine similarity measure.

## 2 Intuitionistic Fuzzy Set and Similarity Measures

Let $X=\{x_1,x_2,...,x_n\}$ be the universe of discrete discourses. Consider two intuitionistic fuzzy sets (IFSs) $\tilde{A}$ and $\tilde{B}$ in $X$. We first describe the aspects of IFSs discussed by Atanassov [2] as follows.

**Definition 1.** (Atanassov [2]) An intuitionistic fuzzy set (IFS) $A$ in $X$ is defined as $A=\{(x,\mu_A(x),\nu_A(x))|x\in X)\}$ where $\mu_A:X\to[0,1]$ and $\nu_A:X\to[0,1]$ with the condition $0\le\mu_A+\nu_A(x)\le 1, \forall x\in X$.

For each IFS $\tilde{A}$ in $X$, the numbers $\mu_A(x)$ and $\nu_A(x)$ denote the degree of membership and non-membership of $x$ to $A$, respectively, and the number $\pi_A(x)=1-\mu_A(x)-\nu_A(x)$ denotes a hesitancy degree of $x$ to $\tilde{A}$. In this paper, we use IFSs(X) to denote the class of all IFSs of $X$.

**Definition 2.** If $\tilde{A}$ and $\tilde{B}$ are two IFSs of X, then

(i) $\tilde{A}\subseteq\tilde{B}$ if and only if $\forall x\in X$, $u_{\tilde{A}}(x)\le u_{\tilde{B}}(x)$ and $\nu_{\tilde{A}}(x)\ge\nu_{\tilde{B}}(x)$.

(ii) $\tilde{A}=\tilde{B}$ if and only if $\forall x\in X$, $u_{\tilde{A}}(x)=u_{\tilde{B}}(x)$ and $\nu_{\tilde{A}}(x)=\nu_{\tilde{B}}(x)$.

Measuring a similarity between IFSs is important in IFSs researches. Some methods had been proposed to calculate the similarity degree between IFSs where Li et al. [14] introduced the following definition.

**Definition 3.** (Li et al. [14]) A mapping S: IFSs(X)×IFSs(X)→$[0,1]$. $S(\tilde{A},\tilde{B})$ is said to be the degree of similarity between $\tilde{A}$ and $\tilde{B}$ in IFSs(X) if $S(\tilde{A},\tilde{B})$ satisfies the following properties:

(S1) $0\le S(\tilde{A},\tilde{B})\le 1$;

(S2) $S(\tilde{A},\tilde{B})=1$ iff $\tilde{A}=\tilde{B}$;

(S3) $S(\tilde{A},\tilde{B})=S(\tilde{B},\tilde{A})$;

(S4) $S(\tilde{A},\tilde{C})\le S(\tilde{A},\tilde{B})$ and $S(\tilde{A},\tilde{C})\le S(\tilde{B},\tilde{C})$ if $\tilde{A}\subseteq\tilde{B}\subseteq\tilde{C}$ in IFSs(X).

(S5) $S(\tilde{A},\tilde{A}^C)=0$ iff $\tilde{A}$ is a crisp set.

Consider two IFSs $\tilde{A}$ and $\tilde{B}$ in IFSs(X), Dengfeng and Chuntian [8] proposed a similarity measure between them as follows:

$$S_{DC}(\tilde{A},\tilde{B}) = 1 - \frac{1}{\sqrt[p]{n}} \sqrt[p]{\sum_{i=1}^{n} | m_{\tilde{A}}(i) - m_{\tilde{B}}(i) |^p}$$

where $m_{\tilde{A}}(i) = (u_{\tilde{A}}(x_i) + 1 - v_{\tilde{A}}(x_i))/2$, $m_{\tilde{B}}(i) = (u_{\tilde{B}}(x_i) + 1 - v_{\tilde{B}}(x_i))/2$, $1 \le p < \infty$. When $p = 1$, $S_{DC}(\tilde{A},\tilde{B}) = S_C(\tilde{A},\tilde{B})$ where $S_C(\tilde{A},\tilde{B})$ was in Chen [15] as follow:

$$S_C(\tilde{A},\tilde{B}) = 1 - \frac{1}{2n}\sum_{i=1}^{n}\left|(u_{\tilde{A}}(x_i) - v_{\tilde{A}}(x_i)) - (u_{\tilde{B}}(x_i) - v_{\tilde{B}}(x_i))\right|$$

Hong and King [16] proposed new similarity measures $S_H(\tilde{A},\tilde{B})$, $S_L(\tilde{A},\tilde{B})$ and $S_O(\tilde{A},\tilde{B})$ (also see Li et al. [14]) as follows:

$$S_H(\tilde{A},\tilde{B}) = 1 - \frac{1}{2n}\sum_{i=1}^{n}\left|u_{\tilde{A}}(x_i) - u_{\tilde{B}}(x_i)\right| + \left|v_{\tilde{A}}(x_i) - v_{\tilde{B}}(x_i)\right|$$

$$S_L(\tilde{A},\tilde{B}) = 1 - \frac{1}{4n}\sum_{i=1}^{n}\left|(u_{\tilde{A}}(x_i) - v_{\tilde{A}}(x_i)) - (u_{\tilde{B}}(x_i) - v_{\tilde{B}}(x_i))\right| - \frac{1}{4n}\sum_{i=1}^{n}\left|u_{\tilde{A}}(x_i) - u_{\tilde{B}}(x_i)\right| + \left|v_{\tilde{A}}(x_i) - v_{\tilde{B}}(x_i)\right|$$

$$S_O(\tilde{A},\tilde{B}) = 1 - \left(\frac{1}{2n}\sum_{i=1}^{n}(u_{\tilde{A}}(x_i) - u_{\tilde{B}}(x_i))^2 + (v_{\tilde{A}}(x_i) - v_{\tilde{B}}(x_i))^2\right)^{1/2}$$

Liang and Shi [9] proposed a similarity measure between $\tilde{A}$ and $\tilde{B}$ as follows:

$$S_e^p(\tilde{A},\tilde{B}) = 1 - \frac{1}{\sqrt[p]{n}} \sqrt[p]{\sum_{i=1}^{n}(\varphi_{t\tilde{A}\tilde{B}}(i) + \varphi_{f\tilde{A}\tilde{B}}(i))^p} \quad \text{where } \varphi_{t\tilde{A}\tilde{B}}(i) = | u_{\tilde{A}}(x_i) - u_{\tilde{B}}(x_i) | / 2,$$

$\varphi_{f\tilde{A}\tilde{B}}(i) = | (1 - v_{\tilde{A}}(x_i)) - (1 - v_{\tilde{B}}(x_i)) | / 2$ and $1 \le p < \infty$. To get more information on IFSs, Liang and Shi [9] gave another similarity measure as follows:

$$S_s^p(\tilde{A},\tilde{B}) = 1 - \frac{1}{\sqrt[p]{n}} \sqrt[p]{\sum_{i=1}^{n}(\varphi_{s1}(i) + \varphi_{s2}(i))^p} \quad \text{where } \varphi_{s1}(i) = | m_{\tilde{A}1}(x_i) - m_{\tilde{B}1}(x_i) | / 2$$

and $\varphi_{s2}(i) = | m_{\tilde{A}2}(x_i) - m_{\tilde{B}2}(x_i) | / 2$ with $m_{\tilde{A}1}(i) = (u_{\tilde{A}}(x_i) + m_{\tilde{A}}(i))/2$, $m_{\tilde{A}2}(i) = (m_{\tilde{A}}(i) + 1 - v_{\tilde{A}}(x_i))/2$, $m_{\tilde{B}1}(i) = (u_{\tilde{B}}(x_i) + m_{\tilde{B}}(i))/2$ and $m_{\tilde{A}2}(i) = (m_{\tilde{B}}(i) + 1 - v_{\tilde{B}}(x_i))/2$. Mitchell [10] interpreted IFSs as ensembles of ordered fuzzy sets from statistical a viewpoint and proposed a similarity measure between $\tilde{A}$ and $\tilde{B}$ as follows:

$S_{HB}(\tilde{A},\tilde{B})=\frac{1}{2}(\rho_u(\tilde{A},\tilde{B})+\rho_v(\tilde{A},\tilde{B}))$ where $\rho_u(\tilde{A},\tilde{B})=1-\frac{1}{\sqrt[p]{n}}\sqrt[p]{\sum_{i=1}^{n}|u_{\tilde{A}}(x_i)-u_{\tilde{B}}(x_i)|^p}$, $\rho_v(\tilde{A},\tilde{B})=1-\frac{1}{\sqrt[p]{n}}\sqrt[p]{\sum_{i=1}^{n}|v_{\tilde{A}}(x_i)-v_{\tilde{B}}(x_i)|^p}$. Hung and Yang [11] proposed several similarity measures of IFSs based on Hausdorff distance. For two IFSs $\tilde{A}$ and $\tilde{B}$ in IFSs(X), they first defined $I_{\tilde{A}}(x_i)=[u_{\tilde{A}}(x_i),1-v_{\tilde{A}}(x_i)]$ and $I_{\tilde{B}}(x_i)=[u_{\tilde{B}}(x_i),1-v_{\tilde{B}}(x_i)], i=1,\dots,n.$ The Hausdorff distance $H(I_{\tilde{A}}(x_i),I_{\tilde{B}}(x_i))$ between $I_{\tilde{A}}(x_i)$ and $I_{\tilde{B}}(x_i)$ was then defined as follows:

$$H(I_{\tilde{A}}(x_i),I_{\tilde{B}}(x_i))=\max\{|u_{\tilde{A}}(x_i)-u_{\tilde{B}}(x_i)|,|1-v_{\tilde{A}}(x_i)-(1-v_{\tilde{B}}(x_i))|$$

They defined the distance $d_H(\tilde{A},\tilde{B})$ between $\tilde{A}$ and $\tilde{B}$ with $d_H(\tilde{A},\tilde{B})=\frac{1}{n}\sum_{i=1}^{n}H(I_{\tilde{A}}(x_i),I_{\tilde{B}}(x_i))$. In Hung and Yang [11], they proposed three similarity measures of $\tilde{A}$ and $\tilde{B}$ as follows:

$$S_{HY}^1(\tilde{A},\tilde{B})=1-d_H(\tilde{A},\tilde{B}),\quad S_{HY}^2(\tilde{A},\tilde{B})=\frac{e^{-d_H(\tilde{A},\tilde{B})}-e^{-1}}{1-e^{-1}},\quad S_{HY}^3(\tilde{A},\tilde{B})=\frac{1-d_H(\tilde{A},\tilde{B})}{1+d_H(\tilde{A},\tilde{B})}.$$

## 3 Modified Cosine Similarity Measure between IFSs

By considering the information carried by the membership and nonmembership degrees in intuitionistic fuzzy sets (IFSs) as a vector representation, Ye [13] proposed a cosine similarity measure between IFSs $\tilde{A}$ and $\tilde{B}$ based on the cosine concept as follows:

$$C_{IFS}(\tilde{A},\tilde{B})=\frac{1}{n}\sum_{i=1}^{n}\frac{u_{\tilde{A}}(x_i)\times u_{\tilde{B}}(x_i)+v_{\tilde{A}}(x_i)\times v_{\tilde{B}}(x_i)}{\sqrt{(u_{\tilde{A}}(x_i))^2+(v_{\tilde{A}}(x_i))^2}\times\sqrt{(u_{\tilde{B}}(x_i))^2+(v_{\tilde{B}}(x_i))^2}}$$

In Ye [13], he also proved that $C_{IFS}(\tilde{A},\tilde{B})$ satisfies the conditions: (P1) $0\le C_{IFS}(\tilde{A},\tilde{B})\le 1$; (P2) $C_{IFS}(\tilde{A},\tilde{B})=C_{IFS}(\tilde{B},\tilde{A})$; (P3) $C_{IFS}(\tilde{A},\tilde{B})=1$ if $\tilde{A}=\tilde{B}$.

However, the above three conditions are only for the correlation coefficient. If we consider the general definition of a similarity between IFSs as shown in Definition 3 of Section 2, Ye's [13] cosine similarity measure $C_{IFS}(\tilde{A},\tilde{B})$ between IFSs does not satisfy. For example, if we give $\tilde{A}=\{\langle x,0.1,0.1\rangle\}$ and $\tilde{B}=\{\langle x,0.4,0.4\rangle\}$, then $C_{IFS}(\tilde{A},\tilde{B})=1$. That is, the condition (S2) in Definition 3 is not satisfied. Furthermore, if $\tilde{A}=\{\langle x,0.0,0.1\rangle\}$ and $\tilde{B}=\{\langle x,0.1,0.0\rangle\}$ are given, then $C_{IFS}(\tilde{A},\tilde{B})=0$. This is also unreasonable. We find that Ye's [13] cosine similarity measure $C_{IFS}(\tilde{A},\tilde{B})$ is only to consider one side of the information between membership and nonmembership degrees, but not consider the middle-side and opposite-side information between them. If we also consider the middle information with $(u_{\tilde{A}}(x_i)+1-v_{\tilde{A}}(x_i))/2$ and $(u_{\tilde{B}}(x_i)+1-v_{\tilde{B}}(x_i))/2$ and the opposite-side information with $(1-u_{\tilde{A}}(x_i)),(1-u_{\tilde{B}}(x_i)),(1-v_{\tilde{A}}(x_i))$ and $(1-v_{\tilde{B}}(x_i))$, we could get a good similarity measure between IFSs $\tilde{A}$ and $\tilde{B}$ through the cosine concept. We next utilize them to propose a new similarity measure between IFSs.

We first define the two items $C^{*}_{IFS}(\tilde{A},\tilde{B})$ and $C^{**}_{IFS}(\tilde{A},\tilde{B})$ as follows:

$$C^{*}_{IFS}(\tilde{A},\tilde{B})=\frac{1}{n}\sum_{i=1}^{n}\frac{(\frac{1+u_{\tilde{A}}(x_i)-v_{\tilde{A}}(x_i)}{2}\times\frac{1+u_{\tilde{B}}(x_i)-v_{\tilde{B}}(x_i)}{2})+v_{\tilde{A}}(x_i)\times v_{\tilde{B}}(x_i)}{\sqrt{(\frac{1+u_{\tilde{A}}(x_i)-v_{\tilde{A}}(x_i)}{2})^2+(v_{\tilde{A}}(x_i))^2}\times\sqrt{(\frac{1+u_{\tilde{B}}(x_i)-v_{\tilde{B}}(x_i)}{2})^2+(v_{\tilde{B}}(x_i))^2}}$$

$$=\frac{1}{n}\sum_{i=1}^{n}\frac{((1+u_{\tilde{A}}(x_i)-v_{\tilde{A}}(x_i))\times(1+u_{\tilde{B}}(x_i)-v_{\tilde{B}}(x_i))+4v_{\tilde{A}}(x_i)\times v_{\tilde{B}}(x_i)}{\sqrt{(1+u_{\tilde{A}}(x_i)-v_{\tilde{A}}(x_i))^2+(2v_{\tilde{A}}(x_i))^2}\times\sqrt{(1+u_{\tilde{B}}(x_i)-v_{\tilde{B}}(x_i))^2+(2v_{\tilde{B}}(x_i))^2}}$$

$$C^{**}_{IFS}(\tilde{A},\tilde{B})=\frac{1}{n}\sum_{i=1}^{n}\frac{(1-u_{\tilde{A}}(x_i))\times(1-u_{\tilde{B}}(x_i))+(1-v_{\tilde{A}}(x_i))\times(1-v_{\tilde{B}}(x_i))}{\sqrt{(1-u_{\tilde{A}}(x_i))^2+(1-v_{\tilde{A}}(x_i))^2}\times\sqrt{(1-u_{\tilde{B}}(x_i))^2+(1-v_{\tilde{B}}(x_i))^2}}$$

We propose a new similarity measure between IFSs $\tilde{A}$ and $\tilde{B}$ as follows:

$$S_{IFS}(\tilde{A},\tilde{B})=\frac{1}{3}(C_{IFS}(\tilde{A},\tilde{B})+C^{*}_{IFS}(\tilde{A},\tilde{B})+C^{**}_{IFS}(\tilde{A},\tilde{B})) \tag{1}$$

As defined by equation (1), we can show that $S_{IFS}(\tilde{A},\tilde{B})$ is a similarity measure.

**Proposition 1.** The measure $S_{IFS}(\tilde{A},\tilde{B})$ is a similarity measure between IFSs $\tilde{A}$ and $\tilde{B}$.

**Proof:** It is trivial to claim $S_{IFS}(\tilde{A},\tilde{B}))=S_{IFS}(\tilde{B},\tilde{A}))$ that is the condition (S3) of Definition 3. To claim the conditions (S1) and (S3) of Definition 3, we first consider the following induction. For any $a\ge 0,b\ge 0,x\ge 0,y\ge 0$, we have that

$(ay-bx)^2\ge 0$ iff $(ax+by)^2\le(a^2+b^2)(x^2+y^2)$ iff

$0\le(ax+by)\le\sqrt{a^2+b^2}\sqrt{x^2+y^2}$

iff $0\le\frac{ax+by}{\sqrt{a^2+b^2}\sqrt{x^2+y^2}}\le 1$. Since $u_{\tilde{A}}(x_i),u_{\tilde{B}}(x_i),v_{\tilde{A}}(x_i)$ and $v_{\tilde{B}}(x_i)$ are all between 0 and 1, it is easy to show that $0\le C_{IFS}(\tilde{A},\tilde{B})\le 1$, $0\le C^{*}_{IFS}(\tilde{A},\tilde{B})\le 1$ and $0\le C^{**}_{IFS}(\tilde{A},\tilde{B})\le 1$ based on the fact $0\le\frac{ax+by}{\sqrt{a^2+b^2}\sqrt{x^2+y^2}}\le 1$ for $a\ge 0,b\ge 0,x\ge 0,y\ge 0$.

We then have $0\le S_{IFS}(\tilde{A},\tilde{B}))\le 1$ that is the condition (S1) of Definition 3.

Similarly, we consider the following induction:

$S_{IFS}(\tilde{A},\tilde{B})=1$ iff $(C_{IFS}(\tilde{A},\tilde{B})+C^{*}_{IFS}(\tilde{A},\tilde{B})+C^{**}_{IFS}(\tilde{A},\tilde{B}))=3$ iff

$C_{IFS}(\tilde{A},\tilde{B})=1$, $C^{*}_{IFS}(\tilde{A},\tilde{B})=1$ and $C^{**}_{IFS}(\tilde{A},\tilde{B})=1$ iff

$u_{\tilde{A}}(x_i)\times v_{\tilde{B}}(x_i)=v_{\tilde{A}}(x_i)\times u_{\tilde{B}}(x_i)$,

$((1+u_{\tilde{A}}(x_i)-v_{\tilde{A}}(x_i))\times v_{\tilde{B}}(x_i)=v_{\tilde{A}}(x_i)\times(1+u_{\tilde{B}}(x_i)-v_{\tilde{B}}(x_i))$,

$(1-u_{\tilde{A}}(x_i))\times(1-v_{\tilde{B}}(x_i))=(1-v_{\tilde{A}}(x_i))\times(1-u_{\tilde{B}}(x_i))$

iff $u_{\tilde{A}}(x_i)=u_{\tilde{B}}(x_i)$ and $v_{\tilde{B}}(x_i)=v_{\tilde{A}}(x_i)$ iff $\tilde{A}=\tilde{B}$.

Thus, we claim $S(\tilde{A},\tilde{B})=1$ iff $\tilde{A}=\tilde{B}$ that is the condition (S2) in Definition 3.

We next prove that $S_{IFS}(\tilde{A},\tilde{B})$ satisfies the condition (S4) in Definition 3 as follows.

If $\tilde{A}\subseteq\tilde{B}\subseteq\tilde{C}$, then for each $x\in X$, we have that $u_{\tilde{A}}(x)\le u_{\tilde{B}}(x)\le u_{\tilde{C}}(x)$ and $v_{\tilde{A}}(x)\ge v_{\tilde{B}}(x)\ge v_{\tilde{C}}(x)$. We consider a function $f$ with

$f(x)=(ax+by)/\sqrt{x^2+y^2}$.

Then, we have that $\frac{d}{dx}f(x)=\frac{y(ay-bx)}{(x^2+y^2)^{3/2}}$. In this case, if $a=u_{\tilde{A}}(x_i)$, $b=v_{\tilde{A}}(x_i)$,

$x=u_{\tilde{B}}(x_i)$ and $y=v_{\tilde{B}}(x_i)$, then $\frac{d}{dx}f(x)<0$. That is, $f(x)$ is a decreasing function of x. Let us consider another function $g$ with $g(y)=(ax+by)/\sqrt{x^2+y^2}$.

Then, we have $\frac{d}{dy}g(y)=\frac{x(bx-ay)}{(x^2+y^2)^{3/2}}$. In this case, if $a=u_{\tilde{A}}(x_i),b=v_{\tilde{A}}(x_i)$, $x=u_{\tilde{B}}(x_i)$ and $y=v_{\tilde{B}}(x_i)$, then $\frac{d}{dy}g(y)>0$. That is, $g(y)$ is an increasing function of y. Hence, we have

$$C_{IFS}(\tilde{A},\tilde{C})=\frac{u_{\tilde{A}}(x_i)\times u_{\tilde{C}}(x_i)+v_{\tilde{A}}(x_i)\times v_{\tilde{C}}(x_i)}{\sqrt{((u_{\tilde{A}}(x_i))^2+(v_{\tilde{A}}(x_i))^2)}\times\sqrt{((u_{\tilde{C}}(x_i))^2+(v_{\tilde{C}}(x_i))^2)}}$$

$$\le\frac{u_{\tilde{A}}(x_i)\times u_{\tilde{B}}(x_i)+v_{\tilde{A}}(x_i)\times v_{\tilde{B}}(x_i)}{\sqrt{((u_{\tilde{A}}(x_i))^2+(v_{\tilde{A}}(x_i))^2)}\times\sqrt{((u_{\tilde{B}}(x_i))^2+(v_{\tilde{B}}(x_i))^2)}}=C_{IFS}(\tilde{A},\tilde{B})$$

Similarly, we can claim that $C_{IFS}(\tilde{A},\tilde{C}) \le C_{IFS}(\tilde{B},\tilde{C})$. We know that $C^*_{IFS}$ and $C^{**}_{IFS}$ have a similar form as $C_{IFS}$ with $\dfrac{ax+by}{\sqrt{a^2+b^2}\sqrt{x^2+y^2}}$ so that, for $\tilde{A} \subseteq \tilde{B} \subseteq \tilde{C}$, we can claim that $C^*_{IFS}(\tilde{A},\tilde{C}) \le C^*_{IFS}(\tilde{A},\tilde{B})$, $C^*_{IFS}(\tilde{A},\tilde{C}) \le C^*_{IFS}(\tilde{B},\tilde{C}), C^{**}_{IFS}(\tilde{A},\tilde{C}) \le C^{**}_{IFS}(\tilde{A},\tilde{B})$, $C^{**}_{IFS}(\tilde{A},\tilde{C}) \le C^{**}_{IFS}(\tilde{B},\tilde{C})$. Hence, we have $S_{IFS}(\tilde{A},\tilde{C}) \le S_{IFS}(\tilde{A},\tilde{B})$ and $S_{IFS}(\tilde{A},\tilde{C}) \le S_{IFS}(\tilde{B},\tilde{C})$ if $\tilde{A} \subseteq \tilde{B} \subseteq \tilde{C}$. This is the condition (S4) in Definition 3. We next prove that $S_{IFS}(\tilde{A},\tilde{B})$ satisfies the condition (S5) in Definition 3 as follows.

$S_{IFS}(\tilde{A},(\tilde{A})^c)=0$ iff $C_{IFS}(\tilde{A},(\tilde{A})^c)=0$, $C^*_{IFS}(\tilde{A},(\tilde{A})^c)=0$, $C^{**}_{IFS}(\tilde{A},(\tilde{A})^c)=0$ iff $u_{\tilde{A}}(x_i)\times v_{\tilde{A}}(x_i)=0$, $((1+u_{\tilde{A}}(x_i)-v_{\tilde{A}}(x_i))\times(1+v_{\tilde{A}}(x_i)-u_{\tilde{A}}(x_i))+4v_{\tilde{A}}(x_i)\times u_{\tilde{A}}(x_i)=0$, $(1-u_{\tilde{A}}(x_i))\times(1-v_{\tilde{A}}(x_i))=0$ iff $u_{\tilde{A}}(x_i)=1$, $v_{\tilde{A}}(x_i)=0$ or $u_{\tilde{A}}(x_i)=0$, $v_{\tilde{A}}(x_i)=1$ iff $\tilde{A}$ is a crisp set. Thus, the proof is completed. □

If we follow the proof of Proposition 1, we can find that the cosine similarity measure $C_{IFS}$ between IFSs proposed by Ye [13] can satisfy (S4), but it cannot satisfy (S2) and (S5). We next use an example to demonstrate this phenomenon and also make comparisons of our modified similarity measure with some existing measures.

***Example 1.*** We consider the data used in Ye [13] where the data were first used in Li et al. [14]. The six data sets are as follows:

$X_1=\{\tilde{A}=(x,\ 0.3,\ 0.3),\tilde{B}=(x,\ 0.4,\ 0.4)\}$, $X_2=\{\tilde{A}=(x,\ 0.3,\ 0.4),\tilde{B}=(x,\ 0.4,\ 0.3)\}$,
$X_3=\{\tilde{A}=(x,\ 1,\ 0),\tilde{B}=(x,\ 0,\ 0)\}$,
$X_4=\{\tilde{A}=(x,\ 0.5,\ 0.5),\tilde{B}=(x,\ 0,\ 0)\}$,
$X_5=\{\tilde{A}=(x,\ 0.4,\ 0.2),\tilde{B}=(x,\ 0.5,\ 0.3)\}$, $X_6=\{\tilde{A}=(x,\ 0.4,\ 0.2),\tilde{B}=(x,\ 0.5,\ 0.2)\}$.

The degrees of some existing similarity measures between the two IFSs $\tilde{A}$ and $\tilde{B}$ are shown in Table 1. We find that, for the 1st data set $X_1$ with $\tilde{A}\ne\tilde{B}$, Ye's [13] similarity measure $C_{IFS}(\tilde{A},\tilde{B})$ is given with $C_{IFS}(\tilde{A},\tilde{B})=1$. Note that similar cases from $C_{IFS}(\tilde{A},\tilde{B})$ are also occurred for the 3rd $X_3$ and 4th $X_4$ data sets. Obviously, $C_{IFS}(\tilde{A},\tilde{B})$ does not satisfy the condition (S2) of Definition 3 for a similarity measure. However, most other similarity measures give the result with $C_{IFS}(\tilde{A},\tilde{B})\ne 1$. On the other hand, by comparing the 5th data set $X_5$ and the 6th data set $X_6$, the similarity $S(X_6)$ for $X_6$ should be larger than the similarity $S(X_5)$ for $X_5$. We find that the similarity measures $S_H$, $S_O$, $S_{HB}$, $S_e^p$ and our $S_{IFS}$ present the correct case with $S(X_6) > S(X_5)$, but others do not. Our modified cosine similarity measure $S_{IFS}$ actually corrects the drawbacks of the Ye's measure $C_{IFS}$.

**Table 1.** Similarity measures between IFSs $\tilde{A}$ and $\tilde{B}$

| Data sets | $X_1$ | $X_2$ | $X_3$ | $X_4$ | $X_5$ | $X_6$ |
|---|---|---|---|---|---|---|
| $S_C$ | 1 | 0.9 | 0.5 | 1 | 1 | 0.95 |
| $S_H$ | 0.9 | 0.9 | 0.5 | 0.5 | 0.9 | 0.95 |
| $S_L$ | 0.95 | 0.9 | 0.5 | 0.75 | 0.95 | 0.95 |
| $S_O$ | 0.9 | 0.9 | 0.3 | 0.5 | 0.9 | 0.93 |
| $S_{DC}$ | 1 | 0.9 | 0.5 | 1 | 1 | 0.95 |
| $S_{HB}$ | 0.9 | 0.9 | 0.5 | 0.5 | 0.9 | 0.95 |
| $S_e^p$ | 0.9 | 0.9 | 0.5 | 0.5 | 0.9 | 0.95 |
| $S_s^p$ | 0.95 | 0.9 | 0.5 | 0.75 | 0.95 | 0.95 |
| $S_{HY}^1$ | 0.9 | 0.9 | 0 | 0.5 | 0.9 | 0.9 |
| $S_{HY}^2$ | 0.85 | 0.85 | 0 | 0.38 | 0.85 | 0.85 |
| $S_{HY}^3$ | 0.82 | 0.82 | 0 | 0.33 | 0.82 | 0.82 |
| $C_{IFS}$ | 1 | 0.96 | 0 | 0 | 0.9971 | 0.9965 |
| $S_{IFS}$ | 0.997 | 0.859 | 0.902 | 0.902 | 0.995 | 0.997 |

## 4 Conclusions

Although Ye's [13] cosine similarity measure has good concept and merit, the measure is not satisfied the definition of a similarity between IFSs. In this paper we analyzed the drawback of Ye's [13] similarity measure and then modify it. We showed that the modified measure satisfies the definition of a similarity between IFSs. The example presented better results of our modified similarity measure than Ye's [13] measure and some other existing measures.

## References


1. Zadeh, L.A.: Fuzzy Sets. Information and Control 8, 338–356 (1965)
2. Atanassov, K.: Intuitionistic Fuzzy Sets. Fuzzy Sets and Systems 20, 87–96 (1986)
3. Atanassov, K.: Intuitionistic Fuzzy Sets: Theory and Applications. Physica-Verlag, Heidelberg (1999)

4. Atanassov, K., Georgeiv, G.: Intuitionistic Fuzzy Prolog. Fuzzy Sets and Systems 53, 121–128 (1993)
5. Pankowska, A., Wygralak, M.: General IF-Sets with Triangular Norms and Their Applications to Group Decision Making. Information Sciences 176, 2713–2754 (2006)
6. De, S.K., Biswas, R., Roy, A.R.: An Application of Intuitionistic Fuzzy Sets in Medical Diagnosis. Fuzzy Sets and Systems 117, 209–213 (2001)
7. Hung, W.L., Yang, M.S.: On the J-Divergence of Intuitionistic Fuzzy Sets with Its Application to Pattern Recognition. Information Sciences 178, 1641–1650 (2008)
8. Dengfeng, L., Chuntian, C.: New Similarity Measures of Intuitionistic Fuzzy Sets and Application to Pattern Recognition. Pattern Recognition Letters 23, 221–225 (2002)
9. Liang, Z., Shi, P.: Similarity Measures on Intuitionistic Fuzzy Sets. Pattern Recognition Letters 24, 2687–2693 (2003)
10. Mitchell, H.B.: On the Dengfeng-Chuntian Similarity Measure and Its Application to Pattern Recognition. Pattern Recognition Letters 24, 3101–3104 (2003)
11. Hung, W.L., Yang, M.S.: Similarity Measures of Intuitionistic Fuzzy Sets Based on Hausdorff Distance. Pattern Recognition Letters 25, 1603–1611 (2004)
12. Xu, Z.S., Chen, J.: An Overview of Distance and Similarity Measures of Intuitionistic Fuzzy Sets. International Journal of Uncertainty, Fuzziness and Knowlege-Based Systems 16, 529–555 (2008)
13. Ye, J.: Cosine Similarity Measures for Intuitionistic Fuzzy Sets and Their Applications. Mathematical and Computer Modelling 53, 91–97 (2011)
14. Li, Y., Olson, D.L., Qin, Z.: Similarity Measures between Intuitionistic Fuzzy (Vague) Sets: A Comparative Analysis. Pattern Recognition Letters 28, 2687–2693 (2007)
15. Chen, S.M.: Measures of Similarity between Vague Sets. Fuzzy Sets and Systems 74, 217–223 (1995)
16. Hong, D.H., Kim, C.: A Note on Similarity Measures between Vague Sets and between Elements. Information Science 115, 83–96 (1999)

# An Equipment Failure Prediction Accuracy Improvement Method Based on the Gray GM(1,1) Model

Liangli Ma[1], Haohan Liu[2], and Zebo Feng[3,4]

[1] College of Electronic Engineering, Naval University of Engineering, Wuhan, China, 430033
[2] 61660 Army, Beijing, China, 100089
[3] Department of Information Security, Naval University of Engineering, Wuhan, China, 430033
[4] Department of Military Affairs, Navy Headquaters, Beijing, China, 100841

**Abstract.** Failure prediction is an important and difficult research aspect. Firstly we introduce the concepts of Gray GM(1,1) model. In this paper, addressed to the existing problems of GM(1,1) in the prediction accuracy aspect, that is often affected by the smoothness of the sequence of collected failure datum. In order to improve the prediction accuracy, we introduce the concept of transforming trying to improve the smoothness of the original failure datum sequence. That is using GM(1,1) model to implement the transforming. And apply it to equipment prediction. Based on many collected failure datum, we use the proposed method. Then MATLAB simulation is applied to implement the residual test and posterior difference test. The results show that the above methods are valid and accuracy test meets the requirements. And the program proposed in this paper is shown to improve accuracy on the failure prediction.

**Keywords:** Equipment Failure Prediction, Gray GM(1,1) model, Residual Test, Posterior Difference Test, Prediction Accuracy.

## 1 Introduction

With the rapid development of equipment complexity, how to improve its failure prediction becomes an important research aspect. The complexity of the equipment makes the equipment failure uncertainty, nonlinear, concurrency. Once in the event of failure, it will result in significant economic loss[1]. Now, many failure prediction methods are presented, but the prediction accuracy didn't meet the needs. So using new methods to improve the equipment prediction accuracy becomes an important content.

Among the fault prediction methods, the one based on gray theory can be very accurate model, which can be effectively predicted. Grey system theory has been proposed by Professor Deng Julong since 1982. The gray prediction thought and relational analysis are widely used, which successfully resolved a large number of practical problems in production, life and scientific research[2,3]. GM (1,1) model is the most basic model of the gray theory, with advantages of modeling simply and accurately. So, it became so a hot topic, that many scholars have been optimized for

J. Lei et al. (Eds.): AICI 2012, LNAI 7530, pp. 294–300, 2012.

this model to apply to the actual [4,5]. Traditional GM(1,1) model can only predict a smooth sequence, and for the volatile sequence, the forecast accuracy is not high. By transforming the original sequence, it can improve the smoothness of the sequence and make high prediction accuracy.

### 1.1 Checking the PDF File

In some cases, it is the Contact Volume Editor that checks all the pdfs. In such cases, the authors are not involved in the checking phase.

## 2 Gray GM (1,1) Model

If a system has the fuzziness of level and structure relations, the randomness of dynamic change, Incompleteness and uncertainty of indicator data, it will be called a gray system. Gray model is to use less or inaccurate data series to represent the system behavior. The original series are transformed and built approximate differential equation, then solving the equation and get the model. The aim of the model is quantitative analysis the system using part incomplete information[6].

The basic idea of the GM(1,1) model is as follow. Firstly, the original data is processed in the form of accumulation to dilute the Influence of random factors of observed data sequence, so that the internal law of observed data sequence is improved. Then, the data sequence is constructed to a gray model with low about differential, difference and similar exponential[7]. The concrete steps of GM (1,1) model's establishment are as follow:

(1) The original sequence

$$X^{(0)} = (x^{(0)}(t_1), x^{(0)}(t_2), \cdots, x^{(0)}(t_n)), \tag{1}$$

(n is the number of samples) is first-order accumulated according in the form of

$$x^{(1)}(t_k) = \sum_{i=1}^{k} x^{(0)}(t_i), \tag{2}$$

and turn into the sequence

$$X^{(1)} = (x^{(1)}(t_1), x^{(1)}(t_2), \cdots, x^{(1)}(t_n)). \tag{3}$$

(2) Establishing differential equation

$$\frac{dX^{(1)}}{dt} + aX^{(1)} = u, \tag{4}$$

where a and u are unknown parameters. Let the equation above to be discretized.

$$dX^{(1)} = x^{(1)}(k+1) - x^{(1)}(k), \tag{5}$$

$$dt = (k+1) - k = 1, \tag{6}$$

where k is the discretization of t.

Set $Z^{(1)}(k+1) = \frac{1}{2}(x^{(1)}(k) + x^{(1)}(k+1))$.

According to the equation (4)

$$x^{(0)}(k+1) + aZ^{(1)}(k+1) = u. \tag{7}$$

(3) Calculate the parameters a and u.

According to least squares theory, the above equation is combined with to get the parameters.

$$[a, u]^T = (B^T B)^{-1} B^T Y_N, \tag{8}$$

Where

$$B = \begin{bmatrix} -Z^{(1)}(2) & 1 \\ -Z^{(1)}(3) & 1 \\ \vdots & \vdots \\ -Z^{(1)}(n) & 1 \end{bmatrix}, Y_N = \begin{bmatrix} x^{(0)}(2) \\ x^{(0)}(3) \\ \vdots \\ x^{(0)}(n) \end{bmatrix}.$$

(4) To solve the differential equation,

Let $x^{(1)}(1) = x^{(0)}(1)$ to be the initial condition, we can get

$$x^{(1)}(k+1) = \left[ x^{(0)}(1) - \frac{u}{a} \right] \exp(-ak) + \frac{u}{a}, \tag{9}$$

from equation (4), where k is the discretization of t.

## 3 Accuracy Test

For the predict using gray model, it need the accuracy test. The test generally includes the residual test and the posterior difference test.

### 3.1 Residual Test

Let the original sequence be $X^{(0)} = (x^{(0)}(1), x^{(0)}(2), \cdots, x^{(0)}(n))$, and the corresponding sequence of model predictions $\hat{X}^{(0)} = (\hat{x}^{(0)}(1), \hat{x}^{(0)}(2), \cdots, \hat{x}^{(0)}(n))$, then the gray model residuals is $\varepsilon(i) = x^{(0)}(i) - \hat{x}^{(0)}(i)$.

Its residual series is $\varepsilon^0 = (\varepsilon^0(1), \varepsilon^0(2), \cdots, \varepsilon^0(n))$.

Therefore, the relative residuals of the model is

$$e(i)=\left|\frac{x^{(0)}(i)-\hat{x}^{(0)}(i)}{x^{(0)}(i)}\right|*100\% \tag{10}$$

The average relative error is

$$\overline{e}=\frac{1}{n}\sum_{i=1}^{n}e(i) \tag{11}$$

The modeling accuracy is

$$p^{0}=(1-\overline{e})*100\% \tag{12}$$

### 3.2 Posterior Difference Test

The original sequences of the mean and mean residuals are

$$\overline{x}=\frac{1}{n}\sum_{i=1}^{n}x^{(0)}(i),\quad \overline{\varepsilon}^{(0)}=\frac{1}{n}\sum_{i=1}^{n}\varepsilon^{(0)}(i) \tag{13}$$

The original sequence of variance and residual variance are

$$s_1^2=\frac{1}{n}\sum_{i=1}^{n}(x^{(0)}(i)-\overline{x})^2,\quad s_2^2=\frac{1}{n}\sum_{i=1}^{n}(\varepsilon^{(0)}(i)-\overline{\varepsilon})^2 \tag{14}$$

Its variance ratio test is

$$c=s_2/s_1 \tag{15}$$

A predictive model for assessing the quality, c values is made as small as possible, generally require c less than 0.35, the maximum does not exceed 0.65.

## 4 Improved GM(1,1) Model

Gray system theory is successful application to the field of failure prediction, but often meet the situation that the prediction accuracy is not high or can not pass the accuracy test. Therefore, the prediction is often not accepted. There are two approaches to improve prediction accuracy in the gray prediction model: one is correcting model to adapt the raw sequence data, the other is transforming the original sequence to improve the smoothness of the sequence, so as to meet the appropriate model. In this paper, the method changing the original sequence is supported to improve the modeling accuracy.

### 4.1 Transformation of the Original Sequence

Let the original sequence be $X^{(0)}=(x^{(0)}(1),x^{(0)}(2),\cdots,x^{(0)}(n))$, the transformed sequence is $V^{(0)}=(v^{(0)}(1),v^{(0)}(2),\cdots,v^{(0)}(n))$, then

$$v^{(0)}(k) = (\ln \frac{x^{(0)}(k)+R}{M})^T \quad (k=1,2,\cdots,n) \tag{16}$$

Where R, M, T are constant value depend on the original sequence. And must have

$$\frac{x^{(0)}(k)+R}{M} > 1 \quad (k=1,2,\cdots,n) \tag{17}$$

### 4.2 Reduction of the Predict Sequence

For the transformation sequence, GM(1,1) model is used to predict, but the values of R, M, T should be continuous adjustment so as to achieve the best accuracy of the test results. Finally, we can get the values of the sequence predicted $\hat{V}^{(0)} = (\hat{v}^{(0)}(1), \hat{v}^{(0)}(2), \cdots, \hat{v}^{(0)}(n))$. With the formula

$$\hat{x}^{(0)}(k) = M * \exp((\hat{v}^{(0)}(k))^{1/T}) - R \quad (k=1,2,\cdots,n) \tag{18}$$

We can restore the value and got the predicted values.

## 5 Case Study

It using the traditional GM(1,1) model and the improved model to predict and compare by an example of system failure. Let R=0, M=11, T=1.05 to transform the original sequence. The original data and the transform data are shown in table 1.

**Table 1.** The original data and the transform data

| Numbers | 1 | 2 | 3 | 4 | 5 | 6 |
|---|---|---|---|---|---|---|
| Original sequence $X^{(0)}$ | 11.266 | 30.864 | 25.822 | 38.121 | 40.193 | 28.947 |
| Transformation sequence $V^{(0)}$ | 0.020 | 1.033 | 0.847 | 1.256 | 1.313 | 0.966 |
| Numbers | 7 | 8 | 9 | 10 | 11 | |
| Original sequence $X^{(0)}$ | 50.944 | 63.820 | 79.213 | 99.817 | 132.842 | |
| Transformation sequence $V^{(0)}$ | 1.566 | 1.808 | 2.043 | 2.294 | 2.608 | |

Using MATLAB and GM(1,1) model to predict two sets of data, we can get the models as follows:

$$\hat{x}^{(1)}(k+1) = 76.9147 * \exp(-0.2069 * k) - 65.6487 \tag{19}$$

$$\hat{v}^{(1)}(k+1) = 6.3542 * \exp(-0.1238 * k) - 6.3344 \tag{20}$$

Two models are used to predict the data and the forecast data with the improve model is restored. We can get the traditional model prediction and the relative residuals, transformation sequences and their relative predictive value of the residuals, improved model final prediction and their relative residuals shown in Table 2.

**Table 2.** Predicted Results

| Nu-mb-ers | Original Sequence | Traditional model | | Transfor mation Sequence | Prediction of Transformation Series | | Improve Model | |
|---|---|---|---|---|---|---|---|---|
| | | Predictive Value | Relative Residuals (%) | | Predictiv e value | Relative residuals (%) | Predictive Value | Relative Residua ls (%) |
| 1 | 11.266 | 11.266 | 0 | 0.020 | 0.020 | 0 | 11.266 | 0 |
| 2 | 30.864 | 17.676 | 42.731 | 1.033 | 0.838 | 18.930 | 25.603 | 17.047 |
| 3 | 25.822 | 21.738 | 15.817 | 0.847 | 0.948 | 11.995 | 28.458 | 10.210 |
| 4 | 38.121 | 26.733 | 29.873 | 1.256 | 1.073 | 14.590 | 32.054 | 15.914 |
| 5 | 40.193 | 32.877 | 18.203 | 1.313 | 1.215 | 7.472 | 36.646 | 8.824 |
| 6 | 28.947 | 40.432 | 39.676 | 0.966 | 1.375 | 42.316 | 42.604 | 47.181 |
| 7 | 50.944 | 49.724 | 2.395 | 1.566 | 1.556 | 0.635 | 50.474 | 0.923 |
| 8 | 63.820 | 61.151 | 4.183 | 1.808 | 1.761 | 2.620 | 61.079 | 4.294 |
| 9 | 79.213 | 75.204 | 5.062 | 2.043 | 1.993 | 2.411 | 75.699 | 4.436 |
| 10 | 99.817 | 92.486 | 7.345 | 2.294 | 2.256 | 1.671 | 96.373 | 3.451 |
| 11 | 132.842 | 113.740 | 14.380 | 2.608 | 2.553 | 2.075 | 126.457 | 4.806 |

When testing the model, the results of residual test and posterior difference test are shown in Table 3.

**Table 3.** The Results of Accuracy Test

| | Traditional Model | Prediction of Transformation Series | Improved Model |
|---|---|---|---|
| Average relative error (%) | 16.333 | 9.520 | 10.644 |
| Modeling accuracy (%) | 83.667 | 90.480 | 89.356 |
| Posterior standard deviation | 0.216 | 0.221 | 0.156 |

According to table 2 and table 3, after transforming, the sequence becomes smooth, which has been well predicted. Compare with the traditional model, the improved one makes greatly improvement on modeling accuracy and the posterior standard deviation. And the prediction of the last three data is more accuracy than using the traditional model, which is more suitable to predict the data.

## 6 Conclusions

In many cases, failure data is very valuable, but often contents not smooth sequence. These data can not meet the requirements of modeling. But we can transform the original sequence to improve the smoothness of the sequence, and create a model of transformation sequence, finally restore the results of the forecast. In this paper, example is used to shows that compare with the traditional model, the improved one makes higher modeling accuracy, higher prediction accuracy and lower ratio of posterior standard deviation. It is of great significance for failure prediction.

## References

1. Zuo, X.-Z., Kang, J., Li, H., et al.: Overview of Fault Prediction Technology. Fire Control & Command Control 35(1), 1–5 (2010)
2. Gopalsamy, B.M., Mondal, B., Ghosh, S.: Optimisation of machining parameters for hard machining: grey relational theory approach and ANOVA. The International Journal of Advanced Manufacturing Technology 45(11-12), 1068–1086 (2009)
3. Li, G.-D., Yamaguchi, D., Nagai, M.: A grey-based decision-making approach to the supplier selection problem. Mathematical and Computer Modeling 46(3-4), 573–581 (2007)
4. Li, G.-D., Yamaguchi, D., Nagai, M.: The development of stock exchange simulation prediction modeling by a hybrid grey dynamic model. The International Journal of Advanced Manufacturing Technology 36(1-2), 195–204 (2008)
5. Guo, R., Cheng, C.Y., Cui, Y.H.: L1-Normed GM(1,1) Models and Reliability Analysis. In: IEEE International Conference on Systems, Man and Cybernetics, pp. 775–779 (2006)
6. Zhao, Q., Gao, J., Wu, T., Lu, L.: The Grey theory and the preliminary probe into information acquisition technology. In: Proceedings of 2004 IEEE International Conference on Information Acquisition, pp. 402–404 (January 2004)
7. Lee, Y.-S., Tong, L.-I.: Forecasting energy consumption using a grey model improved by incorporating genetic programming. Energy Conversion and Management 52(1), 147–152 (2011)
8. Li, S.-F., Forrest, J.: Advances in grey systems theory and its applications. In: Proceedings of 2007 IEEE International Conference on Grey Systems and Intelligent Services, Nan Jin, pp. 1–6 (2007)

# A Group Decision Making Procedure for Selecting Data Warehouse Systems

Santoso Wibowo[1], Hepu Deng[2], and Xibao Zhang[3]

[1] Faculty of Business Informatics, Central Queensland University
Melbourne 3000, Victoria, Australia
s.wibowo1@cqu.edu.au
[2] School of Business IT & Logistics, RMIT University
GPO Box 2476 V, Melbourne 3001, Victoria, Australia
hepu.deng@rmit.edu.au
[3] Department of International Economics and Trade, Qingdao University
7 East Hongkong Road, Qingdao, China 266071
xibao.zhang@qdu.edu.cn

**Abstract.** Evaluating and selecting the most suitable data warehouse system for development is complex and challenging. To effectively solve this problem, this paper presents a group decision making procedure for evaluating and selecting data warehouse systems. The subjectiveness and imprecision of the decision making process are adequately modeled by the use of interval-valued based intuitionistic fuzzy numbers. The concept of ideal solutions is adopted for determining the overall performance of each alternative data warehouse system across all the selection criteria on which the decision is made. An example is presented for demonstrating the applicability of the proposed procedure for solving real world data warehouse system selection problems.

**Keywords:** Multicriteria Analysis, Uncertainty Modeling, Intuitionistic Fuzzy Sets, Data Warehouse Systems.

## 1 Introduction

A data warehouse is a collection of integrated databases in organizations for improving their competitiveness in the marketplace. The popularity of data warehouses is due to their capabilities in providing organizations with accurate, reliable and well-integrated information for supporting decision making in a timely manner [1]. As a result, evaluating and selecting the most suitable data warehouse system (DWS) to develop becomes critical in modern organizations.

The process of evaluating and selecting the most suitable DWS is complex and challenging. The complexity of the selection process is due to (a) the involvement of multiple decision makers (DMs), (b) the multi-dimensional nature of the decision making process, (c) the conflicting nature of the multiple selection criteria, and (d) the presence of subjectiveness and imprecision in the decision making process. The challenge of the selection process comes from the need for making transparent

J. Lei et al. (Eds.): AICI 2012, LNAI 7530, pp. 301–308, 2012.

decisions in a timely manner. To effectively solve this problem, a comprehensive evaluation of the overall performance of alternative DWS is necessary [1, 2].

This paper presents a group decision making procedure for solving the DWS selection problem. The subjectiveness and imprecision of the decision making process are adequately modeled by the use of interval-valued based intuitionistic fuzzy numbers. The concept of ideal solutions is adopted for determining the overall performance of each alternative DWS across all the selection criteria on which the final decision is made. An example is presented for demonstrating the applicability of the proposed procedure for effectively solving the DWS selection problem.

## 2 A Data Warehouse System Selection Problem

Organizations are facing with an increasing pressure to obtain current and accurate information for making better and timely decisions [3]. Traditional databases in these organizations, however, are incapable of effectively handling the large amount of information and the demand for online information access, update, and maintenance [4]. As a result, adopting appropriate DWSs for storing, maintaining, and utilizing the massive data efficiently becomes critical in organizations [1, 3].

There are various factors that affect the performance of DWSs in an organization. Much research has been done on identifying the critical factors for determining the performance of DWSs in an organization [1, 2, 5-10]. Deng and Wibowo [1], for example, show that the success of DWSs is directly related to the support of the DWS to the organizational strategy. Stewart and Mohamed [2] believe that the success of the DWS is dependent on its capability to reduce the intensity of the competition. AbuAli and Abu-Addose [5] state that the success of a DWS is based on the ability of the DWS to improve the organization's profile and support its business partners effectively. Lin et al. [6] state that the performance of a DWS is dependent on its compatibility with regulations and standards. Hwang et al. [7] show that the success of DWS is determined by the flexibility for integration with other software systems and hardware platforms. Rao [8] shows that the vendor reputation plays an important role in the success of DWSs development. Yeh et al. [9] believe that the success of DWS is related to the ability of the DWS to accommodate different organization's requirements. Roper-Lowe and Sharp [10] state that the success of a DWS is based on its ability to reduce the risk.

A comprehensive review of the related literature shows that the DWS evaluation and selection problem can be formulated as a multicriteria analysis problem. Four most important criteria are identified for evaluating and selecting DWSs in an organization including the Financial Costs ($C_1$), the User-friendly Interface ($C_2$), the System Functionality ($C_3$), and the Vendor Support ($C_4$) [1, 4].

The Financial Cost reflects on the economic feasibility of the development and implementation of a DWS with respect to the resource limitations of an organization [2]. It is measured by the initial purchasing cost of the system, and the post implementation cost to maintain the system. The User-friendly Interface of an alternative DWS concerns with the easiness of the DWS to learn and use. The system must be capable of supporting end-users from the organization. This is measured by the ease of operation, the variety of interfaces, and the ease of learning [4].

The System Functionality of a DWS reflects on the capabilities of the DWS to provide necessary functions required by the end-users. This is measured by the variety of access means by end-users, query functionality, database support, data quality check, ease of source data transformation, and ease of source data administration [2]. Vendor Support concerns about the ongoing technical assistance and service support provided by the vendor after the implementation [6]. This is measured by technical support, consultant service, training, quality of service, and delivery time for service.

To effectively evaluate and select the most suitable DWS in a given situation, it is important for the DMs to simultaneously consider the multiple selection criteria discussed as above. To facilitate the evaluation and selection of the most appropriate DWS, an effective group decision making procedure is presented in the following.

## 3 A Group Decision Making Procedure

Multicriteria group decision making involves in evaluating a set of decision alternatives with respect to multiple, often conflicting criteria for selecting the most appropriate alternative in a given situation with multiple DMs [9]. Formulated as a multicriteria group decision making problem, the evaluation and selection of DWSs usually involves in (a) discovering all the alternatives, (b) identifying the selection criteria, (c) assessing the alternatives' performance ratings and the criteria weights by individual DMs, (d) aggregating the alternative ratings and criteria weights for producing an overall performance index for each alternative across all the criteria, and (e) selecting the best alternative in the given situation [9].

Subjectiveness and imprecision are always present in decision making due to incomplete information, abundant information, conflicting evidence, ambiguous information, and subjective information [1, 3]. To adequately model the subjectiveness and imprecision, interval-valued based intuitionistic fuzzy numbers [11] are used for representing the subjective assessment of the DM.

Interval-valued based intuitionistic fuzzy numbers [12] are the generalization of the intuitionistic fuzzy numbers. Their fundamental characteristic is that the values of the membership function and non-membership function are intervals rather than exact numbers. The introduction of intervals for describing the value of membership and non-membership helps to reduce the cognitive demand on the DMs in representing their subjective assessments in the decision making process.

The decision making process starts with the determination of the performance of each alternative $A_i$ ($i$ = 1, 2, ..., $n$) with respect to each criterion $C_j$ ($j$ = 1, 2, ..., $m$) by individual DMs $D_k$ ($k$ = 1, 2, ..., $s$). Intuitionistic fuzzy number $y_{ij}^k = ([a_{ij}^k, b_{ij}^k], [c_{ij}^k, d_{ij}^k])$ can be used for representing the performance rating of alternative, in which $[a_{ij}^k, b_{ij}^k]$ indicates the degree that alternative $A_i$ satisfies the criterion $C_j$ whereas $[c_{ij}^k, d_{ij}^k]$ shows the degree that alternative $A_i$ does not satisfy the criterion $C_j$. As a result, an interval-valued based decision matrix for the multicriteria group decision making problem for each DM can be obtained as

$$y_{ij}^k = \begin{bmatrix} ([a_{11}^k, b_{11}^k], [c_{11}^k, d_{11}^k]) & ([a_{12}^k, b_{12}^k], [c_{12}^k, d_{12}^k]) & \cdots & ([a_{1m}^k, b_{1m}^k], [c_{1m}^k, d_{1m}^k]) \\ ([a_{21}^k, b_{21}^k], [c_{21}^k, d_{21}^k]) & ([a_{22}^k, b_{22}^k], [c_{22}^k, d_{22}^k]) & \cdots & ([a_{2m}^k, b_{2m}^k], [c_{2m}^k, d_{2m}^k]) \\ \cdots & \cdots & \cdots & \cdots \\ ([a_{n1}^k, b_{n1}^k], [c_{n1}^k, d_{n1}^k]) & ([a_{n2}^k, b_{n2}^k], [c_{n2}^k, d_{n2}^k]) & \cdots & ([a_{nm}^k, b_{nm}^k], [c_{nm}^k, d_{nm}^k]) \end{bmatrix} \quad (1)$$

The weightings of the criteria $C_j$ for each DM $D_k$ can be represented as

$$w_j^k = (w_1^k, w_2^k, \ldots, w_m^k) \quad (2)$$

where $w_j^k = ([e_{ij}^k, f_{ij}^k], [g_{ij}^k, h_{ij}^k])$ is the interval-valued based intuitionistic fuzzy number obtained from the DMs for assessing the relative importance of the selection criterion. $[e_{ij}^k, f_{ij}^k]$ indicates the degree where the DM considers the selection criterion $C_j$ to be important whereas $[g_{ij}^k, h_{ij}^k]$ indicates the degree where the DM considers the criterion $C_j$ to be unimportant.

To determine the weighted interval-valued based intuitionistic fuzzy performance matrix for each DM, the interval-valued based intuitionistic fuzzy weighted averaging operator [12] is used for aggregating the interval-valued based intuitionistic fuzzy decision matrix in (1) and the interval-valued based intuitionistic fuzzy criteria weightings in (2). The use of this operator is due to the ability of this operator in dealing with the multicriteria decision making problem in which the performance of each alternative and the relative importance of the selection criteria are represented by interval-valued based intuitionistic fuzzy values [13]. The weighted interval-valued based intuitionistic fuzzy performance matrix for each DM can be determined as

$$Z^k = \begin{bmatrix} z_{11}^k & z_{12}^k & \cdots & z_{1m}^k \\ z_{21}^k & z_{22}^k & \cdots & z_{2m}^k \\ \cdots & \cdots & \cdots & \cdots \\ z_{n1}^k & z_{n2}^k & \cdots & z_{nm}^k \end{bmatrix} \quad (3)$$

where

$$z_{ij}^k = \left[ \frac{\sum_{j=1}^m [a_{ij}^k, b_{ij}^k][e_{ij}^k, f_{ij}^k]}{\sum_{j=1}^m [e_{ij}^k, f_{ij}^k]}, \frac{\sum_{j=1}^m [1 - c_{ij}^k, 1 - d_{ij}^k][1 - g_{ij}^k, 1 - h_{ij}^k]}{\sum_{j=1}^m [1 - g_{ij}^k, 1 - h_{ij}^k]} \right] \quad (4)$$

$[a_{ij}^k, b_{ij}^k], [c_{ij}^k, d_{ij}^k]$ represent the interval-valued based intuitionistic fuzzy numbers for the performance ratings of the alternatives by each DM and $[e_{ij}^k, f_{ij}^k], [g_{ij}^k, h_{ij}^k]$ represents the interval-valued based intuitionistic fuzzy numbers for the relative importance of the selection criteria by each DM.

By aggregating all the weighted interval-valued based performance matrices given in (3) using interval arithmetic [9], the overall weighted interval-valued based intuitionistic fuzzy performance matrix for the problem can be obtained as

$$R = \begin{bmatrix} r_{11} & r_{12} & \cdots & r_{1m} \\ r_{21} & r_{22} & \cdots & r_{2m} \\ \cdots & \cdots & \cdots & \cdots \\ r_{n1} & r_{n2} & \cdots & r_{nm} \end{bmatrix} \tag{5}$$

where $r_{ij} = \frac{\sum_{k=1}^{s} [a_{ij}^k, b_{ij}^k], [c_{ij}^k, d_{ij}^k]}{s}$ and $r_{ij} = [a_{ij}, b_{ij}], [c_{ij}, d_{ij}]$.

The concept of the ideal solution is first introduced as the best decision outcome in a given decision situation [14]. Such a concept is then extended to include the negative ideal solution in order to avoid the worst decision outcome in the decision making process [15]. This concept has since been widely used for solving practical decision problems [16] due to (a) its simplicity and comprehensibility in concept, (b) its computation efficiency, and (c) its ability to measure the relative performance of the decision alternatives in a simple mathematical form [1].

To rank all the alternatives based on the weighted interval-valued based fuzzy performance matrix in (5), the concept of the positive and negative ideal solutions is used. The positive (or negative) ideal solution consists of the best (or worst) criteria values attainable from all the alternatives [14, 16]. The fuzzy positive ideal solution $\alpha^+$ and the fuzzy negative ideal solution $\alpha^-$ can be determined respectively as

$$\alpha^+ = (\alpha_1^+, \alpha_2^+, ..., \alpha_m^+) = \{\Big\langle ([(\max_i a_{ij}, \max_i b_{ij}) \big| j \in J_1, (\min_i a_{ij}, \min_i b_{ij}) \big| j \in J_2)], \\ [(\min_i c_{ij}, \min_i d_{ij}) \big| j \in J_1, (\max_i c_{ij}, \max_i d_{ij}) \big| j \in J_2]) \Big\rangle i = 1, 2, ..., n\} \tag{6}$$

where $\alpha_j^+ = [a_j^+, b_j^+], [c_j^+, d_j^+]$ and ($j$ = 1, 2, …, $m$).

$$\alpha^- = (\alpha_1^-, \alpha_2^-, ..., \alpha_m^-) = \{\Big\langle ([\min_i a_{ij}, \min_i b_{ij}) \big| j \in J_1, ((\max_i a_{ij}, (\max_i b_{ij}) \big| j \in J_2)], \\ [((\max_i c_{ij}, (\max_i d_{ij}) \big| j \in J_1, (\min_i c_{ij}, \min_i d_{ij}) \big| j \in J_2]) \Big\rangle i = 1, 2, ..., n\} \tag{7}$$

where $\alpha_j^- = [a_j^-, b_j^-], [c_j^-, d_j^-]$.

Based on (6) - (7), the Hamming distance between alternative $A_i$ and the positive ideal solution and the negative solution can be calculated respectively as follows

$$S_i^+ = \frac{1}{4} \sum_{j=1}^{m} \left[ \left| a_{ij} - a_j^+ \right| + \left| b_{ij} - b_j^+ \right| + \left| c_{ij} - c_j^+ \right| + \left| d_{ij} - d_j^+ \right| \right] \tag{8}$$

$$S_i^- = \frac{1}{4} \sum_{j=1}^{m} \left[ \left| a_{ij} - a_j^- \right| + \left| b_{ij} - b_j^- \right| + \left| c_{ij} - c_j^- \right| + \left| d_{ij} - d_j^- \right| \right] \tag{9}$$

The most preferred alternative should not only have the shortest distance from the positive ideal solution, but also the longest distance from the negative ideal solution [15]. Following this line, an overall performance index for each alternative $A_i$ across all the criteria can be determined by

$$P_i = \frac{S_i^-}{S_i^- + S_i^+} \qquad i = 1, 2, \quad , n \tag{10}$$

The larger the performance index $P_i$, the more preferred the alternative $A_i$. The group decision making procedure presented above is summarized as

Step 1. Obtain the decision matrix for each DM as expressed in (1).
Step 2. Determine the criteria weighting of each DM as expressed in (2).
Step 3. Obtain the interval-value weighted decision matrix as given in (3).
Step 4. Obtain the overall weighted performance matrix as shown in (5).
Step 5. Determine the fuzzy positive ideal solution and the fuzzy negative ideal solution using (6) and (7) respectively.
Step 6. Calculate the Hamming distance between alternative $A_i$ and the positive ideal solution and the negative solution by (8) and (9) respectively.
Step 7. Compute the overall performance index for each alternative by (10).
Step 8. Rank the alternatives in descending order of their index values.

## 4 An Example

To demonstrate the applicability of the group decision making procedure, an example of evaluating and selecting the most suitable DWS from six available DWSs with respect to multiple DMs and multiple evaluation and selection criteria is presented.

To start with the DWSs evaluation and selection process, the relative performance of all available DWSs in regard to DMs $D_1$, $D_2$, and $D_3$ can be determined by making their subjective assessments using the interval-valued based intuitionistic based fuzzy numbers as shown in Table 1.

**Table 1.** Performance Assessments of Data Warehouse System Alternatives

| Alternatives | | $C_1$ | $C_2$ | $C_3$ | $C_4$ |
|---|---|---|---|---|---|
| $A_1$ | $D_1$ | ([0.6,0.7],[0.2,0.3]) | ([0.3,0.6],[0.3,0.4]) | ([0.4,0.6],[0.2,0.4]) | ([0.5,0.6],[0.1,0.3]) |
| | $D_2$ | ([0.5,0.8],[0.1,0.2]) | ([0.5,0.7],[0.2,0.3]) | ([0.4,0.7],[0.1,0.2]) | ([0.4,0.5],[0.3,0.4]) |
| | $D_3$ | ([0.5,0.6],[0.3,0.4]) | ([0.3,0.7],[0.1,0.3]) | ([0.7,0.8],[0.1,0.2]) | ([0.5,0.6],[0.1,0.3]) |
| $A_2$ | $D_1$ | ([0.3,0.6],[0.2,0.4]) | ([0.5,0.6],[0.1,0.3]) | ([0.1,0.3],[0.5,0.6]) | ([0.5,0.8],[0.1,0.2]) |
| | $D_2$ | ([0.5,0.7],[0.2,0.3]) | ([0.5,0.8],[0.1,0.2]) | ([0.6,0.8],[0.1,0.2]) | ([0.5,0.7],[0.2,0.3]) |
| | $D_3$ | ([0.5,0.6],[0.1,0.3]) | ([0.7,0.8],[0.1,0.2]) | ([0.5,0.6],[0.3,0.4]) | ([0.5,0.6],[0.1,0.3]) |
| $A_3$ | $D_1$ | ([0.4,0.5],[0.3,0.4]) | ([0.1,0.3],[0.5,0.6]) | ([0.3,0.6],[0.2,0.4]) | ([0.7,0.8],[0.1,0.2]) |
| | $D_2$ | ([0.3,0.7],[0.1,0.3]) | ([0.7,0.8],[0.1,0.2]) | ([0.5,0.8],[0.1,0.2]) | ([0.6,0.8],[0.1,0.2]) |
| | $D_3$ | ([0.5,0.6],[0.1,0.3]) | ([0.1,0.3],[0.5,0.6]) | ([0.7,0.8],[0.1,0.2]) | ([0.5,0.6],[0.3,0.4]) |
| $A_4$ | $D_1$ | ([0.5,0.8],[0.1,0.2]) | ([0.6,0.8],[0.1,0.2]) | ([0.1,0.3],[0.5,0.6]) | ([0.3,0.6],[0.2,0.4]) |
| | $D_2$ | ([0.5,0.8],[0.1,0.2]) | ([0.6,0.8],[0.1,0.2]) | ([0.6,0.7],[0.2,0.3]) | ([0.3,0.6],[0.3,0.4]) |
| | $D_3$ | ([0.7,0.8],[0.1,0.2]) | ([0.5,0.6],[0.3,0.4]) | ([0.5,0.8],[0.1,0.2]) | ([0.5,0.7],[0.2,0.3]) |
| $A_5$ | $D_1$ | ([0.1,0.3],[0.5,0.6]) | ([0.3,0.6],[0.2,0.4]) | ([0.7,0.8],[0.1,0.2]) | ([0.5,0.8],[0.1,0.2]) |
| | $D_2$ | ([0.6,0.7],[0.2,0.3]) | ([0.3,0.6],[0.3,0.4]) | ([0.6,0.7],[0.2,0.3]) | ([0.3,0.6],[0.3,0.4]) |
| | $D_3$ | ([0.5,0.8],[0.1,0.2]) | ([0.5,0.7],[0.2,0.3]) | ([0.5,0.8],[0.1,0.2]) | ([0.5,0.7],[0.2,0.3]) |
| $A_6$ | $D_1$ | ([0.5,0.7],[0.2,0.3]) | ([0.5,0.8],[0.1,0.2]) | ([0.6,0.8],[0.1,0.2]) | ([0.5,0.7],[0.2,0.3]) |
| | $D_2$ | ([0.5,0.6],[0.1,0.3]) | ([0.7,0.8],[0.1,0.2]) | ([0.5,0.6],[0.3,0.4]) | ([0.5,0.6],[0.1,0.3]) |
| | $D_3$ | ([0.7,0.8],[0.1,0.2]) | ([0.5,0.8],[0.1,0.2]) | ([0.5,0.8],[0.1,0.2]) | ([0.6,0.8],[0.1,0.2]) |

Similarly, the criteria weights for selecting the DWSs can be obtained directly from the DMs $D_1$, $D_2$, and $D_3$ as shown in Table 2.

**Table 2.** Criteria Weights of Data Warehouse Systems

| | | $C_1$ | $C_2$ | $C_3$ | $C_4$ |
|---|---|---|---|---|---|
| Criteria Weights | $D_1$ | ([0.4,0.5],[0.3,0.4]) | ([0.7,0.8],[0.1,0.2]) | ([0.3,0.6],[0.2,0.4]) | ([0.7,0.8],[0.1,0.2]) |
| | $D_2$ | ([0.5,0.8],[0.1,0.2]) | ([0.6,0.8],[0.1,0.2]) | ([0.6,0.7],[0.2,0.3]) | ([0.3,0.6],[0.3,0.4]) |
| | $D_3$ | ([0.7,0.8],[0.1,0.2]) | ([0.5,0.6],[0.3,0.4]) | ([0.5,0.8],[0.1,0.2]) | ([0.5,0.7],[0.2,0.3]) |

Based on (3) - (5), the overall weighted interval-valued based intuitionistic fuzzy performance matrix for the problem can be obtained. Table 3 shows the results.

**Table 3.** The Overall Weighted Interval-valued based Intuitionistic Fuzzy Performance Matrix

| | $C_1$ | $C_2$ | $C_3$ | $C_4$ |
|---|---|---|---|---|
| $A_1$ | ([0.45,0.62],[0.32,0.43]) | ([0.48,0.67],[0.29,0.36]) | ([0.45,0.69],[0.15,0.35]) | ([0.54,0.68],[0.27,0.38]) |
| $A_2$ | ([0.51,0.87],[0.62,0.77]) | ([0.57,0.72],[0.42,0.53]) | ([0.46,0.67],[0.51,0.68]) | ([0.45,0.57],[0.67,0.72]) |
| $A_3$ | ([0.68,0.89],[0.71,0.82]) | ([0.47,0.68],[0.59,0.64]) | ([0.79,0.87],[0.53,0.67]) | ([0.51,0.66],[0.58,0.69]) |
| $A_4$ | ([0.48,0.65],[0.46,0.69]) | ([0.59,0.64],[0.67,0.72]) | ([0.48,0.62],[0.56,0.68]) | ([0.57,0.69],[0.72,0.87]) |
| $A_5$ | ([0.55,0.73],[0.42,0.57]) | ([0.65,0.81],[0.56,0.62]) | ([0.67,0.88],[0.69,0.75]) | ([0.45,0.62],[0.42,0.53]) |
| $A_6$ | ([0.45,0.62],[0.42,0.43]) | ([0.43,0.72],[0.36,0.58]) | ([0.68,0.83],[0.59,0.62]) | ([0.61,0.87],[0.47,0.59]) |

By using (6) - (10), the overall performance index for each DWS alternative across all the criteria can be calculated in a computational efficient manner. Table 4 shows that alternative $A_1$ is the obvious choice for selection.

**Table 4.** The Performance Index of Data Warehouse System Alternatives and their Rankings

| Alternatives | Index | Ranking |
|---|---|---|
| $A_1$ | 0.78 | 1 |
| $A_2$ | 0.69 | 2 |
| $A_3$ | 0.43 | 6 |
| $A_4$ | 0.50 | 5 |
| $A_5$ | 0.56 | 4 |
| $A_6$ | 0.62 | 3 |

It is evident that the proposed procedure is capable of effectively dealing with the involvement of multiple DMs, the multi-dimensional nature of the decision process, and the presence of subjectiveness and imprecision in the multicriteria group decision making problem. With its simplicity in concept and efficiency in computation, the proposed group decision making procedure is applicable for effectively solving the general multicriteria group decision making problem.

## 5 Conclusion

The DWS evaluation and selection process is complex and challenging due to the involvement of multiple DMs, the multi-dimensional nature of the decision process and the subjectiveness and imprecision inherent in the human decision making process. To effectively solve this problem, this paper has presented a group decision making procedure for solving the DWSs selection problem. A DWS selection problem is presented that shows the proposed group decision making procedure is simple and effective for solving the general DWS selection problem.

## References


1. Deng, H., Wibowo, S.: Intelligent decision support for evaluating and selecting information systems projects. Eng. Lett. 16, 412–418 (2008)
2. Stewart, R., Mohamed, S.: IT/IS projects selection using multi-criteria utility theory. Logistics Inform. Manage. 15, 254–265 (2002)
3. Wibowo, S., Deng, H.: Intelligent Decision Support for Effectively Evaluating and Selecting Ships under Uncertainty in Marine Transportation. Exp. Syst. Appl. 39, 6911–6920 (2012)
4. Wixom, B.H., Watson, H.J.: An empirical investigation of the factors affecting data warehousing success. MIS Q. 25, 17–41 (2001)
5. AbuAli, A.N., Abu-Addose, H.Y.: Data Warehouse Critical Success Factors. Eur. J. Sci. Res. 42, 326–335 (2010)
6. Lin, H.Y., Hsu, P.Y., Sheen, G.J.: A fuzzy-based decision-making procedure for data warehouse system selection. Exp. Syst. Appl. 32, 939–953 (2007)
7. Hwang, H.G., Kua, C.Y., Yen, D.C., Cheng, C.C.: Critical factors influencing the adoption of data warehouse technology: a study of the banking industry in Taiwan. Decis. Support Syst. 37, 1–21 (2004)
8. Rao, S.S.: Enterprise resource planning: business needs and technologies. Ind. Manage. Data Syst. 100, 81–88 (2000)
9. Yeh, C.H., Deng, H., Wibowo, S., Xu, Y.: Multicriteria group decision for information systems project selection under uncertainty. Int. J. Fuzzy Syst. 12, 170–179 (2010)
10. Roper-Lowe, G.C., Sharp, J.A.: The analytic hierarchy process and its application to an information technology decision. J. Oper. Res. Soc. 41, 49–59 (1990)
11. Atanassov, K., Gargov, G.: Interval-valued intuitionistic fuzzy sets. Fuzzy Sets Syst. 31, 343–349 (1989)
12. Chen, S.M., Lee, L.W., Liu, H.C., Yang, S.W.: Multiattribute decision making based on interval-valued intuitionistic fuzzy values. Exp. Syst. Appl. (forthcoming, 2012)
13. Yue, Z.L., Jia, Y.Y., Ye, G.D.: An approach for multiple attribute group decision making based on intuitionistic fuzzy information. Int. J. Uncertain. Fuzziness Knowl. Based Syst. 17, 317–332 (2009)
14. Zeleny, M.: Multiple Criteria Decision Making. McGraw-Hill, New York (1982)
15. Hwang, C.L., Yoon, K.S.: Multiple Attribute Decision Making: Methods and Applications. Springer, Berlin (1981)
16. Yeh, C.H., Deng, H., Chang, Y.H.: Fuzzy Multicriteria Analysis for Performance Evaluation of Bus Companies. Eur. J. Oper. Res. 126, 1–15 (2000)

# An Architecture for Cloud Computing and Human Immunity Based Network Intrusion Detection

Caiming Liu[1,2], Chunming Xie[3], Yan Zhang[2], Qin Li[2,*], and Lingxi Peng[4]

[1] School of Information Science & Technology,
Southwest Jiaotong University, 610031 Chengdu, China
[2] School of Computer Science, Leshan Normal University, 614004 Leshan, China
[3] Teaching Affairs Office, Leshan Normal University, 614004 Leshan, China
[4] School of Computer and Education Software,
Guangzhou University, 510006 Guangzhou, China
liucaiming@gmail.com, wkywawa@tom.com

**Abstract.** Traditional intrusion detection system is confronted with the pressure of processing massive network traffic data which increases sharply. Besides, its feature of static detection causes the weak adaptability for the network environment. To overcome the former problems, an architecture for network intrusion detection based on cloud computing and artificial immune principle is proposed. It consists of local intrusion detection sub-system and cloud computing platform which provides the services of intrusion detection. The local intrusion detection sub-system captures and simply preprocesses the network traffics. The cloud computing platform deals with the true transactions of intrusion detection. It interacts with the local intrusion detection sub-system through standard service interface and responds the intrusion detection requests of the local intrusion detection sub-system. Furthermore, it simulates the good features of artificial immune principle and adopts self-learning mechanism to evolve intrusion detection elements to make the proposed architecture adaptive for the real network environment.

**Keywords:** Intrusion Detection, Cloud Computing, Artificial Immune, Architecture.

## 1 Introduction

Network attacks harm Internet more and more seriously. According to the statistics of the network security company Symantec in "*2011 State of Security Survey*", most of the respondents suffered network attacks [1]. Intrusion Detection System (IDS) is usually adopted to analyze and discover the attacks hid in network traffics. However, along with the gradual increment of network bandwidth, Internet users and

* Corresponding author.

J. Lei et al. (Eds.): AICI 2012, LNAI 7530, pp. 309–316, 2012.

applications, network traffics increase fast. Massive communication data of Internet makes that IDS which is deployed in the network node can not complete analyzing network traffic data in time. It is difficult to detect network attacks in real time. IDS is apt to be flooded by network data and not to perform its function of intrusion detection. Furthermore, traditional IDS relies on the static signature to detect network attacks. However, the environment of Internet is dynamic and changeful. Traditional intrusion detection technology is difficult to adapt the change of Internet environment.

Cloud Computing (CC) [2] makes the idea that the network is the computer [3] be true. It changes the way of traditional disperse service provision. It fuses a lot of resources of computing, storage and software to provide service for network users. It makes the provision way of software service and digital resource become intensive, largescale and specialized [4]. The model of CC integrates massive resources of software and hardware to provide a strong platform for the analysis and storage of largecscale data. This characteristic of CC can remedy the fault of insufficient ability of data processing of traditional IDS. CC platform may render services of virtualized data processing and storage for network intrusion detection. Traditional IDS deployed locally only captures signatures of local network traffics and submit them to CC platform which effectively analyzes network data and sends the detection result to Local IDS.

Artificial Immune System (AIS) [5] has the good features such as self-adaptation, diversity, and etc [6]. Many researchers have paid close attention to it [7, 8]. It became one of the hot spots in the field of intelligent computation. AIS was used to solve the problems of network security. It was especially applied in network intrusion detection effectively [9-12].

The intensive model of computing and storage of CC and the excellent adaptation of AIS are made full use of to propose an architecture for network intrusion detection in this paper. The proposed architecture is expected to provide a new idea to solve the problems of insufficient data processing and weak adaptation to network environment of traditional IDS. In the rest of this paper, section 2 describes the proposed architecture in detail, section 3 concludes this paper.

# 2 Description of Proposed Architecture

## 2.1 General Framework of Proposed Architecture

The general framework of the proposed architecture is shown in Fig. 1.

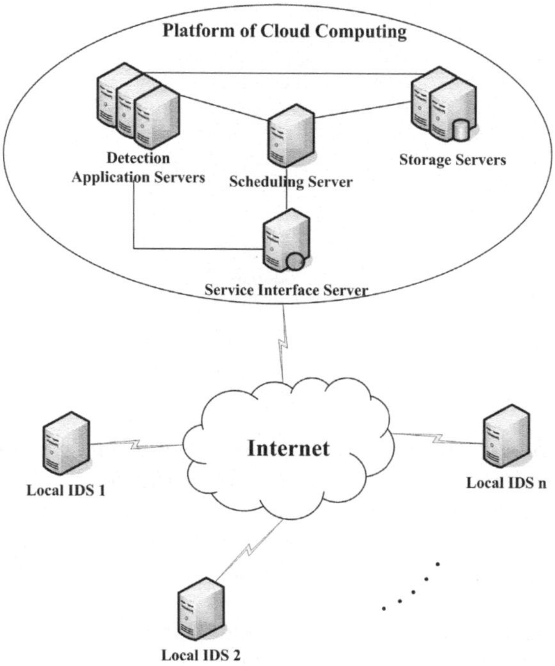


**Fig. 1.** General framework

The proposed architecture consists of Local Intrusion Detection Sub-system (LIDS) and CC Platform (CCP) which provides the services of intrusion detection. LIDS captures network traffics and gets their signatures. It also submits the request of intrusion detection service to CCP and receives the detection result. CCP is formed by service interface server, scheduling server, application severs of intrusion detection and storage servers. It receives the request of LIDS, allocates appropriate resources to process the intrusion detection service for LIDS and returns the detection result to LIDS.

## 2.2 Local Intrusion Detection Sub-system (LIDS)

LIDS is mainly responsible for collecting and preprocessing network traffics. However, it doesn't analyze the signatures of massive IP packets and store abundant static signatures of network attacks. The framework of LIDS is shown in Fig. 2. LIDS is formed by service communication module, alarm module, network traffic collection module and data preprocessing module. The above modules perform lightweight algorithm and processing logic. They utilize few resources of local system to ensure that LIDS can complete its work swimmingly.

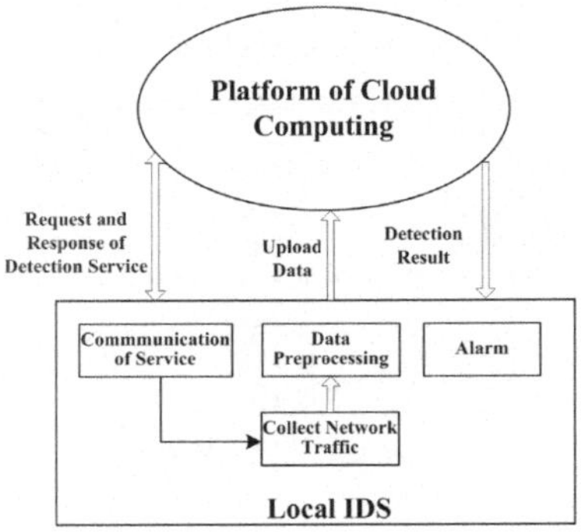


**Fig. 2.** Framework of LIDS

The service communication module requests the intrusion detection service of CCP and receives the response of request success. The alarm module receives the detection result. If local network and host suffer attacks, it raises an alarm. The network traffic collection module captures the initial network IP packets. The data preprocessing module packs the packets' signatures into binary strings and puts the tag of LIDS into the binary strings.

The proposed architecture analyzes the signatures of the IP packets to verify whether there are attacks in the network. Let the data set of network signatures be *SData*. It meets $SData = \{\langle sIP, dIP, \Pr otocol, TTL, Len \rangle\}$, where, *sIP* is the source IP address, *dIP* is the destination IP address, *Protocol* means the protocol type of the packet, it may be ICMP, IGMP, TCP, UDP, and etc, *TTL* and *Len* is the lifecycle and the length of the packet, respectively.

The data preprocessing module deals with the signature data set *SData* and sends the preprocessing data to CCP. Let preprocessing data set be *PData* which is described in Eq. (1).

$$PData = \{\langle IID, sdata \rangle | IID \in N, \forall sd \in SData \wedge sdata = toBinaryString(sd)\} \quad (1)$$

Where, *IID* is the identifier of the LIDS $IDS_{IID}$, it is a long sequence integer and unique in the global scope. *sdata* is a *l*-long binary string. It is transformed into by the function *toBinaryString*().

### 2.3 Cloud Computing Platform for Intrusion Detection (CCP)

1) Summary of CCP

CCP provides a unified intrusion detection service interface for LIDS. The service interface is managed by the service interface server which carries out the following tasks. First, it receives the intrusion detection requests of LIDS. Then, it implements a series of mechanisms to judge whether the requests are reasonable. Second, it accepts the request and responds LIDS. Third, it asks the scheduling server to allocate intrusion detection resources (Including processing and storing) for LIDS. Fourth, it receives the network traffic signatures preprocessed by LIDS and sends them to the detection application servers. Finally, it receives the intrusion detection results and sends them to LIDS.

The scheduling server allocates appropriate resources for intrusion detection according to the request of service interface server. It creates the services of intrusion detection application. The services are run by the intrusion detection application server.

The intrusion detection application server adopts immune mechanisms to learn from and analyze data signatures of network traffics. It aims at recognizing the attacks hid in network traffics. The storage servers are responsible for storing immune detection elements.

2) Allocation of Intrusion Detection Resources

Once the CCP accepts the request of intrusion detection of LIDS $IDS_{IID}$, it uses the initialization operation to create the resource of intrusion detection service for $IDS_{IID}$. Let the two dimension vector of service resource of intrusion detection be $\langle APP_{IID}, STO_{IID}\rangle$ which is shown in Eq. (2).

$$\langle APP_{IID}, STO_{IID}\rangle = Create(IID) \tag{2}$$

Where, $APP_{IID}$ means intrusion detection application service. It is responsible for the transaction of intrusion detection for LIDS $IDS_{IID}$. The scheduling server allocates $APP_{IID}$ to appropriate detection application servers. $STO_{IID}$ is storage service. It is responsible for the storage function for intrusion detection elements. The scheduling server puts $STO_{IID}$ into some storage servers. The function *Create*() is the creation operation of intrusion detection resource.

If the LIDS $IDS_{IID}$ requested the service of intrusion detection before, the operation *Read*() which is shown in Eq. (3) is used to create the service resource of intrusion detection.

$$\langle APP_{IID}, STO_{IID}\rangle = \mathrm{Re}\,ad(IID) \tag{3}$$

3) Simulate of Immune Elements

In the immune system, immune cells constantly evolve to recognize pathogens. To simulate this kind of mechanism, antigen is used to simulate signatures of network traffics in LIDS, detector is used to simulate immune cell in the proposed architecture.

Let the antigen set of intrusion detection service $APP_{IID}$ be $STO_{IID}.Antigen$, where the round dot means ownership. $STO_{IID}.Antigen$ is shown in Eq. (4).

$$STO_{IID}.Antigen = \{ant \mid \forall pd \in PData \wedge ant = pd.sdata, |ant| = l\} \tag{4}$$

Where, *PData* is the signature data of network traffics sent by LIDS $IDS_{IID}$. The antigen *ant* is a *l*-long binary string.

Let the normal antigen set of $APP_{IID}$ be $STO_{IID}.Self$, abnormal antigen set (attacks) be $STO_{IID}.Nonelf$. They meet $STO.Self \cup STO.Nonself = STO.Antigen$. The proposed architecture aims to recognize $STO_{IID}.Nonself$.

Detector is used to discover the attack behavior hid in network IP packets. Let the detector set of $APP_{IID}$ be $STO_{IID}.Detector$ which is shown in Eq. (5).

$$STO_{IID}.Detector = \{\langle pattern, age, count, type \rangle\} \tag{5}$$

Where, *pattern* is the signature pattern of detector. It has the same data style as antigen. It is used to match antigen. *age* means life generation of detection. *count* is the sum of antigen matched by detector. *type* means the class of detector.

4) Evolution of Detector

The detector set of the intrusion detection application service $APP_{IID}$ is classified into immature detector set $STO_{IID}.Immature$, mature detector set $STO_{IID}.Mature$ and remember detector set $STO_{IID}.Remember$. The class of detector set is indicated by the domain *type* of $STO_{IID}.Detector$.

The immature detector set $STO_{IID}.Immature$ simulates the initial stage of immune cells in AIS. It includes initializing detectors newly generated. It is shown in Eq. (6).

$$\begin{aligned} &STO.\mathrm{Im}mature = \{imm \mid imm \in \mathrm{Im}mature, \\ &imm.type = i, imm.age < \alpha, imm.count = 0\} \end{aligned} \tag{6}$$

Where, $\alpha$ is the tolerance threshold of an immature detector. The immature one will evolve to a mature detector until it passes self-tolerance [6]. *count* means the sum of self antigens matched by the immature one. If the immature one recognizes a self antigen, namely $count > 0$, it dies.

The mature detector set $STO_{IID}.Mature$ simulates the middle stage of immune cells in AIS. It is evolved to by an immature one. It is shown in Eq. (7).

$$STO.Mature = \{mat \mid mat \in Mature, mat.age < \lambda, mat.count < \delta\} \tag{7}$$

Where, $\lambda$ means the survival generation of a mature detector. $\delta$ denotes the activation threshold of the mature one. The mature detector will evolve to a remember one when it matches antigens whose sum is equal with or greater than $\delta$ in $\lambda$.

The remember detector set $STO_{IID}.Remember$ simulates the top stage of immune cells in AIS. It is evolved to by a mature one or shifted to by system administrators through the signatures of known network attacks. It is shown in Eq. (8). Its signature domain *pattern* can match the harmful antigens accurately.

$$STO.\mathrm{Re}member = \{rem \mid rem \in \mathrm{Re}member, rem.age \geq \lambda, rem.count \geq \delta\} \tag{8}$$

5) Detection and Alarm of Network Attacks

According to the network environment, the detectors gradually evolve to remember ones which can recognize network attacks exactly. The remember detectors use their signature domain *pattern* to match antigens. Once an antigen is matched by the signature pattern of a remember detector, it is recognized as an attack. Let harmful antigens detected by the intrusion detection application service $APP_{IID}$ be $STO_{IID}.Harm$. The detection process of network attacks is shown in Eq. (9).

$$STO_{IID}.Harm = \{harm \mid harm \in STO_{IID}.Antigen, \exists rem \in STO_{IID}.Remember \wedge f_{match}(rem, harm) = true\} \tag{9}$$

Where, $f_{match}()$ is the match function between detector and antigen. If the remember detector *rem* matches the antigen *harm*, it returns the value *true*. Otherwise, it returns *false*.

When the intrusion detection application service $APP_{IID}$ detects an attack antigen, it forms a piece of alarm information and informs the service interface server which sends the alarm to the $IDS_{IID}$.

# 3 Conclusion

IDS supplements the active defense tools of network security. However, along with the fast development of Internet, it is confronted with more and more complicate network environment. Furthermore, the data processing pressure of IDS increases sharply. The proposed architecture in this paper constructs a cloud computing platform with superior resources to provide intrusion detection services. Moreover, the principles and mechanisms of AIS are applied to intrusion detection service to improve its adaptation ability for network environment. The above measures are expected to resolve the former faults of traditional IDS.

**Acknowledgments.** This work is supported by the National Natural Science Foundation of China (No. 61103249 and 61100150), the Open Fund of Artificial Intelligence Key Laboratory of Sichuan Province (No. 2011RYJ01), the Construction Plan for Scientific Research Innovation Groups of Leshan Normal University and the Scientific Research Fund of Leshan Normal University (No. Z1113 and Z1065).

# References

1. Symantec, State of Security Survey (2011), http://www.symantec.com/content/en/us/about/media/pdfs/symc_state_of_security_2011.pdf
2. Chen, K., Zheng, W.M.: Cloud Computing: System Instances and Current Research. Journal of Software 20, 1337–1348 (2009)
3. Wikipedia, http://en.wikipedia.org/wiki/John_Gage
4. Feng, D.G., Zhang, M., Zhang, Y., Xu, Z.: Study on Cloud Computing Security. Journal of Software 22, 71–83 (2011)
5. Mo, H.W., Zuo, X.Q.: Artificial Immune System. Science Press, Beijing (2009)
6. Li, T.: Computer immunology. Publishing House of Electronics Industry, Beijing (2004)
7. Xiao, R.B., Wang, L.: Artificial immune system: principle, models, analysis and perspectives. Chinese Journal of Computers 25, 1281–1293 (2002)
8. Jiao, L.C., Du, H.F.: Development and Prospect of the Artificial Immune System. Acta Electronica Sinica 31, 1540–1548 (2003)

9. Dasgupta, D.: An immunity-based technique to characterize intrusions in computer networks. IEEE Transactions on Evolutionary Computation 6, 281–291 (2002)
10. Dasgupta, D.: Immunity-based intrusion detection system: a general framework. In: The 22nd National Information Systems Security Conference (1999)
11. Kim, J., Bentley, P.J.: Towards an artificial immune system for network intrusion detection: an investigation of dynamic clonal selection. In: The Congress on Evolutionary Computation, pp. 1015–1020 (2002)
12. Hofmeyr, S.A.: An immunological model of distributed detection and its application to computer security. Ph. D. dissertation, Department of Computer Sciences, University of New Mexico (1999)

# Robust Algorithm for Detection of Copy-Move Forgery in Digital Images Based on Ridgelet Transform

Guorui Sheng[1,2], Tiegang Gao[1], Yanjun Cao[1,2], Lin Gao[1,2], and Li Fan[1,2]

[1] College of Software, Information Security Technology Lab., Nankai University, Tianjin, China, 300071
[2] College of Information Technical Science, Nankai University, Tianjin, China, 300071
{ShengGuoRui001,GaoTiegang,CaoYanjun528411, Gao2689,FanLihlm}@gmail.com

**Abstract.** The rapid development in image processing techniques has enabled people to easily synthesize realistic images; this may result in the social problems when a doctored image cannot be distinguished from a real one by visual examination, so, the digital image authentication becomes more and more important in our daily life. In this paper, a robust algorithm based on ridgelet transform is proposed to detect copy-move forgery which is the most popular method in image tampering. The proposed method has the advantage of simplicity and low complexity of calculating. Simplicity lies in the vectors got by calculating Hu moments of the ridgelet transform domain; vectors with the length of 7 results in low complexity of calculating. The given experimental results show that our algorithm has good performance of detection of copy-move forgery, even the image undergo some kind of JPEG compress.

**Keywords:** copy-move, ridgelet transform, Hu moments, image forensics.

## 1 Introduction

With the development of modern image processing technologies, it's easier for people to form a digital image, and at the same time, one can tamper a digital image easily with the help of any editing software, such as Photoshop, without leaving any visible trace. So people may have any excuse to doubt the truthfulness of any one image, and consequently, it's very important for researchers to develop reliable methods to authenticate digital images.

At present, digital watermarking scheme can be used for fragile authentication [1-2]. However, the watermark must be embedded into the image before the tampering happens. One more attracting method of image authentication, which is called blind detection method has been developed for recent years, the method does not need the embedding of digital watermarks ahead of time, some intrinsic statistical features of image were used to detect the image forgery. Image splicing and copy-move method, may be the most popular kind of image forgery. The purpose of the copy-move forgery is usually to conceal a special people or object in the original image. To achieve this purpose, one region is copied from the original image and pasted onto somewhere in the same image.

J. Lei et al. (Eds.): AICI 2012, LNAI 7530, pp. 317–323, 2012.

Some kinds of copy-move detecting methods have been introduced these years. Fridrich proposed an overlapping block matching method based on DCT [3][4], the proposed method proved to be more effective than exhaustive search. A similar method, which used PCA instead of DCT, was proposed by Popesecu [5]. The method was claimed to be more effective due to the dimension reduction by PCA. The method mentioned above perform well when the copy-move region have not been post processed. But when the copied region has been post processed by some steps (including rotating, adding noise, blurring and so on), the methods does not perform well. In the other hand, these two methods is time consuming. In [6], a kind of copy-move forgery detection method based on dyadic wavelet transform was proposed by Najah Muhammad. The method was claimed to be simple and effective. In [7], the combination of Radon transform and Fourier-Mellin transform was used to detect copy-move forgery.

In this paper, a detection method based on ridgelet transform is proposed. It is well known, wavelets can efficiently represent *pointlike* singularities. But in higher dimensions, there are some situations which wavelets can not deal with efficiently, for instance, singularities along lines and curves. In order to find a way to solve these problems, researchers try to find other ways to offer the representations of images. Minh N. Do proposed a new systems representations named ridgelet, which can deal with the situation of *linelike* phenomena in 2-D [8]. Due to its powerful ability of images representations, it is believed that ridgelet transform can be used in detecting copy-move forgery, and may have better results than other transforms. The experiments results confirm this viewpoint.

The rest of the paper is organized as follows. The next section introduces the ridgelet transform briefly. The section 3 explains our proposed method. Experimental results are shown in the section 4. Finally, conclusion is drawn in the section 5.

## 2 Ridgelet Transform

First the continuous ridgelet transform is reviewed and further more, be compared with the continuous wavelet transform. If $f(x)$ is an integrable bivariate function, the continuous ridgelet transform in $\mathbb{R}^2$ can be defined as:

$$RI_f(a,b,\theta)=\int_{\mathbb{R}^2} \psi_{a,b,\theta}(x) f(x)dx \tag{1}$$

The ridgelet $\psi_{a,b,\theta}(x)$ in 2-D are defined from a function which is wavelet typed in 1-D $\psi(x)$ as:

$$\psi_{a,b,\theta}(x) = a^{-1/2}\psi((x_1 cos\theta + x_2 sin\theta - b)/a) \tag{2}$$

To make the comparison, the continuous wavelet transform in $\mathbb{R}^2$ is:

$$W_f^2(a_1, a_2, b_1, b_2) = \int_{\mathbb{R}^2} \psi_{a_1,a_2,b_1,b_2}(x) f(x)dx \tag{3}$$

It's easy to find that continuous ridgelet transform appears similar to the continuous wavelet transform except that the $(b_1, b_2)$, which is the point parameters, are replaced by the $(b, \theta)$, which is the line parameters. Some conclusions can be drawn from this: the wavelet analysis is very effective at representing images with isolated point singularities. While the continuous ridgelet transform will be effective in representing images which has singularities along lines. Since the singularities are often joined together along edges or contours in images, the ridgelet transform is sure to be powerful in copy-move forgery detecting.

Points and lines can be related by Radon transform, so the wavelet and ridgelet transform can also be linked by Radon transform. The Radon transform is:

$$RA_f(\theta, t) = \int_{\mathbb{R}^2} f(x)\delta(x_1 cos\theta + x_2 sin\theta - t)dx \tag{4}$$

Performing the 1-D wavelet transform to the slices of the Radon transform, the ridgelet transform can be represented precisely:

$$RI_f(a,b,\theta) = \int_{\mathbb{R}} \psi_{a,b}(t) RA_f(\theta, t)dt \tag{5}$$

# 3 Proposed Method

## 3.1 Algorithm Framework

In this section, the proposed method for detecting copy-move forgery and locating the tampered region is presented in detail.

The copy-move forgery is done by copying one region from the image and pasting it to another location. Our method tries to find any copy-move forgery by searching the whole image to find regions that have very similar internal structure. The main framework of the proposed method is as follows:

Step 1: Division: Divide the input image into overlapping sub-blocks.

Step 2: Representation: Performing ridgelet transform on every sub-block, getting the result and compute the Hu-moments of every result to represent every sub-block.

Step 3: Matching: Compute the Euclidean distance of features corresponding to each pair of sub-blocks to find similar pairs.

The following is the detail of each step in our framework.

A square of $B \times B$ pixels slides along the image following the trace from the upper left corner right down to the lower right corner. $(M - B + 1) \times (M - B + 1)$ blocks will be got if the input image is $M \times N$ size. Next the feature of each sub-block will be computed based on ridgelet transform.

### 3.2 Representation Using Ridgelet Transform

Ridgelet transform was performed on each sub-block to extract features. Since the result got from ridgelet transform is too complex to be used in matching similar features, Hu moments [9] for each result from ridgelet transform is calculated.

After computing the Hu moments for every ridgelet transform result, a vector whose length is 7 can be obtained for every sub-block. These $(M\text{-}B+1)(N\text{-}B+1)$ vectors will be used as the feature for sub-blocks and be compared to find similar regions.

### 3.3 Matching

$(M\text{-}B+1)(N\text{-}B+1)$ vectors are obtained after extracting all sub-blocks' features got from computing Hu moments of matrixes which are from performing ridgelet transform on each sub-block. Then the vectors be sorted in a lexicographically order. By doing this, the computing complexity for measure the similarity of each two vectors can be reduced obviously. Let the matrix A_S denotes the sorted vectors, its size will be $(M\text{-}B+1)(N\text{-}B+1)\times 7$.

Let $\vec{a_i}$ denotes the $i$th row of matrix A_S and $(x_i, y_i)$ denotes the coordinate of the top-left corner of the certain sub-block. The Euclidean distance of each consecutive vector $\vec{a_i}, \vec{a_j}$ is computed to find if they two are similar. Let $d$ be the Euclidean distance of two vectors, if $1/(1+d) > r\_threshold$, corresponding vectors are marked as similar. And then, the shift vector between these two vectors is calculated: $s = (s_1, s_2) = (x_i - x_j, y_i - y_j)$. Considering that $s$ maybe a negative value which should be normalized further, which means it needs multiplying by -1 if it's negative. Every $s$ is put into the matrix Q. Let $s^{(1)}, s^{(2)}, s^{(3)}$ … …denote all normalized vectors in Q, the special $s^{(i)}$ whose occurrence in Q exceeds a predefined threshold *o_threshold* will be marked as candidate copy-move blocks. Finally there is another consideration should be focus on. For the copy-move regions are assumed to be not overlapping, and at the same time, the image is segmented into overlapping sub-blocks, a wrong judgment is very likely to be drawn that two close blocks be counted as copy-move region. So an additional computation is needed when a matching is found. That is, compute the distance of blocks corresponding to every marked $s^{(i)}$ in the Q, only the shift vector which representing two blocks not so close be retained, using the threshold *d_threshold* .

Our algorithm finally outputs a black color image, except for regions which are judged to be the copy-move forgery. Special color besides black is assigned to these regions. Researchers can easily distinguish the copy-move forgery and make the judgment of which region is original and which region is tampered.

## 4 Experiment Result

In this section the experiment procedure and related parameters are introduced first, and then the visual results of the experiments are given.

The experiments were performed on the Matlab R2009a. All images are from the data set [10]. These images are all BMP format with 128×128 pixels. By default, the parameters in our experiments are set as follows: *B*=16, *r_threshold*=0.995, *o_threshold*=5, *d_threshold*=40. The images shown in Fig.1 were the results of our copy-move forgery detecting method. Each row includes three images: original image, tampered image and the output of our algorithm. The results show that our algorithm performs well while the duplicated regions are in different positions of the image: vertical, horizontal, diagonal and random.

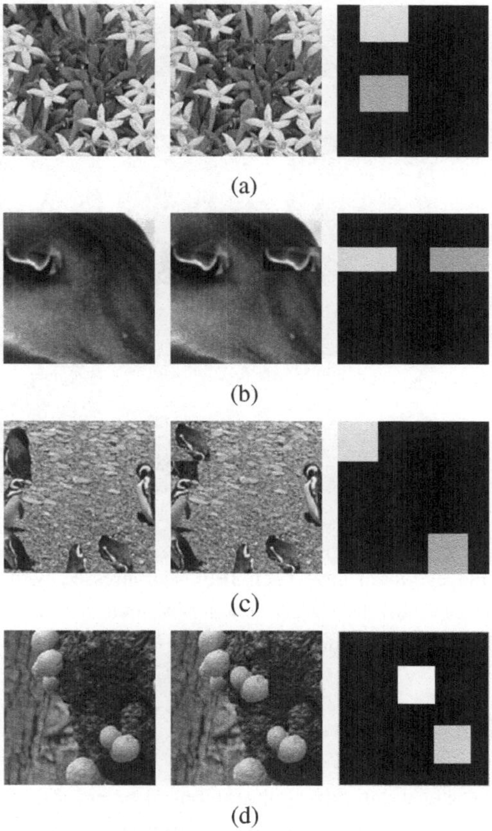


**Fig. 1.** Results of images which have different positions of duplicated regions in (a) vertical (b) horizontal (c) diagonal (d) random directions. Each row includes three images: original image, tampered image and the output of our algorithm.

The robustness of the algorithm was also tested, when the tampered image is compressed by JPEG, the proposed scheme can still detect duplicated regions, the detect outcomes are shown in Fig.2. These show that the proposed algorithm has better robustness against JPEG compress attacks. As many images are in the format of JPEG saving, the proposed detecting method is strongly believed to have wide application in image forensics.

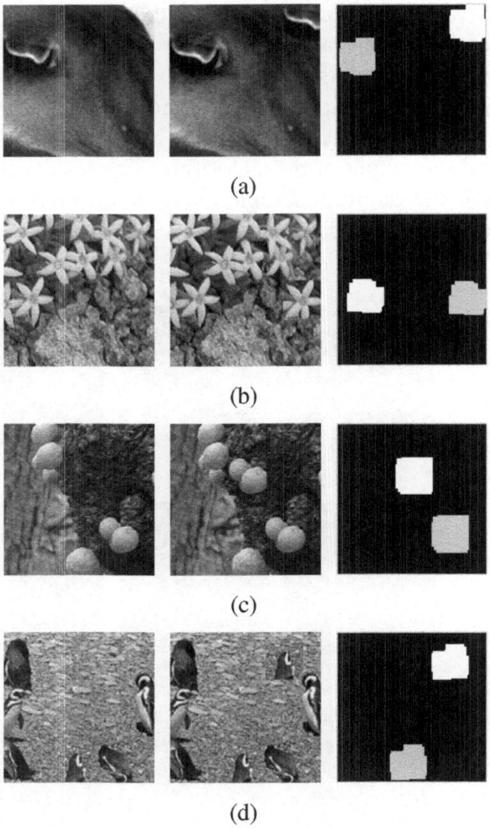


**Fig. 2.** Results of images which have been JPEG compressed with different quality factors: (a)=90dB (b)=80dB (c)=70dB (d)=60dB. Each row includes three images: original image, tampered image and the output of our algorithm.

## 5 Conclusion

A new copy-move forgery detecting scheme based on ridgelet transform is proposed in this paper. The proposed scheme has the advantage of simplicity and low complexity of computing. Experimental results verify the effectiveness of the proposed algorithm to detect duplicated regions. At the same time, the algorithm is robust to JPEG compression, this may makes the algorithm has a wide field of application in image forensics.

**Acknowledgements.** The author would thank the support from National Science Fund of China (60873117), Key Program of Natural Science Fund of Tianjin (Grant #11JCZDJC16000), China.

## References

1. Katzenbeisser, S., Petitcolas, F.: Information Techniques for Steganography and Digital Watermarking. Artec House (2000)
2. Fridrich, J., Goljan, M., Du, M.: Invertible authentication. In: Proc. SPIE, Security and Watermarking of Multimedia Contents, vol. 4313, pp. 197–208 (2001)
3. Fridrich, J., Soukalm, D., Lukáš, J.: Detection of copy-move forgery in digital images. In: Digital Forensic Research Workshop, Cleveland, OH, pp. 19–23 (2003)
4. Cao, Y., Gao, T.: A robust detection algorithm for copy-move forgery in digital images. Forensic Sci. Int. 214, 33–43 (2011)
5. Popescu, A.C., Farid, H.: Exposing Digital Forgeries by Detecting Duplicated Image Regions, Tech. Rep. TR2004-515, Dartmouth College (2004)
6. Muhammad, N., Hussain, M.: Copy-move forgery detection using dyadic wavelet transform. In: Eighth International Conference on Computer Graphics, Imaging and Visualization (CGIV), pp. 103–108 (2011)
7. Jing, T., Li, X.: Image Tamper Detection Algorithm Based on Radon and Fourier-Mellin Transform. In: IEEE International Conference on Information Theory and Information Security (ICITIS), pp. 212–215 (2010)
8. Do, M.N., Vetterli, M.: The Finite Ridgelet transform for Image Representation. IEEE Transactions on Image Processing 12(1) (January 2003)
9. Hu, M.-K.: Visual Pattern Recognition by Moment Invariants. IRE Transactions on Information Theory, 179–187 (1962)
10. Ng, T.-T., Hsu, J., Chang, S.-F.: Columbia Image Splicing Detection Evaluation Dataset, http://www.ee.columbia.edu/ln/dvmm/downloads/AuthSplicedDataSet/

# Function Projective Synchronization and Parameters Identification of Different Hyperchaotic Systems Based on Adaptive Control

Jun Dong[1], Guang-jun Zhang[1,2], Hong Yao[1], Xiang-bo Wang[1], and Jue Wang[2]

[1] College of Science, Air Force University of Engineering Xi'an Shaanxi PRC,
[2] School of Life Science and Technology Xi'an Jiao Tong University Xi'an Shaanxi PRC
Zhanggj3@126.com

**Abstract.** In this paper a novel three dimensions chaotic system with uncertain parameters and lorenz hyperchaotic system are as examples, the function projective synchronization and parameters identification of different hyperchaotic systems are researched. First, based on the Lyapunov theory of stability and adaptive control method, the adaptive nonlinear controller and adaptive identifying rule to uncertain parameter are designed logically. And by the controller and identifying rule, the function projective synchronization of different systems between three dimensions response system with uncertain parameter and drive system is realized. Second, the feasibility of the controller and identifying rule to uncertain parameter designed in this paper is analyzed theoretically, and the function projective synchronization and parameters identification are proved strictly theoretically. Finally, the theoretical results are verified by numerical simulation.

**Keywords:** hyperchaotic system, function projective synchronization, adaptive control method, parameters identification.

## 1 Introduction

In the nonlinear science, the research on chaos is one of the key problems in the field of nonlinear dynamics. Because of the very extensive applied foreground and tremendous potential value in the market in the secure communication, signal processing, and life science etc., chaos control and chaotic synchronization induce the extensive attention and are researched widely[1-5].

Pecora and Corroll advance the concept of chaotic synchronization for the first time in 1990, and the chaotic synchronization in two coupled chaotic systems is realized in electrocircuit experiment[3-4]. Later, different kinds of chaotic synchronization are realized from different aspects, such as general synchronization, pulse synchronization, phase synchronization, anti-synchronization etc.[6-9]. In 1999, Mainieri and Rehacek initially put forward the concept of projective synchronization in the researches on part-linear chaotic system[10]. That is to say, not only the phases of outputs of coupled drive system and response system is locked in definite conditions, but also does the amplitude of each correspondence state evolve by a

J. Lei et al. (Eds.): AICI 2012, LNAI 7530, pp. 324–332, 2012.

certain ratio. This characteristic of proportion is applied to secure communication in which binary system digit is extended to M system digit to the more rapid transmission being actualized[11-12]. This kind of novel chaotic synchronization phenomenon is gradually becoming a novel research hotspot. Recently, Abdurahman etc. put forward the problem of projective synchronization for unified hyperchaotic system with constant as proportional factor, by the Lyapunov theory of stability the adaptive controller is designed and full state hybrid projective synchronization for unified hyperchaotic system with known parameter is actualized[13]. But if constant proportional factor is replaced by proportional function with time as independent variable, the chaotic signal which equal to the output of drive chaotic system multiplying by proportional factor changing with time is obtained. This chaotic signal will synchronize with the output of response system. And so after function projective synchronization the topologic structures of the two chaotic systems are identical in the sense of this proportional function. Because the time-varying proportional function is adopted in function projective synchronization and the intercepting difficulty of chaotic system output increases, the more secure communication is obtained in secure communication. In the practical application of chaotic synchronization, especially in the secure communication, drive system and response system are usually different. So the researches on function projective synchronization of different hyperchaotic systems are of practical important significance.

Therefore, in this paper based on the Lyapunov theory of stability and adaptive control method, the adaptive controller and adaptive identifying rule to uncertain parameter are designed. By the adaptive controller and adaptive identifying rule to uncertain parameter the function projective synchronization between response system with uncertain parameter and certain drive system is achieved. By adjusting the value of $\lambda_i (i = 1,2,3 \cdots 8)$ the speed of error convergence to zero and the effect of parameters identification are improved. This method is robust; the applicable area of the method is extensive and is easily realized. The applicable area of projective synchronization is extended. On the base of theoretically analysis this method is verified by numerical results.

## 2 The Analysis of System and the Description of Function Projective Synchronization

By a state variable is added to Lorenz chaotic system a four dimensions hyperchaotic Lorenz system is presented as follow:

$$\begin{cases} \dfrac{dx_1}{dt} = a \cdot (x_2 - x_1) + x_4 \\ \dfrac{dx_2}{dt} = b \cdot x_1 - x_1 \cdot x_3 - x_2 + x_4 \\ \dfrac{dx_3}{dt} = x_1 \cdot x_2 - c \cdot x_3 \\ \dfrac{dx_4}{dt} = -d \cdot x_2 \end{cases} \tag{1}$$

When $a = 10$ , $b = 28$ , $c = 8/3$ , $d = 5$ , the four Lyapunov exponentials of the system are respectively: $\lambda_{L_1} = 0.7326$ , $\lambda_{L_2} = 1.5169$ , $\lambda_{L_3} = -1.8435$ , $\lambda_{L_4} = -13.7046$ . Two of the four Lyapunov exponentials are positive. The four dimensions Lorenz system in this case is hyperchaotic. The hyperchaotic attractor of system (1) is shown in figure 1.

A novel chaotic dynamical system is presented by Cai etc[14]. It is depicted by the three dimensions autonomous differential equation as follows:

$$\begin{cases} \frac{dy_1}{dt} = a_1 \cdot (y_2 - y_1) \\ \frac{dy_2}{dt} = b_1 \cdot y_1 + c_1 \cdot y_2 - y_1 \cdot y_3 \\ \frac{dy_3}{dt} = y_1^2 - h_1 \cdot y_3 \end{cases} \quad (2)$$

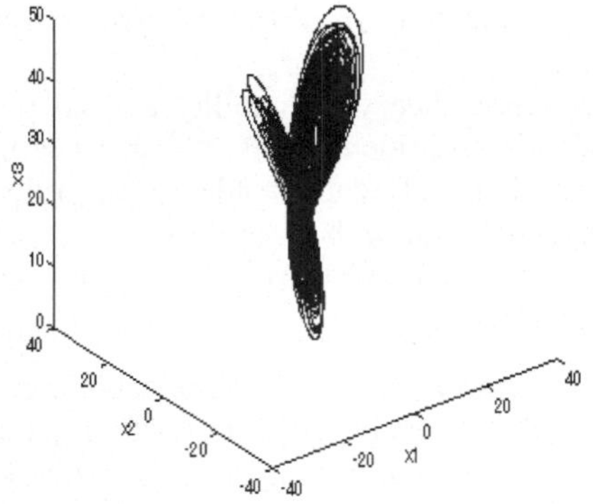


**Fig. 1.** The hyperchaotic attractor of system (1)

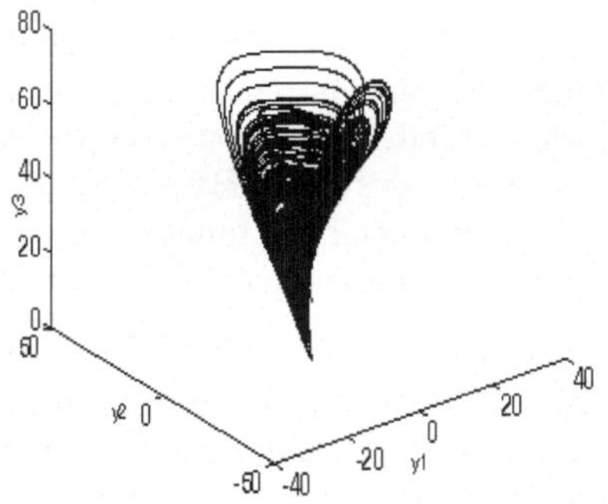


**Fig. 2.** The chaotic attractor of system (2)

When $a_1 = 20$ , $b_1 = 14$ , $c_1 = 10.6$ , $h_1 = 2.8$ , the system (2) is chaotic[14]. The chaotic attractor of system (2) is shown in figure 2.

In this paper, when the parameters of system (2) are unknown the adaptive controller and parameters identification which can make the function projective synchronization between response system and drive system to be realized will presented.

Suppose the drive system and the response system to be respectively as follows:

$$\dot{x} = f(t, x) \quad (3)$$

$$\dot{y} = g(t, y) + u(t, x, y) \quad (4)$$

Where *u(t,x,y)* is nonlinear controller. Suppose $\alpha(t)$ to be proportional function of time, it is continuous and differentiable, and $\alpha(t) \neq 0.0 \quad t \in (0, \infty)$. The error of two systems is defined as: $e(t) = y - \alpha(t) \cdot x$ . If proportional function $\alpha(t)$ make $\lim_{t \to +\infty} \| y(t) - \alpha(t) \cdot x(t) \| = 0.0$ the function projective synchronization between system (3) and system (4) is realized.

## 3 The Function Projective Synchronization of Lorenz Hyperchaotic System and the Novel Chaotic System

In this paper, the Lorenz hyperchaotic system (1) is as drive system, the novel chaotic system (2) presented by Cai etc.[14] is as response system after controlled. That is as follows:

$$\begin{cases} \frac{dy_1}{dt} = a_1 \cdot (y_2 - y_1) + u_1 \\ \frac{dy_2}{dt} = b_1 \cdot y_1 + c_1 \cdot y_2 - y_1 \cdot y_3 + u_2 \\ \frac{dy_3}{dt} = y_1^2 - h_1 \cdot y_3 + u_3 \\ \frac{dy_4}{dt} = 0 + u_4 \end{cases} \tag{5}$$

Where $u_i (i = 1,2,3,4)$ is controller. Let the synchronous error between system (1) and system (5) to be: $e_i(t) = y_i - \alpha(t) \cdot x_i \quad (i = 1,2,3,4)$, where y(4)=0.0. When y(4)=0.0 the system (5) is the same as the system (2). The error system is obtained as follows:

$$\begin{cases} \dot{e}_1 = a_1 \cdot (y_2 - y_1) - \alpha(t) \cdot [a \cdot (x_2 - x_1) + x_4] - \dot{\alpha}(t) \cdot x_1 + u_1 \\ \dot{e}_2 = b_1 \cdot y_1 + c_1 \cdot y_2 - y_1 \cdot y_3 - \alpha(t) \cdot (b \cdot x_1 - x_1 \cdot x_3 - x_2 + x_4) - \dot{\alpha}(t) \cdot x_2 + u_2 \\ \dot{e}_3 = y_1^2 - h_1 \cdot y_3 - \alpha(t) \cdot (x_1 \cdot x_2 - c \cdot x_3) - \dot{\alpha}(t) \cdot x_3 + u_3 \\ \dot{e}_4 = 0 - \alpha(t) \cdot (-d \cdot x_2) - \dot{\alpha}(t) \cdot x_4 + u_4 \end{cases} \tag{6}$$

In order that the function projective synchronization between system (1) and system (5) can be realized the appropriate controller $u_i (i = 1,2,3,4)$ which can make $\lim_{t \to +\infty} \| y(t) - \alpha(t) \cdot x(t) \| = 0.0$ must be designed. The adaptive controller designed is as follows:

$$\begin{cases} u_1 = -\hat{a}_1 \cdot (y_2 - y_1) + \alpha(t) \cdot [a(x_2 - x_1) + x_4] + \dot{\alpha}(t) \cdot x_1 - \lambda_1 \cdot e_1 \\ u_2 = -\hat{b}_1 \cdot y_1 - \hat{c}_1 \cdot y_2 + y_1 \cdot y_3 + \alpha(t) \cdot (b \cdot x_1 - x_1 \cdot x_3 - x_2 + x_4) + \dot{\alpha}(t) \cdot x_2 - \lambda_2 \cdot e_2 \\ u_3 = \hat{h}_1 \cdot y_3 - y_1^2 + \alpha(t) \cdot (x_1 \cdot x_2 - c \cdot x_3) + \dot{\alpha}(t) \cdot x_3 - \lambda_3 \cdot e_3 \\ u_4 = \alpha(t) \cdot (-d \cdot x_2) + \dot{\alpha}(t) \cdot x_4 - \lambda_4 \cdot e_4 \end{cases} \tag{7}$$

Let the parameter estimating error of system to be $\tilde{\theta} = \hat{\theta} - \theta$, the equation (7) is substituted into the equation (6), then we can obtain:

$$\begin{cases} \dot{e}_1 = -\tilde{a}_1 \cdot (y_2 - y_1) - \lambda_1 \cdot e_1 \\ \dot{e}_2 = -\tilde{b}_1 \cdot y_1 - \tilde{c}_1 \cdot y_2 - \lambda_2 \cdot e_2 \\ \dot{e}_3 = \tilde{h}_1 \cdot y_3 - \lambda_3 \cdot e_3 \\ \dot{e}_4 = -\lambda_4 \cdot e_4 \end{cases} \tag{8}$$

With the effect of adaptive controller $u_i (i = 1,2,3,4)$ the synchronous error $e_1, e_2, e_3, e_4$ will decrease gradually, and in the meantime the uncertain parameter $\hat{a}_1, \hat{b}_1, \hat{c}_1, \hat{h}_1$ will vary with the law as follows:

$$\begin{cases} \dot{\hat{a}}_1 = e_1 \cdot (y_2 - y_1) - \lambda_5 \cdot (\hat{a}_1 - a_1) \\ \dot{\hat{b}}_1 = e_2 \cdot y_1 - \lambda_6 \cdot (\hat{b}_1 - b_1) \\ \dot{\hat{c}}_1 = e_2 \cdot y_2 - \lambda_7 \cdot (\hat{c}_1 - c_1) \\ \dot{\hat{h}}_1 = -e_3 \cdot y_3 - \lambda_8 \cdot (\hat{h}_1 - h_1) \end{cases} \tag{9}$$

The uncertain parameters $\hat{a}_1, \hat{b}_1, \hat{c}_1, \hat{h}_1$ are as the estimation of the unknown parameters $a_1, b_1, c_1, h_1$, and then the parameter estimation errors of response system are as follows:

$$\begin{cases} \dot{\tilde{a}}_1 = e_1 \cdot (y_2 - y_1) - \lambda_5 \cdot \tilde{a}_1 \\ \dot{\tilde{b}}_1 = e_2 \cdot y_1 - \lambda_6 \cdot \tilde{b}_1 \\ \dot{\tilde{c}}_1 = e_2 \cdot y_2 - \lambda_7 \cdot \tilde{c}_1 \\ \tilde{h}_1 = -e_3 \cdot y_3 - \lambda_8 \cdot \tilde{h}_1 \end{cases} \tag{10}$$

The variable $\lambda_i$ $(i = 1,2,3,\cdots 8)$ in equation (7) and equation (9) is greater than zero. By the controller designed according to equation (7) and the control law of uncertain parameters from equation (9) the synchronous error between response system (5) and drive system (1) will gradually approach to zero with the time. And in the meantime the uncertain parameters which need estimating will be to the scheduled perfect value according to the equation (9).

Below, the function projective synchronization of two systems is proved.

**Theorem 1.** For arbitrary determinate proportional function $\alpha(t)$, if adaptive controller suffice the equation (7) and the adaptive control law of response system parameters suffice the equation (9), then the function projective synchronization will be realized.

**Prove:** the Lyapunov function is constructed as:

$$V = \frac{1}{2}(e_1^2 + e_2^2 + e_3^2 + e_4^2 + \tilde{a}_1^2 + \tilde{b}_1^2 + \tilde{c}_1^2 + \tilde{h}_1^2)$$

The derivative to time of Lyaponov function along to error orbit is:

$$\begin{aligned}\dot{V} = {} & e_1 \cdot [-\tilde{a}_1 \cdot (y_2 - y_1) - \lambda_1 \cdot e_1] + e_2 \cdot (-\tilde{b}_1 \cdot y_1 - \tilde{c}_1 \cdot y_2 - \lambda_2 \cdot e_2) + e_3 \cdot (\tilde{h}_1 \cdot y_3 - \lambda_3 \cdot y_3) + e_4 \cdot (-\lambda_4 \cdot e_4) \\ & + \tilde{a}_1 \cdot [e_1 \cdot (y_2 - y_1) - \lambda_5 \cdot \tilde{a}_1] + \tilde{b}_1 \cdot (e_2 \cdot y_1 - \lambda_6 \cdot \tilde{b}_1) + \tilde{c}_1 \cdot (e_2 \cdot y_2 - \lambda_7 \cdot \tilde{c}_1) + \tilde{h}_1 \cdot (-e_3 \\ & \lambda_5 \cdot \tilde{a}_1] + \tilde{b}_1 \cdot (e_2 \cdot y_1 - \lambda_6 \cdot \tilde{b}_1) + \tilde{c}_1 \cdot (e_2 \cdot y_2 - \lambda_7 \cdot \tilde{c}_1) + \tilde{h}_1 \cdot (-e_3 \cdot y_3 - \lambda_8 \cdot \tilde{h}_1) \\ = {} & -\lambda_1 \cdot e_1^2 - \lambda_2 \cdot e_2^2 - \lambda_3 \cdot e_3^2 - \lambda_4 \cdot e_4^2 - \lambda_5 \cdot \tilde{a}_1^2 - \lambda_6 \cdot \tilde{b}_1^2 - \lambda_7 \cdot \tilde{c}_1^2 - \lambda_8 \cdot \tilde{h}_1^2 < 0\end{aligned}$$

From the Lyapunov theory of stability, the error system (8) of synchronization and parameter estimating error system (10) are asymptotically stable in origin. That is to say, with the effect of adaptive controller depicted by equation (7) and parameter adaptive law depicted by equation (9) the system (1) and system (5) will reach function projective synchronization.

## 4 The Analysis of Numerical Simulation

In order to verify the validity of the method presented in this paper the hyperchaotic Lorenz system (1) and the novel chaotic system (5) are numerically calculated respectively by Matlab. After the step is chosen as $\Delta t = 0.001$ by four orders Runge-Kutta the system (1), system (5), system (8) and system (9) are respectively numerically simulated. The simulated results of their synchronous error are shown in Figure 3. The simulated results of response system parameter estimation are shown in Figure 4-1, 4-2, 4-3 and 4-4. In these figures, the system (1) are respectively chosen as $x_1(0) = 1, x_2(0) = 2$, $x_3(0) = 3$, $x_4(0) = 4$. The state initial values of system (5) are respectively chosen as $y_1(0) = 5$, $y_2(0) = 6$, $y_3(0) = 7$, $y_4(0) = 0$. In addition, the initial value of response system unknown parameter are chosen as $\hat{a}_1(0) = 22$, $\hat{b}_1(0) = 15$, $\hat{c}_1(0) = 12$, $\hat{h}_1(0) = 4$, the actual value of these parameter are respectively $a_1 = 20$, $b_1 = 14$, $c_1 = 10.6$, $h_1 = 2.8$. The proportional function $\alpha(t)$=0.1+0.25 ·sin(t), the variable $(\lambda_1, \lambda_2, \lambda_3, \lambda_4, \lambda_5, \lambda_6, \lambda_7, \lambda_8)$ =(3,3, 3,3,3,3,3,3). From the Fig.3 it can be seen that the error of system state approaches to zero rapidly with the increase of time. From the Fig.4-1,4-2,4-3 and 4-4 it can be seen that the each uncertain parameter converges to a constant rapidly. And so the function projective synchronization between the Lorenz hyperchaotic system and the novel chaotic system is realized, the uncertain parameters of response system are estimated.

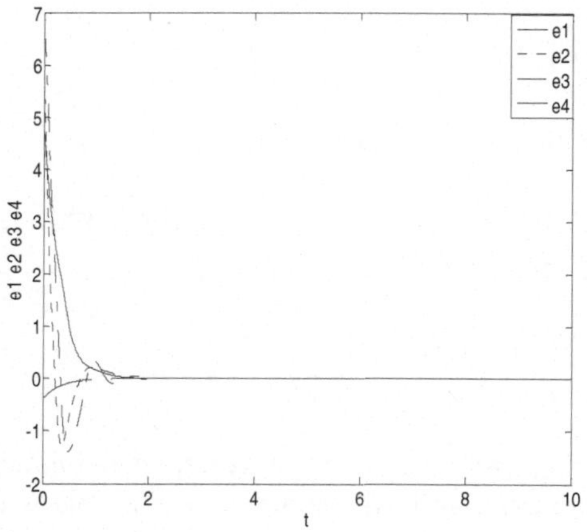


**Fig. 3.** The curve of synchronous error

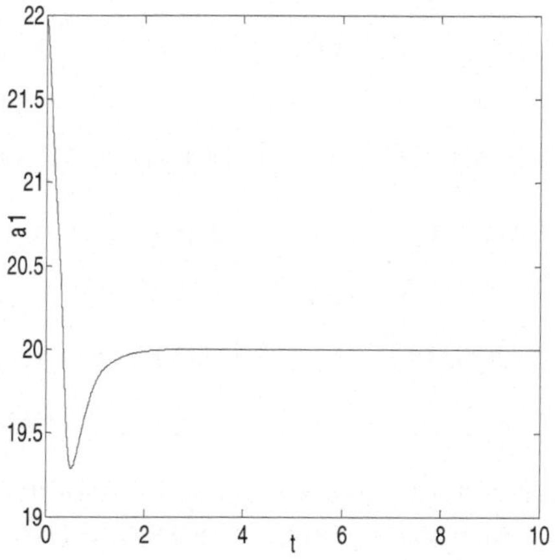


**Fig. 4-1.** The convergent curve of $a_1$ estimation

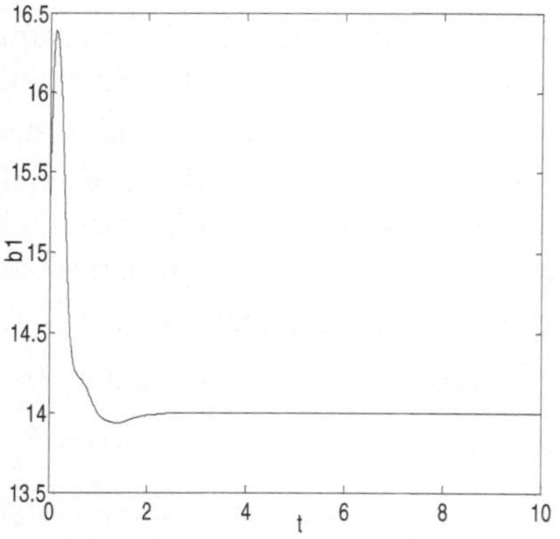


**Fig. 4-2.** The convergent curve of $b_1$ estimation

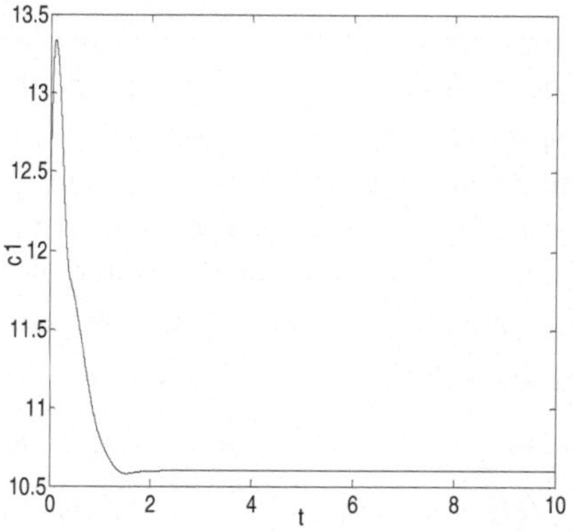


**Fig. 4-3.** The convergent curve of $c_1$ estimation

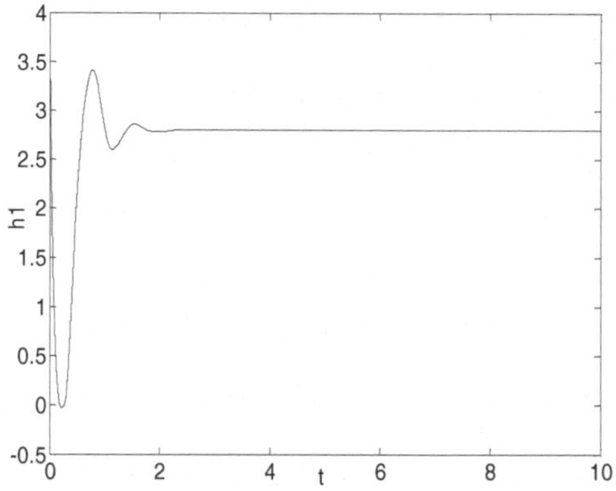


**Fig. 4-4.** The convergent curve of parameter $h_1$ estimation

## 5 Conclusions

In this paper, a novel three dimensions chaotic system and lorenz hyperchaotic systemare as examples, the function projective synchronization between a chaotic system with uncertain parameter and a hyperchaotic system are researched and the value of each uncertain parameter is also estimated. Based on the Lyapunov theory of stability and adaptive control method, the adaptive nonlinear controller and the adaptive identifying rule to uncertain parameter are designed. By the adaptive controller and the adaptive identifying rule to uncertain parameter, the function projective synchronization between a chaotic system with uncertain parameter and a given hyperchaotic system is realized. By adjusting the value of $\lambda_i$ $(i = 1, 2 \cdots 8)$ the speed of error between response system and drive system approaching to zero and the effect of parameter estimation can be improved. The applicable range of this method is wide and it is easy to realize. Because of solving the perturbation problem of response system parameter very well this method is very robust. Finally, on the basis of theoretical analysis, the validity of this method is verified by numerical simulation.

**Acknowledgment.** The National Science Foundation under Grant (10872156, 81071150), Shaanxi province Science Foundation under 2007014 and National post-doctor Foundation under 20080430203 support the work reported in this paper.

## References

1. Wang, G.R., Yu, X.L., Chen, S.G.: Chaotic control, synchronization and utilizing. National Defence Industry Press, Beijing (2001) (in Chinese)
2. Feng, C.W., Cai, L., Kang, Q., Zhang, L.S.: A novel three dimension autonomous chaotic system. Acta Phys. Sin. 60(3), 030503–030507 (2011) (in Chinese)

3. Pecora, L.M., Carroll, T.L.: Synchronization of chaotic systems. Physical Review Letters 64(8), 821–830 (1990)
4. Carroll, T.L., Pecora, L.M.: Synchronizing chaotic circuits. IEEE Transactions on Circuits and Systems 38(4), 453–456 (1991)
5. Dong, J., Zhang, G.J., Yao, H., et al.: The control of complete synchronization and anti-phase synchronization for hyperchaotic systems of different structures. Journal of Air Force Engineering University 13(5), 90–94 (2012)
6. Min, F.H., Wang, Z.Q.: General chaotic synchronization of identical and different novel hyperchaotic system. Journal of Electronics & Information Technology 30(12), 3031–3034 (2008) (in Chinese)
7. Zhang, G., Zhang, W.: Pulse synchronization of complex network. Journal of Dynamics and Control 7(1), 001–004 (2009)
8. Lu, J., Zhang, R., Xu, Z.Y.: Phase synchronization between two adjacent nodes in amplitude coupled dynamical networks. Acta Phys. Sin. 59(9), 5949–5953 (2010) (in Chinese)
9. Cai, N., Jing, Y.W., Zhang, S.Y.: Adaptive synchronization and anti-synchronization of two different chaotic systems. Acta Phys.Sin. 58(2), 0802–0813 (2009) (in Chinese)
10. Mainieri, R., Rehacek, J.: Projective synchronization in three-dimensional chaotic systems. Phys. Rev. Lett. 82(15), 3042–3045 (1999)
11. Li, Z.G., Xu, D.: Chaos. Solitons and Fractals 22, 477 (2004)
12. Chee, C.Y., Xu, D.: Chaos. Solitons and Fractals 23, 1063 (2005)
13. Kadir, A., Wang, X.Y., Zhao, Y.Z.: Projective synchronization for unified hyperchaotic systems. Acta Phys. Sin. 60(4), 040506 (2011) (in Chinese)
14. Cai, G.L., Tan, Z.M., Zhou, W.H., Tu, W.T.: Dynamical analysis of a new chaotic system and its chaotic control. Acta Phys. Sin. 56(11), 6230–6237 (2010) (in Chinese)

# A General Construction for Multi-authority Attribute-Based Encryption*

Guoyan Zhang

School of Computer Science and Technology,
Shandong University, Jinan 250100, China
guoyanzhang@sdu.edu.cn

**Abstract.** An attribute-based encryption scheme is a scheme in which each user is identified by a set of attributes, and some function of those attributes is used to determine decryption ability for each ciphertext. But as an extension for identity-based encryption scheme, the attribute-based schemes are also confronted with the key escrow problem. Furthermore, the attributes belonging to a user usually are monitored by different authorities. One approach to simultaneously resolve the two problems is multi-authority attribute-based encryption schemes, in which the secret keys of the users needed be distributed by different authorities. However, this solution comes at the cost of introducing extra infrastructure and communication.

This paper gives a new approach, in which different attributes sets of a user are still certified by different authorities, but the secret key corresponding to the attributes is generated by the central authority. In order to resolve key escrow problem, different authorities generate secret value for the user, but the central authority cannot obtain the secret value. We give a general construction for multi-authority attribute-based encryption scheme using a general attribute-based encryption scheme. Finally, we present a concrete attribute-based encryption scheme secure against the malicious authorities.

**Keywords:** Attribute-Based Encryption Scheme, Malicious Authorities, Key-Escrow, Multi-Authority.

## 1 Introduction

Sahai and Waters [1]introduced the concept of attribute-based encryption (ABE). In attribute-based encryption schemes, a user's keys and ciphertexts are labeled with sets of descriptive attributes and a particular key can decrypt a particular ciphertext only if there is a match between the attributes of the ciphertext

* This work is supported by the National Natural Science Foundation of China(No.60873232), Open Research Fund from Key Laboratory of Computer Network and Information Integration In Southeast University, Ministry of Education, China, Shandong Natural Science Foundation(No.Y2008A22) and Independent Innovation Foundation of Shandong University(No. 2012TS070).

J. Lei et al. (Eds.): AICI 2012, LNAI 7530, pp. 333–340, 2012.

and the users key. Two variants of ABE were subsequently proposed. In the key-policy variant (KP-ABE) of Goyal, Pandey, Sahai and Waters (GPSW) [2], every ciphertext is associated with a set of attributes, and every user secret key is associated with a threshold access structure on attributes. Decryption is enabled if and only if the ciphertext attributes set satisfies the access structure on the user secret key. In the ciphertext-policy variant (CP-ABE) of Bethencourt, Sahai and Waters (BSW) [3], the situation is reversed: attributes are associated with user secret keys and access structures with ciphertexts. Following, in order to make the access structure more expressive, many schemes have been presented [2–8]. Simultaneously, schemes [6, 9–11] were devoted to get constant-size ciphertexts. Similar to identity-based encryption schemes, the attribute authority is able to compute the private key corresponding to any attribute, and it has to be completely trusted. The KGC is free to engage in malicious activities without any risk of being confronted in a court of law. The malicious activities could include: decrypting and reading messages meant for any user, which is called the key escrow problem. One approach to mitigate the key escrow problem is to employ multi-authority attribute-based encryption, which allows the sender to specify for each authority $k$ a set of attributes monitored by that authority and a number $d_k$ so that the message can be decrypted only by a user who has at least $d_k$ of the given attributes from every authority. Multi-authority attribute-based encryption also allow any number of attribute authorities to be corrupted, and guarantee the security of encryption as long as the required attributes cannot be obtained exclusively from those authorities and the trusted authority remains honest. This is an attractive solution and successfully avoids placing trust in a single entity by making the system distributed. However, this solution comes at the cost of introducing extra infrastructure and communication.

### 1.1 Related Work

Building on the ideas from [12], Chase proposed a solution for multi-authority attribute-based encryption, provided that a trusted central authority is available [13], but a global identifier is a linchpin for tying users keys together. Her system relied on a central authority and was limited to expressing a strict AND policy over a pre-determined set of authorities. Müller, Katzenbeisser, and Eckert [14]give a system with a centralized authority that realizes any LSSS access structure. Their proof is limited to nonadaptive queries. The system achieves roughly the same functionality as the engineering approach above, except one can still acquire attributes from additional authorities without revisiting the central authority. The scheme [15]removed the central authority using a distributed PRF; however, the same limitations of an AND policy of a determined set of authorities remained. Lin et. al. [16] give a threshold based scheme that is also somewhat decentralized. The set of authorities is fixed ahead of time, and they must interact during the system setup. The system is only secure up to collusion of $m$ users, where $m$ is a system parameter chosen at setup such that the cost of operations and key storage scales with m. Scheme [17] proposes a new multi-authority attribute-based encryption system. In their system, any party can become an authority and there is no requirement for any global coordination other than the creation of an initial set of common reference parameters which will be created during a trusted setup). A party can simply act as an authority by creating a public key and issuing private keys to different users that reflect their attributes. Different authorities need not even be aware of each other.

## 2 Preliminaries

### 2.1 Definition

**Definition 1. (Key-Policy Attribute-Based Encryption Scheme).** A generic key-policy attribute-based encryption scheme consists of the following four algorithms:

**-SetUp:** a probabilistic polynomial time (PPT) algorithm run by a key generation center (KGC) given a security parameter $k$ and an universe of attributes $U$ as input which outputs a randomly chosen master secret key $msk$ and master public key $mpk$. The master public key $mpk$ includes a description of the message space $\mathcal{M}$ and ciphertext space $\mathcal{C}$.

**-PrivateKeyExtract:** given the master public key $mpk$, master secret key $msk$, an access tree $T$ over the universe of attributes $U$ and an attributes set $S \in U$ for entity $A$, the KGC runs this PPT algorithm to generate the partial private key $d_A$ for the attributes set $S$. Then the partial private key $d_A$ is transported to entity $A$ over a confidential and authentic channel.

**-Encrypt:** given a plaintext $M \in \mathcal{M}$, master public key $mpk$ as inputs, a sender runs this PPT algorithm to create a ciphertext $C \in \mathcal{C}$ or the null symbol $\perp$ indicating an encryption failure.

**-Decrypt:** given master public key $mpk$, the entity's private key $d_A$, and the ciphertext $C \in \mathcal{C}$ that was encrypted under the access tree $T$ as inputs, the entity as a recipient runs this deterministic algorithm to get a decryption $\sigma$, which is a plaintext message if the set $S$ of attributes satisfies the access tree $T$.

### 2.2 Security Model for Attribute-Based Encryption

We give the security model of the key-policy attribute-based encryption scheme as follows:

Selective-Set Model for $ABE$

**Init.** The adversary declares the set of attributes, $\gamma$ , that he wishes to be challenged upon.

**Setup.** The challenger runs the Setup algorithm of $ABE$ and gives the public parameters to the adversary.

**Phase 1.** The adversary is allowed to issue queries for private keys for many access structures $A_j$, where $\gamma$ cannot satisfy access structures $A_j$ for all $j$.

**Challenge.** The adversary submits two equal-length messages $M_0$ and $M_1$. The challenger flips a random coin $b \in \{0, 1\}$, and encrypts $M_b$ with $\gamma$. The ciphertext is passed to the adversary.

**Phase 2.** Phase 1 is repeated.

**Guess.** The adversary outputs a guess $\acute{b}$ of b.

The advantage of an adversary $A$ in this game is defined as $Pr[\acute{b} = b] - \frac{1}{2}$.

We note that the model can easily be extended to handle chosen ciphertext attacks by allowing for decryption queries in Phase 1 and Phase 2.

**Definition 2.** An attribute-based encryption scheme is secure in the selective-set model of security if all polynomial time adversaries have at most a negligible advantage in the selective-set game.

### 2.3 Complexity Assumptions

The following are the complexity assumption related with our scheme:

**Definition 3. (Decisional Modified Bilinear Diffie-Hellman (MBDH) Assumption).** Suppose a challenger chooses $a, b, c, z \in \mathcal{Z}_p$ at random. The Decisional MBDH assumption is that no polynomial-time adversary is to be able to distinguish the tuple $(A = g^a, B = g^b, C = g^c, Z = e(g,g)^{\frac{ab}{c}})$ from $(A = g^a, B = g^b, C = g^c, Z = e(g,g)^z)$ with more than a negligible advantage.

## 3 A General Construction for Multi-authority Attribute-Based Encryption Scheme

### 3.1 The Construction

We assume that there are $n$ attribute authorities, and every attribute authority $k$ monitors $d_k$ attributes,

**-SetUp:** a probabilistic polynomial time (PPT) algorithm run by a central authority given a security parameter $k$ and an universe of attributes $\mathcal{U}$ as input which outputs a randomly chosen master secret key $msk$ and master public key $mpk$. The master public key $mpk$ includes a description of the message space $\mathcal{M}$ and ciphertext space $\mathcal{C}$. Every attribute authority $k$ generates his own signature secret key $sk$ and verifying secret key $vk$.

**AuthorityCertify:** In order to certify the attribute set $i$ for user $A$, attribute authority $k$ generates signature $Sign_{s_k}(i, A)$ and sends this signature to the central authority with verifying secret key $v_k$.

**-SecretValueExtract:** the $n$ attribute authorities generate a distributed $(d-1)$-degree polynomial $p(x)$ and compute $p(i)$ to users who are in possession of the attribute $i$. The public key $pk$ is published. Any user who has $d$ attributes can recover the partial private key $\alpha = p(0)$.

**-PartialPrivateKey:** the secret key corresponding to the attribute is generated by the central authority normally as in general $ABE$ after he verifies the validity of the signature for the attribute and the user.

**-Encrypt:** The encryption scheme is obtained by encrypting the ciphertext $E_{pk}(m)$, it is to say, $C = E_{ABE}(E_{pk}(m))$.

**-Decrypt:** given master public key *mpk*, the secret value recovered by the user in possession of at least $d$ attributes, partial private key $d_A$, and the ciphertext $C \in \mathcal{C}$, the user as a recipient runs this deterministic algorithm to get a decryption $\delta$, which is a plaintext message if the set $S$ of attributes satisfies the access tree $T$. Otherwise outputs $\perp$.

# 4 Concrete Multi-authority Attribute-Based Encryption Scheme

## 4.1 Construction

Let $(G, G_T)$ be bilinear map groups of order $p > 2^k$ and let $e : G \times G \to G_T$ denote a bilinear map. Let $g$ be a generator for $G$. $k$ is the security parameter. $\Delta_{i,s}(x), i \in \mathcal{Z}_p$ is the lagrange coefficient, and $S$ is a set of elements in $\mathcal{Z}_p$. $\mathcal{U}$ is the universal set of attributes which is associated with a unique element in $\mathcal{Z}^*{}_p$. Our construction is as follows:

**Setup**($d$)**:** First, we take the integers $1, 2, \ldots, |\mathcal{U}|$ to be the universe.

Next, choose $t_1, \ldots, t_{|\mathcal{U}|}$ uniformly at random from $\mathcal{Z}_p$. Finally, choose $y$ uniformly at random in $\mathcal{Z}_p$. The published public keys are:

$$T_1 = g^{t_1}, \ldots, T_{|\mathcal{U}|} = g^{t_{|\mathcal{U}|}}, Y = e(g, g)^y.$$

The master key is:

$$t_1, \ldots, t_{|\mathcal{U}|}, y.$$

Every attribute authority $k$ generates his own signature secret key $s_k$ and verifying secret key $v_k$.

**AuthorityCertify:** in order to certify the attribute $i$ for user $u$, attribute authority $k$ generates signature $Sign_{s_k}(i, u)$ and sends this signature to the central authority with verifying secret key $v_k$.

**SecretValueExtract:** the $n$ attribute authorities generate a distributed $(d-1)$-degree polynomial $p(x)$ and compute $p(i)$ to users who are in possession of the attribute. Any user who has $d$ attributes can recover the secret value $\alpha = p(0)$, publishes the public key $Z = Y^\alpha$.

**PartialPrivateKeyExtract.** After verifying the signature for attributes set $\omega$, the central authority picks $(d-1)$ degree polynomial $q$ randomly so that $q(0) = y$. The partial private key is

$$D_i = g^{\frac{q(i)}{t-i}}, i \in \omega.$$

**Encrypt.** In order to encrypt message $m \in G_T$ with attributes set $\omega$, pick $s \in \mathcal{Z}_P$ and compute the ciphertext as follows:

$$C = (C_0, C_1, C_2) = (\omega', m \cdot Z^s, E_i = T_i^s, i \in \omega').$$

**Decryption.** Suppose that a ciphertext, $C$, is encrypted with a key for attributes set $\omega'$ and we have the partial private key $D_i$ and secret key $f(i)$ for $i \in \omega$, where $|\omega \bigcap \omega'| \geq d$. Choose an arbitrary $d$-element subset, $S$, of $\omega \bigcap \omega'$ and compute

$$\alpha = p(0) = \sum_{i \in S} p(i) \Delta_{i,s}(0),$$

$$m = C_1 / \prod_{i \in S} (((e(D_i, E_i))^{\Delta_{i,s}(0)})^{\alpha}.$$

## 5 The Security Analysis

**Theorem 1.** If the Modified Bilinear Diffie-Hellman problem is hard, then our attribute-based encryption scheme is IND-CPA secure.

Proof. Suppose $\mathcal{A}$ is a polynomial-time adversary, and he can success with advantage $\epsilon$, then we can construct a scheme $\mathcal{B}$ to resolve the Modified Bilinear Diffie-Hellman problem with advantage $\epsilon$ by calling $\mathcal{A}$. The following is the proceed:

$\mathcal{B}$ sets the groups $(G, G_T)$ be bilinear map groups of order $p > 2^k$ and let $e : G \times G \rightarrow G_T$ denote a bilinear map. Let $g$ be a generator for $G$, and he is given the tuple$(A, B, C, Z) = (g^a, g^b, g^c, Z)$ .

$\mathcal{B}$ runs $\mathcal{A}$ to obtain a challenge attributes set $\alpha$.

**Setup.** $\mathcal{B}$ sets the public key as follows. He sets $Y = e(g, A) = e(g, g)^a$. For all attributes $i \in \alpha$, he sets $T_i = C^{t_i} = g^{ct_i}$ for random $t_i$, and for all $i \in \mathcal{U} - \alpha$, he sets $T_i = g^{\omega i}$ for random $\omega_i$.

**phase1**

**PartialPrivateKeyExtract.** $\mathcal{A}$ makes requests for partial private key for attributes set $\gamma$ . If $|\gamma \cap \alpha| < d$. Set three sects $\Gamma, \Gamma_1, S$, where:

$$\Gamma = \gamma \cap \alpha, \Gamma \subseteq \Gamma_1 \subseteq \gamma$$

$$|\Gamma_1| = d - 1,$$

$$andset S = \Gamma_1 \cup \{0\}.$$

We can randomly define the secret key for $\Gamma_1$ as follows:

For $i \in \Gamma$, Set

$$D_i = g_i^s.$$

Where $s_i$ is chosen randomly.

For $i \in \Gamma_1 - \Gamma$, Set

$$D_i = g^{\frac{\lambda_i}{\omega_i}}.$$

Where $\lambda_i$ is chosen randomly.

Then from the definition, $\mathcal{B}$ have chosen a $d-1$ degree polynomial $q(x)$ by choose $(d-1)$ random value and set $q(0)=a$. Especially, $i \in \Gamma, q(i) = ct_i s_i$, and for $i \in \Gamma_1 - \Gamma, q(i) = \lambda_i$.

$\mathcal{B}$ can define the other secret key $D_i$:

$$D_i = (\prod_{j\in\Gamma} C^{\frac{t_j s_j \Delta_{j,S}(i)}{\omega_i}})(\prod_{j\in\Gamma_1-\Gamma} g^{\frac{\lambda_j \Delta_{j,S}(i)}{\omega_i}})Y^{\frac{\Delta_{0,S}(i)}{\omega_i}}.$$

**Secret Value Extract.** $\mathcal{A}$ asks a secret key for attributes set $\beta$, if $|\beta| > d$, Choose a polynomial $f(x)$ and compute $f(i)$ for $i \in \beta$, set $f(0) = x$. compute $Y_1 = Y^x$ and publish it. If $|\beta| < d$, choose $|\beta|$ random values to $\mathcal{A}$.

**Public Key Extract.** Publish $Y_1 = Y^x$.

**Challenge.** $\mathcal{A}$ submit two challenge message $M_1, M_0$ to $\mathcal{B}$, and $\mathcal{B}$ chooses a bit $b \in \{0\}$. The ciphertext is output as:

$$C = (C_0, C_1, C_2) = \alpha, m_b \cdot Y_1, E_i = B_i^t, i \in \alpha).$$

**phase2.** $\mathcal{B}$ acts exactly as it did in phase 1.

**Guess.** If $\mathcal{A}$ correctly guess the bit $b$,$\mathcal{B}$ will decide that the tuple$(A, B, C, Z)$ is the Modified Bilinear Diffie-Hellman tuple, else it is not.

From the above analysis, we find that the advantage of $\mathcal{B}$ is equal to the advantage of $\mathcal{A}$.

**Theorem 2.** If the discrete logarithm problem is hard, then our attribute-based encryption scheme is secure against malicious central authority.

Proof. From the public key $Y, Y_1$, we can see if the central authority want to get the secret key $x$, he must compute the discrete logarithm $\log_y^{Y_1}$, but the discrete logarithm problem is hard, and the central authority cannot obtain the secret key and he cannot obtain any information about the plaintext.

## 6 Conclusion

In order to mitigate the key escrow, this paper gives a new approach which adds new secret value to the user by the attribute authorities, and the central authority don't know the secret value. Compared with the general multi-authority, our approach is simple, and the length of the ciphertext and the public key published to the sender is not increased. Simultaneously, we present an attribute-based encryption scheme secure against the malicious central authority by modifying the scheme [1] using this new technique. Furthermore, compared with the original scheme, our scheme has the same length of the public key and the ciphertext. The security of our scheme is obtained directly from the security of the original scheme. But, our scheme only admits the "threshold" access control, and an efficient scheme admitting expressive access structure secure against malicious is our further research.

## References

1. Sahai, A., Waters, B.: Fuzzy Identity-Based Encryption. In: Cramer, R. (ed.) EUROCRYPT 2005. LNCS, vol. 3494, pp. 457–473. Springer, Heidelberg (2005)

2. Goyal, V., Pandey, O., Sahai, A., Waters, B.: Attribute-based encryption for fine-grained access control of encrypted data. In: Proceedings of the 13th ACM Conference on Computer and Communications Security (CCS 2006), pp. 89–98 (2006)
3. Bethencourt, J., Sahai, A., Waters, B.: Ciphertext-policy attribute-based encryption. In: Proceedings of the 28th IEEE Symposium on Security and Privacy (Oakland), pp. 321–334 (2007)
4. Ostrovsky, R., Sahai, A., Waters, B.: Attribute-based encryption with non-monotonic access structures. In: ACM Conference on Computer and Communications Security, pp. 195–203 (2007)
5. Lewko, A., Sahai, A., Waters, B.: Revocation Systems with Very Small Private Keys. In: IEEE Symposium on Security and Privacy (2010)
6. Attrapadung, N., Libert, B., de Panafieu, E.: Expressive Key-Policy Attribute-Based Encryption with Constant-Size Ciphertexts. In: Catalano, D., Fazio, N., Gennaro, R., Nicolosi, A. (eds.) PKC 2011. LNCS, vol. 6571, pp. 90–108. Springer, Heidelberg (2011)
7. Cheung, L., Newport, C.C.: Provably secure ciphertext policy abe. In: ACM Conference on Computer and Communications Security, pp. 456–465 (2011)
8. Goyal, V., Jain, A., Pandey, O., Sahai, A.: Bounded Ciphertext Policy Attribute Based Encryption. In: Aceto, L., Damgård, I., Goldberg, L.A., Halldórsson, M.M., Ingólfsdóttir, A., Walukiewicz, I. (eds.) ICALP 2008, Part II. LNCS, vol. 5126, pp. 579–591. Springer, Heidelberg (2008)
9. Daza, V., Herranz, J., Morillo, P., Ràfols, C.: Extended access structures and their cryptographic applications. To appear in Applicable Algebra in Engineering, Communication and Computing (2008), http://eprint.iacr.org/2008/502
10. Emura, K., Miyaji, A., Nomura, A., Omote, K., Soshi, M.: A Ciphertext-Policy Attribute-Based Encryption Scheme with Constant Ciphertext Length. In: Bao, F., Li, H., Wang, G. (eds.) ISPEC 2009. LNCS, vol. 5451, pp. 13–23. Springer, Heidelberg (2009)
11. Herranz, J., Laguillaumie, F., Ràfols, C.: Constant Size Ciphertexts in Threshold Attribute-Based Encryption. In: Nguyen, P.Q., Pointcheval, D. (eds.) PKC 2010. LNCS, vol. 6056, pp. 19–34. Springer, Heidelberg (2010)
12. Chase, M.: Multi-authority Attribute Based Encryption. In: Vadhan, S.P. (ed.) TCC 2007. LNCS, vol. 4392, pp. 515–534. Springer, Heidelberg (2007)
13. Müller, S., Katzenbeisser, S., Eckert, C.: Distributed Attribute-Based Encryption. In: Lee, P.J., Cheon, J.H. (eds.) ICISC 2008. LNCS, vol. 5461, pp. 20–36. Springer, Heidelberg (2009)
14. Müller, S., Katzenbeisser, S., Eckert, C.: On multi-authority ciphertext-policy attributebased encryption. Bulletin of the Korean Mathematical Society 46(4), 803–819 (2009)
15. Chase, M., Chow, S.: Improving privacy and security in multi-authority attribute-based encryption. In: ACM Conference on Computer and Communications Security, pp. 121–130 (2009)
16. Lin, H., Cao, Z., Liang, X., Shao, J.: Secure Threshold Multi Authority Attribute Based Encryption without a Central Authority. In: Chowdhury, D.R., Rijmen, V., Das, A. (eds.) INDOCRYPT 2008. LNCS, vol. 5365, pp. 426–436. Springer, Heidelberg (2008)
17. Lewko, A., Waters, B.: Decentralizing Attribute-Based Encryption. In: Paterson, K.G. (ed.) EUROCRYPT 2011. LNCS, vol. 6632, pp. 568–588. Springer, Heidelberg (2011)

# GIS-Based Urban Land Development Intensity Impact Factors Analysis

Minghao Liu[1,2,*], Yaoxin Wang[1], Zhizhong Dai[2], and Qiyuan Li[3]

[1] School of Computer Science, Chongqing University of Posts and Telecoms, Chongqing, 400065, China
[2] College of Architecture and Urban Planning, Chongqing University, Chongqing 400045, China
[3] School of Management China West Normal University, Nanchong, Sichuan, China
1516398568@qq.com

**Abstract.** Urban land development intensity (ULDI) is an important indicator to measure urban livability and sustainable development. Using Jönköping Municipality of Sweden as a case, Geostatistics were explored by integrating methods of GIS spatial and SPSS regression analysis to analyze the correlation between ULDI and driving factors. The results showed that the spatial distribution of all ULDI gradients could well be explained by the selected driving variables as indicated by the high Relative Operating Characteristics (ROC) test statistics (>0.7). The spatial distribution of land use gradient has shown a strong distance decline law and the characteristic of centripetal.

**Keywords:** urban land development intensity (ULDI), neighborhood enrichment index (NEI), regression analysis methods, neighborhood analysis, Sweden, GIS.

## 1 Introduction

Urban land development intensity (ULDI) is an important indicator to measure urban livability and sustainable development. Some debates have arisen on how to understand ULDI since the conception of sustainable development was formed in the late 1980s.

From a regional perspective, land development intensity within a certain area is land relative development degree and its cumulative carrying density under the condition of regional environmental carrying capacity. The purpose of this research is to constrain the developing behavioral norms within the framework of "causing the smallest possible negative feedback". Zhou proposed land development intensity evaluation index consisting of four aspects, namely the conditions of technical support, degree of development, development efficiency, and the feedback effect of resources and ecological environment [1]. Yao Deming (2008) selected 14 indicators to build the evaluation index system of LDI [2]. The economic region (Zhou Bing zhong, 2000) and municipalities (counties) (Yao Deming, 2008) were chosen as the

* Corresponding author.

J. Lei et al. (Eds.): AICI 2012, LNAI 7530, pp. 341–348, 2012.

evaluation unit. Sometimes, according to research needs, a regional LDI refers to the build-up land accounted for a proportion of the total land area of the region.

From the perspective of urban development, LDI usually includes the indicators of development conditions, technical support, development degree, and development efficiency. From the narrow point of view, LDI refer to development degree which usually includes the floor area ratio, building density, building height, and ratio of green space. Three variables, including net employment density, floor space per unit area (floor area ratio), and employment per unit floor space, were used to reveal the spatial patterns of employment and floor space [3]. (John F. McDonald,1985). In order to identify low-density, dispersed settlements, seven building density classes were considered to express ULDI [4] (L. Salvati et al ,2012).

In our research, the building density is also used to measure ULDI according the data availability.

Examining the interdependence between land use and driving factors is a long tradition. Some research focuses on dynamics of land use change and land use pattern conversion from the perspective of land use and land cover change (LUCC). Researchers typically attribute the dynamics of landscape change to five major driving forces: political, economic, cultural, technological, and natural/spatial [5] [6] [7] [8] (e.g., Lambin et al., 2001; Schneeberger et al., 2007; HU Zhao-ling et al,2007; Hersperger and Matthias, 2009). The natural/spatial factors mainly include site and non-site factors [9] (Y. Ye et al.,2011). Site factors, including landscape types and the distances to roads and city centers, had significant impacts on the expansion of construction land.

More recent work has focused on assessing urban land use intensification potential and the bearing capacity of resources and the environment[10] [11] [12] [13] [14] (Hongzeng Lin, et al.,2006; Y. Meng,2008;Chen, Y.2006, Fan, J. et al.,2009; Y. Xu et al.2011). Previous work on LDI, however, limited their focus of the land use changes from the perspective of regional sustainable development to a relatively large scale. Furthermore, though urban area was given much attention, most studies, in recent work, concentrated on assessing urban land use intensification or on a specific driving factor, such as traffic[15], ecological [16] (D. Hu et al.2008),or land price[17] (B.s. Tang, 2010). In contrast, very little work has been reported on how to examine the correlation between ULDI and driving factors from the perspective of Geostatistics.

The objectives of the study are as follows: (1) to explore the spatial characteristics of construction land; (2) to identify the internal drivers of ULDI; and (3) to detect and evaluate the correlation between ULDI and driving factors.

# 2 Materials and Methods

## 2.1 Description of the Study Area

Jönköping Municipality (Jönköpings kommun) is a municipality in Jönköping County, southern Sweden. The city of Jönköping with 89,396 inhabitants (2010) is the municipal seat. Jönköping is situated at the southern end of Lake Vättern in one of Sweden´s economically most prosperous regions. The municipality has a population of 128,305 and is rapidly expanding with an annual population growth of approximately 1,000, and ranks among Sweden´s ten largest municipalities.

The Municipality of Jönköping, which has an area of some 192807hm², includes the local districts of Jönköping, Huskvarna, Gränna, Visingsö, Skärstad, Lekeryd, Tenhult, Barnarp, Norrahammar, Månsarp, Bankeryd, and Norra Mo.

## 2.2 Data and Methods

The land cover database is derived from Corine land cover 2009. It is the year 2009 update of the first CLC database which was finalized in the early 1990s as part of the European Commission programme to COoRdinate INformation on the Environment (Corine). Spatial resolution is 100*100 meter. DEM data and administrative maps come from Corine database. Spatial resolution of DEM data is 1000*1000 meter and is reclassified into 100*100 meter. The traffic data comes from OpenStreetMap (http://download.geofabrik.de/osm/europe/). The roads were divided into trunk roads (railway and expressway) and other roads in this paper. Population data is provided by Eurostat (http://epp.eurostat.ec.europa.eu). Spatial resolution of Population data is 100*100 meter.

In this study, as in the case of Jönköping Municipality of Sweden, the use of regression analysis methods and the neighborhood analysis method were explored to analyze the correlation between ULDI and driving factors.

On the one hand, regression equations were established by using an integrated method of GIS and SPSS. On the basis of ULDI classification and zoning, three ULDI gradients were formed, and seven factors such as elevation, slope, closest distance to main roads (railway and trunk), the other roads, waters, as well as the market were selected from natural environmental factors and socio-economic factors. After collecting spatial data of these factors, a unified spatial database (uniform projection coordinates, resolution and scope of the study) was established with ArcGIS software. The vector data of the ULDI classification and the impact factors data were converted to raster data. Through the process of the raster to points spatial analysis functions of ArcGIS, each factor data was put into the SPSS software for regression analysis. The value of Relative Operating Characteristics (ROC) method was used to test regression analysis results. Based on the regression parameters, the weight sum function was used to calculate the probability maps of the land intensity class.

On the other hand, certain distances from the center of downtown (near the central station of Jönköping) were chosen to measure the enrichment of ULDI gradient by using neighborhood analysis method.

### 2.2.1 ULDI Classification

In this study, urban land in Jönköping Municipality is divided into three types of LDI zones according to the land-use properties and building density.

Various types of land were extracted by attribute in ArcGIS platform. At the same time the different land use types were merged to a high development intensity zone, such as continuous urban areas (SL> 80%) and discontinuous density urban areas (SL: 50% - 80%); discontinuous density urban areas (SL: 30% - 50%) and intermittent low-density urban areas were combined into middle development intensity zones; industrial zone, urban greening, the leisure and entertainment facilities land, intermittent very low density were combined into low intensity of development zones. The merged layers were converted into raster data format for raster spatial analysis.

### 2.2.2 Impacting Factors of ULDI

In this section, the various factors impacting the land development intensity are provided. Two factor groups comprising 6 separate sets account for the land development intensity, which include elevation, slope, population density, distance to road, distance to market, and distance to water (Fig.1).

Topography is an important determinant factor of shaping land development intensity. In general, regions with low elevation are more suitable for human occupation than those at high elevation. A higher slope value indicates a steeper incline. In general, a higher slope indicates a smaller chance to be used by human activities.

Distance parameters were employed to control the impact extent. Distance to road is used to describe the accessibility to service. Distance to market is used to describe the chance or opportunity for urban development.

Distance to water body is also used to describe the accessibility and chance of human recreation. In general, a higher population density indicates higher land development intensity. The different land use patterns are related to different human activities.

### 2.2.3 Logistic Regression Analysis

Geostatistics is a discipline for studying the natural phenomena with characteristics of randomness, structure, spatial correlation, dependence based on regionalized variables, and variogram.

Currently, Geostatistics is widely used in many fields, which has become an important branch of spatial statistics. There are two ways to implement Geostatistics

1) Different index data is chosen to calculate in the mathematical model, and then the results are linked with the GIS spatial data through geocoding to visualize the results.

2) GIS spatial analysis data is imported into other software or mathematical models to be calculated, and then the results or parameters are visualized in GIS.

Here, the second method was adopted by integrating the logistic regression model in SPSS and the GIS analysis model.

In fact, the logistic regression model was used to evaluate the possibility of one land type being converted to another type by constructing the correlation between the various influencing factors and various types of land, namely land suitability for different types. The Logistic regression can be expressed as follows:

$$\log\left(\frac{P_i}{1-P_i}\right) = \beta_0 + \beta_1 X_{1i} + \beta_1 X_{2i} + \cdots + \beta_n X_{ni} \tag{1}$$

Here, $P_i$ is the probability for the occurrence of the considered land use type $i$ for a grid cell where X is the driving factors.

### 2.2.4 Neighborhood Analysis

In addition to regression analysis, this study also introduces the concept of neighborhood enrichment index (NEI) to analyze the spatial correlation between LDI and the driving factors.

NEI is defined by the occurrence of a land use type in the neighborhood of a location relative to the occurrence of this land use type in the study area as a whole following:

$$F_{i,k,d} = \frac{N_{k,d,i}/N_{d,i}}{N_k/N} \tag{2}$$

In which:

$F_{i,k,d}$ characterizes the enrichment of neighborhood d of location i with land use type k;

$N_{k,d,i}$ is the number of cells of land use type k in the neighborhood with size d of cell i ;

$N_{d,i}$ is the total number of cells in the neighborhood d of location i;

$N_k$ is the number of cells with land use type k in the whole raster;

N is the total number of cells in the raster.

## 3 Results and Discussion

### 3.1 The Results of Logistic Regression Analysis

After a series of data processing steps in ArcGIS, including unified geographic coordinates and projections, unified studies scope and resolution, ArcGIS raster data is converted to point data (dbf format), which is output to SPSS software. For each type of land use and its driving factors, logistic stepwise regression statistical method was used and Relative Operating Characteristics (ROC) method put forward by Pontius and Schneider(2001) [18] is used to indicate the validation of the models. Analysis results are shown in Table 1.

**Table 1.** Beta values[a] for regression results of the spatial distribution of different ULDI types in 2009

| | grade 1 | | grade 2 | | grade 3 | |
|---|---|---|---|---|---|---|
| Driving factors | B | Exp(B)-1 | B | Exp(B)-1 | B | Exp(B)-1 |
| elevation | -,061 | 0.059 | -,008 | -0,008 | -,005 | -0,005 |
| slope | ,760 | 1,138 | -,151 | -0,14 | ,051 | 0,052 |
| Population | ,003 | 0,003 | ,001 | 0,001 | ,001 | 0,001 |
| distance to water | -,002 | -0,002 | | | ,000 | 0 |
| distance to market | | | 0 | 0 | ,000 | 0 |
| distance to main roads | 0 | 0 | 0 | 0 | ,000 | 0 |
| Other roads | | | -,002 | -0,002 | -,002 | -0,002 |
| Constant | -6,489 | -0,998 | -3,686 | -0,975 | -1,323 | -0,734 |
| ROC | 0.988 | | 0.868 | | 0.801 | |

[a] All variables (driving forces) significant at $p < 0.01$).

Based on this GIS dataset, logistic regression models were constructed to determine the relations between ULDI gradient and potential driving factors. For each of the gradients, a logistic regression was run in 2009 (Table 1). Seven driving factors, including population density, elevation, slope, distance to main roads, distance to other roads, distance to market, and distance to water body were selected to evaluate the suitability of a certain grid cell to be converted to a development intensity type.

Table 1 shows the results of regression analysis. Exp (B) is the beta coefficient of the natural function base e exponential, and its value is equal to Odds Ratio, Exp (B) values indicate the change in odds upon one unit change in the independent variable. When Exp (B) -1>0, the probability increases upon an increase in the value of the independent variable; when Exp (B) -1 <0 the probability decreases. The driving factors that have no significant contribution to the explanation of the land use distribution are excluded from the final regression equation.

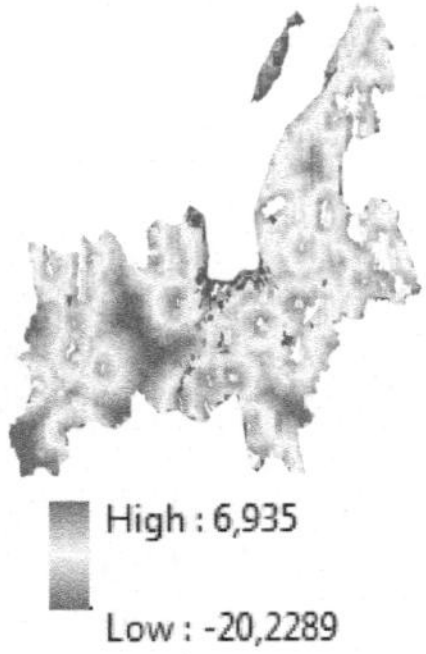


**Fig. 2.** Map of Grade 1

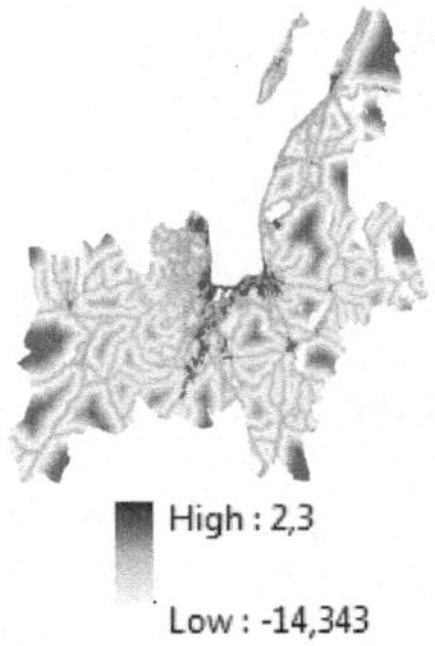


**Fig. 3.** Map of Grade 2

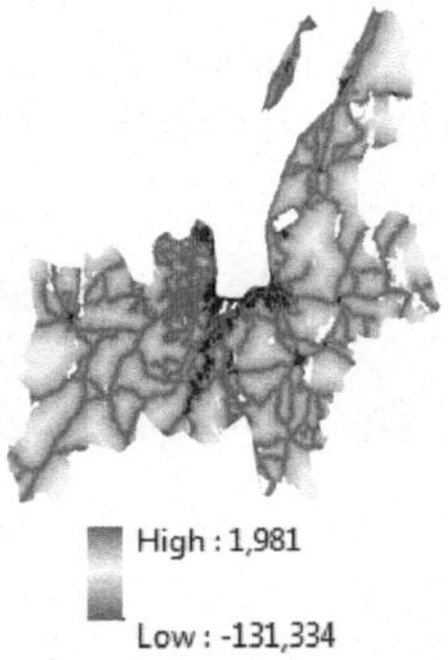


**Fig. 4.** Map of Grade 3

In this research, Exp (B) -1 is a positive value for the selected driving factors such as slope, population density in the grade 1 and 3 columns. When slope values increase in the urban construction land, the probability increases in ULDI. The possible reason is that Jönköping urban construction land is located in south, west and east sides of Vättern. The gentle slope and low-lying places are often occupied by lakes and wetlands, increasing the slope value account for the increase of urban land in grades 1 and 3, but for grade 2, increasing the slope value accounts for the decrease of urban land. The reason is that grade 2 is mainly occupied by industry sectors. The results show that the industry land use is more sensitive to slope.

For the factors of elevation and distance to water body in grade one, the Exp (B) -1 is negative, indicating that they are negatively correlated with ULDI, which is also consistent with the realities.

The spatial distribution of ULDI classification could well be explained by the selected driving variables as indicated by the high ROC test statistics, the ROC parametric test results were 0.988, 0.868, and 0.801. It should be noted that the spatial distribution of factors may not directly result in land development intensity changing, but there is a quantitative relationship between land development intensity and their impacted spaces factors, and those quantitative relationships may change over time.

The probability distribution of each raster was calculate according to the formula (1) through the ArcGIS Spatial Analysis Tool (the overlay/Weight sum) when regression parameters were obtained in the SPSS (table1). The map of probability distribution was formed in ArcGIS (Fig.2, Fig.3, and Fig.4).

### 3.2 The Results of Neighborhood Analysis

In our research, is the raster in the center of Jönköping Municipality (near the central station of Jönköping), 500m and 1000m from were chosen to measure the enrichment of land use types.

Firstly, the center raster of city was extracted based on vector data through raster data conversion in ArcGIS. Secondly, Euclidean distance analysis, re-classification and spatial data plus analysis (spatial analysis tools/math/plus) were performed and parameters data were extracted. Finally, the NEI was calculated within the range of 500m and 1000m according to the formula (2). The greater in number for NEI, the more concentration is indicated for a type of land use in a given space (Tab.2.).

**Table 2.** The results of neighborhood enrichment index

| NEI | $F_{i,1,500}$ | $F_{i,2,500}$ | $F_{i,3,500}$ | $F_{i,1,500\text{-}1000}$ | $F_{i,2,500\text{-}1000}$ | $F_{i,3,500\text{-}1000}$ |
|---|---|---|---|---|---|---|
| The results | 445.25 | 12.28 | 15.15 | 177.22 | 17.46 | 10.58 |
| NEI | $F_{i,1,0\text{-}1000}$ | $F_{i,2,0\text{-}1000}$ | $F_{i,3,0\text{-}1000}$ | $F_{i,1,>1000}$ | $F_{i,2,>1000}$ | $F_{i,3,>1000}$ |
| The results | 242,39 | 15,12 | 11,30 | 0.56 | 0.97 | 0.98 |

The results (Tab2) show that NEI in the grade 1, 2, and 3 of land development intensity types are 445.25, 12.28, 15.15; 177.22, 17.46, 10.58; and 0.56, 0.97, 0.98. Most raster of grade 1 were concentrated within 500 meters from the city center. The NEI reached to 445.25, although the NEI declined in the range of 500-1000m. There are still 177.22, which show obviously the characteristic of centripetal; NEI for grade 2 is slightly larger in the range of 500-1000m, but did not show obviously the characteristic of centripetal. NEI for grade 3 in the 500m, 500-100m, and beyond the range 1000m declined in turn from 15.15, 10.58, to 0.98, which shows the characteristic of layered structure. On the whole, grades 1 and 2 land types have shown a strong distance decline law - farther from the center, smaller of the NEI.

**Acknowledgements.** We thank Neil, Darielle Axe for their linguistic revision. This research was financially supported by Natural Science Foundation Project of CQ CSTC (Grant NO.2011jjA30014), and National Social Science Foundation of China (Grant NO. 11GBL051).

## References

1. Zhou, B.-Z., Bao, H.-S.: Evaluation on Exploitative Intensity of Land Resources in the Yangtze River Delta Region. Scientia Geographic Sinica 20(3), 218–223 (2000)
2. Yao, D.-M., et al.: Research of the Land Developing Intensity Evaluation of Hainan Province. Journal of Hebei Agricultural Sciences 12(1), 86–90 (2008)
3. Mcdonald, J.F.: The Intensity of Land Use in Urban Employment Sectors: Chicago 1956 – 1970. Journal of Urban Economics 18, 261–277 (1985)

4. Salvati, L., et al.: Low-density settlements and land use changes in a Mediterranean urban region. Landscape and Urban Planning 105, 43–52 (2012)
5. Lambin, E.F., Geist, H.J., Lepers, E.: Dynamics of land-use and land-cover change in tropical regions. Ann. Rev. Environ. Resour. 28, 205–241 (2003)
6. Schneeberger, N., Burgi, M., Hersperger, A.M., Ewald, K.C.: Driving forces and rates of landscape change as a promising combination for landscape change research – an application on the northern fringe of the Swiss Alps. Land Use Policy 24(2), 349–361 (2007)
7. Hu, Z.-L., et al.: Analysis of Urban Expansion and Driving Forces in Xuzhou City Based on. Journal of China University of Mining & Technology 17(2), 268–271 (2007)
8. Hersperger, A.M., Matthias, B.: Going beyond landscape change description: quantifying the importance of driving forces of landscape change in a Central Europe case study. Land Use Policy 26(3), 640–648 (2009)
9. Ye, Y., et al.: Research on the influence of site factors on the expansion of construction land in the Pearl River Delta, China: By using GIS and remote sensing. International Journal of Applied Earth Observation and Geoinformation xxx, xxx–xxx (2011)
10. Hong, Z., et al.: Evaluation Index System of Land Use Intensification Potential in Urban Area. Journal of Earth Sciences and Environment 28(1), 106–110 (2006)
11. Meng, Y., et al.: Industrial land-use efficiency and planning in Shunyi, Beijing. Landscape and Urban Planning 85, 40–48 (2008)
12. Chen, Y., Du, P., Zheng, Y., Lin, J.: Evaluation on ecological applicability of land construction in Nanning city based on GIS). Journal of Tsinghua University (Science & Technology) 46, 801–804 (2006) (in Chinese)
13. Fan, J., Li, P.: The scientific foundation of Major Function Oriented Zoning in China. Journal of Geographical Sciences 19, 515–531 (2009)
14. Xu, Y., et al.: Assessing construction land potential and its spatial pattern in China. The assessment of construction land potential. Landscape and Urban Planning 103, 207–216 (2011)
15. Ghen, K., Dai, Z.: Projects Development Intensity Critical Control System Based on Road Traffic. Urban Planning Forum 187(2), 41–47 (2010)
16. Hu, D., et al.: Relationships between rapid urban development and the appropriation of ecosystems in Jiangyin City, Eastern China. Landscape and Urban Planning 87, 180–191 (2008)
17. Tang, B.S., Yiu, C.Y.: Space and scale: A study of development intensity and housing price in Hong Kong. Landscape and Urban Planning 96, 172–182 (2010)
18. Pontius Jr., R.G., Schneider, L.C.: Land-cover change model validation by an ROC method for the Ipswich watershed, Massachusetts, USA. Agriculture, Ecosystems and Environment 85, 239–248 (2001)

# Application of Adaptive Fuzzy Control in the Variable Speed Wind Turbines

Mohamad Eydani Asl, Seyed Hamidreza Abbasi, and Faridoon Shabaninia

School of Electrical and Computer Engineering
Shiraz University, Iran
mohammad.eydani@gmail.com, hamid_abs62@yahoo.com,
shabani@shirazu.ac.ir

**Abstract.** In order to maximize energy capturing in wind turbines, the wind turbine generator needs to tune the speed of the wind turbine according to the wind. As it is obvious from the literatures, conventional controllers have low efficiency to obtain the better dynamic quality and robustness based on the mathematic models due to uncertainty of wind speed and wind direction. In this paper, an adaptive fuzzy controller is proposed for the variable speed wind power system with uncertain model parameters and force disturbances. The control scheme is based on indirect adaptive fuzzy method using feedback linearization. The uncertainties are approximated by a fuzzy approximator and an adaptive law improves the approximation. The stability of the proposed controller is proved using lyapunov method. Simulation results indicate that the strong robustness and better tracking performance can be achieved rather than a conventional PI controller.

**Keywords:** Wind Turbine, Adaptive Fuzzy Control, Feedback Linearization.

## 1 Introduction

Wind power technology has a rapid development in response to the demands for increased use of green energy [1]. The advances in wind power technology made necessary the design of efficient control systems. Thus, in order to improve wind turbines performance, namely to make them more efficient and reliable, a good regulation of the electrical power and a reduction of the loads on the different parts of the wind turbine have been the primary objectives [2- 4].

In recent years, it has shown that the control theory plays an important role in modern wind energy conversion systems (WECS). Thus, in order to improve quality and quantity of produced green energy from the wind turbine, various control structures such as PI regulator [5, 6], optimal control in LQ [7], and LQG form [4,8] have been developed. These control schemes which use the pitch angle as a control input give acceptable performance for rotor speed regulation, but showed poor results in power regulation.

Fuzzy control doesn't need exact model of controlled model and is suitable for strong coupling, time-varying and nonlinear system or control. It was applied in wind

J. Lei et al. (Eds.): AICI 2012, LNAI 7530, pp. 349–356, 2012.

direction measurement because of its simple and good robust [9-10]. The paper [11] proposed a fuzzy controller to change the stator voltage of the wind turbine, and add anti-torque generator to change the output power, but the response of the system to disturbances is not strong. In order to improve the flexibility of controller, an adaptive algorithm which updates the fuzzy sets during operation is useful.

Adaptive fuzzy control theory has many unique merits to solve mentioned problem. There have been a wide variety of applications of adaptive fuzzy control in areas such as power plants and nonlinear systems [12-17]. The most prominent property of the proposed approach is its insensitivity to parameter variations and external disturbances, robustness.

In this paper a fuzzy adaptive controller based on indirect approach is proposed for variable speed wind power system with uncertain model parameters and force disturbances. The uncertainties are approximated by an adaptive fuzzy approximator and the approximated model is used in the feedback linearization scheme. The proposed method is applied to speed tracking control of a variable speed wind power system. Simulation results indicate that the control approach is robust and improves tracking accuracy considerably.

This paper is organized as follows. First, the system model of variable speed wind power system is analyzed in section 2. Then the design of indirect fuzzy adaptive controller with adaptive law is presented in section 3 and the stability of the method is proved. In section 4, the simulation results for variable speed wind power system with proposed control concept are performed. Finally the paper is concluded in section 5.

## 2 Wind Turbine Generator Model

The typical diagram of wind turbine generator system is illustrated in Fig. 1 It is composed of wind turbine, gearbox and generator:

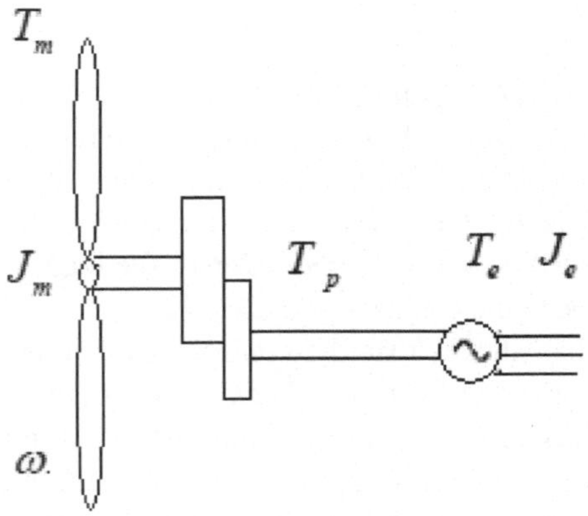


**Fig. 1.** The structure of wind turbine system

Based on Betz theory, the wind torque extracted from the wind can be expressed as the following equation,

$$p_m = k_w \omega^3 . \tag{1}$$

Where

$$k_w = \frac{1}{2} c_p(\lambda) \rho \pi \frac{R^5}{\lambda^3} . \tag{2}$$

$\omega$ is rotor speed, is air density, R is wind wheel radius, $c_p$ is power coefficient (maximum value Betz's limit 59.3%), v is wind speed, is tip speed ratio, the ratio is defined:

$$\lambda = \frac{\omega R}{v} . \tag{3}$$

The dynamics equation of the whole wind turbine generator system is [7],

$$T_m - T = J_m \dot{\omega} + B_m \omega + K_m \theta . \tag{4}$$

$$T_p - T_e = J_e \dot{\omega}_e + B_e \omega_e + K_e \theta_e . \tag{5}$$

Where, $T_m$ and $T_e$ are torque of wind turbine and generator. T is the torque before gearbox and $T_p$ is the torque after gearbox. $J_m$ and $J_e$ are wind turbine rotor inertia and generator inertia. $\omega_e$ is generator speed, $\theta$ and $\theta_e$ are axis angle distance of wind turbine and generator, $B_m$ , $K_m$ , $B_e$ , $K_e$ are coefficient of friction. Speed ratio of gearbox is,

$$\gamma = \frac{\omega_e}{\omega} . \tag{6}$$

Using

$$T_p \omega_e = T\omega . \tag{7}$$

Combining (3) and (4), yields

$$J\dot{\omega} + B\omega + K\theta = \frac{p_m}{\omega} - n\frac{p_e}{\omega_e} . \tag{8}$$

Where

$$\begin{cases} J = J_m + n^2 J_e \\ B = B_m + n^2 B_e \\ K = K_m + n^2 K_e \end{cases} . \tag{9}$$

$p_e$ is the generator energy form the system,

$$p_e = K_\phi \omega_e \, c(I_f) . \tag{10}$$

$K_\phi$ is mechanical correlation coefficient, $c(I_f)$ is the magnetic flux of the generator.

$$J\dot{\omega} + B\omega + K\theta = k_w \omega^2 - \gamma K_\phi c(I_f) . \tag{11}$$

Assuming

$$f(\omega) = \frac{k_w\omega^2 - B\omega - K\theta}{J}\ , A = \frac{\gamma K_\phi}{J}\ , u = c(I_f)\ . \tag{12}$$

Thus, (11) can be rewritten as

$$\dot{\omega} = f(\omega) - Au\ . \tag{13}$$

# 3 Adaptive Fuzzy Control

## 3.1 Control Law

Following the Standard method of feedback linearization, the output of controller is defined:

$$u = \frac{1}{A}\left[f(\omega) - \dot{\omega}_{ref} + Ke\right]\ . \tag{14}$$

Where

$$\omega_{ref} = \frac{v.\lambda_{opt}}{R}\ . \tag{15}$$

K is a positive constant and we define error as

$$e = \omega - \omega_{ref}\ . \tag{16}$$

However, in the practical process, the operation condition is complex and $f(x)$ usually cannot be precisely obtained. So it is difficult to implement the control law (14). In this paper, according to indirect adaptive fuzzy control method, a fuzzy model is utilized to solve above problem.

## 3.2 Fuzzy Approximation Function

In the paper the T-S fuzzy logic system $\hat{f}(x|\Theta_f)$ is used to approximate the unknown function $f(x)$. The fuzzy logic system is briefly described. The basic configuration of the fuzzy logic system includes a fuzzy base, consists of a collection of fuzzy IF-THEN rules is written as:

$$R^i: IF\ x_1\ is\ f_1^i\ and\ x_2 is\ f_2^i \dots and\ x_n is\ f_n^i\ , THEN\ y = B^j\ . \tag{17}$$

By using the singleton fuzzification, product inference and center average defuzzification, the output value of the fuzzy system is [16]:

$$y(x) = \frac{\sum_{j=1}^{m} y^j\left(\prod_{i=1}^{n}\mu_{f_j^i}(x_i)\right)}{\sum_{j=1}^{m}\left(\prod_{i=1}^{n}\mu_{f_j^i}(x_i)\right)}\ . \tag{18}$$

Where, $\mu_{f_j^i}(x_i)$ is the membership function of the linguistic variable $x_i$. By introducing the vector $\xi(x)$, (18) can be rewritten as

$$y(x) = \Theta^T \xi(x) . \tag{19}$$

Where

$$\begin{gathered} y = [y^1 \quad y^2 \ldots \quad y^m] \\ \xi(x) = [\xi^1(\mathrm{x}) \quad \xi^2(\mathrm{x}) \quad \ldots \quad \xi^{\mathrm{m}}(\mathrm{x})]^{\mathrm{T}} \\ \xi_i(x_1, x_2, \ldots, x_n) = \frac{\prod_{i=1}^{n} \mu_{f_j^i}(x_i)}{\sum_{j=1}^{m}(\prod_{i=1}^{n} \mu_{f_j^i}(x_i))} \quad . \end{gathered} \tag{20}$$

Then $\hat{f}(x|\Theta_f)$ can be rewritten as

$$\hat{f}(x|\Theta_f) = \Theta_f^T \xi(x) \ . \tag{21}$$

Where $\xi(x)$ fuzzy vector is like (18) and $\Theta_f^T$ is modified by adaptive law. So the control law (14) is changed as:

$$u = \frac{1}{A}[\hat{f}(\omega) - \dot{\omega}_{ref} + Ke\,] \ . \tag{22}$$

Define the optimal parameter:

$$\Theta_f^* = \arg \min_{\Theta_f \in \Omega_f} [sup\left|\hat{f}\left(x\middle|\Theta_{f_{x \in R^n}}\right) - f(x)\right|] \ . \tag{23}$$

And the minimum approximation error as:

$$\psi = f(x) - \hat{f}(x|\Theta_f^*) \ . \tag{24}$$

We consider the following Lyapunov function:

$$V = \frac{1}{2}e^T P e + \frac{1}{2\beta}\tilde{\Theta}_f^T \tilde{\Theta}_f \ . \tag{25}$$

Where $\tilde{\Theta}_f = (\Theta_f^* - \Theta_f)$ is the error of estimated parameter. P is a positive linear matrix and β is a small positive scalar.
The time derivative of V is:

$$\dot{V} = -\frac{1}{2}e^T Q e + e^T P B \psi + \frac{1}{\beta}\tilde{\Theta}_f^T(\beta e^T P B \xi(x) + \dot{\Theta}_f) \ . \tag{26}$$

We choose the adaptive law as [12]:

$$\dot{\Theta}_f = -\beta e^T P B \xi(x) \ . \tag{27}$$

Where P is a positive linear matrix and β is a small positive scalar.

Based on fuzzy universal approximation theorem, the approximate error can be adequate small by embedding enough many fuzzy system rules, and then lead to derivative of lyapunov function be negative definite. Hence, the system is stable and error will asymptotically converge to zero.

# 4 Simulation Results

Simulations are investigated with a rated 500KW variable speed wind turbine generator. The main parameters of system include: Wind turbine rotor inertia $J_m = 35000\ \mathrm{kg.m^2}$, generator rotor inertia is $J_m = 32\ \mathrm{kg.m^2}$, Wind turbine radius is R = 25m, the speed ratio of gearbox is $\gamma = 28$, coefficient of friction is $B_m = 1000\ \mathrm{Nms^{-1}}$, $K_m = 1000\ \mathrm{Nm}$, $B_e = 100\ \mathrm{Nms^{-1}}$, $K_m = 100\ \mathrm{N}$. Air density is $\rho = 1.225\mathrm{kg/m^3}$. The wind turbine operation is simulated in the time domain (s) and the simulation time used is 10s.

In this paper, the actual wind speed is adopted as input of simulation. The variable curve of wind according to time is shown in Fig. 2 and the speed tracking curve of wind turbine is described as Fig. 3. Also the RMS error of the proposed controller and a conventional PI controller are compared in Fig. 4 and Table I. Using the total RMS errors listed in Table I as comparison criteria, adaptive fuzzy controller shows less total error than a conventional PI controller.

**Table 1.** Comparison between total RMS errors in steady state condition

| **Total RMS error in steady state condition(m/s)** | |
|---|---|
| ***Adaptive fuzzy controller*** | ***PI controller*** |
| 0.013 | 0.062 |

The obtained simulation results shown in Fig. 3 indicate that the proposed controller provide good robustness and transient performance. It is obvious directly from the figure that when the wind speed is above or below of the rated speed and has strong disturbance, the proposed controller can effectively maintain the wind turbine speed at a near desired reference smoothly.

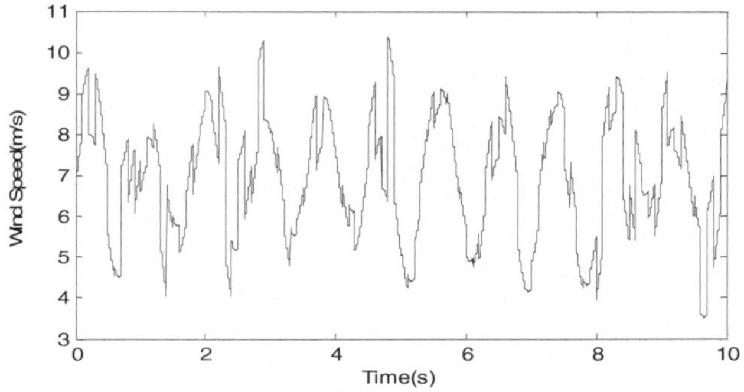


**Fig. 2.** Wind speed curve

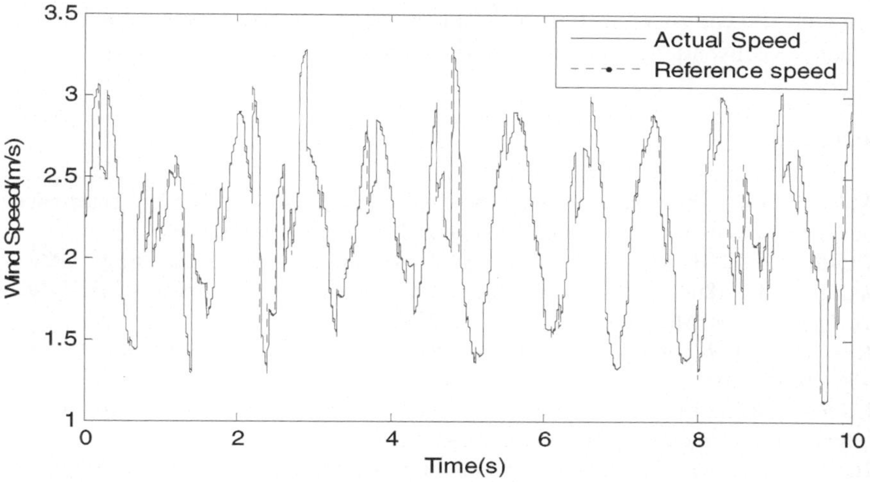


**Fig. 3.** Reference speed and actual speed of wind turbine

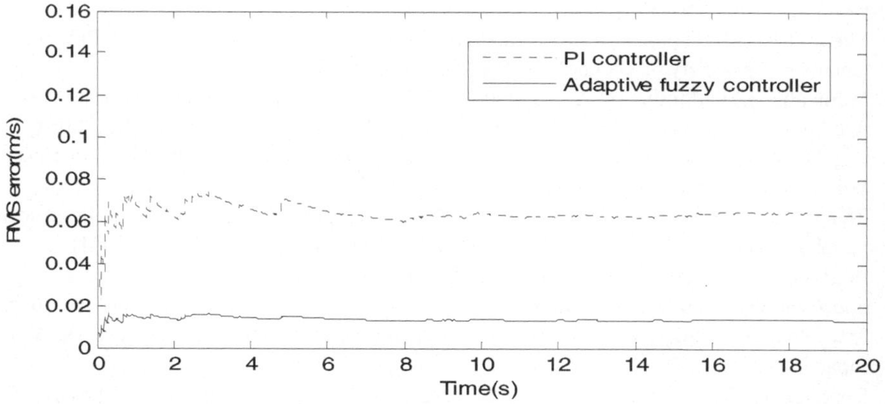


**Fig. 4.** RMS error of the adaptive fuzzy controller and a conventional PI controller

## 5 Conclusions

In this work, an adaptive fuzzy logic application has been developed for improving captured energy from a variable speed wind turbine. This paper presents a dynamic model of a variable speed wind turbine using adaptive fuzzy control. Simulation results show the robustness of the proposed controller in tracking the desired reference speed. The reference speed $\omega_{ref}$ and the actual output speed of the wind turbine $\omega$ were described in Fig. 3. It can be seen when wind speed is continuously variable and has strong disturbances, the proposed controller can effectively maintain good tracking performance. The response speed is fast and the error is small. Also by comparing the total RMS error between the adaptive fuzzy controller and a conventional PI controller in Fig. 5, adaptive fuzzy controller shows smaller total error than the conventional PI controller which indicates better performance.

## References

1. Kaldellis, J.K.: The wind potential impact on the maximum wind energy penetrationin autonomous electrical grids. Renewable Energy 33(7), 1665–1677 (2008)
2. Chadli, M., El Hajjaji, A.: Wind energy conversion systems control using T-Sfuzzy modeling. In: 18th Mediterranean Conference on Control and Automation, IEEE-MED 2010, Marrakech, Morocco, pp. 1365–1370 (2010)
3. Kusiak, A., Li, W., Song, Z.: Dynamic control of wind turbines. Renewable Energy 35(2), 456–463 (2010)
4. Leithead, W.E., Delasalle, S., Reardon, D.: Role and objectives of control of wind turbines. IEEE Proceeding Part C 138, 135–148 (1991)
5. Hand, M.M., Balas, M.J.: Non-linear and linear model based controller design for variable-speed wind turbines. NREL Report No. CP-500-26244, p. 1e8
6. Knudsen, T., Andersen, P., Toffner-Clausen, S.: Comparing PI and robust pitch controllers on a 400 kW wind turbine by full-scale tests. In: Proceeding of the European Wind Energy Conference, pp. 546–550 (1997)
7. Ekelund, T.: Modeling and Linear Quadratic Optimal control of wind turbines. PhD thesis, Chalmers University of Technology, Sweden (April 1997)
8. Munteanu, I., Cultululis, N.A., Bratcu, A.I., Ceanga, E.: Optimization of variable speed wind power systems based an LQG approach. Control Eng. Pract. 13, 903–912 (2005)
9. Xu, D., Zhu, J.: Tracking wind direction fuzzy control system for WTG. Acta Energy Solaris Sinaca 21(2), 187–188 (2000)
10. Zhang, X.-F., Xu, D.-P., Lv, Y.-G., Liu, Y.-B.: Adaptive fuzzy control for large-scale variable speed wind turbines. Journal of System Simulation 16(3), 573–577 (2004)
11. Prats, M.A.M., Carrasco, J.M., Galvan, E., et al.: Improving transition between power optimization and power limitation of variable speed variable pitch wind turbines using fuzzy control techniques. In: IECON 26th Annual Conference of the IEEE, Nagoya, Aichi, Japan (2000)
12. Boukhezzar, B., Siguerdidjane, H.: Nonlinear control of variable speed wind turbines for power regulation. In: Proc. 2005 IEEE CCA, Toronto, Canada (August 2005)
13. Mieczarski, W.: Fuzzy Logic Techniques in Power Systems. Physica - Verlag (1998)
14. Palm, R., Driankov, D., Hellendoorn, H.: Model Based Fuzzy Control. Springer (1997)
15. Song, Y.D., Dhinakaran, B., Bao, X.: Control of Wind Turbines Using Nonlinear Adaptive Field Excitation Algorithms. In: Proceedings of the American Control Conference, Chicago, vol. 3, pp. 1551–1555 (2000)
16. Tong, S.C.: Adaptive fuzzy control for Nonlinear system. Science Pubishing, Beijing (2006)
17. Wang, L.X.: Adaptive fuzzy system and control-design and stability analysis. Guo Fang Industry Pubishing, Beijing (1995)

# Short-Term Traffic Flow Forecasting Based on Grey Delay Model*

Huan Guo[1,**], Xinping Xiao[1], and Yuxiao Tang[2]

[1] College of Science, Wuhan University of Technology, Wuhan 430063, China
[2] School of Finance, Zhongnan University of Economics and Law, Wuhan 430073, China
{guohuan.2007,sunnytyx}@163.com, xiaoxp@263.net

**Abstract.** Under the real circumstances of traffic flow's grey features and traffic system's delay effect, this paper construct a grey delay model ( $GM(1,1,\tau)$ model) and investigate relevant properties, Finally, we complete a traffic experiment on the section of Youyi Avenue, estimate the delay time of the system and establish the delay model. The comparison between the delay model with the $GM(1,1)$ model without considering practical conditions of the traffic system shows that incorporating existing delay effect of the traffic system into the model can reasonably reflect the characteristics of the system and provide satisfactory predictions.

**Keywords:** Grey system feature, Delay time, Traffic flow, $GM(1,1,\tau)$ model.

## 1 Introduction

To resolve traffic congestions and avoid traffic accidents and relevant problems, many countries are paying serious attentions to research on traffic planning methods. Many practical systems for traffic planning and controlling have been developed. However a common prerequisite for operating these systems is that historical data are collected and analyzed to make continuous, simultaneous and exact prediction of future traffic flows. Predicting short-term traffic flow is the premise and key to implement traffic information service, traffic control and induction and is an important component of intelligent transportation systems. Traffic flow prediction is directly related to the quality of traffic information service and the effect of traffic control and induction. Therefore, it is both urgent and practically significant for the current development of intelligent transportation systems to investigate the theory and methodology of effective short-term traffic flow modeling and forecasting and to accurately and rapidly predict and distinguish traffic conditions from accumulated information.

Due to the complexity and uncertainty of traffic flow, it is challenging to make accurate predictions. Many researches in this area have emerged in recent years and

* This work was supported by the National Natural Science Foundation of China under Grant 70971103 and the General Education Program (GEP) Requirements in the Humanities of Social Sciences under Grant 11YJC630155.

** Corresponding author.

J. Lei et al. (Eds.): AICI 2012, LNAI 7530, pp. 357–364, 2012.

significant results have achieved. Current research on predicting traffic flow focuses in three directions: developing models based on statistical theory [1,2], developing models based on nonlinear prediction [3,4], and developing intelligent algorithms [5,6].

Grey generation space model, which studies uncertain systems with small amount of information (or data), sublimates the level of the system's original information through grey generating and achieves effective control of the system by developing a model (called GM model) with the properties of partial difference and partial differential. Traffic flow system has been recently investigated using grey models. Fang et al. [7] developed a distribution model with the largest hidden ability for the military road network traffic flow, based on the idea of grey system and grey information entropy theory. Sun et al. [8] proposed a self-adaptive GM(1,1) model to predict the traffic flow of an intersection without any detector. By analyzing temporal characteristics of short-term traffic flow and applying grey system theory to predict short-term traffic flow, Deng et al. [9] established a rolling GM(1,1) prediction model. Based on residual correction of a grey system GM(1,1) model, Dai and Huang [10] established a traffic flow forecast model. However the use of grey models for predicting traffic flow is restricted to the use of traditional grey models, without considering delayed effects.

Due to delayed effects in traffic time series, the traditional prediction using traffic flow time series does not achieve satisfactory results. This paper presents a grey delay model and investigates relevant properties, and ultimately applying the established model to predict short-term traffic flow.

## 2 Establishment of the Grey Delay Model

### 2.1 Establishment of the Grey Model

Let $X^{(0)} = (x^{(0)}(k))$, $k = 1,2,\cdots,n$, be the original data sequence describing the traffic system's characteristics (e.g. traffic flow) and $X_\tau^{(0)} = (x^{(0)}(k-\tau))$, $k = \tau+1,\cdots,n$, be its delay time sequence, where the difference between $X^{(0)}$ and $X_\tau^{(0)}$ is the delay time $\tau$. Then $X^{(1)} = (x^{(1)}(k))$ and $X_\tau^{(1)} = (x^{(1)}(k-\tau))$ are respectively the first order accumulated generating operation series of $X^{(0)}$ and $X_\tau^{(0)}$, where

$$x^{(1)}(k) = \sum_{i=1}^{k} x^{(0)}(i),\ k = 1,2,\cdots,n\ ;\ x_\tau^{(1)}(k-\tau) = \sum_{i=\tau+1}^{k} x^{(0)}(i-\tau),\ k = \tau+1,\cdots,n$$

According to reference [11], three modeling conditions are required when constructing grey differential equations, and are the configurable condition, material condition and quality condition. Possessing a grey differential coefficient with the most information of data is the configurable condition for constructing the differential equation, and having white background value and grey differentials $x^{(0)}(k)$ and $x^{(0)}(k-\tau)$ is the material condition for constructing the differential equation. The white background values $z^{(1)}(k)$ and $z^{(1)}(k-\tau)$ are in the set of monotonically increasing background, and only data in the background set can ensure the quality of the grey differential equation. Especially in the traffic system, grey differential $x^{(0)}(k)$ (e.g., the traffic flow) is determined jointly by the information of $z^{(1)}(\boldsymbol{k})$ and $z^{(1)}(k-\tau)$. According to the actual situation of the traffic system, we build the following $GM(1,1,\tau)$ model.

**Theorem 1.** Suppose that $X^{(0)}=(x^{(0)}(1),x^{(0)}(2),\cdots,x^{(0)}(n))$ is a nonnegative sequence. We say that $X^{(1)}=(x^{(1)}(1),x^{(1)}(2),\cdots,x^{(1)}(n))$ is said to be AGO series of $X^{(0)}$ if it is modeled by the grey differential equation $GM(1,1,\tau)$ given as

$$x^{(0)}(k)+az^{(1)}(k)=bz^{(1)}(k-\tau) \tag{1}$$

Where $z^{(1)}(k)=0.5x^{(1)}(k)+0.5x^{(1)}(k-1)$. We call this a $GM(1,1,\tau)$ model, with its white form differential equation given as

$$\begin{cases} \dfrac{dx^{(1)}(t)}{dt}+ax^{(1)}(t)=bx^{(1)}(t-\tau) \\ x^{(1)}(0)=x^{(0)}(1) \end{cases} \tag{2}$$

Denote

$$B=\begin{bmatrix} -\frac{1}{2}\left(x^{(1)}(\tau+2)+x^{(1)}(\tau+1)\right) & \frac{1}{2}\left(x^{(1)}(2)+x^{(1)}(1)\right) \\ -\frac{1}{2}\left(x^{(1)}(\tau+3)+x^{(1)}(\tau+2)\right) & \frac{1}{2}\left(x^{(1)}(3)+x^{(1)}(2)\right) \\ \vdots & \vdots \\ -\frac{1}{2}\left(x^{(1)}(n)+x^{(1)}(n-1)\right) & \frac{1}{2}\left(x^{(1)}(n-\tau)+x^{(1)}(n-\tau-1)\right) \end{bmatrix},$$

$$Y=\begin{bmatrix} x^{(0)}(\tau+2) \\ x^{(0)}(\tau+3) \\ \vdots \\ x^{(0)}(n) \end{bmatrix}$$

We then have

$$[a\ b]^T=(B^TB)^{-1}B^TY \tag{3}$$

**Proof.** For the white form, we use $x^{(1)}(k)-x^{(1)}(k-1)=x^{(0)}(k)$ to replace the differential coefficient element, and use $z^{(1)}(k)$ , $z^{(1)}(k-\tau)$ to replace $x^{(1)}(k)$, $z^{(1)}(k-\tau)$, where $k=\tau+1,\cdots,n$. We obtain equations

$$\begin{cases} x^{(0)}(\tau+2)=-\frac{1}{2}\left(x^{(1)}(\tau+2)+x^{(1)}(\tau+1)\right)\cdot a+\frac{1}{2}\left(x^{(1)}(2)+x^{(1)}(1)\right)\cdot b \\ x^{(0)}(\tau+3)=-\frac{1}{2}\left(x^{(1)}(\tau+3)+x^{(1)}(\tau+2)\right)\cdot a+\frac{1}{2}\left(x^{(1)}(3)+x^{(1)}(2)\right)\cdot b \\ \vdots \\ x^{(0)}(n)=-\frac{1}{2}\left(x^{(1)}(n)+x^{(1)}(n-1)\right)\cdot a+\frac{1}{2}\left(x^{(1)}(n-\tau)+x^{(1)}(n-\tau-1)\right)\cdot b \end{cases}$$

In matrix forms, these become $Y = B[a \ b]^T$. By the method of least squares, we have

$$[a \ b]^T = (B^T B)^{-1} B^T Y .$$

Furthermore, set

$$C = \sum_{i=2}^{n-\tau} x^{(1)}(\tau+i)x^{(1)}(i) \, ; E = \sum_{i=\tau+2}^{n} [x^{(1)}(i)]^2 \, ; F = \sum_{i=2}^{n-\tau} [x^{(1)}(i)]^2 \, ;$$

$$H = \sum_{i=2}^{n-\tau} x^{(1)}(i)x^{(0)}(\tau+i)$$

Then

$$a = \frac{CH - GF}{EF - C^2}, b = \frac{EH - CG}{EF - C^2} \tag{4}$$

**Theorem 2.** From the $GM(1,1,\tau)$ model

$$x^{(0)}(k) + az^{(1)}(k) = bz^{(1)}(k-\tau)$$

We derive the $GM(1,1,\tau)$ model of type $x^{(1)}$, namely the $GM(1,1,\tau,x^{(1)})$ model, given by

$$GM(1,1,\tau,x^{(1)}) \, \vdots \; x^{(0)}(k) = \beta\left(x^{(1)}(k-\tau) + x^{(1)}(k-\tau-1)\right) - \alpha x^{(1)}(k-1)$$

$$\beta = \frac{0.5b}{1+0.5a}, \quad \alpha = \frac{a}{1+0.5a}$$

**Proof.** $z^{(1)}(k)$ can be expressed as

$$z^{(1)}(k) = 0.5x^{(1)}(k) + 0.5x^{(1)}(k-1)$$

Substitute $z^{(1)}(k)$ into the model $GM(1,1,\tau)$, we have

$$x^{(0)}(k) + a\left(0.5x^{(1)}(k) + 0.5x^{(1)}(k-1)\right) = bz^{(1)}(k-\tau)$$

$$x^{(0)}(k) + a\left(0.5x^{(1)}(k) + 0.5x^{(1)}(k-1)\right) = b\left(0.5x^{(1)}(k-\tau) + 0.5x^{(1)}(k-\tau-1)\right)$$

$$(1+0.5a)x^{(0)}(k) = b\left(0.5x^{(1)}(k-\tau) + 0.5x^{(1)}(k-\tau-1)\right) - ax^{(1)}(k-1)$$

$$x^{(0)}(k) = \frac{0.5b}{1+0.5a}\left(x^{(1)}(k-\tau) + x^{(1)}(k-\tau-1)\right) - \frac{a}{1+0.5a}x^{(1)}(k-1)$$

$$x^{(0)}(k) = \beta\left(x^{(1)}(k-\tau) + x^{(1)}(k-\tau-1)\right) - \alpha x^{(1)}(k-1) .$$

## 2.2 Verification and Test of the Model

To examine the effect of model simulation and forecasting, we define some indices of prediction error: mean relative error (MRE) and fitting degree (FD) (between the predicted and true values), whose formulas are given as follows:

$$MRE = \frac{1}{n}\sum_{i=1}^{n}\left|\frac{\hat{x}(i) - x(i)}{x(i)}\right| \tag{5}$$

$$FD = 1 - \frac{\sqrt{\sum_{i=1}^{n}[\hat{x}(i) - x(i)]^2}}{\sqrt{\sum_{i=1}^{n}\hat{x}(i)^2} + \sqrt{\sum_{i=1}^{n}x(i)^2}} \tag{6}$$

Smaller values of MRE represent better precision of simulation and forecasting. With regard to FD, any value of FD larger than 0.85 means good prediction, and any value of FD larger than 0.9 indicates satisfactory prediction.

## 3 Analyses of Examples

To solve practical traffic system problems, we conducted experiments using vision trace to photograph the section of Youyi Avenue between Yuanlin Road and Jianshe Road of Wuchang District of Wuhan from 15:00 to 17:00 on May 19, 2010.

The length of the experimental section is $L = 949m$. The traffic capacity of the experimental section is $C = 1600$ vehicles/h, and the design speed is $V_0 = 60km/h$ [12]. In our experiment, we used 5-minute time intervals and so the traffic capacity is $c = 134$ vehicles and the running time is $T_0 = L/V_0 = 56.94s$ if system delay is not considered. Taking 5 minutes as time intervals and using Sava video analysis software to cope with the video information collected from the experiment, the observed traffic flow are listed in Table 1.

In this experiment, we adopted every 5 minutes as a time interval and determined system delay time by using the three-parameter relationship of the traffic system [12]. The system delay time is 11.75 minutes, close to two units of time interval. So we take $\tau = 2$ as the delay factor when modeling with the gray system and construct the $GM(1,1,2)$ model

$$x^{(0)}(k) + a \cdot z^{(1)}(k) = bz^{(1)}(k-2) \tag{7}$$

Due to large values of the traffic flow, it is very likely to result in a singular matrix when solving for the parameter vector $P = [a\ b]^T$. Using Theorem 2 and modeling with the first 13 observed data values, we obtain $a = -0.4965,\ b = -0.4960$. Hence $\alpha = -0.6605,\ \beta = -0.3299$, and the model $GM(1,1,2,x^{(1)})$ is given as

$$x^{(0)}(k) = 0.6605 \cdot x^{(1)}(k-1) - 0.3299 \cdot \left(x^{(1)}(k-2) + x^{(1)}(k-3)\right) \tag{8}$$

Table 1 shows the fitting results from modeling (8) with the first 13 data values. Comparing the fitted results in Fig 1, the delay model $GM(1,1,2)$ is better than $GM(1,1)$ for the fitted curve and fitted values.

**Table 1.** Comparison of fitted results

| Time segment | Observed flow | $GM(1,1)$ model | | $GM(1,1,2)$ model | |
|---|---|---|---|---|---|
| | | Fitted value | Relative error | Fitted value | Relative error |
| 15:00-15:05 | 183 | 183 | 0 | 183 | 0 |
| 15:05-15:10 | 208.5 | 208.5 | 0 | 208.5 | 0 |
| 15:10-15:15 | 184.5 | 184.5 | 0 | 184.5 | 0 |
| 15:15-15:20 | 180 | 199.12461 | 0.106248 | 188.193 | 0.045517 |
| 15:20-15:25 | 219 | 195.3315 | 0.108075 | 189.7808 | 0.133421 |
| 15:25-15:30 | 217.5 | 191.6107 | 0.119031 | 207.7459 | 0.044847 |
| 15:30-15:35 | 192 | 187.9607 | 0.021038 | 210.4523 | 0.096105 |
| 15:35-15:40 | 196.5 | 184.3803 | 0.061678 | 198.6454 | 0.010918 |
| 15:40-15:45 | 163.5 | 180.8681 | 0.106227 | 186.4594 | 0.140424 |
| 15:45-15:50 | 165 | 177.4338 | 0.075356 | 171.6615 | 0.040373 |
| 15:50-15:55 | 181.5 | 174.0431 | 0.041085 | 168.4001 | 0.072176 |
| 15:55-16:00 | 174 | 170.7278 | 0.018806 | 175.185 | 0.00681 |
| 16:00-16:05 | 190.5 | 167.4756 | 0.120863 | 179.7401 | 0.056482 |
| MRE (%) | | 7.7841 | | 6.4707 | |
| FD | | 0.954923 | | 0.961806 | |

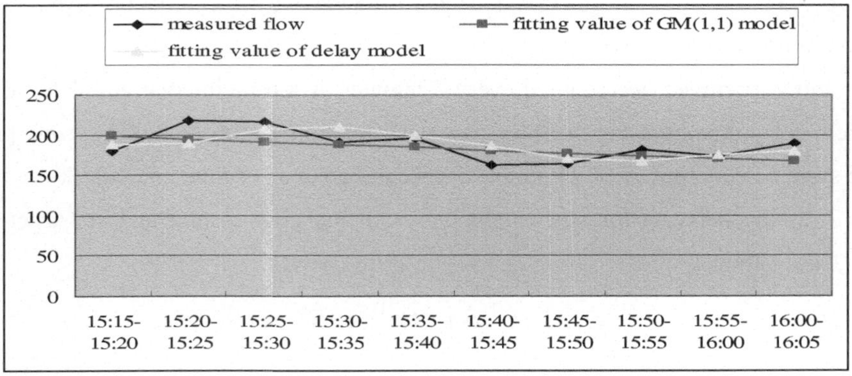


**Fig. 1.** Observed flow and fitted values of the model

Table 2 lists the prediction of the remaining values. Without considering the phenomenon of system delay, the MRE of the $GM(1,1)$ model is 25.924% and the FD is 0.839476(<0.85). However for the $GM(1,1,2)$ model considering the system delay phenomenon, the MRE is 6.0262% and the FD is 0.908177(>0.9). Comparing the predicted results in Fig.2, we see that the delay model $GM(1,1,2)$ reflects the delay phenomenon of the real system and its predicted results are close to observed values, directly showing that the delay model is reasonable.

**Table 2.** Comparison of predicted results

| Time segment | Observed flow | $GM(1,1)$ model | | $GM(1,1,2)$ model | |
|---|---|---|---|---|---|
| | | Predicted value | Relative error | Predicted value | Relative error |
| 16:05-16:10 | 199.5 | 164.2854 | 0.176514 | 188.3936 | 0.055671 |
| 16:10-16:15 | 198 | 161.1559 | 0.186081 | 196.677 | 0.006682 |
| 16:15-16:20 | 249 | 158.0861 | 0.365116 | 210.9248 | 0.152913 |
| 16:20-16:25 | 177 | 155.0748 | 0.123871 | 218.0993 | 0.232199 |
| 16:25-16:30 | 207 | 152.1208 | 0.265117 | 202.5713 | 0.021395 |
| 16:30-16:35 | 202.5 | 149.223 | 0.263096 | 198.5816 | 0.01935 |
| 16:35-16:40 | 205.5 | 146.3805 | 0.287686 | 204.6356 | 0.004206 |
| 16:40-16:45 | 210 | 143.5921 | 0.316228 | 206.2275 | 0.017964 |
| 16:45-16:50 | 211.5 | 140.8569 | 0.33401 | 209.6809 | 0.008601 |
| 16:50-16:55 | 190.5 | 138.1737 | 0.274679 | 206.4334 | 0.08364 |
| MRE (%) | | 25.924 | | 6.0262 | |
| FD | | 0.839476 | | 0.908177 | |

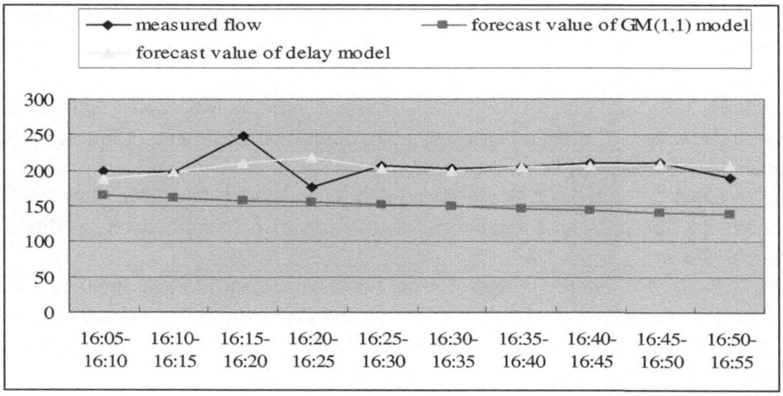


**Fig. 2.** Observed flow and predicted values of the model

## 4 Conclusions

Traffic system is a complex and highly uncertain system due to human participation and dynamic traffic conditions. These are all essential characteristics of the traffic system that are suitable for grey theory modeling. We establish the grey delay model $GM(1,1,\tau)$ and analyze its relevant properties specifically with respect to the delay effect, small sample data and uncertainty of the real traffic system. Finally, we conducted an experiment on the section of Youyi Avenue between Yuanlin Road and Jianshe Road of Wuchang District of Wuhan, analyzed the real condition and constructed the delay model $GM(1,1,2)$ for the purpose of model fitting and prediction. Comparison of the results with the $GM(1,1)$ model without delay effect clearly demonstrates the superiority of the delay model $GM(1,1,2)$ over the $GM(1,1)$ model. To some degrees this study can resolve certain real problems in the traffic system and illustrates that modeling with the consideration of the real situations of the traffic system can better reflect the system's characteristics and provide satisfactory predictions.

## References

1. Queen, C.M., Albers, C.J.: Intervention and Causality: Forecasting Traffic Flows Using a Dynamic Bayesian Network. Journal of the American Statistical Association 104, 669–681 (2009)
2. Tan, M.C., Wong, S.C.: An Aggregation Approach to Short-term Traffic Flow Prediction. IEEE Transactions on Intelligent Transportation Systems, 66–69 (2009)
3. Xie, Y.C., Zhao, K.G.: Gaussian Processes for Short-term Traffic Volume Forecasting. Transportation Research Record, 69–78 (2010)
4. Castro-Neto, M., Jeong, Y.S.: Online-SVR for Short-term Traffic Flow Prediction Under Typical and a Typical Traffic Conditions. Expert System with Applications 36(3), 6164–6173 (2009)
5. Reuter, U., Moller, B.: Artificial Neural Networks for Forecasting of Fuzzy Time Series. Computer-Aided Civil and Infrastructure Engineering 25(5), 363–374 (2010)
6. Huang, S., Sadek, A.W.: A Novel Forecasting Approach Inspired by Human Memory: The Example of Short-term Traffic Volume Forecasting. Transportation Research Part C-merging Technologies 17(5), 510–525 (2009)
7. Fang, Z.G., Liu, S.F.: The Military Traffic Flows Distribute Model Research of Maximum Gray Information Entropy Based on Maximum Concealment. Chinese Journal of Management Science 11(3), 56–61 (2003)
8. Sun, Y., Chen, S.F., Zhou, Z.Q.: Application of Grey Models to Traffic Flow Prediction at Non-detector Intersections. Journal of Southeast University (Natural Science Edition) 32(2), 256–258 (2002)
9. Deng, Z.L., Li, Q., Chen, Q.: Research on Short Time Traffic Flow Prediction Based on Gray System Theory. Technology of Highway and Transport 2(1), 117–119 (2006)
10. Dai, L.J., Huang, H.O.: Traffic Flow Forecast Model Based on Residual Correction of Grey System Theory. China Water Transport 8(7), 127–128 (2007)
11. Deng, J.L.: Grey Theory Foundation. Huazhong University of Technology Press, Wuhan (2002)
12. Dai, J.F., Ma, J.X.: Traffic Engineering Conspectus. China Communications Press, Beijing (2006)

# A Novel Color Image Segmentation Method Based on Improved Region Growing

Jun Liu[1], Yan Ma[1], Kun Chen[1], and Shun-bao Li[2]

[1] Department of Information Mechanical and Electrical Engineering, Shanghai Normal University, Shanghai 200234, China
[2] Department of Mathematic & Science, Shanghai Normal University, Shanghai 200234, China
liujunout@126.com, {ma-yan,lsb}@shnu.edu.cn, 414131298@qq.com

**Abstract.** This paper proposes an improved color image segmentation method based on improved region growing. Firstly, the color image is transformed from RGB to YCbCr color space. Then, seed points are selected automatically and region growing algorithm has been employed for image segmentation under predefined three criterions. Finally, region merging algorithm has been proposed. Smller regions are merged first and larger regions are merged lately. Extensive experiments have been carried out on the random images from the internet by using the proposed algorithm and the results show its effectiveness.

**Keywords:** Color image segmentation, Seed points, Region growing, Region merging.

## 1 Introduction

Image segmentation is the key and the first step in image processing and analysis. Image segmentation is to divide digital images into regions that have same common features. Then it can provide some basis for further computer processing[1]. On the other hand, image segmentation is the classification to image pixels. An image's independent pixels are classified into one class by certain algorithm. So it can reach the purpose of image segmentation. Image segmentation is also an important step in pattern recognition and detection. Up to now, There are a lot of image segmentation methods. Clustering, region-based, boundary-based and fusion methods are main color image segmentation methods. The key of clustering methods is the selection of threshold. Traditional threshold selection method is to select the difference between foreground color and background, then to do clustering[2]. Region-based color image segmentation method is based on the similarity (eg. Luminance, intensity, texture, chroma)[3]. Boundary-based method utilizes the pixel differences among different regions, eg. intensity, luminance and texture[4]. Fusion method is to combine boundary-based and region growing methods to getter better segmentation effect[5].

Seed-based region growing methods is one of the fusion methods. It was proposed by Adams and Bischof [6]. Firstly, select initial seed points by some method. Then combine each pixel to its nearest seed region by growing region.

J. Lei et al. (Eds.): AICI 2012, LNAI 7530, pp. 365–373, 2012.

In this paper, we improved the method proposed by Frank Y.Shih[7]. We optimize the region merging part and consider the unprocessed seed points. We propose new methods and get better results.

## 2 Algorithm

Figure 1 describes the algorithm flow of this paper. Firstly, Transform color space from RGB to YCbCr. Secondly , use auto-selection algorithm to select the seed points. Thirdly, utilize region growing algorithm to segment the image. Finally, use region growing methods to merge the over-segmentation part.

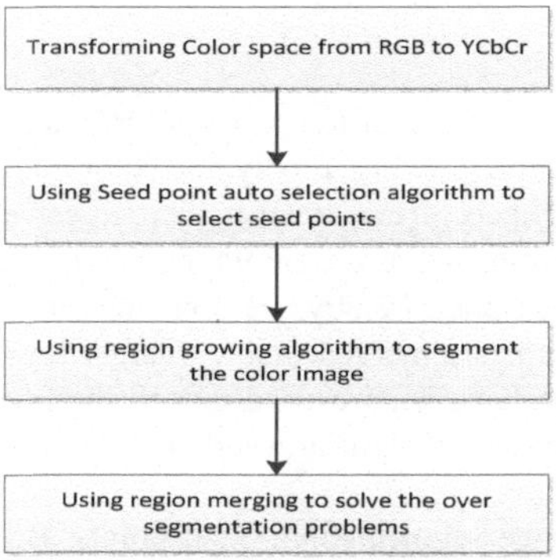


**Fig. 1.** Algorithm flow chart

## 3 Color Space Transform

Basicly, a color image is based on RGB color space. RGB color space has good effect in color display. But it is not suitable for color analysis because of its relationships among three components. Thus, RBG color space is always transformed to other color spaces in image processing and analysis. Cheng et al.[8] proposes and compare several color space effect in image segmentation. Each color space has its advantage and disadvantage. And YCbCr is suitable for image segmentation. There are three main reasons for selecting YCbCr in this paper:

(1) YCbCr is widely applied for image compression standard(eg. MPEG and JPEG.)

(2) Human vision can be easily identified by using Euclidean distance in YCbCr.

(3) Intensity and chroma can be easily separated and processed.

The formula from RGB to YCbCr is listed as following. First quantize r,g,b components to [0,1]. According to formula(1), Y is between [16,255], Cb and Cr are between [16,240].

$$\begin{bmatrix} Y \\ Cb \\ Cr \end{bmatrix} = \begin{bmatrix} 65.481 & 128.553 & 24.966 \\ -39.797 & -74.203 & 112 \\ 112 & -93.786 & -18.214 \end{bmatrix} \times \begin{bmatrix} R \\ G \\ B \end{bmatrix} + \begin{bmatrix} 16 \\ 128 \\ 128 \end{bmatrix} \tag{1}$$

## 4 Seed Points Auto-selection

To region growing color image segmentation, the selection of seed points and region growing criterion are most important. Traditional region growing criterion uses fixed seed points to segment image. But considering the difference among images, using stationary seed points can not get results well. The seed points auto-selection method in this paper satisfy three criterions:

First, the seed pixels must have high similarity with its neighbor pixels.

Second, One region at least has one seed point to form segmentation region ready to be segmented.

Last, seed points among different regions are disconnected.

The algorithm that calculating similarity between a seed point and its neighbor pixels is as following:

1) Calculating neighbor pixels`(3x3) mean values in Y,Cb,Cr components separately. xi denotes Y,Cb,Cr:

$$\bar{x} = \frac{1}{9}\sum_{i=1}^{9} x_i \tag{2}$$

2) Calculating standard deviation in Y,Cb,Cr components of 3x3 neighbor pixels:

$$\sigma_x = \sqrt{\frac{1}{9}\sum_{i=1}^{9}(x_i - \bar{x})^2} \tag{3}$$

3) Calculating total standard deviation:

$$\sigma = \sigma_Y + \sigma_{Cb} + \sigma_{Cr} \tag{4}$$

4) Quantizing the deviation to [0,1]. $\sigma_{\max}$ is the max value among all deviations:

$$\sigma_N = \sigma / \sigma_{\max} \tag{5}$$

5) Getting the similarity of one pixel with its neighbor pixels:

$$H = 1 - \sigma_{\max} \tag{6}$$

According to similarity H, we define first selection condition: the similarity of seed points H must bigger than the threshold.

6) Calculating the Euclidean distances of seed points with its eight neighborhood(based on YCbCr):

$$di = \frac{\sqrt{(Y-Y_i)^2 + (C_b - C_{bi})^2 + (C_r - C_{ri})^2}}{\sqrt{Y^2 + C_b{}^2 + C_r{}^2}} \quad (7)$$

7) Calculating each pixel`s biggest relative Euclidean distance according to step 6.

$$di = \frac{\sqrt{(Y-Y_i)^2 + (C_b - C_{bi})^2 + (C_r - C_{ri})^2}}{\sqrt{Y^2 + C_b{}^2 + C_r{}^2}} \quad (8)$$

According to biggest distance dmax, defining the second selection condition: the Euclidean distance of seed points with its eight neighborhood must smaller than the threshold.

If a pixel satisfies both two selection conditions, it is selected as a seed point. To the first condition, Otus[2] method is used for its threshold selection. To the second condition, we select 0.05 as the threshold according the experiments.

Each connected seed points are considered as one (one seed point). Thus the seed points in a region can be one pixel or a group of connected pixels. The first condition can judge whether seed points have high similarity with its neighbor pixels. The second condition ensure seed pixels aren`t the boundary values between two regions.

We select two images to show the difference of selecting threshold in the second condition. Table 1 and figure 2 show the amounts and difference when selecting threshold as 0.04 and 0.05. The white points in figure 2 denote seed points. According to that, different threshold can result the amounts difference of seed points.

**Table 1.** Connection of threshold selection and seed points amounts

| Images | Amounts | Size | Threshold |
|---|---|---|---|
| peppers.jpg | 369 | 256x256 | 0.05 |
| peppers.jpg | 330 | 256x256 | 0.04 |
| house.png | 255 | 256x256 | 0.05 |
| house.png | 833 | 256x256 | 0.04 |

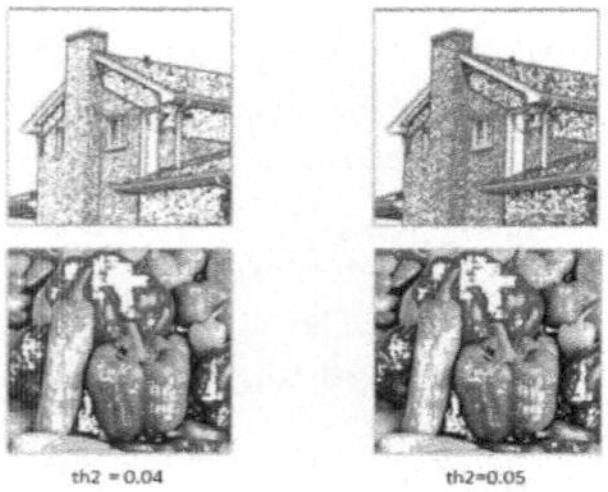


**Fig. 2.** Connection of threshold selection and seed points amounts

## 5 Region Growing

Assumed that S1,S2,….Si denote initial seed points, Ai denotes the seed points region Si`s relevant region. The region growing algorithm is as following:

Calculating all pixels` $\overline{Y}, \overline{C_b}, \overline{C_r}$ in Si.

Marking each seed region with one label.

Calculating all seed region`s four neighbor, and recording the pixels which are not seed points, saving into List T.

If T is not null, removing the first point p and check its four neighbors. If its neighbors all have same labeled label, marking p to that label. If its neighbors have different labels, calculating every distance of p and its different labels. Then marking p to its nearest region and update the mean value of this region. Then checking p`s four neighbors. If its four neighbors are not marked also not in List T, adding its four neighbors into T.

In Step 3, T denotes the pixels that are not classified and at least one region of region Si. In Step 4, p that has the smallest distance is picked. If several pixels have same smallest value, we select the pixel which has the biggest distance of each neighbors. If p has same distance to different label neighbors, we classifying p into the biggest region.

Figure 3 and Figure 4 shows the segmentation results. It also shows the over-segmentation result.

**Fig. 3.** Region grow result

**Fig. 4.** Region grow result

## 6 Region Merging

After region growing algorithm, we get the segmentation result which can initially segment the object. But it exist the problem of over-segmentation, a simple and efficient region merging algorithm is necessary. The region merging strategy is mainly considering the similarity and size of the regions. Figure 5 and Figure 6 show the process.

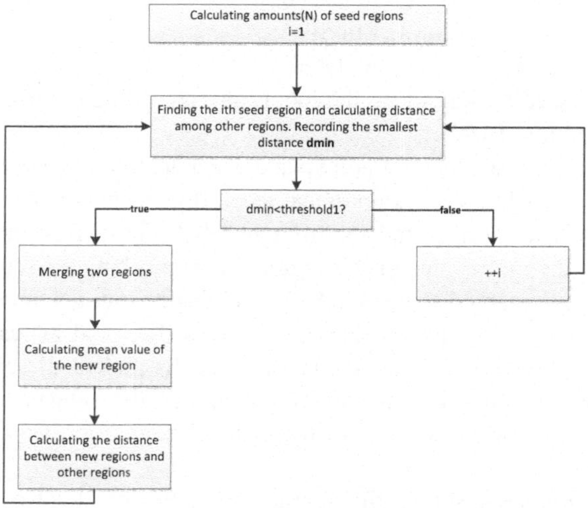


**Fig. 5.** Region merging based on similarity

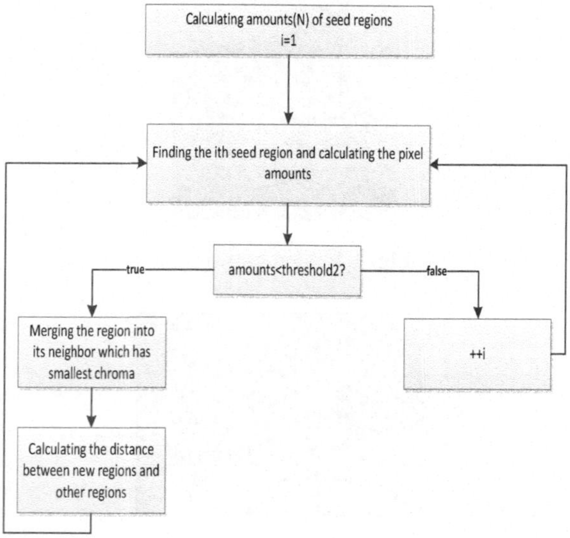


**Fig. 6.** Region merging based on the size

In Figure 5, we set the color similarity thresholding to 0.1 according to the experiment results. Meanwhile, the distance between two regions is calculating as follow:

$$di = \frac{\sqrt{(\overline{Y}_i - \overline{Y}_j)^2 + (\overline{C}_{bi} - \overline{C}_{bj})^2 + (\overline{C}_{ri} - \overline{C}_{rj})^2}}{\min(\sqrt{\overline{Y}_i^2 + \overline{C}_{bi}^2 + \overline{C}_{ri}^2}, \sqrt{\overline{Y}_j^2 + \overline{C}_{bj}^2 + \overline{C}_{rj}^2})} \quad (9)$$

While in Figure 6, the thresholding threshold2 is chosen to 1/150 of the size of the image.

We observe that region merging based on similarity and size to some extent can solve the problem of over-segmentation. But we find some problems after performing the algorithm on 300 images. Some seed points are not calculated, while later exist as independent seed points or seed regions. But the region merging algorithm of figure 5 and figure 6 is based on seed regions. To solve the problem, we improved the region merging algorithm. It can be described as follow:

1) Relabeling regions after segmentation by 8-neighbors, which yields that not adjacent and color-homogeneous regions are marked with different labels.

2) Search adjacent regions after relabeling.

3) Calculate the mean value of Y, Cb, Cr components for relabeled regions.

4) Merge small regions. If the size of region is smaller than threshold $T_1$, it will be merged into its bigger adjacent region with the smallest Euclidean distance in YCbCr color space, and whose size is greater than $T_1$.

5) Merge big regions. If the size of region is smaller than threshold $T_2 (T_1 << T_2)$, we calculate Euclidean distance in YCbCr color space with its adjacent regions whose size is greater than $T_2$. 6. Search the smallest distance dc. If $dc < T_3$, the region will be merged into its adjacent bigger region with the smallest distance, and vice versa.

In this paper $T_1, T_2, T_3$ chosen to 0.003S, 0.05S, 50. (S denotes the size of image).

## 7 Experiment Result

We chose 300 images randomly from the Internet to perform the algorithm, the operating environment is Matlab. Figure 7(a) and Figure 8(a) are original images, Figure 7(c) and Figure 8(c) are images after region merging by the primary method, Figure 7(d) and Figure 8(d) are results by the improved region merging algorithm. From Figure 7(b) and 7(c) we can know that the results are not obvious by the primary region merging method, and exist many independent points. Similarly, Figure 8(c) is result by the original method, there is not any difference. Because the region merging method of the original paper is based on seed points' region merging, it exists many independent points. Figure 7(d) and Figure 8(d) are the effect by the improved region merging, comparing to Figure 7(c) and 8(c) it get rid of the influence of independent seed points and small regions, and get better segmentation results.

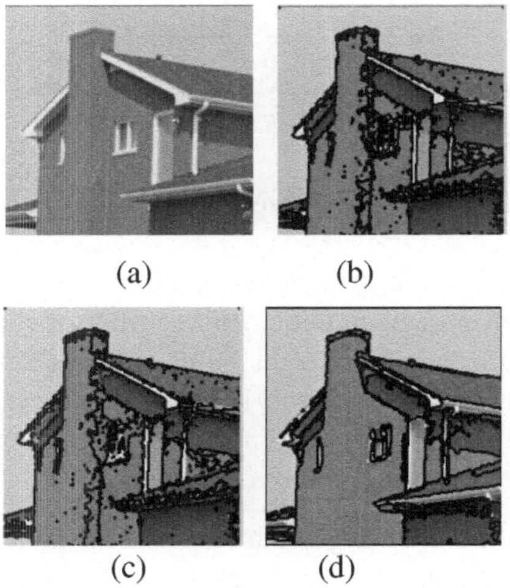


**Fig. 7.** (a)Original (b)Region growing (c) Merging of Paper[7] (d) Merging of the paper

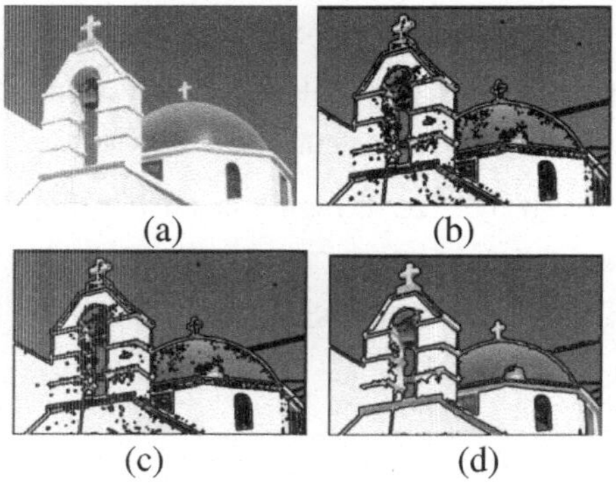


**Fig. 8.** (a)Original (b)Region growing (c) Merging of Paper[7] (d) Merging of the paper

# 8 Summary

This paper proposes an improved color image segmentation method based on region growing. Considering the unsolved seed points merging and get the better result.

# References

1. Freixenet, J., Muñoz, X., Raba, D., Martí, J., Cufí, X.: Yet Another Survey on Image Segmentation: Region and Boundary Information Integration. In: Heyden, A., Sparr, G., Nielsen, M., Johansen, P. (eds.) ECCV 2002, Part III. LNCS, vol. 2352, pp. 408–422. Springer, Heidelberg (2002)
2. Otsu, N.: A threshold selection method from gray-level histogram. IEEE Transactions on Systems, Man, and Cybernetics 9(1), 62–66 (1979)

3. Tremeau, A., Bolel, N.: A region growing and merging algorithm to color segmentation. Pattern Recognition 30(7), 1191–1203 (1997)
4. Basak, J., Chanda, B.: On edge and line linking with connectionist model. IEEE Transactions on System, Man, and Cybernetics 24(3), 413–428 (1994)
5. Pavlidis, T., Liow, Y.T.: Integrating region growing and edge detection. IEEE Transactions on Pattern Analysis and Machine Intelligence 12(3), 225–233 (1990)
6. Adams, R., Bischof, L.: Seeded region growing. IEEE Transactions on Pattern Analysis and Machine Intelligence 16(6), 641–647 (1994)
7. Shih, F.Y., Cheng, S.: Automatic seeded region growing for color image segmentation. Image and Vision Computing 23, 877–886 (2005)
8. Cheng, H.D., Jiang, X.H., Sun, Y., Wang, J.: Color image segmentation: advance and prospects. Pattern Recognition 34, 2259–2281 (2001)

# A Self-adaptive Segmentation Method by Fusion of Multi-color Space Components

Kun Chen[1], Yan Ma[1], Jun Liu[1], and Shun-bao Li[2]

[1] Department of Information Mechanical and Electrical Engineering,
Shanghai Normal University, Shanghai200234, China
[2] Department of Mathematic & Science, Shanghai Normal University, Shanghai200234, China
kun_1949301@126.com, liujunout@126.com,
{ma-yan,lsb}@shnu.edu.cn

**Abstract.** This paper presents a new, simple, and efficient segmentation approach. Firstly, choose the best segmentation components among six different color spaces. Then, Histogram and SFCM techniques are applied for initialization of segmentation. Finally, fuse the segmentation results and merge similar regions. Extensive experiments have been taken on Berkeley image database by using the proposed algorithm. The results show that, compared with some classical segmentation algorithms, our method could achieve better image partitioning and better performance.

**Keywords:** Color image segmentation, Histogram, SFCM, Fusion.

## 1 Introduction

Image segmentation is a popular technique for image processing. We should solve two problems for color image segmentation: (1) choose the right color space; (2) select the appropriate segmentation strategy.

Many methods have been proposed and studied in the last decades to solve the color image segmentation problem. Some researchers prefer to use more complicated feature selection procedures or more elaborate clustering techniques and then improve the final segmentation result by complex optimization method. Such as segmentation based on lossy data compression [1-2], wavelet-domain hidden markov models [3], graph-based [4-5], Mean-Shift [6] and etc. Some researchers also use information fusion strategies to get better performance. They prefer to fuse the results associated with the simple method applied on different color spaces rather than to consider complex segmentation theory or model. Eg. Mignotte [7] proposed a method called FCR by fusion of multi-color spaces based on local histogram and K-means clusters.

Learning from Mignotte's idea, we propose a novel, simple and efficient method by fusion of multi-color space components. First, we choose six different color components elaborately through extensive experiments. Then we propose a peak-finding algorithm to determine cluster number of each component and initialize cluster centroid for SFCM clustering. Then, we propose a clustering method to fuse six different segmentation results. Finally, a region merging method is proposed to

J. Lei et al. (Eds.): AICI 2012, LNAI 7530, pp. 374–382, 2012.

merge the previous segmentation results. The proposed method is tested on Berkeley natural image database. Extensive experiments show that, compared to the state-of-the-art segmentation methods recently proposed in the literature, our method performs competitively in terms of visual evaluations and quantitative performance measures.

## 2 Initial Segmentation

### 2.1 Peak Finding

How to determine initial cluster centroids has always been a problem of clustering. Good initial cluster centroids not only can yield better cluster results but also can make cluster faster. Selecting initial cluster centroids randomly is likely to lead the optimization of the algorithm's objective function to local extreme, therefore the accuracy of the cluster results will be affected. In this paper we utilize histogram technique to find cluster centroid. Here we take gray component as an example to propose the peak finding algorithm. In this way we also can obtain the peaks of other components. The procedure is as follows:

1) Quantize gray component into 0-255 intensity levels, count the frequency, and create the histogram. Let g(i) be the gray component histogram, $x_i$ be the number of pixels associated with ith intensity level in g(i).The histogram of gray component can be represented by the following equation:

$$g(i) = x_i, \qquad 0 \le i \le 255 \tag{1}$$

2) Smooth histogram. Use 1D Gaussian filters with size of $1\times5$ for $g(i)$ to smooth twice, and the result of smoothing depends on Gaussian standard covariance $\sigma_g$ . The histogram is smoother with bigger $\sigma_g$ .We have a new histogram $T_g(i)$ after smoothing.

3) Search for initial peaks. We search turning points on which gradient value varies from positive to negative. We take these turning points as initial peaks and get initial set of peaks $P_1$.

4) Remove small peaks. If the value of peak in set $P_1$ is less than threshold $T_1$, it is removed from $P_1$. So we have new set of peaks $P_2$.

5) Remove adjacent peaks and generate final peaks. If two peaks in $P_2$ are close enough, we think the gray values of the regions represented by the two peaks are similar. Therefore, we remove the smaller one while the distance between two peaks is less than threshold $T_2$. We get final set of peaks $P_3$.

### 2.2 Spatial FCM Clustering

The classical FCM [8] algorithm is to assign pixels to each cluster by using fuzzy memberships. Let $X = (x_1, x_2, \cdots x_n)$ denotes an image with $n$ pixels to be partitioned into $c$ clusters, where $x_i$ represents multispectral (features) data. The result of classification can be represented by a fuzzy membership degree matrix $U=\{\mu_{ik}\}$, where $\mu_{ik}$ represents the membership degree of kth pixel to ith cluster centroid. It is subject to the following constraints:

$$\mu_{ik} \in [0,1], \forall i,k; 0<\sum_k \mu_{ik} < n, \forall i; \sum_i \mu_{ik} = 1, \forall k \tag{2}$$

FCM algorithm is an iterative optimization that minimizes the cost function defined as follows:

$$J=\sum_{k=1}^{n}\sum_{i=1}^{c}\mu_{ik}^{l}\left\|x_k - v_i\right\|^2, \tag{3}$$

Where $U=\{\mu_{ik}\}$ is the membership degree matrix according to Eq.(2), $V=\{v_1,v_2,\cdots v_c\}$ is the set of cluster centroids, $\left\|x_k - v_i\right\|$ represents the distance of pixel $x_k$ to cluster centroid $v_i$, and we use Euclidean distance in initial segmentation. The parameter $l$ controls the fuzziness of the resulting partition, and $l=2$ is used in this study.

The membership functions $\mu_{ik}$ and the centroids $V_i$ are updated iteratively as follows:

$$\mu_{ik}=\frac{\left\|x_k - v_i\right\|^{-2/(l-1)}}{\sum_{j=1}^{c}\left\|x_k - v_j\right\|^{-2/(l-1)}} \tag{4}$$

$$v_i=\frac{\sum_{k=1}^{n}\mu_{ik}{}^{l}x_k}{\sum_{k=1}^{n}\mu_{ik}{}^{l}} \tag{5}$$

The standard FCM algorithms is optimized when pixels close to their centroids are assigned high membership values, while those that are far away are assigned low values.

One of the problems of classical FCM algorithm in image segmentation is the lack of spatial information. Since image noise and artifacts often impair the performance of FCM segmentation, it would be attractive to incorporate spatial information into FCM. Chuang et al. [9] proposed a spatial FCM algorithm in which spatial information can be incorporated into fuzzy membership functions directly using

$$\mu_{ik}^{'}=\frac{\mu_{ik}^{p}h_{ik}^{q}}{\sum_{j=1}^{c}\mu_{jk}^{p}h_{jk}^{q}} \tag{6}$$

Where $p$ and $q$ are two parameters controlling the respective contribution. The variable $h_{ik}$ includes spatial information by

$$h_{ik} = \sum_{j \in N_k} \mu_{ij} \tag{7}$$

Where $N_k$ denotes a local window centered around the image pixel $k$. The weighted $\mu_{ik}$ and the centroid $v_i$ are updated as usual according to Eq. (4) and (5).

## 3 Fusion of Initial Segmentation Results

We get six different initial segmentation results from six different color space components by using the method proposed in section 2. The cluster number of them is different, we record them as $K_i, 1 \le i \le 6$. We use SFCM algorithm again to fuse above six results which with different cluster number and get a new result $I_{fusion}$ after fusion.

### 3.1 Extract Feature Vector

For each initial segmentation result with $K_i (1 \le i \le 6)$ cluster number, considering the squared fixed-size ( $N_W \times N_W$ ) neighborhood centered around the pixel. Let $W_x$ represent the neighborhood of pixel location $x$ .We calculate the normalized local histogram of the class labels for each pixel within $W_x$:

$$h(W_x) = (\frac{n_1}{N_w^2}, \frac{n_2}{N_w^2} \cdots \frac{n_{K_i-1}}{N_w^2}, \frac{n_{K_i}}{N_w^2}) \tag{8}$$

Where $h(W_x)$ represents the feature vector of pixel location $x$ in one of the six segmentation results, $n_j$ denotes the number of pixels whose class labels are $j$ within $W_x$. We do the same process toward six different segmentation results described above. After that, we get six feature vector location in the same place for each pixel. Then combine them in series and normalized. Finally, we get the fused local histogram of the class labels $h^*(W_x)$ with dimension $M = \sum_{i=1}^{6} K_i$ , which is used as feature vector for input in the final clustering.

### 3.2 Fusion of Initial Segmentation by SFCM

We adopt SFCM algorithm (described in Section 2.2) again to partition $h^*(W_x)$ into $N$ classes.

$$N = ceil(\sum_{i=1}^{6} K_i / 6) \tag{9}$$

Where ceil(A) represents round the elements of A to the nearest integers. We get segmentation result $I_{fusion}$ by fusion, in which the distance between two feature vectors from local histogram of the class labels is calculated by Bhattacharya distance:

$$D_B[h_1^*, h_2^*] = \left(1 - \sum_{i=1}^{M} \sqrt{h_1^* \cdot h_2^*}\right)^{1/2} \tag{10}$$

Where $h_1^*, h_2^*$ denote two normalized feature vectors from local histogram of the class labels, $M$ denotes the dimension of feature vector.

# 4 Region Merging

Segmentations with clustering are often featured with numerous discrete small regions. The spatial connectivity between pixels in the same cluster could hardly be guaranteed. These minor regions on one hand preserves the image detail but on the other hand largely affects the segmentation quality. To generate reasonable segmentations, a simple and effective region merging strategy is necessary for this issue. In this paper, the region merging method is presented in LUV color space. The steps are as follows:

1、 Relabel regions after segmentation by 8-neighbors, which yields that not adjacent and color-homogeneous regions are marked with different labels.
2、 Search adjacent regions after relabeling.
3、 Calculate the mean value of L, U, V components for relabeled regions.
4、 Merge small regions. If the size of region is smaller than threshold $T_3$, it will be merged into its bigger adjacent region with the smallest Euclidean distance in LUV color space, and whose size is greater than $T_3$.
5、 Merge big regions. If the size of region is smaller than threshold $T_4 (T_3 << T_4)$, we calculate Euclidean distance in LUV color space with its adjacent regions whose size is greater than $T_4$. Search the smallest distance dc. If $dc < T_5$, the region will be merged into its adjacent bigger region with the smallest distance, and vice versa.

# 5 Experiment Results and Analysis

The proposed algorithm is demonstrated on the computer Inter Core2 Duo CPU T6570 2.10 GHz. We use Matlab R2011a to test the segmentation results on natural images in the Berkeley segmentation database[10]. We have done numerous experiments which show that the results are best when the involved parameters $\sigma_g$ ,$T_1$ ,$T_2$ chosen to 3, 0.001S (S denotes the size of image),15, the window size $N_w \times N_w$ chosen to $5\times5$ and $T_3,T_4,T_5$ chosen to 0.003S, 0.05S, 50. We will analyze our algorithm from the following two aspects: the choice of different color components and compare the algorithm with some state-of-art methods qualitatively and quantitatively. The quantitative comparison is based on the following performance measures, namely a probabilistic measure called PRI [7] (higher probability is better) and three metrics VoI [7], GCE [7], and BDE [7] (lower distance is better).

### 5.1 Choice of Different Color Components

Extensive experiments show that the selection of different color components has important influence on the segmentation result. In order to compare with FCR [7] algorithm, we choose six components to fuse. Which six different color components are the best? We use self-adaptive histogram and SFCM clustering techniques to quantitatively test the components of HSV, YIQ, YCbCr, LAB, LUV color spaces and gray component on randomly chosen images.

Table 1 shows the PRI, VoI, GCE and BDE performance of these 14 components on 100 randomly chosen images in the Berkeley segmentation database. Best performance of each measure is marked with bold. Second best is marked with underline. In PRI indice, V(HSV) component is best, Gray component is second best; In VoI indice, B component is best, I is second best; In BDE indice, Cr component is best, V component is second best. Table 1 also shows that some component is the best in one indice, but worse in other indices. Eg. A component has the best GCE indice, but PRI、BDE is worse. Therefore, we need to consider different performance measures of components together to select the best components. In the analysis, we choose Gray, V(HSV), I, Cr, B, U as six different components to fusion.

**Table 1.** The performance measures of 14 components

| component | PRI | VOI | GCE | BDE |
|---|---|---|---|---|
| Gray | <u>0.7045</u> | 3.0394 | 0.3894 | 10.3919 |
| H | 0.6773 | 2.8031 | 0.3204 | 12.6409 |
| S | 0.6860 | 3.1189 | 0.3909 | 12.2657 |
| V(HSV) | **0.7146** | 3.0790 | 0.3953 | <u>10.1278</u> |
| Y | 0.6980 | 2.9322 | 0.3855 | 10.4577 |
| I | 0.6833 | <u>2.6635</u> | 0.3169 | 11.6272 |
| Q | 0.6511 | 2.9453 | 0.3485 | 11.9675 |
| Cb | 0.6782 | 2.9347 | 0.3612 | 10.2231 |
| Cr | 0.6776 | 2.9786 | 0.3565 | **9.8885** |
| L | 0.6958 | 2.8327 | 0.3282 | 10.7621 |
| A | 0.6190 | 2.6721 | **0.2878** | 14.6134 |
| B | 0.6767 | **2.6045** | <u>0.2952</u> | 11.5503 |
| U | 0.6568 | 2.7091 | 0.3013 | 10.7357 |
| V(LUV) | 0.6901 | 2.8074 | 0.3258 | 10.9033 |

### 5.2 Comparison with State-of-the-Art Methods

We test 300 images on Berkeley image database and compare our method with state-of-the-art methods such as: Mean-shift [6], NCuts [4], FH [5], CTM [1-2] and FCR [7]. Fig.1 shows the segmentation results of FCR, Mean-shift, CTM and our proposed method with 5 randomly chosen images. Fig.1 (a) is original images. Fig.1 (b) shows FCR segmentation results. Fig.1 (c) is Mean-shift results. Fig.1 (d) shows CTM results. Fig.1 (e) is our proposed method. It is obvious that FCR and Mean-shift methods have over-segmentation problem in Fig.1. For certain images, these two methods can only yield small piece regions, and can't generate the right object, especially Mean-shift method. Our method can get better results which are close to human perception and has less over-segmentation problem.

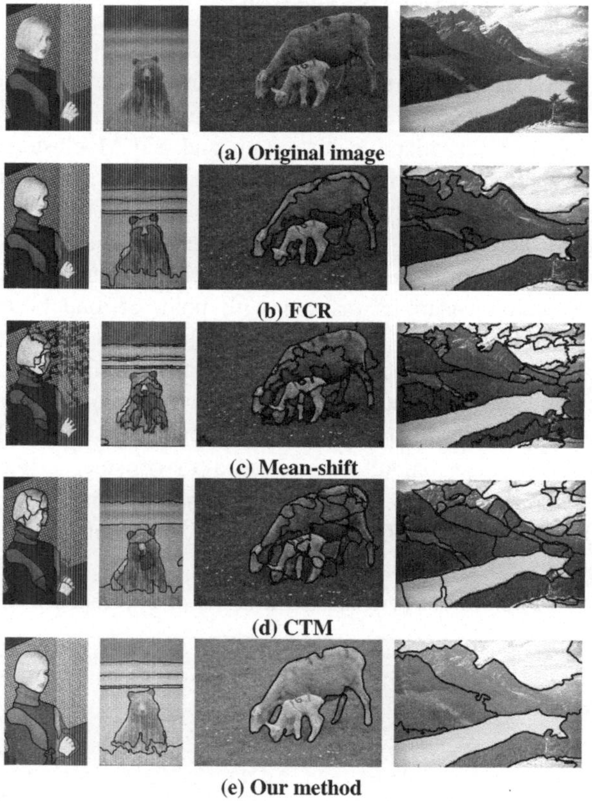


**Fig. 1.** Comparison of FCR, Mean-shift, CTM and our method

Table 2 shows the mean value of performance measures over the 300 images of the Berkeley image database in different methods. Best performance of each measure is marked with bold. Second best is marked with underline. From Table 2 we can see that our method outperforms other methods for several different internal parameters, all the well-known segmentation algorithms presented in Table 2 in terms of PRI and

BDE indices, second best in VoI indice and is obviously better than FCR in PRI, VoI and BDE indices.

**Table 2.** Performance measures comparison to state-of-the-art methods

| Algorithm | PRI | VoI | GCE | BDE |
|---|---|---|---|---|
| Humans | 0.8754 | 1.1040 | 0.0797 | 4.9940 |
| FCR | 0.7842 | 2.3925 | 0.2169 | 9.2463 |
| CTM($\eta$ =0.1) | 0.7561 | 2.4640 | **0.1767** | 9.4211 |
| CTM($\eta$ =0.2) | 0.7617 | **2.0236** | 0.1877 | 9.8962 |
| Mean-shift | 0.7550 | 2.4770 | 0.2594 | 9.7001 |
| NCuts | 0.7229 | 2.9329 | 0.2182 | 9.6038 |
| FH | 0.7841 | 2.6647 | 0.1895 | 9.9497 |
| Our Method | **0.7906** | 2.1395 | 0.2218 | **9.0625** |

Table 3 shows the average runtime of 100 randomly chosen images in the same platform. It is obvious that our method faster than FCR algorithm.

**Table 3.** Runtime of Our Method and FCR

| Algorithms | Our Method | FCR |
|---|---|---|
| Runtime/s | 138.732 | 317.859 |

## 6 Conclusion

This paper proposes a novel, simple and efficient method by fusion of multi-color space components. Results show that the method provides good segmentation on a variety of color images. Histogram and SFCM cluster techniques are used in initial segmentation. The strategy not only can locate initial cluster centroids quickly but also can solve the problem of that clustering number is fixed. Then an effective fusion and region merging strategy is used to make segmentation result more close to human perception. The proposed method has been successfully applied on the Berkeley image database, and performs competitively among the recently reported state-of-the-art segmentation methods in terms of visual evaluations and quantitative performance measures.

## References

1. Yang, A.Y., Wright, J., Sastry, S., Ma, Y.: Unsupervised Segmentation Of Natural Images Via Lossy Data Compression. Comput. Vis. Image Understand. 110, 212–225 (2008)
2. Ma, Y., Derksen, H., Hong, W., Wright, J.: Segmentation Of Multivariate Mixed Data Via Lossy Coding And Compression. IEEE Transactions on Pattern Analysis and Machine Intelligence 29, 1546–1562 (2007)

3. Choi, H., Baraniuk, R.G.: Multiscale Image Segmentation Using Wavelet-Domain Hidden Markov Models. IEEE Transactions on Image Processing 10, 1309–1321 (2001)
4. Shi, J., Malik, J.: Normalized Cuts and Image Segmentation. IEEE Transactions on Pattern Analysis and Machine Intelligence 22, 888–905 (2000)
5. Felzenszwalb, P., Huttenlocher, D.: Efficient Graph-Based Image Segmentation. Int. J. Comput. Vis. 59, 167–181 (2004)
6. Comanicu, D., Meer, P.: Mean shift: A Robust Approach Toward Feature Space Analysis. IEEE Transactions on Pattern Analysis and Machine Intelligence 24, 603–619 (2002)
7. Mignotte, M.: Segmentation By Fusion Of Histogram-Based K-Means Clusters In Different Color Spaces. IEEE Transactions on Image Processing 17, 780–787 (2008)
8. Bezdek, J.C.: Pattern Recognition With Fuzzy Objective Function Algorithms. Plenum Press, New York (1981)
9. Chuang, K.S., Tzeng, H.L., Chen, S., Wu, J., Chen, T.J.: Fuzzy C-Means Clustering With Spatial Information For Image Segmentation. Computerized Medical Imaging and Graphics 30, 9–15 (2006)
10. Martin, D., Fowlkes, C., Tal, D., et al.: A database of human segmented natural images and its application to evaluating segmentation algorithms and measuring ecological statistics. In: Proc. of the 8th IEEE International Conference on Computer Vision, pp. 416–423 (2001)

# Traffic Vehicle Behavior Prediction Using Hidden Markov Models

Jian Wu[1,2], Zhi-ming Cui[1], Peng-peng Zhao[1], and Jian-ming Chen[1]

[1] The Institute of Intelligent Information Processing and Application, Soochow University, Suzhou 215006, China
[2] State Key Laboratory for Novel Software Technology, Nanjing University, Nanjing 210093, China
jianwu@suda.edu.cn

**Abstract.** This paper intends to focus on vehicles steering by analyzing the trajectory. The target values we need to measure are the vehicle's speed and turning angle. These two values are needed to be quantified to certain levels. To create the Hidden Markov Model, HMM learning algorithm and the two values above are used. In HMM, turning left, going straight and turning right are the hidden states and data from the video are used to compute the parameters. HMM can be used to analyze vehicles' driving and to predicate the probable steering in time. The experimental results show that in the case of getting good vehicle trajectory, it is pretty suitable to use HMM to predicate vehicle behavior.

**Keywords:** HMM, Vehicle Behavior, Baum-Welch Algorithm, Forward / Backward Algorithm.

## 1 Introduction

Video-based event detection is an important research in academic fields such as Intelligent Traffic, Security Monitoring, Intelligent Searching. In recent years, vehicle detection, vehicle tracking and vehicle behavior state analysis have already become the focuses of the study[1-3]. Kamijo et al [4,5] proposed an algorithm used in traffic event: Spatio-temporal Markov random field model. Saunier et al [6-8] developed a video-based intersection traffic safety analysis system automatically which firstly takes the feature-based vehicle tracking algorithm into use to track the vehicles within the intersection. As the economic loss due to traffic accidents is serious, taken traffic safety into account, different kinds of analysis and research attach great importance to it. The steering action is one kind of the vehicle behavior which is also an important action in the process of vehicles. If the steering action can be predicated, then illegal behaviors will not appear and traffic accidents will not occur. For example, when the driver's steering action has been predicated in time as the action is forbidden currently, the driver would be warned and the probable illegal behavior can be avoided. When the traffic is heavy, if steering is controlled without delay, traffic accident can be avoided too. The research to steering behavior has not shown a

J. Lei et al. (Eds.): AICI 2012, LNAI 7530, pp. 383–390, 2012.

solution which can be recognized as a mature program for all situations. In this paper, by measuring and classifying the vehicle speed changes and direction changes, training the parameters drawn from sections of multiple videos, the HMM is built and finally real-time prediction to traffic video based vehicle steering is achieved.

## 2 Hidden Markov Model

Hidden Markov Model(HMM) consists of two sequences, one sequence describes the statistical relationship of state transition, the other describes the statistics corresponding relationship between hidden states and observation states. From the observer's perspective, only the observed states' value can be seen which is different from the Markov Chain Model in which observed value and state have the one-to-one relationship. So the states cannot be seen directly, their existence and characteristics are perceived by a random process. The theoretical foundation is setup by Baum et al. HMM has wide application in the field of speech recognition, image process, biomedical and so on [9-12] .

Hidden Markov model is usually described with a five-tuple $\lambda = (N, M, A, B, \prod)$ which is short for $\lambda = (A, B, \prod)$.

(1) $N$ is the number of state, $S = (s_1, s_2, ..., s_N)$ is the collection of states.

(2) $M$ is the number of every state's value, $V = (v_1, v_2, ..., v_M)$ is the collection of values.

(3) Define $Q = (q_1, q_2, ..., q_T)$ as a fixed-length state sequence whose length is $T$, and then $O = (o_1, o_2, ..., o_T)$ is the observed state's value.

(4) $A = \{a_{ij}\}$ is the probability matrix of state transition, $a_{ij} = P(q_{t+1} = s_j \mid q_t = s_i)$ shows the probability of the transition from $i$ to $j$ at time $t$, $1 \le i, j \le N, a_{ij} \ge 0$, $\sum_{j=1}^{N} a_{ij} = 1$.

(5) $B = \{b_i(k)\}$ is the probability distribution matrix of state value, $b_i(k) = P(o_t = v_k \mid q_t = s_i)$ shows the probability of $v_k$'s existence when the state is $i$.

(6) $\prod = \{\pi_i\}$ is the initial state probability distribution vector, $\pi_i = P(q_1 = s_i)$ is the probability that the initial state is $i$, $\pi_i \ge 0, \sum_{j=1}^{N} \pi_i = 1$.

## 3 HMM-Based Vehicle Behavior Description

Hidden Markov Model assumes that the learning processing progress consists of a series of limited and discrete observed states and hidden states. HMM can be described as HMM $=(\prod, A, B)$. $\prod$ is the initial probability; $A$ is the probability matrix of state transition; $B$ is the mixed-state probability matrix, that is the probability matrix of given observed values. In this paper, there are three hidden states: going straight, turning left and turning right. The observed states are vehicle speed changes and direction changes quantified results. After measuring the speed changes and direction changes, they should be quantified suitably according to the values. Probability matrix is got from training the video data.

In this paper, the classic Baum-Welch Algorithm is used to train the model's data by computing with the a series of discrete observed value. Model validation and probability calculations are achieved by Forward/Backward Algorithm. The Model runs like the way in the Fig.1.

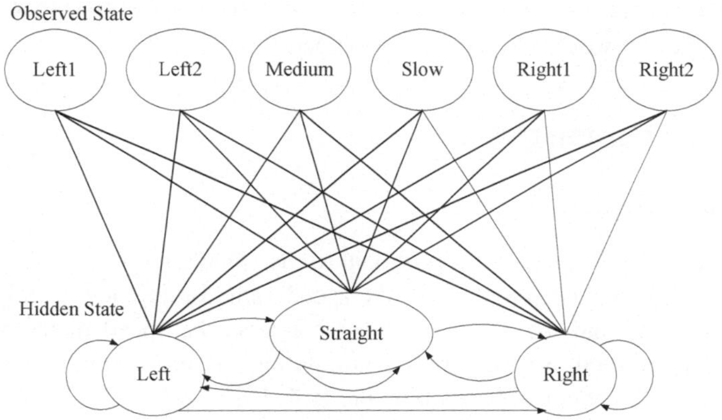


**Fig. 1.** HMM Schematic Diagram

In the experiment, the parameters for HMM according to the chosen video is as follows:

Initial probability matrix which is the $\prod$ vector:

$$\begin{matrix} straight & left & right \\ (1.0 & 0.0 & 0.0) \end{matrix}$$

State transition probability matrix $A$:

| | *straight* | *left* | *right* |
|---|---|---|---|
| *straight* | 0.9 | 0.05 | 0.05 |
| *left* | 0.15 | 0.7 | 0.15 |
| *right* | 0.15 | 0.15 | 0.7 |

Mixed-state probability matrix $B$:

| | *slow* | *medium* | *left* 1 | *left* 2 | *right* 1 | *right* 2 |
|---|---|---|---|---|---|---|
| *slow* | 0.25 | 0.45 | 0.05 | 0.10 | 0.05 | 0.10 |
| *left* | 0.30 | 0.10 | 0.25 | 0.30 | 0.035 | 0.015 |
| *right* | 0.25 | 0.15 | 0.025 | 0.045 | 0.32 | 0.21 |

Initial state shows system's beginning situation which is going through in this paper. The observed states can be achieved by data in the experiment. The observed states' values determine the current hidden state according to the probability which means the current hidden state is turning left, turning right or going through. And the three hidden states can also transfer to another according to the probability. When the times of vehicle's turning state continues reach the pre-set threshold $N$ (if the road or video parameters are different, $N$ would be different, too. In this paper, $N$ =10), the actual turning event would be believed to happen soon.

# 4 Traffic Parameters

## 4.1 Vehicle Speed

As the actual vehicle's speed is proportional to the vehicle's speed in the video on the same road and in the same camera. So the speed value in the video can be chosen as the speed we need and the computation to converse the speed in the video to the actual speed is omitted. So the work is computing the parameters according to the videos which can be obtained by training the HMM. The vehicle speed can be reached according to the difference of different video frames.

In the experiment, C++ and OpenCV are used. By the use of background subtraction method, the background and the foreground can be separated. In the video, vehicle is seen as a white block and then its trajectory can be reached. After computing the centroid's displacement size of the block in the continuous two frames and the frame rate, the vehicle's speed in the video is reached which can represent vehicle's actual speed just as the explanation above. Assume the centroid's pixel difference in adjacent two frames is $\mathrm{PixelDiff}$ , frame rate is $\mathrm{FPS}$, the object's speed $V$ can be computed:

$$\mathrm{V} = \mathrm{PixelDiff}*\mathrm{FPS} \tag{1}$$

If the vehicle's actual speed is too slow, then the pixel difference is not evident and the resulting speed value may has a large deviation. Parameter $\alpha$ is used to adjustment the value:

$$\mathrm{V}' = \alpha*\mathrm{PixelDiff}*\mathrm{FPS} \tag{2}$$

## 4.2 Vehicle Direction

Just as getting the speed value, the direction value is also not the actual vehicle's direction, it is the direction shown in the video. In the same way above, the vehicle in the video is seen as a block. The block's trajectory is tracked to compute the direction angle, which can be used to judge the vehicle's turning degree. As Fig.2 shows:

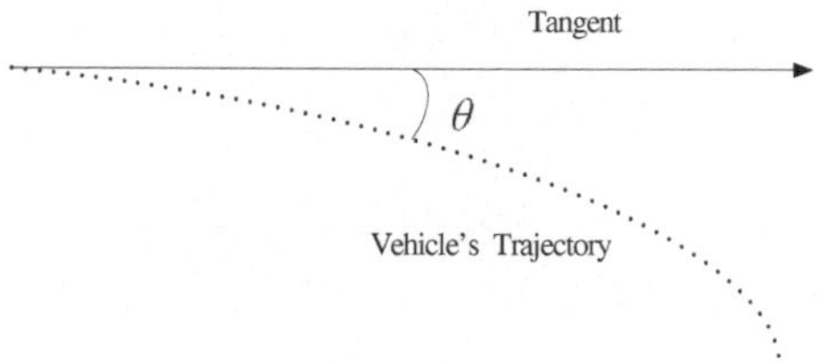


**Fig. 2.** Vehicle's direction

In the real-time tracking process, vehicle's trajectory and its tangent's direction, which is the vehicle's driving direction. The changes of the vehicle's driving direction can be obtained from the value measured by the two adjacent points in the trajectory.

$$\Delta = PixelDiff * (\theta_2 - \theta_1) \tag{3}$$

In the computation and parameter training below, the changes of vehicle direction is quantified.

## 5 Experimental Result and Analysis

In this paper, a video clip of a junction in Suzhou is chosen to carry on the experiment. In the video, there are vehicles going straight and vehicles steering. By the use of the real-time frames of the video and the built HMM, the prediction of steering action can be realized. Some pictures of the video are shown below as Fig.3.

**Fig. 3.** Vehicle's trajectories

(a) Example 1

(b) Example 2 (c) Example 3

**Fig. 3.** (*Continued*)

In Fig.3, vehicle's trajectories are marked with blue lines, which are used to determine the vehicle's current state in the system. Three pictures at the beginning show the vehicle is turning right, three pictures in the middle show that the vehicle is going through and the last three pictures show the vehicle is turning left. In the experiment, if there is only the target vehicle in the frames of the video and the video is clear, the experimental results are better.

The specific experiment is as follows: take the HMM into use, according to the trajectory of vehicle in the video (in the experiment, the vehicle shows as a white block), the system should judge whether the vehicle is going to turning left or turning right and then compare the experimental result with the factual results. The final result shows the correct rate which represents the experimental effect in this paper.

The experimental result contains four parts: ahead, delay, correct and error. The ahead column means the predication occurred $M$ frames earlier; delay column shows the predication has more than $M$ frames of delay; the correct column is the correct rate; and the error column shows that the predication is wrong. In this paper, $M$ =10. The experimental result is shown in Table1.

**Table 1.** Experimental results

| | ahead | delay | correct | error |
|---|---|---|---|---|
| Going through | 2.79% | 7.11% | 89.2% | 10.8% |
| Turning left | 4.6% | 9.56% | 83.65% | 16.35% |
| Turning right | 5.5% | 8.72% | 80.16% | 19.84% |

It can be seen from Table 1 that in the case of less external interference factors, it is pretty precise to predicate the steering action by the use of HMM. The early or delayed situation may have some kind of relationship with the threshold value $N$. The value of parameters is still needed to be discussed. The error happens because of the external interference (such as the vehicle trajectory tracking error, the speed or the direction may have some kind of error) or the defects exist in the system design.

## 6 Conclusions

The experimental result shows that HMM-based vehicle behavior analysis method can predicate vehicle's steering in a good way. In this paper, the number of the observed states can adjusts. The more the number of the states are, the predicated results are more accurate and the process of parameter training and the system's running are time-consuming. As the position and angle of cameras are different, the system's parameter are different, the parameter used should be computed again. And different parameters can lead to different final correct rate, some of which are too low. The method of obtaining the parameter is still to be improved. Whether the parameter can be always a same one is to be studied.

**Acknowledgments.** This research was partially supported by the Natural Science Foundation of China under grant No.60970015, 61003054 and 61170020, the 2009 Special Guiding Fund Project of Jiangsu Modern Service Industry (Software Industry) under grant No.[2009]332-64, the opening project of Jiangsu Province Support Software Engineering R&D Center for Modern Information Technology Application in Enterprise under grant No.SX201102, the Program for Postgraduates Research Innovation in University of Jiangsu Province in 2011 under grant No.CXLX11_0072, and the Beforehand Research Foundation of Soochow University.

## References

1. Li, Z.-M.: Traffic Engineering. China Communications Press, Beijing (2002)
2. Yi, S.: Video-based Traffic Incident Detection Method Research. Southeast University, Nanjing (2007)
3. Shunsuke, K., Masao, S.: Classification of Traffic Events based on the Spatio-Temporal MRF Model and the Bayesian Network. In: Proc. of the 9th World Congress on ITS, Chicago (2002)
4. Shunsuke, K., Katsushi, I., Masao, S.: Event Recognitions from Traffic Images based on Spatio-Temporal Markov Random Field Model. In: Proc. of the 8th World Congress on Intelligent Transport Systems, pp. 1–12 (2001)
5. Kamijo, S., Matsushita, Y., Ikeuchi, K., Sakauchi, M.: Traffic Monitoring and Accident Detection at Intersections. IEEE Trans. on Intelligent Transportation Systems 1(2), 108–118 (2000)
6. Saunier, N.: Automated Road Safety Analysis Using Video Data. In: Proc of the 86th Annual Meeting of the Transportation Research Board (2007)
7. Saunier, N., Sayed, T.: A Feature-based Tracking Algorithm for Vehicles in Intersections. In: Proc. of the 3rd Canadian Conference on Computer and Robot Vision (2006)
8. Saunier, N., Sayed, T.: Clustering Vehicle Trajectories with Hidden Markov Models Application to Automated Traffic Safety Analysis. In: Proc. of International Joint Conference on Neural Networks, pp. 4123–4138 (2006)
9. Bashir, F.I., Khokhar, A.A., Schonfeld, D.: Object Trajectory- Based Activity Classification and Recognition Using Hidden Markov Models. IEEE Trans. on Image Processing 16(7), 1912–1919 (2007)
10. Wang, H.-Q., Peng, J.-X., Yu, Q.-Z.: A Precise Eye Localization via a Dynamic Probability Distribution HMM Model. Journal of Image and Graphics 11(1), 26–32 (2006)
11. Feng, C.-J.: Application on Faults Diagnosis of Rotating Machine in Hidden Markov Models. Zhejiang University, Hangzhou (2002)
12. Shen, Y.-Y., Kang, X., Liu, F.: The Fault Diagnosis of Cars Bogie Based on HMM Method and Its Application. Railway Locomotive & Car 26(4), 24–26 (2006)

# Efficient Visual Tracking by Using LBP Descriptor

Minglei Tong*, Hong Han, and Jingsheng Lei

School of Computer and Information, Shanghai University of Electric Power,
Shanghai 201300, China
tongminglei@gmail.com

**Abstract.** Visual object tracking is a hard problem in many applications for example in video surveillance, human computer interaction(HCI), video communication and compression, augmented reality, traffic control, sports analysis and video editing. The common works towards this task are the ambiguity existing among object and the background because of the moving object and the changing illumination. To track object from cluttered background, LBP descriptor (Local Binary Patterns) is applied in this paper to enable the efficient tracking-by-detection. LBP descriptors are extracted only in region of interest in each frame, to ensure the tracker's high efficiency. After that, tracking is continued using a Bayesian state inference framework in which a particle filter is used for propagating sample distributions over time. The dynamic template updating scheme keeps track of the most representative particles throughout the tracking procedure. Experimental results demonstrate the efficiency of the proposed tracker.

**Keywords:** Visual tracking, Feature extraction ,LBP.

## 1 Introduction

Visual tracking is a critical problem in many computer vision applications such as intelligent surveillance, human computer interaction and intelligent transportation system, etc. Most traditional algorithms are able to track objects in short durations and in well controlled environments. However, these algorithms usually fail to observe the object motion or due to the variation of objects appearance or illumination of the background. The state of this art has advanced significantly during the past several decades.

Tracking can be regarded as the estimation of states for a time series model. In visual tracking context, it generally involves an appearance based model to describe the target, and matching is conducted to differentiate the object from background based on the appearance variety.

Local invariant features exhibit competitive performance in computer vision and pattern recognition tasks due to their attracting representation characteristics. [1] gives a

* This work was supported by the National Natural Science Foundation of China (Grant No. 61105016), the Foundation of Young Teacher in Shanghai University of Electric Power (Grant No. K-2010-16) and the Innovation Program of Shanghai Municipal Education Commission (Grant No. Z2010-062),the Innovation Program of Shanghai Municipal Education Commission(SDL 10026).

J. Lei et al. (Eds.): AICI 2012, LNAI 7530, pp. 391–399, 2012.

thorough empirical study on the performance of various local features. SIFT [2] outstands among many types of local features including shape context [3], PCA-SIFT [4], etc.

In this paper, under the framework of tracking-by-detection, the LBP feature descriptors to represent the target appearance model are proposed. Such localized appearance representation allows a set of features to describe the target in the initial frame, in addition with another set of features to describe the background. The tracker separates target from background in image sequences.

The rest of this paper is organized as follows. Section 2 reviews the related work. Section 3 describes the tracking by detection model built on LBP. Section 4 gives the tracking algorithm. Experimental results are shown in section 5 and Section 6 concludes this paper.

## 2 Related Work

During past decades, many algorithms and systems have been proposed for tracking [5, 6]. Recently, progress among feature representation, object detection and recognition [7], etc. have lead to the new framework namely tracking-by-detection. Shi and Tomasi [8] adopt search methods in a Newton-Raphson manner to handle object tracking under affine transformations. Their tracker monitors the feature qualities during tracking via a measure of feature dissimilarity. Recently, a generative model leveraging on SURF feature is proposed by He et al. [9] to deal with appearance changes, background clutter, illumination changes and occlusion, which is common in visual tracking. Other state-of-the-arts in this field are such as [10]. More related paper regarding with this paper is feature point based approach, such as [11]. In [11], the tracker tracks a single maximally stable extremal region(MSER) feature. In this paper, the proposed tracker is built on LBP, whose computational time is faster than MSER, while maintaining the approximate efficacy.

Our work is motivated in part by the subspace representations as appearance models in conjunction with particle filter framework [12]. In contrast to the eigen tracking algorithm [12], our algorithm does not require a PCA training and appearance model but use a updating LBP descriptor as measurements on-line during the object tracking process. The improvement is the computation of feature extraction is only limited within the region of interest in the temporal tracking context. This improvements significantly speed up the computational efficiency of the tracker.

## 3 The Model

### 3.1 LBP Descriptors

In this section, we have a brief review on the LBP descriptor scheme [13]. The local binary pattern (LBP) operator is defined as a gray-scale invariant texture measure, derived from a general definition of texture in a local neighborhood. Through its recent extensions, the LBP operator has been made into a really powerful measure of image texture, showing excellent results in many empirical studies. The LBP operator can be seen as a unifying approach to the traditionally divergent statistical and structural models of texture analysis. Perhaps the most important property of the LBP operator in real-world applications is its invariance against monotonic gray level changes. Another equally important is its computational simplicity, which makes it possible to analyze images in challenging real-time settings. The LBP originally appeared as a generic texture descriptor. The operator assigns a label to each pixel of an image by a 3x3 neighborhood thresholds with the center pixel value and considering the result as a binary number. In different publications, the circular 0 and 1 resulting values are read either clockwise or counter clockwise. In this research, the binary result will be obtained by reading the values clockwise, starting from the top left neighbor, as can be seen in the following figure. In one image, given a pixel position $(x_c, y_c)$, LBP is defined as an ordered set of binary comparisons of pixel intensities between the central pixel and its surrounding pixels. The resulting decimal label value of the 8-bit word can be expressed as follows:

$$LBP(x_c, y_c) = \sum_{n=0}^{7} s(l_n - l_c)2^n \tag{1}$$

where $l_c$ corresponds to the grey value of the center pixel $(x_c, y_c)$, $l_n$ to the grey values of the 8 surrounding pixels, and function s(k) is defined as::

$$s(k) = \begin{cases} 0, if \text{ k>=0;} \\ 1, if \text{ k<0.} \end{cases} \tag{2}$$

The proposed approach in visual tracking is very robust in terms of grayscale variations caused, e.g., by changes in illumination intensity since the LBP is by definition invariant against any monotonic transformation of the gray scale. This should make it very attractive in situations where nonuniform illumination conditions are a concern, e.g., in visual inspection.

### 3.2 The Tracking Framework of Particle Filter

The particle filter [14] is a Bayesian sequential importance sampling technique for estimating the posterior distribution of state variables characterizing a dynamic system. It consists of essentially two steps: prediction and update. The state variable $\mathbf{X}_t$ the affine motion parameters (and thereby the location) of the target at time t. Given a set of observed images $\mathcal{I}_t = \{\mathbf{I}_1, \cdots, \mathbf{I}_n\}$, we aim to estimate the value of the hidden state variable $\mathbf{X}_t$. Using Bayes' theorem, we have the following result

$$p(\mathbf{X}_t|\mathcal{I}_t) \propto p(\mathbf{I}_t|\mathbf{X}_t) \int p(\mathbf{X}_t|\mathbf{X}_{t-1})p(\mathbf{X}_{t-1}|\mathcal{I}_{t-1})d\mathbf{X}_{t-1} \tag{3}$$

The tracking process is governed by the observation model $p(\mathbf{I}_t|\mathbf{X}_t)$, where we estimate the likelihood of $\mathbf{X}_t$ observing $\mathbf{I}_t$, and the dynamical model between two states $p(\mathbf{X}_t|\mathbf{X}_{t-1})$. The dynamical model we adopt is the same as that in work [1]. The location of a target object in an image frame can be represented by an affine image warp. This warp transforms the image coordinate system, centering the target within a canonical box such as the unit square. The state at time t consists of the six parameters of an affine transformation $\mathbf{X}_t = (x_t, y_t, \theta_t, s_t, \alpha_t, \varphi_t)$ where $x_t$, $y_t$, $\theta_t$, $s_t$, $\alpha_t$, $\varphi_t$, denote x, y translation, rotation angle, scale, aspect ratio, and skew direction at time t. Each parameter in $\mathbf{X}_t$ is modeled independently by a Gaussian distribution around its counterpart in $\mathbf{X}_{t-1}$, and thus the motion between frames is itself an affine transformation.

$$p(\mathbf{X}_t|\mathbf{X}_{t-1}) \propto N(\mathbf{X}_t; \mathbf{X}_{t-1}; \Psi) \tag{4}$$

where $\Psi$ is a diagonal covariance matrix whose elements are the corresponding variances of affine parameters.

However, the observation model is very different with [1]. We model image observations using LBP descriptor.

### 3.3 Algorithm and Implementation

Detailed algorithm and implementation with respect to the tracker is described as follows.

---

**Algorithm 1.** LBP tracking using nearest neighboring classifier.

---

1: Locate the target object in the first frame, either manually or by using an automated detector, and use a single particle to indicate this location.
2: Draw particles from the particle filter, according to the dynamical model.
3: **for** $i$ such that $i = 1 : N$ **do**
4: Uniformly initialize the weight w to 1/N for every particle.
5: **end for**
6: **for** $t$ such that $t \leq$number of frames **do**
7: For each particle, extract the corresponding window from the current frame. calculate the LBP descriptor of every particles.
8: Calculate the likelihood under the observation model.
9: Resample particles according to the dynamical model.
10: Updating the observation model
11: **end for**

---

## 4 Experimental Results

We implemented the proposed tracker in MATLAB based on the implementation of the tracker from the test sequences downloaded from http://www.cs.toronto.edu/ dross/ivt/ [12], namely *david* . Figure 1 shows some typical sampling results of the tracking sequence. The second sequence namely *fish*. In that video, a carved fish on the desk is showed in a moving shot with various illumination. Figure 2 gives some typical samples of the tracking results. The Third sequence namely *sylv*. In that video, a toy is held on the hand moving and rotating in a dim room. The fourth sequence namely *trellis*. In that sequence, a boy is moving in a very cluttered backgroud. Figure 3 and Figure 4 respectively demonstrates the efficacy of the proposed tracker on the third and fourth sequence. One can observe the tracker performs robustly towards different tracking scenarios against noise and variation.

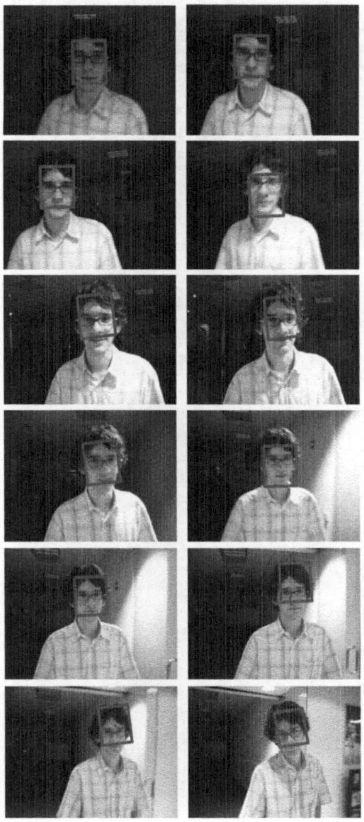

**Fig. 1.** The tracking results of the *davidin* sequence

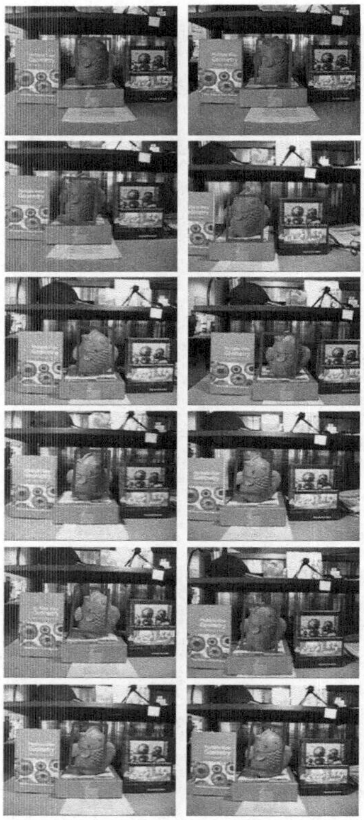

**Fig. 2.** The tracking results of the *fish* sequence

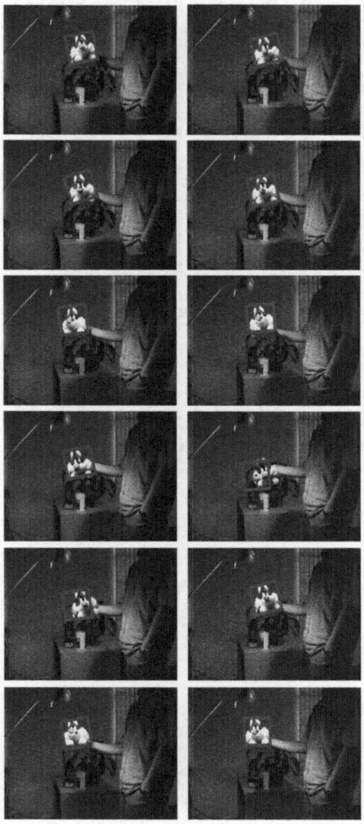

**Fig. 3.** The tracking results of the face *sylv* sequence

**Fig. 4.** The tracking results of the *trellis* sequence

## 5 Conclusion

In this paper we propose using the LBP feature for visual tracking under a tracking-by-detection framework and we improve the efficiency by using LBP feature which has been verified its excellence in visual tracking. On the other hand, we improve the scheme by updating the background appearance model. In experiments involving four challenging sequences, the proposed tracker demonstrates competitive performance. Future work would combine a variety of features representation into a unified framework to compensate current LBP features.

## Acknowledgement

## References

1. Mikolajczyk, K., Schmid, C.: A performance evaluation of local descriptors. In: TPAMI, pp. 257–263 (2003)
2. Lowe, D.G.: Object recognition from local scale-invariant features. In: ICCV (1999)

3. Belongie, S., Malik, J., Puzicha, J.: Shape matching and object recognition using shape contexts. PAMI 24, 509–522 (2002)
4. Ke, Y., Sukthankar, R., Sukthankar, R.: Pca-sift: A more distinctive representation for local image descriptors. In: CVPR, pp. 506–513 (2004)
5. Comaniciu, D., Ramesh, V., Meer, P.: Real-time tracking of non-rigid objects using mean shift. In: CVPR (2000)
6. Sidenbladh, H., Black, M.J., Fleet, D.J.: Stochastic Tracking of 3D Human Figures Using 2D Image Motion. In: Vernon, D. (ed.) ECCV 2000. LNCS, vol. 1843, pp. 702–718. Springer, Heidelberg (2000)
7. Felzenszwalb, P., McAllester, D., Ramanan, D.: A discriminatively trained, multiscale, deformable part model. In: IEEE CVPR (2008)
8. Shi, J., Tomasi, C.: Good features to track. In: CVPR (1994)
9. He, W., Yamashita, T., Lu, H., Lao, S.: SURF Tracking. In: ICCV (2009)
10. Santner, J., Leistner, C., Safiari, A., Pock, T., Bischof, H.: PROST Parallel Robust Online Simple Tracking. In: IEEE CVPR (2010)
11. Donoser, M., Bischof, H.: Efficient Maximally Stable Extremal Region (MSER) Tracking. In: CVPR (2006)
12. Ross, D.A., Lim, J., Lin, R.-S., Yang, M.-H.: Incremental learning for robust vi- sual tracking. International Journal of Computer Vision 77(1-3), 125–141 (2008)
13. Ojala, T., Pietikainen, M., Maenpaa, T.: Multiresolution gray-scale and rotation invariant texture classification with Local Binary Patterns. IEEE Transactions on Pattern Analysis and Machine Intelligence 24(7), 971–987
14. Doucet, A., de Freitas, N., Gordon, N.: Sequential Monte Carlo Methods in Practice. Springer (2001)

# Trilateral Filtering-Based Retinex for Image Enhancement

Li Yang[1], Xiaobo Lu[1], Weili Zeng[2], and Wei Geng[1]

[1] School of Automation, Southeast University, Nanjing, 210096
{yangmismiley,xblu2008,gwahfy}@yahoo.cn
[2] School of Transportation, Southeast University, Nanjing, 210096
zengwlj@yahoo.com.cn

**Abstract.** Retinex is a theory describing the color consistency of human visual system. Its essence is to put aside the influence of illumination image L from image S, and to get reflectance properties R which is the real appearance of the object. Trilateral filter is a recently reported filtering technology which can preserve edges and remove noise as well. In this paper, we propose a Retinex image enhancement method based on trilateral filter. The proposed method enjoys the benefits of trilateral filter, which not only enhance image contrast and aviod "halos" but also preserve edges and texture. Experimental results show the good behavior of our proposed method.

**Keywords:** Retinex, Trilateral Filter, Enhancement, Illumination Image, Reflectance Image.

## 1 Introduction

The basic principle of Retinex theory is separating an image into two parts: illumination and reflectance, and trying to reduce the influence of the illumination to the reflectance, to achieve the purpose of image contrast enhancement [1]. The main goal of Retinex model is image reconstruction, which makes the reconstructed image the same with what observers see on the scene. The foundation of Retinex model is the illumination -reflectance imaging model. The basic principle is very similar to homomorphic filter: illumination, which can be got by using a low-pass filter to the input image, is much smoother than the reflectance, and to divide the input image by the smoothed image we can get the reflectance image [2-3]. Retinex theory explains how the human visual system works: in different brightness environment and light conditions, what people see is mainly concerned with objects reflect and has little to do with the irradiation light. And the human visual system can also remain color constancy [4]. According to the proposed Retinex algorithm by Land, the image can be can be defined as follows:

$$S(x, y) = L(x, y) \cdot R(x, y) \tag{1}$$

$S(x, y)$ represents what human eye sees, and $L$ is the brightness of the surrounding environment which has nothing to do with the object, and $R$ refers to the object

J. Lei et al. (Eds.): AICI 2012, LNAI 7530, pp. 400–407, 2012.

reflective characteristics, it contains a object's detail characteristics. If we let $L$ the radiation intensity of illumination (brightness image), and $R$ the object reflective characteristics, then we can use the former formula to describe the imaging process. The Retinex algorithm flow is illustrated in Fig. 1.

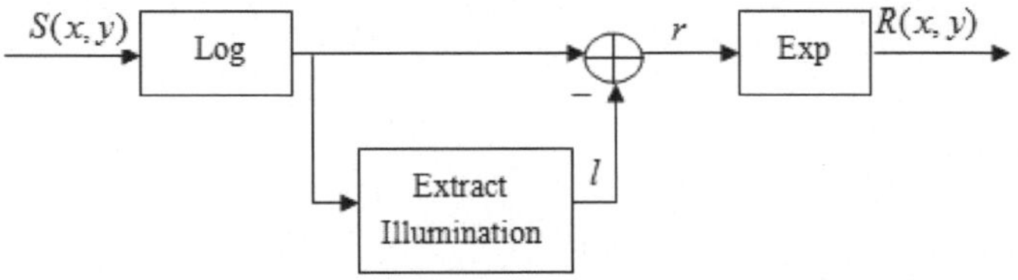


**Fig. 1.** Retinex Algorithm Flow

Smith and Tomasi first proposed the bilateral filter for digital image processing on the basis of Gaussian filter function [5-7]. Bilateral filter can reduce noise and maintain the image edge character at the same time. It contains two weight functions, which can be measured by space distance in space domain and brightness similarity in brightness domain on the neighboring pixel attenuation. The form is also Gaussian function. The definition of bilateral filter is

$$I_{out} = \frac{1}{k(x)} \int_{-\infty}^{\infty} I_{in}(x+\zeta)c(\zeta)s(I_{in}(x+\zeta)-I_{in}(x))d\zeta , \tag{2}$$

$$k(x) = \int_{-\infty}^{\infty} c(\zeta)s(I_{in}(x+\zeta)-I_{in}(x))d\zeta . \tag{3}$$

In the formula above, $k(x)$ is a normalized function, $c(\zeta)$ means domain filter and $s(I_{in}(x+\zeta)-I_{in}(x))$ is range filter. $I_{out}$ is the result of $I_{in}$ by bilateral filtering.

Michael Elad put forward a bilateral filtering based Retinex algorithm. The fundamental principle of bilateral filter Retinex model is as follows: first we convert the input image to logarithmic domain and second we use bilateral filter to estimate the intensity of illumination from the original image and then the image is divided into two components. We can adopt various methods to the two components, and finally we synthesis the two parts together as a new image, and the result image can avoid "halo artifact" phenomenon effectively. On the basis of $0 \le R \le 1$ , we can get $L \ge S$ and $l \ge s$ . In order to restore the real appearance of the object, we shall make sure $\|l-s\|_2$ as small as possible, that is to say the following restrictions shall be met

$$\min_{l \ge s} \lambda\|l-s\|_2^2 + \left\{\|D_x l\|_2^2 + \|D_y l\|_2^2\right\} . \tag{4}$$

Since $r = s - l$ , in order to piecewise smooth the reflectance $r$ , put $r = s - l$ into (4) then we get

$$\min_{l \geq s} \lambda \|l-s\|_2^2 + \left\{\|D_x l\|_2^2 + \|D_y l\|_2^2\right\} + \alpha\left\{\|D_x(s-l)\|_2^2 + \|D_y(s-l)\|_2^2\right\}. \tag{5}$$

To piecewise smooth $l$ and $r$ with bilateral filter

$$B_{W,P}\{x\} = \sum_{m=-P}^{P} \sum_{n=-P}^{P} \left(C_{m,n}x - x\right)^T W_{[m,n]}(s)\left(C_{m,n}x - x\right). \tag{6}$$

The weight function is

$$W_{[m,n]}(k,j) = \frac{\rho'\{s[k,j] - s[k-m, j-n]\}}{s[k,j] - s[k-m, j-n]} \cdot V[m,n]. \tag{7}$$

The model of Retinex by two bilateral filters is

$$\min_{l \geq s, r} \left\{\lambda_l \|l-s\|_2^2 + B_{W_l,P_l}\{l\}\right\} + \alpha\left\{\lambda_r \|r-(s-l)\|_2^2 + B_{W_r,P_r}\{r\}\right\}. \tag{8}$$

The difference between bilateral filter and traditional Gaussian filter is that the former introduced a brightness component to preserve image edges well so that the bilateral filtering Retinex algorithm can keep image characteristics and avoid "halo artifact" while enhancing image contrasts. However, this method also has some deficiencies: (1) Noise reduction ability in high gradient or high curvature area is poor. (2)Smoothing the drastic model features (such as low peak, etc) makes the sharp area weaken too much. (3) Inclined to mix the characteristics of adjacent areas. This paper proposed a Retinex image enhancement algorithm based on trilateral filter, which can not only eliminate halo phenomenon like the bilateral filtering Retinex do but also overcome the deficiency of the previous algorithm. The proposed method can enhance the contrast of image in high gradient or high curvature areas and remain fine image edge and the detail information.

## 2 Retinex by Trilateral Fliter

Considering some defects of the bilateral filtering Retinex algorithm, Choudhury designed a trilateral filter on the basis of bilateral filter. Trilateral filter not only considers space position and brightness like bilateral filter, but also brings the image average gradient into consideration. The filter function tilts along with the direction of local image gradient and has gradient-keep characteristics. The window for bilateral filter is rectangular while that of the trilateral filter is tilting and the slope is $\theta$. First we calculate gradient image $\nabla I_{in}(x)$ and then calculate the mean gradient through bilateral filter:

$$G_\theta(x) = \frac{1}{k_\theta(x)} \int_{-\infty}^{\infty} \nabla I_{in}(x+\zeta) c(\zeta) s(\|\nabla I_{in}(x+\zeta) - \nabla I_{in}(x)\|) d\zeta, \tag{9}$$

$$k_\theta(x) = \int_{-\infty}^{\infty} c(\zeta) s(\|\nabla I_{in}(x+\zeta) - \nabla I_{in}(x)\|) d\zeta. \tag{10}$$

$k_\theta(x)$ is the normalized function, and $c(\zeta)$ is a domain filter, and $s(\|\nabla I_{in}(x+\zeta)-\nabla I_{in}(x)\|)$ is a range filter. For discrete image, we calculate the image gradient with forward difference method:

$$\nabla I_{in}(m,n) \approx (I_{in}(m+1,n)-I_{in}(m,n), I_{in}(m,n+1)-I_{in}(m,n)). \tag{11}$$

Then we calculate the intensity of tangential plane:

$$P(x,\zeta) = I_{in}(x) + G_\theta \cdot \zeta . \tag{12}$$

Then we get the result of $I_{in}(x+\zeta)$ minus $P(x,\zeta)$:

$$I_\Delta(x,\zeta) = I_{in}(x+\zeta) - P(x,\zeta). \tag{13}$$

Adopting bilateral filter to $\Delta I_{in}(x,\zeta)$ to update $I_{in}(x+\zeta)$ .Meanwhile, the trilateral filter brings in a gradient threshold function to divide the pixel in the adjacent domain further. When the difference between $\zeta$ and $x$ exceeds $R$, we ignore $\zeta$ :

$$f_\theta(x,\zeta) = \begin{cases} 1, if \ \|G_\theta(x,+\zeta) - G_\theta(x)\| < R \\ 0, otherwise \end{cases}. \tag{14}$$

$R$ is the threshold parameter. In order to make sure $R$ changes with image texture, we set:

$$R = 0.15(\|\max(G_{avg}(x)) - \min(G_{avg}(x))\|). \tag{15}$$

Then the output image after trilateral filter is:

$$I_{out}(x) = I_{in}(x) + \frac{1}{k_\Delta(x)} \int_{-\infty}^{\infty} I_\Delta(x,\zeta) c(\zeta) s(I_\Delta(x,\zeta)) f_\theta(x,\zeta) d\zeta \cdot \tag{16}$$

$$k_\Delta(x) = \int_{-\infty}^{\infty} c(\zeta) s(I_\Delta(x,\zeta)) f_\theta(x,\zeta) d\zeta . \tag{17}$$

$k_\Delta(x)$ is also a normalized function.

Inspired by the Retinex algorithm based on bilateral, this paper put forward a Retinex algorithm based on trilateral filtering. And bilateral filter Retinex model is similar to the Retinex model based on the trilateral filtering. The basic principle of the proposed algorithm is still transforming the input image to logarithmic domain, and then estimating illumination by applying trilateral filter to the original image to decompose the image into two different components. Dealing with the two components respectively, we can combine the results of each processing as the ultimate result. The difference between our proposed method and former algorithm is taking gradient information into consideration. And namely the filter window is as follows:

$$W_{TF[m,n]}(k,j)=\frac{\rho\{s[k,j]-s[k-m,j-n]\}}{s[k,j]-s[k-m,j-n]}\cdot V[m,n]\cdot\frac{\rho'\{\nabla s[k,j]-\nabla s[k-m,j-n]\}}{\nabla s[k,j]-\nabla s[k-m,j-n]} \quad (18)$$

So the expression of the Retinex algorithm based on trilateral filter is as below:

$$\min_{l\geq s,r}\left\{\lambda_l\left\|l-s\right\|_2^2+B_{TF}\{l\}\right\}+f_s(r). \quad (19)$$

$f_s$ is Sigmoid enhancement function that can enhance the contrast of the reflectance image:

$$f_s(r)=\frac{2}{1+e^{-a*r}}-1. \quad (20)$$

In the above expression, $a$ is control parameter. For the value of the reflectance image is in the logarithm domain, there may be some negative values. The greater $a$ is the steeper the mapping curve will be, and the enhancement of the reflectance will be more significant as well.

In the specific implementation, this paper adopted the following technics: first dividing the image into illumination and reflection through trilateral filtering Retinex algorithm, then applying trilateral filter to illumination again, third adopting Sigmoid function to the reflectance image to enhance its contrast and highlight the details of the image, finally synthesizing the results of former steps as the ultimate enhancement results. The algorithm process is shown in Fig. 2:

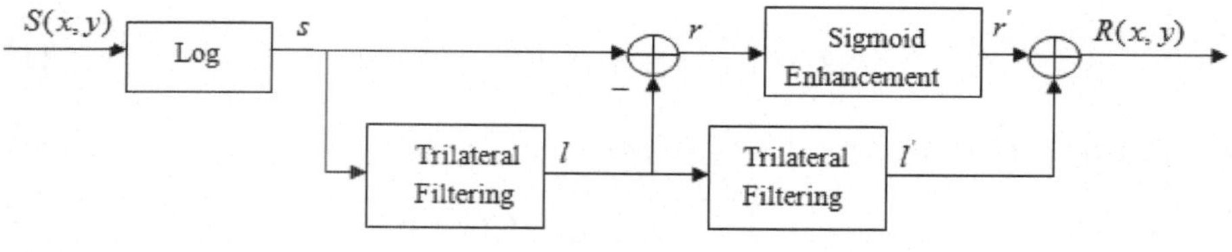


**Fig. 2.** Flow of Retinex Algorithm by Trilateral Filter

## 3 Experimental Results

In order to validate the effectiveness of the proposed trilateral filtering Retinex algorithm, we compared the results of our proposed method in this paper with those of the Retinex algorithm by bilateral filter [12].We evaluated their effectiveness in two aspects, visual effect and the gradient keeping ability. Average gradient is namely the clarity of the image in other words. Since average gradient can reflect the contrast expression ability of images in details, we chose the quantitative evaluation of average gradient keeping ability as an evaluation index. The expression of average gradient is:

$$\bar{g} = \frac{1}{(M-1)(N-1)} \times \sum_{i=1}^{M-1} \sum_{j=1}^{N-1} \sqrt{\frac{(F(i,j)-F(i+1,j))^2 + (F(i,j)-F(i,j+1))^2}{2}} \tag{21}$$

In this paper, we chose eight different low contrast images in specific experiments, the respective average gradient value of the 8 pictures is illustrated in Table 1.

**Table 1.** Average Gradient Value

| Fig. No. | Original Image | Retinex by Two Bilateral Filters | Retinex by Trilateral Filter |
|---|---|---|---|
| 1 | 2.1343 | 2.6768 | 3.6023 |
| 2 | 1.5791 | 5.9727 | 5.1995 |
| 3 | 3.0961 | 3.2820 | 4.2487 |
| 4 | 1.8464 | 3.2560 | 2.7580 |
| 5 | 1.8301 | 5.7101 | 5.7849 |
| 6 | 1.7736 | 3.0954 | 3.4373 |
| 7 | 1.9895 | 2.4509 | 2.5035 |
| 8 | 3.6174 | 3.5838 | 4.4752 |

Fig. 3 gives the results of a low contrast image enhanced by Retinex by two bilateral filters method and by our proposed method. (a) stands for the original image and (b) for the result of Retinex by two bilateral filters and (c) for that of Retinex by trilateral filter. As we can see in Fig. 3, the two algorithms can enhance the image contrasts effectively. Nevertheless, the bilateral filtering Retinex algorithm weakened the boundary and texture information (as shown in the girl's hair, the outline of the trees and the shadow) of the images , while the ability of gradient-keeping of the trilateral filter can avoid this problem effectively, which makes the boundary of the image and texture information better reserved.

**Fig. 3.** Results (a) Original Image, (b) Retinex by Two Bilateral Filters, (c) Retinex by Trilateral Filter

By the above results of the two algorithms, we can see that they both can significantly enhance the image contrasts and make the results more accord with human visual characteristics. At the same time the two algorithms can improve the average gradient of images, and the ability of improving picture average gradient of trilateral filter Retinex algorithm is more significant. Therefore, we can consider our proposed method a kind of effective image enhancement Retinex algorithm with "gradient keep" function.

## 4 Conclusions

Compared with the traditional Retinex image enhancement algorithm, the trilateral filter Retinex algorithm takes image gradient into the calculation so that it has a gradient-keeping ability. In the procession of degraded images, the proposed method can significantly improve the image contrasts to ensure the results more corresponding to the human visual characteristics and more applicative to the computer recognition system. And it can keep image edges and texture information very well which makes the detail information of processed images richer, so we can see it a kind of effective image enhancement algorithm.

The shortcomings of the trilateral filter Retinex algorithm lie in the high complexity of the algorithm and complicated computation, and therefore it cannot be used in real-time image processing. Future research focuses on how to speed up the proposed Retinex algorithm and how to improve the operation efficiency in order to make it suitable for real-time processing system.

**Acknowledgments.** This work was supported by the National Natural Science Foundation of China under grant 60972001, the National Key Technologies R & D Program of China under grant 2009BAG13A06, the Scientific Innovation Research of College Graduate in Jiangsu Province under grant CXZZ_0163, and the Scientific Research Foundation of Graduate School of Southeast University under grand YBJJ1140.

## References

1. Land, E.H.: The Retinex theory of color vision. Sci. Amer. 237, 108–128 (1977)
2. Land, E.H., McCann, J.J.: Lightness and Retinex theory. Journal of Optical Society of America 61(l), 1–11 (1971)
3. Land, E.H.: An alternative technique for the computation of the designator in the Retinex theory of color vision. Proceedings of the National Academy of Science of the United States of America 83(10), 3078–3080 (1986)
4. Rahman, Z., Jobson, D.J., Woodell, G.A.: Retinex Processing for automatic image enhancement. Journal of Electronic Imaging 13(l), 100–110 (2004)
5. Moore, A., Allman, J., Goodman, R.M.: A real-time neural system for color constancy. IEEE Transactions on Neural Networks 2(2), 237–247 (2002)

6. Meylan, L., Isstrunk, S.S.: Color image enhancement using a Retinex-based adaptive filter. Citeseer (2004)
7. McCann, J.: Capturing a black cat in shade: the past and present of Retinex color appearance models. Electronic Imaging 13(1), 36–47 (2004)
8. Ciurea, F., Funt, B.: Tuning Retinex parameters. Journal of Electronic Imaging 13(1), 58–64 (2004)
9. Bertalmio, M., Caselles, V., Provenzi, E.: Issues about Retinex theory and contrast enhancement. International Journal of Computer Vision 83(1), 101–119 (2009)
10. Xiong, W., Funt, B.: Stereo Retinex. Image and Vision Computing 27, 178–188 (2009)
11. Eland, M.: On the original of bilateral filter and ways to improve it. IEEE Transactions on Image Processing 11(10), 1141–1151 (2002)
12. Eland, M.: Retinex by Two Bilateral Filters. In: The 5th international Conference on Scale-Space and PDE in Computer Vision, Hofgeismar, Germany, April 7-9 (2005)
13. Tomasi, C., Manduchi, R.: Bilateral filtering for gray and color images. In: Proc. 6th Int. Conf. Computer Vision, New Delhi, India, pp. 839–846 (1998)
14. Choudhury, P., Tumblin, J.: The Trilateral Filter for High Contrast Images and Meshes. In: Proc. Eurographics Symposium on Rendering, pp. 1–11 (2003)
15. Vaudrey, T., Klette, R.: Fast Trilateral Filtering. Multimedia Imaging Report 40 (2009)
16. Shen, J., Jin, X., Sun, H.: High dynamic range image tone mapping and retexturing using fast trilateral filtering. Visual Compute 23, 641–650 (2007)

# Face Recognition Using a Modified Fuzzy Linear Discriminant Analysis Method

Qianzhou Xu[1], Xiaobo Lu[1], and Weili Zeng[2]

[1] School of Automation, Southeast University, Nanjing, 210096
xuqianzhou@gmail.com, xblu2008@yahoo.cn
[2] School of Transportation, Southeast University, Nanjing, 210096
zengwlj@yahoo.com.cn

**Abstract.** Linear discriminant analysis (LDA) is a simple but widely used algorithm in the area of face recognition. However, it has some shortcomings in which the relationship of each face to a class is assumed to be crisp. This algorithm was modified by incorporating the membership grade of each face pattern into the calculation of the between-class and within-class scatter matrices, which is known as Fuzzy Fisherface. The Fuzzy Fisherface method introduces a gradual level of assignment of each face pattern to a class by using a membership grading based upon the k-Nearest Neighbor (KNN) algorithm, and it obtains an obviously better performance than the LDA method. However, when computing the fuzzy memberships, only the belong-to information is considered while the not-belong-to information is ignored. In this paper, a further modified fuzzy linear discriminant analysis method is proposed to solve this problem. The experiments were performed on the ORL and FERET face databases, and the results show consistent improvement in the recognition rate.

**Keywords:** Linear discriminant analysis, Fuzzy memberships, k-Nearest Neighbor algorithm, Face recognition.

## 1 Introduction

Face recognition has been applied in many applications and has attracted substantial research efforts from the areas of computer vision, bio-informatics and machine learning. With increasing human–machine interaction, many commercial systems are being developed, which attempt to solve various problems through person verification or recognition. The appearance based techniques use the holistic features of the face image whereas the geometrical facial structure features are utilized in the latter [1]. Some researchers have also adopted a hybrid methodology by applying the feature based techniques on the localized regions of the facial image [2]. They all need a subspace projection step that transforms facial images from a high-dimensional space to a more manageable, lower-dimensional space. Principal Component Analysis (PCA) followed by Fisher's Linear Discriminant Analysis (LDA) provides means to reduce the dimension into a subspace where the problem becomes a linear classification. Such methods have been employed extensively and gave the most promising results [3].

J. Lei et al. (Eds.): AICI 2012, LNAI 7530, pp. 408–415, 2012.

LDA is a method to determine the linear combination of features that best separates two or more classes of objects. Its outstanding performance in feature extraction makes it widely used in the area of face recognition. However, it also has some problems which limit its further application. First of all, LDA method is sensitive to outliers, because it is based on data covariance matrices as is principal component analysis (PCA) [4]. These outliers always hinder us from estimating the central of a class accurately, and then the calculated between-class and within-class scatter matrices are not as relevant as we imaged. To reduce the effect of outliers, fuzzy memberships were introduced to robustly estimate the fuzzy scatter matrices, which is called fuzzy LDA (FLDA) [5]. FLDA introduces the k-Nearest Neighbor (KNN) algorithm to find out the closeness between a certain face and its siblings which we call belong-to information, so that it may prevent some outliers from dominating the estimation of scatter matrices. While FLDA method uses the relationship of its close siblings in a class, it ignores every face's inherent alienation to its class, which we call not-belong-to information. Thus, a modified fuzzy linear discriminant analysis method is proposed in this paper, which aims to overcome this shortage.

## 2 Related Algorithm

### 2.1 Principal Component Analysis (PCA)

PCA method is one of the most successful techniques that have been used in face recognition area. It can be used to perform prediction, redundancy removal, feature extraction, data compression, etc.[6]. PCA essentially reduces the large dimensionality of the data space so that the projection of the data is in the direction of the maximum variance of the data used to find the feature space [7]. Large 1-D vectors of pixels are constructed from 2-D face image, by concatenating the columns, and projected onto the eigenvectors of the covariance matrix of the training image vectors. If there are $N$ (no. of images) vectors of size $M$ (rows multiplied by columns of an image) then, the mean vector of all the images is,

$$\overline{\Gamma} = \frac{1}{N}\sum_{i=1}^{N}\Gamma_i \tag{1}$$

where $\Gamma_i$ is the size $M$ vector constructed from face image $i$ .The set of $T$ orthonormal vector $w_i$ are sought which form the projection matrix $W$ of order ( $M \times T$ ) and the feature vectors are then given by the following linear transformation,

$$y_k = W^T\Gamma_k \tag{2}$$

PCA relies on maximizing the total scatter of the training vectors. The total scatter matrix $S_T$ is given by,

$$S_T = \sum_{i=1}^{N} (\Gamma_i - \overline{\Gamma})(\Gamma_i - \overline{\Gamma})^T \tag{3}$$

The scatter of the transformed feature vectors is given by $W^T S_T W$. The projection matrix is

$$W_{PCA} = \arg\max_{W} \left| W^T S_T W \right| \tag{4}$$

where $W$ are the eigenvectors of the covariance matrix,

$$C = P^T P \tag{5}$$

where $P$ is a matrix composed of the mean centered images as the column vectors which are placed side by side. The nonzero eigen-values of the covariance matrix have corresponding orthonormal eigenvectors. We choose the certain largest eigenvectors which have been ranked by their eigen-values for projection. Thus, image vectors are projected onto a subspace formed by the most significant eigenvectors (principal components) of the covariance matrix. When a test image is projected onto the $N$ dimensional subspace, it is classified to the class of the vector that minimizes the Euclidean distance with it [3].

### 2.2 Linear Discriminant Analysis (LDA)

Linear discriminant analysis (LDA) is extension on the basis of the PCA method. Although the PCA work retains the most significant feature of the original face images after projecting them from high- dimensional space to low-dimension space, it ignores the information of class discrimination. Thus, the LDA method is implemented to find a better subspace in which the projected vectors of the different classes are maximally separated.

The between-class scatter matrix $S_B$ and the within-class scatter matrix $S_W$ are defined as [8]:

$$S_B = \sum_{i=1}^{C} n^i (\overline{X}^i - \overline{X})(\overline{X}^i - \overline{X})^T \tag{6}$$

$$S_W = \sum_{i=1}^{C} \sum_{X_k \in n^i} (X_k - \overline{X}^i)(X_k - \overline{X}^i)^T \tag{7}$$

where $n^i$ is the number of training vectors in $i$ ith class, $C$ is the number of distinct classes, $\overline{X}^i$ is the mean of all the vectors belonging to the $i$ th class and $X_k$ represents the set of samples belonging to $i$ th class. $S_W$ represents the scatter of the features around the mean of each class and $S_B$ represents the scatter of features around the overall mean for all the classes.

LDA's aim is to maximize $S_B$ while minimizing $S_W$, that is, the optimal subspace $E_{optimal}$ can be obtained like this [9],

$$E_{optimal} = \arg\max_{E} \frac{\left|E^T S_B E\right|}{\left|E^T S_W E\right|} \tag{8}$$

### 2.3 Fuzzy Linear Discriminant Analysis (FLDA)

Fuzzy Linear Discriminant Analysis (FLDA) [10] is an extension of LDA using fuzzy memberships. In Fuzzy LDA, the basic LDA is changed. The modification is the introduction of fuzziness into the belonging of projected vector to the classes which solves binary classification problems. In conventional LDA approach, every vector is supposed to have a crisp membership. But this does not take into account the resemblance of images belonging to different classes, which may occur under varying conditions. In FLDA, each vector is assigned the membership grades of every class based upon the class label of its *k*-nearest neighbors. This Fuzzy *k*-nearest neighbor method (KNN) is utilized to evaluate the membership grades of all the vectors [11].

$\mu_{ij}$ stands for the membership grade of $j$ th vector in the $i$ th class and it satisfies two obvious properties:

$$\sum_{i=1}^{C} \mu_{ij} = 1 \tag{9}$$

$$0 < \sum_{i=1}^{M} \mu_{ij} < M \tag{10}$$

The class labels of the *k* vectors located in the closest neighborhood of each vector is collected during the training phase. Then the membership grade of $j$ th vector in the $i$ th class is evaluated as follow [11]:

$$\mu_{ij} = \begin{cases} 0.51 + 0.49 n_{ij} / k, \Gamma_j \in C_i \\ 0.49 n_{ij} / k, \Gamma_j \notin C_i \end{cases} \tag{11}$$

where $n_{ij}$ stands for the number of the neighbors of the $j$ th data that belong to the $i$ th class, $\Gamma_j$ stands for the $j$ th data and $C_i$ stands for the $i$ th class.

The moderated membership grades are used in the computations of the statistical properties of the patterns. The mean vector of each class is obtained from the below equation:

$$\tilde{m}_i = \frac{\sum_{j=1}^{M} \mu_{ij} X_j}{\sum_{j=1}^{M} \mu_{ij}} = \frac{XU^T}{\sum_{j=1}^{M} \mu_{ij}}, i = 1, 2, \ldots, C \tag{12}$$

Then, the corresponding fuzzy within-class scatter matrix and fuzzy between-class scatter matrix can be redefined as follow

$$S_{FB} = \sum_{i=1}^{C} n_i (\tilde{m}_i - m)(\tilde{m}_i - m)^T \tag{13}$$

$$S_{FW} = \sum_{i=1}^{C} \sum_{X_k \in x_i} (X_k - \tilde{m}_i)(X_k - \tilde{m}_i)^T \tag{14}$$

## 3 Proposed Approach

In the FLDA method, the scatter matrices with fuzzy set theory are redefined and some nearest membership information is incorporated, thus, we can obtain a better feature extraction effect than LDA. However, it still has some shortcomings. As we can figure out in Eqs.(14), since the grade of membership $\mu_{ij}$ is calculated by weighting the contribution of the k-nearest neighbor vectors, so that if $j$ th data belong to the $i$ th class, $\mu_{ij}$ will always be dominant, which should be dangerous in some cases. As we all know, in the face recognition area, we have to pay a lot attention to deal with interferences bought by some extreme situations such as illumination, attitude, expression, accessories, etc. When projected from a high-dimensional space to a lower-dimensional space, these interferences may turn to some outliers. We can see this situation from Fig. 1 as a schematic diagram.

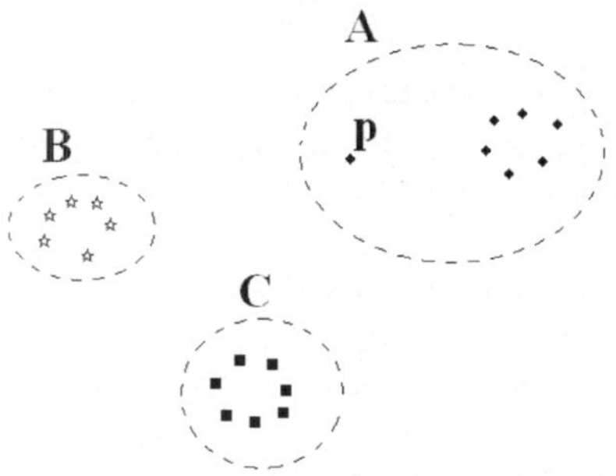


**Fig. 1.** Outliers in a class which can't be weakened by FLDA

Let Fig. 1 be a schematic diagram of three classes which has already been projected from a high-dimensional space. Actually, $p$ th data belongs to the $i$ th class although it is so isolated. When we compute the membership grade of the $p$ th data in class $A$ corresponding to Eqs.(11), we got a topping 1 which means the $p$ th data totally belongs to class $A$ because its $k$-nearest neighbors are all belongs to class $A$ (If $k$ is small enough) and so itself does. Then, we use this membership grade to compute the center of the class and within-class or between-class scatter matrix, which makes the whole within-class and between-class scatter matrix not so appropriate. In other words, this isolated outlier pollutes the class, but the FLDA can't weaken its affect.

We overcome this shortcoming by redefining the membership grade and then the within-class and between-class scatter matrix, in which the not-belong-to information is incorporated. Let $o_{ij}$ be the number of the points having the farthest distance between the $j$ th data which belong to the $i$ th class, and then we can redefine the membership function as below,

$$\xi_{ij} = \begin{cases} 0.51 + (0.49 n_{ij} - 0.49 o_{ij}) / k, \Gamma_j \in C_i \\ (0.49 n_{ij} - 0.49 o_{ij}) / k, \Gamma_j \notin C_i \end{cases} \tag{15}$$

where $\xi_{ij}$ stands for the new membership grade of $j$ th vector in the $i$ th class. To make $\xi_{ij}$ satisfies Eqs.(9), we also need a normalization as below,

$$\xi_{ij}^{'} = \frac{\xi_{ij}}{\sum_{i=1}^{C} \xi_{ij}} \tag{16}$$

The center of each class can be estimated as below,

$$\tilde{m}_i^{'} = \frac{\sum_{j=1}^{M} \xi_{ij}^{'} X_j}{\sum_{j=1}^{M} \xi_{ij}^{'}}, i = 1, 2, ..., C \tag{17}$$

Then, the corresponding fuzzy within-class scatter matrix and fuzzy between-class scatter matrix can be redefined as follow,

$$S_{FB}^{'} = \sum_{i=1}^{C} n_i (\tilde{m}_i^{'} - m)(\tilde{m}_i^{'} - m)^T \tag{18}$$

$$S_{FW}^{'} = \sum_{i=1}^{C} \sum_{X_k \in x_i} (X_k - \tilde{m}_i^{'})(X_k - \tilde{m}_i^{'})^T \tag{19}$$

Finally, we can get the optimal projection matrix based on Fisher's criterion,

$$E_{optimal} = \arg\max_{E} \frac{\left| E^T S_{FB}^{'} E \right|}{\left| E^T S_{FW}^{'} E \right|} \tag{20}$$

## 4 Experimental Results

To investigate the effectiveness of the proposed method, three existing methods (PCA, LDA and FLDA) and the proposed method (M-FLDA) were implemented and tested using Matlab in the FERET database and ORL database which have become a standard database for testing and evaluating face recognition algorithms. Here are some samples of the two databases.

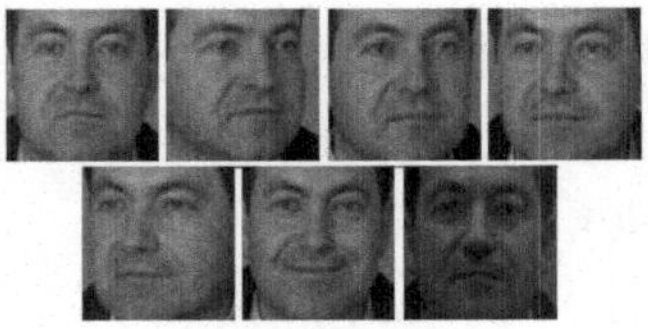

**Fig. 2.** One person's face images in FERET database

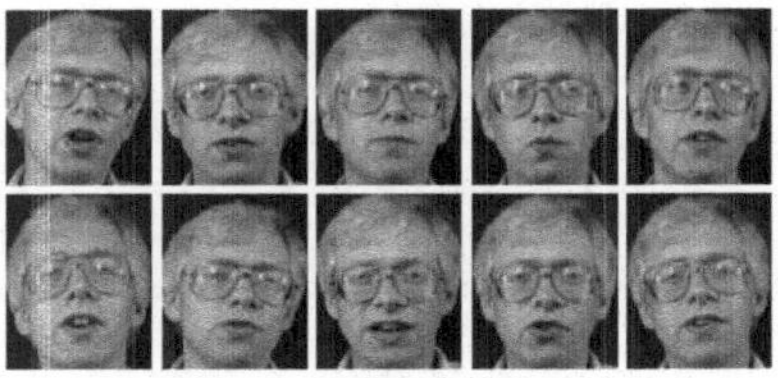

**Fig. 3.** One person's face images in ORL database

The proposed method was tested on a subset of the FERET database. This subset includes 1,400 images of 200 individuals (each individual has 7 images). In order to make full use of the available data and to evaluate the generalization power of algorithms more accurately, we choose 3 (or 4) face images of each individual as training samples randomly in each test while the rest as test samples and then run the test 10 times. The classical PCA, LDA, FLDA and the proposed Fuzzy LDA (M-FLDA) method are respectively used for feature extraction. The KNN parameter k is set as 2. Finally, we adopt Euclidean distance as classifier to test their recognition in FERET database. The recognition rate is as bellow,

**Table 1.** Average recognition rate on FERET database

| | 3 | 4 |
|---|---|---|
| PCA | 34.87 | 41.17 |
| LDA | 60.65 | 73.44 |
| FLDA | 61.63 | 75.66 |
| Proposed Method | 62.56 | 76.25 |

**Table 2.** Average recognition rate on ORL database

| | 4 | 5 |
|---|---|---|
| PCA | 70.89 | 75.08 |
| LDA | 84.46 | 88.22 |
| FLDA | 87.44 | 90.83 |
| Proposed Method | 88.06 | 91.51 |

All the test methods are also performed on ORL database. The ORL database contains 40 persons, each having 10 different images. The images of the same person are taken at different times, under slightly varying lighting conditions and with various facial expressions, which we can figure out from Fig.3. In the experiments, we also split the whole database into two parts evenly. We choose 4 (or 5) images randomly as training samples in each test and also ran the test 10 times. The KNN parameter k is set as 2. Euclidean distance is also chosen as classifier as we did before. Then the recognition rate is recorded in Table 2.

## 5 Conclusions

This paper proposes a modified fuzzy linear discriminant analysis face recognition method. It makes up one of the FLDA's shortcomings that it can't weaken the bad affect caused by isolated outliers. In this method, the not-belong-to information is incorporated. By doing this, it can help to compute better fuzzy between-class scatter matrix and fuzzy within-class scatter matrix, which is important to classification. Experimental results also show that our proposed method is effective.

## References

1. Kresimir, D., Mislav, G., Sonja, G.: Independent Comparative Study of PCA, ICA, and LDA on the FERET. Data Set. 15, 252–260 (2005)
2. Gottumukkal, R., Asari, V.K.: An improved face recognition technique based on modular PCA approach. Pattern Recognition Letters 25, 429–436 (2004)
3. Amar, K., Syed, F.A.: A genetically modified fuzzy linear discriminant analysis for face recognition. Journal of the Franklin Institue 348(10), 2701–2717 (2011)
4. Jayadeva, R.K., Chandra, S.: Learning the optimal kernel for Fisher discriminant analysis via second order cone programming. European Journal of Operational Research 203(3), 692–697 (2010)
5. Wankou, Y., Hui, Y., Jingyu, Y.: Face recognition using complete fuzzy LDA. In: Proceedings of the 2008 International Conference on Pattern Recognition, pp. 1–4 (2008)
6. Keun-Chang, K., Witold: Face recognition using a Fuzzy Fisherface Classifier. Pattern Recognition 38(10), 1717–1732 (2005)
7. Turk, M., Pentland, A.: Eigenfaces for Recognition. Cognitive Neuroscience 3(1), 71–96 (1991)
8. Peter, N.B., Joao, P.H., David, J.K.: Eigenfaces vs. Fisherfaces: recognition using class specific linear projection. IEEE Transactions on Pattern Analysis and Machine Intelligence 19(7), 711–719 (1997)
9. Wendy, S.Y.: Analysis of PCA-based and fisher discriminant-based image recognition algorithms. Computer Science Technical Report CS-00-103 (2000)
10. Chen, Z.P., Jiang, J.H., Li, Y.: Fuzzy linear discriminant analysis for chemical data sets. Chemometrics and Intelligent Laboratory Systems 45, 295–302 (1999)
11. Keller, J.M., Gray, M.R.: A Fuzzy K Nearest Neighbor Classifier Algorithm. IEEE Transactions on Systems, Man and Cybernetics 15(4), 580–585 (1985)

# Progressive Image Registration Based on Probability Boosting Tree

LiTing Guo, JianBing Yi, and Xuan Yang

School of Computer and Software Engineering,
Shenzhen Key Laboratory of Service Computing and Appliction, Shenzhen University,
518060 Shenzhen, China
549878752@qq.com, yijianbing8@163.com, xyang0520@263.net

**Abstract.** This paper proposed a novel progressive image registration method based on landmarks. At first, identification of corresponding control points is implemented by classifying of image features. We constructed a Probability Boosting Tree (PBT) to calculate the matching probability of each pair of corresponding control points and selected the highly matched ones as control points. Next, a progressive image registration is performed by inserting new landmarks selected by PBT one by one, which estimates the deformation model more and more accurate. Experimental results indicate that the proposed algorithm in this paper is feasible and can improve the precision of image registration significantly.

**Keywords:** Probability Boosting Tree, Progressive image registration, Corresponding control points.

## 1 Introduction

Image registration based on landmarks plays an important role in medical image registration. It extracted the corresponding control points from the images and estimated the transformation models between images [1][2]. In order to extract the corresponding control points correctly, D. G. Lowe [3] proposed the SIFT algorithm, which is only suitable for matching of feature points on images with rigid deformation rather than with elastic deformation. Rangarajan [4] proposed extracting the shape feature points and determined the correspondence between the feature points by mutual information. Feng [5] presented hierarchical mutual information to automatically determining the corresponding control points. But this algorithm depends on the statistical information of image intensity and is unable to extract enough points. Actually, it is difficult to ensure that all the corresponding control points are matched correctly, so it is needed to decide which one is benefit to improve the image registration accuracy individually, and estimate the transformation model between two images progressively.

In this paper, we take the problem of extracting the pairs of corresponding control points as a classification problem to deal with. The Probability Boosting Tree, which is good at classifying two different samples, is used to estimate the corresponding

J. Lei et al. (Eds.): AICI 2012, LNAI 7530, pp. 416–423, 2012.

control points. Then, the corresponding control points are added to the control points set optionally in order to estimate the transformation model between two images, and improve the accuracy of image registration progressively.

## 2 Probability Boosting Tree Classifier

The pairs of corresponding points have similar properties in images. On the contrary, the properties of the ones with no corresponding relation are different. So we can structure a suitable classifier to decide whether the control points have the corresponding relation. In this paper, we build a decision classifier based on the Probability Boosting Tree algorithm proposed by Zhuowen Tu [6] to estimate corresponding points in two images. The main idea of PBT classifier is constructing a series of weak classifiers based on the multidimensional features, and the weak classifiers are selected together to construct an AdaBoost strong classifier. Finally, the AdaBoost strong classifiers are cascaded to be a PBT decision classifier, which performs well to classify samples.

One defect of AdaBoost is that samples classified incorrectly will receive more weights during the following iterations. Therefore, after some steps, weak classifiers will become invalid because of the unexpected weights. In order to resolve the issue, a new learning algorithm—probability boosting tree, is applied to find the corresponding control points in images.

The probability boosting tree is trained recursively. At each node the empirical distribution $\hat{q}(y)(y \in \{+1,-1\})$ of the sample is calculated, and if the number of samples at the node is not empty ($0 < \hat{q}(y) < 1$), an AdaBoost strong classifier is trained on the data at the node. Each sample is then passed to the left or right sub-trees, weighted by $q(-1 | x_i)$ and $q(+1 | x_i)$ respectively, where $q(+1 | x_i)$ is the probability that $x_i$ is a positive sample according to the AdaBoost strong classifier, and $q(-1 | x_i)$ is the probability that $x_i$ is a negative sample. The AdaBoost classifier at each node is used not to return the class of the sample but rather to assign the sample to the left or right sub-tree.

Details for training PBT are given below.

Step1: Given a training set with class label $S = \{(x_1, y_1, w_1), \ldots, (x_m, y_m, w_m)\}$; $x_i \in R^n$, $y_i \in \{-1,+1\}$ is the category label of sample $x_i$, $w_i$ is the weight of sample $x_i$ and $\sum_{i=1}^{m} w(i) = 1$.

Step2: Compute the empirical distribution $\hat{q}(+1) = \sum_i w_i \delta(y_i = +1)$ and $\hat{q}(-1) = \sum_i w_i \delta(y_i = -1)$.

Step3: If the current tree depth is *L* then exits; otherwise, continues.

Step4: On training set *S*, train a strong classifier using an AdaBoost algorithm with *T* weak classifiers whose error rate must be smaller than 0.5.

Step5: Initialize two empty sets $S_L$ and $S_R$.

Step6: Use the trained strong classifier to classify the sample and compute the probability for each sample $q(+1 \mid x_i) = \dfrac{\exp\{2H(x_i)\}}{1+\exp\{2H(x_i)\}}, q(-1 \mid x_i) = \dfrac{\exp\{-2H(x_i)\}}{1+\exp\{-2H(x_i)\}}$, where $H(x_i)$ is the decision function obtained by AdaBoost.

Step7: Split the samples into set $S_L$ or $S_R$ using the decision boundary of the trained strong classifier and the tolerance ε. Then:

If $q(+1 \mid x_i) - \frac{1}{2} > \varepsilon$, $(x_i, y_i, 1) \rightarrow S_R$

else If $q(-1 \mid x_i) - \frac{1}{2} > \varepsilon$, $(x_i, y_i, 1) \rightarrow S_L$

else $(x_i, y_i, q(+1 \mid x_i)) \rightarrow S_R$ and $(x_i, y_i, q(-1 \mid x_i)) \rightarrow S_L$

Step8: Normalize all the weights of the samples in $S_L$, and then train the left children recursively using $S_L$ respectively (go to step 2).

Step9: Normalize all the weights of the samples in $S_R$, and then train the right children recursively using $S_R$ respectively (go to step 2).

Once the whole recursion training realized, a PBT is learned. The goal of the PBT algorithm is to learn the probability $P(y \mid x)$ for each sample as formula (1).

$$\tilde{p}(y \mid x) = \sum_{l_1,\cdots,l_n} \tilde{p}(y \mid l_n,\cdots,l_1,x) q(l_n \mid l_{n-1},\cdots,l_1,x),\cdots,q(l_2 \mid l_1,x) q(l_1 \mid x) \quad (1)$$

where $y \in \{+1,-1\}$, $\tilde{p}(y \mid l_n,\cdots,l_1,x)$ is the distribution probability at the leaf node, $q(l_n \mid l_{n-1},\cdots,l_1,x)$ outputs the discriminative probability for each tree node and $l_i$ is the tree level.

# 3 Progressive Elastic Image Registration Based on Probability Boosting Tree

## 3.1 Estimating the Corresponding Control Points

In this paper, the feature points extracted by SIFT algorithm and the edge feature points extracted by Canny operator detection are combined. The feature points extracted by SIFT are invariant to scale and rotation, and are distributed in the area with large change of image intensity. Moreover, the edge feature points are distributed in the area of the image edges. All these feature points are distributed non-uniformly on the whole image.

In order to find the matched points of the edge feature points, the correlation coefficient is maximized to search the corresponding ones. In the floating image *IS*, we select a template named $W_1$ which is centered of the point $p$ and with size $w_1 \times w_1$ , and then move the template $W_1$ on the reference image *IR* and calculate the correlation coefficient between $W_1$ and the window $W_2$ centers on every point $(i, j)$ in the feature points set on *IR*, where $W_1$ is same to $W_2$ .The point where correlation coefficient is maximum on reference image is the corresponding point of point $p$.

By using SIFT and the correlation coefficient maximized, a large number of candidate corresponding points are collected. There are still mismatched points pairs in these candidate point pairs. Next, all these candidate points pairs are put into PBT classifier to eliminate the mismatched and select the ones with high matching precision.

## 3.2 Progressive Elastic Image Registration

When the control points are distributing unevenly on the image, the accuracy of the registration results will not be improved significantly. An example is shown in Fig. 1.

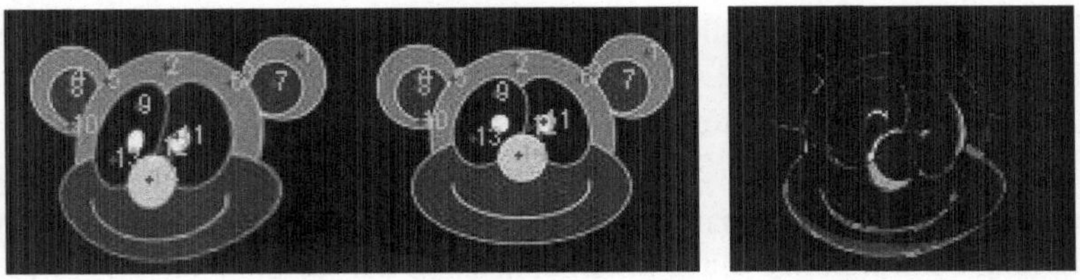

**Fig. 1.** The corresponding control points on the original image and reference image and the difference between registered image and reference image

It can be seen from Fig. 1 that the bottom part of the registered image is not matched well due to lacking necessary control points at the chin and mouth of the monkey. It indicates that when the control points are distributed unevenly on the image, the transformation function obtained based on these corresponding points will reduce the accuracy of image registration. To solve the issue, this paper proposes a method of progressive registration. A suitable control points pairs set which can reflect the spatial transformation relations of images correctly is searched by the way of adding points progressively. These control points make the estimation of transformation model parameters more precise and finally improve the registration results.

The area needed to add control points is determined by comparing the differences of intensity of the two images. Calculate the differences of intensity between the images, the region with large intensity difference is where needs to add control points.

Although the candidate control points pairs have been classified by the PBT, it can't guarantee that all the points pairs are accurately matched and are benefit to improve the registration accurately. If there are incorrect matching point pairs or control points locating in unexpected area, unsatisfied results will be led to.

We add the selected control points pairs one-by-one into the control points set, which is used to estimate the transformation model. Suppose $U=\{(p_{U_1},q_{U_1}),(p_{U_2},q_{U_2}),\cdots,(p_{U_m},q_{U_m})\}$ is the initial control points pairs set, $W=\{(p_{W_1},q_{W_1}),(p_{W_2},q_{W_2}),\cdots,(p_{W_n},q_{W_n})\}$ is the candidate control points pairs set, and Dix is the difference marked map between the two images. If the point pair $(p_{W_i},q_{W_i})$ lies on the region need new control points, we will put it into the set U, and then construct the deformation model based on radial basis function expansion method according to the corresponding relationship of the new control points pairs. Then the image *IS* is warped and interpolated to be a temporary registration image $IS'$. Computing the mutual information between $IS'$ and *IR*, if the mutual information is larger than the one before adding $(p_{W_i},q_{W_i})$, the control point pair $(p_{W_i},q_{W_i})$ will be added on the set *U*. Otherwise, it will be removed out from the sets *U*. Repeat this process until all the candidate control points pairs are processed.

## 4 Experimental Results

In this paper, large and small deformation image registrations are performed in MATLAB. The register images are slices comes from the Brain Web dataset.

(1) Large deformation registration.

The experimental image is $256\times256$ in size. The original one is the reference image, and is deformed manually to be the register image. There are 160 pairs of corresponding control points in the initial control points set and 344 pairs of candidate control points. The CSTPS [9] is adopted to be the radial basis function with support set c=30. In the progressive registration process, up to 59 pairs of corresponding control points are added in the control points set, and registration results are showed in Fig. 2. It can be seen that top right corner area, middle area and the bottom left corner area of the brain image are all got better registration. This suggests that the progressive registration method can effectively improve image registration effect, and is suitable for large deformation image registration.

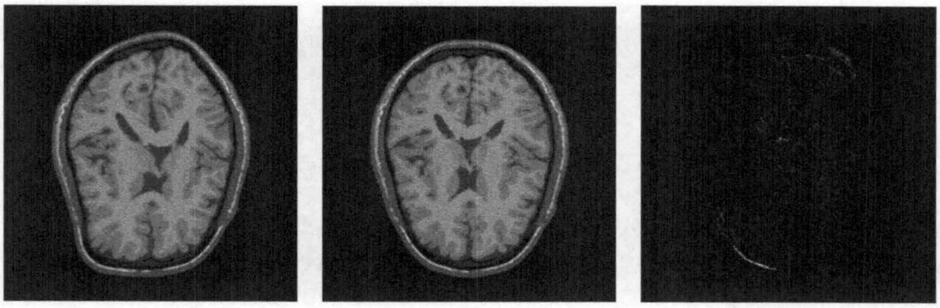

**Fig. 2.** The results of progressive registration process: (a) original image, (b) reference image, (c) the difference between final registered image and reference image.

Mutual information (*MI*), correlation coefficient (*CC*) and mean square deviation (*MSD*) are used to measure the result of registration in this paper. Table 1 lists the measure of registered results during the progressive registration procedure, and the three similarity measure showed in Fig. 3. We can see that the proposed method can improve the precision of image registration step by step.

**Table 1.** Registration results metrics

| Image | *MI* | *CC* | *MSD* |
|---|---|---|---|
| Original image and reference image | 3.6300 | 0.9420 | 19.0947 |
| Initial registration image and reference image | 3.7023 | 0.9634 | 15.1379 |
| The first iteration | 3.7070 | 0.9638 | 15.0442 |
| The tenth iteration | 3.7375 | 0.9750 | 12.4983 |
| The thirtieth iteration | 3.7676 | 0.9775 | 11.8682 |
| The fiftieth iteration | 3.8137 | 0.9884 | 8.5037 |
| Final registered image and reference image | 3.8317 | 0.9935 | 6.3653 |

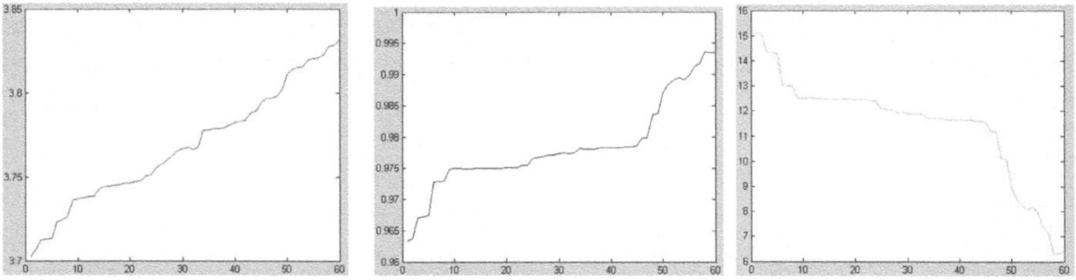

**Fig. 3.** Similarity measure (*MI, CC, MSD*)

(2) Small deformation registration.

The experimental image is a $181 \times 217$, there are 133 pairs of corresponding control points in the initial control points set and 731 candidate pairs of corresponding control points.

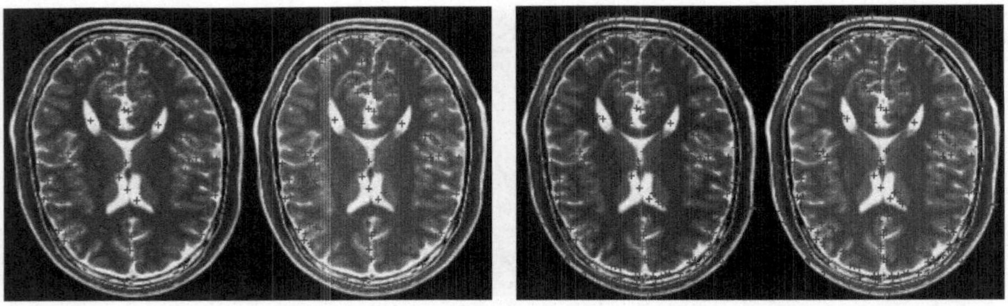

**Fig. 4.** Distribution of corresponding control points on original image and reference image: (a) initial corresponding control points, (b) after adding corresponding control points

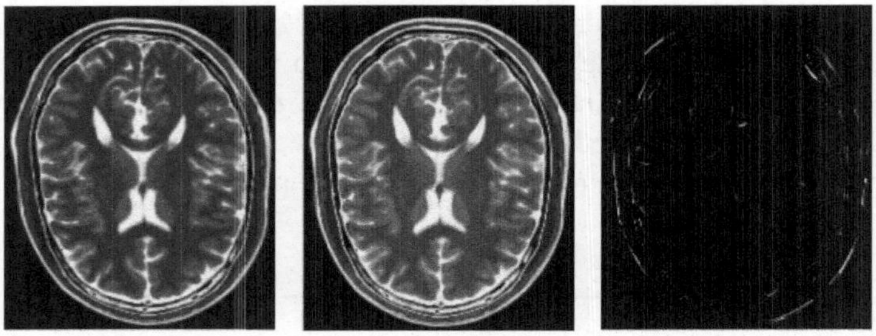

**Fig. 5.** The results of progressive image registration: (a) original image, (b) reference image, (c) the difference between final registered image and reference image

Fig. 4 shows the initial distribution of corresponding control points and the distribution of inserting new corresponding control points on original image and reference image. There are 103 pairs of corresponding control points added in the whole process, which are mainly distributed at the boundary of the image. The registered result is shown in Fig. 5, we can see that differences between two images decreases significantly than before adding points. Table 2 lists the similarity measure data of images. From the table data it can be seen that the final registered image and reference image is the best.

**Table 2.** Registration results metrics.

| Image | *MI* | *CC* | *MSD* |
|---|---|---|---|
| Original image and reference image | 2.3375 | 0.7588 | 45.5147 |
| Initial registered image and reference image | 3.0186 | 0.9079 | 30.3978 |
| Final registered image and reference image | 3.3041 | 0.9622 | 22.6323 |

## 5 Conclusion

Progressive image registration based of probability boosting tree is proposed in this paper. It uses the PBT to classify the feature points of SIFT algorithm and the correlation coefficient maximum algorithm respectively as corresponding ones and non-corresponding ones, in order to select the pairs of corresponding control points with high precision of matching. Through progressively adding points, correctly matched corresponding control points are inserted into the original points set step by step. The progressively estimated deformation model improves the registration result gradually, and enhances the image registration precision. Finally, experiment results on large deformation images and small deformation images show the feasibility of our method.

**Acknowledgments.** This work was supported by the Chinese National Science Foundation under Grant 60972112.

## Reference

1. Brown, L.: A Survey of Image Registration Techniques [J]. ACM Computing Survey 24(4), 325–376 (1992)
2. Elsen, P., Maintz, J., Pol, E., Viergever, M.: Automatic Registration of CT and MR Brain Image using Correlation of Geometrical Features. IEEE Trans. Med. Image 14(6), 384–398 (1995)
3. Lowe, D.G.: Distinctive Image Features from Scale-invariant Interest Points. International Journal of Computer Vision 60(2), 91–110 (2004)
4. Rangarajan, A., Chui, H., Duncan, S.: Rigid points feature Registration using Mutual Information. Medical Image Analysis 3(4), 425–440 (1999)
5. Feng, L., Zhang, M., He, M.: A Non-Rigid Medical Image Registration Approach Based on Hierarchical Mutual Information and Thin-Plate Spline. Journal of Computer-Aided Design and Computer Graphics 17(7), 1492–1496 (2005)
6. Tu, Z.: Probabilistic Boosting-Tree: Learning Discriminative Models for Classification, Recognition and Clustering. In: ICCV, vol. 10(2), pp. 1589–1596 (2005)
7. Freund, Y., Schapire, R.: A Decision-theoretic Generalization of On-line Learning And an Application to Boosting. Journal of Computer and System Science, 119–139 (1997)
8. Zheng, Q., Chellappa, R.: A Computational Vision Approach to Image Registration. IEEE Transactions on Image Processing 2(3) (1993)
9. Zhang, Z., Yang, X.: Elastic image warping using a new radial basic function with compact support. In: Proc. Biosignals 2008, pp. 216–219 (2008)

# Rapid Image Segmentation Using Color, Texture and Syntactic Visual Features

Wei Liu, Jiangxin Wu, Liyuan Zuo, Huai Yuan, and Hong Zhao

Research Academy, Northeastern University,
110179 Shenyang, China
{lwei,wujiangxin,zuoly,yuanh,zhaoh}@neusoft.com

**Abstract.** In this paper, a graph-based hierarchical segmentation algorithm which integrates the color, texture and syntactic visual features is presented. Firstly, it utilizes the color information to conduct coarse segmentation in LUV color space and obtains many color-consistent regions. Next, the texton feature of these regions is extracted and a fine segmentation result can be acquired by merging adjacent regions which have similar texture information. Finally, the syntactic visual processing method is introduced to constrain the small regions. The proposed algorithm is quantitatively and qualitatively evaluated based on a standard image segmentation database. The experiment results demonstrate that this algorithm is efficiently and effectively.

**Keywords:** Image segmentation, Graph theory, Textons, Syntactic visual features.

## 1 Introduction

Image segmentation is the process of partitioning a digital image into multiple coherent and meaningful regions. In computer vision, it is one of the oldest, fundamental studied problems.

This image segmentation algorithm refers many techniques which can simply conclude into three aspects: (1) Graph theoretical framework; (2) Texture segmentation; (3) Syntactic visual features.

Graph-based image segmentation techniques are considered to be one of most efficient segmentation techniques. They explicitly organize the image elements into mathematically structures, making the formulation of image segmentation problem more flexible and the computation more efficient [1].The representative segmentation algorithms underlying this technique contain: Graph cut [2], Local Variation [3], Normalized Cuts (Ncuts) [4], Segmentation by Weighted Aggregation (SWA) [5]. Among these segmentation algorithms, the Local Variation algorithm is one of the most rapid and effectively algorithm.

Texture is a fundamental and significant visual cue in defining visual perception for human beings to recognize objects. It can describe a wide variety of surface characteristics such as plants, minerals, road, sky, fur and skin. The task of texture segmentation is to extract texture features (elements). There are several approaches in

J. Lei et al. (Eds.): AICI 2012, LNAI 7530, pp. 424–434, 2012.

this field, such as gray-level co-occurrence matrices (GLCM) [6], Gabor filters [7], local binary pattern (LBP) [8], textons [9] and many others. Textons is first used by Malik et al. [9] for image segmentation.

Ferran and Casas [10] have proposed some new important information which they call Syntactic Features to image segmentation. Syntactic features, including homogeneity, compactness, inclusion or symmetry, represent geometric properties of image regions and their spatial configurations. These features provide strong evidences to improve the anomalous segmentation results.

In this paper, we present a fast and fully automatic image segmentation method based on graph theoretical framework. It is a hierarchical merging algorithm that integrates the color, texture and syntactic features one after the other. We derive merging schemes from the disjoint-set forest which is used in Local Variation segmentation algorithm and test our method on the Berkeley Segmentation Dataset (BSDS) and compare our results with several existing segmentation algorithms.

The paper is divided as follows. Section 2 summarizes the steps of the proposed segmentation algorithm. Section 3 describes how we improve the Local Variation algorithm to conduct the coarse segmentation. Section 4 introduces the textons technique for fine segmentation. Section 5 adds the syntactic processing. Furthermore, Section 6 provides experiment results of the proposed algorithm. Finally, we give certain future work and draw a conclusion in Section 7.

## 2 Algorithm Overview

This graph-based hierarchical merging algorithm consists of several steps which can be briefly outlined here:

(1).Utilize an improved Local Variation algorithm for coarse segmentation and return some color-consistent regions.

(2).Compute texton of each region and merge two adjacent regions which have similar texture feature. Then obtain a fine segmentation result.

(3).Deal with inclusion according to syntactic visual features to produce the final segmentation results.

Figure 1 demonstrates the details of this segmentation algorithm.

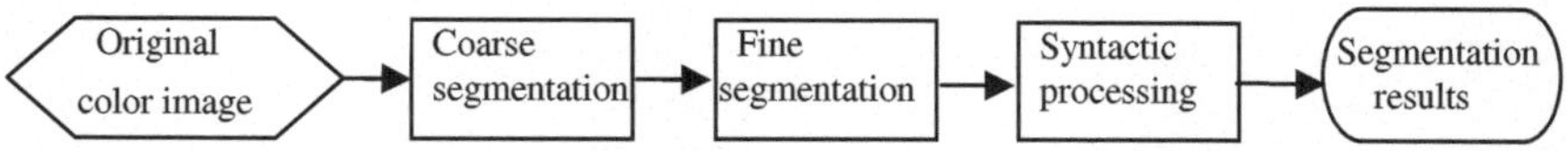


**Fig. 1.** The process of the proposed segmentation algorithm

## 3 Coarse Segmentation Based on Color Feature

The Local Variation algorithm is chose for coarse segmentation due to its efficiency. But this algorithm has three weaknesses. Firstly, Gaussian smoothing would blur the image edges. Secondly, the choice of the color space and weight of edge is unreasonable. Thirdly, some trivial regions are appeared in the segmentation results. So we will improve it from these aspects.

### 3.1 Edge-Preserving Smoothing Filter

Gaussian filter is most widely used for image preprocessing because of it nicely smooth homogeneous areas and efficiently restrains salt & pepper noise. However, it possesses the disadvantage of blurring intensity value details and extending edge region. Many edge-preserving filters which are more suitable for feature extraction had been proposed, including Bilateral filtering [11], SUSAN filtering [12]. An image can be represented as a 2-D lattice where the space of the lattice is known as spatial domain and the gray-level or color information is represented in the range domain. The core idea of those methods is introduced a range filter, videlicet, let domain filter smooth within the similar regions. Inspired by those strategies, we propose a new and simple ternary range filter and add it to Gaussian low-pass domain filter.

We consider an arbitrary pixel $a$ and its 5*5 neighborhood system $S$. First, all these 25 pixels can be classified into three classes: central pixel $a$, 12 primitive pixels and their symmetrical pixels about center $a$. Then we construct a ternary range filter with the initial weight $w(i)=1$ for each pixel $i$. When the intensity difference between the pixel $i(x,y)$ and center $a$ is greater than a given threshold $R$ (set $R=20$), at the same time, the intensity difference between its symmetrical pixel $i'(y,x)$ and $a$ is less than $R$, then its intensity weight $w(i)$ becomes to 0 and its symmetric pixel point's weight $w(i')$ turns into 2 for compensating the unbalance weight. Namely,

$$w(i)=\begin{cases}0 & if\ \ |I(i)-I(a)|>R \text{ and } |I(i')-I(a)|\le R\\ 2 & if\ \ |I(i)-I(a)|\le R \text{ and } |I(i')-I(a)|>R\ .\\ 1 & otherwise\end{cases} \tag{1}$$

Where $I(i)$ is the intensity of pixel $i$ and $i\in S$.

An edge-preserving filter is constructed by combining the ternary range filter and Gaussian low-pass domain filter:

$$Filter(a)=\sum_{i\in S}G_\sigma(a,i)w(i)I_i\ . \tag{2}$$

Where $G_\sigma=e^{-\frac{\|a-i\|^2}{2\sigma^2}}$ is the Gauss function for noise reduction.

In edge regions, the ternary range filter can effectively suppress the smoothing scope of Gaussian filter. Thus, it smoothes the similar region and achieves the purpose of preserving the edge information; When $w(i)=w(i')=1$, the ternary range filter is in an inhibitory state. This case is equivalent to only using Gaussian low-pass domain filter for smoothing. It can remove salt & pepper noise and reserve image original data. Figure 2 demonstrates the way to obtain the weights of the ternary range filter.

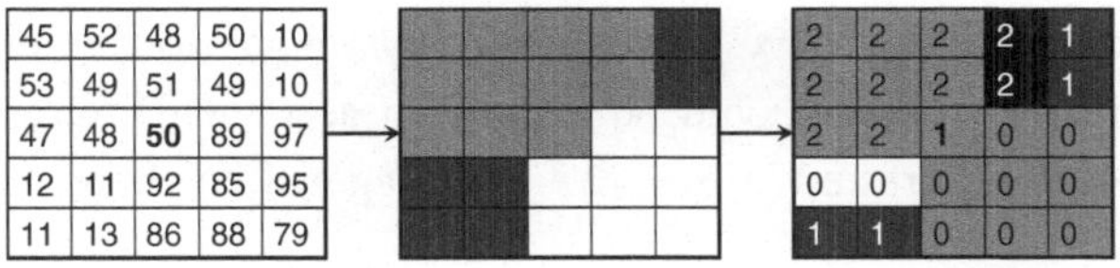


**Fig. 2.** A simple example of how to obtain the weights of the ternary range filter

We use the proposed edge-preserving filter (applied three times corresponding to color space) to preprocess the input color image. The ternary range filter works better than Gaussian and linear range filter. Experiments proved that the smoothing strategy can efficiently preserve the edge as well as reduce the image noise. Figure 3 demonstrates the segmentation results under the Local Variation algorithm which is applied to the Gauss filter and the edge-preserving filter respectively.

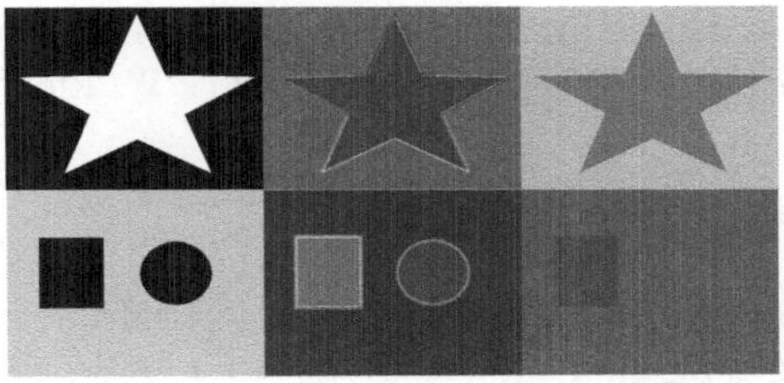

(a) (b) (c)

**Fig. 3.** The segmentation results. From left to right: (a) Synthetic images. (b) Generate 31 regions and 22 regions respectively by using Gauss filter. (c) Generate 2 regions and 3 regions respectively by using the edge-preserving filter.

When we apply Local Variation algorithm with the edge-preserving filter, the results are accurate by comparing the number of regions between (a) and (c). However, when we apply the same algorithm with Gauss filter which dims boundary information and generates many pseudo regions, it results in incorrect segmentation.

### 3.2 Color Model and Edge Weight

Some color models have been widely used in the image segmentation literature, for example, RGB, XYZ and HSV. In the Local Variation algorithm, the authors suggest calculating the color weight with RGB color model. But this model has some well known shortcomings: It is non-intuitive and non-uniform in color separation [13]. LUV color model is considered to be an approximation of the perceptually uniform and referred to as uniform color space. In order to obtain a more meaningful segmentation effect, LUV space is employed to replace RGB space. This is motivated by literature [14] which utilizes such representation in method for natural image segmentation. According to [3], the Euclidean metric is used to express the color difference $d$ between two colors $(L_1,U_1,V_1)$ and $(L_2,U_2,V_2)$:

$$d=\sqrt{(L_1-L_2)^2+(U_1-U_2)^2+(V_1-V_2)^2}\,. \tag{3}$$

When computing the pixel-to-pixel color similarity weight $w(e)$ between 8-connected neighbors, the pixel position should be taken into account. Different from [3], we use $w(e) = d$ for 4-neighbors and $w(e) = \sqrt{2} * d$ for diagonal-neighbors.

### 3.3 Compactness Ratio

A lot of trivial regions will be appeared when using the Local Variation algorithm for coarse segmentation with small similarity threshold $Tc$ [3]. It is well known that real-world objects tend to be compact. So we add a procedure at the end of the Local Variation algorithm to remove those complex regions.

Compactness ratio, sometimes called the shape factor or shape complexity, is a very important concept of the urban spatial morphology to reflect the city shape [15]. It is a numerical quantity representing the degree to which a shape is compact. Therefore, compactness can be introduced into region processing. We calculate it by using the following formula which was proposed by H. W. Richardson in 1961 [16]:

$$Compactness = \frac{2\sqrt{\pi * Area}}{Perimeter} \propto \frac{\sqrt{Area}}{Perimeter}. \tag{4}$$

If the compactness ratio of a region is less than a given threshold (choose 0.15), it must be merged into its adjacent region based on color information.

Figure 4 shows the coarse segmentation results. Both algorithms have the same parameter $Tc = 150$ [3]. By comparing the improved Local Variation algorithm with the Local Variation algorithm, we can find that the former which improves in three aspects can obtain a charming segmentation result.

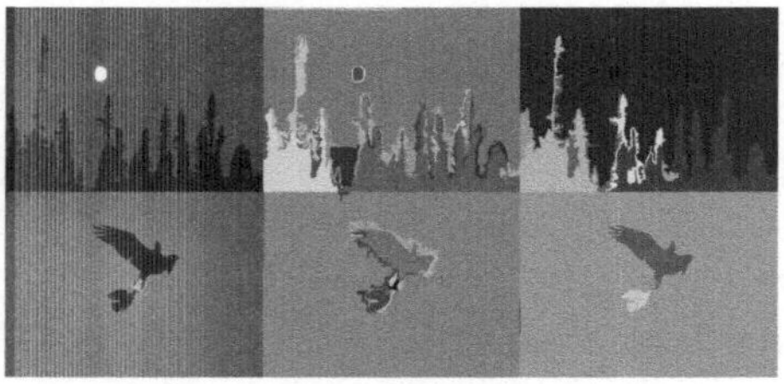

(a) (b) (c)

**Fig. 4.** Experiment results of two segmentation algorithms. From left to right: (a) Original images from BSDS. (b) Generate 57 regions and 27 regions using the Local Variation algorithm respectively. (c) Generate 10 regions and 7 regions using the improved Local Variation algorithm respectively.

# 4 Fine Segmentation Based on Texture Feature

As the analysis in the previous parts, the color cue is not sufficient to capture the range of different appearances of objects in an image. Winn et al. [17] and Shotton et al. [18] have successfully applied textons to image segmentation and image understanding. This technique utilized a 17-dimension filter-bank which performs strong robustness (it added the color cue) and high efficiency (its dimensions are less than Gabor filters [7] and MR8 [9] (Maximum Response filter sets)) to extract texture features of images. We choose it with appropriate improvement to extract texture feature of each region and then merge regions which have similar texture feature.

### 4.1 Texton Filter-Bank and Texton Histogram

In order to upgrade the processing speed, the first three Gaussian smoothing filters are operated on the original color image only at scale $\kappa$ (rather than $\kappa$, $2\kappa$, $4\kappa$ [17][18]; set $\kappa = 1.0$). So the original image is convolved with an 11-dimensional filter-bank. Each pixel is transformed into an 11-dimensional vector of filter responses after filtering. And the filter responses for all pixels within a region (acquired from coarse segmentation) are transformed into normalized histograms. (Each dimensional response has a histogram with $n$ bins, set $n = 16$). Next, all these 11 histograms are cascaded into a texton histogram with $K$ bins ($K = 11*16 = 176$). At last, each region is presented by a texton histogram. It is clear that this statistics method is superior to k-means clustering algorithm which considers the relationship between pixels.

### 4.2 Texture Weight Metric

We compute the texture similarity between adjacent regions by comparing their texton histograms. A number of methods are available for comparing histograms. Among them a simple and effective choice is $\chi^2$ test:

$$\chi^2(h_i, h_j) = \frac{1}{2}\sum_{k=1}^{K} \frac{\left[h_i(k) - h_j(k)\right]^2}{h_i(k) + h_j(k)} . \tag{5}$$

Where $h_i$ is the histogram of region $i$. If the texture of two adjacent regions is more similar, the $\chi^2$ value is smaller.

### 4.3 Regions Merging Based on Texton Feature

The regions obtained by coarse segmentation can be represented by a weighted region adjacency graph $G = (V, E, W)$. We use the $\lambda^2$ value as the weight $w(i, j)$ between regions $i$ and $j$. Obviously, $\lambda^2$ value plays a decisive role in determining the overall performance of texture segmentation process. We apply the disjoint-set forest algorithm which is used in Local Variation segmentation algorithm to partition the graph, namely, grouping the adjacent regions if their texture similarity is less than a threshold $T_t$ (the texture threshold $T_t$ can be set to 0~1).

# 5 Post Processing Based on Syntactic Visual Features

Homogeneity, compactness, regularity, inclusion and symmetry are called syntactic features [10] [19]. This paper adds the "inclusion" into the post processing.

Larger objects may contain certain smaller holes or objects in real world. These less significant parts always have dissimilar color, texture and other basic features with their container (larger objects). Major or total part of a small region included or inside another region provides very strong evidence for merging them [19].

We use the similarly criterion introduced by Adamek [19] to define the inclusion:

$$C_{ij} = \frac{F_{ij}}{\min(L_i, L_j)} . \tag{6}$$

Where $L_i$ is the perimeter of regions $i$, $F_{ij}$ is the length of their common border.

We also introduce the area ratio $R_{ij}$ between neighbor regions $i$ and $j$ :

$$R_{ij} = \frac{\min(S_i, S_j)}{\max(S_i, S_j)} . \tag{7}$$

Where $S_i$ is the area of region $i$. When a relatively small region ( $R_{ij} < 0.1$ ) is major or total included by the other region, they should be merged into a region.

# 6 Algorithm Evaluation

Evaluating the results produced by segmentation algorithms is a challenging task. This is because people often tend to incorporate into their segmentations semantic considerations which are beyond the scope of the image data [23]. In this section, we compared our algorithm with several state of the art algorithms in three aspects: effects comparison, runtime evaluation and quantitative analysis. All segmentation algorithms are running on the world famous image database BSDS.

## 6.1 Effects Comparison

In order to compare the segmentation results fairly, we employ the appropriate parameters for the following state-of-the-art algorithms. (1). SWA [5]. We tested it with the default parameters (used the full range of features) and chose the optimal results among multiple scales for each image. (2). Ncuts [4]. We selected the number of regions that obtained best result for each image. (3). Mean Shift [14] [20].We selected the spatial and spectral bandwidth parameters that obtained the best result for each image. (4). Local Variation [3]. We tested roughly 40 different sets of parameters and selected best result for each image.

Figure 5 shows several samples of segmentation results. The left column are original images, the second column shows the result of our method, the third, fourth, fifth and sixth columns contain results obtained using SWA, Ncuts, Mean Shift and

Local Variation, respectively. The experiment results show the perceived regions are extracted by our method outperforming the other four algorithms.

### 6.2 Runtime Evaluation

We random choose 100 images from BSDS and calculate the average runtime of each algorithms so as to compare the computational cost between the proposed algorithm and other four algorithms which have been made available publicly. Table 1 shows the results. The hardware configuration is 2.93Hz Intel Core2 processors with 2GB of RAM.

Ncuts is quite slow on account of a large amount of calculation. Its runtime is just a reference value which can not comparable with other algorithms. For Mean Shift, the running time rely heavily on its parameters, so we choose the default parameters [14][20]. The default parameters of SWA are also used [5]. For the Local Variation algorithm, we choose the threshold value $T_c = 300$ as suggested by the authors [3]. For the proposed algorithm, the parameters are $(T_c, T_t, C_{ij}) = (150, 0.5, 0.8)$.

From Table 1, we can draw the conclusion that the runtime of our method is significantly faster than Mean Shift, SWA and Ncuts. Although the runtime of our method is slower compared with Local Variation algorithm, the effect of ours which combines texture and syntactic visual features is markedly more accurate.

### 6.3 Quantitative Analysis

In this section, we quantitatively compare our method with other four algorithms. The comparison is based on four quantitative performance measures [21]: (1) BDE (Boundary Displacement Error). (2) PRI (Probabilistic Rand Index). (3) GCE (Global Consistency Error). (4) VoI (Variation of Information).

We report it in Table 2, the average scores over the BSDS for five segmentation methods. The best two results are highlighted in Red and Bold in descending order. The data of Mean Shift and Ncuts which lists below is derived from the literature [22], and the parameters of other algorithms are same as Section 6.2.

Note that, more is greater the PRI measure, better is the segmentation. In the other hand, more are smaller the VoI, GCE and BDE measures, better is the segmentation. We remark that our method achieves the best score for VoI measure and comes in second for PRI and GCE measures compared to the other four algorithms. The BDE measure of our method is not good, since we remove the internal boundaries to form a relatively integrated region which is right in practical application. Though the BDE measure of the Local Variation is superior to the other algorithms, our method wonderfully overcomes the disadvantage that many slender irregular regions emerge between the boundaries of two regions in the Local Variation.

**Table 1.** Comparison of Average Running Time among the Five Algorithms

| Algorithms | Our method | Mean Shift | SWA | Ncuts | Local Variation |
|---|---|---|---|---|---|
| Runtime (s) | 0.877 | 1.532 | 3.859 | 11.768 | 0.276 |

**Table 2.** Comparison of Average Scores among the Five Algorithms

| | BDE | PRI | GCE | VoI |
|---|---|---|---|---|
| Humans | 4.994 | 0.8574 | 0.0797 | 1.1040 |
| Mean Shift | 9.7001 | 0.7550 | 0.2598 | 2.477 |
| Ncuts | **9.6038** | 0.7229 | 0.2182 | 2.9329 |
| Local Variation | 8.3378 | 0.7852 | 0.2341 | 2.4017 |
| SWA | 14.5795 | 0.7076 | 0.1984 | **2.0033** |
| Our Method | 12.5623 | **0.7585** | **0.2031** | 1.8947 |

**Fig. 5.** Experiment results

# 7 Conclusion and Future Work

We have introduced a graph-based hierarchical merging algorithm for image segmentation. In each step, we extract a visual cue and then merge the regions into larger regions. The segmentation results and the computational speed under our algorithm are effectively and efficiently.

Although the running speed of our method is fast, the largest portion of this time is taken by the computation of the 11-dimensional filter responses. In certain circumstances, we can only use 3-dimensional color information or even 1-dimensional lightness information to implement texture merging.

All the parameters of our method have strong robustness except the texture threshold. It is hard to set a fixed value for all images because the contrast ratio is different from each other. As the future work, we aim at reducing the dependence of the texture threshold, for example, according to the image content we can use an adaptive threshold.

**Acknowledgments.** This research is supported by the Fundamental Research Funds for the Central Universities of China under Grant No.N100418001; the National High-Tech Research and Development Plan of China under Grant No. SS20112AA010105.

# References

1. Peng, B., Zhang, L., Zhang, D.: A Survey of Graph Theoretical Approaches to Image Segmentation (2012), `http://www4.comp.polyu.edu.hk/~cslzhang/papers.html`
2. Boykov, Y., Funka-Lea, G.: Graph cuts and efficient n-d image segmentation. International Journal of Computer Vision (IJCV) 70(2), 109–131 (2006)
3. Felzenszwalb, P., Huttenlocher, D.: Efficient graph-based image segmentation. International Journal of Computer Vision (IJCV) 59(2), 167–181 (2004)
4. Shi, J., Malik, J.: Normalized cuts and image segmentation. IEEE Trans. Machine Intell. 22, 888–905 (2000)
5. Sharon, E., Galun, M., Sharon, D., Basri, R., Brandt, A.: Hierarchy and adaptivity in segmenting visual scenes. Nature 442(7104), 810–813 (2006)
6. Haralick, R.M.: Statistical Structural Approaches to Texture. IEEE Proceedings 67(5) (1979)
7. Weldon, T.P., Higgins, W.E., Dunn, D.F.: Efficient Gabor Filter Design for Texture segmentation. Pattern Recognition (1996)
8. Ojala, T., Pietikainen, M.: Unsupervised texture segmentation using feature distributions. Pattern Recognition 32, 477–486 (1999)
9. Malik, J., Belongie, S., Leung, T.K., Shi, J.: Contour and texture analysis for image segmentation. International Journal of Computer Vision (IJCV) 43(1), 7–27 (2001)
10. Bennstrom, C.F.: Casas. Binary-partition-tree creation using a quasi-inclusion criterion. In: Proceedings of the Eighth International Conference on Information Visualization (IV). IEEE Computer Society Press, London, UK (2004)

11. Tomasi, C., Manduchi, R.: Bilateral filtering for gray and color images. In: International Conference on Computer Vision (ICCV), Bombay, India, pp. 836–846 (1998)
12. Smith, S.M., Brady, J.M.: Susan – a new approach to low level image processing. International Journal Computer Vision (IJCV) 23(1), 45–78 (1997)
13. Sotelo, M., Rodriguez, F., Magdalena, L., Bergasa, L., Boquete, L.: A color vision-based lane tracking system for autonomous driving on unmarked roads. Autonomous Robots 16(1), 95–116 (2004)
14. Comaniciu, D., Meer, P.: Mean Shift: A robust approach toward feature space analysis. TPAMI 24(5), 603–619 (2002)
15. Roo, G.D.: Environmental conflicts in compact cities: complexity, decision making, and policy approaches. Environmental and Planning and Design 27(2), 121–162 (2000)
16. Richardson, H.W.: The Economics of Urban Size. Lexington Books (1973)
17. Winn, J., Criminisi, A., Minka, T.: Object Categorization by learned universal visual dictionary. In: International Conference on Computer Vision (ICCV), Beijing, China, vol. 2, pp. 1800–1807 (2005)
18. Shotton, J., Winn, J., Rother, C., Criminisi, A.: TextonBoost: Joint appearance, shape and context modeling for multi-class object recognition and segmentation. In: European Conference on Computer Vision (ECCV), pp. 1–15 (2006)
19. Adamek, T., O'Connor, N., Murphy, N.: Region-Based Segmentation of Images Using Syntactic Visual Features. In: 6th Int. Workshop Image Analysis for Multimedia Interactive Services, pp. 1–4 (2005)
20. Georgescu, B., Christoudias, C.M.: The Edge Detection and Image Segmentation (EDISON) system
21. Doggaz, N., Ferjani, I.: Image Segmentation Using Normalized Cuts and Efficient Graph-Based Segmentation. In: Maino, G., Foresti, G.L. (eds.) ICIAP 2011, Part II. LNCS, vol. 6979, pp. 229–240. Springer, Heidelberg (2011)
22. Yang, A.Y., Wright, J., Ma, Y.: Unsupervised segmentation of natural images via lossy data compression. Comput. Vis.Image Understand 110, 212–225 (2008)
23. Sharon, A., Galun, M., Brandt, A., Basri, R.: Image Segmentation by Probabilistic Bottom-Up Aggregation and Cue Integration. IEEE Transactions on Pattern Analysis and Machine Intelligence 34(2), 315–327 (2012)

# Multi-resolution Image Fusion Algorithm Based on Regional Cross Entropy and Regional Priority

Wen Ge, Peng Li, and Jing Li Xu

School of Electronics and Information Engineering, Shenyang Aerospace University, No.37 Daoyi South Avenue, Daoyi Development District, 110136 Shenyang, China
gewenbox72@sina.com.cn

**Abstract.** Based on image contents, the better to simulate the process pattern of human eyes vision, a fusion algorithm of integration and highlight for the image details is proposed. Through wavelet transform, a regional cross entropy fusion rule is used for the low-frequency component which reflects approximate content, and a region brightness details priority weighted fusion rule is used for the high-frequency component which reflects detail features of image. Finally, the fusion image is reconstructed through an inverse transform of wavelet. Experimental results show that by using this algorithm, the mutual information between the images can be fused organically, the image clarity is raised, the fusion image details and brightness information are enhanced. Strong support for the follow-up information analysis and extractive ability of the images are provided.

**Keywords:** Regional cross entropy, Regional priority, Wavelet transform, fusion image, Region brightness.

## 1 Introduction

Image fusion is the process to comprehend organically the complementary information and redundant data from several sensors images to obtain the image with richer details, more reliable and easier-to-understand and read for the same scene. There are many image fusion methods, among others, the wavelet transform is fully used for image fusion[1-6] because of its good time frequency localization property and multi-scale analytical ability. The multi-resolution wavelet transform is adopted. The wavelet coefficient and scale coefficient on pixel level are analyzed treatment and image fused. The fusion rule based on regional cross entropy is proposed for the low-frequency component, this method is by means of judging the similarity of being fused images local region, and adopts different fusion strategy to fuse, it can reflect the approximate contents of the image. The fusion rule based on regional brightness details priority weighted, it is aimed at the characteristic of image visual information mainly containing brightness and details information, this fusion rule is proposed for the high-frequency component. It obtains the fusion effect with better visual property and richer and obvious details.

J. Lei et al. (Eds.): AICI 2012, LNAI 7530, pp. 435–441, 2012.

## 2 Wavelet Decomposition of Image

Because the wavelet transform is possessed of the good localization property both in frequency domain and time domain simultaneously, it is used widely in image treatment and analyses. Mallat proposed the algorithm of wavelet transform rapid-decomposition and reconstruction, using two one-dimension filters realized the wavelet rapid-decomposition for the two-dimension image, and then using two one-dimension reconstruction filters realized the image reconstruction. Mallat algorithm decomposition equation under the scale $j$ is [7]:

$$\begin{cases} C_j = H_r H_c C_{j-1} \\ D_j^1 = H_r G_c C_{j-1} \\ D_j^2 = G_r H_c C_{j-1} \\ D_j^3 = G_r G_c C_{j-1} \end{cases} \quad (j = 0,-1,\ldots,-J). \tag{1}$$

Where, $H$ and $G$ is one-dimension mirror filter operator respectively, the lower mark $r$ and $c$ represents the line operation and column operation for matrix respectively, and the corresponding two-dimension image reconstruction algorithm is:

$$C_{j-1} = H_r^* H_c^* C_j + H_r^* G_c^* D_j^1 + G_r^* H_c^* D_j^2 + G_r^* G_c^* D_j^3 \\ (j = -J,-J+1,\ldots,-1). \tag{2}$$

Where $H^*$ and $G^*$ is the conjugate transposed matrix of the $H$ and $G$ respectively; and $C_j$ is the approximation of the source image on $2^{-j}$ resolving power (image low-frequency part); $D_j^\varepsilon$ $(\varepsilon = 1,2,3)$ is the error of the approximation (image high-frequency part).

## 3 Image Fusion Rule

In the image fusion process, designing the reasonable image fusion rule is the key to obtain the high quality fusion image. The fusion rule is proposed, which maintains and restores the source images detail characteristic while raises the image clarity. The high frequency information and low frequency information fusion process are described in detail below.

### 3.1 Low-Frequency Information Fusion

According to the low-frequency information containing the most energy of the image, a fusion rule based on regional cross entropy is proposed for low-frequency component. The cross entropy is also called relative entropy, it can be used for measuring the difference between the two images. This is the relative measurement of the containing information of the two images. Because the local property of image is expressed and reflected by many pixels in the region, the adopting region cross entropy reflects the similarity of the two being fused image local region. The region window selects window 3*3, the cross entropy of two being fused images is:

$$C(p,q)=\sum_{i=1}^{M}\sum_{j=1}^{N} p_{ij}\log_2\frac{p_{ij}}{q_{ij}}. \tag{3}$$

Where: $C(p,q)$ is cross entropy; $p_{ij}$ is the regional gradation distribution in image A; $q_{ij}$ is the regional gradation distribution in image B.

$$P_{ij}=B(i,j)/\sum_{i=1}^{M}\sum_{j=1}^{N} B(i,j). \tag{4}$$

Where: $B(i,j)$ is the gradation value of the spot $(i,j)$ in the regional image.

While the difference between the two images is littler, the littler the cross entropy is, while the difference between the two images is larger, the larger the cross entropy is. Supposed, threshold is $T$.

While $C(p,q)<T$, the fusion algorithm for low-frequency is:

$$L_F=\frac{E_{fA}}{E_{fA}+E_{fB}}L_A+\frac{E_{fB}}{E_{fA}+E_{fB}}L_B. \tag{5}$$

Where: $L_A$, $L_B$ and $L_F$ is the low-frequency coefficient of the two being fused images and the after-fused image respectively. $E_{fA}$, $E_{fB}$ is the regional information entropy of image A and image B respectively. According to the principle of Shannon's information entropy theory, the regional information entropy $E_f$ is defined:

$$E_f=-\sum_{i=1}^{M}\sum_{j=1}^{N} P_{ij}\log_2 P_{ij}. \tag{6}$$

While $C(p,q)\geq T$, the fusion algorithm is:

$$L_F=\begin{cases} L_A & E_{fA}\geq E_{fB} \\ L_B & E_{fA}<E_{fB} \end{cases}. \tag{7}$$

Through the above fusion rule, the information supplied by low-frequency component can be fused together effectively.

### 3.2 High-Frequency Information Fusion

Mathematics image provides the visual information mainly including brightness and the detail information, and the high-frequency component contains the important characteristic and detail information in the source images. In order to obtain the fusion effect for better visual characteristic and more rich and prominent details, a fusion rule of the regional brightness and details superior weighted for high-frequency information is proposed. Brightness information is described with the pixels average value, detail information is described with the ratio of the sum for the difference of the 2 pixels and pixels sum[8]. The regional brightness and details of the window, in which $(i, j)$ is the central point, are respectively:

$$I(i,j)=\frac{1}{M\times N}\sum_{m=1}^{M}\sum_{n=1}^{N}f(i+m-\frac{M+1}{2},j+n-\frac{N+1}{2}). \tag{8}$$

$$D(i,j)=\frac{1}{\sum_{i=1}^{M}\sum_{j=1}^{N}f(i+m-\frac{M+1}{2},j+n-\frac{N+1}{2})}\times$$
$$\sum_{m=1}^{M}\sum_{n=1}^{N}\left|f(i+m-\frac{M+1}{2},j+n-\frac{N+1}{2})-f(i+m+1-\frac{M+1}{2},j+n+1-\frac{N+1}{2})\right|. \tag{9}$$

If the brightness and the details of A image are greater or lesser than of B image simultaneously, then takes wavelet coefficient of the great brightness and the details as the high frequency component. Namely,

$$H_F=\begin{cases}H_A & I_A\geq I_B, D_A\geq D_B\\ H_B & I_B\geq I_A, D_B\geq D_A\end{cases}. \tag{10}$$

If the brightness and the details of image A are not greater or lesser than of image B simultaneously, then the high-frequency component is:

$$\text{While } I_A\geq I_B, D_A<D_B \text{ or } I_A<I_B, D_A\geq D_B$$
$$H_F=\alpha(H_A+K_1H_B)-\beta\left|H_A-K_2H_B\right|. \tag{11}$$

Where, $\alpha(H_A+K_1H_B)$ indicates the weighting average value of the image A/B, affects the brightness of fused image. $\beta\left|H_A-K_2H_B\right|$ indicates the weighting differential value of two images, affects the details information of fused image. $H_A$, $H_B$ and $H_F$ is respectively correspondent wavelet coefficient of A/B image and wavelet coefficient after fused. $K_1$, $K_2$ $\alpha$, $\beta$ is the weighting factor respectively, where $K_1$, $K_2$ adjusts the superior ratio of A/B image; $\alpha$ adjusts the image brightness; $\beta$ decides the image edge. Through the adjustment of $K_1$, $K_2$ $\alpha$, $\beta$

may eliminate the fuzzy edge, highlight the details and adjust the image brightness. $K_1, K_2$ $\alpha, \beta$ are calculate by the following formulas:

$$\alpha = \begin{cases} I_A(i,j)/(I_A(i,j)+I_B(i,j)) & I_A(i,j) \ge I_B(i,j) \\ I_B(i,j)/(I_A(i,j)+I_B(i,j)) & I_A(i,j) < I_B(i,j) \end{cases}. \tag{12}$$

$$\beta = \begin{cases} D_A(i,j)/(D_A(i,j)+D_B(i,j)) & D_A(i,j) \ge D_B(i,j) \\ D_B(i,j)/(D_A(i,j)+D_B(i,j)) & D_A(i,j) < D_B(i,j) \end{cases}. \tag{13}$$

$$K_1 = K_I(i,j)/(K_I(i,j)+K_D(i,j)). \tag{14}$$

$$K_2 = K_D(i,j)/(K_I(i,j)+K_D(i,j)). \tag{15}$$

In formula (14), (15), $K_I$ and $K_D$ represents the ratio of regional brightness and details of A/B high-frequency sub-image respectively.

$$K_I = \begin{cases} I_A(i,j)/I_B(i,j) & I_A(i,j) \ge I_B(i,j) \\ I_B(i,j)/I_A(i,j) & I_A(i,j) < I_B(i,j) \end{cases}. \tag{16}$$

$$K_D = \begin{cases} D_A(i,j)/D_B(i,j) & D_A(i,j) \ge D_B(i,j) \\ D_B(i,j)/D_A(i,j) & D_A(i,j) < D_B(i,j) \end{cases}. \tag{17}$$

## 4 Experiment Results and Evaluation

In order to verify the reasonableness effectiveness of the proposed algorithm, a group of fusion experiments for two remote images from Fig. 1 are performed, and by comparison with wavelet transform and PCA. Fig.1(a), (b) is a middle-wave infrared image, long-wave infrared image for the same target area respectively; Fig.1(c) is PCA; Fig.1(d) is wavelet transform algorithm[9]; Fig.1(e) is the proposed algorithm. To view from visual effect of fusion results, the effect of Fig.1(e) fusion image is better, because the detail information is rich, the target is more clear and the edge is continuous. The effect of Fig.1(c) is relatively poor. The effect of Fig.1(d) is better than that of Fig.1(c), but the contract is lost and some interesting targets are not so clear.

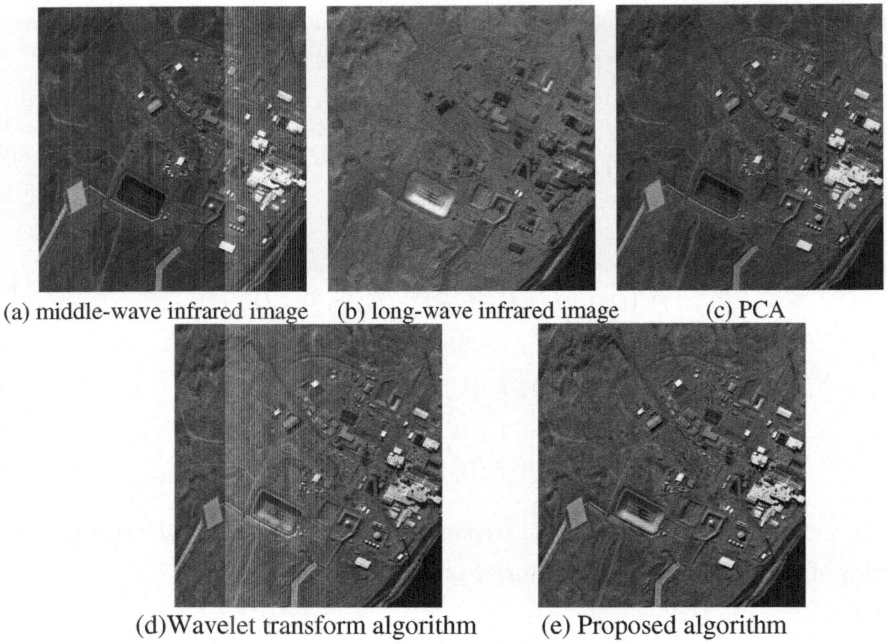


**Fig. 1.** Source images and fused images

Except for the visual comparison, three objective indexes, information entropy, average gradient and standard deviation are adopted to perform quantitative analysis for the quality of the fusion image. The magnitude of the value of information entropy reflects the magnitude of information in the fusion image. The average gradient reflects the clarity of the image. The standard deviation reflects the magnitude of topographical contrast of image. The results of specific evaluation are as shown in table1. From table 1 it is known, the fusion effect of the proposed fusion method is good, its image clarity and the information quantity are improved greatly.

**Table 1.** Objective Evaluation of Both Fusion Algorithms

| Algorithms | Information Entropy | Average Gradient | Standard Deviation |
|---|---|---|---|
| PCA | 8. 9645 | 48. 2546 | 10. 8953 |
| Wavelet Transform | 12. 7845 | 56. 7546 | 14. 4528 |
| Proposed Algorithm | 19. 3667 | 59. 2341 | 18. 5622 |

## 5 Conclusions

For reasons that it has the merits of better time frequency localization property frequency, wavelet transform is applied to the image fusion. Meanwhile the proposed fusion rules for enhancement of the image brightness and details are applied. The source images information is fused together effectively in the fusion image, as well as the clarity of information is applied. The source images information is fused together effectively in the fusion image, the image brightness and texture characteristic are enhanced, and the visual effects is improved obviously. For the follow-up process of the image provides fuller, more reliable information, a wide applied prospect can be expected.

## References


1. Amolins, K., Zhang, Y., Dare, P.: Wavelet-based Image Fusion Techniques: An Introduction, Review and Comparion. Photogrammetry & Remote Sensing 62, 249–263 (2007)
2. Piella, G.: A General Framework for Multiresolution Image Fusion: from Pixels to Regions. Information Fusion 68, 259–280 (2003)
3. Liu, B., Zhu, Q., Deng, J.X.: Fusion Method of Multispectral Image Based on Red-Black Wavelet Transform. Chinese Journal of Scientific Instrument 32, 408–414 (2011)
4. Wu, Z.G., Wang, Y.J., Li, G.J.: Application of Adaptive PCNN Based on Wavelet Transform to Image Fusion, vol. 18, pp. 708–715 (2010)
5. Wu, H., Wang, H.S.: Sobel Operator and Wavelet Transform. Computer Simulation 28, 232–235 (2011)
6. Tao, G.Q., Li, D.P., Lu, G.H.: On Image Fusion Based on Different Fusion Rules of Wavelet Transform. Acta Photonica Sinica 33, 221–224 (2003)
7. Sun, Y.K.: Wavelet Analysis and Application. China Machine Press, Beijing (2005)
8. Zhou, S., Shen, Y., Hao, J.S.: A Daptive Pixel-weighted CT/MRI Fusion Based on Local Priority. Journal of HARBIN Institute of Technology 38, 1314–1317 (2006)
9. Gong, C.L.: A New Wavelet Image Fusion Method Based on Local Energy. Laser & Infrared 38, 1266–1269 (2008)

# Graph-Cut Optimization for Video Moving Objects Detection with Geodesic Active Contour

Chunsheng Guo and Zhiyu Wang

College of Communication Engineering
Hangzhou Dianzi University
Hangzhou, China, (310018)
Guo.chsh@gmail.com, wangzhiyu_1989@126.com

**Abstract.** The traditional graph-cut for video moving objects detection is a global optimization algorithm, the result may be over-smoothing. The lack of local information in graph-cut limits the ability to precisely localize object boundaries. In this paper, moving objects detection algorithm is improved by introducting geodesic active contour. By the Kalman prediction of the number of objectives pixels and objectives-background pixel-pairs, and adaptive updating of the nodes flux with geodesic active contour, the proposed algorithm is successfully applied to video moving objects detection. Though adaptive updating of the nodes flux with geodesic active contour, the proposed algorithm will have a better edge capture ability of moving objects. Experimental results show that the proposed algorithm is more effective than graph-cut for video moving objects detection in complex backgrounds.

**Keywords:** Moving objects detection, Graph-cut, Geodesic active contour, Kalman prediction.

## 1 Introduction

Moving objects detection for a static camera has been extensively studied for many years. Moving objects detection plays a very important role in many vision applications with the purpose of subtracting interesting target area and locating the moving objects from image sequences. It is widely used in vision systems such as traffic control, video surveillance of unattended outdoor environments, video surveillance of objects, etc. Accurate moving object detection is essential for the robustness of intelligent video-surveillance systems.

Recently, the graph-cut algorithm has been used to detect video moving objects though the energy minimization technique under the framework of MAP-MRF (Maximum a posterior-Markov random field). Most graph-cuts algorithms focus on the iterative process and the priori information of moving objects in order to improve the detection accuracy. Garrett and Saito [1] used rectangular region included the moving objects to solve single moving object detection. Nagahashi et al. [2] combined the spatial-temporal volumes with multi-scale smoothing, and reduced the number of nodes and edges in order to reduce the calculations. Wang and Guan [3]

J. Lei et al. (Eds.): AICI 2012, LNAI 7530, pp. 442–449, 2012.

used histograms of oriented gradients (HOG) to express the shape of the moving objects in order to improve the detection accuracy. Fukuchi et al. [4] not only constructed the moving objects prior by saliency map, but also updated priors by the detected results. GUO and Wang [5] proposed an adaptive graph-cut (AGC) algorithm to solve the allocation of node flux in the network.

With the global optimization of the graph-cut algorithm, we can get more precise results of moving object detection. However, due to the approximation of the image prior function; these methods are prone to over-smooth the moving objects. In order to solve these problems, our proposed algorithm first adaptively update the nodes flux by the Kalman prediction of the number of objectives pixels and objectives-background pixel-pairs, and optimize the nodes flux with geodesic active contour; the graph-cut algorithm optimization with active contour was successfully applied to detect video moving objects finally. Geodesic active contour is a dynamic model, which is as a simplification of snake energy model with fewer parameters and less sensitivity to the initial contour. It is based on geodesics in a Riemannian space with a metric induced by the image content. Geodesic active contour model enable the proposed algorithm to have a better edge capture ability, which can get an accurate detection. In this paper, we mainly discuss video moving objects detection using the optimized graph-cut with geodesic active contour. The experimental results are verified to be effective, which compare ours method with traditional graph-cut algorithm.

## 2 Moving Objects Detection Based on Graph-Cut

The flow chart of video moving objects detection using graph-cut with geodesic active contour is given in Fig.1. To use minimum graph-cut solving video moving objects detection, the nodes flux should be determined. But these parameters depend upon the number of foreground pixels and objectives-background pixel-pairs, which could be obtained only after moving objects detection. So the proposed algorithm predicts the number of foreground pixels and objectives-background pixel-pairs by kalman prediction. The key problem is how to optimize the flux of nodes with geodesic active contour while using the proposed algorithm. We will have a detailed discussion about this key problem in next section. In this part, we mainly study the traditional graph-cut algorithm for video moving objects detection based on the frame of MRF.

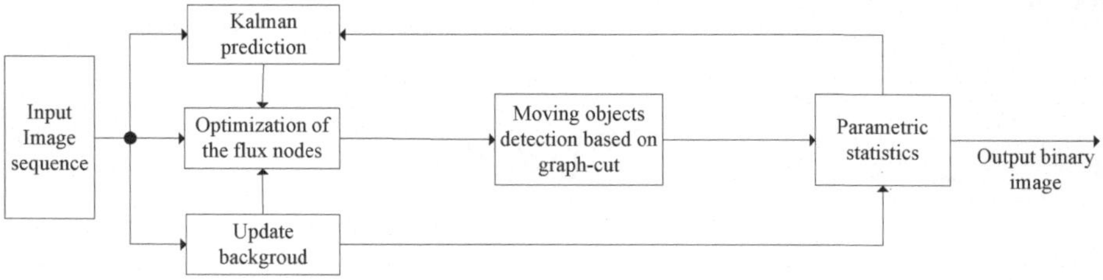


**Fig. 1.** The flow chart of proposed algorithm

Suppose the binary image $U$ of moving objects satisfies Markov random field model. Using first order and second order clique potential to approximate energy function of binary image $U$ , So the prior probability distribution function of $U$ can be simplified as

$$p(U) \propto \exp\left\{ -\gamma \sum_{m \in \Omega} u_m - \beta \sum_{m \in \Omega} \sum_{n \in c_m^2} [u_m u_n + (1-u_m)(1-u_n)] \right\} \tag{1}$$

Where $\Omega$ is the set of pixels in $U$ , $c_m^2$ is a set of second order neighborhood pixels of pixel $u_m$ . $O$ is the number of objective pixels in a video frame. $B$ is the number of objectives-background pixel-pairs in a video frame. Pixel value 1 represents moving objective pixel (or foreground pixel). Pixel value 0 represents background pixel. Then

$$\begin{aligned} \gamma &= -\ln \frac{P(u_m = 1)}{P(u_n = 0)} = -\ln \frac{O}{M \times N - O} \\ \beta &= -\ln \left[ \frac{B}{(N-1)M + N(M-1) - B} \right] \end{aligned} \tag{2}$$

Suppose the background of image follows the Gaussian distribution, and the foreground is consistent with uniform distribution. And all the pixels in $\Omega$ is conditional independence. Therefore, the likelihood function of gray image $Y$ under the condition of giving $U$ is:

$$p(Y|U) = \prod_{m \in \Omega} p(y_m|u_m) = \prod_{m \in \Omega} p(y_m|1)^{u_m} p(y_m|0)^{1-u_m} \tag{3}$$

Where $p(y_m | 0) = \frac{1}{\sqrt{2\pi\sigma_m^2}} \exp\left[ -\frac{(y_m - \mu_m)^2}{2\sigma_m^2} \right]$ , $p(y_m | 1) = \frac{1}{L}$ , $L$ is the number of image gray levels.

Known prior probability and likelihood function, the logarithmic posterior probability can be derived,

$$\begin{aligned} L(U|Y) &= \ln p(U|Y) \propto \ln\left[ p(Y|U) p(U) \right] \\ &= \sum_{m \in \Omega} (\lambda_m - \gamma) u_m + \beta \sum_{m \in \Omega} \sum_{n \in c_m^2} [u_m u_n + (1-u_m)(1-u_n)] \end{aligned} \tag{4}$$

Where $\lambda_m = \ln \frac{p(y_m | 1)}{p(y_m | 0)}$ is log-likelihood ratio.

Now, the problem of video moving objects detection is converted into the problem of maximum a posterior (MAP) estimation. Defined flux $C(U)$ to be

$$C(U)=\sum_{m\in\Omega}u_m\max(0,\gamma-\lambda_m)+\sum_{m\in\Omega}(1-u_m)\max(0,\lambda_m-\gamma)+\beta\sum_{m\in\Omega}\sum_{n\in c_m^2}(u_m-u_n)^2 \tag{5}$$

Where $C(U)+L(U\mid Y)=$ constant, namely the maximum of $L(U\mid Y)$ equals to the minimum of $C(U)$ .Therefore the minimum graph-cut algorithm can be used for video moving objects detection.

# 3 Graph-Cut Optimization for Video Moving Objects Detection with Geodesic Active Contour

In our optimized graph-cut algorithm, the key issue is how to optimize the flux of nodes with geodesic active contour while using graph-cut algorithm. We will have a detailed analysis in this part.

## 3.1 The Define of Geodesic Active Contour

The Geodesic active contour [7] is defined along curve which minimize energy function by the form

$$E(\gamma)=\int_0^{L(\gamma)}\frac{1}{1+b|\nabla u|}ds \tag{6}$$

Where $L(\gamma)$ is the length of curve $\gamma$, $s$ is arc length, $u$ is the observed original image, $\nabla$ is gradient operator, $b$ is an constant.

For simpler, $p=(i,j)$, $p\in N(q)$ then, we have

$$ds=\omega_{pq}=\omega_k \tag{7}$$

Therefore, the discrete form of the length of curve $L(\gamma)=\int_0^{L(\gamma)}ds$ can be expressed as follows:

$$L'(\gamma)=\sum_{p,q\in\Omega}\sum_{p\in N(q)}\omega_k\left((1-u_p)u_q+u_p(1-u_q)\right) \tag{8}$$

Define the gradient of $u(i,j)$ in a 4-neighborhood as

$$\begin{aligned} \nabla u^x &= \left|u(i+1,j)-u(i,j)\right| \\ \nabla u^y &= \left|u(i,j+1)-u(i,j)\right| \end{aligned} \tag{9}$$

Noticing (6)-(9), the discrete representation of the geodesic active contour model to video moving detection can be written as

$$\begin{aligned} E'(\gamma) = &\sum_{(i+1,j)\in\Omega}\sum_{(i,j)\in\Omega}\frac{\omega_k}{1+b\left|\nabla u^x\right|}\left(u_{i+1,j}-u_{i,j}\right)^2 \\ &+\sum_{(i,j+1)\in\Omega}\sum_{(i,j)\in\Omega}\frac{\omega_k}{1+b\left|\nabla u^y\right|}\left(u_{i,j+1}-u_{i,j}\right)^2 \end{aligned} \tag{10}$$

### 3.2 Optimize the Nodes Flux with Geodesic Active Contour

We formulate our algorithm in simplest form as a graph-cut problem using the information of geodesic active contour as one of the unary region terms.

Then the energy function of our algorithm can be written as

$$\begin{aligned} C(U) = &\sum_{(i,j)\in\Omega} u_{i,j}\max\left(0,\gamma-\lambda_{i,j}\right)+\sum_{(i,j)\in\Omega}\left(1-u_{i,j}\right)\max\left(0,\lambda_{i,j}-\gamma\right) \\ &+\sum_{(i+1,j)\in\Omega}\sum_{(i,j)\in\Omega}\beta_1\left(\frac{1}{1+b\left|\nabla u^x\right|}\right)\left(u_{i+1,j}-u_{i,j}\right)^2 \\ &+\sum_{(i,j+1)\in\Omega}\sum_{(i,j)\in\Omega}\beta_2\left(\frac{1}{1+b\left|\nabla u^y\right|}\right)\left(u_{i,j+1}-u_{i,j}\right)^2 \end{aligned} \tag{11}$$

Where $\beta_1 = -\ln\left[\frac{H}{M*(N-1)-H}\right]$, $\beta_2 = -\ln\left[\frac{V}{(M-1)*N-V}\right]$, $H$ is the number of objectives-background pixel-pairs in the horizontal direction, $V$ is the number of objectives-background pixel-pairs in the vertical direction.

Using the minimum graph-cut to get the solution, binary image $U$, of the equation (11), it needs to treat each pixel as a node, additionally, the source node and sink node are added to constitute a nodes set, and a flux is distributed to edge between two neighbor nodes. If $\lambda_{i,j}-\gamma\geq 0$, then a flux $\lambda_{i,j}-\gamma$ is assigned to edge between source node and pixel node, a zero flux is assigned to edge between pixel node and sink node. If $\lambda_{i,j} - \gamma < 0$, then a zero flux is assigned to edge between source node and pixel node, a flux $\gamma - \lambda_{i,j}$ is assigned to edge between pixel node and sink node. The flux of edges between a pixel node and the right adjacent pixel are assigned to $\beta_1 \left( \frac{1}{1+b\left|\nabla u^x\right|} \right)$. The flux of edges between a pixel node and the below adjacent pixel are assigned to $\beta_2 \left( \frac{1}{1+b\left|\nabla u^y\right|} \right)$.

## 4 Results and Analysis

To verify the validity of proposed algorithm, it is compared with adaptive graph-cut algorithm using Fountain sequence. Four performance indicators [8] are used for contrasting the differences between two algorithms, which are Recall $\left( \frac{tp}{tp+fn} \right)$, Precision $\left( \frac{tp}{tp+fp} \right)$, Similarity $\left( \frac{tp}{tp+fn+fp} \right)$ and F-measure $\left( \frac{2tp}{2tp+fn+fp} \right)$, where $tp$ is the total number of true positives(reap. false negatives), ($tp+fn$ ) indicates the total number of items present in the ground truth, $fp$ is the total number of false positives, ($tp+fp$ ) indicates the total number of detected items.

**Table 1.** Detecation Rates of FOUNTAIN Sequence

| | Recall | Precision | F-measure | Similarity |
|---|---|---|---|---|
| proposed algorithm | 0.51 | 0.76 | 0.43 | 0.60 |
| adaptive graph-cut algorithm | 0.32 | 0.74 | 0.28 | 0.43 |

Fountain sequence is 160×128 pixels per frame. The background is dynamic caused by the fountain in this sequence, is classified as complex background condition. This data is usually used for checking the ability of algorithm dealing with complex background. In the experiment , some of the parameters set as follows: $e_k = 20$ , $d_k = 10$ , b=1 .The results of frame 1417,1438,1483 and 1491 are shown in Fig.2 The indicators of quantitative analysis are shown in Table 1. From the results we can see that comparing with adaptive graph-cut algorithm, the proposed algorithm is better than adaptive graph-cut algorithm regarding the ability of dealing with complex background, the proposed algorithm has a good performance in the aspect of quantitative detection indicators, as the introduction of the active contour energy.

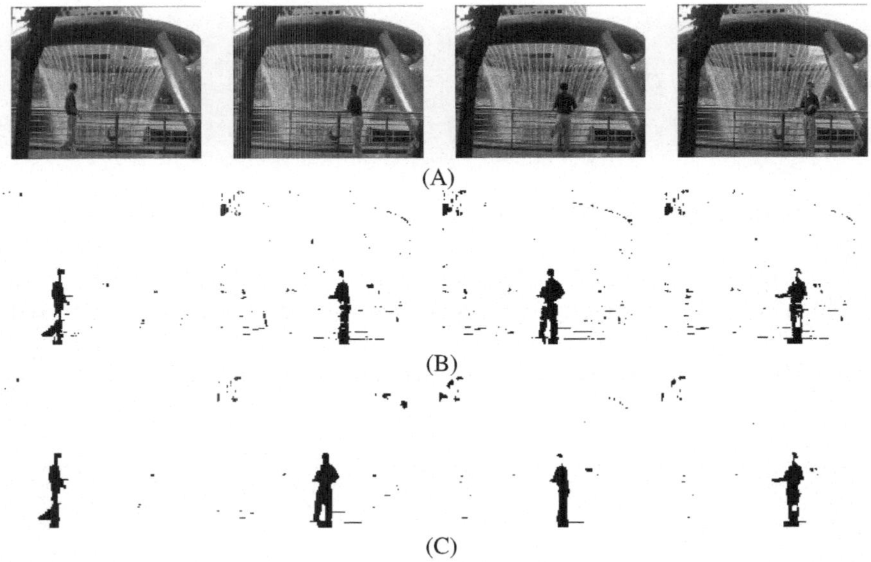


**Fig. 2.** Detection results of Fountain sequence: (A) Original image. (B) Detection results of adaptive graph-cut algorithm. (C) Detection results of proposed algorithm.

## 5 Conclusion

In this paper, graph-cut algorithm optimization with geodesic active contour is exploited to detect video moving objects. The proposed algorithm is successfully applied to video moving objects detection by introducing the kalman prediction of the moving objective pixels number and the foreground-background pixel-pairs number, also by the optimization of nodes flux with geodesic active contour. The proposed method solves the problem of moving objects detection by using optimized graph-cut algorithm very well. Experimental results show that the quantitative detection indicators of the proposed algorithm perform very well in complex background conditions.

**Acknowledgments.** This work is supported by Zhejiang Provincial Natural Science Foundation of China (No. LY12F010003).

## References


1. Garrett, Z., Saito, H.: Live video object tracking and segmentation using graph cuts. In: ICIP, pp. 1576–1579 (2009)
2. Nagahashi, T., Fujiyoshi, H., Kanade, T.: Video Segmentation Using Iterated Graph Cuts Based on Spatio-temporal Volumes. In: Zha, H., Taniguchi, R.-i., Maybank, S. (eds.) ACCV 2009, Part II. LNCS, vol. 5995, pp. 655–666. Springer, Heidelberg (2010)
3. Wang, C.-H., Guan, L.: Graph cut video object segmentation using histogram of oriented gradients. In: ISCAS, pp. 2590–2593 (2008)
4. Ken, F., Kouji, M., Akisato, K., Shigeru, T., Junji, Y.: Saliency-based video segmentation with graph cuts and sequentially updated priors. In: ICME, pp. 638–641 (2009)
5. Guo, C.-S., Wang, P.: Adaptive Graph-cut Algorithm to Video Moving Objects Segmentation. In: The 2nd International Congress on Image and Signal Processing, pp. 1–5 (2009)
6. Boykov, Y., Kolmogorov, V.: An Experimental Comparison of Min-Cut/Max-Flow Algorithms for Energy Minimization in Vision. IEEE Transactions on Pattern Analysis and Machine Intelligence, 1124–1137 (2004)
7. Tao, W.B., Tai, X.C.: Multiple piecewise constant with geodesic active contours (MPC-GAC) framework for interactive image segmentation using graph cut optimization. In: Image and Vision Computing, pp. 499–508 (2011)
8. Guo, C.-S., Gao, H.: Adaptive Graph-cut Algorithm to Video Moving Objects Segmentation Based on Euler's Elastica Model. In: The 4th International Congress on Image and Signal Processing, pp. 404–408 (2011)

# A Video Copy Detection Algorithm Based on Two-Level Features Measure*

Jie Dang, Bei Lu, and Jin-liang Yao

Institute of Computer Application Technology,
Hangzhou Dianzi University, Hangzhou, 310018, China
ericjay_5621@163.com,
{LuBei,YaoJinL}@hdu.edu.cn.com

**Abstract.** Video copy detection is a crucial technique for copyright protection. However, the main disadvantages of most existing approaches are high computational cost and low robustness. In this paper, we consider videos as a set of shots and propose a video copy detection framework that extracts video shots' overall features and spatiotemporal features. To effectively enhance the accuracy of final results, a coarse-to-precise filtration approach is proposed in this paper. In the coarse stage, the video copy shot retrieval is preformed by extracting the features of a video shot based on spatial-chromatic histograms. In the refined stage, the spatiotemporal features improved by quantization encoding are applied to the final verification. The combination of FLANN and "as early as possible to stop" process is adopted to accelerate the detection process in the coarse stage. The experimental results show that the proposed approach is effective in detecting video copies with promising precision and recall rate.

**Keywords:** video copy detection, spatial-chromatic histograms, spatiotemporal feature, FLANN.

## 1 Introduction

### 1.1 The Background of the Research

With the great developments of Internet applications and multimedia technologies, the number of digital video files is increasing fast. As we know, a digital video file can be copied, re-edited easily and distributed over Internet conveniently. Under this situation, the copyright protection and redundancy detection become important problems for users, governors and researchers. A video copy is defined as follows: a video copy is also one video obtained by some editorial operations, which is partly different from the source video in contrast, color, encoding, and so on. Generally, a video copy has identical content with the source one.

* This research is supported in part by National Natural Science Fund (No.61100100) and Zhejiang Provincial Natural Science Fund (No.Y1110232).

J. Lei et al. (Eds.): AICI 2012, LNAI 7530, pp. 450–457, 2012.

### 1.2 State of the Art

How to identify the source video and the copy one is an important problem in video copy detection domain. One solution is to apply watermarking technology [1] in which some additional information is embedded into the media file to be processed. So far no robust watermarking technology is available. Fortunately, content based copy detention (CBCD) presents an alternative to the watermarking approach to solve this problem. The primary advantage of CDBC compared to watermarking technology is the fact that the signature extraction process is not required before the media is distributed. In CBCD approach, the basic idea is to take advantage of a video itself information to determine whether this video is a duplication video or not from a video database. The video itself information can be divided into two categories: spatial features and spatiotemporal features. For spatial features, the purpose of extracting global or local descriptions [2] based on the video frame-level is to perform frames sequence matching. Global descriptions such as texture feature based on an image [3], color histogram intersection [4] and ordinal measure [5] can process the changes of the video encoding, frame resolution and zooming at low computational cost. However, its drawback is ineffective in post-processing, such as rotation, affine and complex editing operations. The other methods based on an image's local features are to make up those disadvantages. Joly et al. [6] presented a technique for content-based video identification based on the points of interest. In Joly' method, the local features are extracted around interest points by using Harris detector, and matched using an approximate nearest neighbors search. Zhao [7] applied PCA-SIFT descriptors for matching with approximate nearest neighbor search and trained SVMs to perform matched patterns learning. Unfortunately, the local features have still some disadvantages, including high complexity of the algorithm, a huge number of feature points, the serious mismatching between feature points. Since a video is a frame sequence that evolves with time, Kim and Vasudev [8] proposed a spatiotemporal matching method to take full advantage of video information and to increase the features' robustness, which combines spatial information by using ordinal signatures of each frame and temporal information which is calculated by the tendency of brightness variation from consecutive video frames. The above technique shows robustness to many distortions caused by digitization and encoding.

## 2 Our Approach

In this paper, we consider videos as a set of shots due to lots of redundant information and propose the CBCD framework which comprises of two main components: an off-line (preprocessing) and an online stage. The spatial-chromatic histograms and spatiotemporal features are combined and integrated into the framework. The whole CBCD framework is shown as Figure 1.

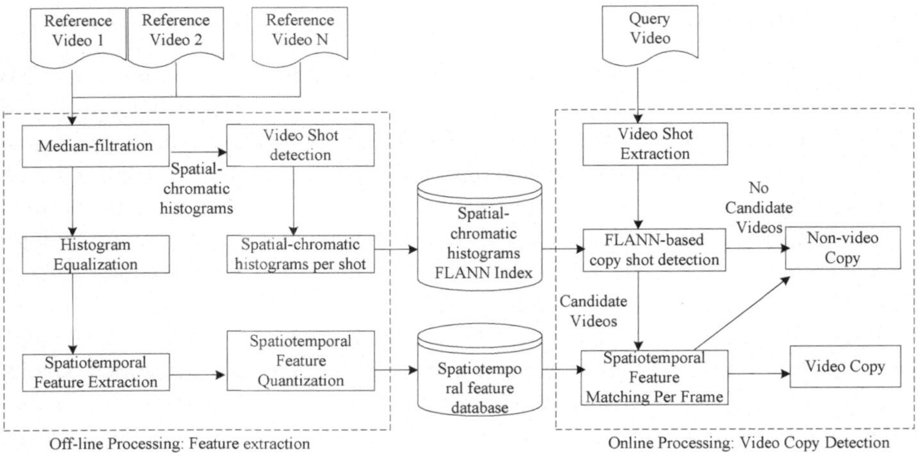


**Fig. 1.** The proposed CBCD framework

In the off-line stage, firstly, a median-filter is used to remove speckle noise. Secondly, the shot-boundaries of a reference video are detected by performing the spatial-chromatic histograms comparison in RGB space. Finally, we use FLANN [9] to store the index of the descriptor of a video shot based on the spatial-chromatic histograms. The second part is the spatiotemporal features extraction based on the frame-level which is stored for the refined filtration step. In the online stage, a coarse-to-precise filtration approach is applied. In the coarse step, the video copy shot retrieval is performed by using FLANN and "as early as possible to stop". If no any results are obtained, then the query process is terminated directly. Otherwise, if candidate videos can be retrieved, then we perform the fine stage using the spatiotemporal features to verify the final results.

## 2.1 Video Copy Shot Detection Based on Spatial-Chromatic Histograms

### 2.1.1 Spatial-Chromatic Histograms

The color histograms are trivial and robust to rotation and small camera viewpoint variations. Unfortunately, the pixels' spatial information has not been considered in color histograms. Therefore, two different images may have similar color distributions. In our approach, we adopt spatial-chromatic histograms (SCH) which integrates color and spatial information. SCH [10] is $c$-entries vectors where each entry $j$ contains the following information: a) the number of pixels with color $j$; b) the information on their spatial arrangement within an image; c) the standard deviation which measures the deviation degree of pixels from the bar-center.

The video shot-boundaries detection process based on SCH can be described as follows. Each detected video shot boundary is considered as dramatic changes of SCH between two adjacent frames. Instead of extracting the first/last frame of a SCH, the average values of SCH during the whole video shot are computed to enhance our approach's robustness.

### 2.1.2 Video Shot Index Storage Using FLANN and Copy Shot Searching

Inspired from the interest point matching technique, we consider every shot feature as a multidimensional point. Then FLANN (Fast Library for Approximate Nearest Neighbors) is applied to perform indexing. In Opencv2.3, we choose the automatic algorithm to return a set of *k* nearest video shots whose similarities are greater than the threshold in reference [10]. We use Jaccard coefficient to measure the similarity of two videos. Let A and B denote a set of video shots of one query video and its reference video respectively. Then the Jaccard coefficient can be computed by following equation.

$$Jaccard\ (A, B) = \frac{A \cap B}{A \cup B} \tag{1}$$

## 2.2 Spatial Temporal Sequence Matching for the Final Detection

The results of transformations and average intensity values of the query frames may be higher or lower than those of the original video, so the histogram equalization is used before the spatiotemporal feature extraction as shown in Fig.1

In fact, the proposed method in reference [8] described the tendency of average gray values of a sub-block which has an important impact on the final results. In this paper, we propose a novel method to improve the precision on the spatiotemporal features.

Defining $V_t = \{V[0], V[1], \ldots\ldots, V[n-1]\}$ as a video sequence with *n* frames and $V_t[i] = \{V^1[i], V^2[i], V^3[i], V^4[i]\}$ as the *i*th frame with *4* partitions, we also define $V_Q = \{p : p+n-1\}$ as a query subvideo in which the number of frames is *n* and the first position is the *p*th frame. The problem is: how to identify whether $V_Q$ is a copy in the V if a query video clip $V_Q$ is given. Firstly, the frames are partitioned into *m* equal-sized blocks and the average values of blocks $G_q^j[i]$ are calculated using following equation.

$$d(G_q^j[i], G_t^j[i]) = \frac{1}{C}\sum_{j=1}^{m}\left|G_q^j[i] - G_t^j[i]\right| \tag{2}$$

According to reference [8], we set *m*=4 and *c*=8. Then the spatial dissimilarity between two sequences is calculated by following equation (3).

$$D_S = (V_q, V_t[p : +n-1]) = \frac{\sum_{i=0}^{n-1} d\left(G_q^j[i] - G_t^j[i]\right)}{n} \tag{3}$$

With regard to temporal feature, a novel method based on reference [8] is proposed in which the temporal feature is improved by quantization encoding. We define $\partial^j$ as an angle which corresponds to the *j*th block and $T_{gap}$ as the time gap:

$$\partial^j = \arctan\left(\frac{G^j[i] - G^j[i-1]}{T_{gap}}\right) \tag{4}$$

Then we adopt quantitative methods to divide the range from -90°to 90°at an interval of 3°, and encode after mapping the result of equation(5) into the divided space. Fig.2 shows the process.

$$\Pi_{t/qi}^{j} = \begin{cases} K & 3n < \partial \leq 3(n+1) \\ 0 & \partial = 0 \\ -K & -3(n+1) \leq \partial < -3n \end{cases} \quad Where \quad n = \{1, \ldots 29\} \quad k = n+1 \tag{5}$$

Therefore, the dissimilarity of temporal features between a query and a reference video is calculated by equation (6).

$$D_T(V_q, V_t) = \frac{1}{4}\sum_{k=1}^{m}\left(\frac{1 - 2\max^{k}(d^{i})}{\left\lfloor \frac{n}{2} \right\rfloor}\right) \tag{6}$$

Where $d_i^j = \sum_{i=1}^{i} J(S_i^j > i)$, $J(B)$ is an indicator function of event B, i.e., $J(B)$ is 1 when B is true and 0 otherwise. $S_i^j = \Pi_{qK}^{j}$, if $\Pi_{ti}^{j} = a, (\Pi_{t-1}^{j})^{a} = i$ when $k = (\Pi_{t-1}^{j})^{i}$. Finally, the proposed methodology for measuring the dissimilarity between two videos relies on the combination of equations (3) and (6).

$$D(V_q, V_t) = \partial D_S(V_q, V_t) + (1-\partial) D_T(V_q, V_t) \tag{7}$$

Where $\partial \in [0,1]$ represents the weight factor balancing spatial and temporal distances. According to reference [8], it is set as 0.5 in our approach.

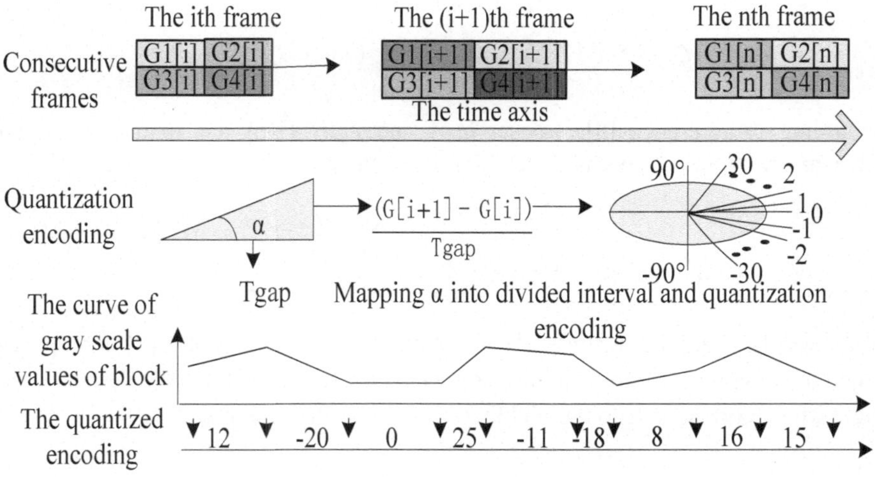


**Fig. 2.** The process of quantization encoding

# 3 Experiment and Analysis

## 3.1 Dataset and Evaluation Criteria

The database in our experiments consists of 60 target videos which are all from Internet with regard to five aspects, including news, sports, commercial advertisement, TV and education. Each video is in AVI format and comprised of 320*240 pixels. The length of every video is about 3-5 minutes. The query database consists of 30 segments in which 20 segments are from the reference database and 10 segments are out of the database. The follows are different transformations: (1) zoom 0.8/1.2; (2) contrast increased by 25% or decreased by 25%; (3) gamma with 5 variation; (4) Gaussian radiue-5 blur; (5) insertion of subtitle; (6) insertion of a logo; (7) crop; (8) combination of some of them (see Fig .3). The precision P and recall rate R are used to evaluate our proposed approach which can be computed by following equation respectively.

$$R = \frac{C_d^r}{C_d}, P = \frac{C_d^r}{C^r} \tag{8}$$

**Fig. 3.** Transformations of query videos

## 3.2 Evaluation of the Coarse Detection

The coarse detection with four different *Fs* values in reference [10] and Jaccard coefficient computation are performed and the results are listed in Table 1.

**Table 1.** The recalls and precisions corresponding to different threshold values

| *Jaccard* | 0.75 | | 0.80 | | 0.85 | | 0.90 | |
|---|---|---|---|---|---|---|---|---|
| *Fs* | *R* | *P* | *R* | *P* | *R* | *P* | *R* | *P* |
| 1.75 | 0.92 | 0.15 | 0.89 | 0.21 | 0.82 | 0.23 | 0.80 | 0.29 |
| 1.80 | 0.89 | 0.16 | 0.85 | 0.28 | 0.78 | 0.29 | 0.72 | 0.31 |
| 1.85 | 0.86 | 0.23 | 0.81 | 0.38 | 0.76 | 0.39 | 0.70 | 0.41 |
| 1.90 | 0.82 | 0.29 | 0.79 | 0.43 | 0.73 | 0.45 | 0.68 | 0.47 |

Because the first filtration is coarse detection, we mainly focus on the value of recall which means the candidate videos should be complete (having the higher recall value). Based on Table 1, we can conclude that the number of candidate videos decreases while the threshold values increases. This is because the higher threshold values exclude more non-copy videos. Note that when Jaccard=0.8, the recall and precision obtained reach the optimal balance. Therefore, in the first detection, the Jaccard parameter has been set as 0.8.

### 3.3 Evaluation of the Precise Detection

In this section, we focus on the precision corresponding to the precise detection. Table 2 displays the results using *Jaccard*=0.8.

**Table 2.** Results of the precise detection

| D | R | P |
|---|---|---|
| 0.1 | 0.79 | 0.82 |
| 0.15 | 0.81 | 0.78 |
| 0.2 | 0.83 | 0.73 |
| 0.25 | 0.89 | 0.69 |

At the same time, we compare the results of our approach with the results in reference [8]. In reference [8], when $\alpha$=0.5 and $D$=0.1, the experimental results are optimal shown in Fig.4.

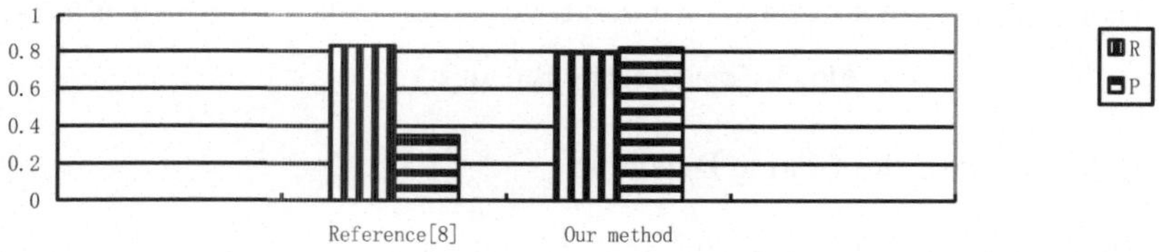


**Fig. 4.** Comparison between our approach and reference [8]

From Fig.4, we can know that our proposed method is more effective than the method in reference [8].

### 3.4 Computational Complexity

The main computational costs are the extractions of SCH and spatial temporal features. For example, a video has $N$ frames, $n*m$ resolution and each frame is quantized into a color space using the palette with 8 colors. We extract SCH and spatial temporal features at a cost of $O\ (8n*m*N)$ and $O\ (32N)$ respectively. Because the off-line method is integrated into the extraction of video database with "as early as possible to stop" strategy, the cost time is saved dramatically.

## 4 Conclusions

In this paper, we proposed a framework for content-based video copy detection. It consists of two parts: the off-line process stage and the online query stage. During the off-line process stage, we extract a video's spatial-chromatic histograms features and spatialtemporal features, and construct two kinds of corresponding feature databases respectively. During the online query stage, the strategy combined FLANN with "as early as possible to stop" is applied to accelerate the process of the overall detection. The experimental results are compared with the results in reference [8]. From the compared results, we can know our proposed approach has better performance and more effective than that of reference [8].

However, the features used in our method are not suitable to some more severe transformations, such as picture-in-picture, the Bokeh. Moreover, the time complexity is still less than the real-time. Therefore, in future work, we will focus on three aspects: (1) Look for better robustness features; (2) Simplify the time complexity; (3) Optimize the detection framework.

## References


1. Langelaar, G., Setyawan, I., Langedijk, R.: Watermarking digital image and video data. A state-of-the-art overview, IEEE Signal Processing Magazine 17(5), 20–46 (2000)
2. Law-To, J., Joly, A.: Video Copy Detection: a Comparative Study. In: Conference on Image and Video Retrieval: Proceedings of the 6th ACM International Conference on Image and Video Retrieval, July 09-11, pp. 371–378 (2007)
3. Wan, H., Chowdhury, M., Hu, H., et al.: Texture feature and its application in CBIR. Journal of Computer- Aided Design & Computer Graphics 15(2), 195–199 (2003) (in Chinese)
4. Satoh, S.: New video analysis based on identical shot detection. In: Proceedings of the IEEE International Conference on Multimedia and Expo (ICME 2002), vol. 1, pp. 69–72 (2002)
5. Bhat, D.N., Nayar, S.K.: Ordinal measures for image correspondence. IEEE Trans. Pattern Anal. Mach. Intell. 20(4), 415–423 (1998)
6. Joly., A., Frelicot., C., Buisson, O.: Robust content-based video copy identification in a large reference database. In: Proceedings of ACM International Conference on Image and Video Retrieval, CIVR 2003, pp. 414–424 (2003)
7. Zhao, W.L., Ngo, W., Tan, H.K., Wu, X.: Near-duplicate keyframe identification with interest point matching and pattern learning. IEEE Transactions on Multimedia 9(5), 1038–1048 (2007)
8. Kim, C., Vasudev, B.: Spatiotemporal sequence matching for efficient video copy detection. IEEE Transactions on Circuits and Systems for Video Technology 15(1), 127–132 (2005)
9. Muja, M., Lowe, D.G.: Fast Approximate Nearest Neighbors with Automatic Algorithm Configuration. In: International Conference on Computer Vision Theory and Applications (VISAPP 2009), pp. 331–340 (2009)
10. Cinque, L., Ciocca, G., Levalldi, S., Pellicano, A., Schettimi, R.: Color-based image retrieval using spatial-chromatic histograms. Image and Vision Computing 19, 879–986 (2001)

# Analyzing Feature Selection of Chromatographic Fingerprints for Oil Production Allocation

Zongrui Yang, Wei Wu*, Mingliang Gao, Qizhi Teng, and YouSong He

Image Information Institute, College of Electronics and Information Engineering, Sichuan University, Chengdu Sichuan 610064, China
wuwei@scu.edu.cn

**Abstract.** Commingling is employed in the petroleum industry to enhance oil recovery and reduce costs. It is of great importance to monitor the production of each oil well oilfields. Nowadays, more and more oilfields use chromatographic fingerprint to estimate single-zone production allocation. However, how to select the features of chromatographic fingerprint remains an unresolved problem. So far, the features of chromatographic fingerprint are still selected by the professional experts. This leads to a certain degree of subjectivity, which easily results in a poor performance of estimation the single-zone production. To our knowledge, there are few researches exploiting the selection of the features of chromatographic fingerprints. In order to select the features of chromatographic fingerprint, principal component analysis (PCA) method, linear correlation method and the variable importance method used in random forest are exploited in this paper. Meanwhile, a joint feature selection method, which combines the linear correlation method and the variable importance method, is proposed. Experimental results with oil samples from an oil field in Hainan offshore basin show that the proposed method can achieve good results.

**Keywords:** Commingled oil well, Single-zone production contribution, Chromatography fingerprinting, Feature selection, Principal component analysis, Random forest.

## 1 Introduction

Commingling has been employed in the petroleum industry to enhance oil recovery and reduce costs in recent years. There are many examples of commingling. For example, we use a single pipeline to transport production from different production zones or use a single separator tank to mix gas/water/oil from different wells [1]. In this case, it is indispensable to monitor dynamically the production of each single-zone in oilfields.

* Corresponding author.
[Supported By Open Fund(PLC200902) of State Key Laboratory of Oil and Gas Reservoir Geology and Exploitation (Chengdu University of Technology) ].

J. Lei et al. (Eds.): AICI 2012, LNAI 7530, pp. 458–467, 2012.

Many traditional methods are proposed to monitor dynamically the production of each single-zone. However, those traditional methods are very expensive and need a long-time cycle to monitor. Furthermore, when those methods are applied, the reservoirs are easily undermined. In addition, some traditional methods even have to shut down the oil wells. Whereas many researches have demonstrated that gas chromatographic fingerprint (also called chromatographic fingerprint) is a powerful tool for monitoring the single-zone production [2, 3, 4, 5]. Compared with traditional methods, the gas chromatographic fingerprint method has many advantages (its cost is very low and its speed is fast) and has a good prospect in production estimation. Meanwhile, it can be carried out in laboratory.

Gas chromatographic fingerprint is a technology to determine production of each single-zone in the commingled oil wells by using oil composition differences [6]. Since the crude oils with different reservoirs or even same reservoir but different positions, their hydrocarbon chromatograms (also called chromatographic fingerprint) may be difference according to organic geochemistry [1, 7]. The difference of the hydrocarbon chromatograms provides a theoretical basis for single-zone production estimation. The chromatographic fingerprint of pure oil from different reservoirs in a certain well is mixed and analyzed to estimate the production of each single-zone. An example of the chromatographic fingerprints of different reservoirs of crude oil is shown in Fig.1.

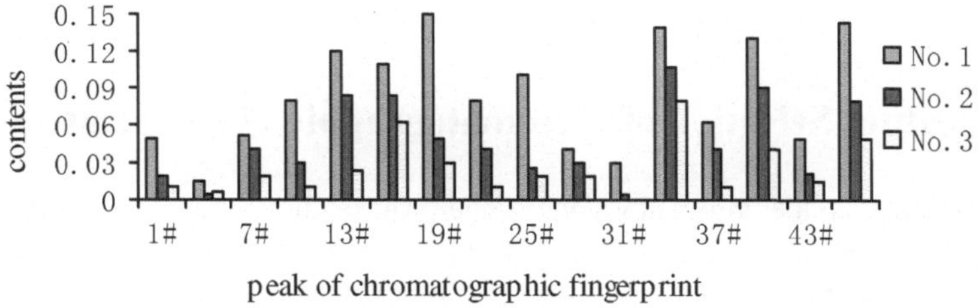


**Fig. 1.** Example of the chromatographic fingerprints of different reservoirs of crude oil

In Fig.1, No.1, No.2 and No.3 represent the different reservoirs and the symbol "#" denotes hydrocarbon compound. It can be seen from Fig.1 that the difference of chromatographic fingerprint in different reservoirs are significantly different.

Reservoir geochemistry have been applied to solve the production allocation of commingled wells [6, 9, 10, 11] since 1987. Production monitoring was applied to manage problem for enhancing oil recovery [11, 12, 13]. The chromatographic fingerprint technology was firstly introduced to the field of single–zone production estimation by Kaufman et al. [6]. This technology has been widely applied in oilfields by many researchers [14, 15, 16] and have achieved good results. Xue and Wu [17] applied the support vector regression to compute the three-zone crude oil production and achieved satisfactory results.

When using chromatographic fingerprint to estimate crude oil production allocation, there are two key technologies. One is to select the features of chromatographic fingerprint, i.e. select applicable hydrocarbon compound; the other is to build a suitable estimation model for crude oil production allocation by using the selection methods of chromatographic fingerprint features. Many researchers [1, 7, 10] focused only on building the estimation model instead of the selection methods of chromatographic fingerprint features. However, the selection of chromatographic fingerprint features plays an important role in the accuracy of the production estimation. Given the same estimation model, the results could be different when using different features.

Crude oil is composed of thousands of compounds. Generally, the professional experts could choose some representative compounds as the features of crude oil. i.e. the chromatographic fingerprint. However, there is certain subjectivity and the selected features may be not suitable for building the estimation model.

To solve these problems, we focus on how to select the features of chromatographic fingerprint in this paper. In order to select such features, this paper adopts the principal component analysis [18], the linear correlation method (PCA) [19] and the variable importance method used in random forest [20] to select the features of chromatographic fingerprint. Finally, we propose a new feature selection method by combining the linear correlation and variable importance method. Experimental results with oil samples from an oil field in Hainan offshore basin show that the proposed method can achieve better results than other methods.

The rest of the paper is organized as follows. In section 2, the details of different approaches are described. In section 3, the experiment results and discussions are presented. In section 4, concluding remarks are presented.

# 2 Feature Selection of Chromatographic Fingerprint

In order to analyze the chromatographic fingerprint features, this paper firstly adopts the principal component analysis (PCA), the linear correlation method and the variable importance method used in random forest to exploit how to select the features of chromatographic fingerprint. Then, a feature selection method by combining the linear correlation method and the variable importance method is proposed.

## 2.1 Feature Selection Based on PCA

PCA is a multivariate statistical analysis method [19] where minority features are selected from a number of features, and these minority features are used to replace the original features. Suppose there are $n$ samples, and each sample has $p$ features. The sample data matrix of $n \times p$ is as follows:

$$X = \begin{bmatrix} x_{11} & x_{12} & \cdots & x_{1p} \\ x_{21} & x_{22} & \cdots & x_{2p} \\ \vdots & \vdots & & \vdots \\ x_{n1} & x_{n2} & \cdots & x_{np} \end{bmatrix} \quad (1)$$

where $x_1, x_2, \cdots, x_p$ and $z_1, z_2, \cdots, z_m (m \le p)$ refer to the original features and the new features respectively. The new features can be obtained by:

The principles of the coefficient $l_{ij}$ $(i = 1, \cdots, m, j = 1, \cdots, p)$ are as follows:

$$\begin{cases} z_1 = l_{11}x_1 + l_{12}x_2 + \cdots + l_{1p}x_p \\ z_2 = l_{21}x_2 + l_{22}x_2 + \cdots + l_{2p}x_p \\ \cdots \\ z_m = l_{m1}x_1 + l_{m2}x_2 + \cdots + l_{mp}x_p \end{cases} \tag{2}$$

(1) $z_i$ is uncorrelated with $z_j$ $(j \ne i, j = 1, 2, \cdots, m)$.

(2) The first principal component $z_1$ is the largest variance in all the linear combination of $x_1, x_2, \cdots, x_p$. The following component $z_i$ $(i = 2, \cdots, m)$ is also the largest variance. However, it must be uncorrelated with the preceding components.

## 2.2 Feature Selection Based on the Linear Correlation

The linear correlation method is a method measuring the linear strength of the linear association between features [18]. The higher the linear correlation coefficient is, the stronger the linearity of features is. The linear correlation between them can be calculated by:

$$r = \frac{\frac{\sum_{i=1}^{n}(x_i' - \overline{x'})(x_i - \overline{x})}{n}}{\sqrt{\frac{\sum_{i=1}^{n}(x_i' - \overline{x'})^2}{n}} \times \sqrt{\frac{\sum_{i=1}^{n}(x_i - \overline{x})^2}{n}}} = \frac{\sum_{i=1}^{n}(x_i' - \overline{x'})(x_i - \overline{x})}{\sqrt{\sum_{i=1}^{n}(x_i' - \overline{x'})^2}\sqrt{\sum_{i=1}^{n}(x_i - \overline{x})^2}} \tag{3}$$

where $x_i$ is the $i$-th feature value of chromatographic fingerprint (also called linear value), which is obtained by using linear correlation method; $x'$ is the actual feature value of chromatographic fingerprint; $\overline{x}$ and $\overline{x'}$ denote the mean value of linear value and actual value respectively.

### 2.3 Feature Selection Based on the Variable Importance Used in Random Forest

Random forest proposed by Leo Breiman [20] in 2001 is a powerful ensemble classification regression algorithm. It consists of a large number of decision trees. So far, random forest has already been applied in many fields, such as ecology, astronomy, drug discovery, etc.

Compared with other classification regression algorithms, random forest has the following characteristics [20, 21]: (1) It has a good estimation performance and it maintains accuracy when some data are missing; (2) It can be used to determine the variable importance of each feature; (3) It has good generalization ability because of not over fitting.

When random forest generates bootstrap training set using the Bagging method, nearly 37% data of original training set do not appear in the bootstrap training set. These data are called out-of-bag (OOB) of the bootstrap training set. This part of data can be used not only to estimate the generalization error of the model, but also to calculate the variable importance and then select the features.

The steps of the variable importance method are as follows:

(1) A model of random forest is created.

(2) Test the model using OOB data and get an OOB estimation accuracy rate $AAC$. Suppose the samples have a total of $M$ features, for the $i$-th feature, randomly disrupt the eigenvalue of the feature in all samples (or randomly add some noises to the feature). Then test the original model using the changed OOB data and get another accuracy rate $AAC_i$. The variable importance is defined as:

$$diff_i = ACC - ACC_i \tag{4}$$

A variable importance set $\{diff_1, diff_2, \cdots, diff_m\}$ can be obtained through the above method. The greater the value $diff_i (i = 1, 2, \cdots, m)$ is, the greater influence the variable has on estimation accuracy rate.

### 2.4 The Proposed Method

Because of the limit of a single method, the features, which are selected by using the single methods, may not be able to achieve satisfactory results. To get a better performance, we propose a feature selection method, which selects the features by combining the linear method and the variable importance method. The reason why we do not select PCA is that the features having been transformed by PCA are different from the original features. So it is difficult to use the intersection operations between the features obtained by PCA and other methods. We select the linear correlation method and the variable importance method as A and B method in the experiment. The feature set $S_A = \{A_1, A_2, \cdots, A_c\}$ and $S_B = \{B_1, B_2, \cdots, B_m\}$ are obtained by A and B method respectively. Then the feature $S_u$ obtained by using the proposed method can be expressed as:

$$S_u = (R_A \cap R_B) \cup R_A^P \cup R_B^P, \quad i = 1, 2, \cdots, I, \quad k = 1, 2, \cdots, K \tag{5}$$

Where

$$R_A \cap R_B = \{t \mid t \in R_A \wedge t \in R_B\} \tag{6}$$

$R_A^P = \{A_1, A_2, \cdots, A_I\}$ and $R_B^P = \{B_1, B_2, \cdots, B_K\}$ are the important features selected by A method and B method respectively. $I$ and $K$ are the numbers of features in feature set $R_A^P$ and $R_B^P$.

## 3 Experimental Results and Discussions

Crude oil samples from Wenchang13-1/2 oilfields in Hainan offshore basin are selected in this paper. These samples are classified into 12 groups, each group having 25 chromatographic fingerprint features numbered 1-25. Experiments are carried out by using the original features, the features obtained by using PCA, the linear correlation method, the variable importance method and the proposed method. The Generalized Additive Model (GAM) [22, 23] is adopted to build the regression model with different features selected by using different feature selection methods. GAM is a nonlinear regression model extension of generalized linear models [24], which has a wide range of applications in the field of economic, medical, etc. It can flexibly characterize the complex relationship among the features to meet the nonlinear production estimation model. Therefore, GAM is used to model the selected features. To verify the performance of the selected features, leave-one-out cross validation is conducted.

The root mean square error is adopted to assess the performance of each methods, the root mean square error is expressed as:

$$RMSE = \sqrt{\frac{1}{N}\sum_{i=1}^{N}\left(y_i - \hat{y}_i\right)^2} \tag{7}$$

Where N is the number of samples and $y_i$, $\hat{y}_i$ is the actual value and the estimation value respectively. Comparison of various feature selection methods with RMSE is shown in Fig. 2.

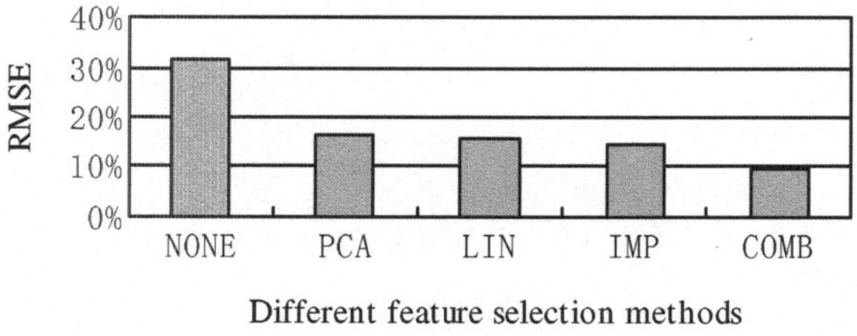


**Fig. 2.** Comparison of various feature selection methods with RMSE

In Fig.2, "NONE" refers to the error rate when using the original features. "PCA", "LIN", "IMP", "COMB" refer to the error rates when using the features obtained by using PCA, the linear correlation method, the variable importance method and the proposed method, respectively. It can be observed from Fig.2 that when using the original features, the RMSE is 32.2% and the result is the worst. When using PCA method, the RMSE is 16.36% and the performance has been greatly improved compared with using the original features. The correlation coefficient of each feature obtained by using the linear correlation method is shown in Table 1. We select those correlation coefficients, which are greater than 0.95(feature number is 2, 3, 5, 8, 9, 10, 13) to model the regression. The RMSE is 15.6% and the performance is slightly better than PCA.

**Table 1.** Linear correlation

| | | | | | | | | | |
|---|---|---|---|---|---|---|---|---|---|
| Feature number | 1 | 2 | 3 | 4 | 5 | 6 | 7 | 8 | 9 |
| Correlation coefficient | 0.918 | 0.973 | 0.956 | 0.938 | 0.968 | 0.944 | 0.935 | 0.958 | 0.957 |
| Feature number | 10 | 11 | 12 | 13 | 14 | 15 | 16 | 17 | 18 |
| Correlation coefficient | 0.965 | 0.727 | 0.952 | 0.961 | 0.946 | 0.937 | 0.938 | 0.942 | 0.921 |
| Feature number | 19 | 20 | 21 | 22 | 23 | 24 | 25 | | |
| Correlation coefficient | 0.956 | 0.955 | 0.927 | 0.902 | 0.914 | 0.907 | 0.849 | | |

The feature importance analysis of the variable importance method is shown in Fig.3. When selecting the important features, the negative features of variable importance are excluded and the high features of variable importance are selected as the final features. Based on the above method, we can obtain the final features, i.e. 1, 2, 3, 4, 9, 11.

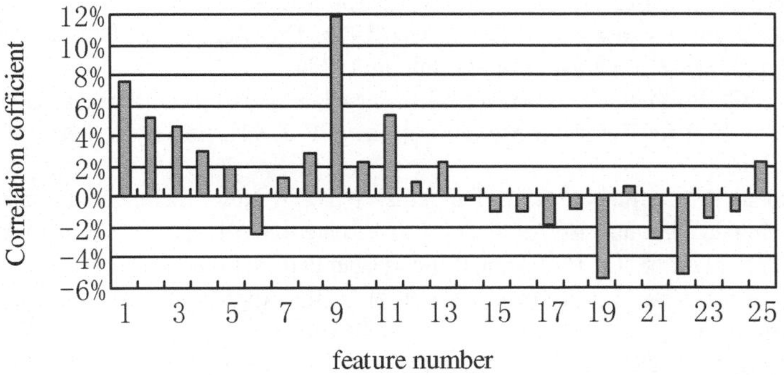


**Fig. 3.** Feature importance analysis of the variable importance method

Finally, we use the proposed method to select features. The number of the important features set $R_A^P$ selected by the linear correlation method is 6, and the number of $R_B^P$ selected by the variable importance method is 3. By formula (5) we can obtain the final features, i.e. 2, 3, 5, 8, 9, 10. When using the proposed method to the regression estimation, the RMSE is 9.96% and the performance is significantly better than other methods.

## 4 Conclusion

This paper introduces the production estimation applications of chromatographic fingerprint and summarizes the advantages and disadvantages of various methods. Then the deficiency of features selection of chromatographic fingerprint technology is analyzed. In order to solve this problem, this paper analyzes several feature selection methods including PCA, the linear correlation method, the variable importance used in random forest method and the proposed method. Experimental results from an oil field in the Hainan offshore basin show that the proposed method achieves better performance than other methods. It is indicated that the proposed method has a great application value in the field of production estimation and crude oil development.

## References

1. Hwang, R.J., Baskin, D.K., Teeerman, S.C.: Allocation of Commingled Pipeline Oils to Field Production. Organic Geochemistry 31, 1463–1474 (2000)
2. Kaufman, R.L., Ahmed, A.S., Elsinger, R.J.: Gas Chromatography as a Development and Production Tool for Fingerprinting Oils from Individual Reservoirs: Applications in the Gulf of Mexico. In: Schumacker, D., Perkins, B.F. (eds.) Proceedings of the 9th Annual Research Conference of the Society of Economic Paleontologists and Mineralogists, New Orleans, pp. 263–282 (1990)
3. Hwang, R.J., Ahmed, A.S., Moldowan, J.M.: Oil Composition Variation and Reservoir Continuity. Organic Geochemistry 21, 171–188 (1994)

4. Nederlof, P.J., Gijsen, M.A., Doyle, M.A.: Application of Reservoir Geochemistry to Field Appraisal. In: Al-Husseini, M.I. (ed.) Geo 1994. The Middle East Petroleum Geosciences, vol. 2, pp. 709–722. Gulf-Petrolink, Bahrain (1994)
5. Nicolle, G., Boibien, C., ten-Haven, H.L., Tegelaar, E., Chavagnac, P.H.: Geochemistry: A powerful Tool for Reservoir Monitoring. In: SPE 37804. Society of Petroleum Engineers, Bahrain (1997)
6. Kaufman, R.L., Ahmed, A.S., Hempkins, W.B.: A New Technique for the Analysis of Commingled Oils and its Application to Production Allocation Calculations. Paper IPA 87-23/21, 247–268. In: 16th Annual Indonesian Petroleum Assoc. (1987)
7. Wen, Z.-G., Zhu, D., Tang, Y.-J., et al.: The Application of Gas Chromatography Fingerprint Technique to Calculating Oil Production Allocation of Single Layer in the Commingled Well. J. Chinese Journal of Geochemistry 24(3), 257–261 (2005) (in Chinese)
8. Bazan, L.W.: The Allocation of Gas Well Production Data Using Isotope Analysis. In: Proceedings of the 1998 Society of Petroleum Engineers Gas Technology Symposium, SPE Paper40032, Calgary, Alberta, Canada, March 15-18, pp. 1–9 (1998)
9. Chang, X.C., Zhang, Z.H., Wang, T.G.: A Study on the Productivity Contribution of Single Horizon in Multiple Horizons Exploitation by Reservoir Geochemistry. Journal of Xi'an Petroleum Institute 15(4), 35–38 (2000) (in Chinese)
10. Chang, X., Li, Z.: Geochemical Surveillance of the Linnan Oil Field with oil Fingerprint. Enegry Exploration & Exploration 28(4), 279–294 (2010)
11. Larter, S.R., Aplin, A.C.: Reservoir Geochemistry: Methods, Applications and Opportunities. In: England, W.A., Cubitt, J. (eds.) The Geochemistry of Reservoirs, vol. 86, pp. 5–32. Geological Society Special Publication, London (1995)
12. Milkov, A.V., Goebel, E., Dzou, L., Fisher, D.A., Kutch, A., McCaslin, N., Bergman, D.F.: Compartmentalization and Time-lapse Geochemical Reservoir Surveillance of the Horn Mountain Oil Field. In: AAPG Bulletin, vol. 91, pp. 847–876. Deep-Water Gulf of Mexico (2007)
13. Weissenburger, K.S., Borbas, T.: Fluid Properties, Phase and Compartmentalization: Magnolia field case study, deepwater Gulf of Mexico, U.S.A. In: Cubitt, J.M., Endland, W.A., Larter, S.R. (eds.) Understanding Petroleum Reservoirs: Towards an Integrated Reservoir Engineering, vol. 237, pp. 231–256. Geological Society Sepcial Publication, London (2004)
14. He, W., Wang, P., Liu, Y., Li, C.: Proportion Fingerprint Palte of Single-source Oil Sample and its Application to Analysis of Commingled wells. J. Petroleum Exploration and Development 28, 82–83 (2001) (in Chinese)
15. Wang, Z., Fu, X., Lu, S., et al.: A Monitoring Technique for Production Allocation of Commingled Production Well with Liquid Chromatography. J. Geochimica 29, 297–301 (2000) (in Chinese)
16. Chen, S., Lin, F., Chen, X., et al.: Calculating Xeparate Zone Peoduction of Commingled Wells Using Oil Chromatography Fingerprint technique. J. Petroleum Exploration and Development 26, 63–71 (1999) (in Chinese)
17. Xue, L., Wu, W., et al.: A Forecasting Model of Productivity Contribution in single Zone of Multi-Zone Production Based on Support Vector regression. Journal of Petrochemical Universities 19(4) (2006) (in Chinese)
18. Wold, S., Exbensen, K., Geladi, P.: Principal component analysis. Chemometrics and Intelligent Laboratory Systems 2, 37–52 (1987)

19. Karl, P.: On Lines and Planes of Closest Fit to Systems of Points in Space. Philosophical Magazine 2(6), 559–572 (1901)
20. Breiman, L.: Random forests. J. Machine Learning 45(1), 5–32 (2001)
21. Richard Cutler, D., Edwards, T.C., et al.: Random Forest for Classification in Ecology. Ecology 88(11), 2783–2792 (2007)
22. Hastie, T.J., Tibshirani, R.J.: Generalized Additive Models. M. Chapman and Hall, London (1990)
23. Stone, C.J.: Consistent Nonparametric Regression. J. The Annals of Statistics 5, 595–620 (1997)
24. Yee, T.W., Mitchell, N.D.: Journal of Vegetation Science 2, 587–602 (1991)

# Studies on Wheat Pests Image Segmentation Based on Variational Level Set*

Jian Li[1], Lijuan Wang[1], and Yi Li[2]

[1] College of Information and Electric Engineering,
Shaanxi University of Science & Technology, Xi'an, China
[2] Unified Communication Development Department,
Huawei Xi'an Institute, Xi'an, China
lijianjsj@sust.edu.cn

**Abstract.** In order to get better wheat pests outline, wheat pests image is segmented by variational level set method in this paper. This kind of level set function corrects the deviation between the level set function and the signed distance function, and it needn't re-initialization during the evolution, so it completely eliminates the procedure of level set re-initialization and improves the efficiency of image segmentation. The effectiveness of this method is verified by MATLAB simulation experiments.

**Keywords:** wheat pests image, image segmentation, variational level set.

## 1 Introduction

Wheat pests are important factors in the wheat production, so it has important practical significance and application value on the identification of wheat pests [1]. Image segmentation is an important part in the identification of wheat pests image [2]. The traditional method of image segmentation uses bottom knowledge of image to realize it simple and quick, but it lacks of robustness, and mostly segmentation results are not ideal because of the influence of noise and boundary contours. Since the geometric active contour model basing on level set method was proposed, it got widely attention of the researchers [3]. In the traditional level set geometry active contour model, it must kept the evolving level set function as an approximate signed distance function during the evolution, and needed to be re-initialized in a certain cycle[4,5]. In this paper ,we use a variational formulation to correct the deviation between the level set function and the signed distance function to realize the level set function without re-initialization during the evolution, and the level set function do not need to be the signed distance function [6].

* This paper is supported by Shaanxi science research project (2009k02-08) and Shaanxi provincial special scientific research project (09JK345).

J. Lei et al. (Eds.): AICI 2012, LNAI 7530, pp. 468–474, 2012.

## 2 Theoretical Basis

Let $I(x,y)$ be an image, $\phi$ be the level set function, and $\Omega$ be the image domain.

The edge indicator function $g$ [7] is defined by:

$$g(x,y) = \frac{1}{1+|\nabla G_\sigma * I|} \tag{1}$$

Where $G_\sigma$ is the Gaussian kernel with standard deviation $\sigma$, $*$ is Convolution molecules, $\nabla$ is Space gradient operator.

Internal energy functional [8] is defined by:

$$\mathrm{P}(\phi)=\int_\Omega \frac{1}{2}(|\nabla \phi|-1)^2 d_x d_y \tag{2}$$

External energy functional [8] is defined by:

$$\mathrm{E}_{g,\lambda,v} = \lambda \int_\Omega g\delta(\phi)|\nabla \phi| d_x d_y + v \int_\Omega g\mathrm{H}(-\phi) d_x d_y \tag{3}$$

Where $\lambda$ and $v$ are weighting coefficients (are constants), $\delta$ is the univariate Dirac function, and H is the Heaviside function.

Energy functional [8] is defined by:

$$\mathrm{E}(\phi) = \mu \mathrm{P}(\phi) + \mathrm{E}_{g,\lambda,v}(\phi) \tag{4}$$

Where the internal energy $\mathrm{P}(\phi)$ use correct the deviation between the level set function and the signed distance function, $\mu > 0$ is weighting coefficients; while the external energy $\mathrm{E}_{g,\lambda,v}$ drives the zero level set toward the object boundaries.

Through minimization the energy functional, we can get the partial differential equations which control the level set evolution [8]:

$$\frac{\partial_\phi}{\partial_t} = \mu\left[\Delta\phi - div\left(\frac{\nabla\phi}{|\nabla\phi|}\right)\right] + \lambda\delta(\phi)div\left(g\frac{\nabla\phi}{|\nabla\phi|}\right) + vg\delta(\phi) \tag{5}$$

## 3 Algorithm Design

Step1: Take $u$ int 8 data in the original image into *double* data.

Step2: Let $\phi$ be the level set initialization function, set the initial contour of image.

Here, the level set initialization function [9] is defined by:

$$\phi(x,y)=\begin{cases}-\rho,(x,y)\in\Omega_0-\partial\Omega_0\\ 0,(x,y)\in\partial\Omega_0\\ \rho,(x,y)\in\Omega-\Omega_0\end{cases} \quad (6)$$

Where $\rho > 0$ is a constant, $\Omega_0$ is the branch area of the image domain $\Omega$, $\partial\Omega_0$ is the boundary of the branch area $\Omega_0$.

Step3: With Gaussian filter function, Convolution molecules and Space gradient operator get edge indicator function $g$.

Step4: Start level set evolution. The process of level set evolution iteration: First, we get a new level set function $\phi$ through calling *EVOLUTION* function; Second, we do evolution iteration for the new level set function $\phi$, and then update the contour line; finally, we get image segmentation results which that required after a certain number of iterations.

In the level set evolution iteration, *EVOLUTION* function main relates with the size of *Dirac* function width $\varepsilon$, time step $\tau$ and weighting coefficients $\mu,\lambda,v$. Here, we use the regularization *Dirac* function $\delta_\varepsilon(x)$ [9], it is defined by:

$$\delta_\varepsilon(x)=\begin{cases}0,|x|>0\\ \frac{1}{2\varepsilon}\left[1+\cos\left(\frac{\pi x}{\varepsilon}\right)\right],|x|\le\varepsilon\end{cases} \quad (7)$$

Where $\varepsilon > 0$ is a constant.

## 4 Experimental Analysis

In this paper, we do simulation experiment based on Matlab [10]. In order to verify the feasibility of the method, some experimental treatments are done under different conditions with three wheat pests images of pleonomus canaliculatus, dolycoris baccarum and chilo suppresslis. First, we make $\rho = 4$ in the level set initialization function, and set the initial contour of image; second, we define the edge indicator function $g$; Third, we make $\tau = 5,\mu = 0.04$ ( Note: The product of $\tau$ and $\mu$ must be less than 0.25 [10] ), and do level set evolution iteration by changing the size of $\varepsilon,\lambda,v$. The purpose is to get better image segmentation results. Initial contour is shown in figure 1.

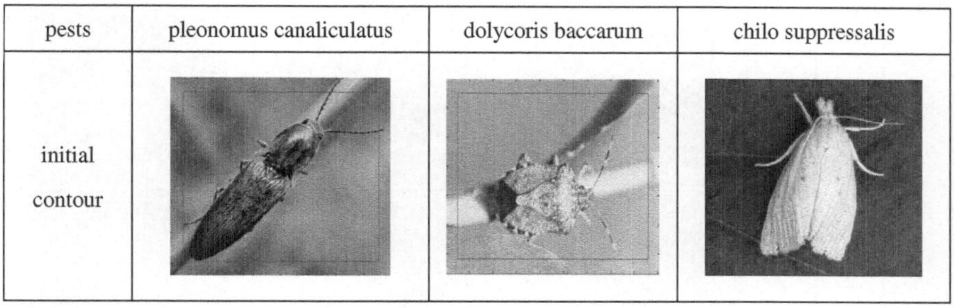


**Fig. 1.** Initial contour

Make $\varepsilon$ fix, change the size of $\lambda$ and $v$, after a certain number of iterations, we get the following segmentation results. Segmentation results are shown in figure 2.

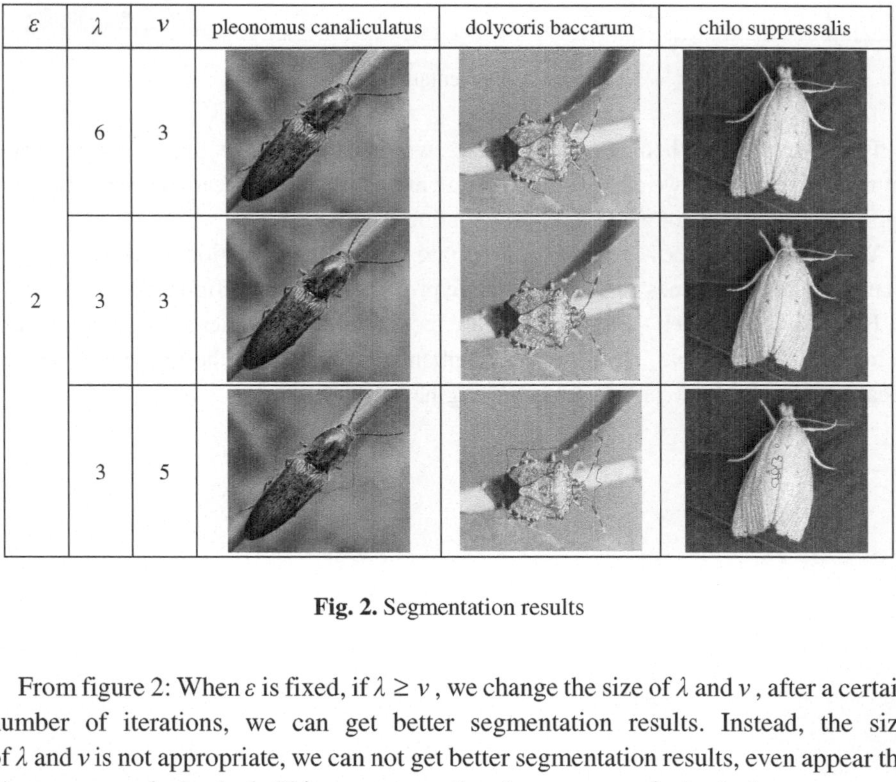


**Fig. 2.** Segmentation results

From figure 2: When $\varepsilon$ is fixed, if $\lambda \geq v$, we change the size of $\lambda$ and $v$, after a certain number of iterations, we can get better segmentation results. Instead, the size of $\lambda$ and $v$ is not appropriate, we can not get better segmentation results, even appear the phenomenon of edge leak. If $\lambda < v$, appear the phenomenon of edge leak.

Make $\lambda$ and $v$ fix, change the size of $\varepsilon$, after a certain number of iterations, we get the following segmentation results. Segmentation results are shown in figure 3.

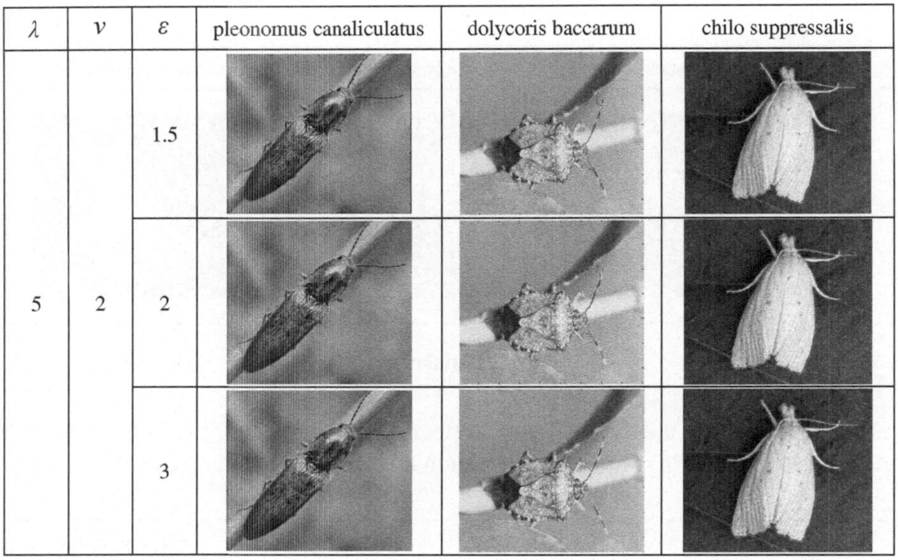


**Fig. 3.** Segmentation results

From figure 3: When $\lambda$ and $v$ are fixed, we change the size of $\varepsilon$ , after a certain number of iterations, we can get better segmentation results. Instead, the size of $\varepsilon$ is not appropriate, we can not get better segmentation results.

When the size of $\varepsilon,\lambda,v$ are more appropriate, we get segmentation results by gradual change of iteration times $iterNum$ . Iterative process is shown in figure 4.

From figure 4: When the size of $\varepsilon,\lambda,v$ are more appropriate, after a certain number of iteration times $iterNum$ , we get better segmentation results. If the iteration times are greater or smaller, we can not get better segmentation results.

| pleonomus canaliculatus ($\varepsilon = 2, \lambda = 3, \nu = 3$) | Dolycoris baccarum ($\varepsilon = 3, \lambda = 6, \nu = 2$) | Chilo suppressalis ($\varepsilon = 1.5, \lambda = 5, \nu = 3$) |
|---|---|---|
| 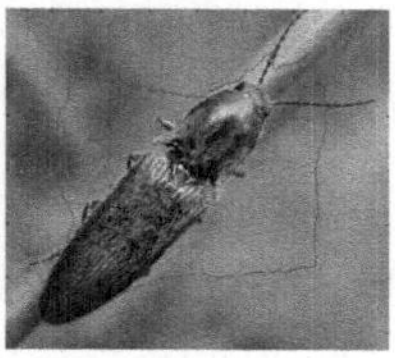 iterate 200 |  iterate 600 |  iterate 300 |
| 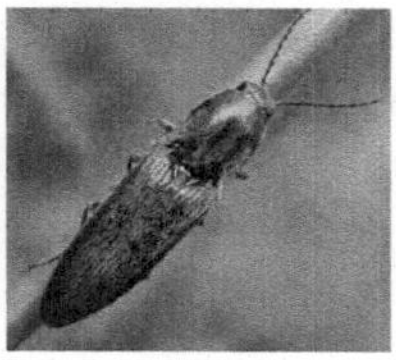 iterate 600 |  iterate 1800 |  iterate 800 |
| 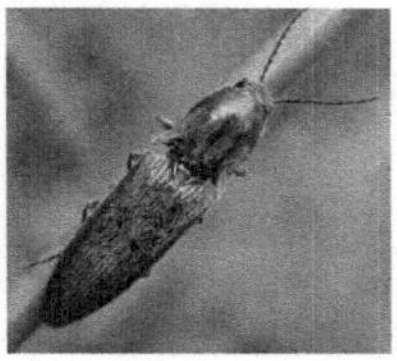 iterate 1400 |  iterate 2400 |  iterate 1000 |

**Fig. 4.** Iterative process

## 5 Conclusions

The variational level set method without re-initiation is a major breakthrough of the traditional level set method in this paper. This method evolution is more faster, numerical computations is more stable, and has effectively improved the efficiency and quality of the wheat pests image segmentation, and it has also provided a strong support for follow-up processing of wheat pests image.

## References

1. Cheng, Y., Hu, X., Zhang, C.: Algorithm for segmentation of insect pests images from wheat leaves based on machine vision. Transactions of the CSAE 23, 187–191 (2007)
2. Bu, T., Lu, C.: Research of Image Segmentation Algorithms. Computer Knowledge and Technology. 6, 1944–1946 (2010)

3. Osher, S., Sethian, J.A.: Fronts propagating with curvature dependent speed: Algorithms based on Hamilton-Jacobi formulation. Journal of Computational Physics 79, 12–49 (1988)
4. Zang, S., Wang, Z.: Image Contour Detecting without Re-initialization Level Set Method. Computer Knowledge and Technology 4, 944–946 (2008)
5. Wang, J., Li, W.: A new method of image segmentation based on level set. Microcomputer Applications 129, 6–10 (2008)
6. Li, C., Xu, C., Gui, C.: Level set evolution without re-initialization: a new variational formulation. Computer Vision and Pattern Recognition 1, 430–436 (2005)
7. Guo, Y.: Segmentation of Images Based on Level Set Methods. Electronic Sci. &Tech. 24, 71–74 (2011)
8. Liu, Z., Luo, W.: The Research about Level Set Function Application in Image Segmentation. Computer and Information Technology 17, 35–38 (2009)
9. Zhang, H., Kang, Z.: Variational level set model integrated with fuzzy clustering for image segmentation. Journal of Electronic Measurement and Instrument 25, 325–330 (2011)
10. Tian, Q., Huang, S., He, C.: Novel adaptive fast level set evolution without re-initialization. Computer Engineering and Applications 46, 174–176 (2010)

# Video Quality Assessment Combining Structural Distortion and Human Visual System

Wei Tu, Zhihua Xie, and Lixin Gan

Key Lab. of Optic-Electronic and Communication,
Jiangxi Sciences and Technology Normal University, Nanchang, Jiangxi, 330013
xie_zhihua@yahoo.com.cn

**Abstract.** This paper presents an improved structural similarity index (SSIM) for video quality assessment based on human visual system (HVS). To integrate visual characteristics to our SSIM, different weighted values are determined by those visual characteristics including contrast sensitivity, multi-channel structure, visual masking and so on. This method has the properties of simple and efficiency as the same of the SSIM method. And it is more suitable for human visual system due to fusing HVS. The experimental results show that the method can reflect people's subjective feelings in a better way and is better than other traditional methods in fitting M2 (Correlation coefficient of Non-linear regression), M3 (Spearman rank), M4 (Outlier Ratio) of VQEG Phase I MOS.

**Keywords:** VS, wavelet transform, SSIM, video quality assessment.

## 1 Introduction

The most commonly method about video quality assessment is composed of two parts, one is subjective measure and the other is objective measure. The subjective measure is the method which under certain conditions different observers are selected to make a different score to different reference videos and distorted videos by a five-mark system which are divided into five equal intervals with the following adjectives from top to bottom: Excellent, Good, Fair, Poor and Bad [1]. Then all scores which different observers gave are weighted and generate the mean opinion score (MOS). Because of different background knowledge, motivation, and different environment of different labs, the entire evaluation process which wastes a lot of time, is very complicated and non-transplant. The objective measure is a method which can give an objective score automatically by a mathematical model which is set up by a variety of influential factors of video quality. According to the dependence on original image, the objective measure is divided into full-reference quality assessment (FR), no-reference quality assessment (NR), and reduced-reference quality assessment (RR) [2]. The commonly full-reference metrics are mean squared error (MSE) and peak signal-to-noise ration (PSNR), which are widely used because they are simple to calculate, have clear physical meanings, and are mathematically easy to deal with for optimization purpose. However, in those methods the image pixels are regarded as independent and the local correlation of inter-pixels is ignored, which is correlated well with perceived quality measurement.

J. Lei et al. (Eds.): AICI 2012, LNAI 7530, pp. 475–483, 2012.

## 2 Human Visual System

Many characteristics of human visual system have closely relation to video quality assessment [3]. At present many characteristics, such as contrast sensitivity, multi-channel structure, visual masking, vision nonlinear law, are researched popularly. But HVS is a very complicated system, to set up simulation model accurately according to characteristics of human physiological and psychological is impossible. However, in recent decade many researchers also have given some simplify models to simulate human visual system.

### 2.1 Luminance Contrast Sensitivity Function (CSF)

The contrast sensitivity function is also named as modulate transform function, which describes the relationship between visual perception and contrast. The different function expression of CSF is generated by different experiments, but all of the functions describe humans' sensitivity to spatial frequency. A model of the CSF for luminance images, originally proposed by Mannos and Sakrison, is given by [4]:

$$CSF(f) = 2.6*(0.192+0.114f)*\exp[-(0.114f)^{1.1}] \tag{1}$$

where spatial frequency is $f=\sqrt{f_x^2+f_y^2}$ with units of cycles/degree. $f_x$ and $f_y$ are the horizontal and vertical spatial frequencies, which are given by:

$$f_x = \sqrt{\frac{1}{M(N-1)}\sum_{m=1}^{M}\sum_{n=2}^{N}(f_{m,n}-f_{m,n-1})^2} \tag{2}$$

$$f_y = \sqrt{\frac{1}{(M-1)N}\sum_{m=2}^{M}\sum_{n=2}^{N}(f_{m,n}-f_{m-1,n})^2}$$

The CSF is a band-pass filter: the HVS is most sensitive to normalized spatial frequencies between 0.025 and 0.125 and less sensitive to very low and very high frequencies [5].

### 2.2 Multi-channels Structure

Visual cortex cells have different sensitivities to different visual information or stimulation, such as color, frequency and direction. Researchers who research visual mask, pattern recognition, the adaptive system consider that the visual sensitivities in human visual system are dealt with independent in different channels. It is early multi-channels theory. However, later through the efforts of researchers they find that the visual mechanism of multi-channels is not isolated. They are contacted with each other, and produce the best vision. To static gray-scale images, the characteristic of multi-channels is described by spatial frequency and orientation. If it has tunable components enough, the entire direction band of image in visual cortex can be masked completely. By this way we can simulate the multi-channels structure completely, but the interaction between inter channels is also not clear [6]. In this paper, we can apply the multi-resolution characteristics of wavelet transform and spatial frequency of image characteristics to simulate the multi- channels structure. In this way, we can get a better video quality metric.

### 2.3 Visual Masking

Visual masking [7, 8] can be described by that the incitement which interacts each other causes the change of the visual threshold when much incitement appears. According to section 2.2, we know that HVS is a multi- channels structure. When an image is inputted it can decompose different perceived components. Every component is its own threshold, and we call it visual threshold. Because of vision threshold being the injury below the threshold will not be perceived. The range of vision threshold is extended when visual masking emerges so more visual injuries can not be perceived. And if our eyes do not perceive the injuries, the influence of injuries can be ignored. Mostly visual masking consists of contrast masking, texture masking, moving masking, switch masking and so on.

### 2.4 Vision Nonlinear Law

Vision nonlinear law is also called as Weber law. To our sense organ when the intensity ($I$) has changed $\triangle I$ which is just notice difference (JND) of vision, the sense difference function is defined in Eq.3:

$$\Delta s = k\frac{\Delta I}{I} \tag{3}$$

$$s = k\log\frac{I}{I_0} \tag{4}$$

Eq.4 is integral of Eq.3 and I0 is an absolute threshold. Eq.4 means that s is in proportion as *logI*, as shown in Fig.1. Fig.1 (b) is a transform of Fig.1 (a) used vision nonlinear law, and I magnify 50 times, which reflects the mapping of image in our visual cortex.

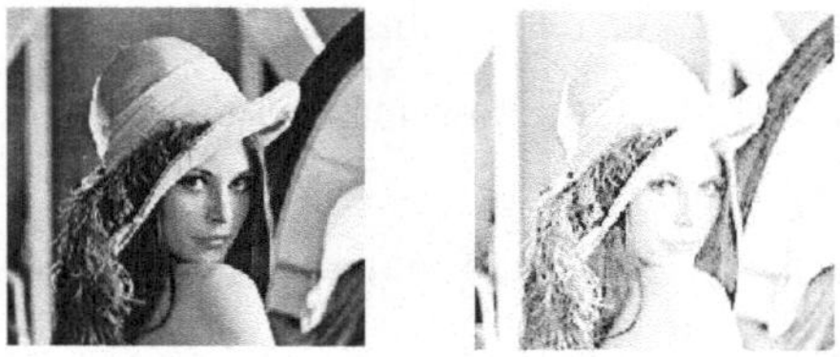

(a) Original image (b) Vision nonlinear transform

**Fig. 1.** Comparison after visual nonlinear transform

## 3 Improved SSIM Index Based on HVS

### 3.1 The Structural Similarity (SSIM) Index

Zhou Wang proposes a new method for video quality assessment based on structural distortion measurement, which is the structural similarity (SSIM) index [9]. Theoretical foundation of this index is that HVS is suitable to extract the structure information of visual scene, so changes between the structure information and quality of perceived image is approximate. If the structure is similar that it has a few changes between reference image and distorted image, which means that the quality is not loss. The structural similarity index is defined the luminance, contrast and structure comparison measures as follows:

$$l(x,y)=\frac{2\mu_x\mu_y+c_1}{\mu_x^2+\mu_y^2+c_1},\quad c(x,y)=\frac{2\sigma_x\sigma_y+c_2}{\sigma_x^2+\sigma_y^2+c_2},\quad s(x,y)=\frac{\sigma_{xy}+c_3}{\sigma_x\sigma_y+c_3} \tag{5}$$

If the two signals are represented discretely as $X\{x_i \mid i=1,2,...N\}$ and $Y\{y_i \mid i=1,2,...N\}$, then the statistical features can be estimated as follows:

$$\mu_x=\bar{x}=\frac{1}{N}\sum_{i=1}^{N}x_i \quad , \quad \mu_y=\bar{y}=\frac{1}{N}\sum_{i=1}^{N}y_i \tag{6}$$

$$\sigma_x=\sqrt{\frac{1}{N-1}\sum_{i=1}^{N}(x_i-\bar{x})^2} \quad , \quad \sigma_y=\sqrt{\frac{1}{N-1}\sum_{i=1}^{N}(y_i-\bar{y})^2} \tag{7}$$

$$\sigma_{xy}=\frac{1}{N-1}\sum_{i=1}^{N}(x_i-\bar{x})(y_i-\bar{y}) \tag{8}$$

Let $\mu_x,\mu_y,\sigma_x^2,\sigma_y^2$ and $\sigma_{xy}$ be the mean of x, the mean of y, the variance of x, the variance of y, and the covariance of x and y.

When $(\mu_x^2+\mu_y^2)(\sigma_x^2+\sigma_y^2)\neq 0$, the similarity index measure between x and y given corresponds to

$$S(x,y)=l(x,y)\cdot c(x,y)\cdot s(x,y)=\frac{(2\mu_x\mu_y+c_1')(2\sigma_{xy}+c_2')}{(\mu_x^2+\mu_y^2+c_1')(\sigma_x^2+\sigma_y^2+c_2')} \tag{9}$$

The SSIM indexing approach is then applied to the Y, Cb and Cr color components independently and combined into a local quality measure using a weighted summation. Let $SSIM_{ij}^{Y}, SSIM_{ij}^{C_b}$ and $SSIM_{ij}^{C_r}$ denote the SSIM index values of the Y, Cb and Cr components of the j-th sampling window in the i-th video frame. The local quality index is given by

$$SSIM_{ij}=W_Y SSIM_{ij}^{Y}+W_{C_b}SSIM_{ij}^{C_b}+W_{C_r}SSIM_{ij}^{C_r} \tag{10}$$

In paper [9] the weights are fixed on $W_Y=0.8$, $W_{C_b}=0.1$, $W_{C_r}=0.1$, respectively.

In the second level of quality evaluation, the local quality values are combined into a frame-level quality index using:

$$Q_i=\frac{\sum_{j=1}^{R_s}w_{ij}SSIM_{ij}}{\sum_{j=1}^{R_s}w_{ij}} \tag{11}$$

Where $Q_i$ denotes the quality index measure of the *i*-th frame in the video sequence, and $w_{ij}$ is the weighting value given to the *j*-th sampling window in the *i*-th frame.

### 3.2 Improved SSIM Index Based on HVS

Because of SSIM algorithm taking into account the spatial structure of image, the performance in consistency of objective measure and subjective measure is much better than traditional method of PSNR or MSE. But SSIM simply simulate the spatial structure of image, other performances of HVS are not utilized absolutely. In response to the lack of SSIM, an improved SSIM index based on HVS is proposed in this paper. In this paper, characteristics of HVS, such as contrast sensitivity, multi-channel structure, visual masking, vision nonlinear law, as the weighting value are fused SSIM algorithm. The detailed flow is given by Figure 2:

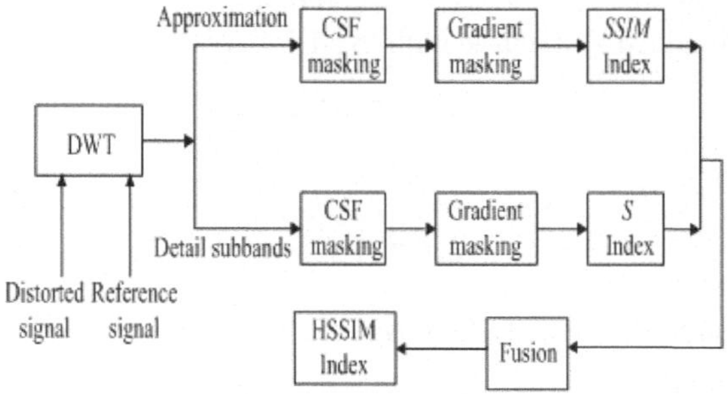


**Fig. 2.** HSSIM index flow which is an improved SSIM based on HVS

The main steps in our method are listed as follow:

Step 1: Simulating Multi-channel Structure of HVS.

Simulating multi-channel structure of HVS is used with multi-resolution characteristics of wavelet transform. Because the multi-resolution characteristics of wavelet transform is similar terribly with multi-channel structure of HVS, to transform with wavelet to reference signal and distorted signal respectively, the multi-channel structure of HVS can be simulated. When signal is transformed with Bi9/7 wavelet, it can be divided into two parts: approximation and details. And approximate signal which concentrates the mainly energy can affect the image mostly. In this paper changes of approximation are computed with SSIM between reference signal and distorted signal. Detail signals describing detail of image are high frequency subbands in which the luminance and contrast can be ignored. In order to describe changes of details structure, comparison measure is given by

$$s'(x,y)=\frac{|\sigma_{xy}|+c}{\sigma_x\sigma_y+c} \tag{12}$$

where c is a small positive number which prevents denominator from being zero.

Step 2: CSF Masking.

As a result of the wavelet decomposition, each frame is divided into several subbands corresponding to spatial frequencies. In this way, this would make it possible to take into account the sensitivity of human visual system to simulate of different spatial frequencies. The DWT CSF making is proposed which is derived from wavelet decompositions of the CSF curve. The weights are determined using different methods and in this paper a peak CSF masking [5] is formed in the following manner.

- Compute the DWT of the CSF in Figure.1. For a three-level wavelet decomposition of the image, the peak CSF mask is a 4-weight, orientation independent perceptual weighting;
- Each weight is computed as the largest value of the CSF curve in its corresponding octave frequency band.
- All of the weights are normalized such that the smallest is one. Figure 4 shows the re-normalized CSF curve with the 4 weights superimposed.
- The weights of CSF in different subbands are given by

$$w_{CSF}^{i} = \frac{CSF_i}{\sum_{i=1}^{N} CSF_i} \tag{13}$$

where $N$ is the sum of subbands, here $N$=10.

Step 3: Gradient masking.

The experimental results show that the performance of HVS is better used SSIM index than PSNR or MSE index. If the image is blurred seriously, the evaluation results used SSIM index will be unreasonable, which are tested using Caps.bmp in the Gaussian Blur of Live Database Release 2 [10]. In Figure 5, Figure 5(a) is an original image, Figure 5(b) is an enhancement image, whose each pixel is added 40 than original that of image, and Figure 55(c) is a Gaussian blur image. The values of PSNR and SSIM to Figure 5(b) and Figure 5(c) are computed, which can be found that the results are deviated from subjective quality assessment. To the defect of SSIM in blurred image, this paper proposed an improvement method of blur masking. According to paper [11], the influence of perceived psychology is decreased systematically to the edge, texture and flat of image and the gradients of these regions are ordinal reduced too. So the performance of quality assessment will be reasonable used gradient weighting to every pixels. In this paper an improved SSIM index used gradient masking is proposed. Improved steps are as follows:

Compute gradients of four directions to every pixel of image:

$$\begin{aligned} grad0(i,j) &= \left| f(i,j) - f(i,j+1) \right| , \\ grad90(i,j) &= \left| f(i,j) - f(i+1,j) \right| , \\ grad180(i,j) &= \left| f(i,j) - f(i,j-1) \right| , \\ grad270(i,j) &= \left| f(i,j) - f(i-1,j) \right| \end{aligned} \tag{14}$$

So, we can get the average of four gradients:

$$grad\_mean(i,j) = mean(grad0, grad90, grad180, grad270) \tag{15}$$

Consider the mean gradient of each pixel as the weight of the image, and HSSIM of different subbands is given by:

$$HSSIM_{approimation}(x,y)=\frac{\sum_{i=1}^{s} grad\_mean * ssim(x,y)}{\sum_{i=1}^{s} grad\_mean} \tag{16}$$

$$HSSIM_{detail}(x,y)=\frac{\sum_{i=1}^{s} grad\_mean * s(x,y)}{\sum_{i=1}^{s} grad\_mean} \tag{17}$$

$$HSSIM=(HSSIM_{detail}+\sum_{i=1}^{N-1} HSSIM^{i}_{approximation})/N \tag{18}$$

Where $N$ is the sum of subbands, here $N$=10, and $s$ is the numb-er of blocks.

# 4 Experiment Results and Analysis

In this paper experimental data is from VQEG Phase•FR-TV test dataset (Including 20 different source sequences, divided into 10 525/60 format sequences and 10 625/50 format sequences, processed by 16 different video systems and evaluated at eight independent laboratories worldwide). The method in VQEG Phase *I* is adopted to evaluate HSSIM index. First, nonlinear fitting is performed using fitting function to HSSIM values and DMOS values of VQEG. Second, take an evaluation with three different metrics which are proposed by [12]:

Metric1: The correlation coefficient between DMOS and model predictions used non-linear regression analysis.

Metric2: The Spearman rank order correlations test for agreement between the rank orders of DMOS and model predictions.

Metric3: The outlier ratio is a percentage of the number of predictions outside the range of $\pm 2$ times of the standard deviations.

**Table 1.** Performance comparison used HSSIM and other models on VQEG Phase I

| Model | Metric 1: Non-linear regression | Metric2: Spearman rank | Metric 3: outline ratio |
|---|---|---|---|
| PSNR | 0.779 | 0.786 | 0.678 |
| KPN/Swisscom CT | 0.827 | 0.803 | 0.578 |
| SSIM | 0.820 | 0.788 | 0.597 |
| HSSIM | 0.840 | 0.836 | 0.574 |

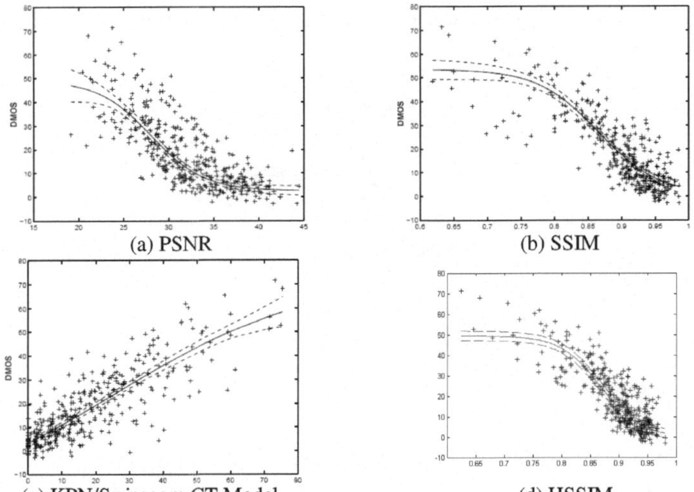


(a) PSNR (b) SSIM

(c) KPN/Swisscom CT Model (d) HSSIM

**Fig. 3.** Scatterplot comparison of different models on VQEG Phase I test dataset. Vertical and horizontal axes are for subjective and objective measurements, respectively. Each sample point represents one test video sequence.

Metric1 is a measure of prediction accuracy, metric 2 is a measure of prediction monotonicity, and metric 3 is a measure of predictions consistency. The detailed define is proposed in VQEG Phase•. Table 1 is performance comparison of different models on VQEG Phase•test data set.

As shown in Fig.3 and Table 1, the performance of proposed method (with HVS adjustment) is better than other models, and the correlation with DMOS is increased more used HSSIM than used other models.

## 5 Conclusions

This paper proposed a new method of objective video quality assessment which based on Human Visual System. It fused human perceived performance into structural similitude algorithm effectively, not only reserves the simply and high efficiency of SSIM algorithm, but also reveals the improvement of human visual characteristic. The experimental results show that it can reflect human subjective perception well, and outperform other video quality assessment methods.

## References


1. ITU-T. Objective perceptual assessment of video quality: Full reference television [EB/OL]. Switezerland: ITU-T Telecommunication Standardization Bureau (TSB) (2004), `http://vqeg.its.bldrdoc.gov`
2. Wang, Z., Sheikh, H.R., Bovik, A.C.: Objective video quality assessment. In: The Handbook of Video Databases: Design and Applications, pp. 1041–1078. CRC Press (2003)

3. Wandell, B.A.: Foundations of vision, pp. 1–10. Sinauer Press, England (1995)
4. Mannos, J.L., Sakrison, D.J.: The effect of a visual fidelity criterion on the encoding of images. IEEE Transactions on Inform. Theory 20(2), 525–536 (1974)
5. Beegan, A.P., Iyer, L.R., Bell, A.E., et al.: Design and evaluation of perceptual masks for wavelet imge compression. Proceedings of 2002 IEEE 10(13-16), 88–93 (2002)
6. Wei, C.K., Chen, L.Z.: An Image Quality Measure Scheme in the Perceptual Field via Masking. Journal of Image and Graphics 9, 690–696 (2004)
7. Come, S., Macq, B.: Human visual quality criterion. In: SPIE Visual Commuication and Image Processing, San Jose, USA, vol. 1360, pp. 2–7 (1990)
8. Ding, X.X., Ding, R.H., Li, J.X.: A Criterion of Image Quality Assessment Based on Property of HVS. Journal of Image and Graphics 9, 190–194 (2004)
9. Wang, Z., Lu, L.G., Bovik, A.C.: Video quality assessment based on structural distortion measurement. Signal Processing: Image Communication 19(2), 121–132 (2004)
10. LIVE Image Quality Assessment Database Release2 [EB/OL], `http://live.ece.utexas.edu/research/quality` (February 19, 2007)
11. Li, J.L., Chen, G., Chi, Z.R., et al.: Image coding quality assessment using fuzzy integrals with a three-component image model. IEEE Transactions on Fuzzy Systems 12(1), 99–106 (2004)
12. Lu, Z.K., Lin, W.S., Yang, X.K., Ong, E.P., Yao, S.S.: Modeling Visual Attention And Motion Effect for Visual Quqlity Evaluation. In: International Symposium on Intelligent Multimedia, Video and Speech Processing (2004)

# A Novel Ocean Wind Field Estimation Method SAR Images Based

Hua Bo, Bin Liu, Shangyan Lv, Haiyun Gu, and Lei Ren

Information Engineering Department, Shnaghai Mairtime University
201306 Shanghai, China
{huabo,syl,hygu,leiren}@shmtu.edu.cn

**Abstract.** Ocean wind field parameters estimation from Synthetic Aperture Radar (SAR) images has been widely accepted. Most wind parameters estimation algorithms use wind direction and corrected $\delta_{VV}$ as priori information before they sent into geophysical model function. Considering same wind direction, different model function would gain different wind speed, so it is crucial to select model function. Moreover tiny errors in wind direction can lead to obvious difference in wind speed. New method based on texture analysis for ocean wind field parameters estimation is discussed and tested. It wills retrieval wind vector from SAR images without priori wind direction information. And simulation results show that this method can give higher accuracy for wind field estimation.

**Keywords:** Ocean Wind Field Estimation, Texture Analysis, SAR Image.

## 1 Introduction

Satellite-based Synthetic Aperture Radar (SAR) sensors can provide high-resolution images of earth surface which also include a lot of ocean information day and night and during most weather condition. Early sensors, such as scattering plan, altimeter and microwave radiometer, are also effective equipment in getting sea surface information [1]. With the development of microwave technology, low spatial resolution of early sensors can't satisfied the observation and information extraction in small areas, then high spatial resolution Synthetic Aperture Radar becomes more and more popular, and it has been widely used in meso- and small-scale ocean movement field, such as wind flow, atmosphere roll, rain roll [2], ocean ship track, oil slick detection as well as ice zone monitor, and especially in wind field estimation of sea surface.

Present, main algorithms of wind field estimation on sea surface include: (1)SWDA (SAR Wind Direction Algorithm) [3-12], it can estimate wind speed under prior wind direction information via GMF-Geophysical model Function which defines relation between wind speed, wind direction, and normalized radar cross section (NRCS); (2)SWA (SAR Wind Algorithm) [13-14], it can estimate wind speed according to the relation among wind speed, cutoff wavelength, and wave height; (3)GM(Gradient Method Model), it can estimate wind field via oil slick information caused by wind field.

J. Lei et al. (Eds.): AICI 2012, LNAI 7530, pp. 484–491, 2012.

Scattering theory shows that, there is certain nonlinear relation between radar backscatter section $\sigma^0$ and sea surface wind speed, which is called "Geophysical model Function" (GMF). Further research suggests, GMF is experiential and theorized model function which contacts normalized radar cross section, sea surface wind speed, wind direction and incident angle together. Under prior wind direction information , it should be first to choose model function before wind speed inversion, and different model function would have dramatic influence in wind speed inversion. Commonly used are CMOD series models, literature [15-18] discuss wind field estimation by CMOD function. Among which, wind direction simulation is mainly through method of wave spectrum inversion by SAR image spectrum put forward by Hasselmanns or through energy extension cord direction in SAR image spectrum; wind speed inversion is mainly through geophysical model function under prior wind direction information.

Both latter two methods don't need wind direction data, and can estimate wind direction without wind stripe. The third one underestimates wind field uniformity, no matter uniform wind field or hurricane wind field.

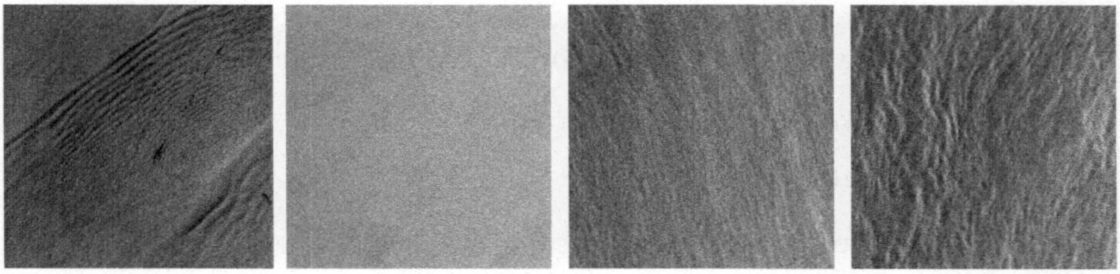

(a) (b) (c) (d)

**Fig. 1.** SAR Marine phenomenon examples: (a) internal waves; (b) wind waves 0-9 knots; (c) wind waves 6-12knots; (d) wind waves 6-24knots

Ocean phenomena such as internal waves, and wind waves would have obvious stripe information in SAR image. As shown in Fig.1, they are from ERS-1 / SAR and ERS-2 / SAR (C band, VV polarization form, Incident Angle≈23°, image resolution for 12.5 m×12.5 m).

Among which, internal waves image was taken from northern south sea of China by ERS-1 (June 1997), rest three wind waves images were from China south sea by ERS-2 (access time was respectively in April 2000, May 1998,and June 1998).

In 1998, Hatten [19], etc. got that Radar scattering section(RCS) were mainly depending on local wind speed and incident angle, and revealed correlation between noise spectrum, wind speed, and wind direction in horizontal polarization x-band radar image. The conclusion can also be applied to c-band radar image. In Fig. 1, wind and waves images (b)-(d) show that different wind speed can create different influence on sea surface movement, produce different RCS, and performance as different texture roughness in SAR image.

Literature[20] have made research on ocean phenomenon detection based on wavelet transform, this paper only researches wind field estimation of waves images. According to corresponding relation between texture roughness and wind speed, we propose a wind speed estimation method based on marine texture. This method will not need phase information of radar, and also doesn't rely on prior wind direction data. Result of this method will be compared with known wind speed information. Texture extraction method will be given in section 2, Wind speed and wind direction estimation algorithm will be discussed in section 3, section 4 is experimental simulation and results analysis, and the last section is conclusion.

## 2 Texture Analysis

Texture analysis is a commonly used image feature extraction method. Gray level co-occurrence matrix (GLCM) is a very good method in texture determining proved by theory and experiment, and is widely used in transforming image gray value into texture information.

### 2.1 Gray Level Co-occurrence Matrixes

GLCM was proposed by Haralick in 1973, it describes frequency related matrix $p_{ij}(s, \theta)$ [21] of pixel i and j in direction angle $\theta$ with distance s. GLCM is a large matrix, we commonly use six eigenvalues of GLCM as texture feature [22]:

Contrast ratio: $Con = \sum_{i,j}(i-j)^2 p_{ij}$ ;

Entropy: $Ent = -\sum_{i,j} p_{ij} \times \log p_{ij}$ ;

Secondary torque of angle or Energy: $Asm = \sum_{ij}(p_{ij})^2$ ;

Equivalence : $Hom = \sum_{ij} p_{ij} / [1+(i-j)^2]$ ;

No similarity : $Dis = \sum_{ij} |i-j| p_{ij}$ ;

Correlation : $Cor = \{\sum_{i,j}(i-\mu_x)(j-\mu_y)P_{ij}\} / (\sigma_x \sigma_y)$

### 2.2 The Texture Cycle

Calculation of texture information in an image need a given information of angle $\theta$ and adjacent pixel distance $s$ , according which we can work out a GLCM matrix $p_{ij}(s, \theta)$, then get a texture value written as $s$-entropy. Texture value can describe image texture discipline , for the smaller the entropy value is, the more "smooth" the texture is, or the smaller the image gray change is; or the greater the entropy value is, the larger the image gray change is, the more "chaos" the image texture is.

$p_{ij}(s, \theta)$ is also a function of $s$, with different spacing adjacent $s$, we get different GLCM value ,then we can get different entropy curve. If image texture is in rule changes, s-entropy curve will also be in rule changes.

Fig.2 shows that there is certain regularity and periodic in texture characteristics of SAR image of waves image, which is just not as obvious as natural optical image. It shows s-entropy curve of (b), and (c) and (d) in Fig. 1 in horizontal direction.

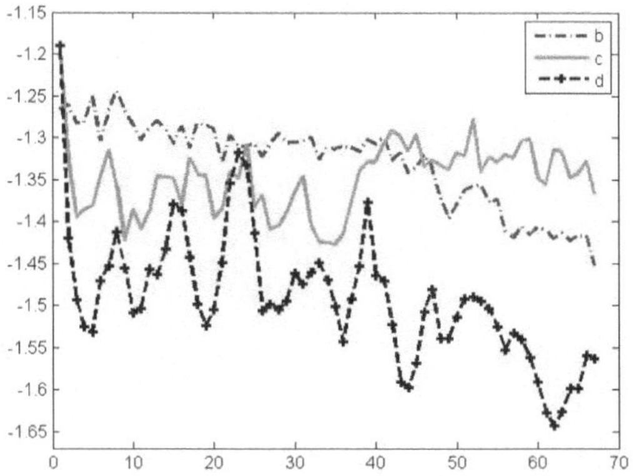


**Fig. 2.** S-entropy curve of wind wave image with different wind speed

We can see that in Fig.2, due to different stripe density of wind and waves, wind waves image will show different texture cycle in s-entropy curve. The corresponding section is, image with smaller wave sessions has a short texture cycle, and image with larger wave sessions has a long texture cycle. Fig.1 (b) has a wind speed of 0-9 knots, its texture entropy values fluctuate most slowly. Fig.1 (d) has a wind speed of 6-24 knots, its texture entropy values fluctuate most frequently. Fig.1 (c) has a wind of 6 - 12 knots, and its texture entropy values fluctuate center. Obviously, there is a corresponding relation between wind speed and fluctuation cycle of texture values. According which, we can estimate wind speed.

Only when there is a consistent between s-entropy curve direction and wind direction, can we correctly estimate wind speed, or there will be a great error. As shown in Fig.3, there are s-entropy curves of Fig.1 (d) in three different directions. Obviously, different direction has different texture periodic.

Analysis of Figure 1 shows that wind direction must have correspondence to direction in which the wind and waves changes most intensively, therefore, wind direction has correspondence to direction in which the s-entropy curve has shortest texture period. In Fig.3, horizontal s-entropy curve has the shortest texture period, so level direction is more close to wind direction. Because GLCM only has three directions, accuracy of wind direction estimation is still very low. At the same time there is 180 ° fuzzy problem in direction. It can be removed by algorithms applied by literature [23-24].

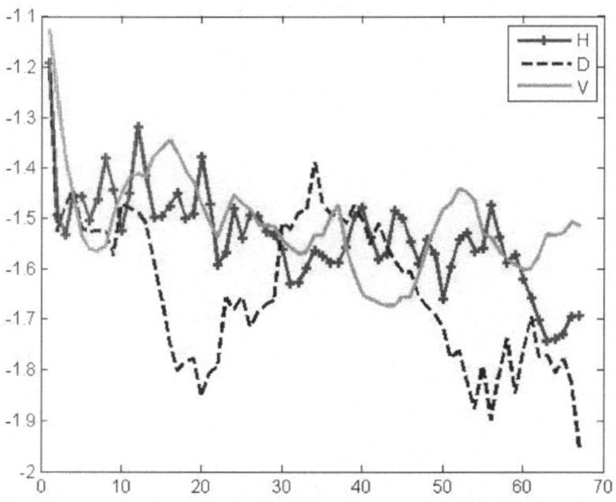


**Fig. 3.** S-entropy curves in three directions

## 3 The Algorithm of Wind Field Estimation

When wind blows across the sea, it causes a certain scope roughness in sea, which causes the change of RCS, and makes wind can be seen in SAR image. Wind vector of local area can be estimated by texture cycle and texture direction.

Because of different imaging time and imaging conditions, SAR image scattering coefficients have great difference throughout whole gray distribution. As shown in Fig.5, it is a SAR image of 1217×1191 pixels, its spatial resolution is 12.5 m ×12.5 m, and it covers about 16 × 15 km. It can be seen from the image, different areas have different gray distribution, and different wind and waves stripe direction.

In order to avoid measurement error induced by gray distribution differences, we will use small size sub-image alternative the whole image to get symbiotic matrix. It can also get wind speed and direction in different region. The specific algorithm procedure is as follows:

(1) Divided the thumper into 25 parts of 5 lines and 5 rows, each branch area has 100× 100 pixels, and each one covers an area of 1.25 × 1.25 kilometers;

(2) Choose a sliding window of N × N size, slip the sliding window in a given direction with pixel units in the sub-image and calculate the entropy value of each slide window;

(3) Draw the s-entropy curve;

(4) Calculate texture cycle of s-entropy curve in different directions, and choose the minimum period direction as wind direction;

(5)There is a 180 ° fuzzy in wind direction, and remove it by algorithm provided in literature [23- 24];

(6)Give corresponding relation between texture cycle and the known wind speed, then according to the relation, estimate unknown wind speed;

(7) Get whole image wind field information by repeating steps (2) to (6) throughout each branch area.

# 4 The Experimental Results and Analysis

Through front experiments and analysis, we can get corresponding relation between texture cycle and wind speed in winds and waves image, as well as method of shortest texture cycle corresponding to wind direction. Both have been tested in ERS-1 and ERS-2 synthetic aperture radar image, and got a good experimental result.

As shown in Fig.4, it reveals comparison between actual wind speed and wind speed estimated by entropy function:

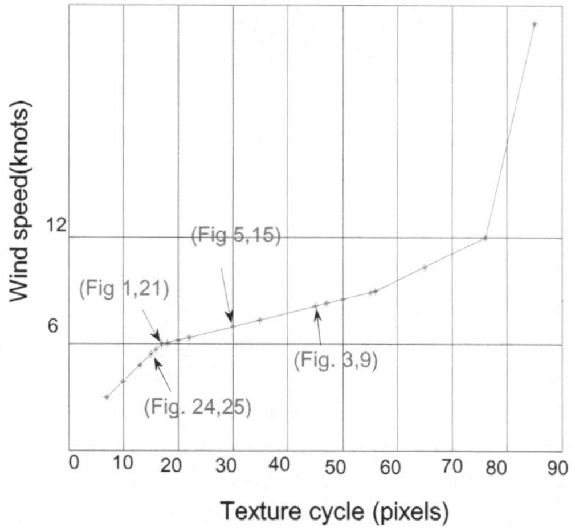


**Fig. 4.** Estimation curve of wind speed

Blue curve represents corresponding relation between texture period and wind speed. Red dots represent information of 25 sub-images of Fig.5; its x-coordinate represents texture cycle, and y-coordinate represents wind speed calculated by the corresponding relation. It has been known that the actual sea surface wind speed in figure 5 is 6~12 knots. Fig.5 shows that, except sub-images 7,12,13,16,17,19 which has no obvious texture information , about 14 out of the rest 19 sub-images have wind speed between 6~12 knots, then there is conclusion that entropy function method has a higher reliability in wind speed estimation.

Comparison between actual wind direction and each calculated sub-images wind direction has been shown as below (Blue arrow represents actual wind direction, and red arrow means wind direction estimated by entropy function).

It can be seen that there is a basic agreement between estimated wind direction and practical wind direction. That is, entropy function method also has high reliability in wind direction estimation.

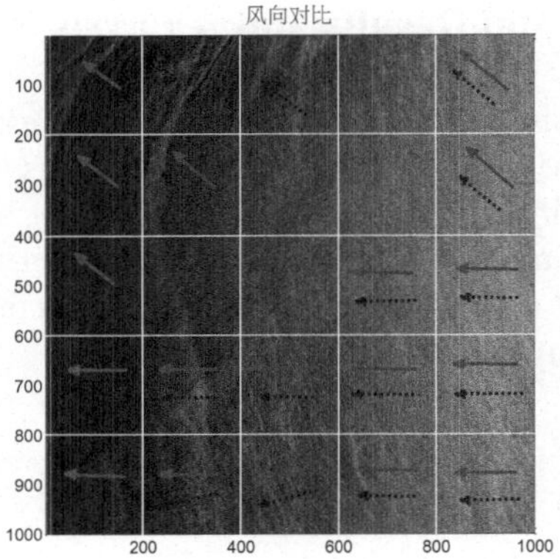


**Fig. 5.** Estimation results of wind direction

# 5 Conclusion

Entropy function based on gray level co-occurrence matrix estimates sea surface wind field data through texture analysis, and does not need priori wind direction information. According to texture analysis on SAR image of sea surface, we can get unknown wind speed by corresponding relation between texture cycle and known wind speed; wind direction must have correspondence to the direction in which wind and waves changes most intensively, that is wind direction corresponds to direction in which s-entropy curve has the shortest texture period. The method has been used on ERS-1 and ERS-2 synthetic aperture radar image for testing and gets good experimental results.

**Acknowledgment.** This work is supported by the science & technology program of Shanghai Maritime University (20120066 and 20110050).

# References

1. Hui, S., Perrie, W., Yijun, H.: Progress in Determination of Wind Vectors from SAR Images. In: Geoscience and Remote Sensing Symposium, pp. 2228–2231 (2006)
2. Katsaros, K., Vachon, P., Lio, W., Black, P.: Microwave remote sensing of tropical cyclones from space. J. Ceanogr. 58, 137–151 (2002)
3. Brown, R.A.: Surface fluxes and remote sensing of air-sea interactions. In: Geernaert, G.L., Plant, W.J. (eds.) Surface Waves and Fluxes, pp. 7–27. Kluwer Acad., Norwell (1990)
4. Etling, D., Brown, R.A.: Roll vortices in the planetary boundary layer: A review. Boundary Layer Meteorology 65, 215–248 (1993)
5. Fetterer, F., Gineris, D., Wackerman, C.C.: Validating a scatter meter wind algorithm for ERS-1 SAR. IEEE Trans. Geosci. Remote Sens. 36, 479–492 (1998)
6. Vachon, P.W., Dobson, F.W.: Validation of wind vector retrieval from ERS-1 SAR images over the ocean. Global Atmos. Ocean Syst. 5, 177–187 (1996)

7. Vachon, P.W., Dobson, F.W.: Wind retrieval from RADARSAT SAR images: Selection of a suitable C-band HH polarization wind retrieval model. Can. J. Remote Sensing 26, 306–313 (2000)
8. Wackerman, C., Rufenach, C., Shuchman, R., Johannessen, J., Davidson, K.: Wind vector retrieval using ERS-1 synthetic aperture radar imagery. IEEE Trans. Geosci. Remote Sensing 34, 1343–1352 (1996)
9. Horstmann, J., Thompson, D.R., Monaldo, F., Graber, H.C., Iris, S.: Can Synthetic Aperture Radars be used to Estimate Hurricane Force Winds? Geophys. Res. Let. (2005)
10. Horstmann, J., Koch, W., Lehner, S., Tonboe, R.: Wind retrieval over the ocean using synthetic aperture radar with C-band HH polarization. IEEE Trans. Geosci. Remote Sens. 38(5), 2122–2131 (2000)
11. He, Y., Perrie, W., Zou, Q., et al.: A New Wind Vector Algorithm for C-band SAR. IEEE Trans. Geosci. Remote Sens. 43(7), 1453–1458 (2005)
12. Vachon, P.W., Dobson, F.W.: Wind retrieval from Radarsat Sar images: Selection of a suitable C-band HH polarization wind retrieval model. Can. J. Remote Sensing 26(4), 306–313 (2000)
13. Chapron, B., Elfouhaily, T.T., Kerbaol, V.: Calibration and validation of ERS wave mode products, DRO/OS/95-02, Inst. Fr. De Rech. Pour l'Exploit. De la mer, Brest, France (1995)
14. Kerbaol, K., Chapron, B., Vachon, P.W.: Analysis of ERS-1/2 synthetic aperture radar wave mode imagettes. J. Geophys. Res. 103(C4), 7833–7846 (1998)
15. Long, A.E.: Towards a C-band radar sea echo model for the ERS-1 scatter meter. In: Proc. 3rd International Colloquium on Spectral Signatures of Objects in Remote sensing, Les Arcs, France, December 16-20, vol. ESA SP-247, pp. 29–34
16. Stoffelen, A., Anderson, D.: Scatter meter data interpretation: Measurement space and inversion. J. Atmos. Oceanic. Technol. 14(6), 1298–1313 (1997)
17. Quilfen, Y.: ERS-1 off-line wind scatter meter products. Technical Report ERS-SCAT / IOA / DOS-01, IFREMER (1998)
18. Vachon, P.W., Dobson, F.W.: Validation of wind vector retrieval from ERS-1 SAR images over the ocean. Global Atmos. Ocean Syst. 5, 177–187 (1996)
19. Hatten, H., Ziemer, F., Seemann, J., Nieto-Borge, J.: Correlation between the spectral background noise of nautical radar and the wind vector. In: Proc. 17th Int. Conf. Offshore Mechanics and Arctic Engineering (OMAE), Lisbon, Portugal (1998)
20. Jie, C., Biao, C., Qin, X.: Detection of ocean features in SAR images with 2D continuous wavelet power spectrum. Chinese Journal of Electronics 38(9), 2128–2133 (2010)
21. Haralick, R.M., Shanmugam, K., Dinstein, I.: Texture Features for Image Classification. IEEE Trans. On Systems, Man and Cybernetics 3(6), 610–621 (1973)
22. April, G.V., Harvey, E.R.: Speckle statistics in four-look synthetic aperture radar imagery. Optical Engineering 30(4), 375–381 (1991)
23. Thompson, D.R., Beal, R.C.: Mapping high-resolution wind fields using synthetic aperture radar. Johns Hopkins Univ. Tech. Dig. 21, 58–67 (2000)
24. Du, Y., Vachon, P.W., Wolfe, J.: Wind direction estimation from SAR images of the ocean using wavelet analysis. Can. J. Remote Sens. 28(3), 498–509 (2002)

# Research of Decentralized Collaborative Target Tracking Architecture in the Sea Battlefield for the Complex Sensor Networks

Li Duan, Kun Feng, Bin Luo, and Ya-Nan Li

College of Electronic and Engineering, Naval University of Engineering, 430033, Wuhan, China
{duanlidragon,dtyyyg,Norosin,Sunny}@126.com

**Abstract.** Considering the issue of decentralized collaborative target tracking architecture in the sea battlefield for the wide perception and complex sensor networks, firstly a new target calculation mechanism of the collaborative target tracking is proposed. To increase the performance of robustness, self-organization and dynamic adaptability for the information dissemination and sharing strategy, research methods and technical route are discussed in detail on the basis of complex network theory. In order to effectively deal with different kinds of information sources in the sensor networks, a generalized fusion machine is presented by way of DSmT model. The proposed architecture is applicable to the further research of collaborative target tracking technologies in the sea battlefield.

**Keywords:** sensor network, decentralized tracking, multi-agent, complex networks, information fusion, DSmT.

## 1 Introduction

Decentralized collaborative target tracking is a key issue of the information fusion in the sea battlefield. Currently, most of the studies are mainly aimed at the specific algorithms for the exploration of sight distance detection and wireless sensor networks. There are few applicable methods which are suitable for wide perception and complex sensor networks. As a result, this makes the decentralized collaborative target tracking application has been severely constrained. With the rapid development of sensor technology, signal detection and processing, and computer science, more and more over-the-horizon sensors are connected into sensor networks. The target tracking system of sea battlefield is developing toward the direction of wide perception, complex network, multi-sources information, and collaborative intelligence [1], which brought out new challenges for target tracking in the following three aspects:

1) With the sensor detection range is constantly expanding, the curvature of the Earth cannot be ignored. Therefore the traditional Under-the-Horizon mechanism of target position information processing which regards the plane Cartesian coordinates as unified

J. Lei et al. (Eds.): AICI 2012, LNAI 7530, pp. 492–499, 2012.

frame of reference is not appropriate, the coordinate transformations among cooperative platforms become complicated and the system capability is badly restricted.

2) Now the scale of sensor network is gradually increasing, and the network topology is more complex. In order to ensure information could be transferred to the correct sensor node at the right time and in the right way, it is the key factor to seek more effective information dissemination and sharing strategies for the target tracking system in the future.

3) Because of the diversity and complexity of information, especially the effect of its uncertainty, incompleteness, inconsistency and impreciseness, the breadth and depth of application for the target tracking system is severely restricted.

In recent years, the concept of generalized fusion, intelligent cooperation technology, and complex network theory have emerged, which provide a new the theoretical foundation for solving the problems of collaborative tracking. The primary goal of this paper is to analyze several key technologies and explore new solutions, which are applicable to the wide-area decentralized collaborative target tracking for the complex sensor network.

## 2 Multi-platforms Collaborative Tracking Architecture in a Wide Sea Battlefield

The emphasis of this paper is to promote the application breadth and depth of the current collaborative target tracking architecture in the sensor networks. The research can be divided into three sections as: collaborative target tracking calculation mechanism, information dissemination and sharing strategies, and the generalized fusion machine. The relationship between them is illustrated as Fig. 1:

The first layer in Fig.1 is for sensors to perform perception in an unknown environment of the wide battlefield according to the unified geodetic coordinates. It simplifies the information processing of collaborative tracking and broadens the coverage area for network perception.

If the platforms on the first layer have effectively completed target tracking, information dissemination and sharing strategies will be provided by the second layer. Within the complex network topology, every node will discover and select the optimal sensor set in a dynamic and intelligent way and form the self-organized area alliance. It ensures that information could be transferred to the correct sensor node in the right time and proper way. Consequently the sensor detection network preserves the relevant information of link assignment, which underlies the information dissemination and sharing in the third layer of fusion.

The third layer performs the multi-source and incomplete information fusion (including homogeneous/heterogeneous information) by means of the generalized fusion machine, which ensuring that more accurate tracking information will be obtained. The obtained information will be resent to the first layer for the target position information revision, which could in return provide more precise information for the next collaborative tracking. The above-mentioned layers are closely related in a progressive manner, forming the multi-platform collaborative tracking architecture in the wide sea battlefield.

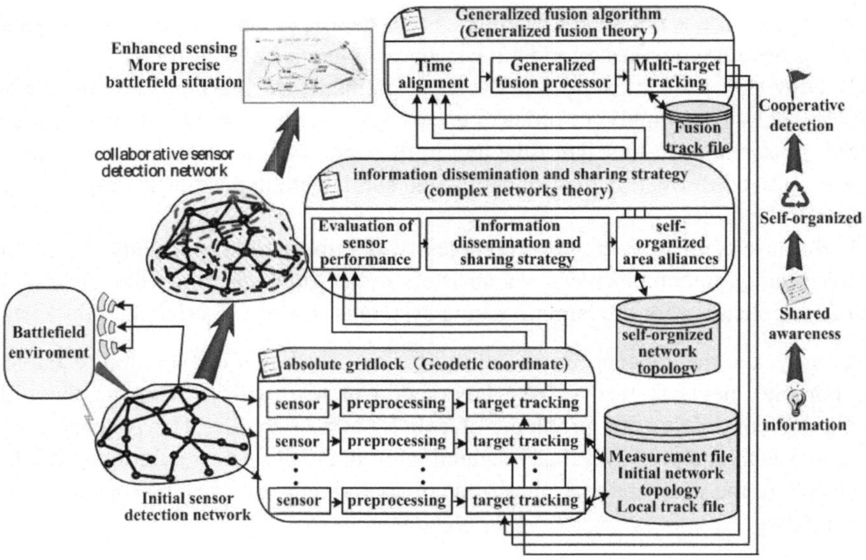


**Fig. 1.** Multi-platform collaborative tracking architecture in sea battlefield

## 3 Collaborative Target Tracking Calculation Mechanism in the Sea Battlefield

Currently, the target tracking and fusion algorithm are completed in the plane Cartesian coordinates system, which has resulted in a series of collaborative tracking problems in the wide-area sensor network. On the one hand, when the distance of detecting target is very far, the effect of the Earth curvature cannot be neglected. Adopting the target tracking in the traditional Under-the-Horizon mechanism will of course create an inaccurate result. On the other hand, in the process of information fusion and collaboration among platforms, the information share between each cooperative-platform need to be repeatedly switched among the relative coordinates system, fusion coordinates system and the absolute coordinates systems. The repetitious conversions between different coordinates are illustrated in the Fig.2.

This conversion process is time consuming and complicated. Besides, there are many conversion errors, such as the navigational system positioning error, the angle measurement and ranging error between each node, the sensor detection error, the information processing delay and so on are inevitably superimposed on the process. The errors will be further highlighted when the scale of system is expanding, which will ultimately have serious impact on the final tracking performance. At present, most of researches tend to using space registration algorithm [2-4] for such problems.

The first step adopted by U.S. army in expanding capabilities of CEC system and successor TCN system is to change relative gridlock to an absolute gridlock co-ordinate system [5]. Over many years of efforts, the China Geodetic Coordinate System 2000 (CGCS 2000) and "Beidou II" navigation and positioning system are established and officially put into service [6], higher requirements are proposed for the similar systems. In recent years, with the deepening study on geodetic coordinates and a new nonlinear filtering method[7-8] (UKF, particle filter, and etc.) is put forward, which makes it possible to perform decentralized target tracking in the geodetic coordinate system. In view of these considerations, it is reasonable to complete multi-platform collaborative target tracking in a unified geodetic coordinate system, shown in Fig. 3:

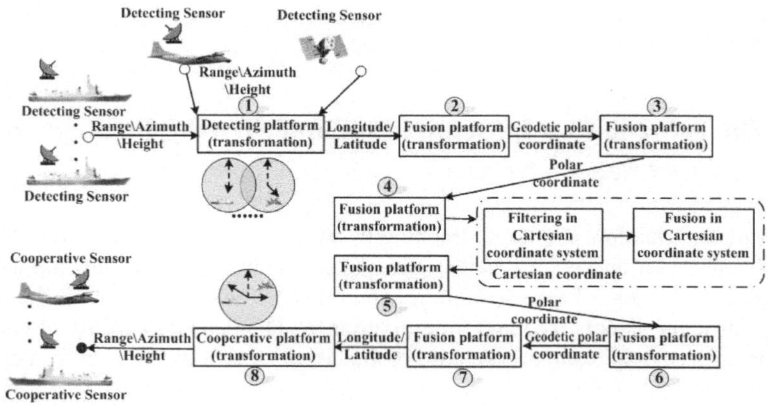


**Fig. 2.** The traditional information processing of multi-platform collaborative target tracking

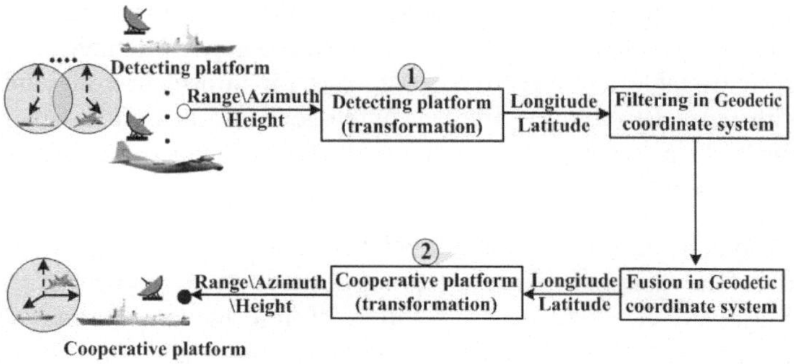


**Fig. 3.** Information processing of multi-platform collaborative tracking in the geodetic coordinate system

If the target calculation is fulfilled in the geodetic coordinate system, the repetitious conversions between different coordinates can be avoided. The system information flow for target calculation can be simplified greatly. On the other hand, it could meet the sensor network's demands of wide perception, expanding the awareness coverage area of the platforms. The collaborative target tracking models applied in short range, intermediate range and long-distance circumstances can be unified, and the precision of target tracking can also be improved.

The specific research method is shown as follows: starting with information processing flow of collaborative target tracking, the research is conducted by analyzing the existing collaborative target calculation process in the plane Cartesian coordinates system and the geodetic coordinates system. Secondly the system error models and precision analysis index are building for the performance evaluation of different linear and nonlinear filtering algorithms. Thirdly, by using different collaborative target tracking scenes generated by simulation testbed, a theoretical analysis of simulation result can be given according to the filtering accuracy, the computing speed and the scope of application. At last, the correctness and effectiveness of the proposed method can be verified by the comprehensive analysis of actual measurement data and the simulation data.

## 4 The Information Dissemination and Sharing Strategy in the Complex Sensor Network

In the realistic situation, the target tracking system of sensor network often work in a complex, uncertain and even hostile environment. In view of the uncertainty of the operation environment, the sensor failure, and other factors like damages, the network topology structure is not fixed and time-invariance. The detection ability of each node is also under dynamic change. Considering the above situations, the sensor network is required to be self-organized in accordance with the specific task and provide the efficient information service through collaborative mechanism in the case of limited system resource.

At present, most collaborative target tracking methods of sensor network are based on the simple and clearly defined network topology structure. As for large-scale networks with unknown and dynamically changed topology structure, there is still not find a more reasonable and effective design rule for information dissemination and sharing. With respect to the effective and intelligent information distribution among multi-platforms in a complex sensor network, the difficulties mainly focus in the following two aspects:

1、NP Problems. As for NP problem, relevant studies are mainly studied under the condition that the general network topology structure is known or simple. For instance, J.Ostwald etc. provided a solution of information dissemination in the decentralized sensor network based on market mechanism [9]. But it is a centralized method in essence, requiring a solution for combinational explosion problem. A.Waldock and M.Ruairı have respectively put forward the dynamics region theories [10] and probability collective methods [11], realizing the self-organization of information dissemination and sharing between network nodes. while the relevant principles need to be preset or reset along with environmental change due to the system's lack of expandability. Besides, there are many scholars present other solutions combined with self-organization theories [12] and multi-intelligent agent technologies [13]. To some extent, these methods have provided a better decentralized solving efficiency for the collaborative target tracking problem, but they are all based on the specific topology structures.

2、The complex network with dynamic topology structure will have influence on the effectiveness and reliability of information dissemination and sharing. Along with the rising of research upsurge on the complex network, many scholars have studied in-depth about the complex network topology structure models and dynamic characteristics [14]. Their findings [15] indicate that information dissemination and the transmission efficiency are closely related with the network topology. In the meantime, a new method based on epidemic spread behavior is proved to be an effective information dissemination strategy in the complex sensor network[16-17], since it is easy to perform in distributed manner and highly adaptable to the failure of communication link and nodes. The researches on the complex network and collaborative theory provide the new theoretical basis and method for the design of information dissemination and sharing mechanism in the complex sensor network. The specific research method is illustrated in Fig. 4:

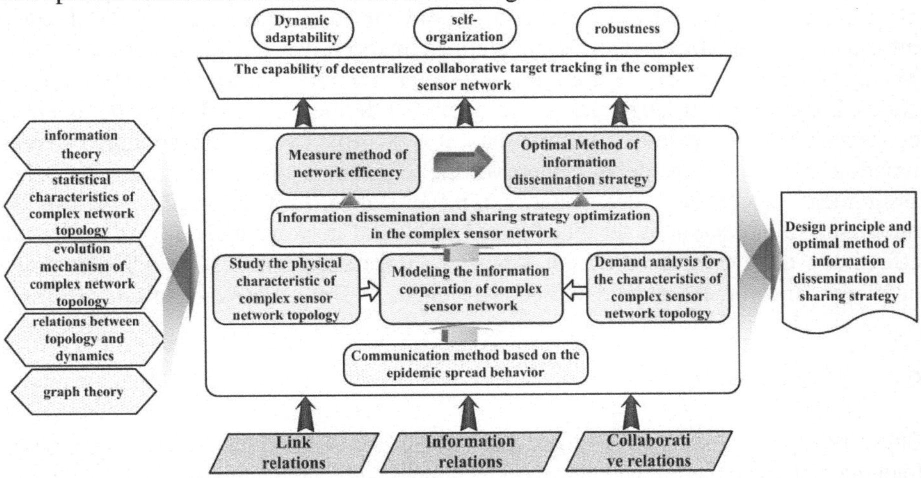


**Fig. 4.** Research method of information distribution and sharing strategy for decentralized collaborative target tracking

## 5 The Generalized Fusion Machine in the Decentralized Collaborative Target Tracking

As more and more geographically distributed homogeneous/heterogeneous sensors are connected into a large tracking network, a large amount of uncertain, incompleteness, inconsistency and imprecise information flood into system. The existing fusion theory and method are unable to cope with the challenges of multi-source, heterogeneous and imperfect information processing during the collaborative target tracking process. The complexity of dealing with the imperfect information in the complex sensor network is mainly represented in the two aspects: the intelligence and the timeliness of information processing.

Earlier researches are focus on a certain feature of the imperfect information with simple treatment. For instance, Rhodes et al. adopted the terminology of imperfect information [18] to describe the uncertain and incomplete information of the linear system in 1969. Dubos and Prade [19] classified the imperfect information into categories as the uncertain, the imprecise, the ambiguous and the inconsistence information in 1988. Recently, J. Dezert et al. present a generalized fusion idea [20], and design a generalized fusion machine to handle imperfect information smartly.

In order to deal with the timeliness of information processing, Xinde Li et al. put forward the hierarchical approximate reasoning method [21]. This method could effectively solve the computing bottleneck caused by the combination explosion problem. Based on the existing generalized fusion research, a generalized fusion machine research method was proposed for the decentralized collaborative target tracking. The specific research method is illustrated as follows:

To begin with, a generalized model, for perceiving, obtaining and representing the imperfect information in network nodes, is built for the decentralized sensor network of collaborative target tracking. And based on the heterogeneity of obtained information, semantic categories are defined in accordance with different criteria. Then, the semantic expression is built after abstract description languages or concretize description languages are respectively defined. By analyzing the essential connection between evidence sources data, it is possible to study the similarity degree function of ESMS in depth, improve the definition of evident sources belief assignment center, and determine the basic standard of threshold. At last, an intelligent preprocessor is designed based on neural network technology, which can effectively deal with the different kinds of information sources including reliable, unreliable, incomplete, redundant, inconsistent, as well as other kinds of information.

## 6 Conclusion

This article puts forwards several key technologies for decentralized collaborative target tracking system applied in wide-area sea battlefield with complex network. By analyzing the development trend of collaborative tracking systems and several new theories and techniques, a new and more effective target calculation mechanism of the decentralized collaborative tracking is presented. In addition, a design method of information dissemination and sharing strategy with merits of strong anti-interference ability, dynamic adaptability and self-organization is studied. By virtue of DSmT theory, this paper also suggests building a new generalized intelligent fusion machine to deal with different kinds of information sources in the sensor networks. The above-mentioned studies of key technologies could bring in new approaches in solving decentralized collaborative target tracking in sea battlefield, which also broaden and deepen the theory, methods and scope of application for decentralized collaborative target tracking.

**Acknowledgments.** This work was supported in part by Hu Bei Province Nature Science Foundation of China under Grant 2009CDB334, 2010CDB01501.

## References

1. Llinas, J.: Studying the complexities in distributed object tracking system. In: Proceeding of Int. Conf. on System, Man and Cybernetics, Washington, vol. 2, pp. 2035–2041 (2003)
2. Lin, X., Kirubarajan, T., Bar-Shalom, Y.: Exact multisensor dynamic bias estimation with local tracks. IEEE Trans. on AES 40(2), 576–590 (2004)

3. Hu, H.-T., Jing, Z.-L., Hu, S.-Q.: An Unscented Kalman Filter Based Multi-Platform Multi-Sensor Registration. J. Shanghai Jiaotong University 39(9), 1518–1521 (2005)
4. Wu, Z.-M., Ren, S.-J., Liu, X.: Research on Collaborative Registration Algorithm for Radar System Error. Acta Armamentarii 29(10), 1192–1196 (2008)
5. Hwang, J.S.: Analysis of Effectiveness of CEC (Cooperative Engagement Capability) Using Schutzer's C2 Theory (Master Thesis). Naval Postgraduate School (2003)
6. Chen, J.-Y.: Chinese Modern Geodetic Datum——Chinese Geodetic Coordinate System 2000(CGCS 2000) and Its Frame. Acta Geodaetica et Cartographica Sinica 37(3), 269–271 (2008)
7. Julier, S., Uhlmann, J., Durrant-Whyte, H.F.: A New method for the nonlinear transformation of means and covariance in filters and estimators. IEEE Trans. Automatic Control 45(3), 477–482 (2000)
8. Gordon, N.J., Salmond, D.J., Smith, A.F.M.: Novel approach to nonlinear-non-Gaussian Bayesian state estimation. IEEE Proceedings Radar and Signal Processing 140(2), 107–113 (1993)
9. Ostwald, J., Lesser, V., Abdallah, S.: Combinatorial auction for resource allocation in a distributed sensor network. In: RTSS (2005)
10. Ruairí, M., Keane, M.: The Dynamics Regions Theory: Role Based Partitioning for Sensor Network Optimization. In: ATSN. AAMAS Workshop, pp. 25–30 (2007)
11. Waldock, A., Nicholson, D.: Cooperative Decentralized Data Fusion Using Probability Collectives. In: ATSN. AAMAS Workshop, pp. 47–53 (2007)
12. Ren, Q.-Q., Li, J.-Z., Gao, H., Cheng, S.-Y.: A Two-Phase Sleep Scheduling Based Protocol for Target Tracking in Sensor Networks. Chinese Journal of Computers 32(10), 1971–1979 (2009)
13. Pavlin, G.: Multi agent systems for flexible and robust Bayesian information fusion. In: 10th International Conference on Information Fusion (2007)
14. Watts, D., Strogatz, S.: Collective dynamics of 'small-world' networks. Nature 393, 440–442 (1998)
15. Xu, Y., Lewis, M., Sycara, K., Scerri, P.: An Efficient Information Sharing Approach For Large Scale Multi-agent Team. In: Proceeding, the 11th Int. Conf. on Information Fusion, Cologne, Germany, pp. 1206–1213 (2008)
16. Moreno, Y., Nekovee, M., Vespignani, A.: Efficiency and reliability of epidemic data dissemination in complex networks. Phys. Rev. E 69, 055101 (2004)
17. Zhang, S.-K., Cui, Z.-M., Gong, S.-R., Sun, Y.: An Investigation on Local Area Control of Compromised Nodes Spreading in Wireless Sensor Networks. Acta Electronica Sinica 37(4), 877–883 (2009)
18. Rhodes, Luenberger, D.: Differential games with imperfect state information. IEEE Transactions on Automatic Control 14(1), 29–38 (1969)
19. Dubois, D., Prade, H.: Possibility Theory: An approach to the computerized processing of uncertain. Plenum Press, New York (1988)
20. Huang, X., Li, X., Dezert, J., Wang, M.: A Fusion Machine Based on DSmT and PCR5 for Robot's Map Reconstruction. International Journal of Information Acquisition 13(3), 201–213 (2006)
21. Li, X., Zhu, B., Dezert, J., Dai, X.: An Improved Fusion Machine for Robot Perception with sonar sensors. Journal of Intelligent and Robotic Systems (2010)

# The Application of Intuitionistic Fuzzy MADM Based on Projection Model in Threat Assessment

Jie Huang, Yongjun Zhao, and Bicheng Li

Zhengzhou Information Science Technology Institute, Zhengzhou 450002, China
Huangjie0922@126.com

**Abstract.** To the question of Threat Assessment, a method of Intuitionistic Fuzzy Multiple Attribute Decision Making based on projection model is proposed. In the method, the intuitionistic fuzzy decision matrix is firstly set up according to the factor values. Then, the weighted score vector of each target is got from the decision matrix and the projection values are calculated between every target and the intuitionistic fuzzy ideal solution. The targets can be ranked by the projection values. Finally, the validity of the method is checked by the threat assessment of aerial defense.

**Keywords:** threat assessment, MADM, intuitionistic fuzzy set, projection model.

## 1 Introduction

Threat Assessment is the level 3 processing in JDL data fusion model. This processing concerns the projection of the current situation into the future to determine the potential impact or threats associated with a current evolving situation. There are many methods for threat assessment, such as MADM [1]-[3], variable weight theory [4], neural network [5], intuitionistic fuzzy reasoning[6] , support vector machine [7] , etc. MADM has become a hotspot of research on threat assessment, because it can synthesize many threat factors and reflect their influence on the final evaluation. But in the traditional MADM, the indefinite of threat attributes can't be reflected. Fuzzy MADM uses fuzzy number to describe their indefinite. Nevertheless, it is difficult to express both the positive and negative information of them.

In this paper, Intuitionistic fuzzy MADM based on projection model is proposed to solve the problem of threat assessment. We consider the threat attributes as intuitionistic number. Decision matrix is an intuitionistic fuzzy decision matrix. Each element of the matrix indicates both the degree that the alternative satisfies the attribute and the degree that the alternative doesn't satisfy the attribute. According to the decision matrix, we can calculate the projection values between every target and the intuitionistic fuzzy ideal solution and use the projection values to describe the threat degree.

This paper is organized as follows: The basic concepts of intuitionistic fuzzy set are first briefly introduced in Section 2. Then, intuitionistic fuzzy MADM based on projection model is given in Section 3. Threat assessment model of aerial defense is described in Section 4. Tests are given in Section 5. Finally, a brief conclusion is drawn.

J. Lei et al. (Eds.): AICI 2012, LNAI 7530, pp. 500–505, 2012.

## 2 Preliminaries

In the following, we will briefly introduce the basic notions of intuitionistic fuzzy set which will be used throughout the paper.

**Definition 1.** Let $X$ be a universe of discourse, then a fuzzy set is defined as:

$$A=\{\langle x,\mu_A(x)\rangle \mid x\in X\} \tag{1}$$

Which is characterized by a membership function $\mu_A : X \to [0,1]$ ,where $\mu_A(x)$ denotes the degree of membership of the element $x$ to the set $A$ [14].

Fuzzy set has been proved to be a suitable tool to model imprecise information.It is widely used in many field nowadays. Nevertheless, it is difficult for fuzzy set to express both the positive and negative information of knowledge. In 1986, Atanassov proposed the concept of an intuitionistic fuzzy set(IFS),shown as follows:

**Definition 2.** An IFS $A$ in $X$ is given by

$$A=\{\langle x,\mu_A(x)\rangle,\nu_A(x) \mid x\in X\} \tag{2}$$

Where $\mu_A : X \to [0,1]$ and $\nu_A : X \to [0,1]$ , with the condition

$$0\le \mu_A(x)+\nu_A(x)\le 1, \quad \forall x\in X$$

The member $\mu_A(x)$ and $\nu_A(x)$ represent, respectively, the membership degree and non-membership degree of the element $x$ to the set $A$ [8][13].

**Definition 3.** For each IFS $A$ in $X$ , if

$$\pi_A(x)=1-\mu_A(x)-\nu_A(x), \quad \forall x\in X \tag{3}$$

Then $\pi_A(x)$ is called the degree of indeterminacy of $x$ to the set $A$ [8][13].

**Definition 4.** Let $\alpha=(\mu_A,\nu_A)$ be an intuitionistic fuzzy number, a score function $S(\alpha)$ of an intuitionistic fuzzy value can be represented as follows:

$$S(\alpha)=\mu_A-\nu_A \quad , \quad S(\alpha)\in[-1,1] \tag{4}$$

The score function $S(\alpha)$ is defined as the difference of the membership function $\mu_A(x)$ and the non-membership function $\nu_A(x)$ [15].

## 3 Intuitionistic Fuzzy MADM Based on Projection Model

In order to represent intuitionistic fuzzy MADM, the assumptions are used as follows:

Let $A=\{A_1, A_2, \cdots, A_n\}$ be a discrete set of alternatives;Let $G=\{G_1, G_2, \cdots, G_m\}$ be a set of attributes; Let $\omega=(\omega_1, \omega_2, \cdots, \omega_m)^T$ be the vector of attributes, where $\omega_j \in [0,1](j=1,2,\cdots m)$ , $\sum_{j=1}^{m}\omega_j=1$ . Suppose that $D=(d_{ij})_{n\times m}$ is the intuitionistic fuzzy matrix, where $d_{ij}=(\mu_{ij}, \nu_{ij})$ , $\mu_{ij}\in[0,1]$ , $\nu_{ij}\in[0,1]$ , $\mu_{ij}+\nu_{ij}\le 1$ , $i=1,2,\cdots,n$ , $j=1,2,\cdots,m$ . $\mu_{ij}$ indicates the degree that the alternative $A_i$ satisfies the attribute $G_j$ , $\nu_{ij}$ indicates the degree that the alternative $A_i$ doesn't satisfy the attribute $G_j$ .

**Definition 5.** [12] A score vector $S(A_i)$ of an alternative $A_i$ can be represented as follows:

$$S(A_i)=(S(d_{i1}), S(d_{i2}), \cdots, S(d_{im}))^T \ (i=1,2,\cdots,n) \tag{5}$$

Where $S(d_{ij})=\mu_{ij}-\nu_{ij}, (j=1,2,\cdots,m)$.

The weighted score vector $S_\omega(A_i)$ of an alternative $A_i$ is given by:

$$S_\omega(A_i)=(\omega_1 S(d_{i1}), \omega_2 S(d_{i2}), \cdots, \omega_m S(d_{im}))^T \ (i=1,2,\cdots,n) \tag{6}$$

**Definition 6.** [12] Let $\alpha_j^+=(1,0)(j=1,2,\cdots,m)$ be the maximal intuitionistic fuzzy numbers, a intuitionistic fuzzy ideal solution $A^+$ can be defined as follows:

$$A^+=(\alpha_1^+, \alpha_2^+, \cdots, \alpha_m^+)^T \tag{7}$$

The score vector $S(A^+)$ of $A^+$ is given by:

$$S(A^+)=(S(\alpha_1^+), S(\alpha_2^+), \cdots, S(\alpha_m^+))^T \tag{8}$$

The weighted score vector $S_\omega(A^+)$ of $A^+$ is given by:

$$S_\omega(A^+)=(\omega_1 S(\alpha_1^+), \omega_2 S(\alpha_2^+), \cdots, \omega_m S(\alpha_m^+))^T \tag{9}$$

**Definition 7.** [12] A projection of $S_\omega(A_i)$ on $S_\omega(A^+)$ can be represented as follows:

$$\Pr j_{S_\omega(A^+)}(S_\omega(A_i)) = |S_\omega(A_i)|\cos\left(S_\omega(A_i), S_\omega(A^+)\right)$$
$$= |S_\omega(A_i)| \frac{\sum_{j=1}^{m} \omega_j^2 S(d_{ij}) S(\alpha^+)}{|S_\omega(A_i)||S_\omega(A^+)|}$$
$$= \frac{1}{|\omega|}\sum_{j=1}^{m} \omega_j^2 S(d_{ij}) \tag{10}$$

The projection $\Pr j_{S_\omega(A^+)}(S_\omega(A_i))$ is defined as the similarity of the alternative $A_i$ and the ideal solution $A^+$. The greater $\Pr j_{S_\omega(A^+)}(S_\omega(A_i))$, the better the alternative $A_i$ will be. We can rank all these alternatives with the projection and then select the most desirable one(s).

## 4 Threat Assessment Model

In this section, we introduce the threat assessment of aerial defense. First, the major factors of attacking targets from air are analyzed. Then, target threat assessment using intuitionistic fuzzy MADM is given.

According to the rules of using weapon system on battles and messages offered by sensors, the following factors will be selected to analyse the target threat degree: the target types( $x$ )、 the distance from target to our party( $r$ )、 the air-raid manner( $h$ )、 the speed of threat target( $v$ )、 the interferential ability of threat target( $g$ ) and the course angle of threat target( $\theta$ ).

As to the threat assessment of aerial defense using the intuitionistic fuzzy MADM, we can regard targets as choice alternatives, and regard the factors which reflect the threat degree of aerial target from different aspects as attributes. Then, quantify the attributes and set up the intuitionistic fuzzy decision imatrix. According to formula (5)-(10), obtain the projection values which can be used to describe the threat degree of each target. Finally, obtain the sequence of targets according to the projection values.

## 5 Tests and Result

Utilizing the data from Ref.[16], we can check the validity of the method. Suppose the enemy party takes a large-scale air formation to attack one area of our party, the detecting equipment has detected 6 groups of targets, and their attributes are list in Table 1. The calculation process of this method works like below.

According to the Table 1, we can build the model with six targets for six attributes. The six possible alternatives $A_i(i=1,2,\cdots,6)$ are to be evaluated using the intuitionistic fuzzy information under the six attributes, as list in the following matrix.

$$D=\begin{pmatrix} (0.625,0.3) & (0.25,0.6) & (0.889,0.0888) & (1,0) & (0.5,0.4) & (1,0) \\ (0.625,0.3) & (0.5,0.4) & (0.889,0.0888) & (1,0) & (0.5,0.4) & (0.875,0.1) \\ (1,0) & (1,0) & (1,0) & (0.25,0.6) & (0.57,0.344) & (0.5,0.4) \\ (1,0) & (0.75,0.2) & (0.889,0.0888) & (0.25,0.6) & (1,0) & (0.875,0.1) \\ (0.625,0.3) & (0.25,0.6) & (0.778,0.1776) & (0.5,0.4) & (1,0) & (0.875,0.1) \\ (0.375,0.5) & (0.125,0.7) & (0.667,0.2664) & (0.25,0.6) & (1,0) & (0.875,0.1) \end{pmatrix}$$

The attribute weighting values can be got by the decision maker as follows:

$$\omega=(0.32,0.1736,0.208,0.1234,0.1206,0.1209)^T$$

We can calculate the threat degree of all targets using formule (5)-(10). The result is $d=\{0.1893,0.2112,0.3731,0.3707,0.1621,0.0163\}$. From this, we can get the sequence of target threat degree. It likes $x_3 > x_4 > x_2 > x_1 > x_5 > x_6$. The sequence accords precisely with the result in Reference [16].

**Table 1.** Target attributes of the enemy part[16]

| | $x$ | $r$ (Km) | $h$ (m) | $v$ (m/s) | $g$ | $\theta$ |
|---|---|---|---|---|---|---|
| $x_1$ | large | 100 | middle | 400 | strong | 50 |
| $x_2$ | large | 150 | middle | 720 | strong | 80 |
| $x_3$ | small | 300 | low | 1600 | none | 30 |
| $x_4$ | small | 260 | low | 1200 | none | 50 |
| $x_5$ | large | 140 | beyond low | 280 | medium | 100 |
| $x_6$ | helicopter | 120 | beyond low | 100 | medium | 150 |

# 6 Conclusions

Aiming at the threat assessment, this paper describes an intuitionistic fuzzy MADM method based on the projection. We use the form of intuitionistic fuzzy numbers to describe the threat of each target according to each factor value and set up the intuitionistic fuzzy decision matrix. The intuitionistic fuzzy MADM based on projection is utilized to aggregate the intuitionistic fuzzy information corresponding to each target. The projection values are calculated between the weighted score vector of each target and the weighted score vector of the intuitionistic fuzzy ideal solution. According to the projection values, the targets can be ranked. Experiment shows that the proposed method can efficiently handle the threat assessment of aerial defense.

## References


1. Qu, C., He, Y.: A Method of Threat Assessment Using Multiple Attribute Decision Making. In: 6th International Conference on Signal Processing, vol. 2, pp. 1091–1095 (2002)
2. Ke, H.-F., Chen, Y.-G.: Threat assessment model of the jamming target of electronic warfare based on multiple attribute and nultiple hierarchy. System Engineering and Electronics 28(9), 1370–1374 (2006)
3. Wang, X.-Y., Liu, Z.-W., Hou, C.-Z., Zhang, C., Yuan, J.-M.: Method of object threat assessment based on fuzzy MADM. Control and Descision 22(8), 859–863 (2007)
4. Cao, K.-J., Jiang, H., Zhao, Z.-G.: Air threat assessment based on variable weight theory. Journal of PLA University of Science and Technology 7(1), 32–35 (2006)
5. Wang, X.-H., Qin, Z., Liu, Y., Shi, Z.-W.: RBF neural network for threat sequencing. Journal of System Simulation 16(7), 1576–1579 (2004)
6. Lei, Y.-J., Wang, B.-S., Wang, Y.: Techniques for threat assessment based on intuitionistic fuzzy reasoning. Journal of Electronics & Information Technology 29(9), 2077–2081 (2007)
7. Yuan, B., Geng, B.-Y., Yang, H.-M.: Radiation source threat assessment based on support vector machine. Fire Control and Command and Control 33(2), 63–65 (2008)
8. Atanassov, K.: Intuitionistic fuzzy sets. Fuzzy Sets and Systems 20(1), 87–96 (1986)
9. Xu, Z.S., Yager, R.: Some geometric aggregation operators based on intuitionistic fuzzy sets. International Journal of General Systems 35, 417–433 (2006)
10. Xu, Z.S., Hu, H.: Projection models for intuitionistic fuzzy multiple attribute decision making.Technical Report (2007)
11. Wei, G., Yi, W.: Method for intuitionistic fuzzy multiple attribute decision making without weight information but with preference information on alternatives. In: 2008 Chinese Control and Decision Conference, pp. 1970–1975 (2008)
12. Xu, Z.S., Hu, H.: Projection models for intuitionistic fuzzy multiple attribute descision making.Techical Report (2007)
13. Atanassov, K.: More on intuitionistic fuzzy sets. Fuzzy Sets and Systems 33, 37-46 (1989)
14. Atanassov, K.: Two theorems for intuitionistic fuzzy sets. Fuzzy Sets and systems 110, 267–269 (2000)
15. Chen, S.M., Tan, J.M.: Handling multicriteria fuzzy decision-making problems based on vague set theory. Fuzzy Set and System 67, 163–172 (1994)
16. Zhang, C., Su, H.-B., Hou, C.-Z., Wang, X.-Y.: An object threat assessment method of indefinite multiple attribute descision making. Journal of China Ordnance 3(1), 38–42 (2007)

# Fault Analysis of Condenser Based on RBF Network and D-S Evidence Theory

Fei Xia[1,2], Hao Zhang[1,2], Wei Liu[3], Daogang Peng[1], Hui Li[1], and Cunmei Xu[1]

[1] College of Electric Power and Automation Engineering, Shanghai University of Electric Power
200090 Shanghai, China
[2] CIMS Research Center, Tongji University
200184 Shanghai, China
[3] Shanghai Chinaust Plastics Corp., Ltd.
201708 Shanghai, China
xiafei@shiep.edu.cn

**Abstract.** A novel Information fusion fault diagnosis method is proposed for condenser fault analysis. Condenser fault diagnoses were analyzed by two algorithms of Radical Basis Function (RBF) neural network. And then the method of information fusion diagnosis was used for improving the results form the two networks. This method has both advantages of the simple features of neural networks and the uncertainty capabilities of information fusion in the application. Through the condenser fault simulation test, it can be verified to improve the accuracy of fault diagnosis, while reducing the complexity of the algorithm.

## 1 Introduction

Condenser fault diagnosis is very important [1]. However the condenser fault diagnosis is more complicated [2]. Although the neural network is a simple and feasible method to achieve a certain degree of condenser fault diagnosis, there is a lower diagnostic accuracy problems [3]. Because the information fusion method has the ability to deal with imperfect knowledge [4], it can improve the accuracy of diagnosis.

This paper focuses on the information fusion technology of the condenser for the fault diagnosis strategy. Although the data fusion method is good, the process of information process is more complex with more time. In most cases, the neural network method is able to correctly judge the failure of condenser. Therefore, the method of fusion decision only is carried for the results which can't be determined correctly by the method of neural network. For the condenser failure, different algorithms of RBF neural network [6] were used respectively for diagnosis. If the two kinds of neural network fault diagnosis results are consistent, the fault type is determined. If the two neural network fault diagnosis results are inconsistent, the integration method of diagnosis will be adopted for getting the final fusion result based on their diagnosis.

J. Lei et al. (Eds.): AICI 2012, LNAI 7530, pp. 506–513, 2012.

# 2 Fundamentals

## 2.1 Radial Basis Function Network

Powell. M.J.D proposed a multi-variable interpolation radial basis function (RBF), Broomhead and Lowe applied RBF to the design of artificial neural network and constructed radial basis function artificial neural network firstly. From then on, people launched a series of RBF neural network research, and also made a lot of encouraging results. Theorem has been proved that RBF network can approximate the nonlinear function and the approximation is unique. RBF network not only has the full power of the comprehensive general network characteristics such as nonlinear mapping, fault tolerance, adaptability, parallel processing and information processing capability etc., which the BP network has, but greater than the BP network capacity of local approximation, not easy to fall into local minimum points, fast training speed and so on.

The RBF network is a three-layer feed forward network with instructors and its structure is similar to BP network: the first layer is input layer, which consists of source nodes; the second layer is hidden layer and the number of elements is selected automatically within the given framework according to the accuracy requirements, and the third layer is output layer, which responds to the role of input mode.

## 2.2 D-S Evidence Theory

Evidence theory [7] is a reasoning method of uncertainty. It is promoted by Dempster initially. He takes the method of many-valued mapping gets the up and down limit of the probability. And then it is promoted and has become evidential reasoning by Shafer in 1976. So it is called D-S theory. D-S evidential theory is used as a reasoning method of uncertainty, has large development these years, and receives more and more concern. It grasp the unknown and uncertain in the problems better than traditional probability theory, and it provides a very useful relative evidence synthesis method. And so it could fuse a number of evidences provided by sources. This method is successfully applied in target recognition and other fields.

The D-S evidence theory provides a very useful synthesis of the formula. It makes us could compose many evidences provided by different evidence sources. The formula is as following.

$$m(H) = \frac{1}{1-k} \sum_{H_i \cap H_j \cdots = H} m_1(H_i) m_2(H_j) \cdots$$

$$\forall H \in 2^{\Theta} \tag{1}$$

And the k represents the evidence's conflict level. The coefficient 1/(1-k) is called normalization factor. Its function is to avoid assigning the non-zero probability to the empty set when coalescence.

## 3 Proposed Methodology

This section mainly described the framework of the proposed method, which was a fault diagnosis method based on information fusion. This method took the advantage of fuzzy logic and neural network. Firstly, the inputs of neural network were fuzzed by the fuzzy membership functions, and the RBF neural network was used to accomplish the initial diagnosis of the fault type. Then, according to the feature of D-S evidence theory which can deal with uncertain information, the initial diagnosis results were fused after further processing.

In fault diagnosis, needing to judge whether a fault has happened and its fault type on the basis of the fault symptoms represented by the system, fault symptoms can be gotten from the data and information provided by some sensors, so each information (Diagnosis parameters) given by sensors used in diagnosing can be regard as the evidence of judging whether some faults has happened. To make evidences provided by a sensor can be combined with others, the key was to construct basic probability distribution function based on the existing evidence, but its general form was not given in D-S Evidence Theory, so its specific form was constructed on the basis of the specific problem. It in comprehensive diagnosis can be constructed according to the reliability of the sensors and diagnostic result of single evidence. The basic probability distribution function of D-S evidence theory diagnosis [9] can be constructed by using local diagnostic results after the local diagnosis of neural network.

Setting I = {diagnostic parameter $i \mid i = 1, 2, \ldots, p$ }, H= {system state $j \mid j = 1, 2, \ldots, q$}. $q$ output nodes of neural network in local diagnosis corresponded to $q$ system states separately, and for the diagnostic parameter $i$, supposing that the output of node $j$ of the network was $C_i(j)$, from the output range of neural network we can know that $1 \geq C_i(j) \geq 0$. On the assumption that the reliability of using parameter $i$ to diagnose locally in neural network.

The basic probability distribution function was defined as follow in the paper:

$$m_i(j) = \frac{C_i(j)}{\sum_j C_i(j)} \times R(i) \tag{2}$$

Where j=1,2,…,q.

$$m_i(\theta) = 1 - R(i) \tag{3}$$

$m_i(j)$ represents the probability of state $j$ occurring when diagnostic parameter $i$ was used to diagnose, and $m_i(\theta)$ was the probability of which state occurred that can't be determined by diagnostic parameter $i$.

After obtaining the output values of neural network diagnosis, from the results obtained by neural network diagnosis, we can see some can determine which fault has happened, but some can't determine the fault type to the end accurately. On this time,

D-S evidence theory [10] was used to fuse the diagnosis results, and then the fusion diagnosis results were obtained.

Because not all the ranges of output value $C_i(j)$ of the trained RBF neural network were between 0 and 1, then using formula (10) to normalize them, the normalized values were obtained represented as $C_i(j)\_GY$ :

$$C_i(j)\_GY = \frac{C_i(j) - \min(C_i(j))}{\max(C_i(j)) - \min(C_j(j))} \tag{4}$$

The range of these data was between 0 and 1, then the value of $m_i(j)$ was obtained by the normalized data $C_i(j)\_GY$ . This formula (5) should be converted into the following form:

$$m_i(j) = \frac{C_i(j)\_GY}{\sum_j C_i(j)\_GY} \times R(i) \tag{5}$$

Where $m_i(j)$ found out represented the possibility of state $j$ was diagnosed by the parameter $i$. The value of $R(i)$ was (1－1/14), $m_i(\theta)$ was obtained according to the formula (6), which represented the possibility can't be determined by the parameter $i$.

According to the formula (1) fusing several information sources, the basic probability distribution provided by two sources were fused, and the results obtained by fusing two information were considered as the basic probability distribution of a signal source, and then fusing the basic probability distribution provided from the rest of the information source in the same way, at last, the fusion of 17 basic probability distribution provided by information sources were realized.

# 4 Experiments Analysis

To validate the algorithm proposed in the paper, the condenser, which is the important auxiliary equipment of steam turbine in power plant, was taken examples to test the novel diagnose fault approach.

## 4.1 Test Data

Due to the complexity of condenser system and particularity of the running environment, manifold fault reasons and fault symptoms were presented. 16 kinds of fault symptoms and 14 kinds of typical fault types from condenser were selected on the basis of the field operating experiences and relevant data in the paper. In order to carry through the simulation experiments, the fault symptom data from 40 groups of known fault modes were selected to diagnose by using Newrb algorithm and K-means algorithm of RBF neural network. Then, the information fusion method proposed in this paper was used for the results. Table 1 was three test data in the stimulation experiments.

### 4.2 Results of Fusion

For the three test data in Table 1, two different algorithms of RBF were adopted respectively. The results were showed in Table 2, Table 3 and Table 4. The analysis is described in the following.

In Table 2, there are the diagnosis results of RBF of test data 3. In the two algorithms, the fault Y4 has the maximum value in the outputs of the both networks. So the fault type in test data 3 is Y4. But the value is only 0.25 in the two algorithms. These results are not obvious bigger than the other outputs of fault type. Then, the proposed fusion method was used for combine the two results in Table 2. After fusion, the output for Y4 increases to 0.3593, which is improved by the approach adopted in this paper.

**Table 1.** The data for test

| Fault symptoms | Test data 1 | Test data 2 | Test data 3 |
|---|---|---|---|
| X1 | 0.75 | 0.75 | 1.00 |
| X2 | 0.50 | 0.56 | 0.50 |
| X3 | 0.50 | 0.50 | 0.45 |
| X4 | 0.50 | 0.75 | 0.50 |
| X5 | 0.50 | 0.50 | 0.50 |
| X6 | 0.50 | 0.50 | 0.55 |
| X7 | 0.55 | 0.55 | 0.50 |
| X8 | 0.75 | 0.75 | 0.25 |
| X9 | 0.75 | 0.75 | 0.75 |
| X10 | 0.50 | 0.50 | 0.70 |
| X11 | 0.75 | 0.70 | 0.50 |
| X12 | 0.45 | 0.50 | 0.48 |
| X13 | 0.50 | 0.50 | 0.50 |
| X14 | 0.50 | 0.50 | 0.50 |
| X15 | 0.50 | 0.50 | 0.53 |
| X16 | 0.50 | 0.50 | 0.55 |
| Fault Type | Circular water shortage | Dirty Condenser Tubes | Rupture of Vacuum system pipe |

Then the same process was done for test data 1 and test data 2. The diagnosis results are showed respectively in Table 3 and Table 4. From Table 3, the maximum value for fault type is nearly 0.35 in two algorithms. And the value in Table 4, for the algorithm of newrb is 0.36, and for algorithm of k-means is 0.20. After combined the results from the two methods, the maximum value for test data 1 is 0.5487 and the one for test data 2 is 0.4288. They are also improved by the fusion method. The clear decision can be made for the two test data. There are fault type Y13 in test data

1 and fault type 9 in test data 2. In order to discriminate the fault type correctly, the difference between the reliability of fault category and the reliability of other categories should be larger than 0.3, and the probability value of uncertainty should be smaller than 0.2 based on test. All examples in this section are satisfied this condition. So the obvious correct fault types were obtained.

**Table 2.** The diagnosis results of RBF (Test Data 3)

| Fault mode | Results of Newrb method | Results of K-meas method |
|---|---|---|
| Y1: Serious accident of circulating pump | 0.0604 | 0.0593 |
| Y2: The interruption of steam supply in real axle seal | 0.0281 | 0.0251 |
| Y3: Condenser filled with water | 0.0580 | 0.0545 |
| Y4: Rupture of Vacuum system pipe | 0.2533 | 0.2263 |
| Y5: Vacuum system is not tight | 0.0719 | 0.0552 |
| Y6:Abnormal working of condensate Pump | 0.0541 | 0.0498 |
| Y7: Rupture of Condenser Tubes | 0.0670 | 0.0613 |
| Y8: Rupture of low pressure copper heater | 0.06287 | 0.0598 |
| Y9: Dirty Condenser Tubes | 0.0618 | 0.0551 |
| Y10: Shortage of cycling water | 0.0741 | 0.0670 |
| Y11: Abnormal working of ejector | 0.0671 | 0.0395 |
| Y12: The overall running state of condenser | 0.0420 | 0.0624 |
| Y13: Circular water shortage | 0.0451 | 0.0550 |
| Y14: Pump serious failure | 0.0540 | 0.1298 |

**Table 3.** The diagnosis results of RBF (Test Data 1)

| Fault mode | Results of Newrb method | Results of K-meas method |
|---|---|---|
| Y1: Serious accident of circulating pump | 0.0504 | 0.0524 |
| Y2: The interruption of steam supply in real axle seal | 0.0512 | 0.0374 |
| Y3: Condenser filled with water | 0.0547 | 0.0520 |
| Y4: Rupture of Vacuum system pipe | 0.0588 | 0.0567 |
| Y5: Vacuum system is not tight | 0.0416 | 0.0492 |
| Y6:Abnormal working of condensate Pump | 0.0514 | 0.0510 |
| Y7: Rupture of Condenser Tubes | 0.0443 | 0.0468 |
| Y8: Rupture of low pressure copper heater | 0.0613 | 0.0563 |
| Y9: Dirty Condenser Tubes | 0.0412 | 0.0492 |
| Y10: Shortage of cycling water | 0.0500 | 0.0563 |
| Y11: Abnormal working of ejector | 0.0654 | 0.0421 |
| Y12: The overall running state of condenser | 0.0389 | 0.0390 |
| Y13: Circular water shortage | 0.3500 | 0.3363 |
| Y14: Pump serious failure | 0.0412 | 0.0810 |

**Table 4.** The diagnosis results of RBF (Test Data 2)

| Fault mode | Results of Newrb method | Results of K-meas method |
|---|---|---|
| Y1: Serious accident of circulating pump | 0.0401 | 0.0539 |
| Y2: The interruption of steam supply in real axle seal | 0.0484 | 0.0226 |
| Y3: Condenser filled with water | 0.0557 | 0.0601 |
| Y4: Rupture of Vacuum system pipe | 0.0474 | 0.1152 |
| Y5: Vacuum system is not tight | 0.0502 | 0.0597 |
| Y6:Abnormal working of condensate Pump | 0.0552 | 0.0617 |
| Y7: Rupture of Condenser Tubes | 0.0425 | 0.0593 |
| Y8: Rupture of low pressure copper heater | 0.0627 | 0.0640 |
| Y9: Dirty Condenser Tubes | 0.3611 | 0.2030 |
| Y10: Shortage of cycling water | 0.0417 | 0.0601 |
| Y11: Abnormal working of ejector | 0.0484 | 0.0472 |
| Y12: The overall running state of condenser | 0.0530 | 0.0275 |
| Y13: Circular water shortage | 0.0452 | 0.0597 |
| Y14: Pump serious failure | 0.0484 | 0.1062 |

## 5 Conclusion

This paper presents a fusion approach for fault diagnosis in the application of the condenser. This method has advantages of neural network and information fusion. Because the information fusion technology has the ability to ameliorate the output of RBF network. It improves the accuracy of fault diagnosis on the one side and reduces the computational complexity on the other side. It can be used as an effective strategy for fault diagnosis applied to online fault diagnosis system in development.

**Acknowledgment.** This work is supported by Project Supported by the State Key Program of National Natural Science Foundation of China (Grant No. 61034004), Leading Academic Discipline Project of Shanghai Municipal education Commission (No. J51301), Shanghai Rising-Star Program (10QA1402900).

## References

1. Chen, C., Mo, C.: A method for intelligent fault diagnosis of rotating machinery. Digital Signal Processing 14(3), 203–217 (2004)
2. Ma, J., Zhang, Y.-S.: Condensation Sets Fault Diagnosis Based on Data Fusion. Turbine Technology 51(2), 141–143 (2009)
3. Qi, Z.-W., Gu, C.-L.: Application of fault diagnosis to equipment based on modified D-S evidential theory. Journal of Naval University of Engineering 20(1), 11–14 (2008)
4. Ye, Q., Wu, X.-P., Song, Y.-X.: Fault diagnosis method based on D-S theory of evidence and AHP. Journal of Naval Universit of Engineering 18(4), 12–17 (2006)
5. Wen, X.-Q., Xu, Z.-M., Sun, Y.-Y., Sun, L.-F.: A Model for Condenser Fault Diagnosis Based on Least Squares-support Vector Machine and Simulation Application. Turbine Technology 30(2), 204–206 (2010)

6. Jiang, J.-G.: Fault Diagnosis of Condenser Based on Wavelet Neural Network. Industrial Engineering Journal 16(4), 40–43 (2010)
7. Ji, D., Wang, X.: Fault Diagnosis of DC Machine Based on D-S Evidential Theory and BP Network. Marine Electric & Electronic Engineering 27(4), 204–206 (2007)
8. Liu, B.-J., Xiao, B., Fu, H.: Discussion of Data Fusion Algorithm Based on D-S Evidential Theory and Fuzzy Mathematics. Microprocessors (3), 70–72 (2007)
9. Xiong, X., Wang, J., Niu, T., Song, Y.: A Hybrid Model for Fault Diagnosis of Complex Systems. Electronics Optics & Control 16(2), 56–59 (2009)
10. Wei, X., Shu, N., Cui, P., Wu, B.: Power Transformer Fault Integrated Diagnosis Based on Improved PSO-BP Neural Networks and D-S Evidential Reasoning. Automation of Electric Power Systems 30(7), 46–49 (2006)

# Simulation Research on Singularity Detection of Transient Power Disturbance

Jian-ping Zhou and Yan Chen

School of Power and Automation Engineering,
Shanghai University of Electric Power, Shanghai, China
zhoujianping@shiep.edu.cn, chenyan1991@yahoo.cn

**Abstract.** According to the actual power system transmission line, the voltage and current transient simulation models are set up. Four different types of transient signals are obtained and they are voltage oscillation, current oscillation, voltage pulse and current pulse. The wavelet and the lifting wavelet are respectively used to detect the above four kinds of transient signals. The simulation results show that both the wavelet and the lifting wavelet can accurately detect the singularity of transient signal. However, the reconstruction error of the lifting wavelet is much less than that of the wavelet. The algorithm of the lifting wavelet is more accurate than that of the wavelet. Therefore, the lifting wavelet is more suitable for the accurate and real-time operating requirements in the power system.

**Keywords:** wavelet, lifting wavelet, transient, detect.

## 1 Introduction

With the continuous improvement of supply reliability, more attention is paid to the new dynamic power quality problems such as dynamic voltage increase, the pulse and the voltage sag [1]. The dynamic power quality includes the short duration voltage change and a variety of transient phenomena. Transient voltage disturbance is the power quality pollution problem of supply voltage sine wave caused by the transient voltage disturbance. Characterized by spectrum and transient duration, transient power quality disturbance can be divided into two types of oscillating transient and pulse transient [1]. At present, there are no unified detection standard and method because disturbance is short and random. Wavelet transform shows the local characteristic of signal both in time and frequency domain, which is suitable to deal with the abrupt signal. Therefore, it is widely used to detect and recognize transient power quality disturbance [2]-[5].

The wavelet transform based on lifting scheme is presented by Sweldens, namely lifting wavelet [6]. The lifting scheme is a simple construction of the second generation wavelet. These wavelets are not necessarily translation and dilation of one fixed function. The lifting scheme allows one to design the filters needed in the transform algorithms. The lifting scheme also leads to a fast in-place calculation of the wavelet transform, i.e., an implementation that does not require additional memory, since fast

J. Lei et al. (Eds.): AICI 2012, LNAI 7530, pp. 514–520, 2012.

calculation is important in power system, especially when a real-time operation is needed. In this paper, simulation is made according to the actual voltage transients and current transients. Lifting wavelet and wavelet transform are adopted respectively to detect the singularity when voltage oscillation, current oscillation, voltage pulse, and current pulse occur in the power grid. Detection performance is evaluated from the modulus maxima and the maximum reconstruction error.

## 2 Modulus Maxima and Maximum Reconstruction Error

Local singularity of the signal can be characterized through the signal's wavelet transform modulus maximum. The modulus maxima of the wavelet transform caused by the signal singularity have different transfer properties in the wavelet transform scale compared to the modulus maxima of wavelet transform caused by detection noise. A small mutation in the waveform can be enlarged and displayed by the wavelet transform. Therefore, modulus maximum is an important indicator to measure whether the transient singular point exists or not.

In addition, both the wavelet transform and the lifting wavelet transform are related to the processes of the signal decomposition and reconstruction. In the process, a small error would be produced and the concept of the reconstruction is introduced based on this small error. The original signal is decomposed into multilayer detail signals and approximate signals, and then the decomposed signals are added together into a sum signal. The difference between this sum signal and the original signal is regarded as the reconstruction error.

As shown in figure 1, the original signal S is decomposed into high-frequency signal D1 and low-frequency signal A1, then low-frequency signal A1 is further decomposed into the second layer of high-frequency signal D2 and the low frequency signal A2, then the low frequency signal A2 further decomposed into the third layer of high-frequency signal D3 and low-frequency signals A3. The last layer of low frequency A3 may be further decomposed according to this principle if necessary. Therefore, the maximum reconstruction error is that the original signal subtracts the highest level of low-frequency signal and three high-frequency signals, i.e. err = max (abs (S-A3-D3-D2-D1)), where abs means absolute value. Therefore, the value of the reconstruction error shows the accuracy of the wavelet function in the processes of signal decomposition and reconstruction.

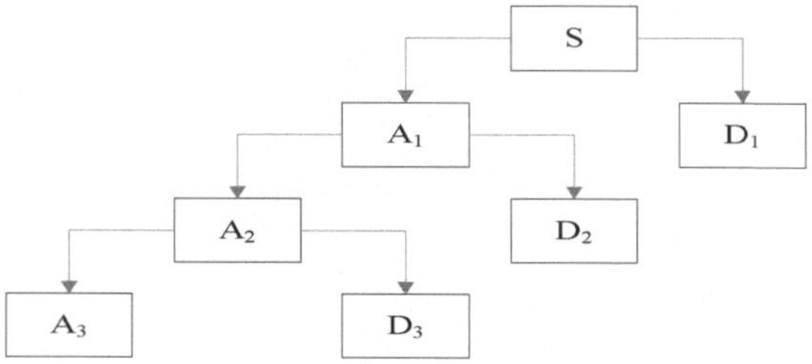


**Fig. 1.** Schematic of the wavelet three layer decomposition

# 3 Oscillation and Pulse Transient Simulation Modeling

## 3.1 Oscillation Transient

Oscillating transient is one of the transient power quality problems. Figure 2 shows the oscillating transient simulation modeling. As can be seen from figure 2, three-phase power supplies to the two three-phase RLC series branch and RLC series branch breaks through single-phase breaker. The two RLC-branch capacitances charge and discharge, which lead to oscillation transient. Breaker is set close and break at 0.02s and 0.04s, respectively. Voltage and current oscillation waveform can be obtained by the oscilloscope display, as shown in figure 3.

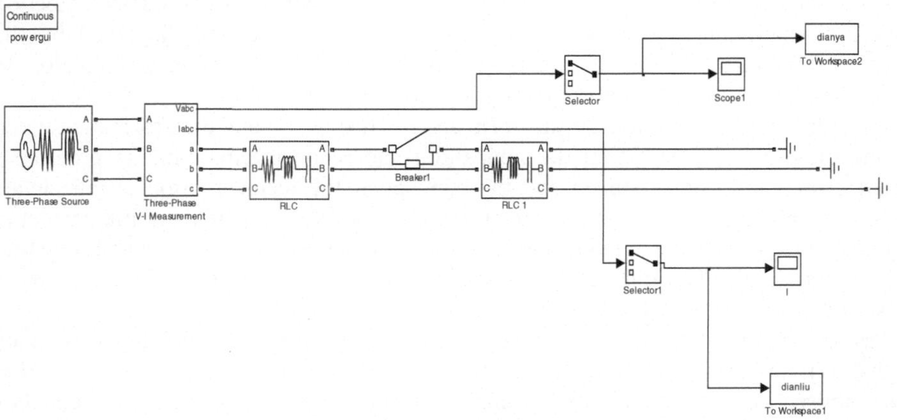


Fig. 2. Oscillating transient signal simulation model

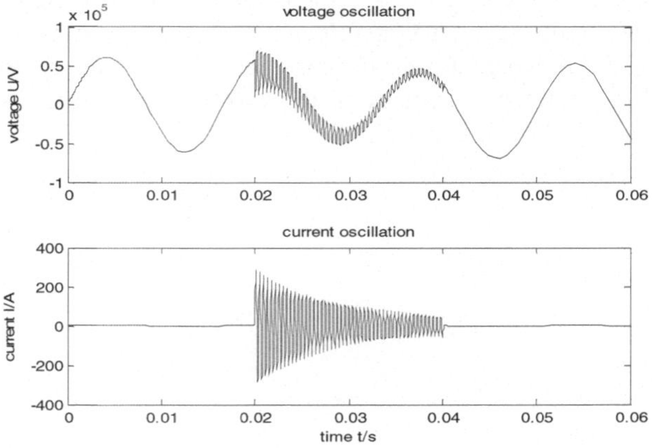


Fig. 3. Voltage and current oscillation waveform

### 3.2 Pulse Transient

Similarly, figure 4 shows the pulse transient simulation modeling. As can be seen from figure 4, a single-phase power supplies to 110MW load after 300km transmission line. The breaker switching leads to a voltage and current pulse waveforms. Set the circuit breaker close at 0.025s. Voltage and current pulse waveform can be shown in figure5.

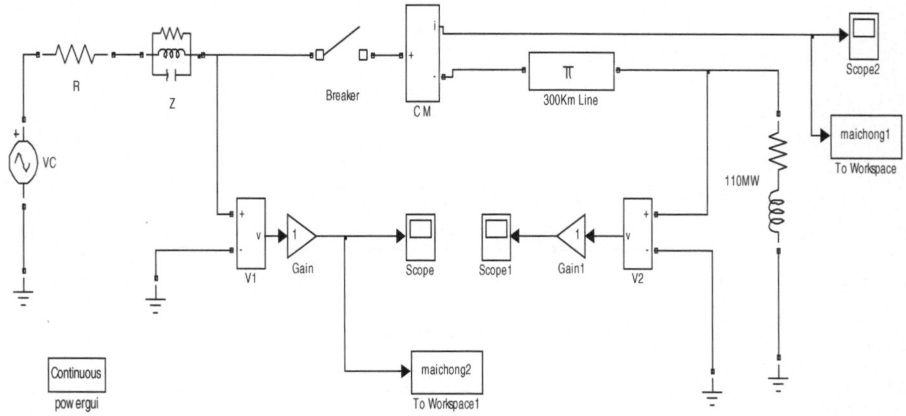


**Fig. 4.** Pulse transient signal simulation model

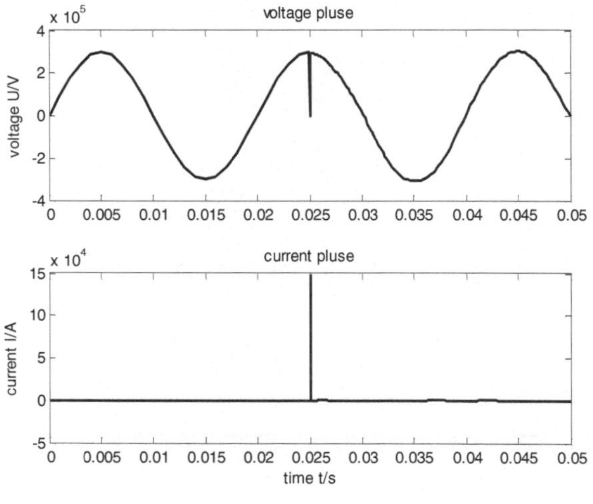


**Fig. 5.** Voltage and current pulse waveform

# 4 Singularity Detection of Oscillating and Pulse Transient

For comparison, db wavelet and its corresponding lifting wavelet are used to detect the above four transient signal. Firstly, db8 wavelet and the corresponding lifting wavelet are used to detect the voltage oscillation and current oscillation. The simulation results suggest that the point of the voltage and current oscillation can be quickly located by db wavelet and lifting wavelet, as shown in figure 6 and figure 7.

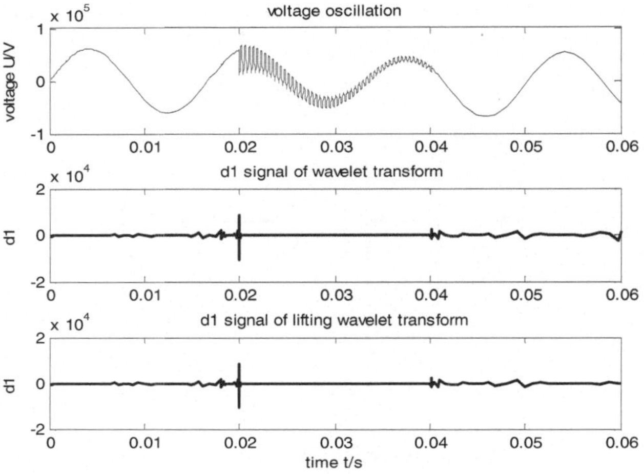


**Fig. 6.** Db wavelet and lifting wavelet used in voltage oscillation

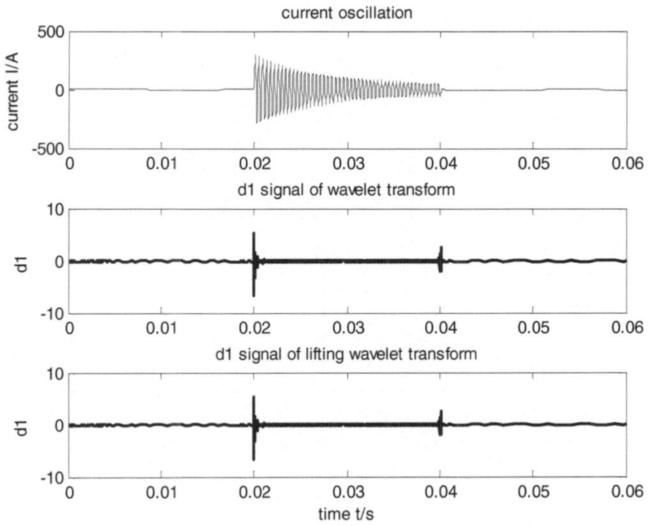


**Fig. 7.** Db wavelet and lifting wavelet used in current oscillation

Analogously, db5 wavelet and the corresponding lifting wavelet are used to detect the voltage pulse and current pulse. The simulation results (not shown) suggest that the point of the voltage and current pulse can be quickly located by db wavelet and lifting wavelet.

To further compare the detection performance of four types of transient singularity using lifting wavelet and wavelet, the modulus maxima and maximum reconstruction errors are respectively compared according to the simulation results, as shown in table 1.

As can be seen from Table 1, the wavelet and lifting wavelet modulus maxima are almost the same, indicating that the wavelet and lifting wavelet can accurately detect transient singular points of voltage oscillation, current oscillation, voltage pulse and current pulse. However, the maximum reconstruction error of lifting wavelet is far less than that of the wavelet, which means that the lifting wavelet in algorithm has more accuracy than the wavelet.

**Table 1.** Detection results of voltage and current transient

| Transient species | modulus maximum | | Maximum reconstruction error | |
|---|---|---|---|---|
| | lifting wavelet | wavelet | lifting wavelet | wavelet |
| voltage oscillation | 10450 | 10450 | 1.3983e-010 | 7.4626e-007 |
| current oscillation | 6.6611 | 6.6611 | 1.0836e-012 | 2.4523e-009 |
| voltage pulse | 105650 | 77787 | 1.9109e-010 | 1.0157e-006 |
| current pulse | 36907 | 38056 | 1.3867e-010 | 2.3705e-007 |

# 5 Conclusion

Based on our analysis and example simulations, the wavelet and lifting wavelet can accurately detect transient singular points of the transient power disturbances. However, the maximum reconstruction error of lifting wavelet is much less than that of the wavelet, suggesting that the algorithm of lifting wavelet is more precise than that of the wavelet, and lifting wavelet is more suitable for fast, accurate and real-time requirements in the electric power system.

**Acknowledgment.** This paper is supported by Leading Academic Discipline Project of Shanghai Municipal Education Commission (J51303).

# References

[1] Cheng, H.-Z., Ai, Q., Zhang, Z.-G.: Power Quality. Tsinghua University Press, Beijing (2006)

[2] Gaouda, A.M., Kanoun, S.H., Salama, M.M.A.: Wavelet-based signal processing for disturbance classification and measurement. IEEE Proceedings-Generation, Transmission and Distribution 149(3), 310–318 (2002)

[3] Chen, X.-X.: Wavelet-based measuring and classification of short duration power quality disturbances. Proceedings of the CSEE 22(10), 1–6 (2002)

[4] Zhang, Y.-H., Chen, X.-D., Wang, H.-Y.: Continuous wavelet-based measuring and classification of short duration power quality disturbances. Electric Power Automation Equipment 24(3), 17–21 (2004)
[5] Wang, C.-S., Wang, J.-D.: Classification on the method of power quality disturbance based on wavelet packet decomposition. Power System Technology 28(15), 78–82 (2004)
[6] Sweldens, W.: The lifting scheme: A construction of second generation wavelet. SIAM J. Math. Anal. 29(2), 511–546 (1998)
[7] Chen, H.X., Chua, P.S.K., Lim, G.H.: Vibration analysis with lifting scheme and generalized cross validation in fault diagnosis of water hydraulic system. Journal of Sound and Vibration 301, 458–480 (2007)
[8] Zhou, W.: MATLAB Wavelet analysis advanced technology. Xidian University Press, Xi'an (2006)

# CMMB Image Sequences Measurement Based on Computation in High-Dimension Space

Jun Yang, Shi-jiao Zhu, and Zhong-qin Bi

School of Computer and Information Engineering,
Shanghai University of Electric Power, Shanghai 200090, China
{yangjun,zhusj,zqbi}@shiep.edu.cn

**Abstract.** At present, video contents from the Internet are accessed with increasing frequency. In this paper, combined with China Mobile Multimedia Broadcasting （CMMB） system characteristics is presented based on high-dimension space computation modal for measurement of quantity of CMMB video sequences. From the relationship between different points, it makes computation in high-dimension space for the measurement of videos. Different with some classic algorithms, such as PSNR, objective model, which discussed for alignment of video sequences and lead to complex computation, the proposed method is based on computation in high-dimension space. Image sequences of original video are classified into different sets. For real quality measurement, a CMMB image is used to find similar among these sets and gave its measurement. Experimental results indicate that the proposed method make the measurement easily and meet the real noised image sequence. The proposed method is constructive and it proves the reliability of this measurement system.

## 1 Introduction

In many countries, network under the home environment has been developed well. However, use rate of mobile device is smaller than home cable network and the broadcast video modal is still play an important role. CMMB (China Mobile Multimedia Broadcasting) is a system for mobile device, such as digital television, digital cinema, wireless handset TV. Its application is mostly about videos.

In this trend, the user experience of CMMB broadcast quality QoE (Quality of Experience) is increasingly highlighted. QoE can be understood and intuitive experience for users to watch programs or perceived effect. It can be quantified to reflect the network's quality about the gap between user's experience and hops.

As we all know, for television programs, an image eventually gives users the perception of influence eventually with audio-visual experience. From the channels between station and end user, source videos are often noised by different disturbed factors before signal reached terminator [1]. Therefore, image quality assessment is an important issue in the real practice and the measurement can be using in communication in order to improve a network's quality [2].

In order to assess image quality, two different categories methods are used in practice: subjective assessment by humans and objective assessment by algorithm

J. Lei et al. (Eds.): AICI 2012, LNAI 7530, pp. 521–529, 2012.

automatically [3]. For human subjectivity assessment, an algorithm is defined by how well it correlates with human perception of quality. Among the algorithms, Automatic Reduced-Reference image quality assessment algorithms from the point of view of image information change are often used. Such changes are measured between the reference-image and natural-image approximations of the distorted image. Algorithms that measure differences between the entropies of wavelet coefficients of reference and distorted images, as perceived by humans, are designed [4]. Feature similarity (FSIM) index for full reference IQA is proposed, which is a dimensionless measure of the significance of a local structure, is used as the primary feature in FSIM [5]. However, these models are algebra method and limited for video sequence or required more precise time. Some application is only make a measure between received image and original one. Meanwhile, real image is often chaotic sequence received or image quality is low. These methods are invalid for this kind of real application.

The paper is organized as follow. Section 2 gives basic definitions and description image calculation in high-dimensional space; section 3 presents an algorithm of image measurement for coving objects in high-dimensional space. Results of different type images from CMMB video and measurement with real-noised images are provided in Section 4, with conclusions in Section 5.

## 2 Application Model

For CMMB application, signals are transmitted for source to user's destinations. Among this way, signals are often noised by out-side noise. The way is shown in Fig.1.

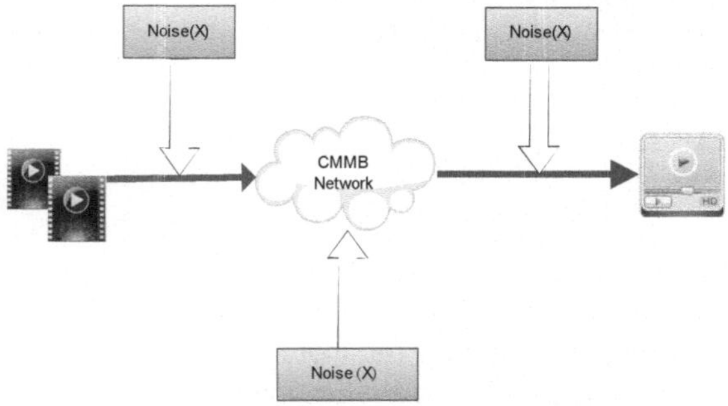


**Fig. 1.** Scheme of CMMB Video Application

For real world, the received video quality is measured by reference objects.

For a digital image which size is M×N, we denote is with $f(x,y)$ as a point in high dimensional space, where $0<x<M, 0<y<N$. It includes original image and noised image. In order to assess the image's quality, we define below scheme as Fig.2.

In Fig 2, images are classified as original, similar, diff and noised images respectively. The main problem is to find sub-set of image in high dimensional space.

Our main idea is to find the points distribution of image in the high dimensional space based on original image. Video quality is decrease when translating in network, such as cable network or wireless network. From the theory of geometrical learning [7], some similar objects in high dimensional space are homeomorphism and can be measured as neighbors [8].

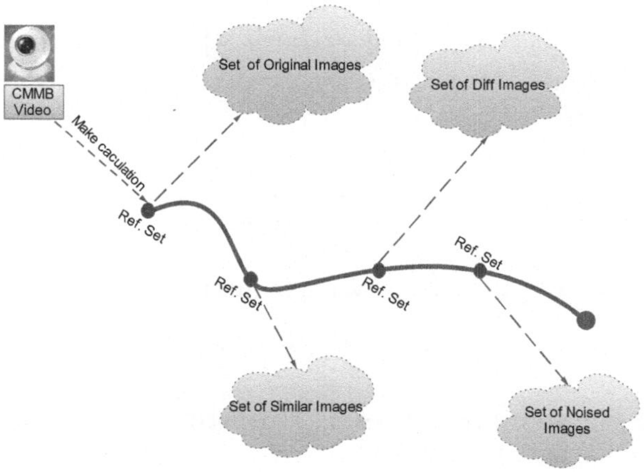


**Fig. 2.** Scheme of point calculation

For the different points in hyper-space, covering entity can be made by adjust parameters [9]. For symbols $X_i \in B_i, X_i \neq Xo_i$, measure distance can be calculated by formula 1 and the shortest distance between current point to center point can be computed by formula 2.

$$X_j = \{X_k \mid \min(\rho(X_k, X_j)), X_k \in B_i\}, X_k \neq Xo_i \tag{1}$$

$$r_x = r - \min(\rho(X_i, Xo_i), \rho(X_j, Xo_i)) \tag{2}$$

When this operating is performed in hype-space for covering vectors of $X_i, X_j$, the result can be got like Fig.3. The covering parameters are shown in two-dimensional space. The distance is described in formula 3.

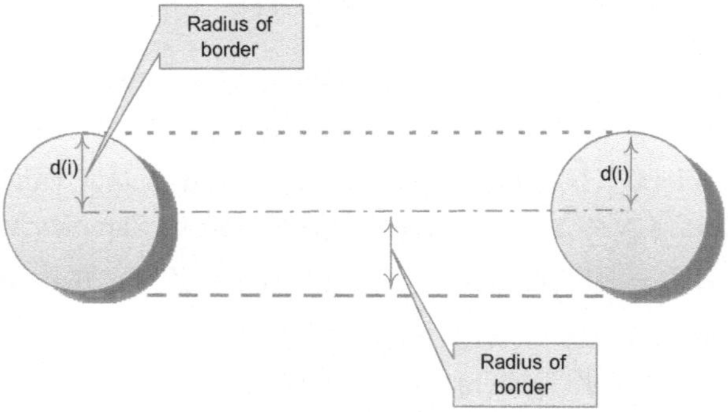


**Fig. 3.** Covering two neighbor points (2-Dim)

Among this expression, where $q(x, x_i, x_j)$ is defined in formula 4.

$$d^2(x,x_i,x_j)=\begin{cases}\|x-x_i\|^2, q(x,x_i,x_j)<0 \\ \|x-x_j\|^2, q(x,x_i,x_j)>\|x_i-x_j\| \\ \|x-x_i\|^2-q(x,x_i,x_j)^2, \text{others}\end{cases} \tag{3}$$

$$q(x, x_i, x_j) = \left\langle x - x_i, \frac{x_j - x_i}{\|x_j - x_i\|} \right\rangle \tag{4}$$

The definition of covering S can be shown in formula 5.

$$S(x,x_i,x_j,r)=\{x \mid d^2(x,x_i,x_j)<r^2\} \tag{5}$$

As for a sample pair, classifier $c:X \rightarrow Y$ is defined as the probability according to an unknown distribution $D$ over $X \times Y$.

## 3 Algorithm

Suppose that in constructive neural network, the whole mapping objects are constructed step by step. Samples covering objects can be generated with different procedure number. The whole architecture is shown in Fig. 4.

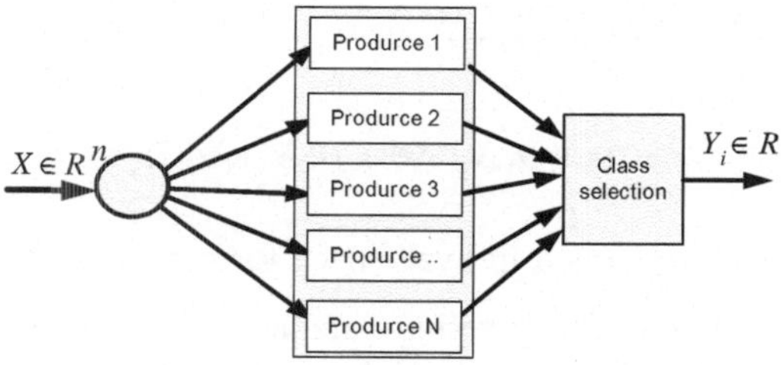


**Fig. 4.** Architecture of Mapping Objects

Let's consider the binary classification problem. The input space is $X$, which an arbitrary subset of is $\mathbb{R}^n$, and the output is $Y=\{0,1\}$ which is a binary classification problem. Therefore a pair of input and output can be symbolized as $z \equiv \{x,y\}$, $x \in X, y \in Y$. The output sequence of the Samples Mapping Objects can be written as $O = \{o_1, o_2, ..., o_p\}$ where $p(p \in Z_+)$ is the produced number which is determined in the process of network construction. The final output of constructive network is $Y_i$ where $i$ class denotation is $Y_i = \min\{i \mid Q(o_i) = 1\}$, $Q$ is a function mapping $y_i \rightarrow \{0,1\}$.

For inside neurons, formula is described as below.

$$\psi = f\left[\sum_{i=1}^{n}\left(\frac{w_j}{|w_j|}\right)^{S}\left|w_j(x_j - w_j^{'})\right|^{P} - \theta\right] \tag{6}$$

where $\theta$ is threshold, $S$ is sign symbol, $P$ is shape symbol, $w^{'}$ is center vector and $w$ is direction vector. For RBF neuron, the parameters are given by $P$ =2, $w_j = 1$, $S = 1$. $w_j^{'}$ is center vector. For hyper-plane neuron, the parameters are given by $P = 1$, $w_j = 1$, $S = 1$. With different parameters, $\psi$ gives different covering shapes in high dimensional space in geometric viewpoint.

For constructive algorithm, let's consider hype-ball neurons in hidden layers where high priority level is characterized by small priority number. The general algorithm can be described as follows:

Input: Feature space data set $X=\{X_1, X_2, ..., X_m\} \subset \mathbb{R}^n$, and the target is $Y \in R$ .Priority number k=1 neuron sets $\{\psi\}$, $D^{`} = \Phi$

Output: Samples Mapping Objects

Algorithm:

While $X \neq \phi$ *do*

For each $X_i$ *in* $X$

→*get max sub set* ($X_{i,sub}$) *of* $X_i$ *and its min Euclid distance (d) of* $X_{j,i \neq j}$

→*use hype-ball neuron(*$\psi$*,d) to cover* $X_{i,sub}$*, and set priority k to neuron*$\psi$

→*add* $X_{i,sub}$ *to temp set* $D^{`}$

End of for

→*set* $X - D^{`}$ *to* $X$, $D^{`} = \Phi$

k = k + 1

End while

For new coming data, the algorithm will update the produce number of neurons or add new neuron with a special value. Here is the algorithm:

Input: new feature training set $X'=\{X'_1, X'_2, ..., X'_m\} \subset \mathbb{R}^n$, $Y \in R$.
Output: Updated Produce Number of Mapping Objects

Algorithm:

```
While X' ≠ φ do
For each X'_i in X'
→Get a consistent subset X'_{i,sub} of X_i randomly and then
do the following produce:
If covered(X'_{i,sub})
do nothing
else
  if(misclassified(X'_{i,sub}))
  Adjust covering space range of hyper-ball neuron ψ
and adjust its priority number to lower one.
else
  add new neuron with higher priority number k.
end if
end if
→add X'_{i,sub} to temp set D`
End of for
→set X-D` to X', D`=Φ
End while
```

When new samples input into constructive neural network, the priority level of hidden neurons is strengthened or weaken. Using this method, the old information is not destroyed (forgot) after new data learning, but can be partly fetched at any latest priority. This produce is similar to man's learning procedure. As this updating and learning method is only to adjust the special priority level of neurons, therefore, it can process large data set with more effective heuristic algorithm.

For sequence video frames, we will split it into some independent images and transform them into eigenvector in feature space using FFT. An image frame is spitted and transformed by FFT on each sub-block. The ultimate identify formula is written as $O_{AVLI} = \{O_{AVLI(1)}, O_{AVLI(2)}, \ldots, O_{AVLI(k)}\}$ where $O_{AVLI(i)}$ is the $i$ piece time quantum. The last result is written as

$$\max\{count(i) \mid O_{AVLI(i)} = O_{AVLI(j)}, i \neq j\} \tag{7}$$

Where $count(i)$ is a function of taking sum number of same class in distinguished sequence and the value $i$ gives the similar measure between the original images and compared one.

## 4 Experimental Results

In order to verify the correctness and effectiveness of our proposed solution, a prototype system is developed by C++ and used in a real project of CMMB signal detecting system. The test video sequence is collected from CMMB source (TV), which is encoded by H.264 with image size of 320*240. Their type is car, animal and people. Transformed sub-block using FFT, an image is split into 12*12 sub-blocks. Noised level is defined from 0 to 5, the smaller is not noised. Video clips are mapping into trends line with different noised level. In Table.1, it gives the comparison between different type videos. The results indicate that the effect of proposed method for measure image based on point calculation in the high dimensional space is coincidence with real condition, and the real life video sequence has some low accurate rate. In Fig.5, it shows the trend of images with different noise.

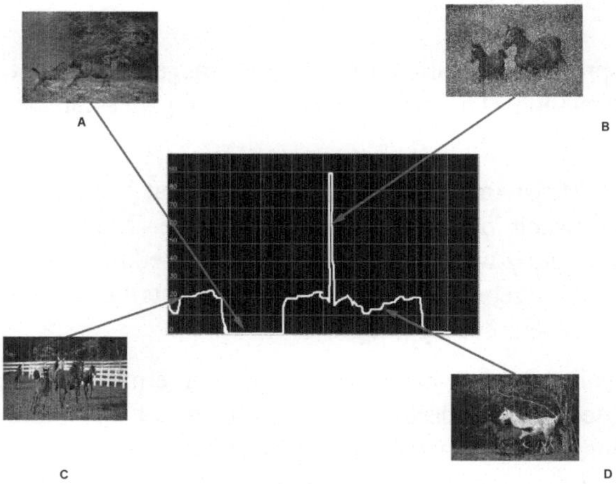


**Fig. 5.** Lines for Different Noised CMMB Images

From the point calculation method, it calculates some relative image points in high dimensional space which provide more information. On the other side, the sequence of image is not very required and its relation-ship can be covered in high-dimensional space. From relative images, relation-ship of noised image and original image can be mapped into trends map which can gives information of image quality measure. Additionally, the computation FFT complexity of point calculation method is about $O(Nlog_2N)$ and this characteristic can be used in some low computing CPU device, such as mobile phone.

From the point calculation method, it calculates some relative image points in high dimensional space which provide more information. On the other side, the sequence of image is not very required and its relation-ship can be covered in high-dimensional space. From relative images, relation-ship of noised image and original image can be mapped into trends map which can gives information of image quality measure. Additionally, the computation FFT complexity of point calculation method is about $O(Nlog_2N)$ and this characteristic can be used in some low computing CPU device, such as mobile phone.

**Table 1.** Comparison Results between images

| Content Type | Video length | Noise Level | Objective --Subjective |
|---|---|---|---|
| Car(400 kbps) | 60 s | High | 0.8% |
| | | Middle | 1.1% |
| | | Low | 0.7% |
| Animal(300 kbps) | 40 s | High | 0.4% |
| | | Middle | 1.0% |
| | | Low | 0.3% |
| People(200 kbps) | 30 s | High | 0.2% |
| | | Middle | 0.4% |
| | | Low | 0.1% |

# 5 Conclusions

In this paper, we present a solution for CMMB image assessment based on high dimensional space method. The contributions of this paper can be summarized as follows:

(1) An idea for CMMB image sequence measurement is proposed. Video image time is not aligned when performing testing produce. Method can be used for different type Video context with noise blind. The proposed method is robust.

(2) The model gives a geometry method. It used points transformed from original image and measured by referencing image.

Experimental results indicate our method is effective in terms of accuracy, robustness, and stability for video quality assessment. The novelty of this solution is the direct embedding of finding reflecting parameters of image using points in high dimensional space. It is a blind solution for video measurements. In addition, the proposed method allows simple and fast implementation for real-time mobile application. More studies in the aspect of geometry in high-dimension space should be invested in future.

**Acknowledgement.** This work is supported by National Natural Science Fund (61073189), Shanghai Technology Innovation Project (09160501700, 10110502200), and Leading Academic Discipline Project of Shanghai Municipal Education Commission (J51303). The authors are grateful for the anonymous reviewers who made constructive comments.

## References


[1] China Multimedia Mobile Broadcasting (EB/OL) (December 04, 2010), http://en.wikipedia.org/wiki/CMMB
[2] Myasnikov, V.V., Ivanov, A.A., Gashnikov, M.V., Myasnikov, E.V.: Computer program for automatic estimation of digital image quality 21(3), 415–418 (2011)
[3] ITU-T Recommendation BT.1788, Methodology for the subjective assessment of video quality in multimedia applications (2007)
[4] Soundararajan, R.: RRED Indices: Reduced Reference Entropic Differencing for Image Quality Assessment. IEEE Transactions on Image Processing 21(2), 517–526 (2012)
[5] Zhang, L., Mou, X., Zhang, D.: FSIM: A Feature Similarity Index for Image Quality Assessment. IEEE Transactions on Image Processing 20(8), 2378–2386 (2011)
[6] Ninassi, A., Meur, O.L., Callet, P.L., Barbba, D.: Does where you gaze on an image affect your perception of quality? Applying visual attention to image quality metric. In: Proc. IEEE Int. Conf. Image Process, ICIP 2007, vol. 2, pp. 169–172 (2007)
[7] Shoujue, W., Jiangliang, L.: Geometrical Learning, descriptive geometry, and biometric pattern recognition. Neuron Computing 67, 9–28 (2005)
[8] Wang, S.J.: Bionic(topological)pattern recognition-A new model of pattern recognition theory and its applications, Acta Electron. Sinica 30(10), 1–4 (2002)
[9] Zhu, S., Wang, Z., Liao, M.: Research on K-classification Covering for PONN. Computer Application 27(2), 330–332 (2007) (in Chinese)

# Two-Dimensional Gibbs Phenomenon for Fractional Fourier Series and Its Resolution*

Meiyu Ding and Hongqing Zhu

Department of Electronics and Communications Engineering,
East China University of Science and Technology, Shanghai 200237, China
hqzhu@ecust.edu.cn

**Abstract.** The truncated Fourier series exhibits oscillation that does not disappear as the number of terms in the truncation is increased. This paper introduces 2-D fractional Fourier series (FrFS) according to the 1-D fractional Fourier series, and finds such a Gibbs oscillation also occurs in the partial sums of FrFS for bivariate functions at a jump discontinuity. In this study, the 2-D inverse polynomial reconstruction method (IPRM) which is a method based on the inversion of the transformation matrix that represents the fraction Fourier space has been used to remove the Gibbs effect. The purpose of this study is to verify the 2-D IPRM has the similar effection for removing the Gibbs oscillation for partial fractional Fourier sums of bivariate functions. Numerical experiments verify the efficiency and accuracy of IPRM.

**Keywords:** Fractional Fourier series, Gibbs phenomenon, inverse polynomial reconstruction method, Gegenbauer polynomials, 2-D.

## 1 Introduction

As it is well known, the representation of a function expanded on Fourier basis functions exhibits evident oscillations where the function has a discontinuity. The oscillations at the neighborhood of discontinuity point or the boundaries do not diminish as the number of terms in the Fourier series is increased [1-2]. The Gibbs phenomenon is considered to be a main drawback which impedes the application of Fourier series in signal processing.

Recently, several available methods had already been addressed to resolve the Gibbs oscillations with one variable. For example, Gottlieb and Shu presented the direct method using Gegenbauer polynomials to overcome the Gibbs effects [1, 3]. Spectral methods can also be applied to the problems because the high frequency components produce the oscillations [4]. Driscoll and Fornberg had discussed the use of a Padé based algorithm for resolution of Gibbs phenomenon [5]. The inverse polynomial reconstruction method had been firstly proposed by Jung and Shizgal in refs. [6-7]. It was proved that the IPRM gave an exact resolution of Gibbs phenomenon [7].

* This work has been supported by National Natural Science Foundation of China under Grant no. 60975004.

J. Lei et al. (Eds.): AICI 2012, LNAI 7530, pp. 530–538, 2012.

Actually, the overshooting property occurs not only for Fourier series, but also for other series expansions. In this paper, we shall investigate a similar Gibbs effect for the truncated sums of series representations employing 2-D fractional Fourier series. The 1-D fractional Fourier series of a finite function, as a generalization of classical Fourier series, had been first proposed by Pei et al. [8]. That is, the ordinary Fourier basis is only a special case of the fractional Fourier basis. In [9], a modified FrFS were proposed to overcome the limitation that original FrFS cannot be applied to chirp signals with arbitrary central frequency.

The current study addresses the 2-D FrFS, and discusses its Gibbs phenomenon. The main purpose is to verify and to expect that 2-D IPRM can also be applied to resolution Gibbs phenomenon of 2-D FrFS. Similar to 2-D classical Fourier series. Firstly, we construct the 2-D FrFS according to the 1-D FrFS. Then, we investigate in detail reconstruction coefficients calculation using the orthogonality of fractional Fourier basis function. Numerical experiments give the reconstructed results by the proposed method.

## 2 Background

### 2.1 One-Dimensional Fractional Fourier Series

Similar to Fourier series expansion, the fractional Fourier series of an aperiodic function $f(x)$, on the finite domain [-*T*/2, *T*/2], can be written as

$$f_N(x) = \sum_{n=-N}^{N} \hat{f}_{\alpha,n}\tilde{\phi}_{\alpha,n}(x) = \sum_{n=-N}^{N} \hat{f}_{\alpha,n}\sqrt{\frac{\sin\alpha + j\cos\alpha}{T}}e^{-j((x^2+(n(2\pi/T)\sin\alpha)^2)/2)\cot\alpha + jnx(2\pi/T)} \tag{1}$$

where $\hat{f}_{\alpha,n}$ are called FrFS expansion coefficients with the parameter $\alpha$ and $\tilde{\phi}_{\alpha,n}(x)$ is an orthogonal basis of FrFS. $\hat{f}_{\alpha,n}$ are computed by the inner product of the signal and chirp basis signals.

$$\hat{f}_{\alpha,n} = \int_{-T/2}^{T/2} f(x)\tilde{\phi}^*_{\alpha,n}(x)dx = \sqrt{\frac{\sin\alpha - j\cos\alpha}{T}}\int_{-T/2}^{T/2} f(x)e^{j((x^2+(n(2\pi/T)\sin\alpha)^2)/2)\cot\alpha - jnx(2\pi/T)}dx \tag{2}$$

where $\tilde{\phi}^*_{\alpha,n}(x)$ is the complex conjugate of $\tilde{\phi}_{\alpha,n}(x)$. Evidently, the classical Fourier series is just a particular case for $\alpha = \pi/2$. Recently, the FrFS has been widely used in signal processing and image processing areas [8-11].

### 2.2 Two-Dimensional Fractional Fourier Series

According to the method of constructing 1-D FrFS orthogonal basis, one could obtain 2-D FrFS basis function, which satisfies orthogonality and separability. For simplicity, this study considers the orders of FrFS be identical (the maximum value of both $m$ and $n$ are $N$). Thus, a 2-D function $f(x, y)$, on the domain [-*T*/2, *T*/2] × [-*T*/2, *T*/2], can be expanded in fractional Fourier series

$$f_N(x,y)=\sum_{m=-N}^{N}\sum_{n=-N}^{N}C_{m,n}\tilde{\phi}_{\alpha,m}(x)\tilde{\phi}_{\beta,n}(y)$$
$$=\sum_{m=-N}^{N}\sum_{n=-N}^{N}C_{m,n}\sqrt{(\sin\alpha+j\cos\alpha)/T}e^{-j((x^2+(m(2\pi/T)\sin\alpha)^2)/2)\cot\alpha+jmx(2\pi/T)}$$
$$\times\sqrt{(\sin\beta+j\cos\beta)/T}e^{-j((y^2+(n(2\pi/T)\sin\beta)^2)/2)\cot\beta+jny(2\pi/T)} \tag{3}$$

where $C_{m,n}$ are FrFS expansion coefficients with the parameters α and $\beta$, which can be obtained as follows

$$C_{m,n}=\int_{-T/2}^{T/2}\left(\int_{-T/2}^{T/2}f(x,y)\tilde{\phi}_{\alpha,m}^{*}(x)dx\right)\tilde{\phi}_{\beta,n}^{*}(y)dy=\sqrt{(\sin\alpha-j\cos\alpha)(\sin\beta-j\cos\beta)/T^2}$$
$$\times\int_{-T/2}^{T/2}\left(\int_{-T/2}^{T/2}f(x,y)e^{j((x^2+(m(2\pi/T)\sin\alpha)^2)/2)\cot\alpha-jmx(2\pi/T)}dx\right)e^{j((y^2+(n(2\pi/T)\sin\beta)^2)/2)\cot\beta-jny(2\pi/T)}dy \tag{4}$$

From Eq. (4), it is clearly observed that the 2-D FrFS can be reduced to 2-D Fourier series in the case of $\alpha=\beta=\pi/2$. Therefore, the fractional Fourier series is only the generalization of classical Fourier series as a result of generalization of projection space.

Similarly to the classical Fourier series expansion, the approximation $f_N(x,y)$ to original function $f(x,y)$ shall exhibit serious oscillations at the interval or discontinuities, called Gibbs phenomenon, which does not disappear as the number of terms in FrFS expansion is increased. As a simple example, we investigate the Gibbs phenomenon of a 2-D step function defined in [-1, 1] × [-1, 1], i.e.

$$f_1(x,y)=\begin{cases}-1 & -1<y<0\\ 0 & x=-1,0,1\\ 1 & 0<y<1\end{cases} \tag{5}$$

Fig.1 plots the sums of the first fifty sums of the FrFS for $f_1(x, y)$ with different parameters $\alpha$ and $\beta$. The errors show Gibbs phenomenon for FrFS always exists nearly to the boundary point and discontinuities regardless of $\alpha$ and $\beta$. From these figures, one can find that the errors between FrFS and the original function are nearly equal under the same condition.

## 3 The 2-D IPRM for FrFS

Taking a cue from the refs [7, 12-13], this paper tries to use the 2-D IPRM to remove the Gibbs oscillation for FrFS of bivariate function. The main difference between the proposed method and previous IPRM is that the Fourier space is substituted by fractional Fourier space. Similarly, here 2-D IPRM also seeks a finite approximation $\tilde{f}_{m_x,m_y}(x,y)$ in Gegenbauer space $G_{m_x,m_y}$ to $f(x,y)$ on the domain [-1, 1] × [- 1, 1],

$$\tilde{f}_{m_x,m_y}(x,y)=\sum_{l=0}^{m_x}\sum_{k=0}^{m_y}\tilde{G}_{l,k}C_k^{\lambda}(y)C_l^{\lambda}(x) \tag{6}$$

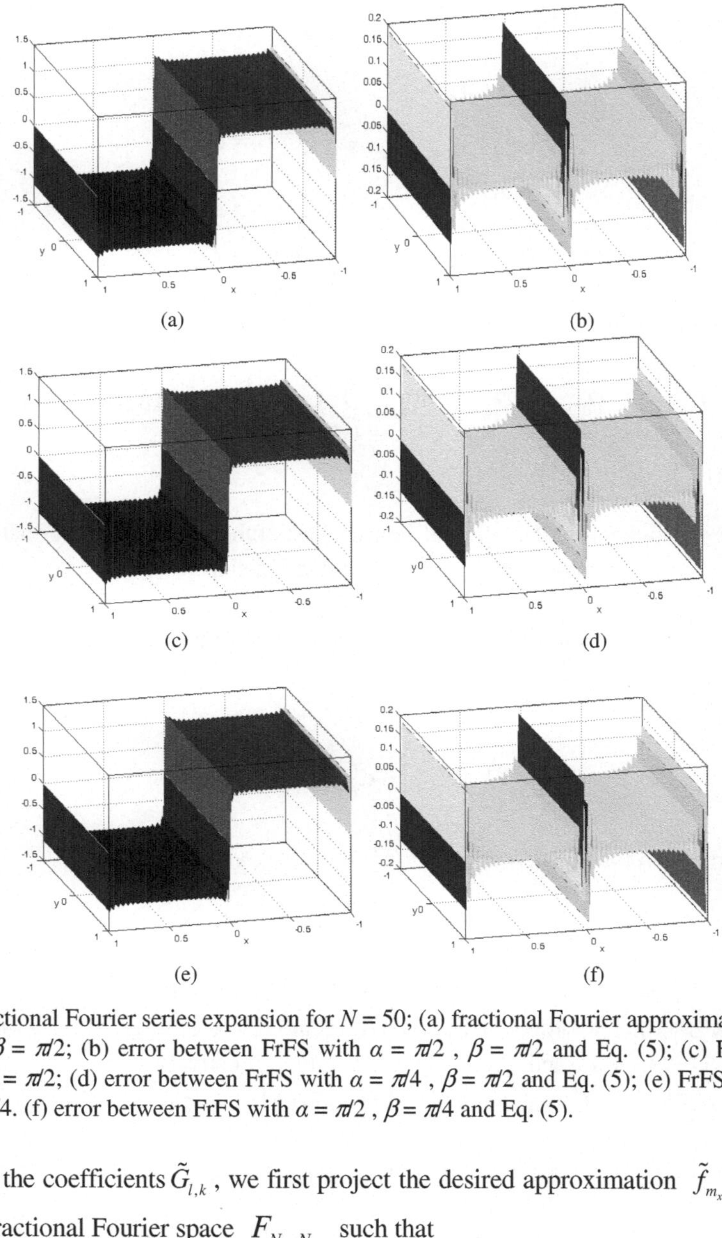


**Fig. 1.** Fractional Fourier series expansion for $N = 50$; (a) fractional Fourier approximation with $\alpha = \pi/2$ , $\beta = \pi/2$; (b) error between FrFS with $\alpha = \pi/2$ , $\beta = \pi/2$ and Eq. (5); (c) FrFS with $\alpha = \pi/4$ , $\beta = \pi/2$; (d) error between FrFS with $\alpha = \pi/4$ , $\beta = \pi/2$ and Eq. (5); (e) FrFS with $\alpha = \pi/2$ , $\beta = \pi/4$. (f) error between FrFS with $\alpha = \pi/2$ , $\beta = \pi/4$ and Eq. (5).

To obtain the coefficients $\tilde{G}_{l,k}$ , we first project the desired approximation $\tilde{f}_{m_x,m_y}(x,y)$ onto the fractional Fourier space $F_{N_x,N_y}$ such that

$$\tilde{f}^{N}_{m_x,m_y}(x,y) = \sum_{n=0}^{N}\sum_{m=0}^{N}((\tilde{f}_{m_x,m_y}(x,y), \tilde{\phi}_{\alpha,m}(x))_{F_x}, \tilde{\phi}_{\beta,n}(y))_{F_y} \tilde{\phi}_{\alpha,m}(x)\tilde{\phi}_{\beta,n}(y) \tag{7}$$

Assuming that the error $E(x, y)$ defined in the following is orthogonal to the fractional Fourier space $\mathbf{F}_{Nx,Ny}$.

$$E(x,y) \equiv \tilde{f}^{N}_{m_x,m_y}(x,y) - f_N(x,y) \tag{8}$$

and

$$((E(x,y), \tilde{\phi}_{\alpha,m}(x))_{F_x}, \tilde{\phi}_{\beta,n}(y))_{F_y} = 0 \tag{9}$$

Thus, we obtain

$$(\tilde{f}^{N}_{m_x,m_y}(x,y), \tilde{\phi}_{\alpha,m}(x))_{F_x}, \tilde{\phi}_{\beta,n}(y))_{F_y} = C_{m,n} \tag{10}$$

and with Eq. (6) and (7), we have

$$\begin{aligned}&[(\tilde{\phi}_{\alpha,m}(x), \sum_{n'=0}^{N}\sum_{m'=0}^{N} \tilde{\phi}_{\alpha,m'}(x)\tilde{\phi}_{\beta,n'}(y)((\sum_{l=0}^{m_x}\sum_{k=0}^{m_y} \tilde{G}_{l,k} C_k^{\lambda}(y) C_l^{\lambda}(x), \tilde{\phi}_{\alpha,m'}(x))_{F_x},\\ &\tilde{\phi}_{\beta,n'}(y))_{F_y})_{F_x}, \tilde{\phi}_{\beta,n}(y)]_{F_y} = C_{m,n}\end{aligned} \tag{11}$$

From the orthogonality of fractional Fourier basis function, the transformation matrix $\mathbf{R}_x$ and $\mathbf{R}_y$ whose elements $R^x_{m',l}$ and $R^y_{n',k}$ are defined as

$$R^x_{m',l} = \int_{-1}^{1} C_l^{\lambda}(x)\tilde{\phi}^*_{\alpha,m'}(x)dx\ , \qquad R^y_{n',k} = \int_{-1}^{1} C_k^{\lambda}(y)\tilde{\phi}^*_{\beta,n'}(y)dy \tag{12}$$

So, Eq. (11) can be rewritten as

$$\sum_{l=0}^{m_x}\left( R^x_{m,l} \sum_{k=0}^{m_y}\left(\tilde{G}_{l,k} R^y_{n,k}\right)\right) = C_{m,n},\ -N \le m,n \le N \tag{13}$$

The matrix form for Eq. (13) is

$$\mathbf{R}_x \tilde{\mathbf{G}} (\mathbf{R}_y)^T = \mathbf{C} \tag{14}$$

where $\mathbf{C}$ and $\tilde{\mathbf{G}}$ are the coefficients matrix of FrFS and reconstruction coefficients matrix by IPRM. The orders of $\mathbf{R}_x$, $\tilde{\mathbf{G}}$, $\mathbf{R}_y$, $\mathbf{C}$ are $(2N + 1) \times (m_x + 1)$, $(m_x + 1)\times (m_y + 1)$, $(m_y + 1)\times (2N + 1)$, and $(2N + 1) \times (2N + 1)$, respectively.

The coefficients $\tilde{\mathbf{G}}$ can be obtained from Eq. (14), that is

$$\tilde{\mathbf{G}} = (\mathbf{R}_x)^{-1}\mathbf{C}\left((\mathbf{R}_y)^T\right)^{-1} \tag{15}$$

The solvability of Eq. (15) demands both $\mathbf{R}_x$ and $\mathbf{R}_y$ are invertible matrix. Therefore, the preconditions of using the above equation are that $m_x$ and $m_y$ are equal to $2N$. However, if the transformation matrix $\mathbf{R}_x$ and $\mathbf{R}_y$ are not ill-posedness, $\tilde{\mathbf{G}}$ can also be obtained according to Eq. (14), and the orders of transformation matrix shall have distinct value.

# 4 Numerical Examples

In this section, we consider two examples to illustrate the effectiveness of the 2-D IPRM for FrFS of the unknown function. Owing to the limit of the space, this study only investigates 2-D IPRM based on the function with single domains, and the IPRM with multiple domains is just the extension of single domains. It should be investigated in our future work.

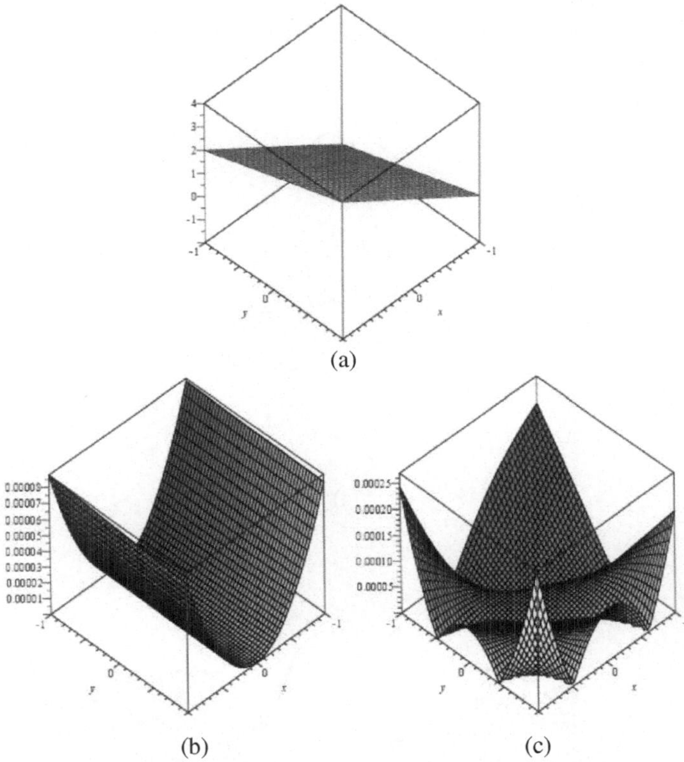

**Fig. 2.** Original function and the absolute error; (a) original function; (b) error between original function and reconstructed function by IPRM with $\alpha = \pi/4$, $\beta = \pi/2$; (c) error between original function and reconstructed function by IPRM with $\alpha = \pi/4$, $\beta = \pi/4$

**Example 1.** In this example, let us first consider a simple function $f(x, y) = 2x + y + 1$ on the domain $[-1, 1] \times [-1, 1]$ (see Fig. 2(a)). Since large $N$ is non-essential for reconstruction by 2-D IPRM, for simplifying the calculation, this study chooses $N = 1$ and $m_x = m_y = 2$. The error curves between reconstructed function and the original function with different $\alpha$ and $\beta$ are shown in Fig. 2(b), (c), respectively. Using 2-D IPRM, we can obtain the nearly exact coefficient matrix of reconstruction $\tilde{\mathbf{G}}$ and the FrFS coefficient matrix $\mathbf{C}$ for the case $(\alpha, \beta) = (\pi/4, \pi/2)$ as follows:

$$\tilde{\mathbf{G}} = \begin{pmatrix} 1 & 0.25 & 0 \\ 0.5 & 0 & 0 \\ 0 & 0 & 0 \end{pmatrix}$$

$$\mathbf{C} = \begin{pmatrix} -0.0412784+0.0490223i & -0.813653-0.685124i & 0.0412784-0.049i \\ 0.141313+0.613509i & 1.92739-0.443948i & -0.141313-0.613509i \\ -0.0412784+0.0490223i & 1.12166+0.944484i & 0.0412784-0.0490223i \end{pmatrix}$$

The reconstructed function by the proposed 2-D IPRM has a high accuracy, which is independent the parameter $\alpha$ and $\beta$.

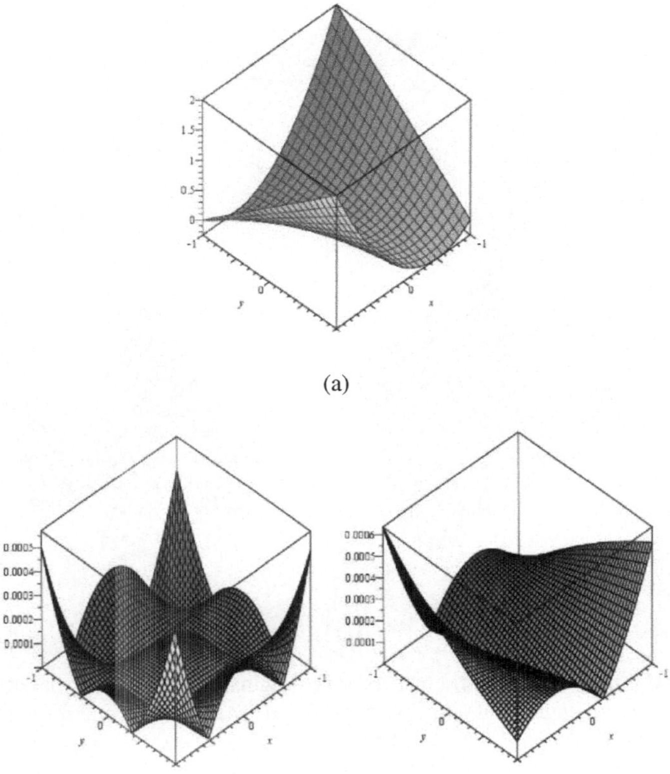

(a)

(b) (c)

**Fig. 3.** Original function and the absolute error; (a)original function; (b)absolute error between original function and reconstructed function by IPRM with $\alpha = \beta = \pi/4$; (c)absolute error between original function and reconstructed function by IPRM with $\alpha = \pi/3$, $\beta = \pi/2$

**Example 2.** We consider another complex function $f(x, y) = x^2 + xy$ to show how the 2-D IPRM works for the reconstruction in domain $[-1, 1] \times [-1, 1]$. Here, we just choose $N = 1$ and $m_x = m_y = 2$ in 2-D IPRM. For the case $\alpha = \beta = \pi/4$, we can obtain the exact coefficient matrix of reconstruction $\tilde{\mathbf{G}}$ and FrFS coefficient matrix $\mathbf{C}$, i.e.

$$\tilde{\mathbf{G}} = \begin{pmatrix} 0.1667 & 0 & 0 \\ 0 & 0.0625 & 0 \\ 0.0833 & 0 & 0 \end{pmatrix}$$

$$\mathbf{C} = \begin{pmatrix} 0.0748348 + 0.198121i & 0.2485992244 - 0.314266i & 0.00671889 - 0.196106i \\ 0.0546410 + 0.0379688i & 0.6206853803 - 0.204971i & 0.0546410 + 0.0379688i \\ 0.00671889 - 0.196106i & 0.2485992244 - 0.314266i & 0.0748348 + 0.198121i \end{pmatrix}$$

The similar results can be obtained in other conditions. Fig. 3 gives original function and the error function with $\alpha = \pi/4$, $\beta = \pi/4$ and $\alpha = \pi/3$, $\beta = \pi/2$. The experimental results of two numerical examples verify that Gibbs' phenomenon for FrFS can be removed by 2-D IPRM.

## 5 Conclusions

In this paper, we proposed 2-D fractional Fourier series according to the definition of 1-D FrFS. It was found that the Gibbs phenomenon can be observed in 2-D FrFS. Therefore, we have applied the 2-D inverse polynomials reconstruction method to resolve the Gibbs oscillations of 2-D FrFS by requiring that the residue is orthogonal to fractional Fourier space. The numerical experiments verified the efficiency of the proposed method.

However, we found the computation of IPRM is very expensive. This is mainly due to the fact that the 2-D FrFS is more complex than the 2-D classical Fourier series. How to decrease the computational complexity will become our future research interests.

## References


1. Gottlied, D., Shu, C.: On the Gibbs Phenomenon and Its resolution. SIAM Rev. 39, 644–668 (1997)
2. Lanczos, C.: Discourse on Fourier Series. Hafner Publishing Company, New York (1996)
3. Gottlieb, D., Shu, C.W., Solomonol, A., Vandeven, H.: On the Gibbs Phenomenon I: Recovering Exponential Accuracy from the Fourier Partial Sum of a Nonperiodic Analytic Function. J. Comput. Appl. Math. 43, 81–92 (1992)
4. Pérez, J.G.W., Rubio, R.G., Moñux, A.O., Fernández, I.M.: Removal of the Gibbs Phenomenon and Its Application to Fast-Fourier-Transform-Based Mode Solvers. J. Opt. Soc. Am. A. 24, 3772–3780 (2007)
5. Driscoll, T.A., Fornberg, B.: A Padé Based Algorithm for Overcoming the Gibbs Phenomenon. Numer. Algorithms 26, 77–92 (2001)

6. Shizgal, B.D., Jung, J.H.: Towards the Resolution of the Gibbs Phenomena. J. Comput. Appl. Math. 161, 41–65 (2003)
7. Jung, J.H., Shizgal, B.D.: Generalization of the Inverse Polynomial Reconstruction Method in the Resolution of the Gibbs Phenomenon. J. Compu. Appl. Math. 172, 131–151 (2004)
8. Pei, S.C., Yeh, M.H., Luo, T.L.: Fractional Fourier Series Expansion for Finite Signals and Dual Extension to Discrete-Time Fractional Fourier Transform. IEEE Trans. Signal Process. 47, 2883–2888 (1999)
9. Barkat, B., Yingtuo, J.: A Modified Fractional Fourier Series for the Analysis of Finite Chirp Signals and Its Application. In: 7th Int. Sym. Signal Process. Appl., pp. 285–288 (2003)
10. Bhandari, A., Marziliano, P.: Sampling and Reconstruction of Sparse Dignals in Fractional Fourier Domain. IEEE Signal Processing Letters 17, 221–224 (2009)
11. Coëtmellec, S., Brunel, M., Lebrun, D.: Fractional-Order Fourier Series Expansion for the Analysis of Chirped Pulses. Optics Communications 249, 145–152 (2005)
12. Jung, J.H., Shizgal, B.D.: Inverse Polynomial Reconstruction of Two Dimensional Fourier Images. J. Sci. Comput. 25, 367–399 (2005)
13. Abdi, A., Hosseini, S.M.: An Investigation of Resolution of 2-Variate Gibbs Phenomenon. Appl. Math. Comput. 203, 714–732 (2008)

# Horizontally UCA DOA Estimation Performance via Spatial Averaging

YaoQing Hou[1,2], Hui Chen[2,3], and Man Zhao[2,3]

[1] School of Electronic and Mechanical Engineering, Wuhan University of Technology
[2] Science and Technology on Electronic Control Laboratory
[3] Key Research Lab. Radar Academy,
288# Huangpu Road Wuhan 430019, China
chhglr@sina.com

**Abstract.** In this paper, a direction-of-arrival (DOA) estimation method for resolving correlation by spatial averaging algorithm based on the horizontally Uniform Circular Array (UCA) is presented. The ambiguity resolution method is proposed for the appearing of the third-order ambiguity on the horizontally UCA. Restructuring the covariance matrix of the received data again, the angle of the sub-arrays will be changed. Finally, the DOA estimation performance of the new algorithm is studied, including the effects of the Signal-to-Noise Ratio (SNR) and the numbers of the snapshots to the performance and further analyzes the effects of imperfections, such as amplitude error and the phase error.

**Keywords:** DOA, Spatial Averaging, UCA, de-correlation, horizontally array.

## 1 Introduction

Direction of arrival (DOA) estimation of multiple narrowband sources is a major research issue in array signal processing. Many high-resolution DOA estimation techniques have been developed over the years. Compared with the Uniform Linear Array (ULA), UCA has many merits, such as 360 azimuthally coverage, almost invariant directional pattern and constant azimuthally resolution. Because of these advantages, the UCA was successfully used in many areas, such as the application in the field of radar, sonar, communication and electronic surveillance. And the discussion about the UCA becomes more and more attractively [1]. However, the structure of the UCA dose not have the Vander monde form, so that many useful techniques based on the ULA cannot be applied directly. For that reason, the DOA estimation for the UCA is more complex especially when the signals are highly coherent. The dominant two kinds of methods for resolving this problem are the phase mode excitation technique and the array interpolation technique, they are all employ beam-space transform to rebuild a manifold with the Vander monde form. Based on these techniques, some methods for de-correlation on the UCA have been proposed. In these methods, the Mode-Space algorithm which is analyzed in the literature [2] [3]

J. Lei et al. (Eds.): AICI 2012, LNAI 7530, pp. 539–546, 2012.

(the literature [4] [5] extend the spatial smoothing to the circular array and further analyze the effects of imperfections, such as mutual coupling, transform error, and directional sensor) and the Virtual Array Transformation Method which is proposed in literature [6-10] are the most representative.

At present, there has not a method that which can estimate the DOA of three coherent sources on the UCA with six sensors. To solve this problem, the paper proposes a new de-correlation algorithm with few sensors on the UCA. Basing on the symmetric configuration of the UCA, it utilizes the spatial averaging algorithm for the de-correlation on the sub-arrays, so the coherent sources can be estimated with few sensors. To resolve the ambiguity angles, the structure of the sub-arrays are rebuild, overlap the two spatial spectrums, the real angles will be separated. Computer simulations are provided to verify the effectiveness and validity of the proposed method.

## 2 Data Model

Consider a circular arrays consisting of $M$ sensors, uniformly spaced on the boundary of a circle with radius $r$ (normalized to wavelength, $\lambda$) and angle $\theta_m$. Each sensor captures narrowband signal transmitted from a stationary source with unknown positions located in the far-field. The azimuth $\theta$ and the elevation $\phi$ as depicted the UCA (H) in Fig. 1 which the UCA placed horizontally.

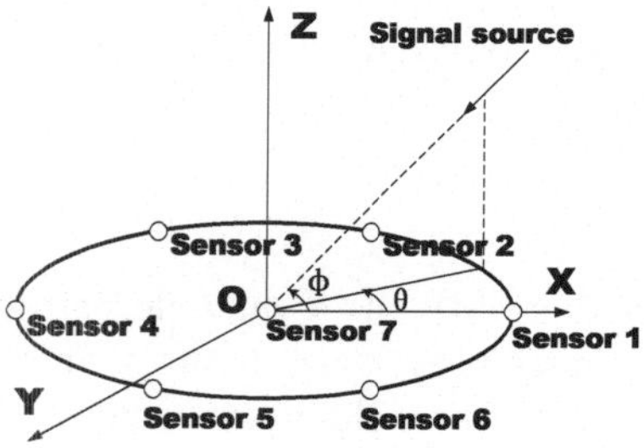


**Fig. 1.** UCA configuration (horizontally)

Assume that there are $k$ narrowband far-field coherent signal sources impinging upon the array. The output of the circular array can be described as:

$$X(k) = AS(k) + N(k) \quad k = 1, 2, \cdots, K \tag{1}$$

where $X(k)$ is the received data, $K$ is the number of snapshots, $A$ is the array manifold matrix, $S(k)$ is a signal vector, $N(t)$ is the noise vector.

$$A = [a_1(\omega_0) \quad a_2(\omega_0) \quad \cdots \quad a_N(\omega_0)] \tag{2}$$

In which signal steering vector is denoted by:

$$a_i(\omega_0) = [\beta_{1i} \quad \beta_{2i} \quad \cdots \quad \beta_{ki}]^{\mathrm{T}} \tag{3}$$

where $\beta_{ki} = \exp(-\mathrm{j}\omega_0 \tau_{Ki})$, and the signal vector is

$$S(k) = [s_1(k) \quad s_2(k) \quad \cdots \quad s_N(k)]^{\mathrm{T}} \tag{4}$$

The formulation of the $\tau$ of the UCA (H) can be expressed as:

$$\tau_{ki} = \frac{r}{c}\left(\cos\left(\frac{2\pi(k-1)}{M} - \theta_i\right)\cos\varphi_i\right) \tag{5}$$

And the $\tau$ can be expressed as:

$$\tau_{ki} = \frac{r}{c}\left(\alpha + \sin\left(\frac{2\pi(k-1)}{M}\right)\sin\varphi_i\right) \tag{6}$$

where $\alpha = \cos\left(\frac{2\pi(k-1)}{M}\right)\cos(\theta_i)\cos\varphi_i$.

## 3 Spatial Averaging De-correlation Algorithm on the UCA

By the analysis aforementioned, to decrease the error of the periodic extension, the number of the sensors should satisfy the formulation:

$$K = \left\lfloor \frac{2\pi r}{\lambda} \right\rfloor \tag{7}$$

where $\lfloor \cdot \rfloor$ represents the bottom integral function. To make the expanding component small enough, the number of the sensors is always: $M > 2K$. For example, if there is a circle with radius $r = 0.5\lambda$, then $K = 3$, basing on the formulation (6), the number of the sensors must be no less than 7. And this is only the situation for the incoherent sources, there are other more sensors needed in the coherent situation.

By the analysis aforementioned, when the number of the sensors is small, there is no way to resolve the coherent sources problem availably at the present time. The main reason is the error of the transformation. To resolve this problem, there should be a spatial averaging algorithm without transformation. The Centro-symmetric structure of the UCA affords the probability for resolving this problem.

Utilizing the averaging algorithm on the UCA, the invariable steering vectors of the sub-array is the most important thing. Taking an example of the six sensors uniformly spaced on the boundary of a circle with the central sensor, basing on the Centro-symmetric structure of the array, it is divided into two sub-arrays, as the Fig. 2.

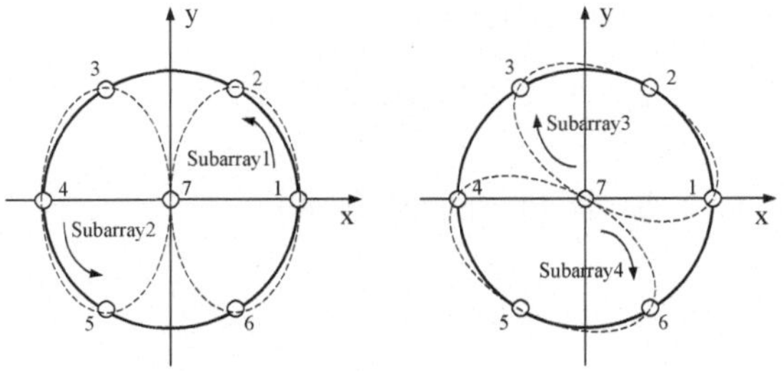


**Fig. 2.** sensor position for smoothing

Considering this structure, there will be four sub-arrays for averaging. More specifically, the two kinds of sub-arrays are composed of the sensors 1、2、7、6 and 7、6、1、2, so do the sensors 4、5、7、3 and 7、3、4、5. Then there are four sub-arrays for averaging, and they are expected to estimate three coherent sources. However, as the Centro-symmetric of the sub-arrays, there may have the ambiguity angles, so there must be a method to overcome this problem. Then, the second structure for resolving ambiguity is proposed, as Fig. 2.

By the assumption aforementioned, the covariance matrix is

$$R_x = E\left\{x(k)x^H(k)\right\} = \mathrm{AR}_s\mathrm{A}^H + \sigma^2\mathrm{I} \tag{8}$$

where $R_x = E\left\{s(k)s^H(k)\right\}$ is the covariance matrix of the signal sources, $I$ is an identify matrix of size $M \times M$, $\sigma^2$ is the noise power, $[.]^H$ represents conjugate transposition, and $E\{.\}$ denotes expectation. Since the sensors are Centro-symmetric, it can be expressed as:

$$\boldsymbol{a}_{ik} = \exp\left(\mathrm{j}\frac{2\pi}{\lambda}(x_k\cos\theta_i + y_k\sin\theta_i)\right) \tag{9}$$

And then the sub-arrays are

$$\text{subarray1:}\begin{cases}\boldsymbol{a} = [a_{i1}, a_{i2}, a_{i7}, a_{i6}]\\ \boldsymbol{a} = [a_{i7}, a_{i6}, a_{i1}, a_{i2}]\end{cases} \tag{10}$$

$$\text{subarray2:}\begin{cases}\boldsymbol{a} = [a_{i4}, a_{i5}, a_{i7}, a_{i3}]\\ \boldsymbol{a} = [a_{i7}, a_{i3}, a_{i4}, a_{i5}]\end{cases} \tag{11}$$

$$\text{subarray3:}\begin{cases}\boldsymbol{a} = [a_{i7}, a_{i3}, a_{i2}, a_{i1}]\\ \boldsymbol{a} = [a_{i2}, a_{i1}, a_{i7}, a_{i3}]\end{cases} \tag{12}$$

$$\text{subarray4:}\begin{cases} \boldsymbol{a} = [a_{i7}, a_{i6}, a_{i5}, a_{i4}] \\ \boldsymbol{a} = [a_{i5}, a_{i4}, a_{i7}, a_{i6}] \end{cases} \tag{13}$$

Basing on those sub-arrays, the three coherent signal sources can be estimated.

# 4 Simulation

## 4.1 Ambiguity Solution Based on Reconstructing Sub-array

In the same situation of the Fig. 3, changing the sub-array from the 1, 2 to 3, 4, the result of the simulation is shown in Fig. 3.

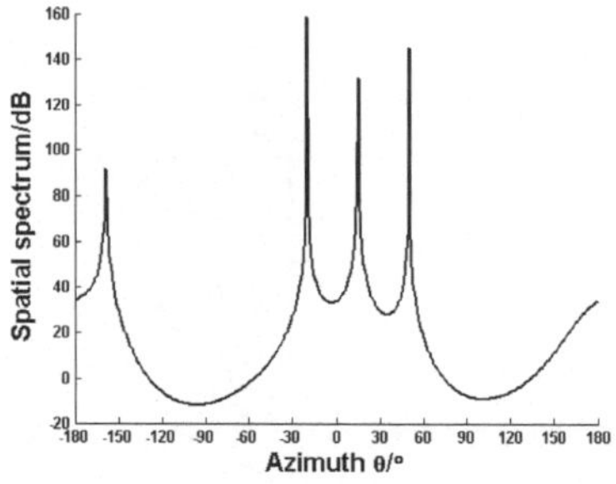


**Fig. 3.** Three coherent signals

Consider a uniform circular array composed of $M = 6$ sensors with one central sensor, which is Centro symmetric. Three far-field coherent sources are impinging on the array with the radius $r = 0.5\lambda$ from $\theta_1 = -20^\circ$ , $\theta_2 = 15^\circ$ and $\theta_3 = 50^\circ$ . For simplicity, the elevation angle is assumed as $\phi_i = 0^\circ$ . The number of snapshots is 200, SNR is 20dB. The simulation results of the proposed algorithm are shown in Fig. 4.

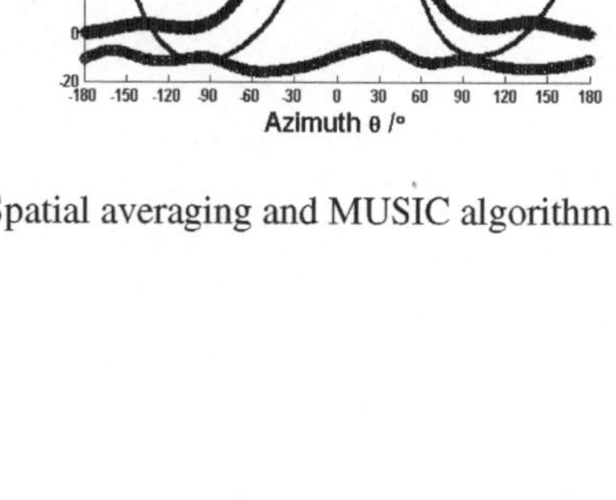

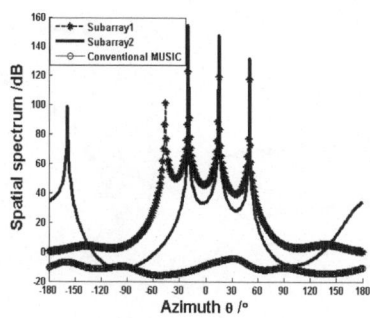


**Fig. 4.** Spatial averaging and MUSIC algorithm (20dB)

In the Fig. 4, (*) painted line describes the spectrum of the Subarray1spatial averaging, (-) painted line describes the spectrum of the Subarray2 spatial averaging, (o) painted line describes the spectrum of the conventional MUSIC algorithm. The conventional MUSIC algorithm can't distinguish the three coherent sources, but in the spectrum of the two sub-arrays spatial averaging, part of the peaks of the two spectrums are overlapping. The coincidence of the three peaks is the real DOA of the target. So there has a way to separate the real DOA angles from the wrong ones.

## 4.2 Performance Analysis of the DOA Estimation on the UCA (H)

In this chapter, by comparing with the emitting signals of coherent and the incoherent, some factors including the SNR, the number of the snapshots and the amplitude and phase errors what can affect the algorithm performance are analyzed. In these simulations, the results are calculated from 1000 Monte Carlo runs.

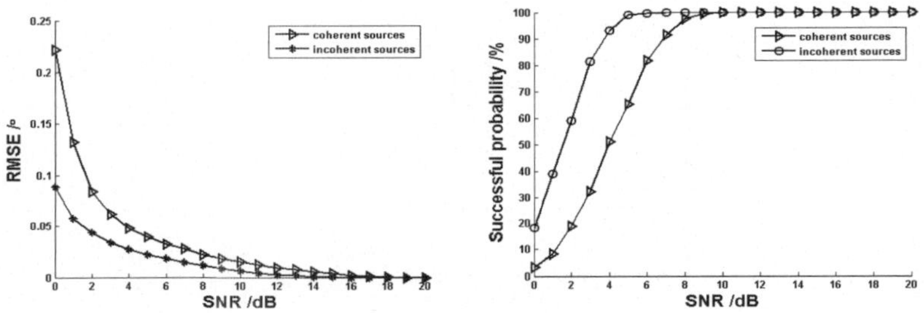


**Fig. 5.** The RMSE versus SNR and the successful probability versus SNR

In the Fig. 5, the results of the simulation show the root mean square error (RMSE) and the successful probability versus SNR from 0dB~20dB. As the increasing of the SNR, the RMSE will decrease, and the successful probability will increase. The results of the simulation show that the performance of the algorithm to the incoherent sources is more effective than coherent sources.

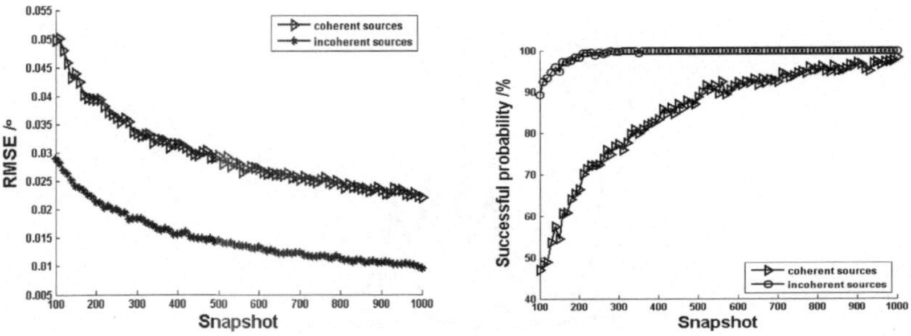


**Fig. 6.** The RMSE and the successful probability versus snapshot

In Fig. 6, the results of the simulation show the RMSE and the successful probability versus the number of snapshots from the 100 to 1000. As the increasing of the number of snapshots, not only the RMSE of the incoherent sources will decrease, but also the coherent sources will decrease. The successful probability will increase as the increasing of the number of the snapshots. The performance in the proposed method to the incoherent sources is better than to the coherent sources.

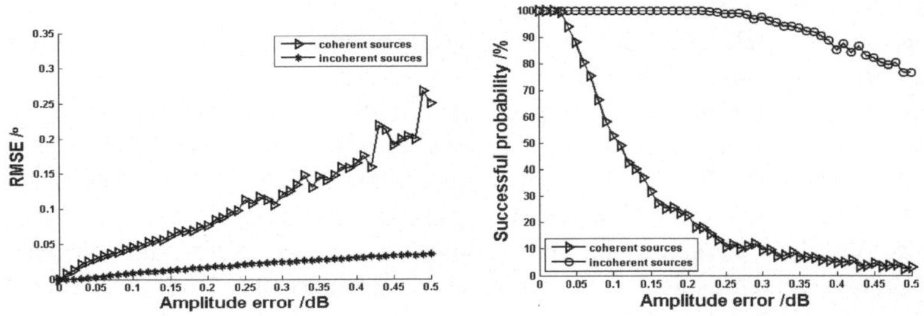


**Fig. 7.** The RMSE and the successful probability versus amplitude error

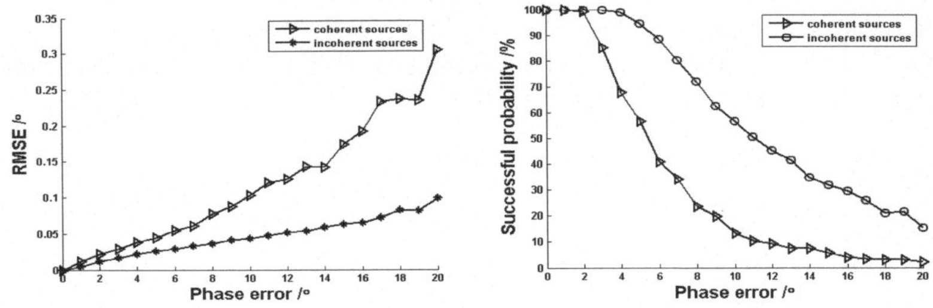


**Fig. 8.** The RMSE and the successful probability versus phase error

At the same situation and algorithm, the imperfections such as the amplitude error and the phase error are analyzed. With the analysis of the amplitude error and phase error, the results of the simulation show in figure 7~8 that the coherent sources are more sensible to the amplitude and phase error.

## 5 Conclusion

In this paper, the spatial averaging algorithm which is used without any transformation on the UCA is proposed. Basing on the Centro-symmetric of the array, covariance matrices of the sub-arrays are averaged to form the spatially averaged covariance matrix. Then make the UCA with six sensors and a center sensor as an example, the performance of the new algorithm on the UCA (H) are analyzed.

## References


1. Joannides, P., Balanis, C.A.: Uniform circular arrays for smart antennas. IEEE Antennas and Propagation Magazine 47(4), 192–206 (2005)
2. Wax, M., Sheinvald, J.: Direction finding of coherent signals via spatial smoothing for uniform circular arrays. IEEE Trans. on Antennas Propagation 42(5), 613–620 (1994)
3. Reddy, K.M., Reddy, V.U.: Analysis of spatial smoothing with uniform circular arrays. IEEE Transactions on Signal Processing 47(6), 1726–1730 (1999)
4. Wang, Y., Chen, H., Wan, S.: An effect DOA method via virtual array transformation. Science in China 44(1), 75–82 (2001)
5. Belloni, F., Koivunen, V.: Beamspace transform for UCA: error analysis and bias reduction. IEEE Transactions on Signal Processing 54(8), 3078–3089 (2006)
6. Ye, Z.F., Xiang, L., Xu, X.: DOA estimation with circular array via spatial averaging algorithm. IEEE Antennas and Wireless Propagation Letters 6(11), 74–76 (2007)
7. Xu, X., Ye, Z.-F., Zhang, Y.-F.: DOA estimation for uncorrelated and coherent signals with centre-symmetric circular array. Journal of University of Science and Technology of China 39(11), 1125–1129 (2009)
8. Wang, Y., Chen, H., Wan, S.: An effect DOA method via virtual array transformation. Science in China 44(1), 75–82 (2001)
9. Abed-Meraim, K., Gazzah, et al.: Optimum ambiguity-free isotropic antenna arrays. IEEE Trans. on Acoust., Speech, Signal Processing, 2157–2160 (2009)
10. Wang, Y., Chen, H., Peng, Y.: Spatial spectrum estimation and its algorithm. Publishing House of Tsinghua university, BeiJing (2004)

# Automatic Object Extraction in Nature Scene Based on Visual Saliency and Super Pixels

Zhaohui Wang, Jinling Su, Yu Chen, and Shengrong Gong

School of Computer Science & Technology, Soochow University, SuZhou, 215006, China
zhhwang@suda.edu.cn

**Abstract.** In this paper we propose an automatic salient object extraction method for nature scene. The proposed method first utilizes an algorithm based on visual attention model to obtain a prior knowledge for Graph Cut, and then constructs the weighted graph of Graph Cut based on super-pixels pre-segmented by the improved watershed algorithm in order to accelerate the speed of proposed method. In this framework, Visual saliency map is obtained using chrominance and intensity features in HSV color space, which provides the approximate region that contains salient object to be segmented. Then the salient object region after extension is cropped as input image, and pre-segmented by the improved watershed algorithm into several regions to construct weighted graph. Finally the salient object is obtained by Graph Cut algorithm. Experiment results show that our algorithm can automatically get salient object without human interactions, and speed up the segmentation without decreasing segmentation accuracy.

**Keywords:** salient object, visual attention, Graph Cut, super pixels.

## 1 Introduction

Considering the diversity and complexity of the natural scene, coupled with the fact that ideal segmentation results always depend on subjective visual perception, interactive segmentation methods are more practical than the fully unsupervised segmentation methods for the optimal result [1]; lazy snapping algorithm [2] get more accurate segmentation results using foreground and background seeds by users.

Graph cut is an efficient framework, which takes image pixels as nodes to construct graph, to segments images by minimizing an energy function over a combination of both region and boundary terms [3]. Now many algorithms are derived from this theoretical framework, M. Ma used Ncut for gray-scale image segmentation[4], but this needs to set the initial threshold manually, and establish gray weight matrix based on image pixels, finally result in complex computation; Y. Zhang et al proposed a coarse-to-fine approach to extract an infrared target[5], they first obtain a confined region containing the entire object, then construct energy function term according to the confined region; J. Zhang et al[6]Proposed a modified graph cut algorithm under the elliptical shape constraint to segment cervical lymph

J. Lei et al. (Eds.): AICI 2012, LNAI 7530, pp. 547–554, 2012.

nodes on sonograms, the disadvantages of these two above algorithms are that they are only available for certain types of image segmentation, lack of versatility; C. Guo and P. Wang[7] proposed an adaptive Graph-cut algorithm to video moving objects segmentation, first do the Kalman prediction of the numbers of objectives pixels and objectives-background pixel-pairs, then adaptive update the nodes flux, however, it is sensitive to noise, which affects final result; C. Rother et al proposed grab cut algorithm[8], which no longer requires users to use brush approach "brush out" foreground and background seeds, but manually select a rectangular box containing the object.

The traditional graph cut algorithms are always pixel-based; they can perform excellent on low-resolution images with fast speed, but slowly on higher resolution images. It would be difficult to get real-time segmentation results for video sequence, moreover, human interaction limits their application, for example, human interaction will greatly slow down the processing speed in some automatic monitoring system or cases, the burden both of the computational complexity and the memory overhead would be heavy severely. Therefore, it is critical to research automatic graph cut segmentation methods with high quality results and fast speed. To address these problems, we propose an automatic salient object extraction method based on visual attention model and super-pixels. We compute the saliency map to obtain regions containing object firstly, then use improved watershed algorithm to pre-segment images into super-pixels that described by Gaussian Mixture distribution to establish weighted graph for Graph Cut.

## 2 Fast Salient Object Extraction Model Based on Graph Cut

In this section, we will introduce our algorithm using saliency map and Graph Cut. We take saliency map as a prior knowledge, combined Graph Cut framework as the extraction model. In order to maximize extraction accuracy and efficiency, we carry out some changes to Graph Cut accordingly.

### 2.1 Salient Object Region Extraction Based on Visual Perception

The interactive of Lazy snapping is prior knowledge of the foreground and background, so the ideal segmentation results depend on the human visual perception [2]. As the visual attention model can extract the most interesting regions of human eyes, which can fast process important content by retaining images useful information, we obtain salient object region using visual saliency model in natural scene automatically.

#### 2.1.1 Generating Visual Saliency Map

Figure 1 shows the framework of computing visual saliency map, it consists of two parts, one is obtaining Saliency Map based on the Hue and Saturation feature in HSV color space, the other is dominant color distribution of background. Warm tones appeal visual attention more effectively. Firsryly we transform images from RGB to HSV color space, two feature map H, SV can be depicted as follows:

$$H=\begin{cases}60\times[(g-b)/(r-min(r,g,b))] & if \quad r=max(r,g,b)\\ 60\times[2+(b-r)/(g-min(r,g,b))] & if \quad g=max(r,g,b)\\ 60\times[4+(r-g)/(b-min(r,g,b))] & if \quad b=max(r,g,b)\end{cases} \tag{1}$$

$$SV(x,y)=\begin{cases}V(x,y).S(x,y).\cos(H(x,y)) & \frac{\sqrt{2}}{2}\le\cos(H(x,y))\le 1\\ 0 & \text{otherwise}\end{cases} \tag{2}$$

Where, S and V are Saturation and Value in position(x, y): $V=max(r,g,b), S=1-min(r,g,b)/V$[9]. I is the intensity feature map, since hue variations are not salient at very low intensity, we set I to zero in the position whose value less than 1/10 of maximum over the entire image.

We subtract mean value from two features for removal some noises of background, then saliency map $S_H$, $S_{SV}$, $S_I$ of three feature map $H$, $SV$, $I$ can be computed using Spectral Residual Approach, more detail refer to reference literature [10]. So we compute Saliency Map ($S_{map}$) from above three saliency feature map as follows:

$$S_{map}=\sqrt{S_H\cdot S_H+S_{SV}\cdot S_{SV}+S_I\cdot S_I} \tag{3}$$

Taking into account that salient region is rare in nature scene, we obtain $BK_{map}$ containing the candidate region of salient object based on the dominate color distribution of background ($BK_{color}$) by color histogram and spatial distribution. The final saliency map (SM) is fused by formula (3) and (5) with the $BK_{map}$ limitation, then we take binarization and open-close operation on SM, Figure1 shows the detail.

$$SM(x,y)=\begin{cases}0 & S_{map}(x,y)<\frac{\min(BK_{map})+\overline{BK_{map}}}{2}\\ 1 & otherwise\end{cases} \tag{4}$$

$$BK_{map}=\begin{cases}r+b+g-BK_{color} & BK_{map}>0\\ 0 & otherwise\end{cases} \tag{3}$$

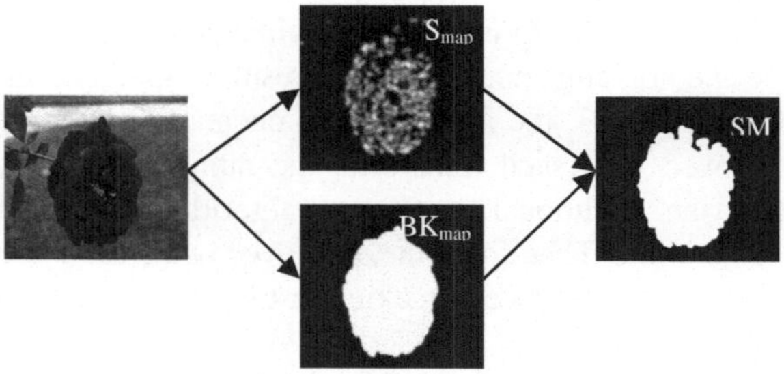


**Fig. 1.** Saliency Map extraction

### 2.1.2 The Extension of the Saliency Region

The enclosing bounding box of the foreground objects (white box) obtained utilizing SM in subsection 2.1.1 may lose the part of target. So we expand the white rectangular bounding box to the black rectangular bounding box according to formula

(6) and (7). Although the expanded bounding box adds some background information, it can be a good solution to the missing of the target part.

$$W_{Black} = W_{white} + \left|1 - W_{Image} / W_{white}\right| \cdot W_{white} \tag{6}$$

$$H_{Black} = H_{white} + \left|1 - H_{Image} / H_{white}\right| \cdot H_{white} \tag{7}$$

Where H and W represent Height and Width, $W_{Black}$, $W_{Image}$, $W_{White}$ are the width of Black rectangle, original image and White rectangle; $H_{Black}$, $H_{Image}$, $H_{White}$ are height of Black rectangle, original image and White rectangle (Figure 2).

**Fig. 2.** The examples of salient region (white and black box is original and extension salient region respectively)

### 2.2 Improved Watershed Algorithm

Since super-pixels often preserve the object boundaries and similar features, we choose super-pixels rather than image pixels as the smallest processing unit for algorithm stability and efficiency. We improve the traditional watershed algorithm to get more accurate segmentation result. Noises from original images bring about many false local minimum in the gradient images, which result in over-segmentation, so we filter the image with mean filter first, then use the sum of Sobel operator and its transposed filter as a new filter to do convolution with images, finally calculate the distance between edge and non-edge pixels, take the result as gradient image.

As can be seen from Figure 3, the image size is decreased by cropping saliency region. After the improved watershed transform, the numbers of pre-segmentation regions are less than the traditional watershed algorithm, but retain adequate information at the same time. Therefore, the improved watershed method greatly reduces the burden of post-merger process, making fast image segmentation feasible.

### 2.3 Graph Cut Model Based on Saliency Map and Super-Pixels

Utilizing visual attention model and improved watershed algorithm described in section 2.1 and 2.2, the extracted significant region is pre-segmented into super-pixels with consistent description.

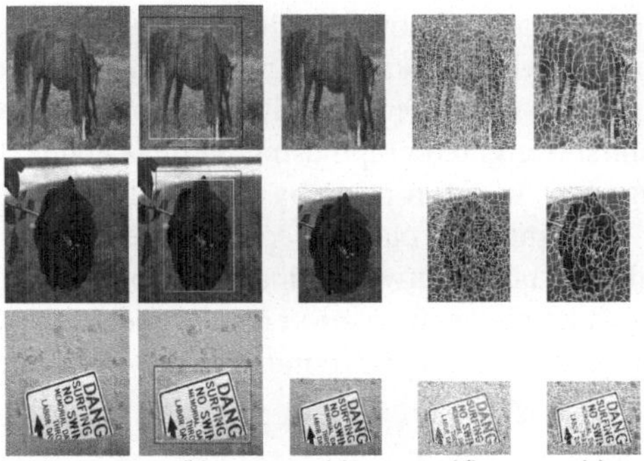


**Fig. 3.** Comparison of super-pixels using traditional and improved watershed algorithm (Column from left to right: (a) original images; (b) labeled images in proposed method; (c) cropped images; (d)pre-segmented images by traditional watershed algorithm; (e) pre-segmented images by improved watershed algorithm respectively.)

Take an image as a weighted undirected graph, assume that a super-pixel block is p, super-pixels set is node set V, and N is the set of all neighboring super-pixel blocks. We assign $A=(A_1,\ldots,A_p,\ldots,A_{|p|})$ is a feasible cut of an image, if $A_p$ belongs to foreground, then $A_p$=1; else $A_p$=0. The energy function is $E(A)$:

$$E(A) = \lambda \cdot \sum_{p \in P} R_p(A_p) + \sum_{(p,q) \in V} B(p,q)\left|A_p - A_q\right| \tag{8}$$

Where $\sum R_p(A_p)$ and B$(p,q)$ are region and boundary terms. λ is a balancing weight between region and boundary cost term. Formula (8) can be rewritten as follows:

$$E(A) = E(A,k,\theta,P) = \lambda \cdot R(A,k,\theta,P) + B(A,P) \tag{9}$$

Where θ is gauss distribution, k presents components of Gaussian Mixture Distribution, P is a super-pixel block.

$$R(A,k,\theta,P) = \sum_{\mathrm{p}} R_p(A_p,k_p,\theta,p_i) \tag{4}$$

$$\begin{aligned} R_p(A_p,k_p,\theta,p) = & -\log[\pi(A_p,k_p,\theta,p) * (\sigma \sum A_p,k_p)/2] \\ & + \frac{1}{2}[p-\mu(A_p,k_p)]^T \sum (A_p,k_p)^{-1} [p-\mu(A_p,k_p)] \end{aligned} \tag{5}$$

$$B(A,P) = \sum_{(p,q) \in V} B(p,q)\left|A_p - A_q\right| = \sum_{(p,q) \in N} \frac{\left|A_p - A_q\right| e^{-\beta Dist(\overline{color(p)},\overline{color(q)})}}{Dist(p,q)} \tag{6}$$

Based on the above definition, we solve the segmentation problem iteratively:

1. Obtain the saliency map according to section 2.1, then get the rectangular bounding box containing foreground according to formula (6) and (7), and take the expanded black rectangle as input image to be segmented.

2. Define the white box as unknown regions to be segmented, image content between black box and white box as background, and initial object set is empty.
3. Utilizing the improved watershed algorithm to pre-segment input image into super-pixels, establish weighted graph using them as nodes.
4. Because the image has been cropped by visual attention model, we use FCM method to cluster them into two classes separately.
5. Calculate the feature distance between super-pixel blocks and each class, then use max flow-min cut to obtain segmentation result, update the Gaussian distribution information of foreground and background, the loop continues until convergence.

# 3 Experimentation

Our experimental dataset is MSRA Salient Object Database with 130,099 high quality images from a variety of sources, mostly from image forums and image search engines [11]. We use F measure as an overall quantitative evaluation [12].

$$F = \frac{precision \cdot recall}{0.5 \cdot precision + 0.5 \cdot recall} \tag{7}$$

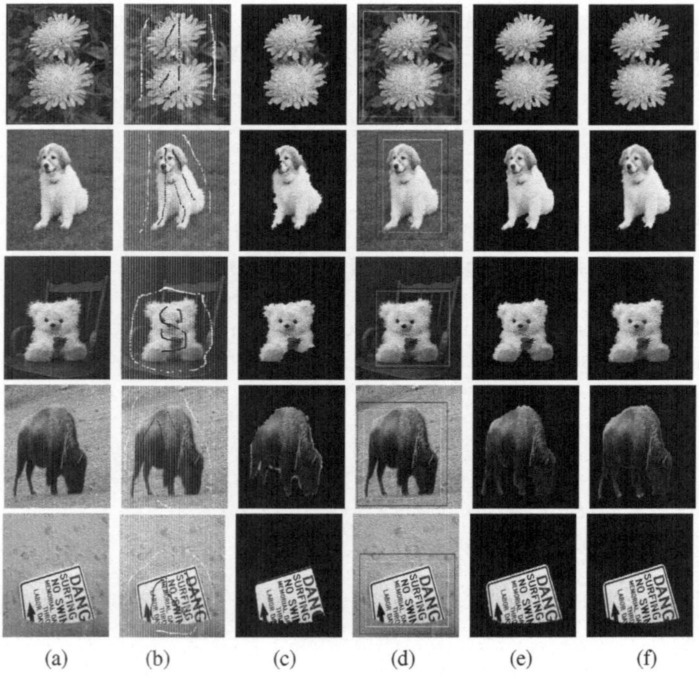

Fig. 4. Comparison of lazy snapping and proposed method From left to right: (a)original images;(b) labeled images;(c) results of lazy snapping; (d)our labeled images; (e)our results; (f) ground truth

It can be seen from figure 4(b) that lazy snapping needs human interaction (black brushes and white brushes are foreground seeds and background seeds labeled by users), so the accuracy and completeness of hand-labeled seeds are important to segmentation results, if labeled points are not sufficient enough, users may hardly obtain a complete segmented object, in other words, its segmentation results rely on users subjective manual interaction more or less. Differently, our labels are obtained by the visual attention model automatically, first calculate the rectangular area containing candidate object to be segmented, and then finish segmentation relatively through the latter iterative processing without interaction. Figure 5(d) shows the quantitative F measure of lazy snapping and our method, we can conclude from Figure 4 and Figure 5(d) that, the proposed algorithm has higher segmentation accuracy than lazy snapping algorithm.

Figure 5(a) shows the comparison between the traditional watershed algorithm and improved watershed algorithm, we can see that the numbers of pre-segmented regions in proposed method are more less than traditional algorithm, accordingly reducing segmentation time and numbers of iterations in the latter iterative processing compared with the original watershed algorithm, as shown in Figure 5(b) and (c). The pre-segmentation stage is 1.3 times than the traditional watershed algorithm due to more complex processing, but it makes average numbers of iterations using improved watershed algorithm only 46.8% of the original watershed algorithm, saving about 34.05% time for follow-up split, as a result leads a faster processing speed.

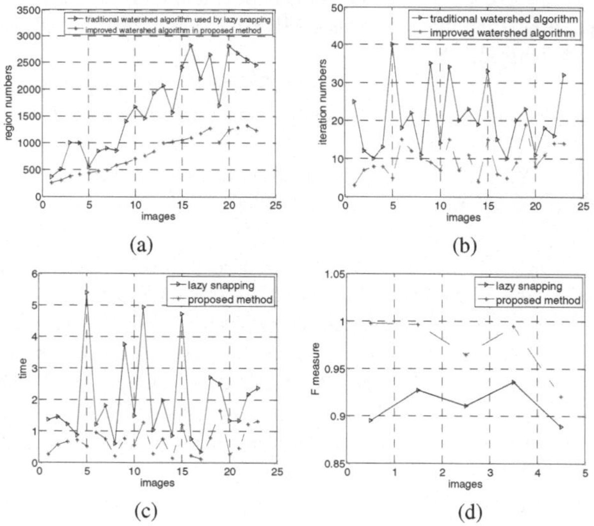


**Fig. 5.** Comparison of lazy snapping and proposed method. From left to right: (a) region numbers pre-segmented by the traditional watershed method and improved method; (b)numbers of iterations using the traditional watershed method and improved method; (c) Graph Cut time of lazy snapping and our method; (d) Quantitative F measure of lazy snapping and our method.

# 4 Conclusion

We propose a new salient object extraction method of nature scene. The method first generates saliency map in HSV color space, then get the rectangle area containing the object, and take regions segmented by the improved watershed algorithm as super-pixels, and super-pixels are clustered by FCM method to build a weighted graph, and finally obtain the salient object by iterative Graph Cut method. Experiment results show that the algorithm can automatically get a prior knowledge of the foreground and background without affecting the accuracy of segmentation.

**Acknowledgments.** This work is supported by National Natural Science Foundation of China (NSFC Grant No. 61170124, 61170020, 61070223), Natural Science Foundation of Jiangsu (Grant No. BK2009116), Science and Technology Support Plan of Jiangsu (Grant No. BE2009048).

## References

1. Xu, N., Ahuja, N., Bansal, R.: Object segmentation using graph cuts based active contours. Computer Vision and Image Understanding 107, 210–224 (2007)
2. Li, Y., Sun, J., Tang, C.K., Shum, H.Y.: Lazy Snapping. In: Proceeding of ACM SIGGRAPH, New York, pp. 303–308 (2004)
3. Boykov, Y., Jolly, M.-P.: Interactive Organ Segmentation Using Graph Cuts. In: Delp, S.L., DiGoia, A.M., Jaramaz, B. (eds.) MICCAI 2000. LNCS, vol. 1935, pp. 276–286. Springer, Heidelberg (2000)
4. Ma, M., He, J., Guo, H., Tian, H.: A New Image Segmentation Method Based on Grey Graph Cut. In: International Joint Conference on Computational Science and Optimization, Huangshan, pp. 477–481 (2010)
5. Yu, S.Y., Zhang, Y., Mao, X.N., Yang, J.: Accurate extraction of infrared target based on graph cut. Electronics Letters 44, 100–101 (2008)
6. Zhang, J., Wang, Y., Shi, X.: An improved graph cut segmentation method for cervical lymph nodes on sonograms and its relationship with node's shape assessment. Computerized Medical Imaging and Graphics 33, 602–607 (2009)
7. Guo, C., Wang, P.: Adaptive Graph-cut Algorithm to Video Moving Objects Segmentation. In: International Conference on Image and Signal Processing, TianJin, pp. 1–5 (2009)
8. Rother, C., Kolmogorov, V., Blake, A.: Grab Cut - Interactive Foreground Extraction using Iterated Graph Cuts. ACM Transactions on Graphics 23, 309–314 (2004)
9. Tian, M.H.: Research on visual attention mechanism modeling and its applications. University of Science and Technology of China (2010)
10. Hou, X.D., Zhang, L.: Saliency detection: A spectral residual approach. In: IEEE Computer Society Conference on Computer Vision and Pattern Recognition, Los Alamitos, pp. 1–8 (2007)
11. Yuan, Z.J., Sun, J., Wang, J.D., Zheng, N.N., Tang, X.O., Shum, H.Y.: Learning to Detect a Salient Object. IEEE Transactions on Pattern Analysis and Machine Intelligence 33, 353–367 (2011)
12. Han, S.D., Tao, W.B., Wu, X.L.: Texture segmentation using independent-scale component-wise Riemannian covariance Gaussian mixture model in KL measure based multi-scale nonlinear structure tensor space. Pattern Recognition 44(3), 503–518 (2011)

# Learning a Ground Object Manifold for Interpreting High-Resolution Sensor Image

Xiangjun Zhao[1,2] and Jianghua Gong[1]

[1] Institute of Remote Sensing Applications, Chinese Academy of Sciences, 100101 Beijing, China
[2] School of Computer Science and Technology, Jiangsu Normal University, 221116 Xuzhou, China
xjzhao@jsnu.edu.cn, jhgong@irsa.ac.cn

**Abstract.** In recent years, the spatial resolution of a remote sensing image becomes much higher than ten years ago. The research of image processing and analyzing based on traditional low resolution image has already not satisfied the need for getting more accurate information. Identifying particular objects from remote sensing image become more important to Digital City and real-time monitoring. The paper proposes a novel semantic manifold interpretation method of high-resolution sensor image, which uses semantics associated with ground object images to improve object recognition works. Our approach first learns the multiple semantic classes by using a semi-supervised manifold learning algorithm to produce a "semantic manifold" of the ground object, and then the RF(Relevance Feedback) iteration based on manifold ranking algorithm is then run on the semantic manifold. The methods are applied to several high-resolution example images, and some buildings as test objects in images are recognized. Those examples illuminate that the method proposed in this paper is effective and accurate, especially for multi-view, multi-spectral, all-weather remote images.

**Keywords:** Semantic Space, Manifold Learning, Sensor Image Interpretation.

## 1 Introduction

Recently, new satellite and unmanned aerial vehicle sensors acquire images with high spectral and spatial resolution, and revisiting time is constantly reduced. Remote sensing images allow Earth observation with unprecedented accuracy. Based on conventional remote sensing image processing method, we can classify different type of large terrain, such as city and farmland. When the resolution of remote sensing image approaches to 1 meter or even less, we can see most small objects on the ground clearly, such as houses, vehicles, and so on. It is difficult to distinguish those small objects from the image background by conventional remote sensing image processing methods. Processing those data is becoming more complex in such situations and how to identify the same ground object from different spectral images and different view images is a vital problem. However, different images of the same ground object have large difference since their different spectral, very different

J. Lei et al. (Eds.): AICI 2012, LNAI 7530, pp. 555–562, 2012.

viewpoint, and different observation time. The traditional image registering method based on computer vision usually fails, and more and more researchers appeal for machine learning method. This paper attempts to construct a manifold of ground object of high-resolution sensor image, and using the semantic information of the ground object, improves the performance of the identification method.

In this paper, we propose a novel ground object identification method that takes into account both a semantic class that reflects user intention for the session, and shared and well-established multiple semantic classes for an improved identification performance. To improve the identification accuracy, a relevance feedback framework based on the manifold learning algorithm is run on the semantic manifold.

The contribution of this paper is the proposal of a novel ground object identification method that combines Relevance Feedback and manifold learning for an improved identification performance. Fig. 1 shows the flowchart of our method. And experimental evaluate the proposed method by using several actual remote sensing images by the aerial vehicle. Evaluation showed that manifold identification method significantly outperforms the same in the original feature space.

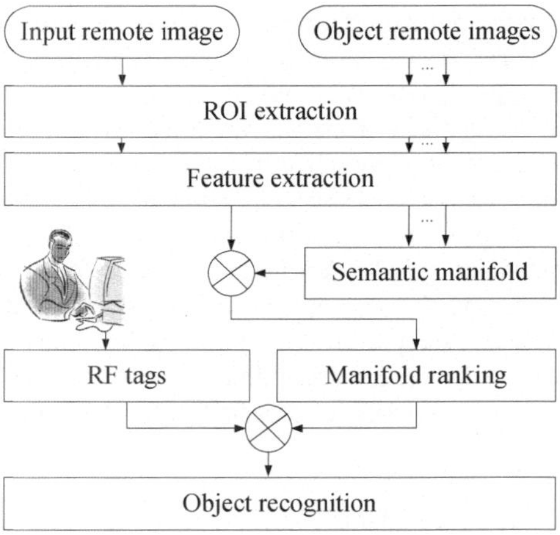


**Fig. 1.** Flowchart of the algorithm that is used to recognize the object in the remote sensing image based on manifold ranking

The following of this paper is structured as follows. The next section will introduce the related work. Section 3 will propose a novel ground object manifold identification method, followed by Section 4 on experiments and their results. We will summarize the paper in Section 5.

## 2 Related Work

Based on conventional remote sensing image, there are many studies on man made object (roads, houses, vehicles) recognition in the high-resolution image[1, 2]. Object recognition algorithm in optical camera image processing are applied to the remote sensing image because of the improvement of the spatial resolution. But there are several different problems faced for remote sensing image processing and optical camera image processing. Remote sensing images are acquired with different viewpoint angle and view field, and the object in remote sensing images usually has scale, translation and distortion. Because the object on the earth is quite variety, from a large city to a small vehicle, and one object has a different appearance on different remote sensing image because of different view point, view field, view angle and weather condition. The algorithm based on supervised or unsupervised methods have good performance to classify large terrain objects. But for small objects in the remote sensing image, the algorithm for recognition should be paid more attention to study.

Content-Based Image Retrieval[3-5] is a long standing research problem in computer vision and ground object recognition. Most of previous ground object recognition techniques build on the assumption that the image space is Euclidean. However, in many cases, the image space might be a non-linear sub-manifold which is embedded in the ambient space. One possible solution to these two problems is to learn a mapping function from the low-level feature space to the high-level semantic space. The former is not always consistent with human perception while the latter is what ground object recognition system desires to have. Specifically, if two object images are semantically similar, then they are close to each other in semantic space. In this paper, our approach is to recover semantic structures hidden in the image feature space.

Many researchers have discussed the relevance feedback method in content-based image recognition from the perspective of machine learning[6, 7], yet most learning methods only take into account current recognition session and the knowledge obtained from the past user interactions with the system is forgotten. To compare the effects of different learning techniques, a useful distinction can be made between short-term learning within a single recognition session and long-term learning over the course of many recognition sessions[8]. Both short- and long-term learning processes are necessary for an image recognition system though the former has been the primary focus of research so far.

As we point out, the choice of the similarity measure is a deep question that lies at the core of image retrieval. In recent years, manifold learning[9-11] has received lots of attention and been applied to face recognition[12], graphics[13], document representation[14], etc. These research efforts show that manifold structure is more powerful than Euclidean structure for data representation, even though there is no convincing evidence that such manifold structure is accurately present. Based on the assumption that the images reside on a low-dimensional sub-manifold, a geometrically motivated relevance feedback scheme is proposed for image object ranking, which is naturally conducted only on the image manifold in question rather than the total ambient space.

## 3 Relevance Feedback on Ground Object Manifold

Our method uses the manifold ranking[15] algorithm for relevance feedback learning. Manifold ranking is a graph-based learning algorithm that can be used in unsupervised or supervised mode. We used the manifold ranking algorithm by Zhou, et al., among others. It is worthwhile to highlight that recovering the image manifold need lots of images. Fortunately, usually there are enough images to recover the ground object manifold. Fig. 2. shows the sub-manifold in the feature space, and two meshes each one corresponds to a sub-manifold.

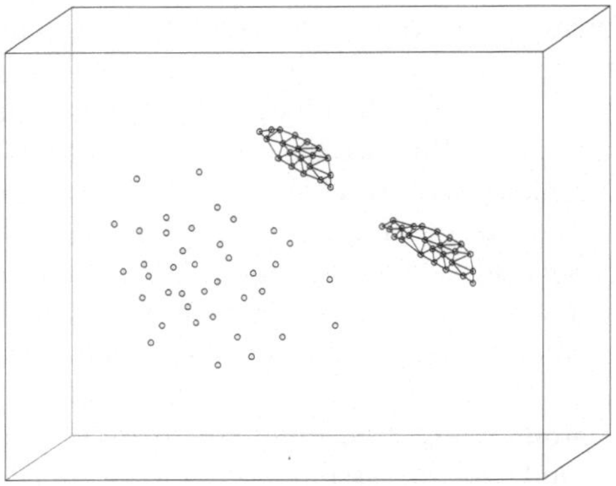

**Fig. 2.** Sub-manifold of ground object

### 3.1 Manifold Ranking

The manifold ranking algorithm tries to similarity of points from a given sample on the manifold of all the samples, taking global distribution of the samples into account.

Let

$$\chi = \{x_1, \ldots, x_q, x_{q+1}, \ldots x_n\}, \tag{1}$$

be a m-dimensional feature vector, in which first $q$ points from $x_1$ to $x_q$ are the recognitions and the rest are the points we want to rank according to their similarity to recognitions .

Let $d: \chi \times \chi \rightarrow R$ denote a distance metric on $\chi$ , e.g., $L^1$ norm distance, that assigns a pair of points $\chi_i$ and $\chi_j$ a distance d($\chi_i$, $\chi_j$). Let $f : \chi \rightarrow R$ be a ranking function that assigns each $\chi_i$ a ranking score $f_i$, forming a rank vector $f = [f_1,\ldots,f_n]^T$. Let the n-dimensional vector $y = [y_1,\ldots,y_n]^T$ be a label vector.

The same manifold ranking algorithm may be used in unsupervised and supervised modes. In the unsupervised mode, the goal is to compute the similarity ranks of the image object in the database to the identification. So we set $y_i = 1$ for the image object to be identified only and $y_i = 0$ for the rest, i.e., all the image objects in the database. In the supervised 1-class learning mode under the relevance feedback framework, the goal is to compute the similarity ranks of the image objects in the database to the expanded unidentified set containing more than one points. The expanded unidentified set consists of the initial image object to be identified and all the image objects in the database that are tagged as activity at current relevance feedback iteration. Thus, we set $y_i = 1$ for the image object to be identified and the image objects that are tagged as relevant during the RF iterations. The initial value $y_i = 0$ is given to the other image objects in the database without the relevant tag.

Intuitively, the process of manifold ranking resembles the process of solving a diffusion equation on an irregular mesh. The manifold is generated by connecting feature points in the high dimensional input features space based on their mutual similarity. The similarity is determined by using a distance measure, e.g., $L^1$-norm, in the feature space. The manifold ranking algorithm iteratively diffuses the initial value of $y_i = 1$ given of the unidentified set to its global distribution of the samples.

The higher the diffused value, the higher the similarity rank of the points in the database to the unidentified set. As the diffusion occurs along the manifold, similarity rank thus computed are better than those computed directly in the input feature space.

Create the similarity matrix **W** where $W_{ij}$ indicates the similarity between samples $x_i$ and $x_j$

$$w_{ij} = \begin{cases} \exp\left(-\dfrac{d(x_i, x_j)}{\sigma}\right), & \text{if } i \neq j; \\ 0, & \text{otherwise.} \end{cases} \tag{2}$$

The distance metric $d(x_i, x_j)$ used for the similarity matrix affects the ranking. The positive parameter $\sigma$ defines the radius of influence. Note that $W_{ii} = 0$ since there is no arc connecting a point with itself. The matrix **W** is positive symmetric. Then, we form a normalized graph Laplacian **L**,

$$\mathbf{L} = \sqrt{\mathbf{D}^{-1}}(\mathbf{D} - \mathbf{W})\sqrt{\mathbf{D}^{-1}}, \tag{3}$$

Where **D** is a diagonal matrix in which $D_{ij}$ equals to the sum of the $i$-th row of **W**, that is, $D_{ij} = \sum_j W_{ij}$ Then the ranking vector $f=[f_1,\ldots,f_n]^T$ can be estimated by iterating the following until convergence

$$f^{t+1} = \frac{1}{1+\mu}(1-\mathbf{L})f^t + \frac{1}{1+\mu}\mathbf{Y}. \tag{4}$$

The parameter $\mu>0$ is a regularization parameter, and affects retrieval performance and the convergence of the iteration above.

Let $f^*$ be the limit of the above iteration. Rank each point $x_i$ as a label $y_i = \arg\max_{j \le c} f^*_{ij}$. In the case above, $f^*$ has a close form solution

$$f^* = (\mathbf{I} + \frac{1}{\mu}\mathbf{L})^{-1}\mathbf{Y}. \tag{5}$$

We use the closed-form solution, for it was faster or our set of parameters when implemented.

### 3.2 Distance Measures

We can use several distance measures for retrieval, and for forming the affinity matrix **W** to be used for manifold ranking. Let $\mathbf{x}=(x_i)$, $\mathbf{y}=(y_i)$ are the feature vectors and $n$ is the dimension of the vectors. Depending on the value of $k$, $d_k(\mathbf{x}, \mathbf{y})$ is the Manhattan distance if k=1.0, or the Euclidian distance if $k$=2.0.

$$d_k(\mathbf{x},\mathbf{y}) = \left[\sum_i^n \left(\|x_i - y_i\|^k\right)\right]^{1/k}. \tag{6}$$

According to Aggarwal, et al[16], $k$<1.0, e.g., $k$=0.5 is expected to perform better for higher dimensional features. As the *cosine* measure is a measure of similarity in the range [0, 1], we converted it to a distance using the following equation

$$d_{\cos}(\mathbf{x},\mathbf{y}) = 1 - \frac{\mathbf{x}\cdot\mathbf{y}}{\|\mathbf{x}\|\cdot\|\mathbf{y}\|}. \tag{7}$$

Another distance measure is the *Kullback-Leibler* divergence (KLD). The KLD is sometimes referred to as information divergence, or relative entropy, and is not a distance metric, for it is not symmetric,

$$d_{KLD}(\mathbf{x},\mathbf{y}) = \sum_{i=1}^{n}(y_i - x_i)\ln\frac{y_i}{x_i}. \tag{8}$$

**Fig. 3.** Building image sequence for constructing a sub-manifold

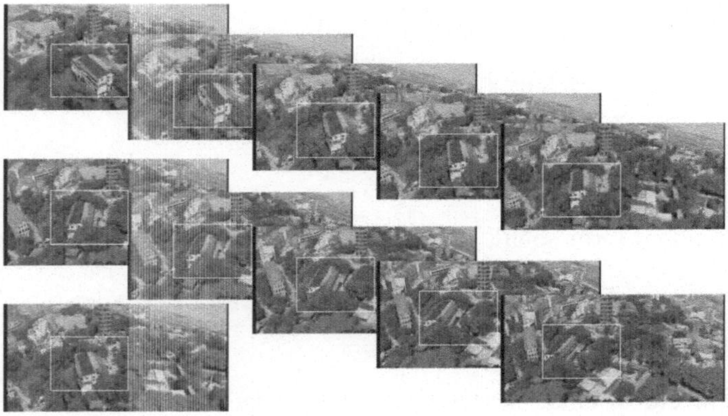

**Fig. 4.** House image sequence for constructing a sub-manifold

## 3.3 Long-Term Relevance Feedback

After finished a remote sensing mission, recognition modular of the same ground object will be used repeatedly. Hence, feedback should have a long-term mechanism. Each recognition result and its original image should be incorporated into the identification database, and the next time those data will be able to play a role.

Especially, the new data should be used to construct sub-manifold. With the increase of the number of sub-manifold and the improvement of manifold shape, the recognizer can get a better and better performance, and higher recognition accuracy and level of automation.

## 4 Experiments and Results

Experiments have been carried out on remote images of the city of Chongqing. The images are acquired from a low-altitude airborne video recordings. Software system is implemented in MATLAB.

The experiment images contain two groups and each group has 272 images, among which, 256 images are used for constructing sub-manifold and the other 16 images are used to identify. In order to verify long-term identification capability, we adopt a gradual experimental method. To speed up the progress of the experiment, we segment the image interactively by the user. At the beginning, there is a low recognition error. At last, classification result is entirely correct. Fig. 1. shows an example, the building image sequence in the figure can generate a sub-manifold. Although the building image in the first and the last picture are quite different, but have a little difference in the adjacent picture. Accordingly, we can construct a triangle mesh, which is the so-called sub-manifold. The picture at the lower left corner of Fig. 1 is to be identified, and experiment shows that the pictures can be accurately linked to the corresponding sub-manifold. Fig. 2 is the other example, and is similar to Fig. 1. The object to be identified in Fig. 2 is the house at the top of the mountain.

## 5 Conclusion

In this paper, we proposed a novel sensor image object identification algorithm that takes into account both long-lived semantics learned from multiple semantic classes and a short-lived, within-session user intention learned by using relevance feedback. The former is accounted for by learning multiple semantic classes at once. The latter is accounted for by using a relevance-feedback framework based on the manifold ranking algorithm. The method performs manifold ranking-based relevance feedback on the semantic-manifold that embodies both feature distribution and multiple semantic classes. Experimental evaluation shows that the proposed approach is quite effective. The use of semantic-manifold instead of the original, ambient feature space improved the effectiveness of the manifold ranking algorithm significantly.

## References

1. Selvarajan, S., Tat, C.W.: Extraction of man-made features from remote sensing imageries by data fusion techniques. In: The 22nd Asian Conference on Remote Sensing, Singapore (2001)

2. Moser, G., Serpico, S.B.: Automatic parameter optimization for support vector regression for land and sea surface temperature estimation from remote sensing data. IEEE Transactions on Geoscience and Remote Sensing 47(3), 909–921 (2009)
3. Zhu, L., Rao, A., Zhang, A.: A theory of keyblock-based image retrieval. ACM Trans. on Information Systems 20(2), 224–257 (2002)
4. Chang, E., Goh, K., Sychay, G., et al.: Cbsa: Content-based soft annotation for multimodal image retrieval using bayes point machine. IEEE Trans. on Circuits and Systems for Video Technology 13(1) (2003)
5. Nizar, B., Nath, G.M.: Discrete visual features modeling via leave-one-out likelihood estimation and applications. Journal of Visual Communication and Image Representation 21(7), 613–626 (2010)
6. Wang, J., Li, J.: Learning-based linguistic indexing of pictures with 2-d mhmms. In: Proc. ACM Multimedia, Juan Les Pins, France, pp. 436–445 (2002)
7. Tong, S., Chang, E.: Support vector machine active learning for image retrieval. In: Proc. ACM Multimedia 2001, Ottawa, Canada (2001)
8. He, X., King, O., Ma, W.-Y., et al.: Learning a semantic space from user's relevance feedback for image retrieval. IEEE Trans. on Circuit and System for Video Technology (1) (2003)
9. Tenenbaum, J.B., Silva, V.D., Langford, J.C.: A global geometric framework for nonlinear dimensionality reduction. Science 290(22) (2000)
10. Belkin, M., Niyogi, P.: Laplacian eigenmaps for dimensionality reduction and data representation. In: Advances in Neural Information Processing Systesms (2001)
11. He, X., Ma, W.-Y., Zhang, H.-J.: Learning an image manifold for retrieval. In: ACM Conference on Multimedia 2004, New York City (2004)
12. He, X., Yan, S., Hu, Y., et al.: Learning a locality preserving subspace for visual recognition. In: IEEE Conf. on Computer Vision, Nice, France (2003)
13. Matusik, W., Pfister, H., Brand, M., et al.: A data-driven reflectance model. In: Proc. of SIGGRAPH 2003 (2003)
14. He, X., Cai, D., Liu, H., et al.: Locality preserving indexing for document representation. In: ACM SIGIR Conference on Information Retrieval, Sheffield (2004)
15. Zhou, D., Bousquet, O., Lal, T.N., et al.: Learning with local and global consistency. In: Proc. NIPS 2003 (2003)
16. Aggarwal, C.C., Hinneburg, A., Keim, D.A.: On the Surprising Behavior of Distance Metrics in High Dimensional Space. In: Van den Bussche, J., Vianu, V. (eds.) ICDT 2001. LNCS, vol. 1973, pp. 420–434. Springer, Heidelberg (2000)

# Robust Adaptive Beamforming with Null Widening under Quadratic Constraint

Yunxiao Jiang

Key Laboratory of Electronic Restriction of AnHui Province,
Electronic Engineering Institute, Hefei 230037, China
j_yunxiao@sohu.com

**Abstract.** The adaptive beamforming algorithm can aim at the direction-of-arrival of jammer automatically. But when the antenna platform vibrates or interference moves quickly, it is possible that the mismatching occurs between adaptive weight and data due to the perturbation of the interference location. To solve these problems above, a novel algorithm for null widening is presented in this paper. By using quadratic constraint on the interference integration matrix, the proposed approach can control the depth and width of null with different user parameters, so the interferences can be suppressed effectively in such case and the robustness of this adaptive beamforming algorithm is improved. The formulas of this algorithm are deduced and the ranges of Lagrange multiplier value are computed. Simulation results show the effectiveness of the proposed algorithm.

**Keywords:** Adaptive beamforming, Null widening, Quadratic constraint, Lagrange multiplier.

## 1 Introduction

Robust adaptive beamforming is a focusing problem in the field of beamforming research. Li[1] and Voyobrov[2,3] had done much constructive work for it. However, their work are fundamentally to correct the steering vector of desired signal so as to improve the robustness of beamformer. If interferences move quickly or the antenna receiver platform shocks frequently, the data mismatch will occur because of the low speed of beamformer output weights' renewing, the interference would move out of null location and cause interference under-nulling. The algorithm for nulling widening can solve such problems hereinabove and improve the robustness. Mailloux[4] and Zatman[5] independently proposed different solution for nulling widening. R.F. LI [6] deduced the null widening method when interference is Gaussian distributed, and proved that the method when interference evenly distributed is equivalent to Zatman's algorithm. S.J. WU [7] and Z.K. YU [8] all use a covariance matrix taper technique to achieve null widening with a matrix T related to the nulling width, these three methods are somewhat same in essence, and attain similar performance in output signal-to-noise-plus-interference (SINR). J.B. WANG [9] proposed a novel method that attains output weights related to the optimal weight using the least squares criteria.

J. Lei et al. (Eds.): AICI 2012, LNAI 7530, pp. 563–570, 2012.

This paper proposed a novel adaptive beamforming algorithm based on interference matrix quadratic constraint which goal is to minimize the square of the norm between the output weight and the optimal weight. The formula of this algorithm are deduced and the scope of Lagrange multiplier value are computed. Simulation results show the validity of the proposed algorithm.

## 2 Array Signal Model

Let us consider a linear array of M sensors that receive signals from multiple narrowband sources. The complex vector of array observation at time can be modeled as

$$x(t) = x_s(t) + x_i(t) + n(t) \tag{1}$$

where $x_s(t)$ , $x_i(t)$ and $n(t)$ are the statistically independent components of the desired signal, inference, and sensor noise, respectively, here $x_s(t) = s(t)a$ , and $s(t)$ is the desired signal waveform, and $a$ is the steering vector associated with the desired signal.

The adaptive beamformer output is given by

$$y(t) = w^H x(t) \tag{2}$$

where $w = [w_{1,} \dots w_M]^T \in C^M$ is the complex vector of beamforming weights, and $(\cdot)^T$ and $(\cdot)^H$ stand for the transpose and hermitian transpose, respectively.

Based on the output power minimization criteria, the adaptive beamforming problem can be described as following optimization problem:

$$\begin{aligned} &\min_w w^H R_x w \\ &\text{s.t.} \quad w^H a = 1 \end{aligned} \tag{3}$$

The Capon beam former gives the solution:

$$w_{capon} = \frac{R_{i+n}^{-1} a}{a^H R_{i+n}^{-1} a} \tag{4}$$

where $R_{i+n}$ is the interference plus noise covariance matrix, for the $R_{i+n}$ is difficult to obtain in practice, we often use its estimation to replace it by using multiple snap method in sense of time and then average the $X(i)X^H(i)$ , $N$ is snap number, the estimate of $R_x$ is defined as

$$R_x = \frac{1}{N}\sum_{i=1}^{N} X(i)X^H(i) \tag{5}$$

The beam former output SINR is defined as

$$SINR = \frac{\sigma_s^2 \,|\, w^H a \,|^2}{w^H R_{i+n} w} \tag{6}$$

where $\sigma_s^2$ is the signal power.

# 3 The Proposed Algorithm

## 3.1 Algorithm Description

The standard Capon beamformer places deep and narrow null in interference location, which will cause under-nulling while data mismatched. Consider a relatively wider null in interference location interval and make the algorithm more robust. Assume the $W_{opt}$ is the optimal beamformer weight , we can refer to Capon beamformer as a comparison, suppose the direction of the interference are $\theta_j$ ( $j = 1, 2, ..., J$ ) , $J$ is the number of interference. Then the formula $\| w_{opt}^H a(\theta_j) \| = 0$ is theoretically right, where the $\| \cdot \|$ represents the European norm. If the interference shake within an interval, we definite a new weight $W$ and expect $\| W^H a(\tilde{\theta}_j) \| = 0$, where $\tilde{\theta}_j \in [\theta_j - \Delta\theta_j / 2, \theta_j + \Delta\theta_j / 2]$ ,the $\Delta\theta_j$ is the pre-set null width which are determined by practical demand. Determine the optimum weight vector $W$ that is the solution to the following quadratic function constrained problem:

$$\begin{aligned} &\min f(W) = \| W - W_{capon} \|^2 \\ &s.t \qquad W^H Q W \le \varepsilon \end{aligned} \tag{7}$$

Where $W_{capon}$ is the Capon beamformer output weight and $Q$ is the interference signal steering vector inegration matrix, $\varepsilon > 0$ is user parameter.

$$Q = \sum_{j=1}^{J} \int_{\theta_j - \frac{\Delta\theta_j}{2}}^{\theta_j + \frac{\Delta\theta_j}{2}} a(\theta) a^H(\theta) d\theta \tag{8}$$

Note that the width of every null can be adjusted according to demand in practice. The problem (7) can be solved by using Lagrange multiper methodology, which is based on the functin

$$f(W)=(W-W_{capon})^H(W-W_{capon})+\lambda(W^HQW-\varepsilon)$$
$$=W^HW-W^HW_{capon}-W_{capon}^HW+W_{capon}^HW_{capon}$$
$$+\lambda(W^HQW-\varepsilon) \quad (9)$$

Differentiation of (9) and make it equal to zero,the above equation yields

$$W=(I+\lambda Q)^{-1}W_{capon} \quad (10)$$

### 3.2 The Range of Lagrange Multiper

Note that weight $W$ is directly related to the Lagrange multiper, so the appropriate Lagrange multiper is vital to the beamformer performance, we will discuss the range of Lagrange multiper in this section.

Using the condition $W^HQW=\varepsilon$ in (7) gives

$$g(\lambda)=W_0^H(I+\lambda Q)^{-1}Q(I+\lambda Q)^{-1}W_0=\varepsilon \quad (11)$$

Suppose the eigendecomposition of the matrix $Q$

$$Q=U\Lambda U^H=\sum_{m=1}^{M}\gamma_m u_m u_m^H \quad (12)$$

where the column vectors of $U=(u_1,u_2,...,u_M)$ contain the eigenvectors of $Q$ and the diagonal matrix $\Lambda=diag(\gamma_1,\gamma_2,...,\gamma_M)$ with $\gamma_1\geq\gamma_2\geq\cdots\geq\gamma_M$ are the corresponding eigenvalues. Then (11) can be written as

$$g(\lambda)=W_0^HU(I+\lambda\Lambda)^{-1}U^HU\Lambda U^HU(I+\lambda\Lambda)^{-1}U^HW_0$$
$$=W_0^HU(I+\lambda\Lambda)^{-2}\Lambda U^HW_0$$
$$=(W_0^HU)(I+\lambda\Lambda)^{-2}\Lambda(W_0^HU)^H$$
$$=\sum_{m=1}^{M}\frac{\gamma_m(W_0^HU)(W_0^HU)^H}{(1+\lambda\gamma_m)^2} \quad (13)$$

Let

$$z_m=|W_0^Hu_m| \quad (14)$$

Then (13) can be rewritten as

$$g(\lambda)=\sum_{m=1}^{M}\frac{\gamma_m(W_0^HU)(W_0^HU)^H}{(1+\lambda\gamma_m)^2}=\sum_{m=1}^{M}\frac{\gamma_m|z_m|^2}{(1+\lambda\gamma_m)^2} \quad (15)$$

(1). Note that $g(\lambda)$ is a monotonically decreasing function of $\lambda > 0$ .According to(11) and (15), $g(0) > \varepsilon$ , and hence , $\lambda \neq 0$ .From (15), it is clear that $\lim_{\lambda \to +\infty} g(\lambda) = 0 < \varepsilon$ .Hence, there is a unique solution within $(0, +\infty)$ to (15). Moreover, $g(\lambda)$ is a monotonically decreasing function of $\gamma_m$ , by replacing the $\gamma_m$ in (15)with $\gamma_1$ and $\gamma_M$ , respectively, we can obtain the following tighter upper and lower bounds on the solution $\lambda > 0$ to (15).

$$\frac{\| W_0 \| \sqrt{\gamma_1} - \sqrt{\varepsilon}}{\gamma_1 \sqrt{\varepsilon}} \leq \lambda \leq \frac{\| W_0 \| \sqrt{\gamma_M} - \sqrt{\varepsilon}}{\gamma_M \sqrt{\varepsilon}} \tag{16}$$

(2). Similarly, note that $g(\lambda)$ is a monotonically increasing function of $\lambda < 0$, From (15), it is clear that $\lim_{\lambda \to -\infty} g(\lambda) = 0 < \varepsilon$ . Hence, there is a unique solution within $(-\infty, 0)$ to (15). By replacing the $\gamma_m$ in (15)with $\gamma_1$ and $\gamma_M$ , respectively, we can obtain the following tighter upper and lower bounds on the solution $\lambda < 0$ to (15).

$$\frac{-\| W_0 \| \sqrt{\gamma_M} - \sqrt{\varepsilon}}{\gamma_M \sqrt{\varepsilon}} \leq \lambda \leq \frac{-\| W_0 \| \sqrt{\gamma_1} - \sqrt{\varepsilon}}{\gamma_1 \sqrt{\varepsilon}} \tag{17}$$

Obviously, for the value of $\varepsilon$ is very small, so the $\lambda$ that satisfying $g(\lambda) = \varepsilon$ is placed in two intervals which is approximately symmetric about zero.

C. Flow of proposed algorithm

The flow of the algorithm can be summarized as follows:

Step1) Compute the constraint matrix $Q$ according to $\Delta\theta_j$ ;

Step2) Compute the range of Lagrange multiper according to $\varepsilon$ ;

Step3) Choose a $\lambda$ , then compute weight $W$ of the beamformer.

## 4 Simulation Results

In the following simulations, we assume a uniform linear array of $M = 10$ spaced half a wave length apart, the additive noise is modeled as a complex Gaussian zero-mean white process. We also assume there interfering sources with plane wavefronts and directions of $30^\circ$, $70^\circ$ and $-30^\circ$, respectively. The interference-to noise ratio( INR) is equal to 30dB, the desires signal is assumed to be a plane-wave that impinges on the array from direction $0^\circ$ .The number of snapshot is 300. In all examples, the experiment results are from 100 Monte Carlo simulations.

In the first example, we simulate the effect on the null depth of user parameter $\varepsilon$. The SNR is 0 dB, the widths of interference null are set as $2^\circ$, $3^\circ$ and $4^\circ$, respectively. Figure 1 shows array beampattern comparisons when $\varepsilon = 10^{-7}$ and $\varepsilon = 10^{-9}$. It is clear that the smaller $\varepsilon$ result in the deeper null.

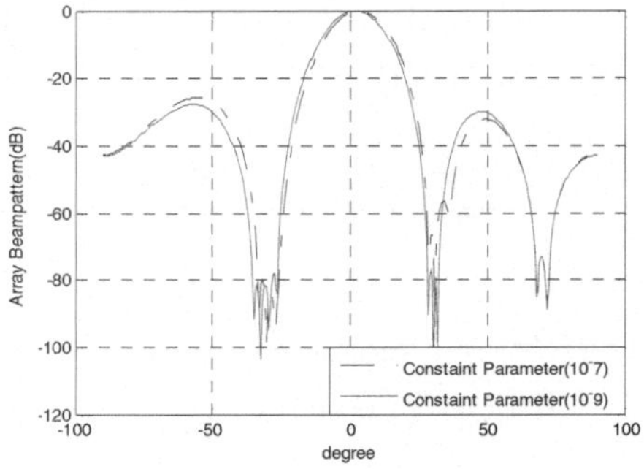


**Fig. 1.** The beamforming performances with different $\varepsilon$

In the second example, call the positive loading or negative loading respectively while $\lambda > 0$ or $\lambda < 0$, the value $\lambda$ is set as the average of the every interval. Figure 2 demonstrates that either the positive loading or negative loading can widen the null effectively, and the the positive loading priors to negative loading in sidelobe control.

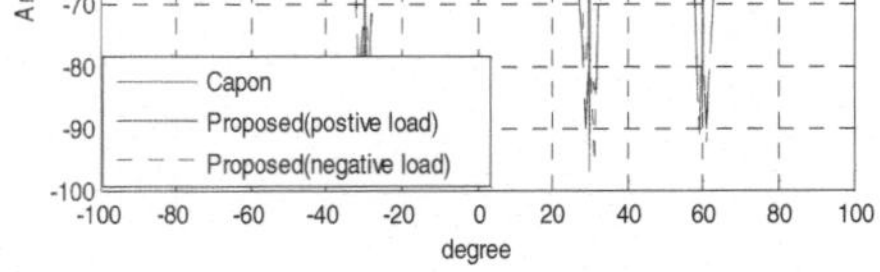

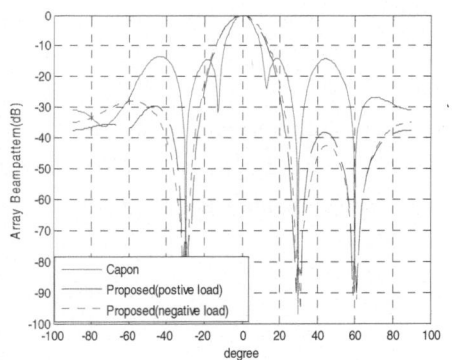


**Fig. 2.** The array beampattern with positive and negative loading $\lambda$

In the third example, the proposed beamforming algorithm is compared to LI's method [6], YU's method [8] and WANG's method [9]. In all methods, SNR= 10dB, SIR=30dB, the widths of interference null are set as $2^\circ$, $3^\circ$ and $4^\circ$. The interference signal steering vector mismatch in LI's method is Gussian distribution modeled

$N(0,4)$, and the $\varepsilon = 10^{-8}$ are set in WANG's and proposed method. Figure 3 shows the array beampattern comparisons. Because the LI's and YU's method all taper covariance matrix with a matrix T, so it is similar in beampattern for two methods. Moreover, the proposed method widened effectively the null to match the width set primarily and has deeper null and lower sidelobe than WANG's method.

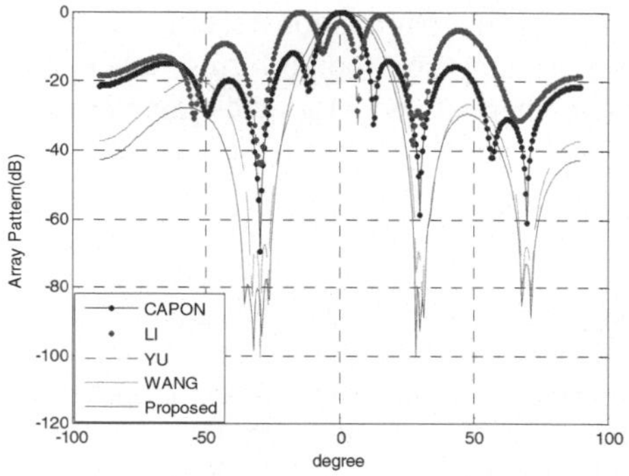


**Fig. 3.** The beamforming peformance comparsion with null width as $2^\circ$, $3^\circ$ and $4^\circ$ respectively

In the fourth example, experiment simulated the relationship between the output signal-to-noise-plus-interference versus SNR, the performance of all methods is shown in figure 4 which depicts that the proposed algorithm outperforms other beamformers. Furthermore, the methods in reference [6] and [8] are almost same, which also verifies the analysis in introduction hereinabove.

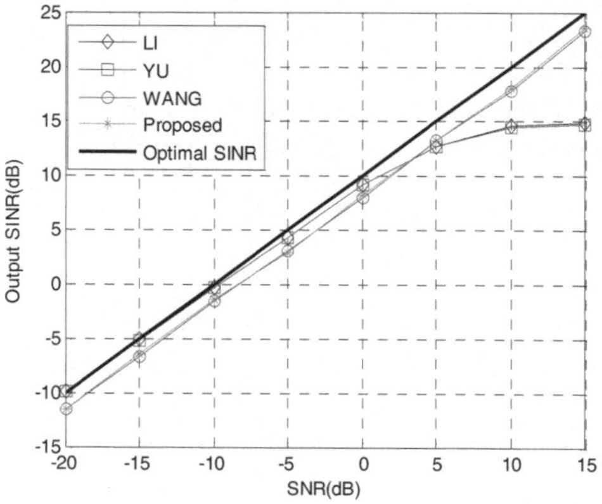


**Fig. 4.** Output SINR versus SNR

## 5 Conclusion

A new algorithm for null widening based on quadratic constraint is presented in this paper according to the data mismatch suffered from interference quick movement or antenna platform vibration. The description of the algorithm and theoretical deduction are proposed, and the range of Lagrange multiper is computed. The presented method can provide an improved robustness against the interference angle shaking and suppress the interference signals. Simulation results demonstrate its validity. It is shown that the proposed approach outperforms other existing null widening algorithms and provides another new way for quick jammer suppression.

## References

1. Li, J., Stoica, P., Wang, Z.: On robust capon beamforming and diagonal loading. IEEE Transactions on Signal Processing 51(7), 1702–1715 (2003)
2. Vorobyov, S.A., Gershman, A.B., Luo, Z.Q.: Robust adaptive beamforming using worst-case performance optimization: A solution to the signal mismatch problem. IEEE Transactions on Signal Processing 51(2), 313–324 (2003)
3. Vorobyov, S.A., Chen, H., Gershman, A.B.: On the relationship between robust minimum variance beamformers with probabilistic and worst-case distortionless response constraints. IEEE Transactions on Signal Processing 56(11), 5719–5724 (2008)
4. Mailloux, R.J.: Covariance matrix augmentation to produce adaptive array pattern thoughs. Electron. Letters 31(10), 771–772 (1995)
5. Zatman, M.: Production of adaptive array troughs by dispersion synthesis. Electron Letter. 47(4), 2141–2142 (1995)
6. Li, R.F., Wang, Y.L., Wan, S.H.: Research on adapted pattern null widening techniques. Modern Radar (2), 42–45 (2004)
7. Wu, S.J., Zhang, J.Z., Zhang, S.: Research on beamforming of the wide nulling algorithm. Journal of Harbin Engineering University 25(5), 658–661 (2004)
8. Yu, Z.K., Yang, S.Y.: Research on the wide nulling algorithm. Applied Science and Technology 33(3), 1–3 (2006)
9. Wang, J.B., Sun, Q., Tang, H.: A novel quiescent pattern control method to widen interference nulls. Journal of Microwaves 23(suppl.), 230–233 (2007)

# Conceptual Analysis of Chinese Query of the Interrogative Sentence

Dani Bi and Yuquan Chen

Department of Computer Science and Engineering
Shanghai Jiao Tong University
800, Dongchuan Rd., Shanghai, China
{bidani,yqchen}@sjtu.edu.cn

**Abstract.** Nowadays, almost every search engine use "Discrete" models, such as boolean logic model. it processes user queries and documents, in the way that will break key words into discrete, unrelated words. It will lose the semantic relation between words, and bring noise in search results. This paper presents a conceptual analysis method, replacing the interrogative word in the query with focus word that is extracted from the query, in order to construct a concept map expressing the connotation semantic information of the query. From the view of Chinese conceptual connotation, we analyze user demands, and restore their retrieval intention, in order to guide the next search, to improve the accuracy of the retrieval system.

**Keywords:** query, interrogative sentence, conceptual analysis, retrieval.

## 1 Introduction

The retrieval model used by most current search engines is still Boolean model. It uses the Boolean model to retrieve results, and then uses a variety of sorting techniques to optimize the retrieval quality. Information retrieval system, which is based on the Boolean model, decomposes the user query into the query keywords, and also the index of documents is generally a discrete word for the object to be processed. This kind of approach has lost most of semantic information, making queries and documents no longer a complete concept. The returned results are only candidates of web pages that contain these keywords. However, what users really want is the information that can answer the queries. For example, the query:"*上海有多大?*" In the top ten of google search results, only three candidates are correct, while the three are all users answers in forums, so the credibility may be not high.

The most important issue of information retrieval is the judgment of the users' information demands, which is called demand analysis. Information demand[1] refers to the desire of the individual/collective orientation and information obtaining, this information can meet the user conscious or unconscious demands. The user is the object of the retrieval system services, thus user demands play a guidance role in the information retrieval system. There are three forms of user query: keyword, phrase, and sentence. Keyword form, including a single keyword, multiple keywords connected by

J. Lei et al. (Eds.): AICI 2012, LNAI 7530, pp. 571–579, 2012.

space; phrase is the form constituted by two or more words; sentence form, including interrogative, declarative and imperative sentences. The query in keyword form is demand representation supported by the major search engines, which is the most suitable for the Boolean model. However, the query of keyword form requires the user to transfer query in the mind into a number of keyword, this transformation may lead to the loss of conceptual information. Moreover, according to the user query log released by Sogou, the query in interrogative form occupies a certain proportion in the collection of user query in the actual network use. Therefore, the analysis of interrogative sentence is necessary. In addition to the analysis of surface structure, this paper focused on the analysis of the user demands in their concepts. The goal of the conceptual analysis of demand is to put user demand into a conceptual structure, so that the retrieval system may be better understand user demands. The conceptual structure, was first raised by John F. Sowa [5], it is developed on the basis of the existential graphs of C.S.Peirce[6] and knowledge representation[7]. Concept map [4] used in this paper is different from the one proposed by John F. Sowa. John F. Sowa only concern whether there is an association between two words, while [4] also concerned about the semantic types of relation between the two words, it would be more effective processing Chinese semantic information.

## 2 Conceptual Analysis

### 2.1 Approach of Conceptual Analysis

Demand analysis of the interrogative sentence can not use "literal translation" to label the concept map of the demand, while we should pay attention to focus of user demand. For example, the query:" *上海哪里能买到火车票?*" figure 1 is parsing results, and figure 2 is its corresponding conceptual structure.

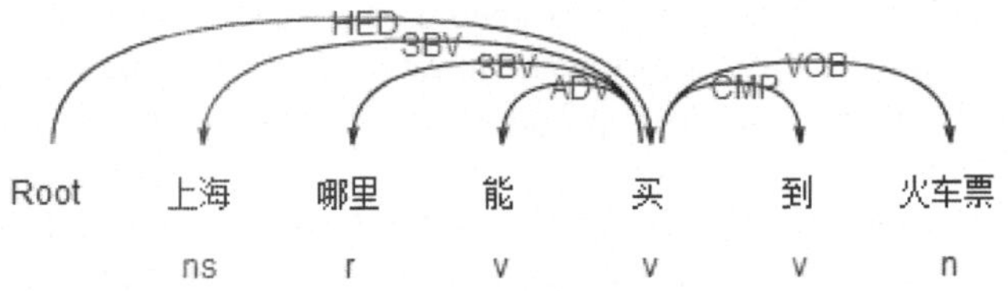


**Fig. 1.** Parsing Results

It can be seen from figure 2, the result of dependency parsing is verb-centered, but the characteristic of Chinese is noun-centered, and its expression form is direct coupling of the concept. If search as the figure 2, the returned results may include information about news stories of buying tickets, these are different from the information user wanted. In fact, what this query want is the information about an entity E of a specific address. According to the result of dependency parsing, building the concept map with <E, A, V> triples for element and noun for center node[4]. Therefore, the correct conceptual analysis map is shown as figure 3, the focus of its inquiry is entity E.

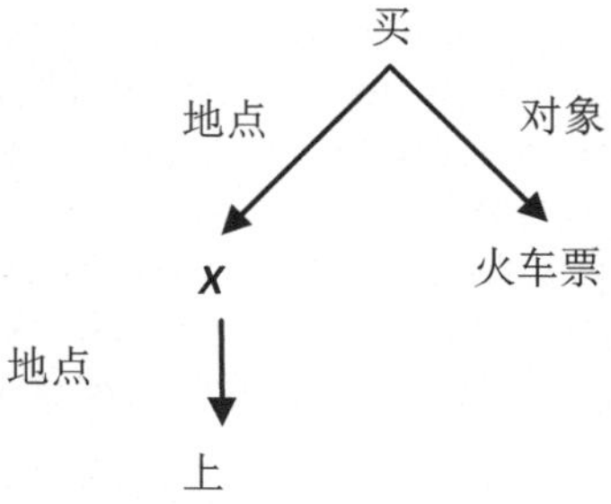


**Fig. 2.** Conceptual Structure

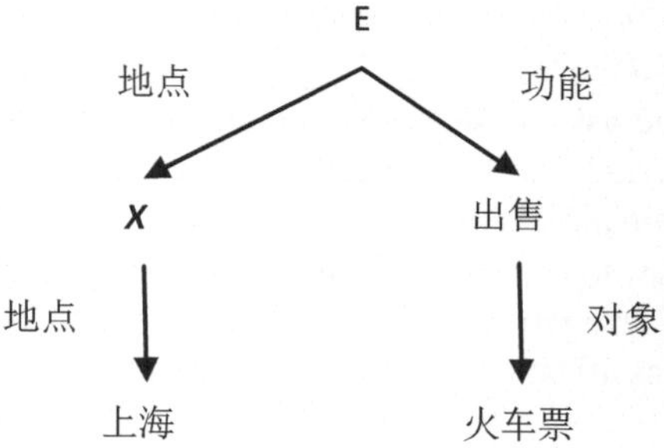


**Fig. 3.** Conceptual Map

The key to analysis of interrogative sentence is to refer the focus of user demands as center, distinguish what the user asked, whether it is entity E, attribute name A, or attribute value V. It can be got from analysis of interrogative sentence, for the interrogative word largely express the focus of user query. If the interrogative word is just replaced by a focus word that user want to inquiry, the accuracy of the retrieval system would be improved. For example, the query sentence:"*上海有多大？*" Its focus information is "*上海占地面积是多少*", an attribute value V. However, original interrogative does not contain the words "*占地*" , "*面积*", make returned results contain a lot of noise, and reduce the accuracy. Table 1 shows the accuracy of search engine with sentences before and after the replacement of focus information, while the accuracy here only calculate the correctness of first ten search engine returned.

**Table 1.** The accuracy before and after replacement of focus information

| | **before replacement** | **after replacement** |
|---|---|---|
| **query** | *上海有多大* | *上海 面积* |
| **accuracy** | 0.3 | 0.6 |

**Table 2.** The information of interrogative classification

| **CLASS** | **SUBCLASS** |
|---|---|
| 人物 | 人名等。 |
| 地点 | 城市，国家，省，河流，具体地址等。 |
| 数字 | 时间点，时间范围，数量，年龄，百分比，速度，重量等。 |
| 实体 | 动物，植物，颜色，娱乐等。 |
| 描述 | 原因，意义，方法等。 |

Interrogative words, such as "*谁*", "*哪里*", can be directly used as focus information. That is, "*谁*" represents information of demand about a person, "*哪里*" represents the information about place. It is relatively easy to get the interrogative word in the sentences, but it is not practical to determine the focus information only by these interrogative words. For instance, the word "*多少*", expresses different focus information with the different followed nouns. Such as the following two sentences: "你到这里要多少时间" and "这个东西要多少钱". They both have the word "多少", and ask the same type focus: attribute value V, but the former query want to get time period, the latter want to get price, they are totally different. Obtaining correct user demands, requires to use syntactic structure analysis. Before studying the method of how to determine the focus information by interrogative words, we should first research interrogative types[3]. Referring to the question answering system, we get the classification information shown in table 2.

## 2.2 Extract Focus Information

Syntactic analysis is to analyze the hierarchical structure of natural language in a given grammar, it is one of the important issues in natural language processing. The main elements of the syntactic structure of dependency grammar is dependency relationship, that is, the binary relation of two words in the sentence, one is called head, and the other is called dependent word. Dependency relationship reflects the dependency on the semantics of the head and dependent word. Figure 1 shows the result of dependency parsing of the query "上海哪里能买到火车票". An arc connected between the two words, indicating the existence of dependency relationship. The word (dependent word) which the arc point to is dependent on the word (head) that the arc initiated. As shown in figure 1, "火车票" is dependent on "买", and "上海" is dependent on "买". The marks on the arc HED, SBV, ADV, CMP, VOB tell the types of relations. HED said the core of the sentence, SBV said the subject-verb relation, ADV said the adverbial-verb structure, CMP said the verb-complement structure, VOB said the verb-object relations.

For the interrogative sentence type classification, we use the approach as [3] described, that is, extracting the classification features from syntactic structure, analyzing the dependency relationship and its type between words pair in the sentence, removing the words pair that has no effect on classifying query, selecting the keywords and relationships as feature to classify. The steps of feature extraction as follows:

1) Find the interrogative words in the query.

Collecting interrogative words to make a table T, such as "谁", "什么" and so on. Segmenting the query, and then matching the table T to find interrogative words.

2) Syntactic analyze the query to get dependency arcs and types of relationship.

Parsing the query, and getting a number of dependency relationships, and to prepare for the next step of feature extraction.

3) Extract the trunk, the interrogative words and its subsidiaries compositions as the classification features.

The subject, predicate and object of a sentence can be seen as the backbone of the sentence. For the query classification, we can assume that the backbone and interrogative words and its subsidiaries compositions (the words dependent on and dependent by interrogative words) of the query will be able to determine the type of the query. According to the dependency arcs and types of relationship such as HED, SBV, VOB, we can extract subject H, predicate S and object O of the query. Assuming the interrogative words extracted from the first step as word C, we can get words D connected to word C according to the dependency relationships extracted in the second step, these words H, S, O, C, D constitute the features of query classification.

Use Bayesian classifier to classify, assuming that the word is independent of each other. Assuming that the user input is Q, and its features extracted are $Q_1$, $Q_2$, …$Q_n$, $C_i$ is the type of the query. The objective of Bayesian classifier is to find a $C_i$, to satisfy $P(C_i|Q) \geq P(C_j|Q)$. The equivalent mathematical formula can be expressed as follows:

$$\arg_{c_i} \max P(C_i|Q) = \arg_{c_i} \max P(Q_1C_i) \times P(Q_2C_i) \times \ldots \times P(Q_nC_i)$$

For the query "*上海哪里能买到火车票*", "*买*" is the head word, "*上海*" and "*哪里*" are both dependent on "*买*", belongs to SBV type relationship, "*火车票*" are dependent on "*买*", belongs to VOB type relationship. Therefore, extracted sentence trunks are "*上海 哪里 买 火车票*". The interrogative word of this query is "*哪里*", its subsidiary composition is "*买*", belongs to sentence trunk. Thus, the features extracted from the whole query are "*上海 哪里 买 火车票*". Through the three steps above, extracting features that express interrogative focus. And with the query classification, this query assigned to a class of address.

## 2.3 Replace the Interrogative Words with Focus Information

After determining the type of query, in order to improve the retrieval accuracy, the interrogative word in the query need to be replaced by the focus word, which represent the sentence type. For example, the focus word of "*上海哪里能买到火车票*" is "*地址*". Because the majority of the pages contain the answer are generally as following forms:"*上海火车票代售点地址……*" or "*上海火车票售票点地址大全……*". They all contain the keywords "*地址*", or "*地点*" or others. Therefore, it has to collect focus words that answers may contain, and corresponding to each type of interrogative sentences. The table is shown as Table 3.

**Table 3.** Focus words corresponded to each type

| TYPE | FOCUS WORDS |
|---|---|
| 地址 | 地址，坐落在，地点 |
| 时间 | 年，月，日 |
| 时间范围 | 天，小时，分钟 |
| 面积 | 占地，面积 |
| 速度 | 速度 |
| 原因 | 因为，所以 |
| …… | …… |

After obtaining the focus words, replacing the interrogative word in the query with them. It should be noted that the focus words may already exist in the query, and it need to eliminate duplication.

## 2.4 Conceptual Analysis, Building Concept Map of Demands

After parsing, with some pairs of dependency relationship (word pairs connected by an dependency arc) of the query, we do semantic analysis, and building user demands concept map. Such as the query "*上海哪儿能买到火车票*", the pairs of dependency relationship got by parsing are *(上海, 买), (哪儿, 买), (能, 买), (到, 买), (火车票, 买)*. The syntactic relation type of (能, 买) is ADV (adverbial-verb structure), (到, 买) is CMP (verb-complement structure). For the Chinese conceptual connotation of the query, these two relationship types are kind of noise. Removing them, we get pairs of dependency relationship as follows: *(上海, 买), (哪儿, 买), (火车票, 买)*. The concept map analyzed now is shown as figure 2, and then replace the interrogative word with focus word, finally get pairs of dependency relationship *(上海, 买), (地址, 买), (火车票, 买)*, the concept map is shown as figure 4.

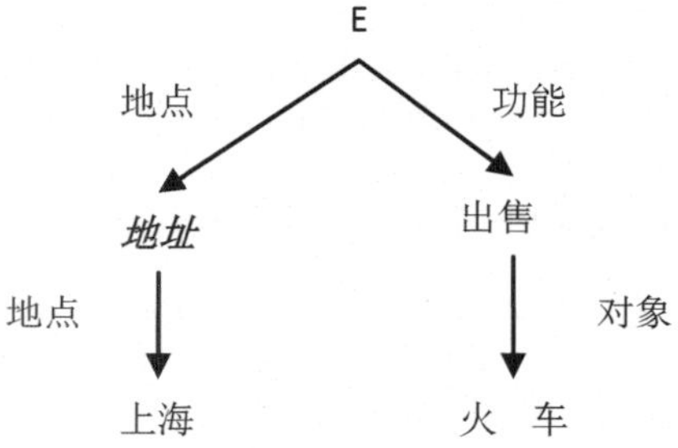


**Fig. 4.** The Finally Concept

We can discover from comparing the actual data, that although the query is generally short, and its pairs of dependency relationship are relatively less, not all relationship can be used to construct concept map, it would bring noise into the concept map with secondary relationships. The dependency relationships that should be kept are ATT(attribute-center structure), VOB(verb-object relation), SBV(subject-verb relation), VV(serial verb structure), CMP(verb-complement structure), POB(prepositional relation), DE(word "的" structure), DI(word "地"structure), DEI(word "得" structure), BA(word "把" structure), BEI(word "被" structure).

The dependency relationship of "*红色的花*" is DE relationship, its pairs of dependency relationship are *(红色, 的), (的, 花)*. If put these two pairs into concept map, it may bring some noise, but they can't be discarded, we should merge this kind of relationship. That is, combining two words each from each words pair into a relationship pair, such as two consecutive pairs connected by DE and ATT relationship is shown as figure 5. It can apply to similar situations such as DEI and DI, which can also merge two consecutive pairs of relationship into one, and the situations are CMP and DEI, DI and ADV. The examples are shown as figure 5. DE relationship merge into (*红色, 花*), DI relationship merge into (*费力, 抬*) and (*抬, 水*), DEI relationship merge into (*跑, 快*).

For instance, the query "*北京奥运会的吉祥物有什么意义*", its syntactic analysis is shown as figure 6. The pairs of dependency relationship are *(北京, 奥运会), (奥运*

*会, 的), (的, 吉祥物), (吉祥物, 有), (意义, 有), (什么, 意义)*. After query classification, the focus word got is "*意义*", replacing the interrogative word, and the pairs of relationship now are *(北京, 奥运会), (奥运会, 的), (的, 吉祥物), (吉祥物, 有), (意义, 有)*. According to the merge rule, after merging pairs *(奥运会, 的), (的, 吉祥物)*, the pairs of relationship are *(北京, 奥运会), (奥运会, 吉祥物), (吉祥物, 有), (意义, 有)*. To construct concept map with these pairs of dependency relationship, the map is shown as figure 7.

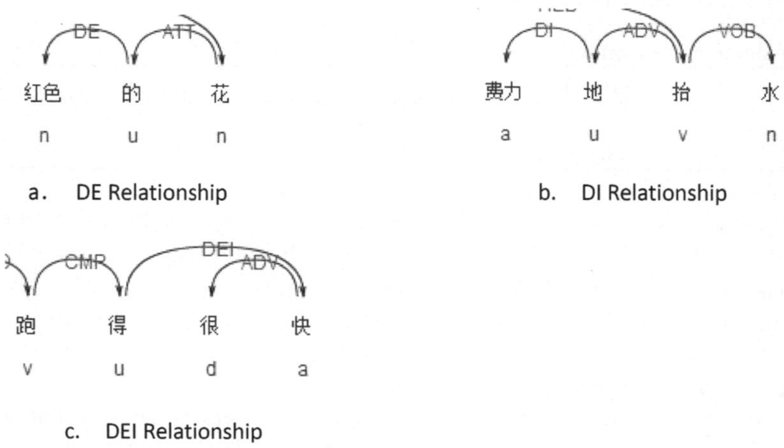


**Fig. 5.** DE、DI、DEI Relationship

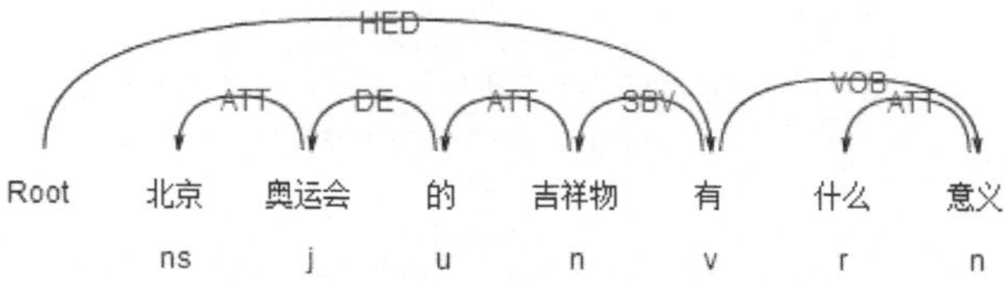


**Fig. 6.** Dependency syntactic analysis

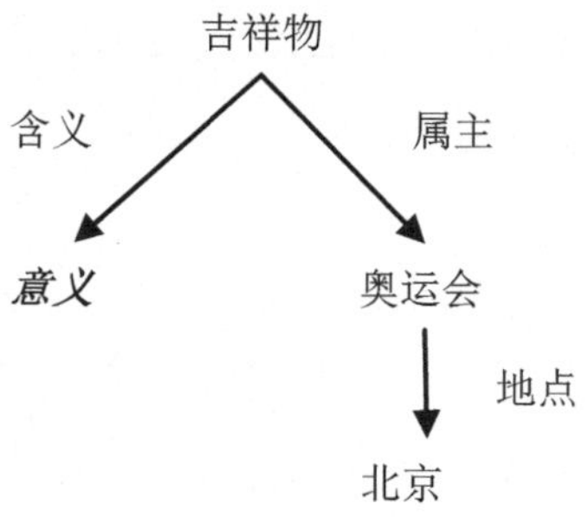


**Fig. 7.** Concept map

# 3 Experimental Results

## 3.1 Data Sources

1) The interrogative classification training data is question sets of information retrieval laboratory of Harbin Institute of Technology, a total of 240 question instances.
2) The concept map building training data is from our laboratory.

## 3.2 Experiment Results and Analysis

**1) The accuracy of query classification**

| classify | accuracy |
|---|---|
| **5 classes** | 77.46% |
| **60 subclasses** | 64.08% |

The semantic category of some queries is rather vague, it is difficult to determine which type it is. Such as "*狗喜欢吃什么*", it is difficult to classify to "*动物*" or "*植物*". This kind of query would bring a loss of accuracy to classification. Also, the inaccuracy of syntactic analysis would lead to the inaccuracy of classification.

**2) The accuracy of concept map**

A map consist of a set of triples $t = < c_1, c_2, r >$, $c_1$ and $c_2$ are concept, $r$ is the relationship. Evaluation indicators are accuracy (P), recall rate (R), F test ($F_{\beta=1}$, F), and accuracy of concept map (C).

$$\mathrm{P} = \frac{\text{the correct number of tuples in result set}}{\text{all tuples in result set}}$$

$$\mathrm{R} = \frac{\text{the correct number of tuples in result set}}{\text{all tuples in training set}}$$

$$\mathrm{F} = \frac{2\mathrm{PR}}{\mathrm{P} + \mathrm{R}}$$

$$\mathrm{C} = \frac{\text{the correct number of concept map in result set}}{\text{all concept map in result set}}$$

| P | R | F | C |
|---|---|---|---|
| **0.7048** | 0.7253 | 0.7197 | 0.4948 |

# 4 Conclusions

The experimental study found that if the query asking specific entity, there is no having to replace the interrogative word with focus word. Because most queries, which ask animals, plants, colors, entertainment tools and others similar, would use the word "*什么*" as interrogative word. And there must exist words represent the object that may near the word "什么". In this case, it's not necessary to determine the object of the query, we can only filter the interrogative word and construct concept map of user demands.

This paper uses the approach of question answering system to classify queries, to get the type of user demands, and the focus information, which is a certain guarantee to the accuracy of the retrieval system. From a semantic point, dependency parsing based concept map construction, represents the user demand information, it is more effective to process Chinese, and improve the accuracy of the retrieval system. But the time spent by the semantic search is larger than general search, applying this method to the actual product may affect the user experience, it should be further optimized.

## References

1. Taylor, R.S.: The process of asking questions. American Documentation, 13(4), 391–396
2. Hui, L.: Conceptual Analysis of User Queries in Information Retrieval (信息检索中用户需求的概念分析研究) (2008)
3. Wen, M., Zhang, Y., Liu, T., Ma, J.: Syntactic structure analysis based Chinese question classification (基于句法结构分析的中文问题分类). Journal of Chinese Information Processing (2005)
4. Lu, R.: Chinese Retrieval and Chinese Semantic Concept Map Construction (中文检索与汉语语义概念图表示). In: The 11th Chinese National Conference on Computational Linguistics (2009)
5. Sowa, J.F.: Conceptual structures: information processing in mine and machine. Addison-Wesley Longman Publishing Co., Inc., Bostom (1984)
6. Roberts, D.D.: The existential graphs of Charles S. Peirce. Approaches to semiotics, p. 168. The Hague, Mouton (1973)
7. Sowa, J.F.: Knowledge representation: logical, philosophical, and computational foundations, vol. XIV, p. 594. Brooks/Cole, Pacific Grove (2000)

# Measuring Information Transmission in Izhikevich Neuron

Zhijun Yang, Li Guo, and Qingbao Zhu

School of Computer Science, Nanjing Normal University,
Nanjing 210046, China
zjyang@njnu.edu.cn

**Abstract.** Izhikevich neuron is a relatively new neuronal model, which has found extensive applications in modeling neuron due to its strong biological plausibility and computational effectiveness. In this work we use the information theoretic method to measure the ability of information transmission of this neuron model. We find that Izhikevich neuron shows low sensitivity to high frequency of random noise; and appropriate noise level can help information transmission through the neuron.

**Keywords:** Neuronal model, Izhikevich neuron, information transmission, mutual information.

## 1 Introduction

Izhikevich's spiking neuron model [1], since being presented some 10 years ago, has been used widely in computational neuroscience to simulate the functionalities in visual [2], auditory [3], tactile [4] signal processings, and to mimic the dynamical behaviours of the large-scale neocortical neuronal network [5]. It has also been implemented with the mixed-signal VLSI process [6,7] and digital techniques [8] for potential real-time applications. Unlike the classical Hodgkin-Huxley spiking neuron model which has strong biological plausibility but weak computational efficiency in, e.g., organising a spiking network, and the leaky integrate-and-firing (LIF) neuron which in contrast has strong computational efficiency but usually weak biological plausibility, Izhikevich neuron can display about 20 different types of neuronal activities by simply modifying a group of 4 parameters of a nonlinear ordinary differential equation (ODE) model [1,9]. It has thus achieved a balance between the computational efficiency and biological plausibility, and has provided an appropriate tool for modeling the individual spiking neurons and neural populations, for a review see [10].

Neurons communicate with action potentials, or spikes. Despite the fact that it is agreed that the spike trains emitted by a neuron usually contain the information of the input signals, the mechanism underlying this encoding is unclear. Generally we have two explanations to the information encoding (and hence decoding accordingly) mechanisms. One is to encode the stimulus with the mean firing rate of the spike trains, and the other is to use the precise times at which spikes occur. The later method has attracted much attention recently with the theoretic and experimental

J. Lei et al. (Eds.): AICI 2012, LNAI 7530, pp. 580–587, 2012.

development of some new spike timing dependent neuronal learning algorithms. In information theoretic analysis, the information transmitted between the stimuli to and outputs of a neuron can be expressed in terms of the interspike interval (ISI) distributions of the relative spike trains. A further mathematical operation based on these distributions result in the measurements of the reliability and variability (corresponding to the entropy and conditional entropy, respectively) of the neuron responses with respect to different stimuli. The shared information between the input and output of a spiking neuron can thus be computed as the mutual information (MI) representing the information transmission capability of that neuron. This method, namely the direct method of maximum likelihood estimation [11], has been used to measure the information transmitted through an LIF neuron [12], and an MNTB neuron with the calyx of Held synapse [13].

Although the dynamics of a single Izhikevich neuron has been well studied, its capability of information transmission is rarely investigated. In this work, we use the direct method to study the efficiency of information transmission through a type of Izhikevich neuron, namely the regular spiking neuron which shows a commonly observed excitatory activity, and the effect of noise on this neuron.

# 2 The Deterministic and Stochastic Regular Spiking Neuron

## 2.1 The Deterministic Model

The Izhikevich spiking neuron [1] is represented by the following differential equations,

$$V' = 0.04V^2 + 5V + 140 - u + I \tag{1}$$

$$u' = a(bV - u) \tag{2}$$

after the spike reaches its apex (here $+30mV$), the membrane voltage and the recovery variable are reset according to the following equation,

$$\text{if } V \geq 30mV \text{, then } V \leftarrow c \text{, } u \leftarrow u + d \text{.} \tag{3}$$

where $V$ is the membrane potential of the neuron, $u$ is a recovery variable and $I$ the synaptic current. By choosing different configurations of the model parameters $\{a, b, c, d\}$ different kinds of neuronal dynamics can be obtained. In our study the neuron is an excitatory one with $\{a, b, c, d\} = \{0.02, 0.2, -65, 8\}$, which delivers regular spikes upon a constant input current. The original Izhikevich neuron is a deterministic model in the sense that it issues reliable, same output spike trains when stimulated by a repeated input spike train.

## 2.2 The Stochastic Model

Since the noise is ubiquitous in the world, we suppose that the input spike train is contaminated by the additive white Gaussian noise (AWGN) such that the amplitude of each spike in the train is different, and therefore each spike can have different impact on the membrane potential fluctuation. Meanwhile, the membrane potential and the recovery variable are both contaminated by the AWGN as well. The degree of contamination is adjustable in terms of the signal noise ratio (SNR) in dB. To investigate the effect of noise in the information transmission of the model, we choose $SNR = 5,10,100dB$, respectively. We thus re-write the original Izhikevich model in the following form,

$$V_{SNR}' = 0.04V_{SNR}^2 + 5V_{SNR} + 140 - u_{SNR} + I_{SNR} \tag{4}$$

$$u_{SNR}' = a(bV_{SNR} - u_{SNR}) \tag{5}$$

$$\textbf{if } V_{SNR} \geq 30mV \textbf{, then } V_{SNR} \leftarrow c, u_{SNR} \leftarrow u_{SNR} + d \tag{6}$$

where,

$$I_{SNR} = f(I) = AWGN(I, SNR), \tag{7}$$

$$V_{SNR} = f(V) = AWGN(V, SNR), \tag{8}$$

$$u_{SNR} = f(u) = AWGN(u, SNR). \tag{9}$$

The input and output waveforms of the model when SNR=5dB for the input spike amplitude and two variables of Izhikevich neuron, and the model performance in absence of the noise, are shown in Fig.1.

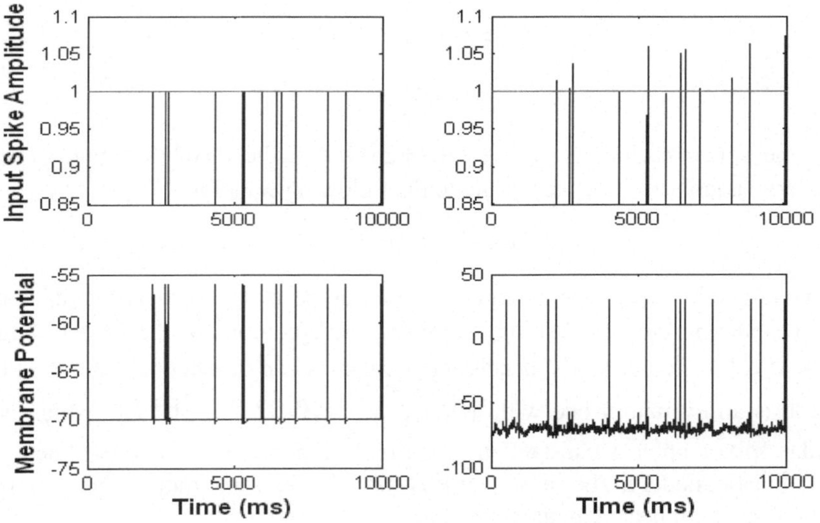


**Fig. 1.** The episodes of the input and output spike trains of the deterministic and stochastic Izhikevich's regular spiking neuron. Top and bottom left: the original regular spiking neuron without noise, the same pattern of spike train is issued on repeated input spike train. Top and bottom right: the revised regular spiking neuron with SNR=5, the pattern of spike train changes on repeated input spike train. The horizontal line in the top figures represents the normalised input spike amplitude.

## 3 Information Theory

We use a direct method [11] to measure the extent of information content contained in the postsynaptic ISI with respect to an independent homogenous Poisson spike train $X$. This involves computing the MI between pre- and post-synaptic ISI to quantify the common information content in both. Suppose the sequence of ISI of an input spike train is $X = \{x_1, x_2, \cdots, x_n\}$, where n is the number of the presynaptic ISIs. $X$ represents a Poisson process conveying temporal information. The greatest value of the ISI, produced by a neuron, when a same spike train of a mean firing rate is repeatedly used to stimulate the neuron, is obtained and taken as a reference. The first percentile of this ISI value is defined as a bin resolution, whose precision can keep the information finite [11,12]. The ISI values produced by the model are then discretised and distributed into the correct bins of the percentile of the greatest ISI value, hence we have the sequence of ISI of the output spike train as $Y = \{y_1, y_2, \cdots, y_m\}$, where $m$ is the number of the postsynaptic ISIs. The probability distribution of the response $P(Y)$ over a long time course can thus be estimated with the maximum likelihood direct estimation method. The total entropy, $H(Y)$ in Shannon's theory [14], is a quantity measuring the amount of variability of the postsynaptic response $Y$ to the ensemble of different inputs, without being constrained by input conditions.

$$H(Y) = -\sum_{i=1}^{100} p(y_i) \log_2 p(y_i) \tag{10}$$

where $p(y_i)$ is the probability of the ISI value $y_i$ which fall in the $i^{th}$ percentile with a value between $y_i$ to $y_{i+1}$. The conditional entropy, $H(Y \mid X)$, is a quantity that measures the reliability of the postsynaptic response $Y$ to the repeated presentations of the same inputs,

$$H(Y \mid X) = -\sum_{i=1}^{100} p(y_i \mid X) \log_2 p(y_i \mid X) \tag{11}$$

where $p(y_i \mid X)$ is the conditional probability of the model responses which fall in the $i^{th}$ percentile, conditioned on the appearance of presynaptic stimulation sequence $X$. In numerical experiments, the stochastic model was tested with different groups of Poisson spike trains with mean frequencies $f \in [1Hz, 100Hz]$. We repeated a particular Poisson spike train of mean rate $f$ as the input to the model for 200 times. Due to the stochastic nature, the model responded differently in its ISI values at each time. We thus account for the reliability of a synapse conditioned on a specific input with an alternative calculation of the conditional entropy,

$$H(Y \mid X) = \frac{\sum_{j=1}^{n}(-\sum_{i=1}^{100} \hat{p}(y_i)\log_2 \hat{p}(y_i))}{n} \tag{12}$$

where $n$ is the total number of input spikes in a spike train inducing the postsynaptic ISIs, $\hat{p}(y_i)$ is the probability of all ISI values in \$200\$ trials which are induced by the same input spike and fall in the $i^{th}$ percentile.

The MI $I(X;Y)$ quantifying the common information between the presynaptic ISI sequence, $X$, and the postsynaptic ISI sequence, $Y$, can be expressed as,

$$I(X;Y) = H(Y) - H(Y \mid X) \tag{13}$$

It is notable that, in order to use the entropy of the discrete ISI distribution upon all possible input stimuli, $p(y_i)$, to represent the variability of the neuron response, ISIs must be independent (Zador, 1998; Stevens, 1996). As Izhikevich neuron will reset to a fixed membrane potential after firing a spike, its ISIs between the neighbouring spike pairs are independent, like the leaky integrate-and-fire neuron.

## 4 Results

In Monte Carlo simulations, Poisson spike trains of 15 different mean firing rates ranging from 1Hz to 100Hz are used to stimulate the neuron. Each input spike train is used repeatitively for 200 times, and the corresponding output spike train is recorded for 400 spikes (and hence the same amount of ISIs). The entropy and conditional entropy of the neuron response are then computed according to the numerical information theoretic method, and shown in Fig.2.

The plot shows that, in the low frequency range, the neuron with little noise has a large variability (expressed in entropy) while its variability to Poisson spike train decreases when mean input frequency is above 50Hz. This confirms that, like Hodgkin-Huxley neuron, Izhikevich neuron (here the regular spiking type) shows low sensitivity to high frequency of random noise as it uses a slow adapting variable and negative feedback to adapt the firing rate of the neuron. The conditional entropy for the (nearly) deterministic neuron shows that, in the full range of stimulus frequency, the neuron is very reliable to reflect the input spikes as its conditional entropy is zero, which is usually not the case in biologically *in vivo* or *in vitro* neurons.

To make the neuron more biological plausible in responding to its external stimuli, different levels of additive Gaussian white noise is incorporated into the neuron's spike generation mechanism. The simulation results show that the additive noise can improve the sensitivity of the neuron to the input signals as it results in larger variability. This effect is more obvious in the higher frequency range. It is also natural to see that the higher noise level can lead to the larger variability (larger entropy value), and less reliability (larger conditional entropy value).

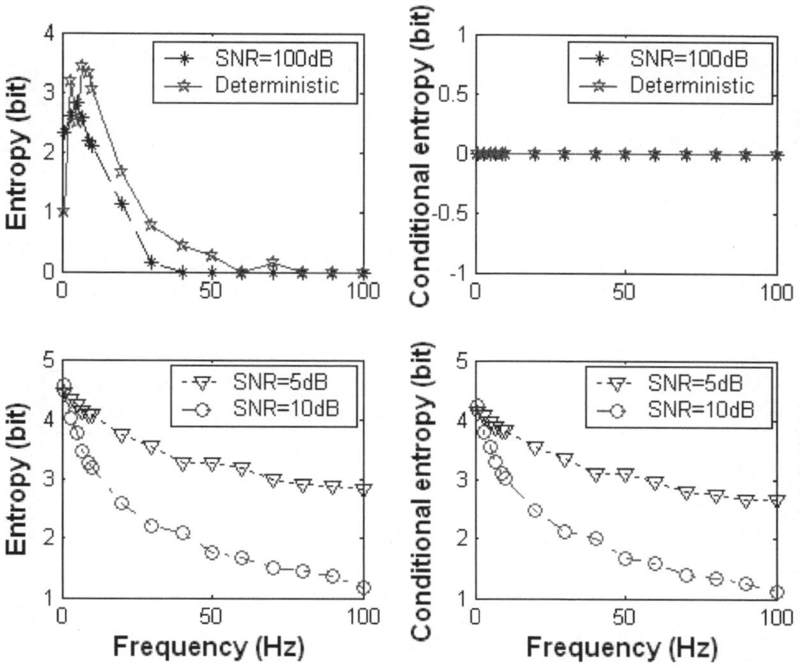


**Fig. 2.** The entropy and conditional entropy of the postsynaptic response, in the form of ISI distribution, to different and repetitive Poission spike trains. Top and bottom left are the entropy for 4 groups of experimental protocols with different noise levels. Top and bottom right are the conditional entropy for the same 4 protocols.

Does added noise have any impact on the neuron's capability for information transmission? Some interesting phenomena can be viewed in Fig. 3. For the (nearly) deterministic model, the MI, which is the measure of common information in both input and output sides, are simply the entropy as the conditional entropy is zero. It is interesting to see that the MI curves display a strong frequency band feature, which fits into the rhythmic patterns in the electroencephalogram. In both delta and theta rhythms (from 1Hz to 8Hz), the deterministic and slightly noisy MI values alternatively become larger. In both alpha and beta rhythms (from 8Hz to 40Hz), the deterministic MI value is consistently larger than the slightly noisy MI value, signifying that the deterministic neuron can transmit more information. In the gamma rhythm (above 40Hz) both MI values approach zero, which means the neuron is not sensitive to the input information.

If more noise is used, we can see from Fig.2 that both entroy and conditional entropy will increase. Qualitatively we cannot determine if MI value will increase or decrease from MI equation in the previous section. However, our quantitative analysis tells that for almost all frequencies the MI value of the more noisy neuron is consistently greater than that of the less noisy neuron. With all other conditions the same, obviously the noise plays a role in helping the information transmission through the Izhikevich neuron.

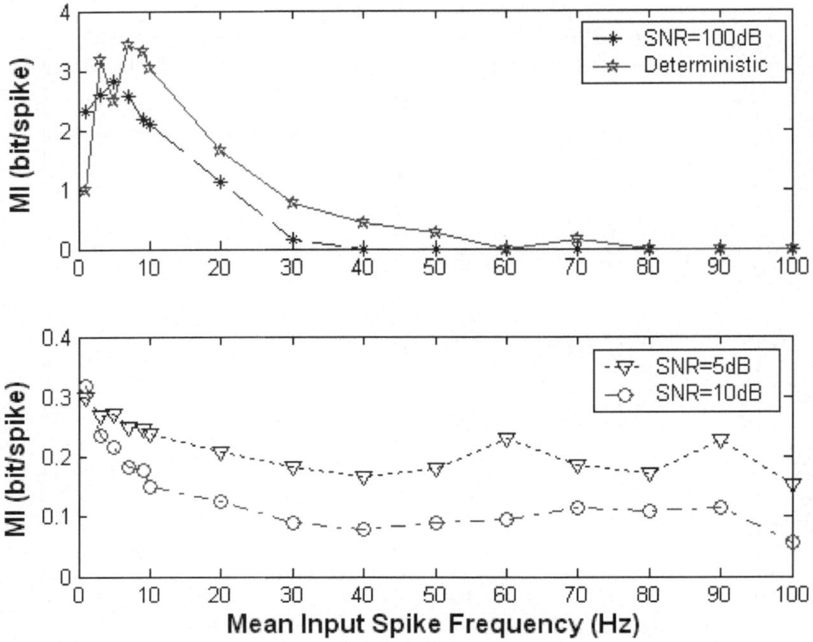


**Fig. 3.** The mutual information of the deterministic and stochastic Izhikevich neurons computed on 4 experimental protocols

# 5 Concluding Remarks

In this work we use the information theoretic analysis method to estimate the entropy for all stimulation patterns, and the conditional entropy conditioned on a specific input. MI is then obtained to measure the ability of information transmission for the regular spiking Izhikevich neuron in a frequency range from 1Hz upto 100Hz. We find that the deterministic neuron is sensitive to the low frequency spiking stimulation (in the delta and theta domain) but much less sensitive to the high frequency stimulation (in the gamma domain).

Through the numerical experiments, we have demonstrated that the additive Gaussian noise may help improve the information transmission capability of a stochastic Izhikevich neuron, as in the gamma high frequency range, the deterministic and slightly noisy neurons have little information transmission capability, while those with higher noise level perform better. This numerical evidence further confirms that noise, unlike what people might believe, may play some role in the spiking communications among neurons.

**Acknowledgments.** This work is supported by the National Natural Science Foundation of China under grant number 61073118.

## References


1. Izhikevich, E.M.: Simple model of spiking neurons. IEEE Transactions on Neural Networks 14(6), 1569–1572 (2003)
2. Super, H., Romeo, A.: Rebound spiking as a neural mechanism for surface filling-in. Journal of Cognitive Neuroscience 23(2), 491–501 (2011)
3. Macherey, O., Carlyon, R.P., Wieringen, A.V., Wouters, J.: A dual-process integrator-resonator model of the Electrically stimulated human auditory nerve. Journal of the Association for Research in Otolaryngology (JARO) 8(1), 84–104 (2007)
4. Arena, P., Patane, L.: A spiking network for object detection in roving robots via a bionic Antenna. To appear in Proceedings of the International Joint Conference of Neural Networks (IJCNN 2012), Brisbane, Australia (2012)
5. Izhikevich, E.M., Edelman, G.M.: A large-scale model of mammalian thalamocortical Systems. PNAS 105, 3593–3598 (2008)
6. Demirkol, A.S., Ozoguz, S.: A low power VLSI implementation of the Izhikevich neuron Model. In: Proceedings of the 9th IEEE International Conference on New Circuits and Systems (NEWCAS 2011), Bordeaux, France, pp. 169–172 (2011)
7. Mizoguchi, N., Nagamatsu, Y., Aihara, K., Kohno, T.: A two-variable silicon neuron circuit based on the Izhikevich model. Artificial Life and Robotics 16(3), 383–388 (2011)
8. Fidjeland, A.K., Shanahan, M.P.: Accelerated simulation of spiking neural networks using GPUs. In: Proceedings of the International Joint Conference of Neural Networks (IJCNN 2010), Barcelona, Spain, pp. 1–8 (2010)
9. Izhikevich, E.M.: Which model to use for cortical spiking neurons? IEEE Transactions on Neural Networks 15(5), 1063–1070 (2004)
10. Herz, A.V.M., Gollisch, T., Machens, C.K., Jaeger, D.: Modeling single-neuron dynamics and computations: A balance of detail and abstraction. Science 314, 80–85 (2006)
11. Zador, A.: Impact of synaptic unreliability on the information transmitted by spiking neurons. Journal of Neurophysiology 79, 1219–1229 (1998)
12. Stevens, C.F., Zador, A.: Information through a spiking neuron. In: Advances in Neural Information Processing Systems, vol. 8, pp. 75–81. MIT Press (1996)
13. Yang, Z., Hennig, M., Postlewaite, M., Forthyse, I., Graham, B.P.: Wide-band information transmission at calyx of Held. Neural Computation 21, 991–1017 (2009)
14. Shannon, C.E., Weaver, W.: The mathematical theory of communication. Univ. of Illinois Press (1949)

# Color Image Segmentation Using Centroid Neural Network

Do-Thanh Sang, Dong-Min Woo, and Dong-Chul Park

Dept. of Electronics Engineering, Myongji University, Yongin, Korea 449-728
sang.dothanh@gmail.com, {dmwoo,parkd}@mju.ac.kr

**Abstract.** Color image segmentation has been attracting more and more attention, mainly because color images can provide more information than gray level images. Many methods have been proposed so far to deal with the problem. However, most methods require fine tuning of parameters, which can be attained after repetitive trial and error. This paper discusses unsupervised learning in terms of Centroid Neural Network (CNN). In fact, CNN is the crucial algorithm to diminish the empirical process of parameter adjustment required for color image segmentation. The simulation results indicate that the proposed technique yields the reasonably segmented images in comparison with other conventional algorithms.

**Keywords:** image segmentation, color image, neural network, unsupervised learning.

## 1 Introduction

The problem of image segmentation plays a key role in many fields of analysis, recognition, classification and description. The goal of image segmentation is the division of an image into a set of disjointed areas with uniform and homogeneous attributes such as intensity, color, tone or texture. Good color segmentation enables the object recognition and localization system to execute more accurately. A large number of different segmentation techniques have been developed so far and have been detailed in the literature. Most published results of color image segmentation are based on gray level image segmentation approaches with different color representations.

The commonly used method in segmenting monochrome images is thresholding, which assigns a pixel to one class if its gray level is less than a specified threshold, and otherwise assigns it to the other classes [1]. This is called bi-level thresholding. Generally, one can select more than one threshold, and use these thresholds to separate the whole range of gray values into several sub ranges. This process is called multilevel thresholding. This method does not require prior information of the image and also has a low computation complexity. Nevertheless, it does not work well for an image without any obvious peaks or with broad and flat valleys. Moreover, it does not use any spatial information at all.

J. Lei et al. (Eds.): AICI 2012, LNAI 7530, pp. 588–595, 2012.

Compared to monochrome images, color images are useful or even necessary in computer vision, because they provide additional information such as intensity. Color image processing thus is becoming more practical nowadays. Among many existing methods of color image segmentation, four main categories can be distinguished: pixel-based techniques, region-based techniques, contour-based techniques, and hybrid techniques. Unsupervised learning is widely applied in some applications, where the image features are unknown, such as nature scene understanding, satellite image analysis, etc. Many algorithms impose spatial constraints on clustering algorithms for segmenting image data. One of the most widely used algorithms employing fuzzy clustering techniques is the Fuzzy c-Means (FCM) [2], which has been proposed as an improvement on earlier clustering algorithms such as the Self-Organizing Map (SOM) [3] and the k-Means. Nevertheless, the FCM has the problem of exhaustive computational burden in classifying each pixel based on color feature space, especially for large images. Recently, Guo and Ming [4] have developed a hybrid technique that incorporates SOM and Simulated Annealing (SA) into color image segmentation. The SA is used to find optimal clusters from SOM prototypes; however, its drawbacks include the need for a lot of trial and error to obtain the optimal parameters, and it is very hard to implement such extensive trial and error.

In order to reduce the influence of the variables' initialization, we offer the idea of using Centroid Neural Networks (CNN) [5] to construct a "natural grouping" for an image without using any prior knowledge. The CNN algorithm does not require a predetermined schedule for a learning coefficient or a total number of iterations for clustering. The effect of color image segmentation is dependent not only on the algorithm but also on the color coordinate, and hence the paper surveys RGB and L*u*v* coordination to obtain the best color segmentation. We compare the segmentation results for natural scene images extracted from the Berkeley Database to show the effectiveness of the proposed method.

The paper is organized as follows. Section 2 elucidates the choosing of color information. Section 3 describes the applicable features of CNN in comparison with FCM and SOM-SA. Experimental results are shown in section 4. Finally, our conclusion is given in section 5.

## 2 Choosing Color Information

Color is discerned by humans as a combination of tristimuli R (red), G (green), and B (blue), which are usually called the three primary colors. We can derive other kinds of color representations (spaces) by using either linear or nonlinear transformations from RGB representation. Several color spaces, such as RGB, CIE XYZ, HSI, L*u*v* are utilized in color image segmentation, but none of these can transcend the others for all kinds of color images. Selecting the best color space for all cases is still one of the difficulties in color image segmentation [6]. The RGB color space can be geometrically illustrated in a 3-dimensional cube (Fig. 1 (a)). The coordinates of each point inside the cube represent the values of red, green and blue constituents, respectively.

RGB is suitable for color display, but not good for color segmentation and analysis because of the high correlation among the R, G, and B components. High correlation means that if the intensity changes, these three components will change accordingly. Moreover, the measurement of a color in RGB space does not represent color differences in a uniform scale. Hence, it is impossible to evaluate the similarity of two colors from their distance in RGB space.

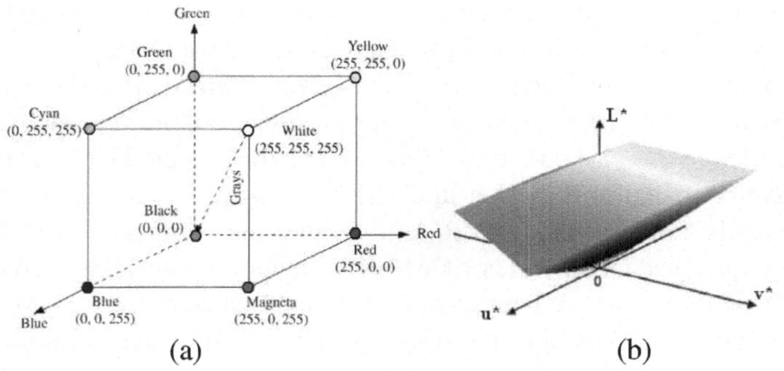


**Fig. 1.** Color space represented in a 3-dimensional cube. (a) RGB color space (b) L*u*v* color space.

Fig. 1 (b) depicts color distributions in L*u*v*. It is concluded that the Modified CIE L*u*v* performs better than other color spaces. The CIE L*u*v* color space is defined from the CIE standard color model XYZ. L* is the luminance component, u* and v* are color components, the u* axis varies from green to red, and the v* axis changes from blue to yellow. The conversion from RGB to Modified CIE L*u*v* is a nonlinear transformation that includes the following two steps.

## 2.1 RGB to CIE XYZ

CIE XYZ can be transmuted from RGB by a 3x3 matrix transformation using Equation (1).

$$\begin{bmatrix} X \\ Y \\ Z \end{bmatrix} = \begin{bmatrix} 0.412453 & 0.357580 & 0.180423 \\ 0.212671 & 0.715160 & 0.072169 \\ 0.019334 & 0.119193 & 0.950227 \end{bmatrix} \begin{bmatrix} R \\ G \\ B \end{bmatrix} \tag{1}$$

## 2.2 CIE XYZ to Modified CIE L*u*v

Intermediate quantities u' and v' can be computed using Equation (2).

$$u' = \frac{4X}{X + 15Y + 3Z} \qquad v' = \frac{9Y}{X + 15Y + 3Z} \tag{2}$$

In [4], researchers used Modified CIE L*u*v* in lieu of standard CIE L*u*v* because the brightness L is proportional to $\sqrt{Y}$ rather than $\sqrt[3]{Y}$ for a complex viewing environment. Then, we have the relations, shown in Equation (3).

$$L^* = 10\sqrt{Y} \quad u^* = 13L^*(u' - u_n') \quad v^* = 13L^*(v' - v_n') \tag{3}$$

where $(u_n', v_n') = (0.1978, 0.4683)$ is used by default.

In an attempt to prove that Modified CIE L*u*v* performs better than RGB acquisition color space, we conduct segmentation using CNN on different images. Fig. 2 depicts the segmented results of the original image and a brighter one. The number of clusters is three, as presented by different colors in the resulting images. Because of the luminance information $Y = [0.2126 \quad 0.7151 \quad 0.0722]\,[R \quad G \quad B]^T$, L*u*v* gives results that are smoother than those of RGB, which has much noise.

(a) (b) (c)

(d) (e) (f)

**Fig. 2.** Effect of image segmentation using RGB and L*u*v*. (a) Original image (b) Brighter image (c) Segmented original image using RGB (d) Segmented brighter image using RGB (e) Segmented original image using L*u*v* (f) Segmented brighter image using L*u*v*.

## 3 Centroid Neural Network

Unsupervised learning is the process of partitioning a set of objects (pattern vectors) into subsets of similar objects called clusters. The purpose is that from a limited set of learning data patterns, we establish a set of clusters in which data patterns in a cluster are as similar as possible and data patterns in distinct clusters are as different as possible. Pixel clustering in three-dimensional color space on the basis of color similarity is one of the popular approaches in the field of color image segmentation. Clustering is often seen as an unsupervised classification of pixels. Generally, a priori knowledge of the image is not used during the clustering process. Colors, dominated in the image, create dense clusters in the color space in a natural way.

The CNN algorithm originated from the conventional *k*-means algorithm finds the centroid of data in corresponding clusters at each presentation of the data vector. In lieu of calculating the centroids of the clusters while every piece of data is being presented, the CNN algorithm updates data weights only when the status of the output neuron for the presenting data has changed: that is, the weights of the winner neuron in the current epoch for the data change only when the winner neuron did not win the data in the previous presentation and the weights of the winner neuron in the previous epoch for the data change only when the neuron does not win the data in the current epoch. We call the former one a "winner neuron" and the latter one a "loser neuron". When an input vector x is applied to the network at time *n*, the weight update equations for winner neuron *j* and loser neuron *i* in CNN can be written as in Equations (4) and (5).

$$w_j(n+1) = \frac{1}{N_j+1}\left[N_j w_j(n) + x(n)\right] = w_j(n) + \frac{1}{N_j+1}\left[x(n) - w_j(n)\right] \quad (4)$$

$$w_i(n+1) = \frac{1}{N_i-1}\left[N_i w_i(n) + x(n)\right] = w_i(n) + \frac{1}{N_i-1}\left[x(n) - w_i(n)\right] \quad (5)$$

In Equations (4) and (5), $w_j(n)$ and $w_i(n)$ represent the weight vectors of the winner neuron and the loser neuron, respectively, while $N_i$ and $N_j$ denote the number of data vectors in cluster $i^{th}$ *and* $j^{th}$ at the time of iteration, respectively.

The learning rule for CNN is based on the following theorem and on the condition for minimum energy clustering:

- **Theorem 1**: The centroid of data in a cluster is the solution that gives minimum energy in $L_2$ norm.
- **Minimum Energy Condition:** The weights for a given output neuron should be chosen in such a way as to minimize the total distance in $L_2$ norm from the vectors in its class, such as

$$w_j = \min_{w} \sum_{i=1}^{N_j} \left\| x_j(i) - w \right\|^2 \quad (6)$$

Using Theorem 1, Equation (6) can be expressed as

$$w_j = \frac{1}{N_j} \sum_{i=1}^{N_j} x_j(i) \quad (7)$$

where $N_j$ is the number of members in cluster j.

When CNN is compared with other conventional competitive learning algorithms, the CNN produces very comparable results with less computational effort. That is, the CNN requires neither a predetermined schedule for learning gain nor a total number of iterations for clustering; it converges stably to suboptimal solutions, while the conventional algorithms, including the Self Organizing Map (SOM), may give unstable results depending on the initial learning gain and the total number of iterations.

Other modifications of this clustering technique are realized by increasing the dimensions of the feature space by introducing additional features, such as the geometrical coordinates of a pixel in the image. By visually comparing the segmented images, the image segmented by using CNN is more stable than that done by FCM and SOM, because the two other algorithms yield different results depending on the initial conditions.

# 4 Experiments

We use the evaluation function in [7] to measure the segmentation results quantitatively as follows:

$$Q(I) = \frac{\sqrt{R}}{10000(W \times H)} \times \sum_{i=1}^{R} \left[ \frac{e_i^2}{1 + \log A_i} + \left( \frac{R(A_i)}{A_i} \right)^2 \right] \tag{8}$$

where W x H is the image size, R is the number of regions of the segmented image, $A_i$ is the area of the $i^{th}$ region, $e_i$ is the sum of the Euclidean distance between the L*u*v* color vectors of the pixels of region $i$ in original image and the color vector attributed to region $i$ in the segmented image. The $R(A_i)$ represents the number of regions having an area equal to $\tilde{A}_i$

In the experiments, we carry out the segmentation of natural scene images from the Berkeley Database by using three clustering algorithms: CNN, FCM and SOM-SA. FCM and SOM-SA require an intensive tuning of parameters. After numerous repetitive trials, we attained the optimal parameters with regard to the evaluation function shown in Equation (8). As a result, the configuration of SOM-SA was determined such that the initial learning rate is 0.7, the total number of iterations is 10000, and the number of neurons is 256. The configuration of FCM was set up such that the exponent weight is 2, and the learning coefficient is 0.3.

The optimally fine tuned FCM and SOM-SA algorithms are compared with the proposed CNN algorithm using the evaluation function shown in Equation (8). Even though the evaluation function depends on the colors set in the segmented images, the function generates the quantitative assessment of segmentation results. The quantitative measurements of the two kinds of natural scene images are shown in Table 1.

Table 1 indicates that the evaluated values of CNN are comparably shown in both images in comparison with FCM and SOM-SA. Since the evaluated values of FCM and SOM-SA are attained after many executions to get the reasonable parameters, above, the proposed CNN algorithm is considered as very efficient.

**Table 1.** Quantitative comparison of segmentation results using different approaches

| | No. of clusters | CNN | FCM | SOM-SA |
|---|---|---|---|---|
| House | 4 | 0.3376 | 0.3376 | 0.335 |
| Flower | 3 | 0.8526 | 0.8166 | 0.8433 |

Fig. 3 and Fig. 4 show the segmentation results of two kinds of natural scene images. The visual comparison of the segmented images indicates that the image segmented by using CNN is very comparable to those of the optimally fine tuned FCM and SOM-SA. In this regard, the proposed CNN algorithm is considered as an easily implemented and effective segmentation method.

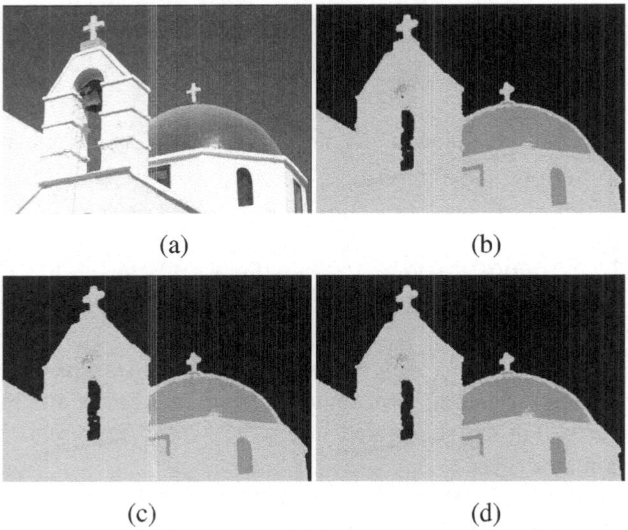


**Fig. 3.** Segmentation results of house image. (a) Original image; images segmented by using (b) CNN (c) FCM (d) SOM-SA.

**Fig. 4.** Segmentation results of flower image. (a) Original image; images segmented by using (b) CNN (c) FCM (d) SOM-SA.

# 5 Conclusions

This paper presents an alternative approach for color image segmentation by using an unsupervised learning algorithm called Centroid Neural Network. The selection of proper parameters in FCM and SOM has been left in state of the art methods until now. Instead, the influences of location of initial cluster centers, number of iterations, etc., of some conventional algorithms are significantly diminished when we apply CNN. In the color image segmentation, we employ L*u*v* space, which gives relatively better results. Since FCM and SOM-SA require an intensive parameter tuning process to get the reasonable parameters, it can be concluded that the proposed CNN algorithm is a very efficient segmentation method.

**Acknowledgments.** This work was supported by National Research Foundation of Korea Grant funded by the Korean Government (2012-004950).

## References

1. Huang, L.K., Wang, M.J.: Image Thresholding by Minimizing the Measures of Fuzziness. Pattern Recognition 28, 41–51 (1995)
2. Yang, J.F., Hao, S.S., Chung, P.C.: Color Image Segmentation Using Fuzzy C-means and Eigenspace Projections. Signal Processing 82, 461–472 (2002)
3. Kohonen, T.: Self-Organizing Maps. Springer, Germany (1995)
4. Dong, G., Xie, M.: Color Clustering and Learning for Image Segmentation Based on Neural Networks. IEEE Trans. on Neural Networks 16 (2005)
5. Park, D.C.: Centroid Neural Network for Unsupervised Competitive Learning. IEEE Trans. on Neural Network 11, 520–528 (2000)
6. Cheng, H.D., Jiang, X.H., Sun, Y., Wang, J.: Color Image Segmentation: Advances and Prospects. Pattern Recognition 34, 2259–2281 (2001)
7. Borsotti, M., Campadelli, P., Schettini, R.: Quantitative Evaluation of Color Image Segmentation Results. Pattern Recognition Letter 19, 741–747 (1998)

# Zhang Fractals Yielded via Solving Time-Varying Nonlinear Complex Equations by Discrete-Time Complex-Valued ZD

Yunong Zhang, Long Jin, Zhijun Zhang, Lin Xiao, and Senbo Fu

School of Information Science and Technology
Sun Yat-sen University, Guangzhou 510006, China
zhynong@mail.sysu.edu.cn, {spaformind,983979597}@qq.com,
iloveyouzhijun@126.com, xiaolin860728@163.com
http://sist.sysu.edu.cn/~zhynong

**Abstract.** In this paper, new fractals, termed Zhang fractals, are yielded by using the discrete-time complex-valued Zhang dynamics (DTCVZD) to solve time-varying nonlinear equations in the complex domain. Such a DTCVZD model is designed by zeroing an indefinite complex-valued error function. Such Zhang fractals generated by the DTCVZD model are quite different from the famous fractals generated by Newton iteration (i.e., Newton fractals) which solves static equations. The presented DTCVZD model with different types of activation functions usable can be seen as a new iterative algorithm to produce fractals. In addition, by comparing the area and degree of blue color in Zhang fractals under the same conditions, the effectiveness of the DTCVZD model using different activation functions for solving time-varying nonlinear complex equations is reflected.

**Keywords:** Time-varying nonlinear equations, Complex-valued, Zhang dynamics (ZD), Discrete-time, Zhang fractals.

## 1 Introduction

The problem of generating fractals is an important and interesting issue in the fields of signal recovery [1] as well as computer geometry and graphics [2], where fractal theories and techniques have developed since 1970s. The fractals may show and reveal the chaotic phenomena [3] appearing in the natural or artificial architectures of microscopic self-computing and/or self-organization. Generation of many fractals is closely related to the Newton iteration method for solving certain types of nonlinear equations in the complex domain, e.g., complex roots finding, complex polynomials solving, and complex-exponential functions solving. Then, by the authors' thoughts, time-varying nonlinear equations solving may output very different figures while generating new fractals.

A special kind of recurrent neural network termed Zhang neural network (ZNN) has been proposed since March 2001 for various time-varying problems solving, e.g., time-varying linear equations solving, time-varying matrix

J. Lei et al. (Eds.): AICI 2012, LNAI 7530, pp. 596–603, 2012.

inversion, time-varying linear matrix equations solving and time-varying matrix square-roots finding [4]. As its reduced or generalized form of ZNN, Zhang Dynamics (ZD) has performed efficiently in the area of time-varying nonlinear equation solving, including the time-varying roots finding. The main difference between ZNN and ZD lies only in the dimensions of neural states.

In this paper, by dicretizing the continuous-time complex-valued Zhang dynamics (CTCVZD) model for time-varying nonlinear equations solving in the complex domain, the corresponding discrete-time complex-valued ZD (DTCVZD) model can be obtained. In addition, such a DTCVZD model is applied to generating new fractals (i.e., Zhang fractals), which are clearly different from the fractals generated by the Newton iteration method (i.e., Newton fractals).

## 2 Complex-Valued Zhang Dynamics

In this paper, we consider the following time-varying nonlinear equation in the complex domain for generating Zhang fractals:

$$f(z(t),t) = g(z(t),t) - a(t) = 0 \in C, t \in [0, +\infty) \tag{1}$$

where $g(\cdot) : C \to C$ denotes a nonlinear complex mapping, $a(t) \in C$ is a smoothly time-varying scalar in complex domain, and $z(t) \in C$ denotes the unknown time-varying complex-valued scalar to be solved for.

The well-known Newton iteration method in the complex domain has been applied to computing the roots of the complex-valued nonlinear equation $f(z) = g(z) - a = 0 \in C$ which is clearly time-invariant (or termed static). The corresponding fractals can thus be generated by the Newton iteration method at the same time. Different from the conventional Newton iteration method, we develop and exploit a DTCVZD model in this paper to find the roots of the complex-valued time-varying nonlinear equation (1), and thus obtain new fractals. Specifically, the time-varying nonlinear equations considered to be solved for yielding fractals in this paper is $f(z(t),t) = z^N(t) - a(t)$, where $N \geqslant 3$ is a positive integer parameter and $a(t)$ is a complex-valued time-varying scalar [5].

### 2.1 Continuous-Time Complex-Valued Zhang Dynamics

By following Zhang *et al.*' design method, a CTCVZD model is firstly constructed to solve time-varying nonlinear equation (1). To monitor and control the solution process of time-varying nonlinear equation (1), the following indefinite lower-unbounded complex-valued error function is defined:

$$e(t) = f(z(t),t) \in C, \tag{2}$$

where, evidently, $z(t) \in C$ converges to a theoretical solution $z^*(t)$ of time-varying nonlinear equation (1) as error function $e(t)$ converges to zero. To make error function $e(t)$ diminish to zero, the following Zhang *et al.*' formula is employed (which is also termed ZNN formula or ZD formula):

$$\dot{e}(t) = -\gamma\phi(e(t)), \tag{3}$$

where design parameter $\gamma > 0$ is the reciprocal of a capacitance parameter in the hardware implementation of the ZD model, which should be set as large as the hardware would permit, or set appropriately for simulative purposes. Meanwhile, $\phi(\cdot) : C \to C$ denotes the activation function. In this paper, five types of activation functions (i.e., the linear, power-sigmoid, power-sum, hyperbolic-sine, and sigmoid activation functions) are introduced and investigated. Other types of activation functions can then be generalized from these five basic types of activation functions.

Expanding the CTCVZD design formula (3), we can have the following complex-valued differential equation for the so-called CTCVZD model:

$$\frac{\partial f(z(t),t)}{\partial z(t)}\frac{dz(t)}{dt} + \frac{\partial f(z(t),t)}{\partial t} = -\gamma\phi(f(z(t),t));$$

and, with $\partial f(z(t),t)/\partial z(t) \neq 0$ assumed, we further have

$$\dot{z}(t) = \frac{-\gamma\phi(f(z(t),t)) - f_t'(z(t),t)}{f_z'(z(t),t)}, \tag{4}$$

where $f_t'(z(t),t) = \partial f(z(t),t)/\partial t$ and $f_z'(z(t),t) = \partial f(z(t),t)/\partial z(t)$, with $z(t) \in C$ denoting the neural state of the ZD model corresponding to a solution of (1).

## 2.2 Discrete-Time Complex-Valued Zhang Dynamics

In this subsection, we present the discrete-time form of the aforementioned CTCVZD model to solve time-varying nonlinear equation (1), which can be viewed as the essential step for the new fractals' generation. By applying Euler forward difference rule to the CTCVZD model, the corresponding DTCVZD model is obtained via the intermediate result

$$\frac{z_{k+1} - z_k}{\tau} = \frac{-\gamma\phi(f(z_k,k\tau)) - f_t'(z_k,k\tau)}{f_z'(z_k,k\tau)}, \text{ with } f_z'(z_k,k\tau) \neq 0;$$

that is,

$$z_{k+1} = z_k - \frac{h\phi(f(z_k,k\tau)) + \tau f_t'(z_k,k\tau)}{f_z'(z_k,k\tau)}, \text{ with } f_z'(z_k,k\tau) \neq 0, \tag{5}$$

where $k = 0, 1, 2, \cdots$ denotes the iteration index, and design parameter $h = \gamma\tau \in C$ denotes the step-size. Note that different choices of step-size $h$ and activation function $\phi(\cdot)$ lead to different convergence performances of DTCVZD (5).

Additionally, let us show the link of DTCVZD iteration (5) to Newton iteration. For simplicity, $\tau f_t'(z_k,k\tau)$ is omitted temporarily, and (5) reduces to

$$z_{k+1} = z_k - h\frac{\phi(f(z_k,k\tau))}{f_z'(z_k,k\tau)}, \text{ with } f_z'(z_k,k\tau) \neq 0, \tag{6}$$

which is a simplified discrete-time complex-valued ZD (S-DTCVZD) model. It follows from S-DTCVZD model (6) that, if we utilize linear activation function $\phi(u) = u$, iteration (6) becomes

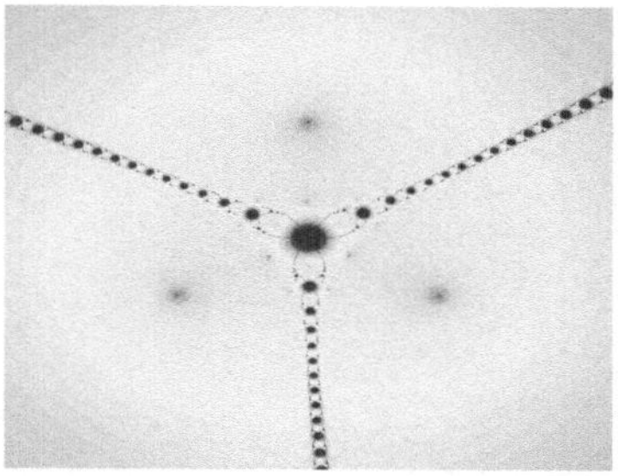

(a) Using linear activation function

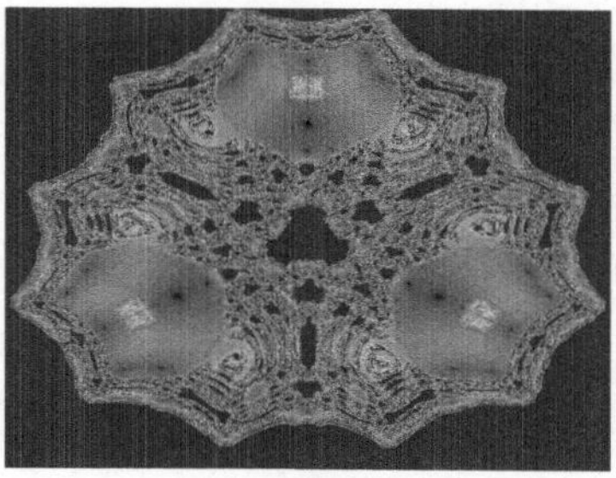

(b) Using power-sigmoid function

**Fig. 1.** Zhang fractals generated by solving time-varying complex-valued nonlinear equation (10) via DTCVZD (5) with $h = 0.2$ and using different activation functions, where horizontal and vertical axes correspond to real and imaginary parts of initial state $z_0 \in [-2, 2] + i[-2, 2]$, respectively, while the color degree corresponds to the MNI required by the DTCVZD model starting with such an initial state $z_0$.

$$z_{k+1} = z_k - h\frac{f(z_k, k\tau)}{f'_z(z_k, k\tau)}. \tag{7}$$

Furthermore, if $h \equiv 1$ and $f(z(t), t) \equiv f(z)$ (i.e., static), iteration (7) becomes

$$z_{k+1} = z_k - \frac{f(z_k)}{f'(z_k)}, \tag{8}$$

which is exactly the Newton iteration for nonlinear equation solving in the complex domain. This implies that Newton iteration can be viewed as a special case of the general DTCVZD model (5). Newton iteration (8) is applied to generating different fractals by solving different types of nonlinear equations, and the fractals yielded are thus called Newton fractals. In the ensuing section, we show that the fractals generated by the DTCVZD model (5) for solving time-varying nonlinear equations are quite different from those generated by Newton iteration.

*Remarks.* The corresponding residual error of the DTCVZD model [which is derived from error function (2)] is defined as

$$E_k = |f(z_k, k\tau)|,$$

where symbol $|\cdot|$ denotes the modulus of a complex number and also the absolute value of a real number. So, the minimum number of iterations (MNI) for a model's residual error converging to its steady state can be defined and used to determine the color degree corresponding to an initial state in the generated Zhang fractals. The MNI decision is based on the following practical criterion:

$$|E_{k+1} - E_k|/E_k \leqslant 10^{-3}, \tag{9}$$

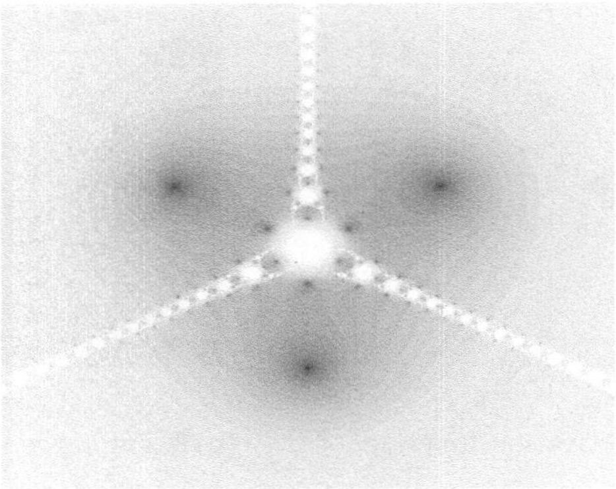

**Fig. 2.** Newton fractal generated by solving time-invariant complex-valued nonlinear equation $z^3 - i = 0$ [which is a very simple one of (10)] via Newton iteration (8)

where we have MNI $= k + 1$ when the criterion (9) is satisfied for the first time during the solving process. For the purpose of better demonstration, the upper limit of MNI is set to 80; i.e., MNI $= 80$ if it is larger than 80. Note that calculating MNI is the key step to produce Zhang fractals.

## 3 Fractals Generation

In this section, some representative and illustrative examples are shown to verify DTCVZD model (5) for generating fractals, which solves time-varying nonlinear equations in the complex domain. We exploit the DTCVZD model with $h = 0.2$ and using different activation functions to generate Zhang fractals.

### 3.1 Fractals Generated by Solving $z^3(t) - a(t) = 0$

Let us consider the following complex-valued time-varying nonlinear equation:

$$z^3(t) - \sin(t) - i\cos(t) = 0. \tag{10}$$

We exploit the DTCVZD model activated by the power-sigmoid activation function to generate new fractals, where the power-sigmoid activation function is

$$\phi(e) = \begin{cases} e^p, & \text{if } |e| \geqslant 1, \\ \frac{1+\exp(-\xi)}{1-\exp(-\xi)} \frac{1-\exp(-\xi e)}{1+\exp(-\xi e)}, & \text{otherwise,} \end{cases}$$

with parameters $\xi = 4$ and $p = 3$. For demonstrating the effectiveness of using different activation functions for generating different fractals, Fig. 1 shows two different Zhang fractals generated by solving (10) via the DTCVZD model with $h = 0.2$. In the figure, the horizontal and vertical axes correspond to the real and imaginary parts of the initial state $z_0 \in [-2, 2] + i[-2, 2]$, respectively, while the color degree corresponds to the MNI required by DTCVZD model (5) starting with such an initial state $z_0$.

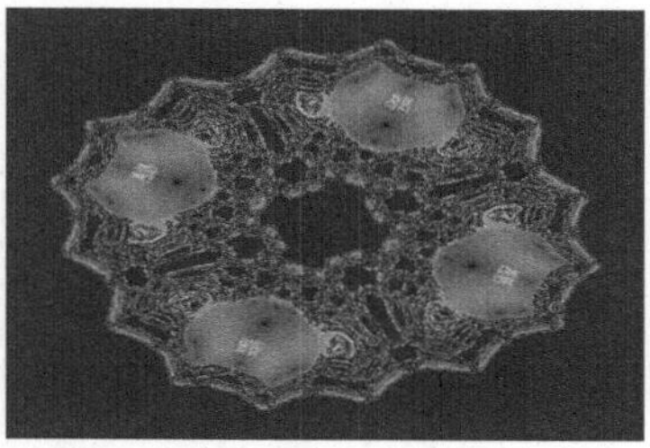

(a) Using power-sigmoid function

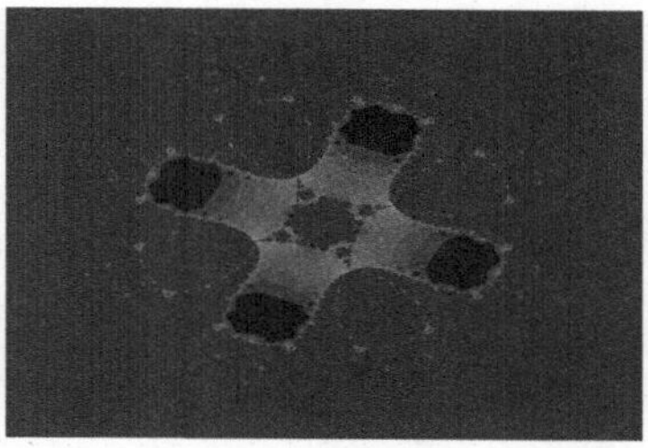

(b) Using power-sum activation function

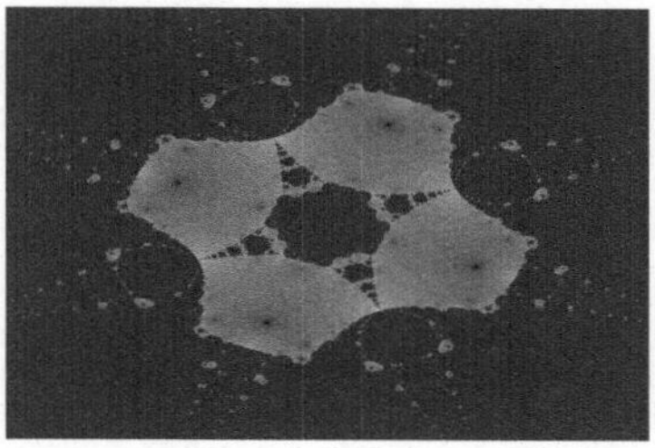

(c) Using hyperbolic-sine function

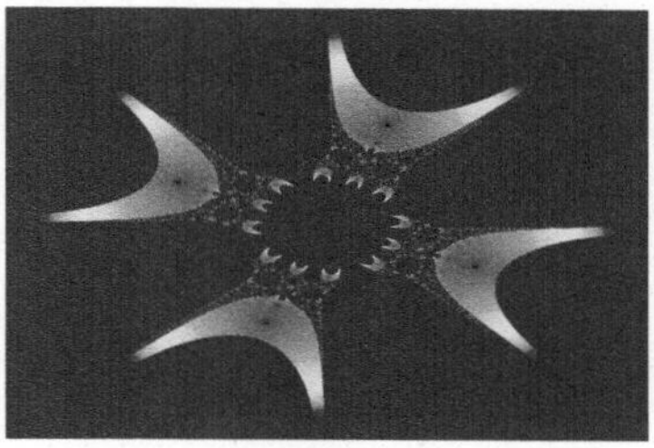

(d) Using sigmoid activation function

**Fig. 3.** Zhang fractals generated by solving (11) via the DTCVZD model with $h = 0.2$ and using different activation functions, where $z_0 \in [-2, 2] + i[-2, 2]$

For comparison, let us consider $f(z(t), t) \equiv z^3 - i$ [i.e., the static case which can be view as a very simple problem of (10) with $t = 0$], where Newton iteration (8) is used. Newton fractal is thus generated as a special case of Zhang fractals: the graphical result is shown in Fig. 2.

### 3.2 Fractals Generated by Solving $z^4(t) - a(t) = 0$

Let us consider the following complex-valued time-varying nonlinear equation:

$$z^4(t) - \sin(t) - i\cos(t) = 0. \tag{11}$$

For the purpose of generating more Zhang fractals and also for comparison, another three nonlinear activation functions (i.e., power-sum, hyperbolic-sine, and sigmoid activation functions) are applied to the DTCVZD model, where
1) the power-sum activation function is

$$\phi(e) = \sum_{k=1}^{M} e^{2k-1}, \text{with } M \geqslant 1;$$

(a) $z_0 \in [-2, 2] + i[-2, 2]$ (b) $z_0 \in [-0.6, -0.1] + i[-0.2, 0.1]$

**Fig. 4.** Zhang fractal generated by solving (12) via the DTCVZD with $h$=0.2

2) the hyperbolic-sine activation function is

$$\phi(e) = \frac{\exp(\zeta e) - \exp(-\zeta e)}{2}, \text{ with } \zeta = 2;$$

3) the sigmoid activation function is

$$\phi(e) = \frac{1 + \exp(-\xi)}{1 - \exp(-\xi)} \frac{1 - \exp(-\xi e)}{1 + \exp(-\xi e)}, \text{ with } \xi = 4.$$

Seeing Fig. 3 and comparing the four Zhang fractals which are generated by solving the time-varying nonlinear equation (11) with different activation functions, we get a general conclusion that the power-sigmoid activation function and power-sum activation function have relatively better effects on solving time-varying nonlinear equation (11) via DTCVZD (5) (note that the better efficacy is reflected by the darker degree and area of blue color in the fractals).

### 3.3 Fractal Generated by Solving $z^5(t) - a(t) = 0$

Let us consider the following complex-valued time-varying nonlinear equation:

$$z^5(t) - \sin(t) - it = 0. \tag{12}$$

For further demonstrating the effectiveness and the wide applications of the proposed model, we use DTCVZD model (5) with a linear activation function and $h = 0.2$ to solve equation (12). The corresponding result is shown in Fig. 4.

## 4 Conclusions

In this paper, some interesting and beautiful Zhang fractals are yielded by the presented discrete-time complex-valued Zhang dynamics (DTCVZD) for solving the time-varying nonlinear equations in the complex domain. In addition, the darker degree and area of blue color in Zhang fractals which are generated by using different activation functions can be used to compare and reflect the solution effect of the nonlinear time-varying complex equations via the DTCVZD model. The provided graphical results demonstrate that the DTCVZD model can be viewed as a new iterative algorithm to generate many different new fractals.

**Acknowledgements.** This work is supported by the National Natural Science Foundation of China under Grants 61075121 and 60935001, and also by the Fundamental Research Funds for the Central Universities of China. Besides, the corresponding author, Yunong, would like to thank and encourage the coauthors with the following thoughts: 1) "Generally speaking, the students without teachers' advising are short thinkers who randomly walk, while the students with teachers' advising appear to be long thinkers who have clear directions", 2) "The painter has already drawn the picture in mind for many times before assigning students to draw on paper", 3) "Around success, there are full of local minima of failure", and 4) "The river said to the streams: you and I will forget all our ways when we reach the heart of the sea" (Gibran). Sincere thanks and best regards.

## References

1. Zhai, M.: Signal Recovery in Power-Line Communications Systems Based on the Fractals. IEEE Trans. Power Delivery 26(3), 1864–1872 (2011)
2. Singh, G.: Fun with Fractal Art. IEEE Comput. Graph. Appl. 29(1), 4–5 (2009)
3. Singh, G.: Beauty in Chaos. IEEE Comput. Graph. Appl. 27(5), 4–5 (2007)
4. Zhang, Y., Ge, S.S.: Design and Analysis of a General Recurrent Neural Network Model for Time-Varying Matrix Inversion. IEEE Trans on Neural Netw. 16(6), 1477–1490 (2005)
5. Zhang, Y., Ke, Z.: Solving for Time-Varying and Static Cube Root via Discrete-Time Real-Valued ZD Models and Complex-Valued ZD Model. Neural Computing and Applications (2011)

# Sub-optimal Multiuser Detector Using a Chaotic Neural Network with Decaying Chaotic Noise

Yunxiao Jiang

Key Laboratory of Electronic Restriction of AnHui Province,
Electronic Engineering Institute, Hefei 230037, China
j_yunxiao@sohu.com

**Abstract.** This paper proposes a sub-optimal multiuser detector (MUD) algorithm for CDMA system based on the neural network with a novel Chaotic Neural Network with Decaying Chaotic Noise, and gives a concrete model of the MUD after appropriate transformations and mappings. On the basis of the Hopfield neural network with transient chaos and time-varying gain (NNTCTG), the proposed chaotic noised Hopfield neural network(CNHNN) introduces the decaying chaotic noise to each neuron, and the noise is gradually reduced to zero. The proposed CNHNN has richer and more complex dynamics than NNTCTG, and the transient chaos enables the network to escape from local energy minima and to settle down at the global optimal solution, so that it can be expected to have much powerful ability to search for globally optimal or sub-optimal solutions ,and can refrain from the serious local optimal problem of Hopfield-type neural networks. Simulation experiments have been performed to show the effectiveness and validation of the proposed method for MUD problem.

**Keywords:** MUD, Hopfield neural network, Chaotic Noise.

## 1 Introduction

MUD came to the focus in CDMA systems as traditional single user receivers have shown poor performance due to uncorrelation produced by multipath propagation. In Verdu's opinion[1], minimizing the objective function of the Optimal multisuer Detector (OMD) is a NP-complete problem, however the OMD has been shown to have the exponential computation complexity in the number of users, and can not satisfy the real time demand. Consequently, many authors proposed different sub-optimal solutions. The first Hopfield type of receiver was introduced by Varanasi and Aazhang [2], They defined a multi-stage detector. However, due to local optimization, this structure does not provide optimal detector. Lots of articles have discussed modified recurrent neural network structures for multiuser detector to achieve performance improvement. Yoon,Chen, et al.[3] proposed an annealed neural network multiuser receiver. Kechriotis and Manolakos introduced another modified structure which was named as hybrid detector[4]. Wang, et al. considered a transiently chaotic neural network based multiuser receiver scheme[5] which originates from chaos theory. In[6], a Hopfield neural network with transient chaos and time-varying gain

J. Lei et al. (Eds.): AICI 2012, LNAI 7530, pp. 604–611, 2012.

(NNTCTG) was proposed. However, the optimality of all the previously listed techniques has not yet been proven theoretically, but rather tested by simulations. Therefore, developing optimal detector for MUD still remains an open question.

In this paper, on the basis of NNTCTG, the neural network with a novel Chaotic Neural Network with Decaying Chaotic Noise was proposed and used to deal with MUD. The proposed CNHNN has richer and more complex dynamics, and the transient chaos enables the network to escape from local energy minima and to settle down at the global optimal solution. Unlike the conventional networks detectors, the detector based on proposed nerual network has richer and far-from equilibrium dynamics with various coexisting attractors, not only of fixed points and periodic points but also of strange attractors, to avoid getting stuck in local minima.

The paper is organized as following: The proposed neural model and its neural dynamics analysis are presented in Section 2, the proposed neural model based MUD in CDMA system is given in Section 3. In Section 4, some simulation experiments are performed for the comparison of our model with the existing detector models. Finally, conclusion remarks are given in Section 5.

## 2 Chaotic Neural Network Models

It is well known that Hopfield network with continuous-time or asynchronously discrete-time state transitions guarantee convergence to a stable equilibrium solution but suffer from local minimum problems. Since the chaotic neural network has richer and more flexible neural dynamics, therefore, it can be used to efficiently escape from local minima problem in chaotically. In this section, two chaotic neural network models are given. The first NNTCTG is presented by Y.tan[6], the second is the proposed CNHNN with Decaying Chaotic Noise which makes use of the decaying chaotic noise parameters of the recurrent neural network to control the evolving behavior of the network

### 2.1 Neural Network with Transient Chaos and Time-Varying Gain (NNTCTG)

In order to obtain a global optimal or sub-optimal convergent solution for nonlinear optimization, Y.tan, et al.[6] proposed a neural network with transient chaos and time-varying gain(NNTCTG), as defined below:

$$x_i(t) = \frac{1}{1+e^{-y_i(t)/(\varepsilon_i(t)+1)}} \tag{1}$$

$$y_i(t+1) = ky_i(t) + \alpha\Big(\sum_{j=1, j\neq i}^{n} w_{ij}x_j(t) + I_i\Big) - z_i(t)(x_i(t) - I_0) \tag{2}$$

$$z_i(t+1) = (1-\beta)z_i(t) \tag{3}$$

$$\varepsilon_i(t+1) = (1-\gamma)\varepsilon_i(t) \tag{4}$$

where $x_i$ is output of neuron i, $y_i$ is internal state of neuron i, $w_{ij}$ is connection weight from neuron j to neuron i, $I_i$ is input bias of neuron i, $I_0$ is a positive parameter, $\alpha$ is positive scaling parameter for inputs, $k$ is damping factor of nerve membrane ($0 \le k \le 1$), $z_i(t)$ is self-feedback connection weight( $z_i(t) \ge 0$ ), $\beta$ is damping factor of the time-dependent $z_i(t)$ ,($0 \le \beta \le 1$), $\varepsilon_i(t)$ is gain parameter of the output function ( $\varepsilon_i(t) \ge 0$ ), $\gamma$ is damping factor of the time-dependent $\varepsilon_i(t)$ ,($0 \le \gamma \le 1$).

The term $z_i(t)(x_i(t) - I_0)$ in (2) is related to inhibitory self-feedback or refractoriness and is the main factor generating chaotic phenomenon. NNTCTG actually has transiently chaotic dynamics which eventually converges to a stable equilibrium point through successive bifurcations with the temporal evolution of $z_i(t)$ and $\varepsilon_i(t)$ in (3). The time evolutions of self-feedback connection weight $z_i(t)$ is shown in Fig.1 with respect to damping parameter $\beta$.

## 2.2 The Proposed CNHNN

In order to prevent the chaotic dynamics of NNTCTG from vanishing so quickly to realize sufficient chaotic searching, this paper proposed the chaotic neural network with decaying noise on the base of NNTCTG, described as follows:

$$x_i(t) = \frac{1}{1 + e^{-y_i(t)/(\varepsilon_i(t)+1)}} \tag{5}$$

$$y_i(t+1) = ky_i(t) + \alpha\left(\sum_{j=1, j\ne i}^{n} w_{ij}x_j(t) + I_i\right) - z_i(t)(x_i(t) - I_0) \tag{6}$$

$$z_i(t+1) = (1-\beta)z_i(t)(1 - z_i^2(t)) \tag{7}$$

$$\varepsilon_i(t+1) = (1-\gamma)\varepsilon_i(t) \tag{8}$$

where $x_i$, $y_i$, $w_{ij}$, $I_i$, $\alpha$, $k$, $z_i(t)$, $\beta$, $\varepsilon_i(t)$, $\gamma$ are the same with the above. The term $(1-\beta)z_i(t)(1 - z_i^2(t))$ in (7) is the decaying chaotic noise which reduced to each neuron. The time evolutions of self-feedback connection weight $z_i(t)$ is shown in Fig.2 with respect to damping parameter $\beta$.

### 2.3 Application to Traveling Salesman Problem (TSP)

The coordinates of 10-city is as follows:

(0.4, 0.4439)、( 0.2439, 0.1463)、( 0.1707, 0.2293)、( 0.2293, 0.716)、(0.5171,0.9414)、( 0.8732, 0.6536)、( 0.6878, 0.5219)、( 0.8488, 0.3609)、( 0.6683, 0.2536) 、( 0.6195, 0.2634). The shortest distance of the 10-city is 2.6776.

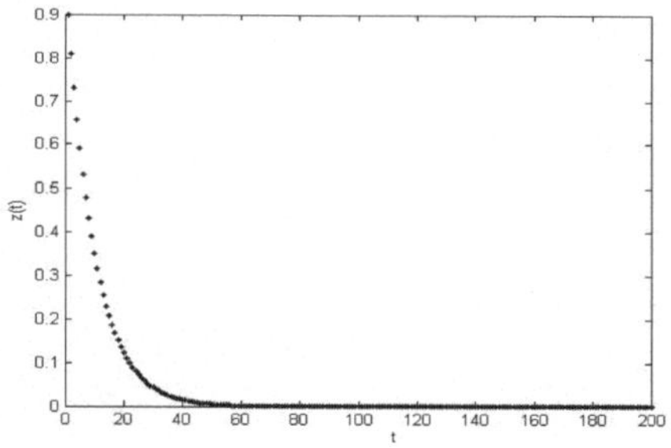


**Fig. 1.** The time evolutions of $z_i(t)$ of NNTCTG

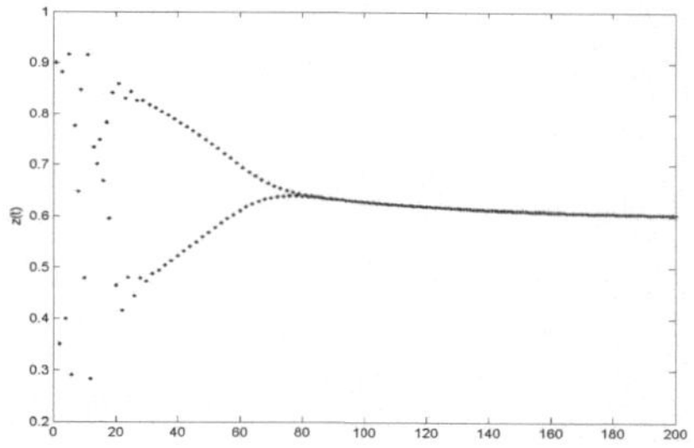


**Fig. 2.** The time evolutions of $z_i(t)$ of CNHNN

Here are the results of the test about Y.tan's NNTCTG model and CNHNN. The objective function we adopt is that provided in the reference [7]. The parameters of the objective function are set as follows: A=2.5, D=1.

The parameters of NNTCTG are set as follows :

$$\alpha = 0.5, k = 1, I_0 = 0.5, \varepsilon(0) = [0.5, 0.5], \gamma = 0.3, z(0) = [0.5, 0.5].$$

The parameters of WCSANN are set as follows :

$$\alpha = 0.5, k = 1, I_0 = 0.5, \varepsilon(0) = [0.5, 0.5], \gamma = 0.3, z(0) = [0.5, 0.5]$$

We make the test for 1000 iterations in different β, as is shown in Table 1. (VN=valid number; GN=global number; VP=valid percent; GP=global percent.)

**Table 1.** Test result of two chaotic neural network

| $\beta$ | Reference | VN | GN | VP | GP |
|---|---|---|---|---|---|
| 0.035 | CNHNN | 961 | 938 | 96.1% | 93.8% |
| | NNTCTG | 891 | 823 | 89.1% | 82.3% |
| 0.011 | CNHNN | 965 | 950 | 96.5% | 95% |
| | NNTCTG | 893 | 830 | 89.3% | 83% |
| 0.008 | CNHNN | 980 | 970 | 98% | 97% |
| | NNTCTG | 911 | 903 | 91.1% | 90.3% |

The time evolution figures of the energy function of CNHNN and NNTCTG in solving TSP are respectively given in Fig.3 and Fig.4 when β =0.008.By comparison, it is concluded that CNHNN is superior to NNTCTG model. From the Fig.3, Fig.4,we can see that the velocity of convergence of CNHNN is much faster than that of NNTCTG in solving TSP.

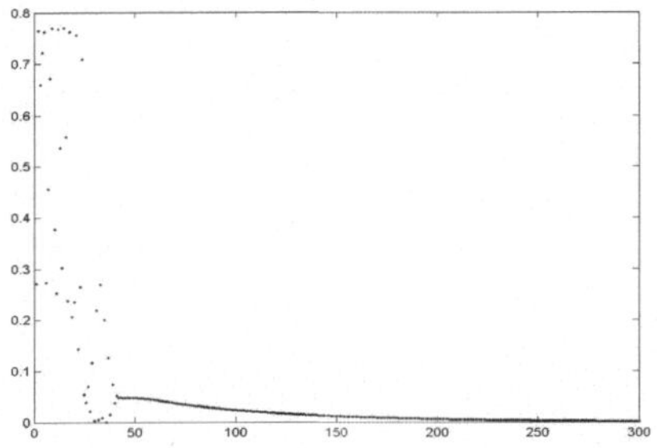


**Fig. 3.** Energy time evolution figure of CNHNN

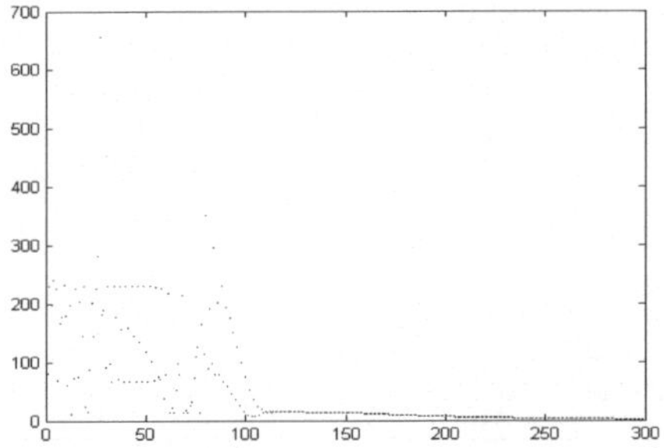


**Fig. 4.** Energy time evolution figure of NNTCTG

The superiority of CNHNN contributes to several factors: First, because of the quality of Morlet wavelet function, the activation function of CNHNN has a further performance in solving combinatorial optimization problems than NNTCTG . Second, it is easier to produce chaotic phenomenon [8] in that the activation function is non-monotonic.

## 3 CNHNN-Based Multiuser Detector in CDMA System

Let us assume that K users are transmitting (2M+1) symbol long packets on the same channel in CDMA system. Applying the complex baseband equivalent model, the transmitted signal of the kth user can be written in the following form:

$$q_k(t) = \sqrt{E_k} \sum_{i=-M}^{M} b_k(i) s_k(t - iT) \tag{9}$$

where $T$ denotes the symbol length, $s_k(t)$ denote the signature waveform related to the kth user, $b_k(i)$ be the ith symbol of the kth user and $\sqrt{E_k}$ is the symbol energy of the kth user.

In[1],Verdu derive the optimal MUD as a maximum likelihood sequence estimation(MLSE) problem which can then be obtained in the following fashion:

$$\hat{b}^{opt} = agr \min_{x \in \{-1,+1\}^k} [x^T Rx - 2x^T b] \tag{10}$$

where $R$ is the signature waveform cross-correlation matrix, $b$ is the symbol of the user. The optimal solution can be found, for example, by exhaustive search using (10), however, it implies exponentially increasing computational complexity as the number of users grows. We cannot afford to use such a time wasting mechanism in real life implementation, thus many sub-optimal MUD algorithms have been researched in the recent past.

In [2][3][4], Hopfield neural network(HNN) is used for the purpose of MUD. The energy function of HNN can be written as:

$$E = -\frac{1}{2}x^T Wx - x^T I \tag{11}$$

where $x$ is the output of HNN, and $W$ is neural weight matrix. $I$ is the input biases. It is apparent from (10) that the objective function of the MLSE is very similar to the HNN energy function (11). With the replacement $W \rightarrow R$ and $x \rightarrow x$, and mapping the HNN parameters to (5)-(8),then we can get the CNHNN-based Multiuser detector.

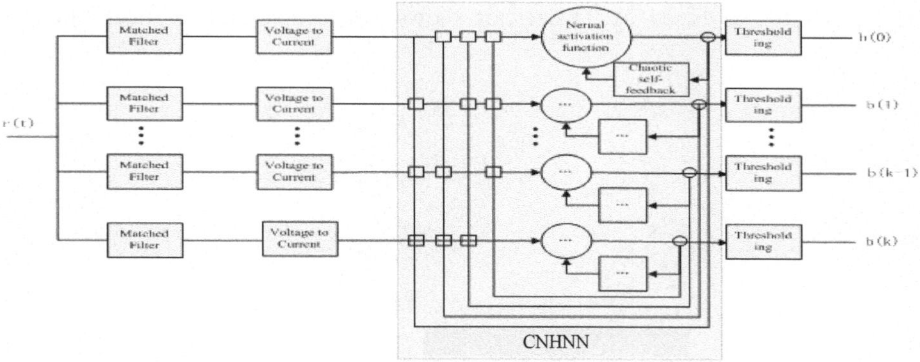


**Fig. 5.** CNHNN -based multiuser detector

The structure of the CNHNN-based MUD receiver is shown in Fig.5. CNHNN-based MUD has richer and far-from equilibrium dynamics with various coexisting attractors to avoid getting stuck in local minima, especially when the number of the users or the packet length becomes very large(in the order of hundreds of users) .

## 4 Experimental Results

The simulations were performed in a $K = 33$ user environment with 31 length Gold codes, which entails that the channel were overloaded. The number of symbols in one block was chosen to be $2M + 1 = 51$, chip level synchronization was assumed, samples were taken in the middle of chip intervals. For all users a $L_k = 5$ path propagation model was used, where the attenuation was generated subjected to Rayleigh distribution. The delay parameters of each path were chosen to be uniformly distributed over one symbol length.

Since the quality of digital communication link can be measured by Bit Error Ratio(BER) in all the figures on

the vertical axis the simulated BER is depicted. The performance of the optimal detector based on MLSE can not be simulated due to its computational complexity (there are $2^{(2M+1)K} = 2^{1683}$ possible sequences), thus the theoretical BPSK AWGN bound is depicted in the figures, which is expected to be close to the curve of optimal detector[9].It is given as the function of bit energy per noise variance ratio( $E_b / N_o$ )in the following form:

$$BER = Q(-\sqrt{\frac{E_b}{N_o/2}}) = \int_{-\infty}^{-\sqrt{\frac{E_b}{N_o/2}}} \frac{1}{\sqrt{2\pi}} e^{-\frac{x^2}{2}} dx \tag{12}$$

In Fig.6 the performance versus bit energy per noise variance ratio $E_b / N_o$ is depicted. CNHNN-based MUD detector shows satisfactory performance, which is

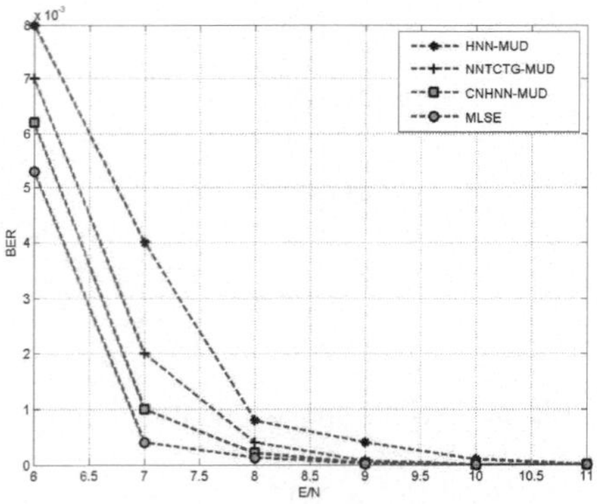


**Fig. 6.** BER vs. Bit Energy per Noise Variance

very close to the optimal detector. CNHNN-based MUD detector results in better performance than NNTCTG-based MUD detector, namely 1..2dB gain in performance can be achieved.

Although we have seen that CNHNN-based MUD detector outperforms the NNTCTG-based MUD detector, but it is still a question how much additional iteration is needed in exchange for better performance. It is worth comparing this in Table 2, where the number of iterations need to reach the steady state by NNTCTG and CNHNN are shown from $E_b / N_o = 6$ to 11dB. It is very clear that CNHNN needs 7~8 additional iterations to exchange for her better performance.

**Table 2.** Number of Iterations needed by NNTCTG-MUD and CNHNN-MUD

| $E_b / N_o$ | Average Iterations | | BER($10^{-3}$) |
|---|---|---|---|
| | NNTCTG | CNHNN | |
| 6 dB | 14.997 | 21.663 | 5.9 |
| 7 dB | 11.5 | 19.6 | 2.1 |
| 8 dB | 10.1 | 18.8 | 0.71 |
| 9 dB | 9.9 | 16.9 | 0.25 |
| 10 dB | 9.7 | 16.0 | 0.17 |
| 11 dB | 9.2 | 15.5 | 0.06 |

## 5 Conclusions

In this paper we have proposed a novel multi-user detection scheme, which makes use of CNHNN optimization algorithm in MUD. The new method resulted in better performance than NNTCTG detector algorithms, namely 1..2dB gain in performance can be achieved, only 7..8 additional iteration is needed. In the future, we would like to develop a more sophisticated, adaptive version of the algorithm.

## References


1. Verdu, S.: Minimum probability of error for asynchronous Gaussian multiple-access channels. IEEE Trans. Inform. Theory IT-32, 85–96 (1986)
2. Varanasi, M.K., Azhang, B.: Multistage Detection in Asynchronous Code-Division Multiple Access Communications. IEEE Trans. on Comm. 38, 509–519 (1990)
3. Yoon, S.H., Rao, S.: Multiuser detection in CDMA based on the annealed neural network. In: IEEE Int. Conf. Neural Networks, vol. 4, pp. 2124–2129 (1996)
4. Kechriotis, G., Manolakos, E.S.: A Hybrid Digital Signal Processing-Neural Network CDMA Mulituser Detector Scheme. IEEE Trans. Cicuits and Systems 43(2), 96–104 (1996)
5. Wang, B., He, Z., Nie, J.: To Implement the CDMA Multiuser Detector by Using Transiently Chaotic Neural Network. IEEE Trans. Aerospace and Elec. Sys. 33, 1068–1071 (1997)
6. Tan, Y., Deng, C., Wang, B., He, Z.: A Neural Network with Transient Chaos and Time-Varying Gain and its Application to Optimization Calculation. Acta Electronica Sinica 26(7), 123–127 (2005)
7. Sun, S., Zheng, J.: A Kind of Improved Algorithm and Theory Testify of Solving TSP in Hopfield Neural Network. Acta Electronica Sinca 1(23), 73–78 (1995)
8. Potapove, A., Kali, M.: Robust Chaos in Neural Networks. Physics Letters A 277(6), 310–322 (2000)
9. Teich, W.G., Seidl, M.: Code Division Multiple Access Communications: Multi-user Detection based on a Recurrent Neural Network Structure. IEEE Trans. Veh. Technol. 46, 979–984 (1996)

# A Formal Method for Testing Reactive System from Symbolic Model

Yongbing Wan, Zhongwei Xu, and Meng Mei

School of Electronics & Information Engineering, Tongji University, Shanghai, China
{wybingsh,xuzhongweish,mei_meng}@163.com

**Abstract.** Testing is one of the most well-established techniques for the verification and validation of systems. Since success or failure verdicts are emitted with respect to the test case execution results, proper test case selection activities need to be performed. In this paper, we propose an approach to address the problem of generating symbolic test cases to test whether an implementation conforms to its specification, given in terms of reactive systems. This approach is based on a unified symbolic semantic model which can unify data operation and abstract behavior. We adapt the sioco conformance relation to deal with this symbolic model, and then describe a test case generation process, as well as a symbolic execution algorithm based on on-the-fly strategy. The approach has been illustrated on a simple example while the soundness and completeness of the symbolic notions is demonstrated.

**Keywords:** symbolic model, conformance testing, symbolic execution, test case generation.

## 1 Introduction

Reactive systems are systems whose role is to maintain an ongoing interaction with their environment rather than produce a final value on termination. Typical examples of reactive systems are traffic control system; programs controlling mechanical and electronic devices in a train or a robot; and system controlling ongoing processes in a nuclear reactor [1]. Behavior of reactive systems is usually very complex. How to ensure the correctness and reliability of the reactive systems, has given rise to considerable attention in the literature. More recently, formal verification and conformance testing are two well established techniques for validating reactive systems. Both approaches consist in comparing the conformance between two representations of the system: Formal verification compares a formal specification of the system with respect to some higher-level requirements that the system should satisfy, while Conformance testing compares the observable behavior of a black-box implementation of the system with that described by its formal specification, according to a conformance relation [2].

Since testing is one of the most important techniques to validate the quality of software systems, and if used in an effective way, it can provide important evidences of product reliability. In this paper, we adopt the latter approach and propose a formal

J. Lei et al. (Eds.): AICI 2012, LNAI 7530, pp. 612–625, 2012.

framework for the conformance testing to reactive systems. The high-level specification is first translated into a formal model, namely an Extended Input-Output Symbolic Transition System (EIOSTS). In the second step, a symbolic implementation relation sioco, with an explicit notion of data and behavior, is well defined, which avoids the state-space explosion during test generation. After that, symbolic test cases can be derived directly from EIOSTS with a given test purpose. Finally, those test cases are executed on a System Under Test (SUT) to rule out errors in the implementation, for which it used to verdict whether or not the SUT conformance to its specification.

### 1.1 Comparison to Related Work

Since formal testing has become a topic of interest to the validation of software systems, it has arisen a few of interesting works in conformance testing for reactive system from researchers [3], [10], [11], [12], some of them have even started to penetrate the industry [13], [14], [18]. Several other works with respect to conformance testing are related to our work.

The approach described in [4], [5] presentes a model based testing theory for Labeled Transition Systems (LTS). As a model for specifications, implementations and test cases, LTS were represented to the complex transition systems. The testing mechanism defines the implementation relation ioco between a specification and an implementation, additionally uses a test generation algorithm, which can exactly test for ioco conformance.

A selective testing method for reactive systems was proposed in [6]. The SUT is modeled by the UML statecharts and the functional property is specified by the temporal logic. The test sequences are derived from the model according to the given property. Such method can also be applied to real-time systems [15].

The approach described in [7] uses symbolic test selection techniques to extract test cases from a specification, which under some sufficient conditions, can be used to perform a compositional validation between the implementation and specification. However, the test selection algorithm is not related to the specification in a formal way.

Wilkerson and Patricia in [8], [16] propose an approach to modeling and testing from STS or IOLTS models of reactive system with interruptions, which make it possible the automatic test generation through direct manipulation of high-level specification.

In [9] the authors propose an approach to test reactive systems specified as IOSTS, and use symbolic execution techniques to extract IOSTS behaviors to be tested in the role of test purposes and to go ground an algorithm of test case generation.

Compared to our approach, these approaches are limited to deterministic finite-state systems, and without considering the relativity between data and behavior. In this work, we propose a unified symbolic semantic model, called EIOSTS, which can unify data operation and abstract behavior, also provides us with an appropriate level of abstraction used to avoid the classical state explosion problem. In order to capture conformance of the SUT to its specification, we define a formal relation sioco, and propose an algorithm to generate test cases on-the-fly.

### 1.2 Paper Organization

In Section 2, we present the model of EIOSTS, and some of its symbolic operations, which used for conformance testing, are also given. In Section 3, we define the symbolic implementation relation sioco with an explicit notion of data and behavior for generating symbolic test cases from EIOSTS. In Section 4, we present the conformance testing with EIOSTS, which describes a test case generation process, as well as a symbolic execution algorithm based on on-the-fly strategy. In Section 5, we prove the soundness and completeness of the generated test cases by the method. Finally, Section 6 presents conclusion and future work.

## 2 Model: Extended Input-Output Symbolic Transition System

This section presents a new symbolic model, named EIOSTS, which is made of variables, input and output actions may carry communication parameters, guards and assignments. The presence of variables is used as parameters in actions, guards and assignment expressions, so as to improve the ability to specify the behaviors and the data of reactive systems.

### 2.1 Syntax of EIOSTS

**Definition 1 (EIOSTS).** An EIOSTS is a tuple $< Q, V, D, C, q_0, \Sigma, T >$, where:

- $Q$ is a nonempty, countable set of states and $q_0 \in Q$ is the initial state;
- $V$ is a countable set of typed variables;
- $D$ is a countable set of data, and for $x \in V \cup D$, *type(x)* denotes the type of $x$;
- $C$ is a set of communication channel names, which carries a tuple of communication parameters $p= <p_1,...,p_k>$;
- $\Sigma = \Sigma_? \cup \Sigma_! \cup \Sigma_\tau$ is a nonempty, countable alphabet, and $\Sigma_?$ represents a countable set of input actions, $\Sigma_!$ represents a countable set of output actions, $\Sigma_\tau$ represent a countable set of internal actions, respectively. For each actions $a \in \Sigma$ has a signature $sig(a) = < t_1, \cdots, t_k >$, that is a tuple of distinct parameters. The signature of internal actions is the empty tuple;
- $T$ is a set of transitions, where each transition is a tuple $< q, a, G, A, q' >$ consists of: a state $q \in Q$, called the origin of the transition and $q' \in Q$ called the destination of the transition; an action $a \in \Sigma$, called the action of the transition; a predicate $G$ with variables in $V \cup sig(a)$, called the guard; an assignment $A$, such that for each variable $x \in V$, there is exactly one assignment in $A$, of the form $x := A_x$, where $A_x$ is an expression on $V \cup sig(a)$.

A simple example of EIOSTS is depicted in Fig.1, and it will serve as the running example in the context. It represents the specification of cash withdrawal of a coffee machine, with inputs are followed by the "?" symbol and outputs are followed by "!" symbol. These symbols are not part of an action's name, but are used only as notations. The machine starts in state $q_0$, after inserting coins (Coin?(m)) moves to $q_1$ from the environment, and there are two possible transitions: one is the user choose cancel (Cancel?), the machine moves to $q_2$ and then returns the paid amount (Return!(Paid)) and moves back to $q_0$, another is the user chooses the beverage (Select?(Beverage)), the machine moves to $q_3$. There are another three possibilities: if there is no cup (Print!(No cup)), the machine moves to $q_4$ then returns the paid amount and moves back to $q_0$, or there is not enough money, then the machine moves back to $q_0$ and returns the difference between the amount already paid and the price of beverage (Print!(Price-Paid)), or the machine delivers the beverage (Deliver! (Beverage)). Lastly, if the amount inserted is more than the price of the beverage, the machine return back to the start state $q_0$.

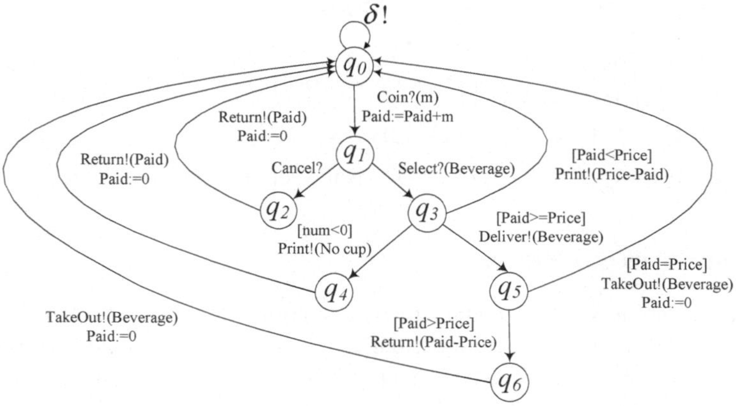


**Fig. 1.** A simple EIOSTS modeling the specification of a Coffee machine

## 2.2 Semantics of EIOSTS

**Definition 2 (IOLIS semantics of an EIOSTS).** The semantics of an EIOSTS $S= < Q,V,D,C,q_0,\Sigma,T >$ is an IOLTS $[[S]] =< S,s_0,\Sigma,T >$, defined as follows:

- $S = Q \times V$ is a finite non-empty set of states and $s_0 \in S$ is an initial state ;
- $\Sigma = \Sigma_I \cup \Sigma_O \cup \Sigma_\tau$ is a finite set of actions, where $\Sigma_I$ is a finite set of input actions, $\Sigma_O$ is a finite set of output actions, and $\Sigma_\tau$ is a finite set of internal actions;
- $T$ is the smallest relation in $S \times \Lambda \times S$ defined by the following rule:

$$\frac{< q,v >,< q',v' >\in S,< a,\pi >\in \Lambda, t =< q,a,G,A,q' >\in T, G(v,\pi) = true, v' = A(v,\pi)}{< q,v > \xrightarrow{<a,\pi>} < q'v' >}$$

The rule says that a transition $< q,a,G,A,q' >$ of an EIOSTS is fireable from the current state $< q,v >$ to another state $< q',v' >$, if there exists a valuation $\pi$ of the communication parameters $p$, carried by the action $a$, satisfies the guard $G$. Then, the assignment $A$ of the transition maps the pair $(v,\pi)$ to $v'$.

Most properties and operations of EIOSTS are defined in terms of their underlying IOLTS semantics. For the sake of convenience, we can use *traces* to denote the behaviors of reactive system. Given $S =< Q, V, D, C, q_0, \Sigma, T >$ be an EIOSTS whose semantics are defined by the IOLTS $[[S]] =< S, s_0, \Sigma, T >$. Let $s, s', s_i \in S$, $a_1, a_2, \cdots, a_n \in \Sigma$, $\varepsilon$ is the empty sequence, $\varepsilon \in \Sigma \cup \{\tau\}, \sigma \in (\Sigma)^*, \Rightarrow$ to denote the observable behavior, defined as follows:

1) $s \xrightarrow{\varepsilon} s' =_{def} s = s'$,
2) $s \xrightarrow{\mu} s' =_{def} (s, \mu, s') \in T$, where $\mu$ is used to denote a traces,
3) $s \xrightarrow[a]{\mu_1 .. \mu_n} s' =_{def} \exists s_0, \cdots s_i : s = s_0 \xrightarrow[\varepsilon]{\mu_1} s_1 \xrightarrow[\varepsilon]{\mu_2} \cdots \xrightarrow{\mu_i} s_i = s'$,
4) $s \Rightarrow s' =_{def} \exists s_1, s_2 : s \Rightarrow s_1 \xrightarrow[\sigma]{a} s_2 \Rightarrow s'$ or $s \xrightarrow{\tau ... \tau} s'$,
5) $s\ after\ \sigma =_{def} \{s' \in S : s \Rightarrow s'\}$,
6) $out(s) =_{def} \{a \in \Sigma^! \cup \{\tau\} \mid \exists s' : s \xrightarrow{a} s'\}$,
7) $trace(s) =_{def} \{\sigma \in (\Sigma \cup \{\tau\})^*_{\sigma} \mid \exists s' : s \xrightarrow{\sigma} s'\}$,
8) $Straces(s) =_{def} \{\sigma \in \Sigma^*_{\delta} \mid s \Rightarrow \delta\}$, is the set of suspension traces.

### 2.3 Operations on EIOSTS

We now define a generic composition operation for EIOSTS, which we then specialize to define a few operations that are needed in conformance testing. These operations are the synchronous product of two EIOSTS, the suspension and the determinization of an EIOSTS.

Given two EIOSTS $S_1 =< Q_1, V_1, D_1, C_1, q_1^0, \Sigma_1, T_1 >$ and $S_2 =< Q_2, V_2, D_2, C_2, q_2^0, \Sigma_2, T_2 >$, we assume that $S_1, S_2$ are compatible if $V_1 \cap V_2 = \varnothing$, $D_1 = D_2$, $\Sigma_{!1} = \Sigma_{!2}$, $\Sigma_{?1} = \Sigma_{?2}$ and $\Sigma_{\tau 1} \cap \Sigma_{\tau 2} = \varnothing$. Note that $S_1$ and $S_2$ may share some parameters, and that a variable of one system may be a parameter of the other.

**Definition 5 (Synchronous Product).** The synchronous product $S{=}S_1 \| S_2$ of two compatible EIOSTS $S_1$, $S_2$ is the EIOSTS $S =< Q, V, D, C, q_0, \Sigma, T >$ defined by: $V = V_1 \cup V_2, D = D_1 = D_2, C = C_1 \wedge C_2, Q = Q_1 \times Q_2, q_0 = q_1^0 \times q_2^0, \Sigma_! = \Sigma_{!1} = \Sigma_{!2}, \Sigma_? = \Sigma_{?1} = \Sigma_{?2}, \Sigma_\tau = \Sigma_{\tau 1} \cup \Sigma_{\tau 2}$. The set $T$ of symbolic transitions of the composed system satisfying the following rules:

1) For $a \in \Sigma_1^\tau$ *and* $q_2 \in Q_2 :< q_1, a, G_1, A_1, q_1' > \in T_1 \Rightarrow << q_1, q_2 >, a, G_1, A_1, < q_1', q_2 >> \in T$;
2) For $a \in \Sigma_2^\tau$ *and* $q_1 \in Q_1 :< q_2, a, G_2, A_2, q_2' > \in T_2 \Rightarrow << q_1, q_2 >, a, G_2, A_2, < q_1, q_2' >> \in T$;
3) For $a \in \Sigma_! \cup \Sigma_? :< q_1, a, G_1, A_1, q_1' > \in T_1 \wedge < q_2, a, G_2, A_2, q_2' > \in T_2 \Rightarrow << q_1, q_2 >, a, G_1 \wedge G_2, A_1 \cup A_2, < q_1', q_2' >> \in T$.

In conformance testing, it is assumed that the environment may observe not only outputs of the implementation, but also absence of outputs. Such absence of output is called quiescence, which contains a self-loop transition [10]. Here, we describe quiescence by adding a special output action $\delta$ that manifests itself in quiescent states.

**Definition 6 (Suspension).** Given an EIOSTS $S= < Q,V,D,C,q_0,\Sigma,T >$, with $\Sigma = \Sigma_? \cup \Sigma_! \cup \Sigma_\tau$ and $\delta \notin \Sigma$, the suspension EIOSTS $S^\delta =< Q^\delta, V^\delta, D^\delta, C^\delta, q_0^\delta, \Sigma^\delta, T^\delta >$, where $Q^\delta = Q, V^\delta = V, D^\delta = D, C^\delta = C, q_0^\delta = q_0, and\ \Sigma_!^\delta = \Sigma_! \cup \{\delta\}, T^\delta = T \cup \{< q, \delta, G_{\delta,q}, (x := x)_{x\in V}, q > \mid q \in Q\}$, also the traces of $S^\delta$ are the suspension traces of *S*.

**Definition 7 (Determinism).** An EIOSTS *S* is deterministic if:

- it has no internal actions: $\Sigma^\tau = \varnothing$,
- it has at most one initial state, and
- for all $q \in Q$ and for each pair of distinct transitions with origin in $q$ and labeled by the same action $a$.

# 3 The Implementation Relation for EIOSTS

## 3.1 The Conformance Relation Sioco

Conformance testing aims at checking a conformance relation between a specification and its implementation, which the former is a formal model of a system, the later is a physical process that can only be controlled and observed at its interfaces. Thus the major issue of conformance testing is to verdict whether an implementation is correct with respect to a specification. This requires a notion of correctness, which is covered by defining an implementation relation.

Our testing theory is close to the conformance testing framework described in [12], and here we propose to adapt a symbolic input-output conformance relation (sioco), as implementation relation.

**Definition 8 (sioco).** Let *F* be a set of subset of suspension traces for an initialized specification EIOSTS $Spec =< Q,V,D,C,q_0,\Sigma,T >$, satisfying $F \in Straces(q_0)$, and given an input-enabled EIOSTS $SUT =< Q,V,D,C,s_0,\Sigma,T >$, *SUT* is symbolic input-output conform to *Spec*, denotes by *SUT sioco Spec*, if and only if:

$$SUT\ sioco\ Spec \Leftrightarrow_{def} \forall \sigma \in F : out\ (s_0\ after\ \sigma) \subseteq out\ (q_0\ after\ \sigma)$$

Intuitively, this relation assumes that a specification *Spec* and a corresponding *SUT* can also be modeled by an EIOSTS, and *SUT sioco Spec* means that one cannot differentiate *SUT* form *Spec* as long as one stimulates *SUT* with sequences of inputs/outputs specified in *Spec*. This is done by running test cases on the *SUT* and obtaining verdicts with respect to the conformance relation.

### 3.2 Testing for Sioco

A test case is used to check the conformance between the implementation and its specification. In the context, a test case is an EIOSTS implementation a strategy for satisfying a give test purpose (typically, staying within a set of traces). The test case takes into account output choices of the specification and expects incorrect outputs of the implementation.

**Definition 9 (Test case).** A test case for a specification *Spec* is a deterministic, input-complete EIOSTS $TC =< Q^{TC}, V^{TC}, D^{TC}, C^{TC}, q_0^{TC}, \Sigma^{TC}, T^{TC} >$, where $\Sigma_?^{TC} = \Sigma_!^S$ and $\Sigma_!^{TC} = \Sigma_?^S$, equipped with three disjoint subsets of states *Pass, Fail, Inconclusive*.

Intuitively, when the state *Fail* is reached, it means rejection, the state *Pass* means that some targeted behaviors have been reached, and *Inconclusive* means that targeted behaviors cannot be reached anymore.

The specification *Spec* contains all the information relevant to conformance, and its mirror followed by input-completion constitutes a test case by itself. However, such a test case is typically too large and is not focused on any part of the system. It is more interesting in practice to test what happens in the course of a given scenario, and if no error has been detected, to end the test successfully when the scenario is completed. This is exactly the reason for introducing test purpose.

A test purpose describes some expected behaviors that we intend to check on the implementation during the test generation, for which they are used to select test cases in order to check specific scenarios. In the context, a test purpose is a specific EIOSTS TP formally defined as follows:

**Definition 10 (Test Purposes).** Let $S =< Q, V, D, C, q_0, \Sigma, T >$ be a specification EIOSTS, a test purpose of *Spec* is a deterministic, complete EIOSTS $TP =< Q^{TP}, V^{TP}, D^{TP}, C^{TP}, q_0^{TP}, \Sigma^{TP}, T^{TP} >$ together with a subset $Accept \in Q^{TP}$ of accepting state, which implies that the states sequence from the initial state to the state marked *Accept*.

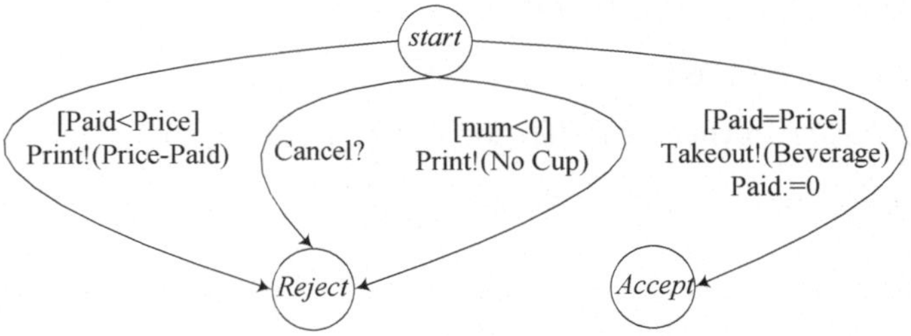


**Fig. 2.** Test Purpose

Fig.2 provides one possible test purpose from the Fig.1, which specifies the behavior that the machine delivers coffee and the user dose not introduce coins more than once and does not cancel. Note that an accepted behavior is indicated by arrival at state Accept. The test purpose rejects behaviors that correspond to pressing the cancel choice, or purchasing the beverages less than the price. However, the rejected behaviors are not necessarily erroneous, they are just behaviors that are not targeted by the test purpose.

# 4 Conformance Testing with EIOSTS

Currently the conformance testing activity consists of: generating test cases from the specification, and executing them on the *SUT* [20]. In this context, we assume that the conformance testing for reactive system based on *sioco* relation, and use the strongly-connected graph to generate test cases. By virtue of the data transaction will direct effect the test selection and execution, we introduce on-the-fly testing approach, including symbolic test generation and test execution. Our test method combines and thus extends, two complementary test method presented in [11] and [12], respectively. It includes four steps outlined in Fig.3 and described in subsections step 1 to step 4.

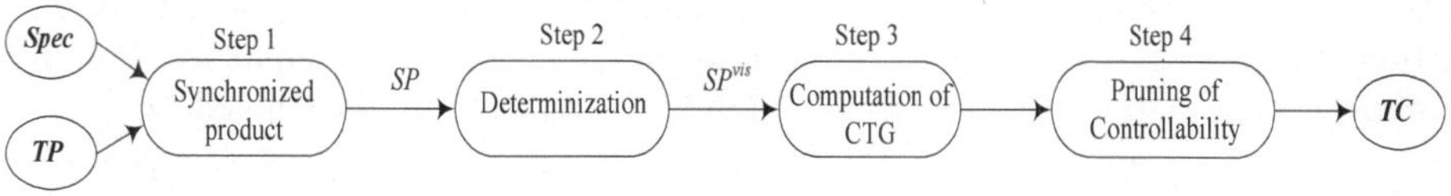


**Fig. 3.** Steps of the test method

## 4.1 Test Generation

In the following, we present a procedure for generation symbolic test case. The first operation is to compute a synchronous product between Spec and TP, and then we will extract the visible behavior from the computed result, after that, we will describe the computation of an EIOSTS complete test graph (CTG). Finally, it produces one test case TC by pruning controllability conflicts.

**Step 1: Synchronous Product**

In this step, it takes a specification Spec and a test purpose TP as inputs. According to the definition 5, we assume the synchronous products *Spec*||*TP* is an EIOSTS $SP =< Q^{SP}, V^{SP}, D^{SP}, C^{SP}, q_0^{SP}, \Sigma^{SP}, T^{SP} >$, equipped with two disjoint sets of states *Accept* and *Refuse*. If the rules satisfied, $SP = Spec \otimes TP$, such that: $Q^{SP} = Q^{Spec} \cup Q^{TP}, V^{SP} = V^{Spec} \cup V^{TP}, D^{SP} = D^{Spec} \wedge D^{TP}, q_0^{SP} = q_0^{Spec} \times q_0^{TP}, \Sigma_!^{SP} = \Sigma_!^{Spec} = \Sigma_!^{TP}$, $\Sigma_?^{SP} = \Sigma_?^{Spec} = \Sigma_?^{TP}, \Sigma_\tau^{SP} = \Sigma_\tau^{Spec} = \Sigma_\tau^{TP}$, and the set of transitions $T$ is defined as follows: $(q^{Spec}, q^{TP}) \xrightarrow{a} (q'^{Spec}, q'^{TP}) \Leftrightarrow q^{Spec} \xrightarrow{a} q'^{Spec} \wedge q^{TP} \xrightarrow{a} q'^{TP}$, $Accept^{SP} = Q^{SP} \cap (Q^{Spec} \times Accept^{TP})$ and $Refuse^{SP} = Q^{SP} \cap (Q^{Spec} \times Refuse^{TP})$.

Here, we assume that *Spec* is input-complete and *TP* is complete, which support that *SP* is input-complete. The effect of the synchronous product is to determine in Spec all the executions that correspond to state Accept and Refuse, respectively. *SP* is built by the next operation but could be built by any traversal. The result of synchronous product of two EIOSTS *Spec* (in Fig.1) and *TP* (in Fig.3) is shown in Fig.4.

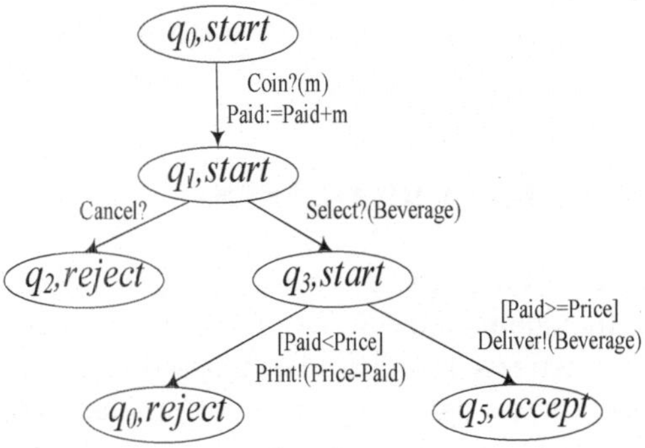


**Fig. 4.** Synchronous Product of Spec and TP

**Step 2: Extracting the Visible Behavior**

The second operation is constructing the visible behavior from *SP*, which internal actions should be eliminated by projection into the observable alphabet. For this purpose, we can use a procedure proposed in [11], and the result is denoted $SP^{vis}$.

According the definitions 7, we assume the EIOSTS $SP^{vis} =< Q^{vis}, V^{vis}, D^{vis}, C^{vis}, \Sigma^{vis}, T^{vis} >$, such that $SP^{vis}=det(\delta(SP))$. $SP^{vis}$ is equipped with *Accept* and *Refuse* states: $Refuse^{vis}=\{P\in Q^{vis} \mid P\cap Refuse^{SP}\neq\varnothing\}$, which means that we ignore every execution if it can correspond to a sequence not to be tested, and $Accept^{vis}=\{P\in Q^{vis} \mid P\cap Accept^{SP}\neq\varnothing\}\setminus Refuse^{vis}$, note that it means we accept an execution only when it cannot correspond to a sequence not to be tested and can correspond to a sequence to be tested.

**Step 3: Computing a Complete Test Graph**

After the results of the $SP^{vis}$ which denotes all visible behaviors of *Spec*, we will extract a test case by selection of accepted behaviors. This operation is a bit more complex as, to compute a test case. We must perform a mirror image, complete it for inputs in all states where an input is possible, ensure controllability, and define verdicts by set *PASS, INCONCLUSIVE* and *FAIL.*

We construct a Complete Test Graph (CTG) proposed in [17] instead of dealing with controllability. CTG is not an interesting EIOSTS as it contains all test cases corresponding to the test purpose, but also easier to explain separately how controllability conflicts are solved. We describe the computation as followings:

1) Let L2A $=\{q\in Q^{vis} \mid \exists\sigma\in A^{vis}, v \xrightarrow{\sigma} Accept^{vis}\}$ be the set of states of $SP^{vis}$ from which an Accept is reachable.
2) Let PASS= $Accept^{vis}$ represents the set of states Accept of $SP^{vis}$.
3) Let Fail = {fail}, where fail$\notin Q^{vis}$, consists of a new state that is reached by every non-specified output transition of $SP^{vis}$ executable from L2A.
4) Let INCONC be the set of states of $SP^{vis}$ that are not in L2A$\cup$PASS and are accessible from L2A by a single output transition of $SP^{vis}$.

5) CTG can be obtain from $SP^{vis}$ by:
   - adding state FAIL and its incoming transitions,
   - removing every state $\notin$ L2A $\cup$ PASS $\cup$ INCONC $\cup$ FAIL, and
   - removing outgoing transitions of every state $\in$ PASS $\cup$ INCONC.

Let $SP_A$ represents the part of SP that leads to a state Accept (L2A), and $SP_A^{vis}$ represents the part of $SP^{vis}$ that leads to a state Accept. A verdict Pass, Inconclusive or Fail, generated after the execution by SUT of a trace $\sigma$ , is interpreted as follows:

Pass means that $\sigma$ conforms to $SP_A^{vis}$.
Fail means that $\sigma$ does not conforms to $SP_A^{vis}$.
Inconclusive means that we cannot determine whether $\sigma$ conforms to $SP_A^{vis}$ or not.

**Step 4: Pruning of controllability**
The graph CTG contains all the behaviors a test case might have, except for controllability [19]. According to the test case definition (Definition 9), now we have to deal with controllability conflicts. In fact some conflicts can be solved during the computing of the CTG by forward DFS [21]. Since some states $q$ of CTG may have an option between inputs and outputs or between outputs. Solving these conflicts consists in extracting a controllable sub-graph of CTG while preserving the required behaviors.

In a conflict state, some transitions must be pruned, that is one output is kept, while all other outputs and inputs are pruned, or all inputs are kept, while all outputs are pruned. We know that a controllability conflict in a state lead to an Accept or Pass state is preserved by backtracking a transition t between a source state, such that its L2A is still false and a target state such that its L2A is true. In this context, we adopt a breadth first traversal to remove all the transitions from the source state in conflict with *t*, because of it is able to select short-cut transition from Pass.

Fig.5 shows an test case that automatically generated by the procedure defined in this section, which covers all the behaviors of the specification come from Fig.1, that are targeted by the test purpose of Fig.2.

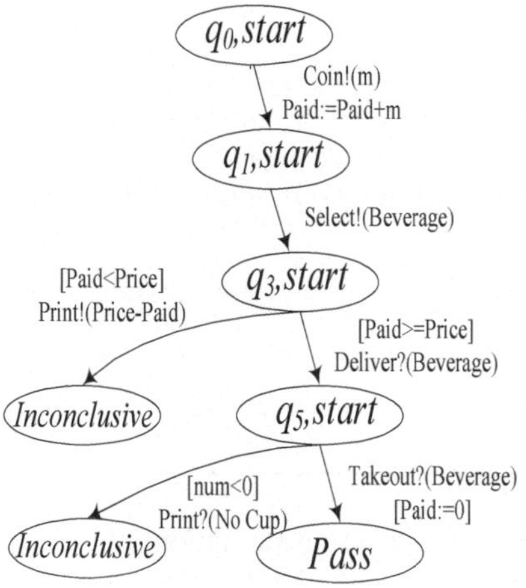


**Fig. 5.** Test Case

### 4.2 Test Execution

In most test generation techniques based on symbolic model, generated test cases are abstract. To obtain an executable test, the symbolic test case should be completely instantiation in the form of sequences (in the deterministic case), trees or graphs where actions carry some valued parameters. In this paper, test cases obtained after symbolic test generation is translated into a concrete test program capable of interacting with an implementation interface-compatible with the original specification. The test program is then ready to be complied and linked with the implementation for test execution. The results of a test execution are: "***Pass***", which means no errors were detected and the test purpose was satisfied, "Inconclusive" means no errors were detected but the test purpose was not satisfied, "***Fail***" means an error was detected and a non-conformance between SUT and its specification.

**Algorithm 1.** on-the-fly testing

**Input:** symbolic test case $TC_s$ (generated by the process of section 4.1)
**Output:** executable test case $TC_d$

```
begin:
T = empty; t = emptyTree; Current_P = t;
Add (Current_P, Adj_Set (Current_P), T); // Adj_Set(Current_P)={(a,q)|(p,a,q)∉ T}
while not empty T do
  Delete ((Source_P, a, Target_P), T)
  if (Target_P == (Accept||Refuse)) then
    switch (a) do
      case ? :
         DataSelect (a, MSG);
         Send (MSG);
         if (Target_P == Refuse) then
           AddVerdict ((Current_P, a, Inconclusive),t);
      case ! :
         Receive (MSG);
         if (Target_P == Accept) then
           AddVerdict ((Current_P, a, Pass), t);
         if (Target_P == Refuse) then
           AddVerdict ((Current_P, a, Fail), t);
      otherwise:
         return;
  end
  Delete (T);
  Add (Target_P, Adj_Set (Target_P), T);
end
```

According to the symbolic data instantiation, the executable test case finally obtained from the algorithm 1, with the symbolic test case $TC_s$ as input and the instantiated test case $TC_d$ after data selection as output, and where *Adj_Set(P)* denotes the set of states and edges which connect with state *p*. Data variables instantiated mainly through the function *DataSelect()*.

There are two strategies in the process of data instantiated, one method is first instantiated all the possible values by using data algebraic and then select the data values. Another is realizing the data instantiated by some certain techniques in the process of test case instantiated, in order to avoid the later procedure of test selection. In this paper, we adopt the second one because of the environment, such as *Receive (MSG), Send(MSG), etc,* are needed to sustain the dynamic test generation and execution. As for the limitation of literature, we do not make a detail introduction for this.

# 5 Properties of the Test Cases

This section consists in discussing properties of the test cases obtained from the procedures presented in section 4.1. However, the generated EIOSTS TC has all the properties required from a test case. It is sound, and conclusive.

In this context, we formalize qualitative properties of TC relative to their ability to satisfy the given TP during test execution. Considering the defined model of test case execution, each trace $\sigma \in Traces(TC \parallel SUT)$ is associated with one of the following scenarios: (1) If all steps of TC are executed and all expected outputs are observed, then the result is Pass, i.e. $Verdict(\sigma) = Pass \cong s_0\ after\ \sigma \subseteq Pass$ ; (2) If, at any moment, any unspecified output is emitted by the *SUT*, the execution is stopped and the result is Fail, i.e. $Verdict(\sigma) = Fail \cong s_0\ after\ \sigma \subseteq Fail$ ; (3) If the SUT, at any moment, blocks to emit an output, or if the output of the SUT are specified by the specification, but the behavior specified by a TP is not exhibited, the result is Inconclusive, i.e. $Verdict(\sigma) = Inconclusive \cong s_0\ after\ \sigma \subseteq Inconclusive$ .

Then, under those uniformity hypotheses, the rejection of *SUT* by a test case *TC* is formally defined as follows:

**Definition 5.1 (may reject):** TC may reject SUT $\cong \exists \sigma \in Traces(TC \parallel SUT):$ $verdict(\sigma) = Fail$

The next definition formally relates the previously defined sioco (Definition 8) to the verdicts considering some properties of test cases.

**Definition 5.2 (Soundness and Conclusiveness):** A test case *TC* is sound for a specification *Spec* and if $\forall$ SUT, *SUT* sioco *Spec* $\Rightarrow \neg$(TC may reject SUT). A test cases is exhaustive of *Spec* and if $\forall$ SUT, $\neg$(*SUT* sioco *Spec*) $\Rightarrow \exists$ TC: *TC* may reject *SUT*. Finally, a test case is complete if it is both sound and exhaustive.

Intuitively, soundness means that the TC rejects only non-conformance implementations. Conclusiveness means that the TC can detect all implementations, which are non-conformance with the specification relative to the TP. Furthermore, a test case that can identify all conforming and non-conformance implementations is called completeness. A complete test case is a very strong requirement for practical testing, that sound test cases are more commonly accepted in practice. In this context, the test cases generated by our approach have the properties stated in Theorem 5.1

**Theorem 5.1:** For any specification *Spec*, and test purpose *TP*, the test cases *TC* derived from *Spec* and *TP* by test generation are sound. Moreover, test generation can be considered as being Conclusive in the following sense: for each non-conformant implementation, one can design a *TP* such that a *TC* generated from this *TP* may reject the *SUT*.

The main ideas of the theorem may discuss as following: for soundness, we need to prove that if a *TC* may reject a *SUT*, then $\neg$*(SUT sioco Spec)*. In this case, we only need to prove that a Fail verdict of a *TC* only occur if the *SUT* emits an unexpected output. In our approach, test cases are generated based on symbolic execution of specifications. This approach allows us to identify all possible traces of a specification. Thus, the unique case where a Fail verdict is obtained during test execution is exactly when the *SUT* emits an unexpected output. For Conclusiveness, we need to prove that for any non-conforming *SUT*, there exist a *TP* and a way of generating a *TC* from *Spec* and a given *TP*, such that *TC* may reject *SUT*. Given that $\neg$*(SUT sioco Spec)*, then there is a trace $\sigma$ of *Spec* such that an output of *SUT* after $\sigma$ is not allowed by *Spec*. Furthermore, the trace $\sigma$ can be used to define a *TP*, and this *TP* can be used to generate test cases where the *SUT* may be rejected.

## 6 Conclusion and Future Work

This contribution has presented a formal approach for testing reactive system based on symbolic execution. In this paper, we present some basic steps towards of the generation of symbolic test cases in the form of extended input-output symbolic transition system with parameters and variables. The approach generalizes existing work on test generation using symbolic execution from abstract model. The test cases generated by our method satisfy some complete properties.

Since the test cases behaviors depend on their parameters and message, to be useful in practice, an effective strategy for data selection should be taken into account in the process of data instantiated. This is the first direction for continuing work. Since the quality of a test suite is directly related to a measure of system performance, the second future work direction is how to make the automatically generated test suite can be expressed, measured, and controlled. In fact, the approaches described in this paper can be used in techniques of other domains. So, an alternative work direction is going to be extended to support time constraints for extending its application to real-time systems.

## References

1. Manna, Z., Pnueli, A.: Temporal verification of reactive systems: safety. Springer-Verlag, New York, Inc. (1995)
2. Constant, C., Thierry, J., Marchand, H.: Integration Formal Verification and Conformance Testing for Reactive System. IEEE Transactions on Software Engineering 33(8), 558–574 (2007)

3. Frantzen, L., Tretmans, J., Willemse, T.A.C.: Test Generation Based on Symbolic Specifications. In: Grabowski, J., Nielsen, B. (eds.) FATES 2004. LNCS, vol. 3395, pp. 1–15. Springer, Heidelberg (2005)
4. Frantzen, L., Tretmans, J., Willemse, T.A.C.: A Symbolic Framework for Model-Based Testing. In: Havelund, K., Núñez, M., Roşu, G., Wolff, B. (eds.) FATES 2006 and RV 2006. LNCS, vol. 4262, pp. 40–54. Springer, Heidelberg (2006)
5. Tretmans, J.: Model Based Testing with Labelled Transition Systems. In: Hierons, R.M., Bowen, J.P., Harman, M. (eds.) FORTEST. LNCS, vol. 4949, pp. 1–38. Springer, Heidelberg (2008)
6. Shuhao, L., Ji, W., Wei, D.: A Framework of Property-Oriented Testing of Reactive Systems. Acta Electronica Sinica 32, 222–225 (2004)
7. Rusu, V.: Combining Formal Verification and Conformance Testing for Validation Reactive System. Software Testing, Verification and Reliability 13(3), 157–180 (2003)
8. Wilkerson, L., Patricia, D.L.: Modeling and Testing Interruptions in Reactive System Using Symbolic Models. In: Proc. of the 2nd Brazilian Workshop on Systematic and Automated Software Testing, pp. 34–43 (2009)
9. Gaston, C., Le Gall, P., Rapin, N., Touil, A.: Symbolic Execution Techniques for Test Purpose Definition. In: Uyar, M.Ü., Duale, A.Y., Fecko, M.A. (eds.) TestCom 2006. LNCS, vol. 3964, pp. 1–18. Springer, Heidelberg (2006)
10. Tretmans, J.: Testing Concurrent Systems: A Formal Approach. In: Baeten, J.C.M., Mauw, S. (eds.) CONCUR 1999. LNCS, vol. 1664, pp. 46–65. Springer, Heidelberg (1999)
11. Claude, J., Thierry, J.: TGV : theory, principles and algorithms: A Tool for the Automatic Synthesis of Conformance Test Cases for Non-Deterministic Reactive Systems. Software Tools for Technology Transfer 7(4), 297–315 (2002)
12. Clarke, D., Jéron, T., Rusu, V., Zinovieva, E.: STG: A Symbolic Test Generation Tool. In: Katoen, J.-P., Stevens, P. (eds.) TACAS 2002. LNCS, vol. 2280, pp. 470–475. Springer, Heidelberg (2002)
13. Koopman, P., Plasmeijer, R.: Testing Reactive System with GAST. In: Gilmore, S. (ed.) Trends in Functional Programming. Trends in Functional Programming, vol. 4, pp. 111–129 (2003)
14. Cartaxo, E.G., Andrade, W.L., Neto, F.G.O.: LTSBT: A tool to generate and select functional test cases for embedded systems. In: Proceedings of the 2008 ACM Symposium on Applied Computing, vol. 2, pp. 1540–1544 (2008)
15. Li, S., Ji, W., Wei, D.: Property-oriented testing of real-time system. In: Proc. of the 11th Asia-Pacific Software Engineering Conference, pp. 358–365. IEEE Computer Society (2004)
16. Andrade, W.L., Machado, P.D.L.: Interruption Testing of Reactive Systems. In: Oliveira, M.V.M., Woodcock, J. (eds.) SBMF 2009. LNCS, vol. 5902, pp. 37–53. Springer, Heidelberg (2009)
17. Ahmed, K.: Complete Test Graph Synthesis for Symbolic Real-time Systems. Electronic Notes in Theoretical Computer Science, 79–100 (2003)
18. Tretmans, J., Brinksma, E.: TroX: automated model-based testing. In: Proc. First European Conference on Model-Driven Software Engineering, pp. 31–43 (2003)
19. Simao, A., Petrenko, A.: Generating asynchronous test cases from test purposes. Information and Software Technology, 1252–1262 (2011)
20. Thierry, J.: Symbolic Model-based Test Selection. Electronic Notes in Theoretical Computer Science, 167–184 (2009)
21. Jéron, T., Morel, P.: Test Generation Derived from Model-Checking. In: Halbwachs, N., Peled, D.A. (eds.) CAV 1999. LNCS, vol. 1633, pp. 108–121. Springer, Heidelberg (1999)

# Study on the Medium and Long Term of Fishery Forecasting Based on Neural Network

Hongchun Yuan[1,*], Yiting Gu[1], Jintao Wang[1], Ying Chen[2], and Xinjun Chen[3]

[1] College of Information Technology, Shanghai Ocean University,
999 Hucheng Huan Road, Shanghai 201306, China
[2] School of Computing and Information Systems,
University of Tasmania, Hobart, Tasmania, Australia
[3] College of Ocean Science, Shanghai Ocean University,
999 Hucheng Huan Road, Shanghai 201306, China
hcyuan@shou.edu.cn, thct@live.cn, wangjintao0510@163.com,
ying.chen@utas.edu.au, xjchen@shou.edu.cn

**Abstract.** The forecasting system for medium to long term fishery resources is based on historical production data of specified fish types and those marine environmental factors. As these systems give a macro level prediction of fishery resources in the coming years they provide indispensable references for the planning and management of catching seasons. This paper introduces a new model for the prediction using Windows XP platform and Visual Studio 2010 development environment with C# programming language. Combining correlation analysis and BP neural network, the new model analyzes marine environmental data and fishery historical production data to forecast fisheries in medium to long terms. Experiments applying this model to forecast the squid production in the Pacific Northwest result in an average relative error of about 13.5% as compared with 23.2% error using linear regression analysis. This result proves that the new model has the potential to provide better forecasts for fisheries.

**Keywords:** Pelagic Fisheries, Correlation Analysis, BP Neural Network, Medium to Long Term Forecasting.

## 1 Introduction

Fishery forecasts play a significant role in pelagic fisheries. Environmental factors in the form of marine environmental data must be considered to improve forecasting accuracy in the hope of providing better reference to fishery departments in their deployment and management in catching seasons[1][2].

Oceanic squid species have short life cycles, usually less than one year, spawn only once a lifetime and die after spawning. So the resource amount of each generation depends entirely on the replenishment of the previous generation and survival rate of replenishment before fishery activities. This life cycle is completely different from fish with a long life cycle. If the existing traditional model is adopted to forecast the

J. Lei et al. (Eds.): AICI 2012, LNAI 7530, pp. 626–633, 2012.

replenishment of oceanic squid, and the relationship between its parent and replenishment, the accuracy can be greatly reduced. There are two reasons for this: (1) the changing environment has a great influence on all stages of their lives (hatching, larvae, adult fish and spawning); (2) the squid has no residual groups but contemporary supplementary groups. Consequently the oceanic squid species are highly sensitive to the changes of marine environment, which calls for research to discover the impact of marine environment changes on their distribution and resource replenishment. This is important to the fishery industry when it comes to assess and manage fishery resources. Nowadays both domestic and international researches are conducting experiments with environmental data in relation to Argentina's Illex argentinus fish. These data are generated with marine remote sensing devices that are new development products combining information and space technology. They use regression methods [4-8] and habit index model to forecast fishery resources. However, the authors of this paper have identified issues with this research approach: (1) most of these methods use linear regression and are unable to deal with the dynamic relationship between marine environmental factors and squid resources; (2) the methods need heavy and complicated calculations because of the need for time consuming manual interventions. So there is an urgent need to create an intelligent information processing technique in data analysis to solve the problems of linear regression and to achieve a higher level of automation.

In this study, SST (sea surface temperature) stands for the key environmental factors of squid resources and correlation analysis will be applied to find significant geographical locations. Then SST and CPUE (catch per unit effort) of those locations will be the input data of BP (Back Propagation) neural network to get the final forecasting model.

## 2 Theory and Methods

### 2.1 Data Sources

In this paper, the environmental data is taken from the marine environment database of U.S. National Oceanic and Atmospheric Administration (NOAA). From this database, the type of the satellite, the intervals and the start time of environmental data can be obtained. Historical fisheries production data is supplied by the squid fishing group of Shanghai Ocean University, including the daily catch, the number of operating days, the number of working ships daily and the operating areas.

### 2.2 Correlation Analysis

The correlation analysis is to analyze the correlation between two or more variables on measuring how closely the variables are related.

Based on the production data, the CPUE will be first preprocessed for the annual catches (t/y). Then calculating the correlation between sea surface temperatures of $1°*1°$ each grid point and the CPUE of the following year to determine the distribution of fishing grounds. Finally, analysis of variance will be performed to find the statistically significant related grid points.

### 2.2.1 Calculating Correlation Coefficient

Let SST and CPUE be represented by two random variables X and Y, then the overall correlation coefficient is:

$$\rho = \frac{\mathrm{cov}(X,Y)}{\sqrt{\mathrm{var}(X)}\sqrt{\mathrm{var}(Y)}} \tag{1}$$

Where cov(X, Y) is the covariance between two variables while var(X) and var(Y) is the variance of the variables X and Y.

As the overall correlation coefficients is generally unknown, the coefficient of a sample will be estimated to represent the correlation coefficient of SST and CPUE. Let $X = (x_1, x_2, ..., x_n)$ and $Y = (y_1, y_2, ..., y_n)$ be two samples of X and Y, then the correlation coefficient of the samples can be calculated as follows:

$$\gamma = \frac{\sum_{i=1}^{n}(x_i - \bar{x})(y_i - \bar{y})}{\sqrt{\sum_{i=1}^{n}(x_i - \bar{x})^2 (y_i - \bar{y})^2}} \tag{2}$$

Then the sample correlation coefficient and overall correlation coefficient are consistent.

### 2.2.2 Test of Significance

(1) Establishing assumptions

$H_0 : \mu_1 = \mu_2 = \cdots \mu_r$, SST has no significant effect on CPUE

$H_1 : \mu_1 \neq \mu_2 \neq \cdots \neq \mu_r$, SST has a significant impact on CPUE

To determine the significance level, presented by $\alpha$, usually taking 0.05 or 0.1

Constructed to test F-statistics, the formula is as the following table:

**Table 1.** F Analysis of variance table

| Variance sources | Sum of squares of deviations | df |
|---|---|---|
| Variance between the group | SSA | r-1 |
| Variance within the groups | SSE | n-r |
| **Mean square** | **F** | |
| MSA = SSA /(r-1) | MSA/MSE | |
| MSE = SSE /(n-r) | | |

Where SSA represents the sum of squares for factor A, while SSE is the sum of squares for errors, the formula is as followed:

$$SSA = \sum_{i=1}^{r} n_i (\bar{x}_{i\bullet} - \bar{\bar{x}})^2 \tag{3}$$

$$SSE = \sum_{i=1}^{r}\sum_{j=1}^{n_i}(x_{ij} - \bar{x}_i)^2 \tag{4}$$

**(2) Judgments and Conclusions**

To achieve the assumed condition, the F-statistics should satisfy the F-distribution that the first degree of freedom ( $df_1$ ) is r-1 and the second degree of freedom ( $df_2$ ) is n-r. The F-statistics will be compared with the threshold $F_\alpha(r-1, n-r)$ which is defined by the given significant level $\alpha$ . The comparison can be made to determine whether the assumption $H_0$ will be rejected or not.

If F≥ $F_\alpha$ ,then the assumption $H_0$ will be rejected which means SST has a significant impact on CPUE.

If F< $F_\alpha$ , then the assumption $H_0$ will not be rejected which means SST has no significant effect on CPUE.

## 2.3 BP Neural Network Training

In this section, SST and CPUE of those grid points, which are significant by calculating the correlation, will be trained as samples of BP neural network. The network will be trained and verified after setting the parameters, and then the mean square error (MSE) will be obtained.

BP neural network is a unidirectional transmission of multilayer feed-forward network with three or more network structures, including the input layer, middle layer (hidden layer) and output layer. There are full connections between the upper and lower layers of the network while there is no connection between each layer of neurons. The transfer function of BP neural network must be differentiable, usually taking the S-type function:

$$f(x) = \frac{1}{1 + e^{-x/Q}} \tag{5}$$

The learning process is divided into two parts, which are the forward propagation and back propagation. On the forward process, a sample (Xp,Yp) will be taken into the network, then the actual output Op will be calculated by the transfer function. On the back propagation process, a difference will be calculated between the actual output Op and the corresponding output Yp , then the weight matrix will be adjusted to minimize the error until $\sum E_p < \varepsilon$ .

The error on the p-th sample and the error of the entire network are calculated as followed:

$$Ep = \frac{1}{2}\sum_{j=1}^{m}(y_{pj} - o_{pj})^2 \tag{6}$$

$$E = \sum E_p \tag{7}$$

# 3 Experiments and Results

The system is developed on the Windows XP operating system platform, and its development environment is Visual Studio 2010 by using the C # language. This system can run immediately without installation, but some of the features need the support of MapObject2.3. In this system, the environmental data is stored in .txt format and its time scales is month, while the historical production data is CPUE with a unit of year in one fishing ground. The SST and the CPUE of northwest Pacific squid are in the range of 130 ° E - 170 ° E, 20 ° N - 30 ° N in the year of 1996 to 2006 from January to May. These are taken as original data to provide medium to long term forecast for the short lifetime of the fish. Two main modules of this system are described in Sections 3.1 and 3.2.

## 3.1 Significant Grid Points by Correlation Analysis

In step one, after selecting the fishing ground (in this case the northwest Pacific squid fishing grounds), the CPUE of the Edit will display with the corresponding data (step 2). The Correlation coefficient(R) and PFSST, which stands for the most suitable sea surface temperature, will be calculated in step three. Step four will find significant grid points by F-statistics. In the last step, the SST and the CPUE of those significant grid points will be stored as BP neural network training and validation data.

The map in Fig. 1 shows nine significant grid points after the calculation above.

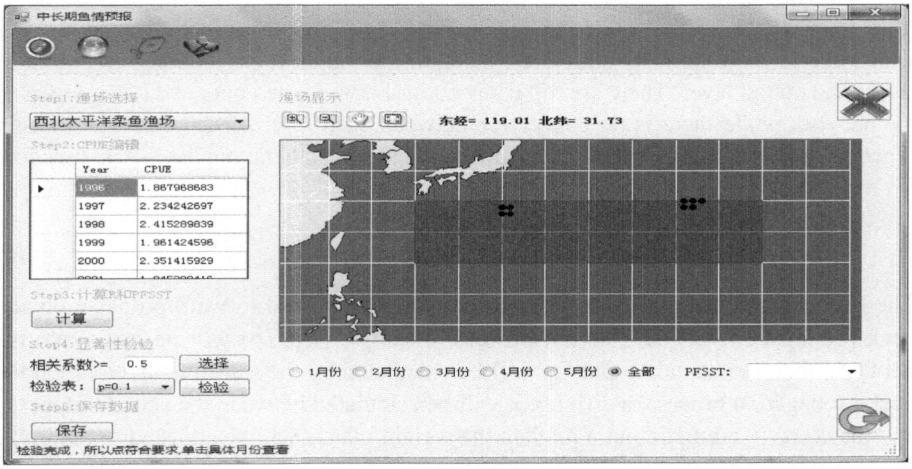


**Fig. 1.** The sample of fisheries correlation analysis

## 3.2 BP Neural Network Prediction Model

Before testing BP neural network, some basic parameters should be set, such as the number of hidden layer nodes, training times, minimum precision, number of training samples and testing samples. The MSE (Mean Square Error) will be generated after training and verifying the network.

Fig. 2 and Fig. 3 show that the MSE is 13.5% according to the data generated by section 3.1.

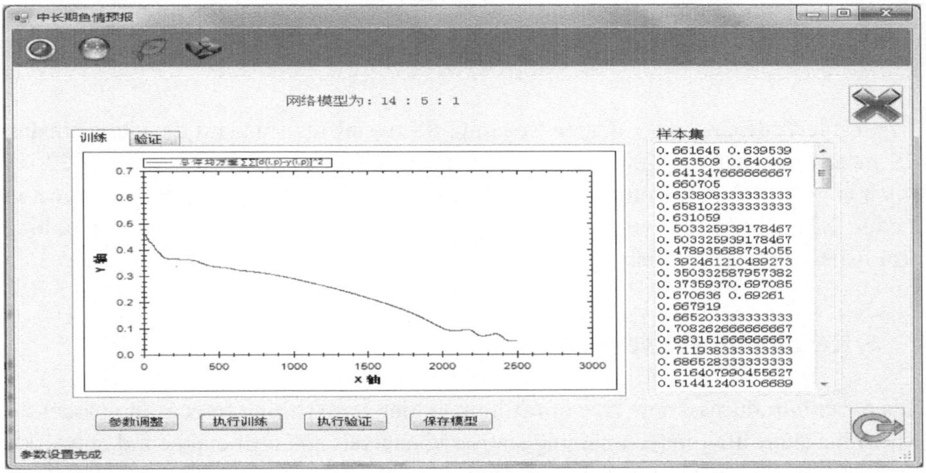


**Fig. 2.** Neural network training sample

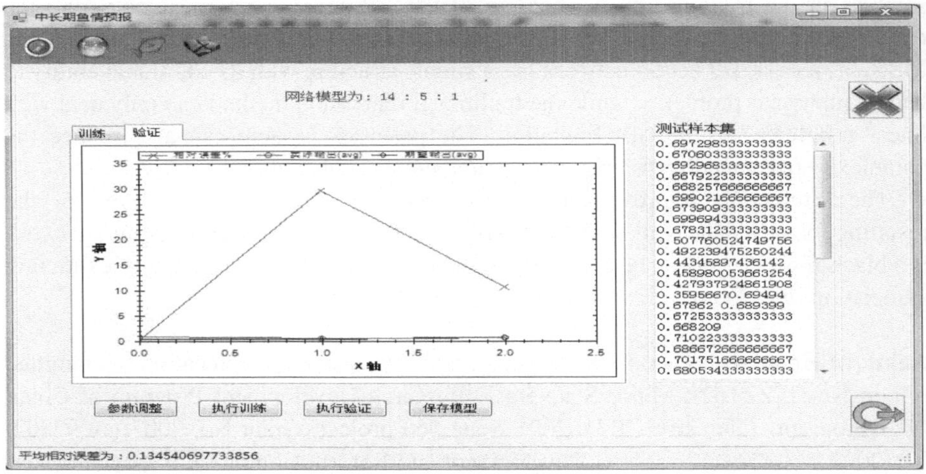


**Fig. 3.** Neural network validation sample

In order to compare the results of this system, the SST and CPUE of nine grid points after correlation analysis are used to calculate the regression analysis with Microsoft Excel, the results are as followed:

**Table 2.** Results of regression analysis

| | df | SS | MS |
|---|---|---|---|
| **Regression analysis** | 14 | 0.232805 | 0.025867223679601 |
| **Residual error** | 0 | 0 | 65535 |
| **Total** | 14 | 0.232805 | |

In Table 2, df means the degree freedom, SS means the standard deviation square, MS means the mean square, and the total error of the linear regression was 23.2% which is greater than the error calculated by our system (which is 13.5% as shown in Section 3.2). Obviously, our system provides a higher accuracy in the medium to long term fishery forecast for fish types with a short life time.

## 4 Results and Discussion

This paper introduces a new type of medium to long term fishing forecasting model for fish with a short life time, combining statistical analysis and artificial neural networks. In this paper, SST is used as the main environmental factors that affect fish resources and correlation analysis is used to find those grid points which also have a significant impact on fish resources. SST and CPUE of those significant grid points are used as samples in the training and verification of the BP neural network. Comparing the results of the new model and the traditional linear regression, the new model has a less MSE. Moreover, BP neural network has a simple structure with its additional ability to handle nonlinear problems, while the traditional regression method can only deal with linear problems, an obvious limitation. The prototype system can also reduce the complexity of manual interventions and cumbersome calculations.

The foreseeable improvement for the model is the comprehensibility of the resulting forecast data to all potential users since the process of training neural network is a black-box operation. The next stage of the research will focus on ways of extracting association rules from the data obtained by neural network.

**Acknowledgments.** This work is supported by Shanghai Education Committee (Grant No. 12ZZ162); Major State Basic Research Development Program of China (973 Program) (No. 2012CB316200); State 863 project (Grant No. 2007AA092202, No. 2007AA092201); National Development and Reform Commission special (Grant No. 2060403).

## References

1. Yang, Y.B., Yuan, H.C., Chen, Y.: A Forecasting Model of WCPO Katsuwonus Pelamis Purse Seine Catch Based on Rough-SVR. In: Conference on Intelligent System and Knowledge Engineering (ISKE 2007), pp. 1341–1346 (2007)

2. Arkhipkin, I.: Intrapopulation Structure of Winter-Spawned Argentine Shortfin Squid, Illex Argentinus (Cephalopoda: Ommastrephidae), During its Feeding Period over the Patagonian Shelf. Fish. Bull. 98, 1–13 (2000)
3. Cao, J., Chen, X.J., Liu, B.L., et al.: Review on the Relationship Between Stock Recruitment of Squid and Oceanographic Environment. Journal of Shanghai Ocean University 19, 232–239 (2010)
4. Feng, S., Tian, Q., Chen, X.J.: The Habitat Suitability Index Of Illex Argentinus By Using Quantile Regression Method In The Southwest Atlantic. Transactions of Oceanology and Limnology 1, 15–22 (2010)
5. Efthymia, V.S., Christos, D.M., John, H.: Modeling and Forecasting Pelagic Fish Production using Univariate and Multivariate ARIMA Models. Fish Sci. 5, 979–988 (2007)
6. Cao, J., Chen, X., Chen, Y.: Influence of Surface Oceanographic Western Winter-Spring Cohort of Neon Flying Squid Omma Pacific Ocean. Mar. Ecol. Prog. Ser. 381, 119–127 (2009)
7. Zheng, X.Q., Li, G., Chen, X.J.: Application of Environmentally Dependent Surplus Production Model For Scomber Japonicus in the East China Sea And Yellow Sea. Transaction of Oceanology and Limnology 3, 41–48 (2010)
8. Feng, B., Chen, X.J., Xu, L.X.: Study on Distribution of Thunnuns Obesus in the Indian Ocean Based on Habitat Suitability index. Journal of Fisheries of China 31, 805–812 (2007)
9. Sakurai, Y., Kiyofuji, H., Saitoh, S., Goto, T., Hiyama, Y.: Changes in Inferred Spawning Sites of Todarodes Pacificus (Cephalopada: Ommastrephidae) Due to Changing Environmental Conditions. ICES J. Mar. Sci. 57, 24–30 (2000)
10. Basson, M., Beddington, J.R., Crombie, J.A., et al.: Assessment and Management Techniques for Migratory Annual Squid Stocks: the Illex Argentinus Fishery in the Southwest Atlantic as an Example. Fish Res. 28, 3–27 (1996)

# Application of Wavelet Neural Network in the Fault Diagnosis of Turbine Generator Unit

Chunmei Xu[1,2], Hao Zhang[1,2], Daogang Peng[2], and Yuliang Qian[1]

[1] School of Electronic and Information, Tongji University, Shanghai 201804 China
[2] School of Electric Power and Automation Engineering,
Shanghai University of Electric Power, Shanghai 200090 China
chunmeixu@yeah.net

**Abstract.** Wavelet neural network(WNN) is a type of feedforward network which is designed by using wavelet function as the activation functions in neural networks. Based on the technique of WNN, a diagnostic method is presented for turbine generator unit. The simulation results show that the proposed method can effectively diagnose the vibration fault of turbine generator, can overcome the random noise disturbance and has good application prospects.

**Keywords:** Turbine, Wavelet Neural Network, Fault Diagnosis.

## 1 Introduction

Turbine generator unit is a very complex system, and plays an important role in electricity production. The running status of turbine generator unit reflects its own safety and stability of operation. If the fault state is not timely collected, follow-up production may produce fault, which can cause significant economic losses in severe, even lead to destroying the machine and causing casualties. The fault type of turbine generator unit has a wide variety and the cause is complex, which increases the difficulty of accurate diagnosis for the fault of turbine generator unit, and therefore looking for a fast and accurate diagnostic method for the safe operation of the turbine generator unit has a very important practical significance.

WNN is a new type of neural network which is constructed based on wavelet analysis theory. It not only makes full use of the localized features of wavelet transform, but also inherits the self-learning of neural networks. Thus, wavelet neural network can overcome the above defects of neural network has a strong approximation and pattern recognition capabilities[1-3].In this paper, The diagnostic method based on WNN is presented for the turbine generator unit. The simulation results show that the proposed method can effectively diagnose all the vibration fault of turbine generator even in the random noise disturbance.

J. Lei et al. (Eds.): AICI 2012, LNAI 7530, pp. 634–641, 2012.

# 2 Wavelet Neural Network

## 2.1 Wavelet Transform

Suppose $\forall f(t) \in L^2(R)$, definition (1) is a continuous wavelet transform of the signal $f(t)$.

$$W_f(a,b) = < f, \psi_{\theta,\gamma} > = \frac{1}{\sqrt{a}} \int_{-\infty}^{+\infty} f(t) \psi^* \left( \frac{t-\gamma}{\theta} \right) dt \tag{1}$$

Where, $\theta > 0$ is scale factor, which determines the changes of $\psi\left(\frac{t-\gamma}{\theta}\right)$ spectral; $\gamma$ is displacement factor, and its value can be positive or negative; the tag "*" is taken conjugate. The wavelet transform is to filter the signal $f(t)$ by using the filter with the same shape, the different bandwidth and frequency. If the scale factor $\theta$ is changed, the time resolution and the frequency resolution of the signal $f(t)$ filtered will be affected, and the analysis results of the signal $f(t)$ around the point $\gamma$ will be affected as the displacement factor $\gamma$ is changing. Wavelet transformation is a time-frequency localization analysis method that the window area is fixed, the time and frequency window can be adjusted. It has high frequency resolution ratio and low time resolution ratio in low frequency part, and high time resolution ratio and low frequency resolution ratio in high frequency. And then the feature allows the wavelet transform with self-adaptive for signal.

## 2.2 The Structure of WNN

The structure of WNN[1,4] is very similar to three-layer BP network. The wavelet function is assigned as activation function for each neuron of hidden layer in conventional BP neural network. The structure of WNN is shown in Fig.1. As is shown in Fig.1, the structure of WNN is made up of input layer, hidden layer and output layer. The number of nodes in the input layer is $p$, hidden layer is $m$ and output layer is $q$.

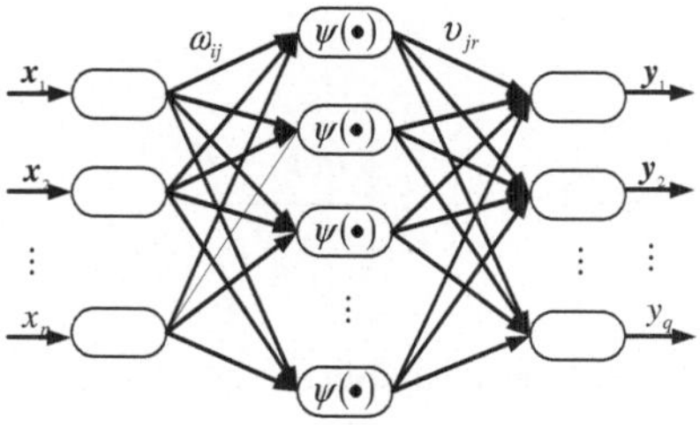


**Fig. 1.** WNN topology structure

### 2.3 The Training Algorithm of WNN

The paper adopts conjugate gradient descent algorithm in WNN. Its basic idea is that square sum of the error, which is the difference between the target and real outputs of the WNN, is used as goal function and the weights of network are automatically adjusted according to the minimization principle of the error square sum. The process of weights adjustment rule is divided into two phases: the first is feed-forward propagation process, which is calculated forward from the input layer to the outputs of each layer of WNN according to the input sample, so the output of the output layer is ultimately obtained; the second is back propagation process, in which the weights is calculated and adjusted backward from the output layer. These two processes are repeated alternately until convergence[1,5-6].

In order to describe the training rule of the WNN better, the network symbol can be set as follows: the input vector of WNN is $X_k = \lfloor x_1^k, x_2^k, \cdots, t_p^k \rfloor$; the input vector of hidden layer is $S_k = \lfloor s_1^k, s_2^k, \cdots, s_m^k \rfloor$ and the output vector is $B_k = \lfloor b_1^k, b_2^k, \cdots, b_m^k \rfloor$; the actual output vector of output layer is $Y_k = \lfloor y_1^k, y_2^k, \cdots, y_q^k \rfloor$ and the target output vector is $T_k = \lfloor t_1^k, t_2^k, \cdots, t_q^k \rfloor$; where, $p$、 $m$ and $q$ are the number of neurons in the input layer, the hidden layer and output layer in WNN respectively, $k = 1,2,\cdots N$, and $N$ in $k = 1,2,\cdots N$ is the number of WNN fault samples. The connective weight from the input layer to the hidden layer is $\{\omega_{ij}\}$, where, $\omega_{ij}$ is the connective weight value from NO. $i$ node in the input layer to NO. $j$ node in the hidden layer, and the connective weights from the hidden layer to the output layer is $\{\upsilon_{jr}\}$, where $\upsilon_{jr}$ is the connective weight value from NO. $j$ node in the hidden layer to NO. $r$ node in the output layer, where $i = 1,2,\cdots,p$ ; $j = 1,2,\cdots,m$ ; $r = 1,2,\cdots,q$ .

The input of each neuron in the hidden layer is shown as (2). The output of each neuron in the hidden layer is shown as (3).

$$net_j^k = \sum_{i=1}^{p} \omega_{ij} x_i^k \tag{2}$$

$$b_j^k = \psi\left(net_j^k\right) = \psi\left(\frac{net_j^k - \theta_j}{\gamma_j}\right) \tag{3}$$

Where, $\theta_j$ is scale factor and $\gamma_j$ is displacement factor of NO. $j$ node in hidden layer; $\psi(\bullet)$ is Mexican Hat wavelet function, shown as (4).

$$\psi(x)=\frac{2}{\sqrt{3}}\pi^{-0.25}\left(1-x^2\right)e^{-\frac{x^2}{2}} \tag{4}$$

The output of each neuron in the output layer is shown as (5)

$$y_r^k=\sum_{r=1}^{q}\upsilon_{jr}b_j^k \tag{5}$$

The trained algorithm of WNN is the same with BP. During the training phase wavelet parameter $\theta_j$, $\gamma_j$ and weight $\omega_{ij}$, $\upsilon_{jr}$ are adjusted to the rule of minimizing the square error sum. The square error of the sample $k$ is written as (6). $E$ is the square error sum of all samples, shown as(7). In (7), $N$ is the number of inputted sample.

$$E_k=\frac{1}{2}\sum_{r=1}^{q}\left(t_r^k-y_r^k\right)^2 \tag{6}$$

$$E=\sum_{k=1}^{N}E_k=\frac{1}{2}\sum_{k=1}^{N}\sum_{r=1}^{q}\left(t_r^k-y_r^k\right)^2 \tag{7}$$

The parameter $\upsilon_{jr}$、$\omega_{ij}$、$\theta_j$ and $\gamma_j$ need to be adjusted. Their partial derivatives are shown as (8), according to the conjugate gradient descent algorithm[1]. Then they are adjusted according to(9). In (9), $\eta_1$、$\eta_2$、$\eta_3$ and $\eta_4$ are learning rate.

$$\begin{cases}\dfrac{\partial E}{\partial \upsilon_{jr}}=-\sum_{k=1}^{N}\left(t_r^k-y_r^k\right)\psi\left(net_j^k\right)\\ \dfrac{\partial E}{\partial \omega_{ij}}=-\sum_{k=1}^{N}\sum_{r=1}^{q}\left(t_r^k-y_r^k\right)\psi'\left(net_j^k\right)\dfrac{x_i^k}{\theta_j}\\ \dfrac{\partial E}{\partial \theta_j}=\sum_{k=1}^{N}\sum_{r=1}^{q}\left(t_r^k-y_r^k\right)\psi'\left(net_j^k\right)\left(\dfrac{net_j^k-\gamma_j}{\theta_j}\right)\dfrac{x_i^k}{\theta_j}\\ \dfrac{\partial E}{\partial \gamma_j}=\sum_{k=1}^{N}\sum_{r=1}^{q}\left(t_r^k-y_r^k\right)\psi\left(net_j^k\right)/\theta_j\end{cases} \tag{8}$$

$$\begin{cases} \upsilon_{jr}(t+1) = \upsilon_{jr}(t) - \eta_1 \dfrac{\partial E}{\partial \upsilon_{jr}} \\ \omega_{ij}(t+1) = \omega_{ij}(t) - \eta_2 \dfrac{\partial E}{\partial \omega_{ij}} \\ \theta_j(t+1) = \theta_j(t) - \eta_3 \dfrac{\partial E}{\theta_j} \\ \gamma_j(t+1) = \gamma_j(t) - \eta_4 \dfrac{\partial E}{\gamma_j} \end{cases} \tag{9}$$

# 3 Application in the Fault Diagnosis for Turbine Generator Unit

## 3.1 Description of the Turbine Fault

The fault type of turbine generator unit has a wide variety and the cause is complex. The signals reflecting fault features are many, including temperature, pressure, vibration, etc. Among all the fault feature signals of turbine generator unit, the vibration signals have the most fault features, which can reflect the operating condition of mechanical equipments more rapidly and directly and also can be monitored and analyzed more easily. In order to verify the fault diagnosis method based on WNN in the paper, three kinds of common faults of turbine generator unit have been chosen: oil whipping, unbalance and no orderliness. As is shown in table1, there is the elements of feature vector components of six different frequency ranges including <0.4f, 0.4f-0.5f, 1f, 2f, 3f,> 3f (f for the rotation frequency) [7]. In the table 1, Type1 is the fault mode of oil whipping, Type2 is unbalance, and Type3 is no orderliness. Among the samples adopted in this paper, (1 0 0) denotes oil whipping, (0 1 0) denotes unbalance and (0 0 1) denotes no orderliness. Table 2 is test sample.

**Table 1.** Training Sample Data

| NO. | <0.4f | 0.4f-0.5f | 1f | 2f | 3f | >3f | Target output | Fault mode |
|---|---|---|---|---|---|---|---|---|
| 1 | 3.35 | 46.60 | 12.15 | 1.94 | 2.30 | 1.67 | 1 0 0 | Type1 |
| 2 | 4.43 | 51.00 | 11.02 | 3.02 | 1.30 | 2.43 | 1 0 0 | Type 1 |
| 3 | 3.24 | 50.00 | 11.61 | 1.24 | 0.90 | 1.30 | 1 0 0 | Type 1 |
| 4 | 5.72 | 46.30 | 12.31 | 3.62 | 1.50 | 0.59 | 1 0 0 | Type 1 |
| 5 | 6.32 | 45.80 | 15.23 | 3.56 | 2.30 | 3.19 | 1 0 0 | Type 1 |
| 6 | 1.51 | 3.29 | 52.92 | 6.59 | 2.5 | 2.54 | 0 1 0 | Type 2 |
| 7 | 2.43 | 1.19 | 54.49 | 4.64 | 0.80 | 1.78 | 0 1 0 | Type 2 |
| 8 | 0.54 | 2.92 | 48.82 | 6.64 | 3.90 | 1.51 | 0 1 0 | Type 2 |
| 9 | 0.81 | 1.73 | 52.00 | 6.43 | 3.60 | 1.89 | 0 1 0 | Type 2 |
| 10 | 1.24 | 1.35 | 49.79 | 4.64 | 1.00 | 2.27 | 0 1 0 | Type 2 |
| 11 | 1.78 | 1.46 | 22.46 | 23.8 | 19.0 | 8.59 | 0 0 1 | Type 3 |
| 12 | 0.92 | 1.24 | 30.38 | 22.0 | 16.0 | 5.67 | 0 0 1 | Type 3 |
| 13 | 0.65 | 2.11 | 21.98 | 26.2 | 18.0 | 11.10 | 0 0 1 | Type 3 |
| 14 | 1.13 | 0.92 | 24.46 | 22.3 | 15.0 | 15.80 | 0 0 1 | Type 3 |
| 15 | 0.92 | 1.40 | 26.08 | 26.0 | 20.0 | 11.40 | 0 0 1 | Type 3 |

**Table 2.** Test Sample

| NO. | <0.4f | 0.4f-0.5f | 1f | 2f | 3f | >3f | Target output | Fault mode |
|---|---|---|---|---|---|---|---|---|
| 1 | 3.01 | 40.60 | 11.15 | 2.94 | 2.30 | 1.07 | 1 0 0 | Type 1 |
| 2 | 0.50 | 2.22 | 48.82 | 6.54 | 2.90 | 1.61 | 0 1 0 | Type 2 |
| 3 | 3.29 | 47.00 | 1105 | 1.44 | 0.80 | 1.01 | 1 0 0 | Type 1 |
| 4 | 2.43 | 1.18 | 53.40 | 4.02 | 0.35 | 1.68 | 0 1 0 | Type 2 |
| 5 | 1.03 | 0.82 | 24.56 | 21.33 | 14.08 | 14.8 | 0 0 1 | Type 3 |
| 6 | 0.89 | 1.04 | 29.38 | 22.00 | 15.80 | 5.47 | 0 0 1 | Type 3 |
| 7 | 0.79 | 1.09 | 28.38 | 21.98 | 15.60 | 5.07 | 0 0 1 | Type 3 |

### 3.2 Design of WNN

According to the WNN, the neuron number of the input layer is equal to the number of the fault feature parameter of the inputted sample and the neuron number of the output layer is equal to the number of fault type, that is, each output neuron is corresponding to a kind of fault. Thus, the construction of WNN network can be established with 6 neurons in input layer, 8 neurons in hidden layer and 3 neurons in output layer.

### 3.3 Diagnostic Results and Analysis

Firstly, training samples and test samples are normalized. Secondly, training samples are input into the network to train WNN network. The test samples is shown in table 2 and input into the trained WNN network for fault diagnose. The diagnosis results are obtained as table 3 shown. The table shows that the proposed method can effectively diagnose the vibration fault of turbine generator, i.e. the accuracy of the fault diagnose is 100%.

**Table 3.** The Diagnosis Results

| NO. | The output of wavelet neural network | | | Diagnosis Results |
|---|---|---|---|---|
| | Type1 | Type2 | Type3 | |
| 1 | 0.9226 | 0.1064 | -0.0083 | Type 1 |
| 2 | -0.0219 | 0.9989 | -0.0009 | Type 2 |
| 3 | 1.0368 | -0.0011 | -0.0066 | Type 1 |
| 4 | 0.0137 | 0.9940 | -0.0034 | Type 2 |
| 5 | -0.0129 | 0.0288 | 0.9899 | Type 3 |
| 6 | 0.0750 | 0.0800 | 1.0187 | Type 3 |
| 7 | 0.0721 | 0.0685 | 1.0239 | Type 3 |

**Table 4.** The Diagnosis Results When Adding 5% Random Noise

| NO. | The output of wavelet neural network | | | Diagnosis Results |
|---|---|---|---|---|
| | Type1 | Type2 | Type3 | |
| 1 | 1.1710 | -0.1841 | 0.0259 | Type 1 |
| 2 | -0.0194 | 0.8728 | 0.0907 | Type 2 |
| 3 | 0.8718 | 0.1914 | -0.0275 | Type 1 |
| 4 | 0.0160 | 1.0120 | -0.0069 | Type 2 |
| 5 | 0.0852 | -0.0707 | 1.0223 | Type 3 |
| 6 | 0.0398 | 0.1807 | 0.8936 | Type 3 |
| 7 | 0.1253 | 0.1550 | 1.0009 | Type 3 |

The test samples are added to 5% random noise and input into the trained WNN network for fault diagnose. The diagnostic results are obtained as table 4 shown. The table shows that the proposed method can effectively diagnose the vibration fault when the test samples are disturbed by the random noise.

## 4 Conclusion

Wavelet neural network is a new type of neural network which is designed by introducing the wavelet analysis theory to neural network. It not only makes full use of the localized features of wavelet transform, but also inherits the self-learning of neural networks. Thus, WNN has a strong approximation and pattern recognition capabilities. The fault diagnosis technology based on WNN can effectively diagnose the vibration fault of turbine generator even in the random noise disturbance, can meet the requirements of the speed and accuracy for fault diagnosis and is easy to implement. So it has good application prospects.

**Acknowledgments.** This work was supported by the State Key Program of National Natural Science Foundation of China (Grant No. 61034004), Shanghai Science and Technology Commission Key Program (No. 10250502000) and Innovation Key Program of Shanghai Municipal Education Commission (No. 12ZZ177). The authors are grateful for the anonymous reviewers who made constructive comments.

## References

[1] Tian, J., Gao, M.: Artifical neural network and its application, vol. 7. Beijing Institute of Technology Press, Beijing (2006)
[2] Zhong, L., Rao, W., Zou, C.: Artificial neural network and its fusion applications, vol. 3. Science Press, Beijing (2008)
[3] Banakar, A., Azeem, M.F.: Generalized Wavelet Neural Network Model and its Application in Time Series Prediction. In: International Joint Conference on Neural Networks, pp. 882–886 (2006)
[4] Matlab Chinese Forum. The Study of 30Ccases of Neural Network Based on Matlab, vol. 4. Beijing University Press, Beijing (2010)

[5] Peng, T., Ma, Q.: Application of wavelet neural network on rolling element bearings fault diagnosis. Computer Engineering and Applications 46(4), 213–215 (2010)
[6] Zhu, J.-M., Jiang, L.-X., Rao, K.-K.: Research of Power Network Fault Diagnosis Based on Wavelet Neural Network. Electric Switchgear (6), 23–25 (2011)
[7] Zhang, B.-D., Sun, C.-X., Ou, J., et al.: A Fuzzy Clustering Method for Turbo-generator Vibration Faulty Diagnosis. Turbine Technology 24(5), 289–291 (2002)

# Density Based Grid Clustering Partition of the Input Space for RBF Neural Network*

Ping Wei, Wenbo Liu, Xin Zuo, and Xionglin Luo

Department of Automation, China University of Petroleum, Beijing 102249, P.R. China
mail_weiping@yahoo.com.cn, liuwb99@163.com,
{zuox,luoxl}@cup.edu.cn

**Abstract.** Selecting proper centers is important for constructing a radial basis function (RBF) neural network. Motivated by the idea of clustering according to density for data mining applications, the algorithm of unevenly partition of the input space is proposed in the paper. By combining the neighboring subsets with low density data, the ultimate clustering centers are selected as the hidden layer centers in RBF neural network. An example is presented to demonstrate the method proposed, and the results illustrate the comparative high accuracy RBF neural network with comparative short training time can be created by selecting the initial partition set and the upper limit for the number of data in one subinterval properly.

**Keywords:** Neural Network, Radial Basis Function, Density Based Grid Clustering, Partition, Modeling.

## 1 Introduction

Radial basis function (RBF) network, comparing with other types of artificial neural networks, has simple network structures, better approximation capabilities, faster learning algorithms, etc. Therefore it is often used for modeling complex systems or identifying the model parameters, and so on. RBF network has three layers, namely, input layer, hidden layer and output layer. Receiving the input information from the environment is the main function of the input layer. The hidden layer is in charge of filtering the input vector into the high-dimension space through radial basis function. Computing the network output by linear combinations of the hidden node responses is in output layer. The relation of input and output in RBF neural network is as follows:

$$\hat{y}_m(k) = \sum_{l=1}^{L} \omega_{m,l} f_l \left( \sqrt{\sum_{n=1}^{N} (x_n - \hat{x}_{l,n})^2} \right) . \tag{1}$$

Where $x_n$ denotes input vector, $\hat{x}_{l,n}$ denotes the centers of the hidden node, $\omega_{m,l}$ is the weights between the hidden layer and the output layer, $\hat{y}_m(k)$ is output vector, $f_l$

* This work is supported by Science Foundation of China University of Petroleum, Beijing (No. KYJJ2012-05-31).

J. Lei et al. (Eds.): AICI 2012, LNAI 7530, pp. 642–649, 2012.

is the radial basis function. There are something need to be defined when constructing a radial basis function network, that is, radial basis function, the parameters of hidden nodes, the weights between the hidden layer and the output layer. The parameters of hidden nodes include the number, width and location, which have great effect on the convergence rate, model accuracy and generalization ability of neural network.

A number of methods, which aim to determine the parameters of hidden nodes, have been proposed by many researchers. Broomhead and Low selected randomly one subset of the training data set as the hidden node centers [1]. Moody and Darken determined the hidden node centers by K-means clustering methodology [2]. Chen et al. proposed the algorithm based on Orthogonal Least Square for selecting the centers which have most great influence on the network output [3]. The support vector was presented as hidden node centers using support vector machine by Schokopf et al. [4]. The optimal subset of training data was selected automatically as the hidden node centers with Genetic Algorithm by Steve et al. [5]. Based on a fuzzy partition of the inputs space, the hidden node centers are determined (to see Ref. [6, 7]). Afterward, non-symmetric partition of the input space was considered additionally to improve the effect of modeling [8]. Improving or combining the methods in existence to optimize the number and location of the hidden node centers have become an important task for constructing RBF neural networks.

The algorithm presented in Ref. [6, 7] for determining the number and locations of hidden nodes requires only one pass of the training data, and decreases the training time obviously. However, the even partition of the input space leads to distribution imbalance of data in every subset. Every subset has the equal width though it is likely that some subsets have lots of data and some have few. For improving the quality of the fuzzy subset, the non-symmetric partition of the input space was proposed [8]. Every variable of input space was partitioned evenly but assigned a different number of fuzzy sets. Motivated by the idea of clustering according to density for data mining applications in Ref. [9], the algorithm of grid partition input space based on the density of the data is proposed, which can find the more proper clustering centers and improve the accuracy and generalization ability of the network.

## 2 Grid Partition and Combination of the Input Space

Finding the hidden node centers is the mainly task when training RBF neural network. Determining the number and location of hidden nodes using proper clustering centers in input space is studied in the paper. Firstly partition the input space evenly. Then combine the neighboring subspaces if the sum of data in the two subspaces is not more than the number defined ahead. Iterating the combination until the number of data in arbitrary two neighboring subspaces is more than the defined number, and therefore every subspace has nearly equal number of data.

Suppose there are $N$ input variables in a neural network, and the $i$th variable as a one-dimension space is partitioned to $d_i$ subsets, $i=1,\cdots,N$. Let $a_{i,j}$ and $\delta_{i,j}$ $(i=1,\cdots,N, j=1,\cdots,d_i)$ denote the centers and the width of the corresponding subset respectively. An arbitrary subset can be described by $A_{i,j}=\{a_{i,j},\delta_{i,j}\}$. Therefore the

$N$-dimension input space is partitioned into $S$ subspaces, $S=\prod_{i=1}^{N} d_i$, and denote subspaces by $A^l, l=1,\cdots,S$. Suppose the $i$th $(i=1,\cdots,N)$ input interval is $[a,b]$, and there are $K$ input date in the interval. It is required that the number of the input date in every interval is not more than $K/m$ ($m$ is a positive integer) after partition and combination. Firstly partition the interval $[a,b]$ into $d_i (d_i > m)$ subintervals, $A_{i,1}, A_{i,2}, \cdots, A_{i,d_i}$, and denote the initial width and center of the $j$th subinterval are $\delta_{i,j}=(b-a)/d_i$ and $a_{i,j}=(j-1/2)\times\delta_{i,j}, j=1,\cdots,d_i$. The steps on how to combine the intervals is as the following Fig.1.

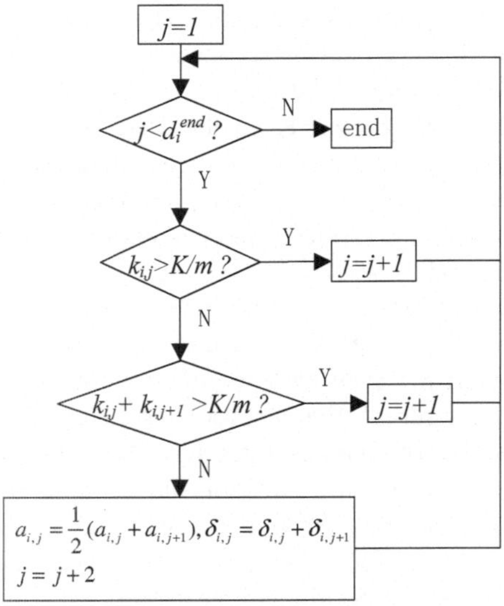


**Fig. 1.** The steps of combining the intervals

After the first combination, there are $d_i'(d_i' \le d_i)$ subintervals in the input interval, [a, b]. Continue the combination of current subintervals until the sum of data in arbitrary two neighboring subintervals is more than *K*/*m*. Finally the ith $(i=1,\cdots,N)$ input interval is partitioned into $d_i^{end}(d_i^{end} \le d_i)$ subintervals. An example of the density based grid partition is displayed in Fig.2 using a one-dimension case. Suppose there are 100 date in the interval [0, 1], and partition the interval into 10 subintervals evenly. It is required that the number of data in every interval is not more than 25 after combination. The centers and the width of every interval are marked in Fig.2, and the numbers above the intervals denote how many data are in the corresponding subintervals. Fig.2(a) is the initial partition. After the first combination, the second partition can be described as Fig.2(b). Continue the combination and get the third partition as Fig.2(c). Now the sum of data in arbitrary two neighboring subintervals is more than 25, and the partition ends.

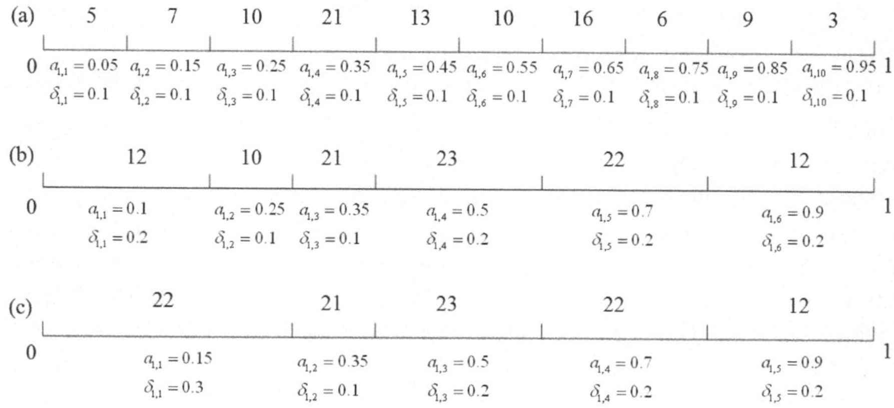


**Fig. 2.** The example of density based grid partition using a one-dimension case

## 3 Selecting the Hidden Nodes of the RBF Neural Network

After grid partition and combination, the input space is partitioned into $S'$ subspaces, $S' = \prod_{i=1}^{N} d_i^{end}$ .Selecting the proper subspaces which uniformly cover the input data distribution, and regarding the subspace centers as the hidden nodes of the RBF neural network. The notion of the multidimensional memberships function is introduced for getting the proper subspaces as usual [6, 7], that is,

$$\mu_{A^l}(x(k)) = \begin{cases} 1 - rd^l(x(k)), rd^l(x(k)) \le 1, \\ 0, \qquad \textit{otherwise}. \end{cases} \tag{2}$$

Where $rd^l(x(k))$ is the Euclidean relative distance of the subspace $A^l$ and input vector $x(k)$, as the following expression,

$$rd^l(x(k)) = \sqrt{\frac{\sum_{i=1}^{N}(a_{i,l} - x_i(k))^2}{N(\delta_{i,l})^2}} \ . \tag{3}$$

The steps for selecting the hidden nodes centers as follows.

(1) Initial clustering number *L*=1 and the training data $x(k), (k=1)$. Compute the membership of input vector $x(1)$ in every subspace and select the subspace which has the maximal membership as the first clustering center $A^l, l=1$. Let $k=2$.

(2) Compute the Euclidean relative distance, $rd^l(x(k))$, of $x(k)$ and every generated clustering center respectively.

(3) Denote $\min(rd^l(x(k))) < 1$ as the minimal value of the $rd^l(x(k))$. If $rd^l(x(k)) \le 1$, it shows there exists a clustering covering $x(k)$, then let $k = k+1$ and

go to Step (2). Else, it shows there are no clustering covering $x(k)$. It need generate another new clustering, then let $L = L+1$.

(4) Compute the membership of $x(k)$ in every subspace and select the subspace which has the maximal membership as the $L$th clustering center $A^l$. $k = k+1$, go to Step (2) until the computation about all training data have been finished.

There are $L$ clustering being generated after the above steps. For arbitrary $x(k)$, there exists a clustering center from which the Euclidean relative distance is no more than 1. Therefore the clustering centers can be selected as the hidden node centers in RBF neural network. The width of the hidden nodes can be computed by P-nearest neighbor heuristic. The valves of P are selected to be equal to about 50 percent of the number of hidden nodes in this paper.

$$\sigma_l = \left( \frac{1}{\mathrm{P}} \sum_{\mathrm{j}=1}^{P} \left\| \hat{x}_l - \hat{x}_j \right\| \right)^{\frac{1}{2}} . \tag{4}$$

## 4 Example and Simulation

Here chose a non-isothermal continuous stirred tank reactor (CSTR) in chemical engineering as an example to demonstrate the method proposed in the paper. The process is described by the following equations which have been discussed in Ref [6].

$$\begin{aligned} \frac{dC_A}{dt} &= \frac{F}{V}\left(C_{A,in} - C_A\right) - 2k_0 \exp\left(-\frac{E}{RT}\right) C_A^{\,2}, \\ \frac{dT}{dt} &= \frac{F}{V}\left(T_{in} - T\right) + 2\frac{(-\Delta H)_R}{\rho c_P} k_0 \exp\left(-\frac{E}{RT}\right) C_A^{\,2} - \frac{UA}{V\rho c_P}\left(T - T_j\right). \end{aligned} \tag{5}$$

Where $C_A$ and $T$, as the output variables of the system, denote the concentration and temperature of the reactant inside the reactor respectively. $F$, $C_{A,in}$, $T_{in}$ and $T_j$ are the input variables of the system, and they are the flow rate into the reactor, the inlet concentration of the reactant, the inlet temperature of the reactant and the temperature of the coolant respectively. $V$ is the volume of the CSTR. $(-\Delta H)_R$ is the heat of the reaction, and so on. The parameters in the system are valued as Table 1.

**Table 1.** Parameter values of CSTR.

| Parameter | $V$ | $UA$ | $P$ | $c_P$ | $(-\Delta H)_R$ | $k_0$ | $E$ | $R$ |
|---|---|---|---|---|---|---|---|---|
| Value | 100 | 20000 | 1000 | 4.2 | 596619 | $6.85\times10^{11}$ | 76534.704 | 8.314 |
| Unit | L | J/s·K | g/L | J/g·K | J/mol | L/s·K | J/mol | J/mol·K |

Construct the model of the reactor by simulink in Matlab. The disturbance is added to the steady state of input variables. And suppose the range of the disturbance is 5 percent of the input variables. The steady state values of input variable are described by Table 2. The input variable $F$ with disturbance can be demonstrated as Fig. 3, and other input variables with their disturbances are similar.

**Table 2.** Steady state values of input variables

| Variable | $F$ | $C_{A,in}$ | $T_{in}$ | $T_j$ |
|---|---|---|---|---|
| Value | 20 L/s | 275 K | 250 K | 1 mol/L |

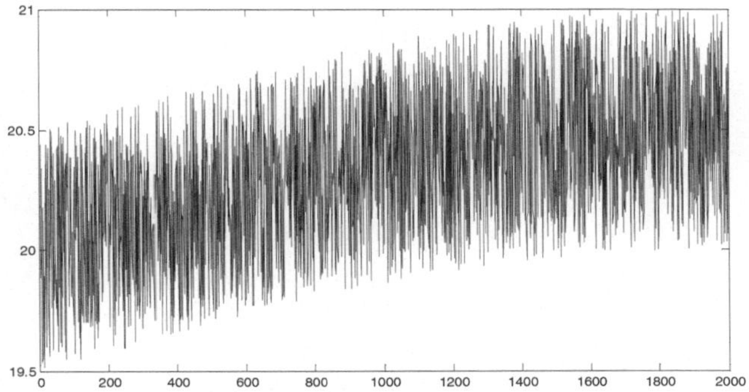


**Fig. 3.** Input variable F with disturbance

Take sample periods, $T_a = 1s$, and compute 2000 output data from the system equation, the former 1000 samples are used to train the neural network, and the other 1000 data are used to test the accuracy of the neural network. The sums of squared errors (SSE) for testing data can be computed by $SSE = \sum_{i=1}^{N} (y_i - \hat{y}_i)^2$. $y_i$ is the value of real data from the system, and $\hat{y}_i$ is the estimated value from the neural network.

Training the RBF neural network by the evenly partition methods proposed in Ref. [6]. The results are described by Table 3. $C$ is the number of fuzzy partition sets. $L$ is the number of hidden nodes. $T$ is the training time. $SSE1$ is the sums of squared errors for concentration. $SSE2$ is the sums of squared errors for temperature. The results of training the RBF neural network by grid clustering partition in the previous sections is as Table 4. $d_i$ is the number of initial partition sets. $K/m$ is the number of data allowed in one subinterval.

From Table 3 and 4, the accuracy and training time of neural network are changing along with the number of partition sets and the hidden nodes. Using the method proposed in the paper, the comparative high accuracy RBF neural network with comparative short training time can be created by selecting the initial partition set and the upper limit for the number of data in one subinterval properly. It is usually that the higher the accuracy is, the longer the training time is. Therefore it is a trade-off between the model accuracy and training time when training a neural network.

**Table 3.** Results of evenly partition

| $C$ | $L$ | $SSE1$ | $SSE2$ | $T$ |
|---|---|---|---|---|
| 4 | 12 | 0.018 | 26349 | 0.55 |
| 5 | 285 | 0.026 | 3710 | 6.95 |
| 6 | 898 | 0.014 | 9508 | 67.52 |
| 7 | 999 | 0.015 | 8988 | 76.48 |
| 8 | 1000 | 0.013 | 10028 | 75.04 |

**Table 4.** Results of density based partition

| $d_i$ | $K/m$ | $L$ | $SSE1$ | $SSE2$ | $T$ |
|---|---|---|---|---|---|
| 12 | 142 | 72 | 0.010 | 21579 | 3.38 |
| 19 | 167 | 177 | 0.009 | 9514 | 11.08 |
| 37 | 200 | 214 | 0.020 | 2943 | 10.01 |
| 17 | 143 | 185 | 0.007 | 7351 | 7.52 |
| 17 | 167 | 165 | 0.003 | 7802 | 7.52 |

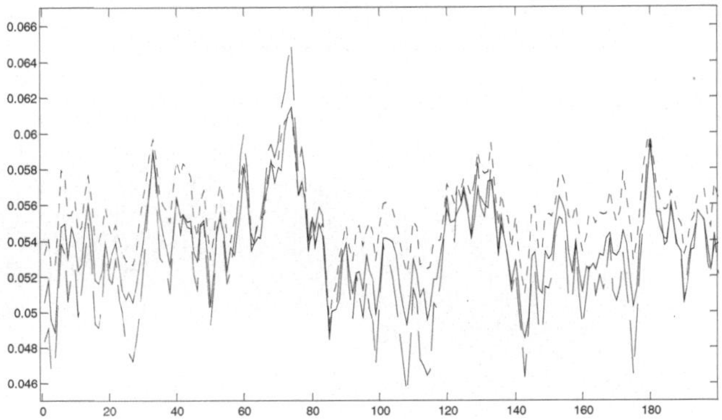


**Fig. 4a.** Real value and RBF neural network predictions of concentration

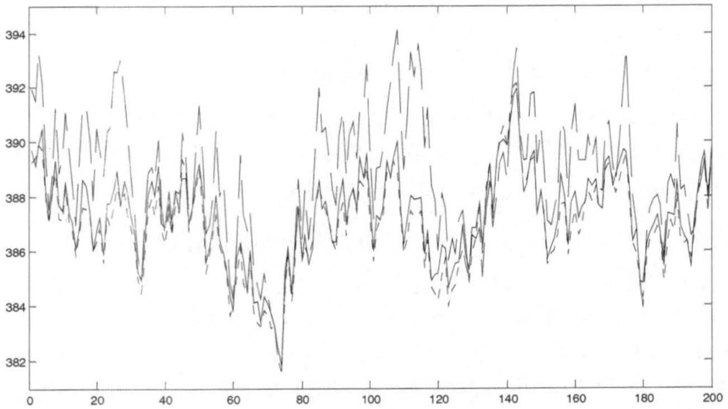


**Fig. 4b.** Real value and RBF neural network predictions of temperature

The RBF neural network model is simulated by selecting the optimal partition for concentration and temperature of reactant. The simulation curve for concentration generated with the method proposed in Ref. [6] is figured by dashed line, and $c$=5. The simulation curve for concentration generated with the method proposed here is figured by real line, and $d_i$=37, $K/m$=200. For temperature, $c$=8 , $d_i$=17, $K/m$=167. The curve of real values for concentration and temperature are figured by segment line respectively.

From Fig. 4, the RBF neural network model can predict the output values basically when the curves of concentration and temperature inside the reactor change sharply. Comparing the simulation curves to the real value, the results generated by the proposed method are closer to the real values.

## 5 Conclusion

The algorithm of grid partition input space based on the density of the data is proposed in the paper. With the method, the comparative high accuracy RBF neural network with comparative short training time can be created by selecting the initial partition set and the upper limit for the number of data in one subinterval properly. Comparing with the partition method in Ref. [6], density based grid clustering partition is more efficient when the data of input space distribute unevenly. That is, it can improve the generalization ability of the RBF neural network.

## References


1. Broomhead, D.S., Lowe, D.: Multi-variable functional interpolation and adaptive networks. Complex System 2, 321–355 (1988)
2. Moody, J., Darken, C.J.: Fast learning in networks of locally-tuned processing units. Neural Computation 1, 281–294 (1989)
3. Chen, S., Cowan, C.F.N., Grant, P.M.: Orthogonal least squares learning algorithm for radial basis function networks. IEEE Transactions on Neural Networks 2, 302–309 (1991)
4. Schokopf, B., Sung, K., Burges, C., et al.: Comparing support vector machines with Gaussian kernels to radial basis function classifiers. IEEE Transactions on Signal Processing 45(11), 2758–2765 (2000)
5. Billings, S.A., Zheng, G.L.: Radial Basis Function Network Configuration Using Genetic Algorithms. Neural Networks 8, 877–890 (1995)
6. Sarimveis, H., Alexandridis, A., Tsekouras, G., Bafas, G.: A fast and efficient algorithm for training radial basis function neural networks based on a fuzzy partition of the input space. Industrial Engineering Chemistry Research 41, 751–759 (2002)
7. Aggelogiannaki, E., Sarimveis, H.: Nonlinear model predictive control for distributed parameter systems using data driven artificial neural network models. Computers & Chemical Engineering 32, 1225–1237 (2008)
8. Alexandridis, A., Sarimveis, H., Ninos, K.: A Radial Basis Function network training algorithm using a non-symmetric partition of the input space-Application to a Model Predictive Control configuration. Advances in Engineering Software 42, 830–837 (2011)
9. Agrawal, R., Gehrke, J., Gunopulos, D., Raghavan, P.: Automatic subspace clustering of high dimensional data for data mining applications. In: International Conference on Management of Data, pp. 94–105. ACM Press, New York (1998)

# Research of the Urban Air Quality Forecast Method Based on Resource Allocation Network

Zhifang Jiang, Shanxiang Zhang, Ruobo Xin, Shenghui Cheng, and Ning Li

School of Computer Science and Technology of Shandong University
Jinan 250010, China
zfjiang@sdu.edu.cn

**Abstract.** An air quality forecast model based on the resource allocation network has been established in consideration of the time-varying characteristics of the urban air quality and the effects of a variety of nonlinear factors to the prediction accuracy. We have used the distance criteria and error criteria to allocate hidden layer nodes dynamically or adjust network parameters. In this way, we have got the minimum neural network structure to meet the error requirements and avoid solving the problem of selecting the initial neural network structure and parameters.

**Keywords:** Air Quality, Resource Allocation Network, Neural Network, Forecast Model, Pollution Sources, Monitoring.

## 1 Introduction

Urban air quality has becoming an important factor restricting the development of the city and it is affected by urban geography, meteorology, pollution distribution and other factors. Its variation with time presents the typical non-linear characteristic[1]. Artificial neural network can be considered as a tool with strong learning capabilities to analyze and predict non-linear objects[2,3] and has been widely used in various science and engineering area. In recent years, artificial neural networks have made great progress in air pollution index forecast and pollutant concentration forecast[1,2].

### 1.1 Related Works

Wu Xiaohong has introduced adaptive momentum adjusting algorithm to traditional BP model to predict air pollution index and both the learning rate and forecast precision are better than traditional algorithm[4]. Mikk has used SOM algorithm and the fuzzy distance matrix to classify the combination of air quality data and meteorological data and then used multi-layer perceptron to these data to improve air quality forecast precision[5]. Hoi,k.I made a comparison between air quality artificial neural network forecast models with more hidden layer nodes and that with less ones and pointed out that the error in forecast period is not sensitive to changes of hidden layer nodes[6]. Du has proposed an air quality forecast model based on the BP neural

J. Lei et al. (Eds.): AICI 2012, LNAI 7530, pp. 650–657, 2012.

network of the samples self-organization clustering[7]. The model forecasts daily pollutant concentrations according to its class categorized by intrinsic characteristics and regularity of pollutant concentration changes and the model is obviously superior to the ordinary BP model not only in the forecast accuracy but also in the stability and the time-consuming.

### 1.2 Main Work of This Paper

This paper presents a forecast model based on RAN which allocates hidden layer nodes dynamically or adjust parameters of the network according to the distance criteria and error criteria. In this way, we can not only get the minimum neural network structure to meet the error requirements but also avoid solving the problem of selecting the initial neural network parameters and thus a faster training speed and a better predict result has been got.

## 2 RAN Structure and Learning Algorithm

### 2.1 RAN Structure

Radial Basis Function Neural Network is a kind of commonly used three-layer feed-forward neural network. Its basic idea is to make radial basis function affiliate with hidden layer nodes to form the hidden layer and maps the input vector directly to it, which means there are no weights between the input layer and hidden layer. However, the mapping between hidden layer and output layer is linear.

Resource Allocating Network (RAN) proposed by Platt is a kind of single hidden layer neural network based on radial basis function which can implement online studying[8]. The network allocates hidden layer nodes dynamically by novelty conditions or adjusts network parameters. In this way, we can get the minimum neural network structure to meet the error requirements and avoid solving the problem of selecting the initial neural network structure and parameters. Its structure is shown in Fig.1.

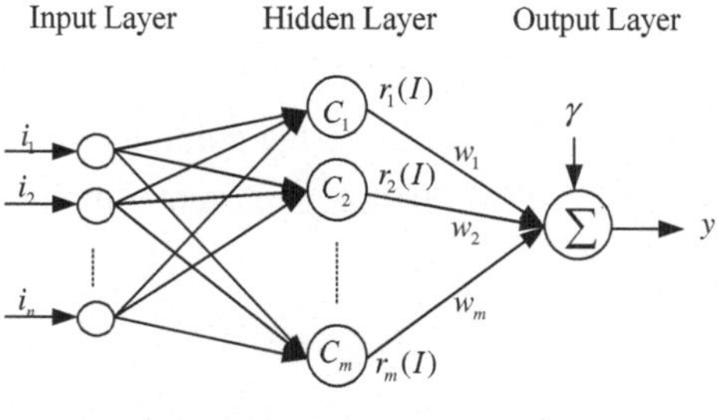


**Fig. 1.** RAN structure diagram

In the figure above, activation function afflicted with each hidden layer neuron node (hereinafter referred to as hidden node) is radial basis function. In this paper, Gaussian function[9] is adopted. For the j-th hidden node, its local mapping result can be got from formula (1).

$$r_j(I) = \exp(-\frac{\left\|I - C_j\right\|^2}{2\sigma_j^2}), j = 1, 2, \cdots, m \tag{1}$$

Here $I = (i_1, i_2, ..., i_j, ..., i_n)^T$ is the input vector, $C_j = (c_{1j}, c_{2j}, \cdots, c_{nj})^T$ is the j-th unit center vector of the network hidden node and $\sigma_j$ is the width of this node, $w_j$ is the weight between this hidden node and the output node, $\gamma$ is the output offset, $y$ is the output of the network calculated by equation (2).

$$y = \sum_{j=1}^{m} r_j w_j + \gamma \tag{2}$$

### 2.2 RAN Learning Algorithm

RAN starts with no hidden nodes and allocates hidden node dynamically or adjust network parameters based on hidden node allocation criteria and parameter adjustment criteria.

Suppose $\{I_i, T_i\}_{i=1}^N$ represent the training sample pairs set and in this notion, $I_i$ is the $i$-th input sample in the set, $T_i$ is the corresponding output and $N$ is the number of the training sample pairs. RAN algorithm traverses the training sample pairs set and checks the novelty conditions for each sample pair $(I_i, T_i)$. If satisfied, the network can have a new hidden node; otherwise, we just adjust network parameters.

For the current input sample $I_i$, calculate the distance between $I_i$ and hidden nodes already generated, that is $\| I_i - C_j \|$, where $C_j, j = 1, 2, \cdots, l (1 \le l \le m)$ represents hidden node center already calculated. For $I_i$, we could then find a unit center $C_j$ denoted as $C_{nearest}$ which meets $\| I_i - C_{nearest} \| = \min\{\| I_i - C_1 \|, \| I_i - C_2 \|, \cdots, \| I_i - C_l \|\}, 1 \le l \le m$

For the $i$-th sample pair $(I_i, T_i)$, novelty conditions are as follows:

Condition 1(distance criteria): Distance between current input sample $I_i$ and unit center $C_{nearest}$ is greater than value $\delta(t)$

$$\| I_i - C_{nearest} \| > \delta(t) \tag{3}$$

$\delta(t)$ is the distance resolution or distance, t is the training time.

Condition 2(error criteria): Error between current output sample $T_i$ and network output $y$ is greater than value $\varepsilon$

$$\| T_i - y \| > \varepsilon \tag{4}$$

Case 1: When condition 1 and condition 2 are both satisfied, we can increase a hidden node for hidden layer. After that, network output should be equal to the output sample. Suppose the new hidden node is numbered as k, then the unit center $C_k = I_i$, the weight between this hidden node and the output node $w_k = T_i - y$, the width of this node $\sigma_k = \tau \| I_i - C_{nearest} \|$, where $\tau$ is the overlap factor which indicates the overlap degree of each hidden node center responding region.

At the beginning, $\delta(t)$ is initialized as network parameter $\delta_{\max}$. With the learning process, it is decreased to $\delta_{\min}$.

$$\delta(t) = \max\{\beta^t \delta_{\max}, \delta_{\min}\} \tag{5}$$

Here $\beta$ is the attenuation constant.

Case 2: When condition 1 and condition 2 are not both satisfied, we then adjust network parameters, such as $w_j$, $\gamma$ and $C_j$, based on Widrow-Hoff LMS algorithm[10]. The error is calculated by following formula.

$$e = \|y - T_i\|^2 \tag{6}$$

Weights between the output node and hidden nodes already generated are adjusted as follows

$$\Delta w_j = \alpha(T_i - y)\exp(-\frac{\|I_i - C_j\|^2}{\sigma_j^2}), j = 1, 2, \cdots, l \tag{7}$$

where $\alpha$ is the learning rate.

Output offset $\gamma$ is adjusted by formula (8).

$$\Delta\gamma = \alpha(T_i - y) \tag{8}$$

By gradient descent method, we can regulate unit centers.

$$\Delta C_j = 2\frac{\alpha}{\sigma_j}(I_i - C_j)r_j[(T_i - y)w_j], j = 1, 2, \cdots, l \tag{9}$$

# 3 Samples Selection and Network Initialization

## 3.1 Samples Selection

As the city thermoelectric and central heating company's production facilities have been incorporated into the basic environment and pollution sources online monitoring system and in winter these production facilities are all in operation, so the pollution monitoring data is more accurate and complete in winter. In this paper, we have chosen 36 real-time emission pollution sources showed in Table 1 and 1 air quality monitoring station in the winter of Jinan, 2010. We selected item $SO_2$ as the object of study. As is listed in Table 2, there are 7 air quality monitoring stations in Jinan and Station "Kegan institute" was chosen as our study object. We selected the $SO_2$ monitoring data per day from Dec.1,2010 to Feb.20,2011, 82 days in total. Among them, monitoring data from the first 72 days can be considered as training sample pairs $\{I_i, T_i\}_{i=1}^{72}$ and the last 10 days as the testing samples.

To increase the convergence rate and the convenience when processing data, we have normalized input samples with the function mapminmax() included in Matlab neural network toolbox. As for output samples, the maximum value is 0.037 and the minimum value is 0.019. They are both in interval [0,1] and so there's no need to normalize them.

**Table 1.** Emission Pollution Sources in Jinan

| | | | |
|---|---|---|---|
| 1Northern thermoelectricity | 2Eastern thermoelectricity | 3Gaoxin heating | 4Gengchen steel 1 |
| 5Gengchen steel 2 | 6Huangtai power 7 | 7Huangtai power 8 | 8Machine plant 2 |
| 9Jigang sintering 1 | 10Jigang shaft furnace 2 | 11Jigang sintering 2 | 12Jigang coking 7 |
| 13Jigang coking 1-2 | 14Jigang coking 3-4 | 15Jigang coking 5 | 16Jigang coking 6 |
| 17Jinan Univ. | 18Jinan Fertilizer | 19Architecture Univ. | 20Jinji heating |
| 21Economic college | 22Kuangshan heating | 23Minghu thermoelectricity | 24Southern thermoelectricity |
| 25Southern thermoelectricity 1 | 26Southern thermoelectricity 2 | 27Qilu Medicine | 28Xiban heating |
| 29Yanshan heating | 30Eastern school of Shandong Univ. | 31New region of Shandong Univ. | 32Ductile of Shandong |
| 33 Shandong Normal Univ. | 34Tianlong Motion | 35Railway Hospital | 36Xiaoya Group |

**Table 2.** Air Quality Monitoring Stations

| | | | |
|---|---|---|---|
| 1Changqing Zone | 2Chemical plant | 3Machine tool plant | 4Monitoring house |
| 5Seed station | 6 Kegan institute | 7Development zone | |

### 3.2 Network Initialization

According to RAN structure diagram, there should be 36 nodes in the input layer corresponding to the stations listed in Table 1; for simplicity, we set only one output node which represents Station "Kegan institute" in Table 2;for the hidden layer, the network starts with no nodes and new node is added with the training process.

Network parameters are initialized as follows: the maximum distance resolution $\delta_{max}$ =0.6, the minimum distance resolution $\delta_{min}$ =0.01, attenuation constant $\beta$ =0.5, desired error accuracy $\varepsilon$ =0.01, overlap factor $\tau$ =0.9, learning rate $\alpha$ =0.01, the maximum training times maxepoch=900.

## 4 RAN Air Quality Forecast Model Implementing

After training the network with the training sample pairs, the number of hidden nodes of the network is increased to 32. All the unit centers of the 32 hidden nodes constitute a matrix showed in Fig.2.

The weight vector between hidden nodes and the output node $w$ is:

( -0.0120 0.0180 -0.0128 0.0186 0.0280 -0.0142 -0.0124 0.0200 -0.0173 0.0102 0.0103 0.0130 -0.0146 -0.0122 0.0114 -0.0140 0.0219 0.0116 0.0659 0.0729 -0.0507 -0.0204 -0.0187 -0.0144 0.0133 0.0134 0.0180 -0.0204 0.0415 0.0271 0.0346 -0.0444)

Output offset $\gamma$ is 0.0310

Width vector of the basis function $\sigma$ is:

$( 0.5400\ 4.6531\ 1.8615\ 2.3996\ 2.4731\ 2.4528\ 1.9209\ 2.2973\ 2.1550\ 2.5245\ 2.2644\ 1.2897\ 2.5458\ 2.7080\ 2.0933\ 1.5326\ 1.8606\ 2.1456\ 2.2011\ 1.9383\ 1.3210\ 1.7238\ 1.4848\ 1.7682\ 1.1759\ 1.7531\ 2.6415\ 2.0485\ 1.3174\ 1.4259\ 1.6584\ 1.2014)^T$

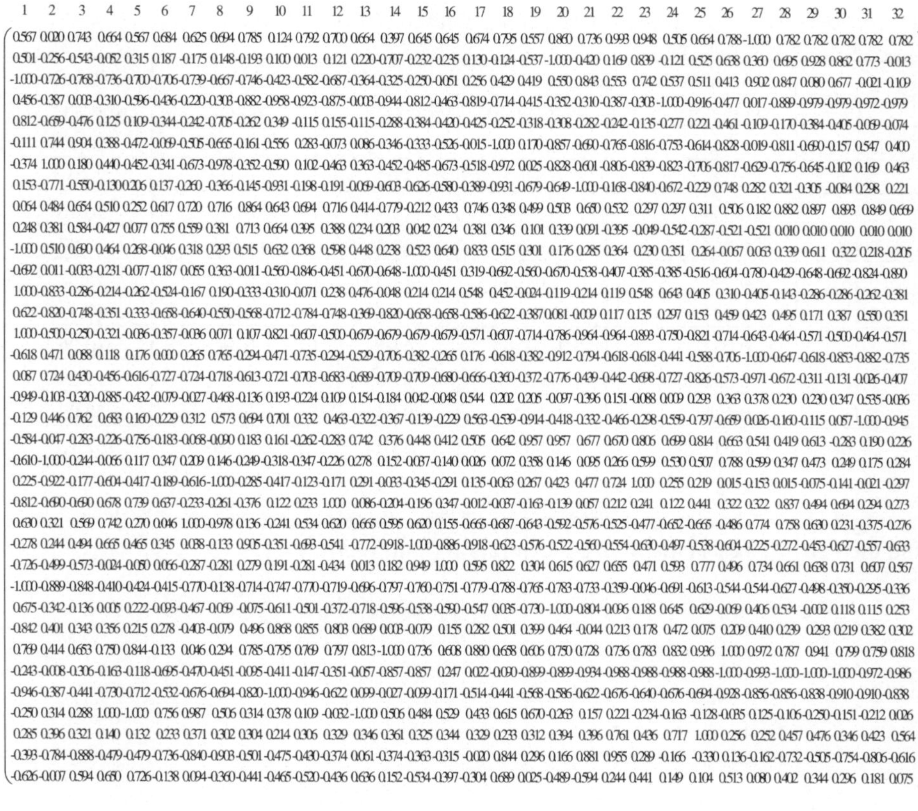

**Fig. 2.** Unit centers of the 32 hidden nodes

# 5 Forecast Result and Analysis

## 5.1 Forecast Result

After training, we have established an air quality forcast model based on RAN and determined the network structure and parameters. Then we tested this model with the input data from Feb.11, 2011 to Feb.20, 2011. The forcast results (unit: mg/m$^3$) are shown in Table 3.

**Table 3.** RAN Model Forcast Results (unit: mg/m3)

| **Date** | 2.11 | 2.12 | 2.13 | 2.14 | 2.15 |
|---|---|---|---|---|---|
| **Forcast result** | 0.0514 | 0.0494 | 0.0496 | 0.0454 | 0.0469 |
| **Date** | 2.16 | 2.17 | 2.18 | 2.19 | 2.20 |
| **Forcast result** | 0.0317 | 0.031 | 0.0321 | 0.0326 | 0.0322 |

Besides, we also established a BP network prediction model with the same data. And the comparison among RAN forcast result, BP forcast result and the actual monitoring data is shown in Table 4.

**Table 4.** Comparison Between BP and RAN

| Date | Monitoring data(mg/m3) | RAN (mg/m3) | Relative error(%) | BP (mg/m3) | Relative error(%) |
|---|---|---|---|---|---|
| 2.11 | 0.043 | 0.0514 | 19.53 | 0.0517 | 20.17 |
| 2.12 | 0.036 | 0.0494 | 37.22 | 0.0392 | 8.89 |
| 2.13 | 0.059 | 0.0496 | 15.93 | 0.0257 | 56.36 |
| 2.14 | 0.057 | 0.0454 | 20.35 | 0.0154 | 73.07 |
| 2.15 | 0.041 | 0.0469 | 14.39 | 0.0183 | 55.38 |
| 2.16 | 0.033 | 0.0317 | 3.94 | 0.0271 | 17.75 |
| 2.17 | 0.033 | 0.0310 | 6.06 | 0.0296 | 10.26 |
| 2.18 | 0.051 | 0.0321 | 37.06 | 0.0607 | 19.09 |
| 2.19 | 0.036 | 0.0326 | 9.44 | 0.0461 | 28.07 |
| 2.20 | 0.045 | 0.0322 | 28.44 | 0.0729 | 62.10 |

Fig.3 shows the intuitive comparison of these three values. Abscissa is the timeline and the vertical axis is the SO2 concentration (unit: mg/m$^3$).

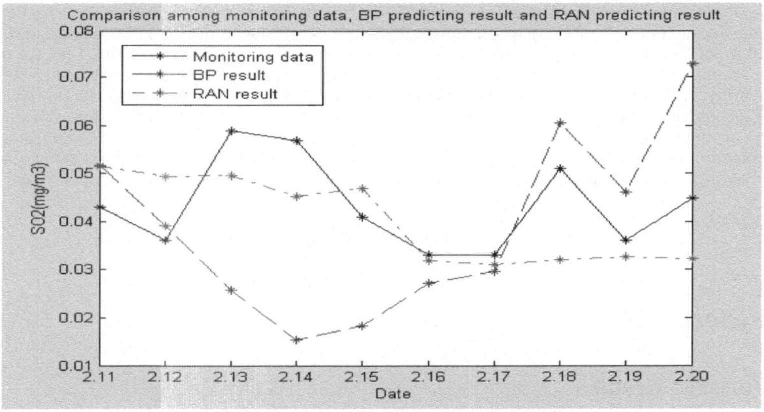


**Fig. 3.** Comparison among monitoring data, BP forcast result and RAN forecast result

## 5.2 Result Analysis

As can be seen form Table 4 and Fig.3, the RAN air quality forcast model can allocate hidden node dynamically according to training samples and therefore weaken the effect caused by samples selection to the network structure. So the RAN results are more accurate than BP. What's more, the BP results fluctuate dramatically, which means it's unstable.

In the training time aspect, we have trained these two models on a PC with Windows XP operating system, Intel(R) Core(TM) 2 Duo CPU E6550@2.33GHz processor, 2GB memory and Matlab 7.11 for win-32. The time cost by RAN model is 0.04688 second which is lower than 0.214068second cost by BP model. With all of the advantages, the RAN model has higher availability.

## References

1. Zhao, H., Liu, A.-X., Wang, K., Bai, Z.: Study on GA_ANN based on ambient air SO2 and NO2 concentration forcasting mode. Computer Engineering and Applications 46(8), 199–201 (2010)
2. Zhang, Y., Yang, Y., Li, W.: Weights Direct Determination of Neural Networks. Zhongshan University Press, Guangzhou (2010)
3. Yang, X.: Neural Network and Its Application in Control. Zhejiang University, Hangzhou (2004)
4. Wu, X., Kang, H., Ren, D.: Forecast of Air Pollution Index in One City Based on the Artificial Neural Net. Mathematics in Practice and Theory 35(2), 87–91 (2005)
5. Kolehmainen, M., Martikainen, H., Hiltunen, T., Ruuskanen, J.: Forecasting Air Quality Parameters Using Hybrid Neural Network Modelling. Environmental Monitoring and Assessment 65(1), 277–286 (2000)
6. Hoi, K.I., Yuen, K.V., et al.: Is a Complex Neural Network Based Air Quality Prediction Model Better Than a Simple One? In: A Bayesian Point of View. AIP Conference Proceedings, vol. 1233(1), p. 764 (2010)
7. Du, X., Jiang, Z., Meng, X., Liu, S., Li, S.: An Air Quality Forecast Model Based on the BP Neural Network of the Samples Self-Organization Clustering. In: The 6th International Conference on Natural Computation (ICNC 2010), pp. 1523–1527 (2010)
8. Platt, J.: A resource-allocating network for function interpolation. Neural Computing 3(2), 212–225 (1991)
9. Zhang, Y., Leithead, W.E., Leith, D.J.: Time-series Gaussian process regression based on Toeplitz of O(N2) operations and O(N)-level storage. In: Proceeding of the 44th IEEE Conference on Decision and Control, pp. 3711–3716 (2005)
10. Ge, L., Huo, A.-Q.: Application research of Widrow-Hoff neural network learning rule. Electric Design Engineering (6) (June 2009)
11. Dang, K., Yang, L., Lin, T.: A New Type of General RBF Neural Network and Its Training Mehtod. Computing Technology and Automation 26(1) (March 2007)

# A Novel Neural Network Approach for Computing Eigen-Pairs of Real Antisymmetric Matrices

Hang Tan[1], Xianhe Huang[1], Huachun Tan[2], and Ying Tang[1]

[1] School of Automation and Engineering,
University of Electronic Science and Technology of China, Chengdu 610054, China
[2] School of Mechanical and Vehicular Engineering,
Beijing Institute of Technology, Beijing 100081, China
tanhang2011@yahoo.com.cn

**Abstract.** In the present paper, we focus on the problem how to compute all eigen-pairs of any real antisymmetric matrix by the conventional neural network approach without modification the original structure of the neural network. Given any $n$-dimensional real antisymmetric matrix, our proposed method is based on a $n$-dimensional ODEs and the preprocessing become comparatively easy. The contributions of this paper are mainly come from two aspects, on the one hand, we constructed the eigen-pairs relationship between those of symmetric matrix and anti-symmetric matrix; on the other hand, we presented a simple method to compute all eigen-pairs of any antisymmetric matrix. Simulations verify the computational capability of the proposed method.

**Keywords:** Neural network, Real antisymmetric matrix , Eigen-pairs.

## 1 Introduction

In many practical fields such as image processing and primary component analysis or minor component analysis, we not only expect to compute the largest or smallest eigenvalues and the corresponding eigenvectors, but also need more eigenvalues and the corresponding eigenvectors due to some important information can't be given only by the largest or smallest eigen-pairs. Many neural network based methods have been proposed to solve the largest or smallest eigen-pairs problems [1,3,7,8]. Two excellent reviews can be found in [2,8]. However, most of those studies only focused on computing largest or smallest eigenpairs of real symmetric matrices. In [8], the following ordinary differential equation (ODE) was proposed:

$$\frac{dx(t)}{dt} = Ax(t) - x(t)^T Ax(t)x(t) \tag{1}$$

and the authors of [7,9] suggested the following ODE:

$$\frac{\mathrm{d}x(t)}{\mathrm{d}t} = x(t)^T x(t)Ax(t) - x(t)^T Ax(t)x(t). \tag{2}$$

J. Lei et al. (Eds.): AICI 2012, LNAI 7530, pp. 658–665, 2012.

Note that $A$ in (1) and (2) must be real symmetric matrix.

The ODEs from (1) to (2) are all efficient ways to compute the eigenvector corresponding to the largest eigenvalue. Moreover, they can also succeed to compute the eigenvector corresponding to the smallest eigenvalue by simply replacing $A$ with $-A$.

However, following the relevant research we only find three neural network based methods for computing eigenpairs of $n$-dimensional real antisymmetric matrix [4,5,6]. Two of them were based on $2n$-dimensional ODEs[4,5], and the other one needs the time-consuming preprocessing such as household transformation although it is based on $n$-dimensional ODEs[6]. To avoid the overlarge network scale and complex preprocessing, this paper presents a neural network based method with a comparatively easy preprocessing that can be summarized by $n$-dimensional ODEs. In the proposed method, the problem of computing eigen-pairs of real antisymmetric matrices is translated into that of the related real symmetric matrix. Then the translated problem is to be solved using (1) or (2). We note that all of this methods only can compute the largest or smallest eigen-pairs problem, in order to compute all eigen-pairs of antisymmetric matrix, we propose a simple method to compute all eigen-pairs of symmetric matrix at first, then we could compute all eigen-pairs of antisymmetric matrix by the relationship of eigen-pairs between antisymmetric matrix and symmetric matrix.

## 2 Main Results

In the following, $A$ is always a real antisymmetric matrix. Thus, $A^T A$ is a real symmetric matrix whose eigenvalues are all nonnegative real numbers because for $\forall x \in R^n$, there exists $x^T A^T A x = \| Ax \|_2^2 \geq 0$, i.e., $A^T A$ is a positive semidefinite matrix. For $\forall x = a + b\mathrm{i} \in C$, we denote the conjugate of $x$ as $\overline{x} = a - b\mathrm{i}$. We cited the following lemma 1 and lemma 2 from [10] at first.

**Lemma 1.** The eigenvalues of $A$ are zeros or pure imaginary numbers.

**Lemma 2.** Let $\lambda\mathrm{i}$ ($\lambda \in R$) and $u$ be an eigenvalue and the corresponding eigenvector of $A$, respectively. Then, $-\lambda\mathrm{i}$ is another eigenvalue of $A$ corresponding to the eigenvector $\overline{u}$.

**Theorem 1.** Let $\lambda\mathrm{i}$ ($\lambda \in R$) and $u$ be an eigenvalue and the corresponding eigenvector of $A$, respectively. Then, $\lambda^2$ is an eigenvalue of $A^T A$. Moreover, both of $u$ and $\overline{u}$ are eigenvectors of $A^T A$ corresponding to the eigenvalue $\lambda^2$.

**Proof.** Following the conditions, we have $Au = \lambda\mathrm{i}u$, which leading to $A^T Au = -AAu = -\lambda\mathrm{i}Au = \lambda^2 u$. Thus, $u$ is an eigenvector of $A^T A$ corresponding to the eigenvalue $\lambda^2$. By the lemma 2 and the fact that both of $A$

and $\lambda$ are real, we have $\overline{Au} = A\overline{u} = \overline{\lambda i u} = -\lambda i\overline{u}$, so $A\overline{u} = -\lambda i\overline{u}$, i.e., $A^T A\overline{u} = \lambda^2\overline{u}$. Thus, $\overline{u}$ is another eigenvector of $A^T A$ corresponding to the eigenvalue $\lambda^2$.

**Theorem2.** Assume that $\lambda^2, \lambda \in R,$ is the eigenvalue of $A^T A$ corresponding to the eigenvector $u$, where $A$ is skew symmetric matrix. If $\lambda = 0$, 0 is the eigenvalue of $A$ corresponding to the eigenvector $u$. If $\lambda \neq 0$, let $v = \frac{Au}{|\lambda|}$. Then, $\pm|\lambda|\mathrm{i}$ are two eigenvalues of $A$ corresponding to two eigenvectors $v \pm u\mathrm{i}$, respectively.

**Proof.** If $\lambda = 0$, we have $A^T Au = 0$.
So $\| Au \|_2^2 = u^H A^H Au = u^H A^T Au = 0$, i.e., $Au = 0$. Therefore, 0 is the eigenvalue of $A$ corresponding to the eigenvector $u$. Next, assume $\lambda \neq 0$. Following the conditions, we have $A^T Au = \lambda^2 u$ and

$$Au = |\lambda| v \tag{3}$$

Hence, $\lambda^2 u = A^T Au = -AAu = -|\lambda| Av$. Since $\lambda \neq 0$, we have

$$v = -|\lambda| u \tag{4}$$

Then it is straightforward to verify

$$A(v + u\mathrm{i}) = -|\lambda| u + |\lambda| v\mathrm{i} = |\lambda| \mathrm{i}(v + u\mathrm{i}) \tag{5}$$

$$A(v - u\mathrm{i}) = -|\lambda| u - |\lambda| v\mathrm{i} = -|\lambda| \mathrm{i}(v - u\mathrm{i}) \tag{6}$$

Thus proving the theorem.

If we can get eigenvalues and the corresponding eigenvectors of the real symmetric matrix $A^T A$, we can get those of $A$ using the theorem 2. As we have introduced before, following those neural network based methods such as (1) and (2) for extracting largest or smallest eigenpairs of real symmetric matrices, in this paper we replace $A$ with $A^T A$ in (2), which reads

$$\frac{\mathrm{d}x(t)}{\mathrm{d}t} = x(t)^T x(t) A^T Ax(t) - x(t)^T A^T Ax(t) x(t). \tag{7}$$

Next, we refer to the following results on convergence of (7) introduced in [10].

**Lemma 3**. (Theorem 4 in [9]) Given any nonzero vector $x(0) \in R^n$, if $x(0)$ is not orthogonal to the eigensubsapce corresponding to the largest eigenvalue of $A^T A$, then the solution of (7) starting from $x(0)$ converges to an eigenvector corresponding to the largest eigenvalue of $A^T A$ that is equal to $\lim_{t\to+\infty} \frac{x(t)^T A^T Ax(t)}{x(t)^T x(t)}$.

**Lemma 4.** (Theorem 5 in [9]) Replacing $A^T A$ with $-A^T A$ in (7), we get

$$\frac{\mathrm{d}x(t)}{\mathrm{d}t} = -x(t)^T x(t)A^T Ax(t) + x(t)^T A^T Ax(t)x(t). \tag{8}$$

Suppose $x(0)$ is a nonzero vector in $R^n$ which is not orthogonal to the eigensubspace corresponding to the smallest eigenvalue of $A^T A$, then the solution of (8) starting from $x(0)$ converges to an eigenvector corresponding to the smallest eigenvalue of $A^T A$ that is equal to $\lim_{t\to+\infty} \frac{x(t)^T A^T Ax(t)}{x(t)^T x(t)}$.

If we randomly choose initial $x(0)$, the projection of $x(0)$ on the eigensubsapces corresponding to the largest or smallest eigenvalues is nonzero with probability one since the dimension of eigensubspaces is generally less than the dimension of $R^n$. Hence, the lemma 3 and 4 almost hold with randomly generated $x(0)$.

We denote $n$ eigenvalues of the $n$-dimensional real antisymmetric matrix $A$ as $\lambda_j$i, $j = 1,\cdots,n$, satisfying $|\lambda_1| \geq |\lambda_2| \geq \cdots \geq |\lambda_n|, \lambda_{2s-1} = -\lambda_{2s}$ and we denote the corresponding eigenvectors as $u_1,\cdots,u_n$, respectively.

In many practically case such as image fusion, we need more eigen-pairs information to describe the real scene; and perform primary component analysis (PCA) to some high dimension data, we also need to compute more than one eigen-pair for more completely representing the high dimension space information with this important low dimension space data. So it it can't avoid to compute more eigen-pairs in some practical fields. However, by the lemma 3 and 4, we can only obtain the largest or smallest eigen-pairs. If we can obtain all eigen-pairs of $A^T A$ according to (7), we can also obtain all eigen-pairs of $A$ by the theorem 2.

Given an eigenvalue $\lambda^2$ of $A^T A$, denote the eigenspace corresponding to $\lambda^2$ by $V_{\lambda^2}$. Let $\lambda_1^2,\cdots,\lambda_n^2$ be all the eigenvalues of $A^T A$ ordered by $\lambda_1^2 \geq \cdots \geq \lambda_n^2$. Suppose that $S_i$ $(i = 1,\cdots,n)$ is an standard orthonormal basis in $R^n$ such that each $S_i$ is an eigenvector of $A^T A$ corresponding to the eigenvalue $\lambda_i^2$. Let $\delta_i$ $(i = 1,\cdots,m)$ be all the distinct eigenvalues of $A^T A$ ordered by $\delta_1 > \cdots > \delta_m$. For any $i, 1 \leq i \leq m$, denote the algebraic sum of themultiplicity of $\delta_1,\cdots,\delta_i$ by $k_i$. Clearly, $k_m = n$. For convenience, denote $k_0 = 0$, it is easy to see that $\lambda_i^2 = \delta_r$, for all $i \in [k_{r-1}, k_r]$, and $S_i \in V_{\delta_r}$, for all $i \in [k_{r-1}, k_r]$.

**Lemma 5.** Let $S_i, i = 1, \cdots, n$ is a standard orthonormal eigenvectors group, then the eigenvalues of symmetric matrix $B = (I - \sum_{j<i} S_j S_j^T) A^T A$ are $[0, \cdots, 0, \lambda_i^2, \cdots, \lambda_n^2]$.

**Proof.** It is very simple to verify that the matrix $(I - \sum_{j<i} S_j S_j^T) A^T A$ is symmetric matrix, for $\forall k < i$, $BS_k = 0$ due to $S_i, i = 1, \cdots, n$ is a standard orthonormal basis, so $S_k (k < i)$ are the eigenvectors of matrix $B$ corresponding to the eigenvalues 0. So the eigenvectors corresponding to the eigenvalues $[0, \cdots, 0, \lambda_i^2, \cdots, \lambda_n^2]$ are $S_1, \cdots, S_n$ respectively. Using this lemma, we could directly using equation (7) to compute all eigen-pairs of symmetric matrix $B = (I - \sum_{j<i} S_j S_j^T) A^T A$, furthermore, through using theorem 2, we can obtain all eigen-pairs of any antisymmetric matrix. Matrix $A^T A$ corresponding to the largest eigenvalue $\lambda_1^2 = \frac{V_1^T A^T A V_1}{V_1^T V_1}$. In order to continue, we need to obtain the standard eigenvector $S_1 = \frac{V_1}{V_1^T V_1}$. If $i = 2$, we replaced the matrix $A^T A$ in the equation (7) with matrix $B = (I - S_1 S_1^T) A^T A$, and then we can obtain the largest eigenvalue of $B$, which are just the secondly largest eigenvalue of $A^T A$, can be denote as $\lambda_2^2$, the corresponding eigenvector canbe denote as $V_2$, so $\lambda_2^2 = \frac{V_2^T A^T A V_2}{V_2^T V_2}$. Now, we have two eigenvectors, which can be performed Smith Transformation.

## 3 Simulations

The following invertible and real antisymmetric matrix $A$ was used in our experiment to verify the proposed method.

$$A = \begin{pmatrix} 0 & 2.9067 & -0.2510 & 2.8864 & 0.5446 & 0.3084 \\ -2.9067 & 0 & 0.6021 & 2.3001 & 2.4557 & 0.9741 \\ 0.2510 & -0.6021 & 0 & -0.2838 & -0.6846 & 2.7128 \\ -2.8864 & -2.3001 & 0.2838 & 0 & -3.1545 & 1.5631 \\ -0.5446 & -2.4557 & 0.6846 & 3.1545 & 0 & -0.9693 \\ -0.3084 & -0.9741 & -2.7128 & -1.5631 & 0.9693 & 0 \end{pmatrix}$$

The symmetric matrix $A^T A$ as below:

$$A^T A = \begin{pmatrix} 17.2349 & 8.1257 & -2.1055 & -7.9928 & 1.4964 & -6.1344 \\ 8.1257 & 21.0812 & -0.4210 & 2.3369 & 8.3067 & -1.9519 \\ -2.1055 & -0.4210 & 8.3340 & 7.0604 & -2.1829 & 0.2891 \\ -7.9928 & 2.3369 & 7.0604 & 26.0965 & 5.8995 & -0.6969 \\ 1.4964 & 8.3067 & -2.1829 & 5.8995 & 17.6861 & -4.2279 \\ -6.1344 & -1.9519 & 0.2891 & -0.6969 & -4.2279 & 11.7861 \end{pmatrix}$$

Using lemma 5 we could obtain all eigenvalues of $A^T A$ as following ,we always use random initial $x_0 = [0.1133, -0.2070, -0.8768, 0.5604, -0.3248, 0.2157]^T$ for running equation (7)): $\lambda_1^2 = \lambda_2^2 = 33.8446$ , $\lambda_3^2 = \lambda_4^2 = 12.9831$ , $\lambda_5^2 = \lambda_6^2 = 4.2817$ , and the corresponding eigenvectors as below:

$$[S_1, \cdots, S_6] = \begin{pmatrix} -0.6011 & 0.1024 & -0.4525 & -0.1440 & 0.2697 & 0.5744 \\ -0.5466 & -0.3951 & -0.0959 & 0.6133 & -0.0596 & -0.3953 \\ 0.1936 & -0.1677 & -0.4885 & 0.0641 & -0.7969 & 0.2379 \\ 0.3877 & -0.7510 & -0.2729 & -0.0750 & 0.4427 & 0.0978 \\ -0.2897 & -0.4764 & 0.6445 & -0.2624 & -0.2803 & 0.3553 \\ 0.2613 & 0.1193 & 0.2398 & 0.7243 & 0.1189 & 0.5668 \end{pmatrix}$$

According to theorem 2, we can obtain the eigenvalues of antisymmetric matrix $A$ as following: $\lambda_1 \mathrm{i} = \sqrt{33.8446}\mathrm{i}$ , $-\lambda_2 \mathrm{i} = -\sqrt{33.8446}\mathrm{i}$ , $\lambda_3 \mathrm{i} = \sqrt{12.9831}\mathrm{i}$ , $-\lambda_4 \mathrm{i} = -\sqrt{12.9831}\mathrm{i}$ , $\lambda_5 \mathrm{i} = \sqrt{4.2817}\mathrm{i}$ , $\lambda_6 \mathrm{i} = \sqrt{4.2817}\mathrm{i}$ . Using the MatLab fuction $[V, D] = eig(A)$ , we got the eigenvalues of $A$ : $\pm 5.8176\mathrm{i}$ , $\pm 3.6032\mathrm{i}$ , $\pm 2.0692\mathrm{i}$ (i.e., $\lambda_1 = -\lambda_2 = 5.8176$, $\lambda_3 = -\lambda_4 = 3.6032$, $\lambda_5 = -\lambda_6 = 2.0692$ ), and the corresponding eigenvectors $V_1, \cdots, V_6$ as following, where $V_2 = \overline{V_1}$ , $V_4 = \overline{V_3}$ , $V_6 = \overline{V_5}$ ,

$$V_1 = \begin{pmatrix} -0.2593 - 0.3445\mathrm{i} \\ 0.0710 - 0.4716\mathrm{i} \\ 0.1682 + 0.0673\mathrm{i} \\ 0.5976 \\ 0.2054 - 0.3365\mathrm{i} \\ 0.0098 + 0.2029\mathrm{i} \end{pmatrix}, \quad V_3 = \begin{pmatrix} -0.1972 - 0.2717\mathrm{i} \\ 0.3904 - 0.2007\mathrm{i} \\ -0.0656 - 0.3422\mathrm{i} \\ -0.1110 - 0.1665\mathrm{i} \\ -0.0330 + 0.4910\mathrm{i} \\ 0.5395 \end{pmatrix}, V_5 = \begin{pmatrix} 0.0665 - 0.4438\mathrm{i} \\ 0.0396 + 0.2799\mathrm{i} \\ -0.5881 \\ 0.2802 - 0.1558\mathrm{i} \\ -0.2618 - 0.1841\mathrm{i} \\ -0.0341 - 0.4081 \end{pmatrix}.$$

As for the eigenvectors of $A$, for example, we can use $u = S_1$ or $u = S_2$ based on the theorem 2 to compute the eigenvectors $\frac{AS_1}{\lambda_1} \pm S_1 \mathrm{i}$ or $\frac{AS_2}{\lambda_1} \pm S_2 \mathrm{i}$ , which corresponding to the eigenvalues $\pm \lambda_1 \mathrm{i}$ of the antisymmetric matrix $A$ . In this paper, we choose $S_1, S_3, S_5$ to compute all eigenvectors $U_1, \cdots, U_6$ of $A$, where $U_2 = \overline{U_1}$ , $U_4 = \overline{U_3}$ , $U_6 = \overline{U_5}$ ,

$$U_1 = \begin{pmatrix} -0.1024-0.6011i \\ 0.3951-0.5466i \\ 0.1677+0.1936i \\ 0.7510+0.3877i \\ 0.4764-0.2897i \\ -0.1193+0.2613i \end{pmatrix} = [1.256+0.648i]V_1, U_3 = \begin{pmatrix} -0.1440-0.4525i \\ 0.6133-0.0959i \\ 0.0641-0.4885i \\ -0.0750-0.2729i \\ -0.2624+0.6445i \\ 0.7243+0.2398i \end{pmatrix} = [1.342+0.444i]V_3$$

$$U_5 = \begin{pmatrix} 0.5744+0.2697i \\ -0.3954-0.0596i \\ 0.2379-0.7969i \\ 0.0978+0.4427i \\ 0.3553-0.2803i \\ 0.5669+0.1189i \end{pmatrix} = [-0.4046+1.3551i]V_5.$$

From the results, we can see that $U_1, U_3, U_5$ are just constant multiple of $V_1, V_3, V_5$ respectively. Thus we verify the proposed method based on the theorem 1, theorem 2 and the lemma 5. In this paper, we only present the transient behavior of all eigenvalues of symmetric matrix $A^T A$ based on lemma 3 and lemma 5, the results can be seen in Fig.1.

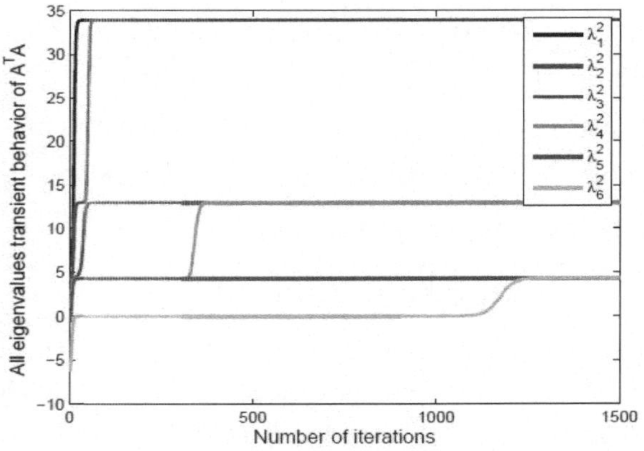


**Fig. 1.** Transient behavior of all eigenvalues of $A^T A$ based on (7) with initial $x_0$

# 4 Conclusion

In this paper we proposed another neural network based method to compute the eigenpairs problem of any antisymmetric matrix, which can be summarized by $n$-dimensional ODEs and avoid too complex preprocessing, which is surely more efficient in practice. Moreover, we extended our method to compute all eigenvalues and corresponding eigenvectors of any real antisymmetric matrix. Experimental results demonstrate the validity of the proposed algorithms.

## References

1. Huang, D., Yi, Z., Pu, X.: A New Incremental PCA Algorithm with Application to Visual Learning and Recognition. Neural Processing Letters 30(3), 171–185 (2009)
2. Martinez, A.M., Kak, A.C.: PCA versus LDA. IEEE T. Pattern Anal. 23(2), 228–233 (2011)
3. Tang, H., Tan, K.C., Yi, Z.: Neural Networks: Computational Models and Applications. Springer (2007)
4. Liu, Y., You, Z.S., Cao, L.P.: A functional neural network for computing the largest modulus eigenvalues and their corresponding eigenvectors of an anti-symmetric matrix. Neurocomputing 67, 384–397 (2005)
5. Liu, Y., You, Z.S., Cao, L.P.: A concise functional neural network computing the largest modulus eigenvalues and their corresponding eigenvectors of a real skew matrix. Theoretical Computer Science 367, 273–285 (2006)
6. Tang, Y., Li, J.P.: Another neural network based approach for computing eigenvalues and eigenvectors of real skew-symmetric matrices. Computers & Mathematics with Applications 60, 1385–1392 (2010)
7. Luo, F.L., Li, Y.D.: Real-time neural computation of the eigenvector corresponding to the largest eigenvalue of positive matrix. Neurocomputing 7(2), 145–157 (1995)
8. Oja, E.: Principal components, minor components, and linear neural networks. Neural Networks 5, 927–935 (1992)
9. Zhang, Y., Yan, F., Tang, H.J.: Neural networks based approach for computing eigenvectors and eigenvalues of symmetric matrix. Comput. Math. Appl. 47, 1155–1164 (2004)
10. Wang, C.: Several Properties of the Real Anti-symmetric Matrix. Journal of Gansu Normal College 15(2), 8–9 (2010)

# Pavement Distress Image Recognition Based on Multilayer Autoencoders

Lukui Shi, Chunying Gao, and Jun Zhang

School of Computer Science and Engineering, Hebei University of Technology, Tianjin 300401, China
shilukui@scse.hebut.edu.cn, gao_chunying@163.com, zhangjun@scse.hebut.edu.cn

**Abstract.** Pavement distress images are typical high dimensional nonlinear data. Manifold learning algorithms can find the intrinsic characteristic hidden in the distress images, which helps to better recognize them. Unlike most of manifold learning algorithms, multilayer autoencoders have solved the data reconstructed problem through building a bi-directional mapping between the high dimensional data and the low dimensional data. An automatic pavement distress image recognition method based on multilayer autoencoders was proposed, which combined the image processing method and multilayer autoencoders. In the method, the distress images were firstly processed with the image processing method. Then the images were reduced dimensions and reconstructed with multilayer autoencoders. Lastly, the distress type was recognized through the network. Experiments showed that the recognition accuracy with the proposed method was great higher than that with the BP neural network.

**Keywords:** Pavement distress image recognition, Manifold learning, Multilayer autoencoders, Image processing, BP neural network.

## 1 Introduction

In order to effectively maintain the damaged pavement, the road management departments need to regularly check road conditions. To timely maintain the damaged road, it is very crucial to fast and accurately obtain a variety of road information. Pavement distress data is one of the important road information need to be acquired. The traditional manual inspection method is low efficient, time-consuming and dangerous [1]. And such method is great subject. Therefore, it is difficult to gain objective and accurate evaluation of pavement damaged conditions with the manual way. With the fast development of the computer technologies and the image processing technologies, many automatic checking systems for pavement distress images have been designed to help people detect pavement distress conditions.

The collection of pavement distress images earliest appeared in the late 1970s. Later the automatic detection systems for pavement distress features gradually emerged with the rapid development of the highway construction. Nowadays, there

J. Lei et al. (Eds.): AICI 2012, LNAI 7530, pp. 666–673, 2012.

are many detecting systems, such as Komatsu system, PCES system, PAVUE system, CREHO system, WDM' road crack detection vehicle, CSIRO's road crack detection vehicle and so on [2, 3, 4]. Although these systems have very mature and complete techniques and functions, the post-data processing, especially pavement distress image automatic recognition, needs to be further improved. How to implement automatic recognition of pavement distress images is yet a technical problem need to be solved. To do this, the median filtering method was used to process pavement distress images in [5]. The histogram analysis method was applied in pavement distress image recognition in [6]. Canny regular was employed in pavement distress image recognition in [7]. A recognition method based on distress density factor was proposed in [8]. An automatic recognition algorithm of pavement distress image based on OTSU and maximizing mutual information was presented in [9].

However, it is very hard to get satisfactory results with the existing image processing technologies for pavement distress images are typical large-scale high dimensional nonlinear data. Contrarily, it has proved that manifold learning algorithms can better project the high dimensional nonlinear data into a low dimensional space and discover the physical meaning behind them. Currently, there are many manifold learning method to be developed, such as isometric feature mapping (ISOMAP) [10], diffusion maps (DM) [11], kernel principal component analysis (KPCA) [12], locally linear embedding (LLE) [13], Laplacian Eigenmaps [13], Hessian LLE [15], local tangent space analysis (LTSA) [16], multilayer autoencoders [17,18] and so on. Although they can guarantee to find the nonlinear manifold hidden in the high dimensional data, most of them don't build a bi-directional mapping between the original data and the corresponding low dimensional data except multilayer autoencoders. So we cannot rebuild the original data from the low dimensional data with these algorithms. It leads to the result that these methods cannot project out of samples into the low dimension space. Contrarily, multilayer autoencoders overcome the problem and can better reconstruct the original data.

An automatic pavement distress image recognition method based on multilayer autoencoders was proposed in the paper, which reduced the dimension of pavement distress images after being processed and reconstructed them with multilayer autoencoders. Finally, the type of pavement distress can be recognized. Experiments illustrated that the recognition accuracy from multilayer autoencoders greatly outperformed that from the BP neural network.

## 2 Related Works

### 2.1 Pavement Distress Characteristics and Classifications

Factors affecting pavement distress include traffic conditions on the road, pavement types, the roadbed, pavement location and even the local climate. Pavement damage can be classified into two categories: functional distress and structural distress [19]. The functional distress mainly deals with ride quality and safety of pavement surface. The common phenomena of the functional distress are waves, popouts, holes, patching, rutting and weeping, etc. The structural distress is associated with the ability of the pavement to carry the design load. Cracking and joint deterioration are typical structural distresses. The structural distress is divided into alligator cracks, longitudinal cracks, transverse cracks and block cracks. The structural damage belongs to the early damage. Once there are structural distresses on the road, it is impossible to discuss other performances of the road. Therefore, we only consider the detection and recognition of the structural distresses here.

## 2.2 Multilayer Autoencoders

Multilayer autoencoders can project the high dimensional data into the low dimensional space with an adaptive multilayer encoder and reconstruct the original data from the low dimensional embedding with a similar decoder. The common part of the encoder and the decoder, namely the coder layer, is just the required low dimensional data. A better training result will be gained with the gradient descent method if the initial weights in the autoencoders are near the optimal solutions. For the rand initial weights, the multilayer autoencoders often trap in local minimal. To do this, Hinton and Salakhutdinov [16] use a two layer network called the restricted Boltzmann machine (RBM) [20] to train the initial values of the network. The procedure is called as pre-training of the network. Multilayer autoencoders, which is shown in Fig. 1, contain three procedures: pre-training, unrolling and fine-tuning [17].

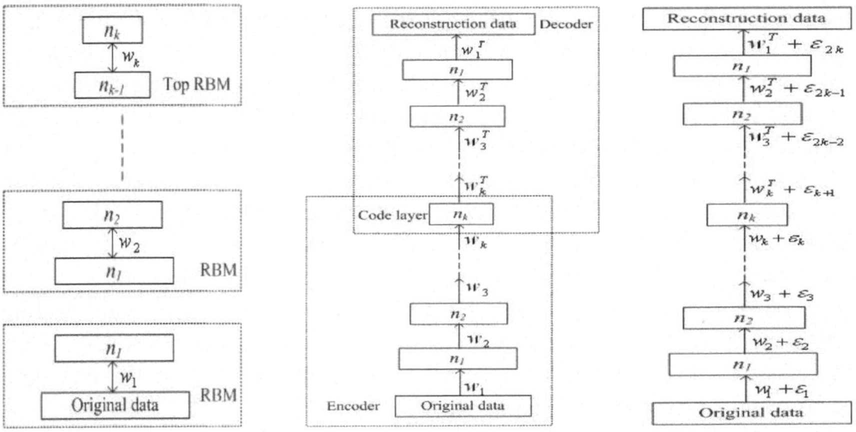


(a) Pre-training (b) Unrolling (c) Fine-tuning

**Fig. 1.** The structure of multilayer autoencoders

In the multilayer autoencoders, one can get better training results with the gradient descent method only if the initial weights are close to the optimal solutions. The network usually traps in poor local minimum while the initial values are randomly selected. The goal of pre-training is to obtain a better initial weight matrix. The process of pre-training utilizes RBM to solve the initial weights of multiple unconnected bi-directional deep neural networks. The network's learning can be executed in single layer each time while the initial weights are computed with RBM.

Namely, we can separately initialize each bi-directional deep neural network. The initial values through such means are very near to the optimal solutions. The single bi-directional deep neural networks obtained from the pre-training procedure are unrolled to multiple bi-directional deep neural networks. After unrolling the network, it is divided into two intersecting components: encoder and decoder. The weight $W$ gained from the pre-training procedure is taken as the initial weight of the encoder and the transposition $W^T$ is taken as the initial weight of the decoder. The encoder is the part of finding the low dimensional manifold and the decoder is the part of reconstructing the original data. Their common part called the coder layer is the required low dimensional data, which reflects the essence of the input data. During the fine-tuning, the weights of each layer are continuously adjusted with the gradient descent method until the error reaches the minimum.

# 3 Pavement Distress Image Recognition Based on Multilayer Autoencoders

If each pavement distress image is taken as a point in the image sample space, the number of pixels contained in each image is the dimension of the whole sample space. Apparently, pavement distress images have higher dimension. It is very hard to gain satisfactorily results with the existing processing methods. Manifold learning methods can find the essential characteristic hidden behind pavement distress images, which can help better recognize and detect distress images. Multilayer autoencoders build a bi-directional map between the high dimensional data and the low dimensional representation, which can directly classify and recognize data instead of executing the classifying algorithms. Therefore, we built a recognition model based multilayer atuoencoders for pavement distress images, which was displayed in Fig 2.

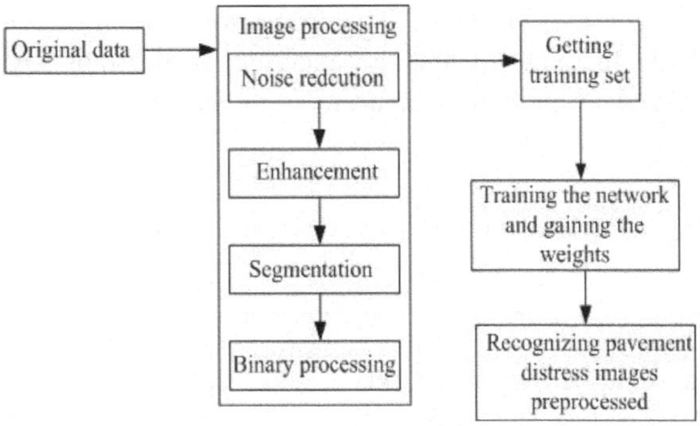


**Fig. 2.** Pavement distress image recognition model based on multilayer autoencoders

In the model, the original pavement images are firstly treated through image noise reduction, enhancement, segmentation and binary processing. Then the structure of multilayer autoencoders is built through training the network on the processed data set. Finally, we can recognize the type of pavement distress images with the trained network.

# 4 Experiments

To validate the effectiveness of the proposed model, we executed the algorithm on the data set self-collected. The data set consists of 400 original pavement images, which contains four categories: alligator cracks, longitudinal cracks, transverse cracks and no distress. Each pavement image has 100x100 pixels.

In experiments, we used two methods to process pavement distress images. One was the method adapted in [21]. In the method, the images were firstly removed the uneven illumination and the shadows with the gray-scale transformation method. Then the image noise was eliminated with the stationary wavelet denoising algorithm. And the images were enhanced with the lateral inhibition method. Next, the images were segmented with the block threshold combining P-tile and FCM histogram. Later, the images were removed speckles with the mathematical morphology method. Lastly, the images were binary processed. Another processing procedure contained several steps. At first, color images were transformed into gray scale images. Then the gray scale images were filtered with a median filter. Thirdly, the images were sharpened with Laplacian method. Next the images were segmented with Otsu's threshold method. Finally, the images were processed with the morphological method. The above two image processing procedures were respectively denoted as process-one and process-two.

We obtained two data sets through processing the original pavement images with the above two processing methods. Each set contained 400 image samples. Each set was divided into two subsets. One was the training set containing 300 samples; the other was the testing set including 100 images. In the training set and the testing set, each class contained the same sample number. At the same time, the class files corresponding to the training set and the testing set were created. We respectively ran two algorithms on two data sets: multilayer autoencoders and BP neural network. Each method was executed five times. The recognized rate for 100 testing samples was given in table 1. In experiments, the encoders in the multilayer autoencoders contained five layers. The number of neurons included in each layer is respectively 10000, 6000, 3000, 1500 and 900.

**Table 1.** The recognized accuracy of different algorithms combined with various processing procedures

| Algorithm | 1 | 2 | 3 | 4 | 5 |
|---|---|---|---|---|---|
| Process-one + multilayer autoencoders | 0.76 | 0.75 | 0.79 | 0.78 | 0.78 |
| Process-two + multilayer autoencoders | 0.63 | 0.63 | 0.65 | 0.66 | 0.66 |
| Process-one + BP neural network | 0.68 | 0.65 | 0.66 | 0.67 | 0.69 |
| Process-two + BP neural network | 0.60 | 0.64 | 0.63 | 0.62 | 0.63 |

From the experimental results, the recognition accuracy of multilayer autoencoders was much higher than that of BP neural network on the data set. Simultaneously, the accuracy of combining multilayer autoencoders with process-one was superior to that of combining multilayer autoencoders with process-two. However, the whole accuracy was not high for all cases. The recognition accuracy from combining multilayer autoencoders with process-one is highest.

For 100 testing samples, the concrete classified results were shown in table 2 and table 3 with multilayer autoencoders combined with two different image processing procedures. In tables, the row expressed the true class of samples and the column was the class obtained from the network. Data in tables were the mean value of five times, which was the number of samples in the true class wrongly classified into the class located in the column.

**Table 2.** The average classified results of combining multilayer autoencoders with process-one

| Type | Longitudinal cracks | Transverse cracks | Alligator cracks | No distress |
|---|---|---|---|---|
| Longitudinal cracks | 16.8 | 5.6 | 1.2 | 1.4 |
| Transverse cracks | 5.4 | 17.6 | 0.8 | 1.2 |
| Alligator cracks | 0.4 | 0.2 | 23.8 | 0.6 |
| No distress | 1.6 | 1.8 | 0.6 | 21 |

**Table 3.** The average classified results of combining multilayer autoencoders with process-two

| Type | Longitudinal cracks | Transverse cracks | Alligator cracks | No distress |
|---|---|---|---|---|
| Longitudinal cracks | 13.2 | 7 | 1.2 | 3.6 |
| Transverse cracks | 6.4 | 15.2 | 0.8 | 2.6 |
| Alligator cracks | 1.8 | 1.4 | 19.6 | 2.2 |
| No distress | 3.2 | 3.8 | 1.4 | 16.6 |

As shown in table 2 and table 3, the number of samples belonging to longitudinal cracks as transverse cracks and the number of samples belonging to transverse cracks as longitudinal cracks were most in all samples recognized wrongly. The sum of the two cases was 11 for 100 testing samples in combining multilayer autoencoders with

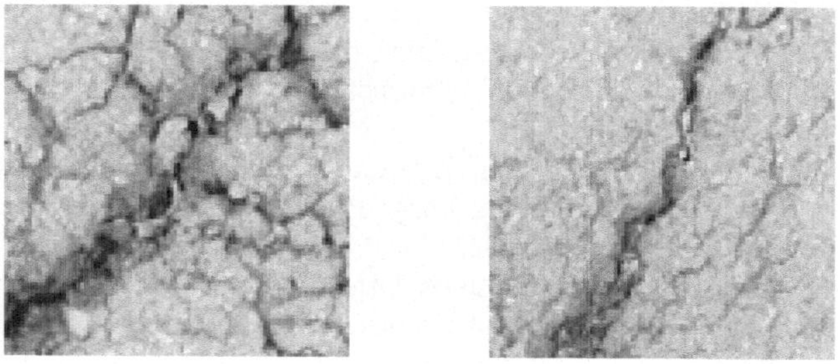

(a) Longitudinal cracks (b) Transverse cracks

**Fig. 3.** Special longitudinal cracks and transverse cracks process-one and the sum was 13.4 in combining multilayer autoencoders with process-two. It was the most principal reason that leads to the lower whole accuracy. And the images recognized wrongly in transverse cracks and longitudinal cracks mainly gathered the two types of images shown in Fig 3. The error can be degraded through increasing the number of distress image's types.

## 5 Conclusions

Pavement distress images are large scale high dimensional nonlinear data. Manifold learning algorithms have a unique advantage in dealing with such data, which can find the intrinsic structure hidden in these high dimensional data. However, most of manifold learning algorithms cannot construct an inverse mapping. Multilayer autoencoders, which overcome the problem, can better reconstruct the original data. An automatic pavement distress image recognition method based on multilayer autoencoders was proposed in the paper, which combined the image processing methods with multilayer autoencoders. In the method, pavement distress images were decreased dimensions and reconstructed with multilayer autoencoders after they were processed with the image processing methods. Then the class was recognized with the network. In experiments, we compared the results from multilayer autoencoders with the BP neural network. Experiments displayed that the accuracy from the proposed method greatly outperformed that from the BP neural network. And the results from combining multilayer autoencoders with process-one were best. In the future, we will increase the types of distress images and check the proposed algorithm on more pavement distress image sets.

**Acknowledgments.** This work was supported by Tianjin Research Project of Application Foundation and Advanced Technology (No. 10JCZDJC16000).

## References

1. Kelvin, C.P.W.: Designs and Implementations of Automated Systems for Pavement Surface Distress Survey. Journal of Infrastructure Systems 6(1), 24–32 (2000)
2. Kelvin, C.P.W., Robert, P.E.: Investigation of Image Archiving for Pavement Surface Distress Survey. Mack-Blackwell Transportation Center, University of Arkansas, Fayetteville (1999)
3. Zhang, J., Sha, A., Gao, H., Sun, Z.: Automatic Pavement Crack Recognition and Evaluation System Based on Digital Image Processing. Journal of Chang'an University (Natural Science Edition) 24(2), 18–22 (2004)
4. Sun, Y.: Automated Pavement Distress Detection Using Advanced Image Processing Techniques. M.S. Thesis, College of Engineering, The University of Toledo (December 2009)
5. SiriPhan, J.: Development of a New Digital Pavement Image Processing Algorithm for Unified Crack Index Computation. A Dissertation Submitted to the Faculty of the University of Utah (1997)

6. Velisky, S.A., Kirsehke, K.R.: Design Considerations for Automated Pavement Crack Sealing Machinery. In: 2nd International Conference on Applications of Advanced Technologies in Transportation Engineering, pp. 77–80 (1991)
7. Peng, H., Li, J., Mu, J.: A Method of Edge Detection Based on Canny Regulation for Detecting Road Surface Image. Journal of Xi'an Institute of Technology 22(4), 322–325 (2002)
8. Xiao, W., Zhang, X., Huang, W.: A New Method for Distress Automation Recognition of Pavement Surface Based on Density Factor and Image Processing. Journal of Transportation Engineering and Information 2(2), 82–89 (2004)
9. Li, G., He, Y., Zhao, Y.: Automatic Recognition Algorithm of Pavement Defect Image Based on OTSU and Maximizing Mutual Information. Microelectronics & Computer 26(7), 241–243, 247 (2009)
10. Tenenbaum, J.B., de Silva, V., Langford, J.C.: A Global Geometric Framework for Nonlinear Dimensionality Reduction. Science 290(5500), 2319–2323 (2000)
11. Nadler, B., Lafon, S., Coifman, R.R., Kevrekidis, I.G.: Diffusion Maps, Spectral Clustering and the Reaction Coordinates of Dynamical Systems. Applied and Computational Harmonic Analysis 21, 113–127 (2006)
12. Schölkopf, B., Smola, A.J., Müller, K.R.: Nonlinear Component Analysis as a Kernel Eigenvalue Problem. Neural Computation 10(5), 1299–1319 (1998)
13. Rowei, S.T., Saul, L.K.: Nonlinear Dimensionality Reduction by Locally Linear Embedding. Science 290(5500), 2323–2326 (2000)
14. Belkin, M., Niyogi, P.: Laplacian Eigenmaps for Dimensionality Reduction and Data Representation. Neural Computation 15(6), 1373–1396 (2003)
15. Donoho, D.L., Grimes, C.: Hessian Eigenmaps: New Locally Linear Embedding Techniques for High-Dimensional Data. Proceedings of the National Academy of Sciences 102(21), 7426–7431 (2005)
16. Zhang, Z., Zha, H.: Principal Manifolds and Nonlinear Dimensionality Reduction via Tangent Space Alignment. SIAM Journal of Scientific Computing 26(1), 313–338 (2005)
17. Hinton, G.E., Salakhutdinov, R.R.: Reducing the Dimensionality of Data with Neural Networks. Science 313(5786), 504–507 (2006)
18. Hu, Z., Song, Y.: Dimensionality Reduction and Reconstruction of Data Based on Autoencoder Network. Journal of Electronics & Information Technology 31(5), 1189–1192 (2009)
19. Al-Mansour, A.: Flexible Pavement Distress Prediction Model for the City of Riyadh. Emirates Journal for Engineering Research 9(1), 8–88 (2004)
20. Hinton, G.E.: Training Products of Experts by Minimizing Contrastive Divergence. Neural Computation 14(8), 1771–1800 (2000)
21. Li, G.: Pavement Distress Recognition Based on Image. M.S. Thesis, School of Computer Science and Engineering, Hebei University of Technology (2008)

# MPSO-Based Operational Conditions Optimization in Chemical Process: A Case Study

Lirong Xia, Jizheng Chu, and Zhiqiang Geng*

School of Information Science and Technology,
Beijing University of Chemical Technology, Beijing 100029,China
xlrmaomao@163.com, {chujz,gengzhiqiang}@mail.buct.edu.cn

**Abstract.** A multi-swarm PSO (MPSO) was proposed, with which the whole swarm is divided into by K-means clustering algorithm randomly to accelerate searching process of global optimum. The big swarm clustering will obey the standard PSO principle to search the global optimal result, which the number of particle is more than a threshold. The small swarm clustering will search randomly inner neighborhood of the global optimal value, and then the outlier particle does not care about the optimal result but flies freely according to themselves velocities and positions. The proposed algorithm enhances its global searching space, and enriches particles' diversity in order to let particles jump out local optimization points. Testing and comparing results with standard PSO and linearly decreasing weight PSO using several benchmark functions show the proposed algorithm is better than other algorithms. Furthermore, the MPSO algorithm is used to optimize the operational conditions in a chemical process case for an ethylene cracking furnace.

**Keywords:** PSO, Ethylene cracking furnace, Operational optimization.

## 1 Introduction

The Particle swarm optimization (PSO) algorithm [1,2] is put forward by Kennedy and Eberhart at 1995, which is based on society behavior simulation of bird swarms. Recently, the PSO algorithm is studied deeply and used widely in many fields. Due to its advantages of parallelism, adaptability, intelligence and so on, PSO has been applied to multi-objective optimization, model identification, intelligent control and decision support, etc. but it is easily to immerse the local optimal value and the precision is not high. The convergence speed and premature convergence have been the most important factors that affect PSO optimization performance [3,4]. In order to stimulate continued evolution of swarms, avoid premature convergence and stagnation, many researchers have pointed out state variables of the whole swarm or some particles could be re-valued according to certain standards to keep the diversity of the population and the evolution of the algorithm [5,6]. Shi and Eberhart had

* Corresponding author: This work is partly supported by the fundamental research funds for the central universities (ZZ1136).

J. Lei et al. (Eds.): AICI 2012, LNAI 7530, pp. 674–681, 2012.

studied the weight which affects the optimization performance, at the same time they put forward an improved PSO by linearly decreasing weight, which the bigger weight can jump out local extremum point, and the smaller weight is easy to converge[7,8]. Suganthan put forward a PSO model based on dynamic neighborhood [9] to improve the PSO algorithm convergent performance, that is to say, the global extremum is substituted by the neighborhood extremum of every particle. Some authors used the multi-swarm competitive and cooperative method to improve the PSO algorithm effectively [10,11]. Several papers studied the Fuzzy c-mean clustering using original PSO algorithm, which the velocity equation of PSO is verified based on FCM cluster analysis of the current particles' positions [12-15].

On the basis of researching PSO's advantages and disadvantages, a multi-swarms parallel PSO algorithm is proposed in this paper. It can overcome the disadvantage about the local optimization and prematurity problem. It is used to optimize the widely used benchmark testing functions, and compare with Linearly Decreasing Weight PSO (LDWPSO) and standard PSO, the results show the proposed MPSO has better optimization performance. Furthermore, it is used to optimize the multi-state of ethylene cracking furnace, and the application results are satisfying in industrial process.

## 2 A Multi-swarm PSO

The whole swarm is clustered using K-means algorithm, and split into different sub-swarms randomly, the bigger particle group obeys the standard PSO principle to search the optimal result, when the number of particle is more than a threshold, that is to say, the particle changes itself position by variant velocity to get the optimization. The smaller clustering searches randomly inner neighborhood of the optimal result to enhance the searching probability for the optimization so as to improve the probability to jump out from the local optimization. The outlier particle does not care about the optimal result and flies freely everyone according to oneself velocity and position, namely is the velocity updating equation doesn't include the third part which is substituted by rand()*Vmax, where Vmax is the maximal velocity. So the outlier particle becomes a single searching body to update oneself, and increases remarkably the diversity of swarm in searching process in order to decrease the clustering of particles.

The basic running process of MPSO is described as follows.

Step1: Initiate the parameters. Set the accelerating constant $c_1, c_2$ and evolution iterative times $T_{\max}$. Let T=1 and produce randomly m particle $X_1, X_2, \ldots, X_m$ in definite space R.

Step2: Call K-means clustering algorithm with minimum total value function $J = \sum_{i=1}^{k}( \sum_{j, x_j \in C_i}(d(x_j - c_i))$, divide $X(t)$ automatically into $X_1(t), X_2(t), \ldots, X_n(t)$. Notice that the clustering radius needs to be selected appropriately, of which size determines the number of the clustering. Furthermore, clustering too little will reduce the diversity of swarms, while too much clustering can also lead to loss of the significance of multi-swarms algorithm.

Step3: Set the threshold. Let $Q_1$ be the big cluster, $Q_2$ be the small cluster, $Q_3$ be the outlier. $Q_1, Q_2, Q_3$ compose the initial swarm $X(t)$, and produce the initial speed of every particle $v_1, v_2, ..., v_m$ to compose the matrix $V(t)$.

Step4: Estimate swarm $X(t)$, namely, calculate the estimating value of every particle in every dimension.

Step5: Compare the current estimating value with the optimization value $P_{best}$, if the current value is better than $P_{best}$, then let $P_{best}$ be the current value in $n$ dimensions.

Step6: Compare the estimating value of every particle with the optimization value of whole swarm, if the current value is better than the global optimal value $gbest$, then let $gbest$ be the estimating value of current particle.

Step7: Update particles.

(1) If $x_i$ belongs to $Q_1$, namely the big cluster, then update the speed and position of every particle according to equation (1) and equation (2).

$$\mathrm{v}_{\mathrm{id}}^{\mathrm{k+1}} = \mathrm{w} \times \mathrm{v}_{\mathrm{id}}^{\mathrm{k}} + c_1 \times rand(\ ) \times (\mathrm{P}_{\mathrm{id}} - \mathrm{x}_{\mathrm{id}}^{\mathrm{k}}) + c_2 \times rand(\ ) \times (\mathrm{P}_{\mathrm{gd}} - \mathrm{x}_{\mathrm{id}}^{\mathrm{k}}) \tag{1}$$

$$\mathrm{x}_{\mathrm{id}}^{\mathrm{k+1}} = \mathrm{x}_{\mathrm{id}}^{\mathrm{k}} + \mathrm{v}_{\mathrm{id}}^{\mathrm{k+1}} \tag{2}$$

(2) If $x_i$ belongs to $Q_2$, then update the position of every particle according to equation (3) and produce a new particles $X(t+1)$. Where $X_{d\max}$ is the searching space of $d$ dimensions.

$$\mathrm{x}_{\mathrm{id}}^{\mathrm{k+1}} = \mathrm{P}_{\mathrm{gd}} + \mathrm{c}_3 \times \mathrm{rand}() \times \mathrm{x}_{\mathrm{dmax}} \tag{3}$$

(3) If $x_i$ belongs to $Q_3$, then update the speed and position of every particle according to equation (4) and (5), and produce new particles $X(t+1)$.

$$\mathrm{x}_{\mathrm{id}}^{\mathrm{k+1}} = \mathrm{x}_{\mathrm{id}}^{\mathrm{k}} + \mathrm{v}_{\mathrm{id}}^{\mathrm{k+1}} \tag{4}$$

$$\mathrm{v}_{\mathrm{id}}^{\mathrm{k+1}} = \mathrm{w} \times \mathrm{v}_{\mathrm{id}}^{\mathrm{k}} + c_1 \times rand(\ ) \times (\mathrm{P}_{\mathrm{id}} - \mathrm{x}_{\mathrm{id}}^{\mathrm{k}}) + rand(\ ) \times \mathrm{V}_{\max} \tag{5}$$

Step8: Judge the end condition. If the result is satisfying then stop, otherwise let T=T+1 and turn to step2. Where, the end condition is maximal iterative times $T_{\max}$, or the estimation value is smaller than the giving precision.

To compute inertial weights $w$, we use the LDWPSO method proposed by Shi, namely use equation (6) to change the weights.

$$w = (w_1 - w_2) \times \frac{MAX_{iter} - iter}{MAX_{iter}} + w_2 \tag{6}$$

Where, $w_1$ and $w_2$ are the initial and end value of inertial weight, respectively. Normally, $w_1$ =0.9 and $w_2$ =0.4, $MAX_{iter}$ and $iter$ are the maximal and current iterative times, respectively.

# 3 Experiments and Testing

For analyzing and testing the proposed MPSO, we use five Benchmark functions by other researcher common used. The functions are in the following.

(1)Girewank $f_1(x) = \frac{1}{4000}\sum_{i=1}^{n} x_i^2 - \prod_{i=1}^{n}\cos(\frac{x_i}{\sqrt{i}}) + 1$ $[-600,600]^n$

(2) Rosenbrock $f_2(x) = \sum_{i=1}^{n-1}[100(x_{i+1} - x_i^2)^2 + (x_i - 1)^2]$ $[-30,30]^n$

(3) Rastrigin $f_3(x) = \sum_{i=1}^{n}[x_i^2 - 10\cos(2\pi x_i) + 10]$ $[-5.12,5.12]^n$

(4) Sphere $f_4(x) = \sum_{i=1}^{n} x_i^2$ $[-100,100]^n$

The four functions have the global minimal value 0 at interval [0,…,0]. Now we use the PSO, LDW PSO and MPSO to optimize the four functions. Three algorithms parameters are set in the following. PSO's parameters are set as follows, c1=c2=2, w=0.7. LDWPSO's parameters are set, c1=c2=2, w1=0.9, w2=0.4. MPSO's parameters are set, let particle swarm scale be n, separate into three groups [n-20, 10, 10] respectively, c1=c2=2, w1=0.9, w2=0.4, c3=0.2. Every algorithm iterates 50 times during the experimental process, and each average optimal value is gained in Table 1.

**Table 1.** Results comparisons of algorithm testing

| Funcitons | Algorithms | Min. | Max. | Avg. | Stand Var. |
|---|---|---|---|---|---|
| | PSO | 0.1008 | 0.3779 | 0.2370 | 0.0129 |
| Griewank | LDWPSO | 0 | 0 | 0 | 0 |
| | MPSO | 0 | 0 | 0 | 0 |
| | PSO | 5.8154 | 14.3469 | 9.8555 | 12.9331 |
| Rosenbrock | LDWPSO | 7.8590 | 8.1224 | 8.0586 | 0.0125 |
| | MPSO | 8.0289 | 8.7280 | 8.3300 | 0.1318 |
| | PSO | 10.9446 | 18.9045 | 13.7305 | 9.6030 |
| Rastrigin | LDWPSO | 0 | 0 | 0 | 0 |
| | MPSO | 0 | 0 | 0 | 0 |
| | PSO | 1.25E-05 | 2.51E-04 | 1.01E-04 | 8.29E-09 |
| Sphere | LDWPSO | 9.89E-31 | 6.56E-28 | 1.35E-28 | 8.49E-56 |
| | MPSO | 3.18E-43 | 2.13E-35 | 4.31E-36 | 8.98E-71 |

Table 1 shows that the searching optimal value gained by MPSO algorithm proposed in the paper is better than the values gained by PSO & LDWPSO in the same condition, that is to say MPSO has the stronger global searching performance than the other two algorithms. Fig.1 –Fig.4 show the iterative process of the proposed multi-swarm PSO.

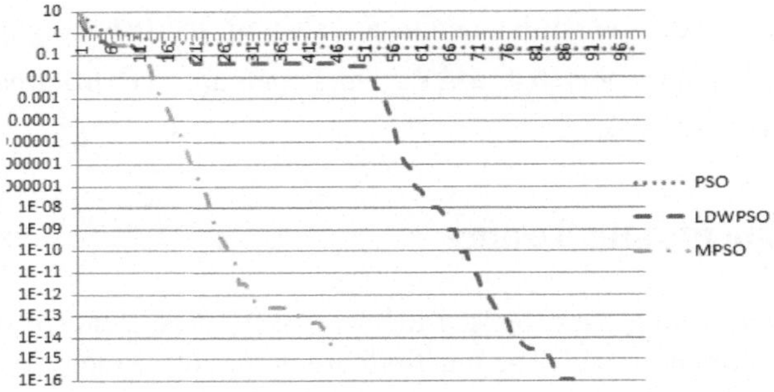


**Fig. 1.** Griewank's optimizaton iterative curves

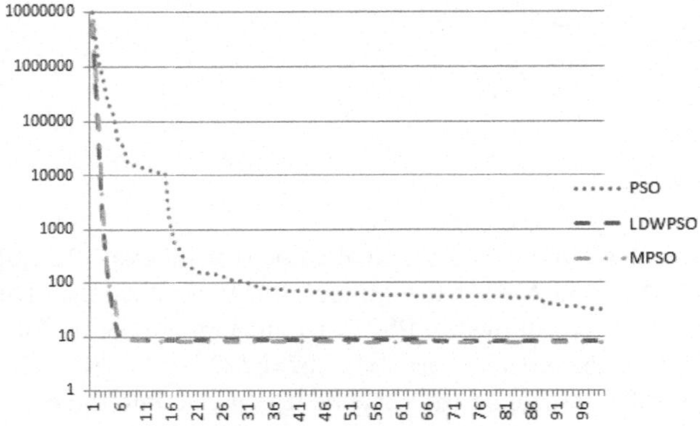


**Fig. 2.** Rosenbrock's optimization interative curves

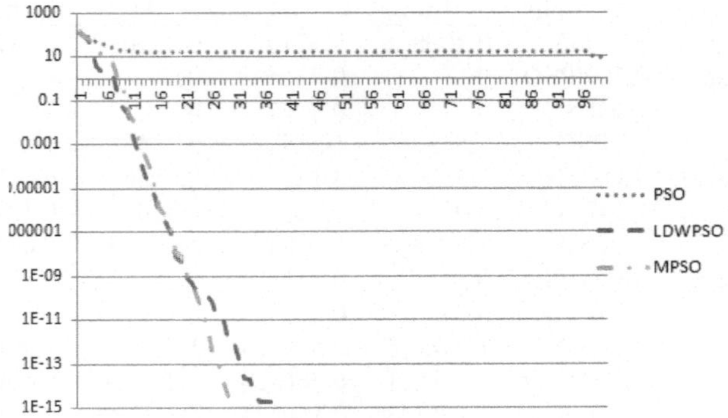


**Fig. 3.** Rastrigin's optimization iterative curves

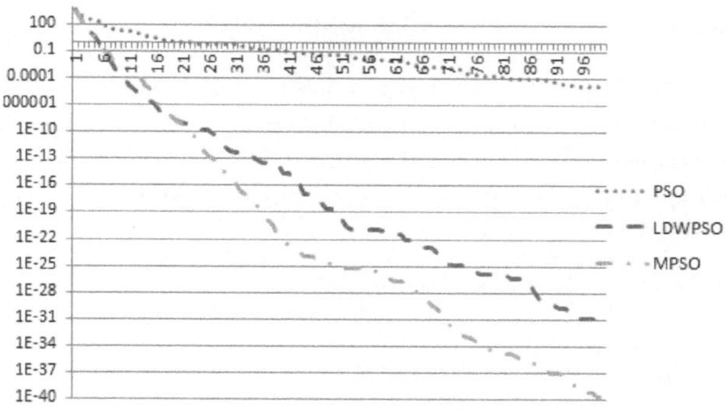


**Fig. 4.** Sphere's optimization iterative curves

From the four functions' optimization results and iterative processes, we can see the performance of proposed MPSO is overwhelming the PSO and LDWPSO.

# 4 A Case Study

## 4.1 Product Model

The cracking furnace reactor system consists of a number of parallel tubular reactors, residing in and sharing a common furnace. In this paper, the furnace structure model is 2-1 type in radiation tubes, including 6 groups radiation tubes and every group includes 8 tubes, the total length of tubes is 28.6 meter.

We use the software SPSS12.0 to build the nonlinear regress model of 4 products according to naphtha samples obtained by experiments and running data in practical industry of one plant.

The model type is $Y_i = f(x_1, x_2, x_3, x_4, x_5, x_6, x_7)$, $x_1$ is specific gravity under 289.6K, $x_2$ is middle average boiling point, $x_3$ is BMCI value, $x_4$ is cracking temperature, $x_5$ is export pressure, $x_6$ is sticking time, $x_7$ is diluting ratio. The 4 detail cracking product models are described as follows:

$Y_1$(hydrogen)=2.999-5.196$X_1$+0.001$X_2$+0.013$X_3$-0.001$X_4$+3.29E-06$X_4$^$^2$-0.198$X_5$+0.559$X_6$ +0.063$X_7$

$Y_2$(methane)=-134.068+9.627$X_1$-0.007$X_2$-0.059$X_3$+0.278$X_4$-0.00014$X_4$^$^2$+3.892$X_5$+10.986$X_6$ -1.164 $X_7$

$Y_3$(ethylene)= -728.722+236.661$X_1$-0.076$X_2$-0.954$X_3$+1.389$X_4$-0.0008 $X_4$^$^2$-11.832 $X_5$ +10.563$X_6$ +2.366$X_7$

$Y_4$(propylene)=-820.679+0.514$X_1$-0.027$X_2$-0.351$X_3$+1.916$X_4$-0.00119$X_4$^$^2$-1.469$X_5$ +3.0198$X_6$ -0.418$X_7$

The values in Table 2 are regression coefficient of each variable. The variables rang are showed as follows.

0.789≤X1≤10.843,187≤X2≤300,19≤X3≤30,760≤X4≤845,0.17≤X5≤0.25, 0.16≤X6≤0.25, 0.6≤X7≤1.

### 4.2 Operational Optimization of Cracking Model

According to operational requirement, cracking furnace can be implemented operational optimization schemes, which are the single product maximal yield operating optimization, the maximal sum of multi-product yields operating optimization. In the paper, the optimization objective is selected by maximal sum of ethylene and propylene yields as follows.

Objective: $maxf(t) = \sum (y_{C_2H_4} + y_{C_3H_6})$

S.t. $760 \le X4 \le 845$.
$0.17 \le X5 \le 0.25$.
$0.16 \le X6 \le 0.25$.
$0.6 \le X7 =$ (diluting vapor (t/h)/oil feed(t/h)) $\le 1$.
$32.17 \le$ oil feed (t/h) $\le 50.55$.
$24.13 \le$ diluting vapor (t/h) $\le 37.91$.

We use the proposed MPSO to apply in the practical cracking furnace in an actual refinery. According to operating requirements, the second optimization object is selected, namely, the maximal sum of ethylene and propylene. The algorithm parameters are set as follows, particle number m = 50, $c_1=c_2=2, w_1=0.9$, $w_2=0.4$, $c_3=0.2$, $V_{max}=0.2$ and $T_{max}=200$. The yields of ethylene and propylene increase by 0.59% and 0.84% on the average, as diagramed in Fig.5 and Fig.6, thus to bring about significant economic benefits for ethylene cracking equipment.

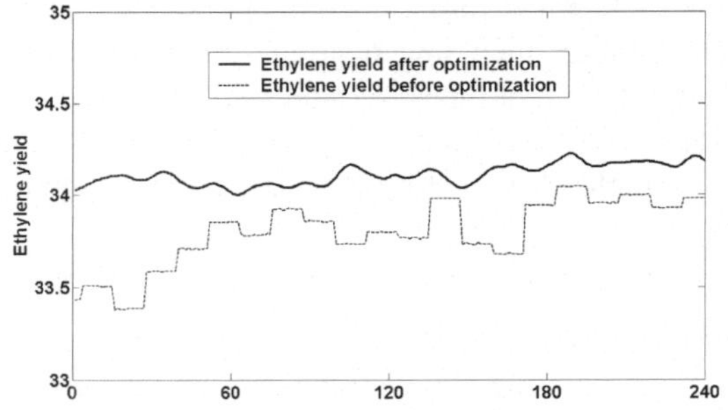


**Fig. 5.** The ethylene yield between after optimization and before

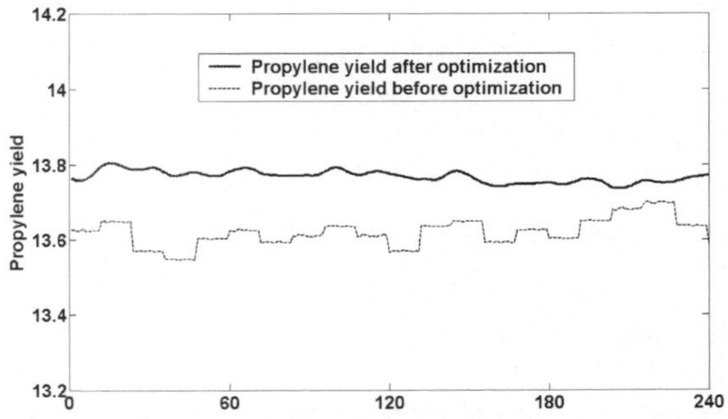


**Fig. 6.** The propylene yield between after optimization and before

From the Fig.5 and Fig.6, we can find easily the optimization operational conditions can improve the products yields. The practical applications of MPSO in ethylene cracking furnace show the proposed algorithm is effective.

## 5 Conclusions

This paper proposes a multi-swarm PSO algorithm based on K-means clustering, which enhances the diversity of the swarms. Different swarms adopt different flight strategies with strong global search abilities. The testing results based on Benchmark functions show the effectiveness and superiority of this proposed algorithm. The practical applications of operational optimization in ethylene cracking furnace show that MPSO is effective and satisfying to apply further in industrial process. Furthermore, this proposed method will be added cooperating and co-evolution mechanism into swarms to improve the global searching abilities.

## References


1. Kennedy, J., Eberhart, R.C.: Proceedings of the IEEE International Conference on Neural Network, pp. 1942–1948. IEEE, Piscataway (1995)
2. Eberhart, R.C., Shi, Y.: Proceedings of 2001 Congress on Evolutionary Computation, pp. 81–86. IEEE, Piscataway (2001)
3. Du, W.L., Li, B.: Multi-strategy ensemble particle swarm optimization for dynamic optimization. Information Sciences 178, 3096–3109 (2008)
4. Niu, B., Zhu, Y.L., He, X.X., Shen, H.: A multi-swarm optimizer based fuzzy modeling approach for dynamic systems processing. Neurocomputing 71, 1436–1448 (2008)
5. Knnk, T., Vesterstr, J.S., Riget, J.: Particle swarm optimization with spatial particle extension. In: Proc. of the 2002 Congress on Evolutionary Computation, pp. 1472–1479 (2002)
6. Xie, X.F., Zhang, W.J., Yang, Z.L.: Adaptive particle swarm optimization individual level. In: 6th International Conference on Signal Processing, pp. 1215–1218 (2002)
7. Shi, Y.H., Eberhart, R.C.: 1998 Annual Conference on Evolutionary Programming, San Diego (March 1998)
8. Shi, Y.H., Eberhart, R.C.: Proceedings of the Congress on Evolutionary Computation, Seoul, Korea (2001)
9. Suganthan, P.N.: Proc. of the Congress on Evolutionary Computation, Washington D.C, pp. 1958–1962 (1999)
10. Goh, C.K., Tan, K.C., Liu, D.S., Chiam, S.C.: A competitive and cooperative co-evolutionary approach to multi-objective particle swarm optimization algorithm design. European Journal of Operational Research 202, 42–54 (2010)
11. Niu, B., Zhu, Y.L., He, X.X., Wu, H.: MCPSO: A multi-swarm cooperative particle swarm optimizer. Applied Mathematics and Computation 185, 1050–1062 (2007)
12. Izakian, H., Abraham, A.: Fuzzy c-means and fuzzy swarm for fuzzy clustering problem. Expert Systems with Application 38, 1835–1838 (2009)
13. Wang, L., Liu, Y.S., Zhao, X.X., Xu, Y.Q.: Particle swarm optimization for fuzzy c-means clustering. In: Proceedings of the Sixth World Congress on Intelligent Control and Automation, Dalian, China, pp. 6055–6058 (2006)
14. Ichihashi, H., Honda, K., Notsu, A., Ohta, K.: Fuzzy c-means classifier with particle swarm optimization. In: Proceedings of the IEEE International Conference on Fuzzy Systems, HongKong, China, pp. 207–215 (2008)
15. Mei, C.L., Zhou, D.W.: An improved particle swarm optimization with fuzzy c-means clustering algorithm. In: Proceedings of the International Conference on Intelligent Human-Machine Systems and Cybernetics, Hanzhou, China, pp. 118–122 (2009)

# Velocity-Free Multi-Objective Particle Swarm Optimizer with Centroid for Wireless Sensor Network Optimization

Ying Gao, Lingxi Peng, Fufang Li, MiaoLiu, and Xiao Hu

Department of Computer Science and Technology, Guangzhou University
Guangzhou, 510006, P.R. China
falcongao@sina.com.cn

**Abstract.** A velocity-free multi-objective particle swarm optimizer with centroid is proposed and applied to optimization of wireless sensor network. Different from the standard PSO, particles in swarm only have position without velocity in the algorithm. Besides, not only the personal best position and the global best position but also the centroid is considered to update the particle position. The initial swarm is generated using the opposition-based learning, and an archive with maximum capacity is used to maintain the non-dominated solutions. The global best solution is selected from the archive on the basis of the diversity of the solutions, and the crowding-distance measure is used for the diversity measurement. The archive gets updated with the inclusion of the non-dominated solutions from the combined population of the swarm and current archive, and the archive which exceeds the maximum capacity is cut using the diversity consideration. The proposed algorithm is applied to some well-known benchmark and optimization of wireless sensor network by maximizing network coverage and lifetime. The relative experimental results show that the algorithm has better performance and is effective.

**Keywords:** Multi-objective particle swarm optimizer, Centroid, Opposition-based learning, Non-dominated solution, Wireless sensor networks.

## 1 Introduction

Multi-objective optimization problems (MOOPs) are widely encountered in various fields of science and technology. In these problems there is a group of conflicting objectives to be simultaneously optimized. Several computational intelligence based approaches, namely, evolutionary computation[1] and swarm intelligence have been used for solving MOOPs. Particle swarm optimization (PSO) [2] belong to the swarm intelligence domain of computational intelligence. And it has proved to be very efficient and capable of providing competitive solutions in many application domains. A few researchers[3-4] have studied the extension of PSO to solve multi-objective optimization problem in continuous domain.

Wireless sensor networks (WSNs) have gained increasing attention from both the research community and actual users in the last years. In many applications, the deployment of WSNs system has generated the WSNs optimization problem that needs to be solved in order to operate the large-scale sensors in an optimal fashion [5-6]. In general, it aims to optimize a set of objectives (coverage, lifetime, etc.) simultaneously by

J. Lei et al. (Eds.): AICI 2012, LNAI 7530, pp. 682–689, 2012.

adjusting the control variables(the coordinates of the sensors, the number of the sensors, and other parameters, etc.) of the system. As a result, in the large-scale deployment environment, the problem is a high-dimensional non linear multi-objective optimization problem with a large number of variables and uncertain parameters. However, existing approaches generally focus on a single objective (ignoring the others), or combine multiple objectives into a single function to be optimized, to facilitate the application of classical optimization algorithms.

The standard PSO uses update rules including velocity, personal best position and global best position. The personal best position visited by itself (i.e. its own experience) and the global best position (i.e. the experience of swarm particles) are two very important position. They influence the position of a particle greatly. In fact, the swarm's centroid[7] is also an important position and includes more information of the swarm. It has more opportunities to guide the search of the whole swarm. In this paper, a velocity-free PSO with centroid is proposed for MOOPs. Different from the standard PSO, particles in swarm only have position without velocity in the algorithm. Besides, not only the personal best position and the global best position but also the centroid is considered to update the particle position. The initial swarm is generated using the opposition-based learning, and an archive with maximum capacity is used to maintain the non-dominated solutions. A crowding-distance measure is used to select the global best solution from the archive based on the diversity of the solutions. The archive gets updated with the inclusion of the non-dominated solutions from the combined population of the swarm and current archive, and the archive which exceeds the maximum capacity is cut using the diversity consideration. Because of discarding the particle velocity and using swarm's centroid information, the algorithm is the simpler and effective. The proposed algorithm is applied to some well-known benchmark and optimization of WSN. The relative experimental results show that the algorithm has better performance and is effective.

## 2 Multi-Objective Optimization Problems

The general MOOPs can be defined as follows:

$$\begin{aligned} \min\ & F(\mathbf{x}) = \{f_1(\mathbf{x}), f_2(\mathbf{x}), \cdots, f_k(\mathbf{x})\} \\ \text{s.t.}\ & g_i(\mathbf{x}) \le 0, \quad i = 1,2,...,m \\ & h_i(\mathbf{x}) = 0, \quad i = 1,2,...,p \end{aligned}$$

The constrains define the feasible region $\Omega$ and any point in $\Omega$ defines a feasible solution. The $k$ components of the vector $F(\mathbf{x})$ are the criteria to be considered. The concept of optimum commonly adopted in MOOPs is Pareto optimality. Pareto optimality is defined as:

A point $\mathbf{x}^* \in \Omega$ is Pareto optimal if $\forall\, \mathbf{x} \in \Omega$ and $\mathrm{I} = \{1,2,\cdots,k\}$

Either: $\forall i \in \mathrm{I}\ f_i(\mathbf{x}^*) \le f_i(\mathbf{x})$

And, there is at least one $i \in \mathrm{I}$ such that $f_i(\mathbf{x}^*) < f_i(\mathbf{x})$

A vector $\mathbf{x} = (x_1, x_2, \cdots, x_n)$ is said to dominate $\mathbf{y} = (y_1, y_2, \cdots, y_n)$, denoted by $\mathbf{x} \prec \mathbf{y}$, if and only if $\mathbf{x}$ is partially less than $\mathbf{y}$, i.e., $\forall i \in \{1,2,\cdots,k\}\ \ x_i \le y_i$ and, at least for one $i$, $x_i < y_i$.

For a given MOOPs $F(\mathbf{x})$, Pareto optimal set $P^*$ is defined as:

$$P^* = \{\mathbf{x} \in \Omega \mid \neg\exists \mathbf{x}' \in \Omega \quad F(\mathbf{x}') \prec F(\mathbf{x})\}$$

Pareto front $PF^*$ is defined as:

$$PF^* = \left\{F(\mathbf{x}) = (f_1(\mathbf{x}), f_2(\mathbf{x}), \cdots, f_k(\mathbf{x})) \mid \mathbf{x} \in P^*\right\}$$

## 3 Opposition-Based Learning

Opposition-based learning[8] can be utilized in a wide range of learning and optimization field to make algorithms faster. Opposite numbers are defined as follows:

Let $\mathbf{x} = (x_1, x_2, ..., x_n)$ be an $n$-dimensional vector, where $x_i \in [a_i, b_i], i = 1,2,...,n$. The opposite vector of $\mathbf{x} = (x_1, x_2, ..., x_n)$ is defined by $\mathbf{x}' = (x_1', x_2', ..., x_n')$ where $x_i' = a_i + b_i - x_i$.

Assume $F(\mathbf{x})$ is a fitness function vector which is used to measure candidate's Pareto optimality, and $\mathbf{x}' = (x_1', x_2', ..., x_n')$ is the opposite of $\mathbf{x} = (x_1, x_2, ..., x_n)$. Now, if $F(\mathbf{x}') \prec F(\mathbf{x})$, then point $\mathbf{x}$ can be replaced with $\mathbf{x}'$; otherwise we continue with $\mathbf{x}$. Hence, the point and its opposite point are evaluated simultaneously to continue with the fitter one.

## 4 Standard Particle Swarm Optimization

In standard PSO, the position vector and the velocity vector of $i$th particle in m-dimensional search space can be represented as $\mathbf{x}_i (i = 1,2,\cdots,N)$ and $\mathbf{v}_i (i = 1,2,\cdots,N)$ respectively, N is the number of particle. In standard PSO, the swarm is updated by the following equations:

$$\mathbf{v}_i(t+1) = w\mathbf{v}_i(t) + c_1 r_1 (\mathbf{p}_i(t) - \mathbf{x}_i(t)) + c_2 r_2 (\mathbf{p}_g(t) - \mathbf{x}_i(t)) \quad (1)$$

$$\mathbf{x}_i(t+1) = \mathbf{x}_i(t) + \mathbf{v}_i(t+1) \quad (2)$$

Where $\mathbf{p}_i(t) (i = 1, \cdots, N)$ and $\mathbf{p}_g(t)$ are given by the following equations, respectively:

$$\mathbf{p}_i(t+1) = \begin{cases} \mathbf{p}_i(t), & f(\mathbf{x}_i(t+1)) < f(\mathbf{p}_i(t)) \\ \mathbf{x}_i(t+1), & f(\mathbf{x}_i(t+1)) \geq f(\mathbf{p}_i(t)) \end{cases} \quad (3)$$

$$\mathbf{p}_g(t) \in \{\mathbf{p}_1(t), \mathbf{p}_2(t), \cdots, \mathbf{p}_N(t) \mid f(\mathbf{p}_g(t)) = \min\{f(\mathbf{p}_1(t)), f(\mathbf{p}_2(t)), \cdots, f(\mathbf{p}_N(t))\}\} \quad (4)$$

$w$ is inertia weight. $c_1$ and $c_2$ are acceleration coefficients. $r_1$ and $r_2$ are two uniform random numbers in the range $[0,1]$. $f(\mathbf{x})$ is minimum objective function.

## 5 Velocity-Free Multi-Objective PSO with Centroid

In standard PSO, the personal best position and the global best position are two very important positions. Besides, the swarm's centroid is also a very important position, and includes the more information of the swarm. It has more opportunities to guide the search of the whole swarm, and influences the performance greatly. In the proposed algorithm, a particle position update without velocity is modified as:

$$\mathbf{x}_i(t+1)=c_1r_1(\mathbf{p}_i(t)-\mathbf{x}_i(t))+c_2r_2(\mathbf{p}_g(t)-\mathbf{x}_i(t))+c_3r_3(\mathbf{p}_c(t)-\mathbf{x}_i(t)) \quad (5)$$

Where $\mathbf{p}_c$ is the swarm's centroid, and given by the following equations,

$$\mathbf{p}_c(t)=\frac{1}{N}\sum_{i=1}^{N}\mathbf{x}_i(t) \quad (6)$$

$c_1, c_2$ and $c_3$ are acceleration coefficients. $r_1, r_2$ and $r_3$ are three independent uniform random sequences distributed in the range $[0,1]$.

Using Pareto-based multi-objective optimization methodology, our implementation of the velocity-free multi-objective PSO with centroid is described as follows:

```
t=0, initialize swarm S(0)
    FOR  i=1 to N
        Initialize x_i(0)
        p_i(0) ← x_i(0)
    END FOR
    EVALUATE(S(0))
    A(0) ← NON_DOMINATED(S(0))
WHILE  t < T  DO
    FOR  i=1 to N
        p_i(t) ← Get_PBest()
        p_g(t) ← Get_GBest()
        p_c(t) = (1/N) Σ_{i=1}^{N} x_i(t)
        x_i(t+1) = c_1 r_1 (p_i(t) − x_i(t)) + c_2 r_2 (p_g(t) − x_i(t)) + c_3 r_3 (p_c(t) − x_i(t))
    END FOR
    EVALUATE(S(t+1))
    A(t+1) ← NON_DOMINATED(S(t+1) ∪ A(t))
    IF |A(t+1)| > M THEN CUT_ARCHIVE(A(t+1))
    t ← t+1
END WHILE
Output the obtained Pareto optimal front
```

In the initialization phase, the position of $i$th particle $\mathbf{x}_i(0)$ is generated by opposition-based learning method. Initial value for the personal best position of the *i*th particle is set to $\mathbf{p}_i(0) \leftarrow \mathbf{x}_i(0)$. The archive $A(0)$ has been initialized to contain the non-dominated solutions from S(0). NON_DOMINATED(S(0)) returns non-dominated solutions from the swarm S(0).

EVALUATE() is used to evaluate fitness of particle in swarm.

Get_PBest() returns the personal best solution. The personal best position of the *i*th particle is updated according to the following equation:

$$\mathbf{p}_i(t+1) = \begin{cases} \mathbf{p}_i(t) & \text{if } F(\mathbf{p}_i(t)) \prec F(\mathbf{x}_i(t+1)) \\ \mathbf{x}_i(t+1) & \text{otherwise} \end{cases}$$

Get_GBest() returns the global best position of swarm. In MOOPs, the choice of a single optimum solution is difficult. To resolve this problem, the concept of non-dominance is used and an archive of non-dominated solutions is maintained, from which a solution is picked up as the global best solution. The proposed algorithm maintains an archive $A(t)$ with maximum capacity M. The global best solution is selected from the archive $A(t)$ on the basis of the diversity of the solutions as in [4], and the crowding-distance measure in [1] is used for the diversity measurement. At iteration, the archive gets updated with the inclusion of the non-dominated solutions from the combined population of the swarm and the archive. If the size of the archive exceeds the maximum capacity M, it is cut using the diversity consideration. CUT_ARCHIVE() is used to cut the archive, and the most sparsely spread M solutions are retained in the archive.

## 6 Bi-objective Optimization Problem for WSN

Consider a 2D static WSN. WSN deployment problem can be formulated as a bi-objective optimization problem:

*Given*: ① H :sink node.
② A: 2D rectangular sensing area.
③ N: number of sensors to be deployed in A.
④ E: initial power supply, the same for all sensors.
⑤ Rs: sensing range, the same for all sensors.
⑥ Pmax: maximum transmission power level, the same for all sensors.
⑦ $P_j$: the transmission power level of sensor j.

Decision variables of a network design X are the horizontal and vertical coordinates of the sensors. We only consider WSN with a predetermined number *N* of sensors, so that the vector of design variables **X** is of constant size 2n.

$$\mathbf{X} = [x_1, y_1, x_2, y_2, \cdots, x_n, y_n]$$

Coverage objective and lifetime objective of the network are considered. Only the sensors that are connected to the H are taken into account in the calculation of these objectives. Coverage objective is equal to the area of the union of the disks of radii Rs centered at each connected sensor, normalized by the total area.

$$\max\ f_C(\mathbf{X}) = \left[\bigcup_{i=1}^{N} R_s^2(x_i, y_i)\right] \Big/ A$$

Lifetime objective is defined as the ratio of the time to first sensor failure (no more energy) and the maximum lifetime of a sensor.

$$\max f_L(\mathbf{X}) = \min_{i=1...N} \{T_{f,i}\} / T_{\max}$$

We assume that all sensors gather data at the same time and then relay it to the H (this is called a sensing cycle). Before the data reaches the H, it may need to be relayed by several sensors. Therefore, at every sensing cycle the sensor nodes need to transmit their own data and possibly the data from other sensors. The data from every sensor needs to be routed to the H in a way that will maximize the remaining energy in the nodes. In order to find these routes, the outgoing edges of every node are weighted by the inverse of the node's remaining energy, and then the Dijkstra algorithm is used to find the route of minimum weight[9]. Repeating this calculation until the energy of at least one node is depleted gives the maximum number of sensing cycles a particular WSN layout can perform. This number is then normalized by *Tmax*, the maximum number of sensing cycles possible (obtained when all sensors are directly connected to the H, so that none act as a communication relay).

The overall optimal solution for WSN deployment problem is represented by a bi-objective optimization problem $\max F(\mathbf{X}) = \{f_C(\mathbf{X}), f_L(\mathbf{X})\}$. The proposed algorithm is used to solve the bi-objective optimization problem.

## 7 Results from Simulations

In the section, performance of the proposed algorithm is compared with that of some multiobjective optimization algorithms. Convergence metric $\gamma$ and diversity metric $\Delta$ measures [1] were used. The smaller value of these metrics is, the better performance of the algorithm is. Population size $N$=150. All experiments were repeated for 20 runs. The maximum number of iterations is set to 2000 in each running. $c_1, c_2$ and $c_3$ are set to 0.5. Multi-objective benchmark functions that show in Table1 are used to test.

Tables2 listed the mean and variances of the convergence and diversity metrics obtained using NSGA-II[1], $\sigma$-MOPSO[3], MOPSO[4] and the proposed algorithm (VFMOPSOC). It shows that VFMOPSOC outperforms NSGA-II, $\sigma$-MOPSO, MOPSO algorithms in both aspects of convergence and distribution of solutions.

The algorithm is also applied to coverage problem of WSN. $F(\mathbf{X}) = \{f_C(\mathbf{X}), f_L(\mathbf{X})\}$ is used as multi-objective function. Simulation experiments are executed to investigate the effect of the proposed algorithm in increasing coverage. Two tests of different number of sensors and areas are conducted. All the sensors used have sensing range Rs = 9. Table3 shows the parameters of these tests. The number of particles in the algorithm is set to 200. The algorithm runs to a maximum number of 5000 iterations. A sample of solutions of the coverage ratio is listed in Tables3. The table shows that the initial coverage of the networks is significantly improved.

**Table 1.** Multi-objective benchmark functions

| Problem | Objective functions | Dim and Bounds |
|---|---|---|
| ZDT1 | $f_1(\mathbf{x}) = x_1$<br>$f_2(\mathbf{x}) = g(x)(1-\sqrt{x_1/g(x)})$<br>$g(\mathbf{x}) = 1+9(\sum_{i=2}^{D} x_i)/(D-1)$ | Bounds : [0,1] |
| ZDT2 | $f_1(\mathbf{x}) = x_1$<br>$f_2(\mathbf{x}) = g(x)(1-(x_1/g(x))^2)$<br>$g(\mathbf{x}) = 1+9(\sum_{i=2}^{D} x_i)/(D-1)$ | Bounds : [0,1] |
| ZDT3 | $f_1(\mathbf{x}) = x_1$<br>$f_2(\mathbf{x}) = g(x)(1-\sqrt{\frac{x_1}{g(x)}}-\frac{x_1}{g(x)}\sin(10\pi x_1))$<br>$g(\mathbf{x}) = 1+9(\sum_{i=2}^{D} x_i)/(D-1)$ | Bounds : [0,1] |
| ZDT4 | $f_1(\mathbf{x}) = x_1$<br>$f_2(\mathbf{x}) = g(x)(1-\sqrt{x_1/g(x)})$<br>$g(\mathbf{x}) = 1+10(D-1)$<br>$+\sum_{i=2}^{D}(x_i^2-10\cos(4\pi x_i))$ | Bounds :<br>$x_1 \in [0,1]$<br>$x_i \in [-5,5]$,<br>$i=2,\ldots,D$ |
| ZDT6 | $f_1(\mathbf{x}) = 1-\exp(-4x_1)\sin^6(6\pi x_1)$<br>$f_2(\mathbf{x}) = g(x)(1-(f_1(x)/g(x))^2)$<br>$g(\mathbf{x}) = 1+9(\sum_{i=2}^{D} x_i/(D-1))^{0.25}$ | Bounds : [0,1] |

**Table 2.** Mean and variances of the convergence and diversity metrics for benchmark functions

| Problem | Algorithm | Convergence γ | Diversity Δ |
|---|---|---|---|
| ZDT1 | NSGA-II | 0.01024± 0.00453 | 0.38096±0.00189 |
| | σ -MOPSO | 0.01408± 0.00051 | 0.40005±0.00582 |
| | MOPSO | 0.00201± 0.00002 | 0.51941± 0.01008 |
| | VFMOPSOC | 0.00069±1.9603e-8 | 0.10821±4.0071e-5 |
| ZDT2 | NSGA-II | 0.070001± 0.02331 | 0.40021± 0.00411 |
| | σ -MOPSO | 0.00399±0.00001 | 0.27632±0.00411 |
| | MOPSO | 0.00088±0.00002 | 0.49989±0.00116 |
| | VFMOPSOC | 0.00043±1.3652e-8 | 0.18794± 1.10002e-6 |
| ZDT3 | NSGA-II | 0.11962± 0.00475 | 0.59732± 0.02005 |
| | σ -MOPSO | 0.01342± 0.00301 | 0.60864± 0.00398 |
| | MOPSO | 0.00402± 0.00001 | 0.70235± 0.00597 |
| | VFMOPSOC | 0.00047±4.4863e-9 | 0.30064±1.2806e-6 |
| ZDT4 | NSGA-II | 0.37926± 0.09735 | 0.59167±0.04871 |
| | σ -MOPSO | 2.03762±1.27158 | 0.74865±0.00038 |
| | MOPSO | 5.87424±3.84769 | 0.76829±0.00096 |
| | VFMOPSOC | 0.11548±0.01514 | 0.16245±4.3776e-6 |
| ZDT6 | NSGA-II | 0.19937± 0.01132 | 0.55768±0.01008 |
| | σ -MOPSO | 0.02114±0.00001 | 0.53454±0.00138 |
| | MOPSO | 0.62863±0.00908 | 0.54896±0.00053 |
| | VFMOPSOC | 0.00094±5.1143e-7 | 0.36313±0.00006 |

**Table 3.** A sample of solutions of the coverage ratio

| | A | N | Initial Coverage % | Final Coverage % |
|---|---|---|---|---|
| Test1 | 50×50 | 25 | 71. 83 | 94.21 |
| Test2 | 100×100 | 100 | 72.38 | 90.87 |

# 8 Conclusions

A Pareto-based velocity-free multi-objective PSO with centroid is proposed and applied to WSN optimization. In the algorithm, particles only have position without velocity, and the personal best, the global best and the centroid are used to update the particle position. Non-dominated sorting and ranking selection procedure are applied to constitute next archive from current population and current archive. A crowding-distance measure is used to select the global best solution from the archive based on the diversity of the solutions. Coverage objective and lifetime objective of WSN are modeled as a bi-objective optimization problem. The proposed algorithm is applied to some benchmark functions and the bi-objective optimization problem. The experimental results show that the algorithm has better performance than NSGA-II, $\sigma$-MOPSO, MOPSO and is effective.

**Acknowledgment.** This work is supported by the Scientific and Technological Innovation Projects of Department of Education of Guangdong Province, P.R.C. and Guangzhou Science and Technology Projects under Grant No. 12C42011563, 11A11020499.

# References

[1] Deb, K., Agrawal, S., Pratap, A., Meyarivan, T.: A fast and elitist multi-objective genetic algorithm: NSGA-II. IEEE Trans. on Evolutionary Computation 6(2), 182–197 (2002)
[2] Kennedy, J., Eberhart, R.C.: Particle Swarm Optimization. In: Proc. IEEE International Conference on Neural Networks, Perth, Australia, pp. 1942–1948 (1995)
[3] Mostaghim, S., Teich, J.: Strategies for Finding Good Local Guides in Multi-objective Particle Swarm Optimization (SIS 2003), pp. 26–33. IEEE Service Center, Inidanapolis (2003)
[4] Coello, C.A.C., Pulido, G.T., Lechuga, M.S.: Handling multiple objectives with particle swarm optimization. IEEE Trans. on Evolutionary Computation 8(3), 256–279 (2004)
[5] Ngatchou, P., Fox, W., Sharkawi, M.: Distributed sensor placement with sequential particle swarm optimization. In: Proc. IEEE Swarm Intelligence Symp., pp. 385–388 (June 2005)
[6] Seah, M., Tham, C., Srinivasan, K., Xin, A.: Achieving coverage through distributed reinforcement learning in wireless sensor networks. In: Proc. 3rd Int. Conf. Intelligent Sensors, Sensor Network. Inf. Proc. (2007)
[7] Gao, Y.: No Velocity Particle Swarm Optimiser with Forgetting Factor and Center. In: ICNC 2009-FSKD 2009, pp. 537–541 (August 2009)
[8] Tizhoosh, H.R.: Opposition-Based Learning: A New Scheme for Machine Intelligence. In: Int. Conf. on Computational Intelligence for Modelling Control and Automation, Vienna, Austria, vol. I, pp. 695–701 (2005)
[9] Cormen, T.H., et al.: Introduction to algorithms. MIT Press, Cambridge (2001)

# The Improved Particle Swarm Optimization Based on Swarm Distribution Characteristics

Wang Hu[1], Jun-jie Hu[2], and Xin Zhang[1]

[1] School of Information and Software Engineering,
University of Electronic Science and Technology of China
Chengdu 610054, China
[2] School of Information science & Technology,
Chengdu University of Technology
Chengdu 610059, China
scuhuwang@126.com

**Abstract.** Due to the deficiency of characteristics of objective function, such as the function derivative, the solutions can only be iterated according to the evolutionary equations of Particle Swarm Optimization (PSO) with the finite information about current swarm state. But in the evolutionary process of PSO, the distribution characteristics of solutions of the objective function are hidden in the many and many fitness evaluations while the evolutionary equations are iterating. The evolutionary strategies, including the balance strategy between the exploration and exploitation, the re-initialization strategy and the generation strategy of new solution from the elite particles, are designed innovatively according to the distribution characteristics of the swarm solutions extracting statistically from the historical evaluations. The experimental results show that these strategies are effective for the optimization precise and efficiency in the early evolutionary process although the complexity of time and space are increased lightly than that of the standard PSO.

**Keywords:** Particle Swarm Optimization, Distribution Characteristics, Evolutionary Strategies, Distribution Entropy.

## 1 Introduction

Particle Swarm Optimization (PSO), inspired by social behavior of bird flocking or fish schooling, is a population based stochastic optimization technique developed by Dr. Eberhart and Dr. Kennedy in 1995[1]. For more than a decade, PSO has been successfully applied in many research and application areas. It is demonstrated that PSO gets better results in a faster, cheaper way compared with other methods for its simple formalization, less parameters and easy realization. But PSO is known to suffer from premature convergence prior to discovering the true global minimizer. Also the evolutionary process of PSO is easily trapped into a long period of stagnation [2] for the best global fitness $P_g$ of the swarm and the best individual fitness $P_{id}$ of the particle can't be improved by the iterative equations of PSO in the late period of evolutionary.

J. Lei et al. (Eds.): AICI 2012, LNAI 7530, pp. 690–697, 2012.

The standard PSO and its varients only employ the finite information, such as the velocity $\boldsymbol{v}(t)$, the location $\boldsymbol{x}(t)$, the individual extremum $P_{id}$ of the particle and the global extremum $P_g$ of the swarm at the prior time t. The evolutionary process of PSO is controlled completely by the two motion equations which are linear iterative function system with two random factors. Due to the deficiency of characteristics of objective function, the particle might fly over some important regions where the better solutions maybe locate, for the influences of the population initialization, motion trajectory and stochastic oscillations. The information share between particle and swarm were well studied to improve the efficiency of information and the performance of optimization in paper [3-6,9]. But in the evolutionary process of PSO, the distribution characteristics of solutions of the objective function are hidden in the many and many fitness evaluations while the evolutionary equations are iterating. However, this rich and useful information of objective function characteristics, which can be extracted statically from the iterations, is ignored in PSO and its varients up to the present.

In this paper, we collected the distribution characteristics of solutions of the objective function and defined the entropy of the swarm, which are the basis of evolutionary strategies, including the balance strategy between the exploration and exploitation, the re-initialization strategy and the generation strategy of new solution from the elite particles. The new PSO with these strategies is proposed to improve the optimization precise and efficiency.

In PSO, $\boldsymbol{x}_{id}$ represents the location of the $i^{th}$ particle in the swarm at the $d^{th}$ dimension among the search space $\Omega_D$. $\boldsymbol{v}_{id}$ is the velocity of particle $i$ at the $d$ dimension. The $r_1$ and $r_2$ are the random factors with the uniform distribution $U(0,1)$. The $c_1$ and $c_2$ are the accelerative factors to adjust the self- cognition and the social impact, respectively. The algorithm initializes the location and velocity of each particle in the population with the random method. Then the particles update their locations and velocities according to the iterative equations (1) and (2) with their best individual solution $P_{id}$ and the global best solution $P_{gd}$ until the stop condition is met.

$$v_{id}^{t+1} = v_{id}^{t} + c_1 r_1 (p_{id} - x_{id}) + c_2 r_2 (p_{gd} - x_{id}) \qquad \text{......(1)}$$

$$x_{id}^{t+1} = x_{id}^{t} + v_{id}^{t+1} \qquad \text{......(2)}$$

In order to improve the intelligence of the particles, all the solutions during the iterations are analyzed statistically after the continuous variables are discretized into the intervals. The distribution characteristics are extracted from the frequency in the intervals. At the same time, the diversity of the population is measured by the entropy of solutions distribution.

## 2 Distribution Characteristics of Solutions

### 2.1 Distribution of Solutions

The solutions of objective function $f$ are always continuous in the solution space $\Omega$. The complex density function is necessary for charactering the distribution of solutions. For simplicity, the range of variable $\boldsymbol{x}$ at the $d^{th}$ ($d \in D$) dimension in $\Omega_{D=d}$

is divided into $K$ intervals, and the label of the $x_d$ falling into the interval, $Seq_{x_d}$, can be computed by the formula (3) defined as below:

$$Seq_{x_d} = \left\lceil \frac{x_d - L_d^a}{L_d^b - L_d^a} \times K \right\rceil \quad \text{......(3)}$$

In (3), $L_d^a$ and $L_d^b$ is the minimum and maximum of variable boundary at dimension $d$. $\lceil\ \rceil$ is the operator to return an integer which is no less than itself.

Each component $x_d$ of solution $\boldsymbol{x}$ can be mapped into an label $Seq_d^k$. The frequency of interval labeled $Seq_d^k$ is the ratio defined as formula (4), in which *Card*() is a statistic function counting the members in a set.

$$Q_d^k = \frac{Card(Seq_d^k)}{\sum_{k=1}^{K} Card(Seq_d^k)} \times 100\% \quad \text{......(4)}$$

$Q_d^k$ is a distribution characteristics of solutions of objective function at the dimension d. The bigger the $Q_d^k$, the more the times of all particles mapped in the interval $k$, also the higher the probability of the extremum in the interval. For an intelligent particle, the most important mission is to find, as early as possible, those intervals which might include the component of the best solution to exploit deeply these regions around these intervals and to prevent itself to fly over these regions. On the contrary, if the $Q_d^k$ is very little or approximate to zero, the interval labeled by $Seq_d^k$ is possible in the unvisited region. The intelligent particles should explore these unvisited regions in a low probability to increase the chance to find the global extremum. So, $Q_d^k$ related to $Seq_d^k$ is an available index for particles to make the intelligent evolutionary decisions. Figure 1 illustrates the solutions distribution, in which the x-axis represents intervals at a certain dimension and the y-axis represents the frequency $Q_d^k$.

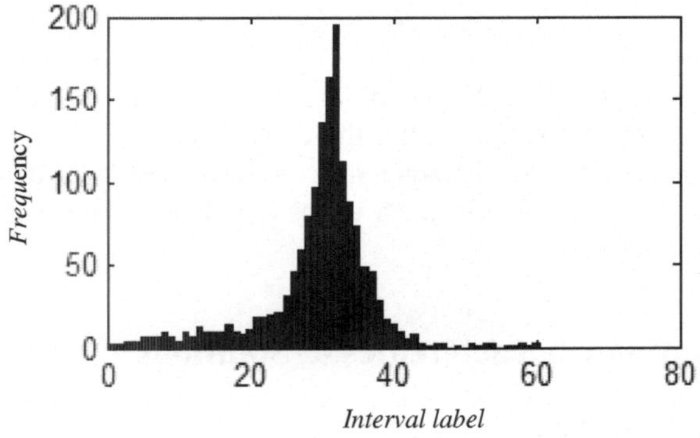


**Fig. 1.** The frequency in interval labels

### 2.2 Diversity of Population

To character the distribution of the whole swarm, the distribution entropy, $H(\Gamma)$, is introduced to describe diversity of the solution series, $\mathbf{\Gamma}=\{\boldsymbol{X}(1), \boldsymbol{X}(2), \ldots, \boldsymbol{X}(t),\ldots,\boldsymbol{X}(T)\}$. The definition of $H(\Gamma)$ is presented as formula (5).

$$H(\Gamma)=\sum_{d=1}^{D}\sum_{k=1}^{K}Q_d^k(\Gamma)*\log(\frac{1}{Q_d^k(\Gamma)}) \qquad \ldots\ldots(5)$$

The bigger the $H(\Gamma)$, the wider the swarm explored in the evolutionary, and the better the diversity of the population. In order to improve the $H(\Gamma)$ of PSO, many methods, such as population re-initiation, parameters regulation and solution mutation, etc., can be employed to scatter the particle in stagnation. On the contrary, the smaller the $H(\Gamma)$, the more concentrative the particles. The swarm is exploiting deeply or stagnating deceptively if the $H(\Gamma)$ is low relatively. So, $H(\Gamma)$ is a vane which can be used to switch the modes of exploration and exploitation. The curve of $H(\Gamma)$ is illustrated in figure 2.

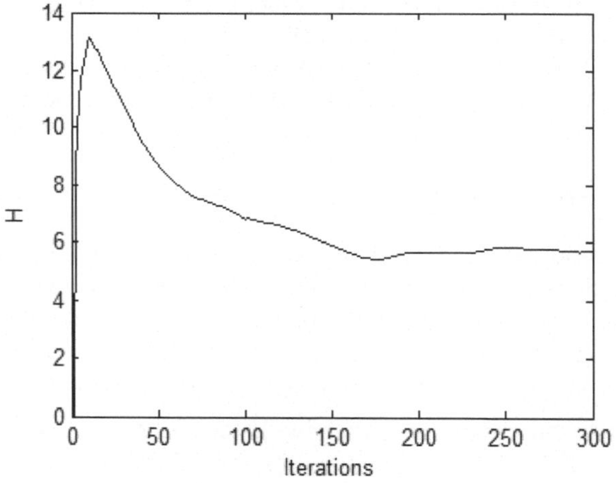


**Fig. 2.** The distribution entropy curve of Ackley

From formula (4) ~ (5), the statistic variables, $Q_d^k$ and $H(\Gamma)$ are accumulative amount which can be computed while the evolutionary process is iterating if the mediate variable $Seq_d^k$ is stored to reduce the operations.

## 3 The Evolutionary Strategies

In this section, three evolutionary strategies based on these characteristics are proposed for intelligent particles to make a wiser decision towards the better situation in probability.

### 3.1 Switch Strategy between Exploration and Exploitation

Particles will stop at the location, determined by the individual extremum $P_{id}$ and the swarm extremum $P_g$, according to the iterative equations (1) and (2) described in section 1. If the swarm is in the stationary state, or the particles are in less activity, the components of solutions will fall into the intervals densely, and the distribution entropy, H(Γ), charactering the uniformity of solutions distribution, will reduce with the decrease of activity. So, the intelligent particles can switch the evolutionary mode of exploration and exploitation, controlled by different parameters and strategies respectively, according to the variety of $H(\Gamma)$.

The absolute values of $H(\Gamma)$ can't be used to indicate the evolutionary situation because its threshold, enslaved to the number of intervals, is hard to determine in different objective function. But the variety trend of $H(\Gamma)$ is a good index to estimate the activity of swarm. If $H(\Gamma)$ is diminished continuously during $T_1$ iterations, and if the individual extremum $P_{id}$ and the swarm extremum $P_g$ are in stagnation during $T_2$ iterations, then the intelligent particles should iterate according to exploration mode to increase the activity of swarm, or else they should take the action of exploitation mode to improve the global extremum. On other conditions, the particles iterate their solutions by the motion equations (1) and (2) commonly.

The values of $T_1$ and $T_2$ will affect the efficiency of the evolutionary strategy for PSO. The values of $T_1$ and $T_2$ between 8 and 15 are recommended based on many experiments.

### 3.2 Re-initialization Strategy

The diversity of population can be increased by re-initialization of the locations and velocities of particles in the mode of exploration. Many re-initialization methods were proposed to avoid the particles falling into the same location [7-8]. In our re-initialization strategy, the particle will be thrown into the unvisited regions according to the distribution of $Q_d^k$ computed from $Seq_d^k$. In the unvisited regions, the global extremum can be found equally in probability if the particle has no knowledge about characteristics of the objective function. So the intelligent particle needs more chances to sample the unvisited regions other than the visited regions in the exploration mode.

The intervals, whose $Q_d^k$ is less than $q$ ($q$ is near to zero), are defined as the unvisited regions. The location of particle to re-initiate can be chosen from the unvisited regions by a certain random probability function.

### 3.3 New Solution Generation Strategy

The important key problem in optimization algorithm is that how to find the potential elite particles, which the global optimum of objective function is near to, in the exploitation mode. Those intervals, visited frequently by particles, with the bigger $Q_d^k$ are the regions where the local extremum and the global extremum maybe locate. So those intervals whose $Q_d^k$ are more than a threshold $Q$ ($Q$ can also be a percentage) are the candidates from which the component of the new solution $\boldsymbol{x}'(t+1)$ can be crossover chosen. The probabilistic model, by which the component of new solution at dimension $d$ is hit from interval $k$, can be defined as the formula (6) as below:

$$P(Seq_d^k \mid K) = Q_d^k \quad \text{......(6)}$$

The potential elite solution $\boldsymbol{x}'(t+1)$ will be composed by all components from the D dimensions, then the elite $\boldsymbol{x}'(t+1)$ is iterated to $\boldsymbol{x}(t+1)$ according to the formula (7).

$$x_{id}^{t+1} = \omega x_{id}^{t} + c_1 r_1 (p_{id} - x_{id}^{t}) + c_2 r_2 (p_{gd} - x_{id}^{t}) \quad \text{......(7)}$$

By the way, this new solution strategy improves the optimization precise and efficiency because the better solutions among particles are fused to the elite solutions $\boldsymbol{x}'(t+1)$ at high probability.

# 4 The Improved Algorithm

The new PSO based evolutionary strategies (ES-PSO) is improved from the classic PSO (C-PSO) with two extra steps, distribution characteristics computation and mode switch between exploration and exploitation. The re-initialization strategy for adding diversity of population and the new solution strategy are employed in the exploration mode and exploitation mode, respectively. ES-PSO can be summarized as follows.

Step 1: Initialize and evaluate population with random locations.

Step 2: Compute the distribution characteristic values of $Q_d^k$ and $H(\Gamma)$ according to formula (3) ~ (5);

Step 3: Determine the evolutionary mode of exploration and exploitation by $H(\Gamma)$ and its threshold;

Step 4: Iterate the new solutions of particles from the common mode according to (1) and (2), the exploration mode with re-initialization strategy and the exploitation with the new solution generation strategy according to equation (6) and (7) respectively by the result of switch strategy.

Step 5: Evaluate the new fitness of each particle in the population again.

Step 6: Go to step 3 if the stopping criterion isn't met, or else exit the algorithm.

The time complex and space complex are increased in ES-PSO due to the computation of solution distribution features at step 3. But the improvement of optimization efficiency leads to the reduction of the iterations in the evolutionary process. According to the no free lunch theorem, the better optimization efficiency of ES-PSO is at cost of algorithm complex. For the complex objective function, whose computations are more than those of the algorithm itself, the total time of computation may decrease for the reduction of iterations.

# 5 Experiments

In this paper, to validate the performance of the evolutionary strategies, the experiment of ES-PSO was executed only comparably with C-PSO because the ES-PSO can be hybridized to PSO and its variants.

The test functions are described as formula (8) and (9).

$$\text{F1:} \quad f_1(x) = 20 + e - 20e^{-0.2\sqrt{\frac{1}{N}\sum_{i=1}^{N} x_i^2}} - e^{\frac{1}{N}\sum_{i=1}^{N}\cos(2\pi x_i)} \quad \ldots \quad \ldots (8)$$

F2: $$f_2(x) = \sum_{i=1}^{n-1} (100(x_{i+1} - x^2)^2 + (x_i - 1)^2) \quad \cdots \quad \ldots(9)$$

In (8) and (9), referring to many experiments setting in other papers, the dimensions $D$ are set to 20, and the range of each dimension is restricted to [-100,100]. For displaying the evolutionary curve completely, the fitness of objective function are transformed by the function *log*() at the base of 10 to shrink the range. The common experimental parameters in ES-PSO and C-PSO are set as follows: the population size $N$ is 20, the max iteration $G$ is 300, the dimension $D$ of objective function is 20, the accelerative constants, $c_1$ and $c_2$, are set to 1.429, and the inertial coefficient $\omega$ is 0.729. In ES-PSO, the special parameters, $T_1$ and $T_2$ are set to 10 respectively. At the same time, in order to reduce the influence by random factors $c_1$ and $c_2$, the fitness is the mean result by 20 runs. The evolutionary curves of $F_1$ and $F_2$ are illustrated in figure 3 and figure 4 respectively.

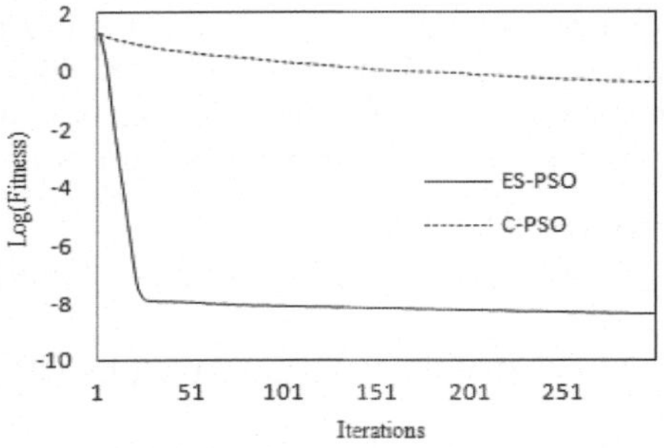


**Fig. 3.** The mean fitness curve of Ackley

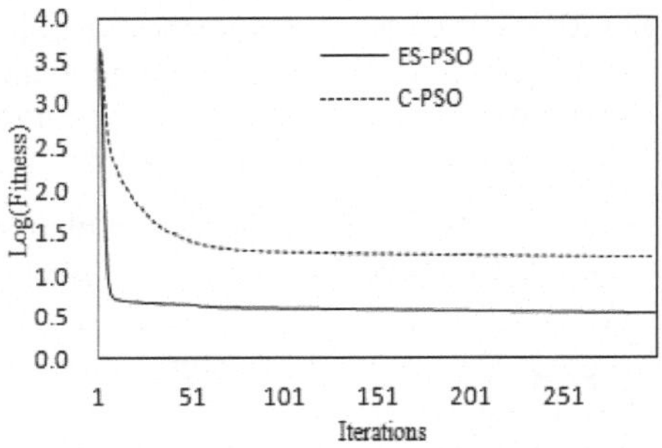


**Fig. 4.** The mean fitness curve of Rosenbrock

From figure 3 and figure 4, the comparative information of ES-PSO and C-PSO can be perceived from the fitness logarithm curve of $F_1$ and $F_2$. Firstly, the mean fitness of ES-PSO is much lesser than that of C-PSO at all iterations in the evolutionary process. This experiment shows that the new method, ES-PSO, can improve the optimization precise in the evolutionary process because the evolutionary strategies proposed in this paper are very effective in different objective functions. Secondly, at the early stage (almost the first 50 iterations within 1000 evaluations of fitness) in evolutionary process, the convergence velocity of ES-PSO is much faster than that of C-PSO. This is a prominent merit of ES-PSO because the optimization time can be reduced for those objective functions, such as online optimization for real applications and image optimization with many computations, that their efficiency is much more important on the condition of their precise is acceptable to a certain extent.

Figure 5 illustrates the comparative oscillation curve of $x_d$, the component of solution at certain dimension, during the evolutionary process of $F_1$ optimized by ES-PSO and C-PSO respectively.

From figure 5, the particle is nearly stationary gradually after some iterations in the evolutionary process of C-PSO, however, the particle is active still at the end of the evolutionary process of ES-PSO. So the diversity, implied by activity of particle, of population in ES-PSO is richer than that in C-PSO. The probability escaping from the local extremum is increased in ES-PSO for the switch strategy and re-initialization strategy.

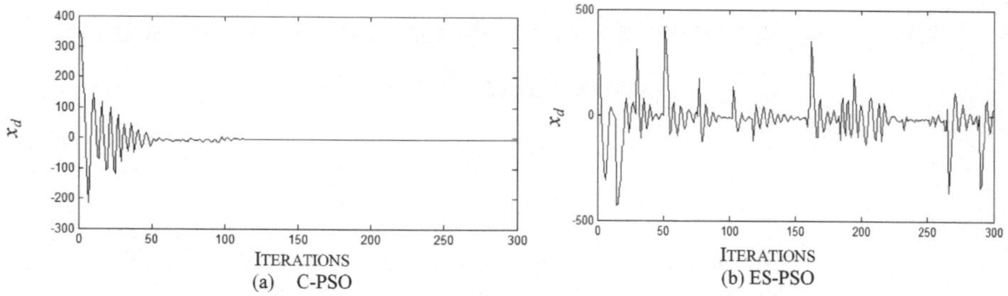


**Fig. 5.** The oscillation curve of particle location at some dimension $x_d$

# 6 Conclusions

The richer information about objective function and evolutionary process leads to the better performance of optimization algorithm although the complex of time and space is increased lightly. Three evolutionary strategies, based on the distribution features of solutions of objective functions, are designed to improve the intelligence of particles in ES-PSO proposed in this paper. The experimental results show ES-PSO is effective for the optimization precise and efficiency in the early evolutionary process. This idea can be hybridized also into the varients of PSO and other optimization methods based on population.

# References


1. Kennedy, J., Eberhart, R.: Particle swarm optimization. In: Proceedings of IEEE International Conference on Neural Networks, pp. 1942–1948. IEEE, Perth (1995)
2. Clerc, M.: Stagnation analysis in particle swarm optimization or what happens when nothing happens, `http://clerc.maurice.free.fr/pso/stagnationanalysis`
3. Liang, J., Qin, A., Suganthan, P., Baskar, S.: Comprehensive learning particle swarm optimizer for global optimization of multimodal functions. IEEE Transactions on Evolutionary Computation 10(3), 281–295 (2006)
4. Kennedy, J., Mendes, R.: Neighborhood topologies in fully informed and best-of-neighborhood particle swarms. IEEE Transactions on Systems, Man, and Cybernetics, Part C: Application and Reviews 36(4), 515–519 (2006)
5. Mendes, R., Kennedy, J., Neves, J.: The fully informed particle swarm: simpler, may be better. IEEE Transactions on Evolutionary Computation 8(3), 204–210 (2004)
6. Hsieh, S., Sun, T., Liu, C., Tsai, S.: Efficient population utilization strategy for particle swarm optimizer. IEEE Transaction on Systems, Man and Cybernetics, Part B: Cybernetics 39(2), 444–456 (2009)
7. Helwig, S., Wanka, R.: Theoretical Analysis of Initial Particle Swarm Behavior. In: Rudolph, G., Jansen, T., Lucas, S., Poloni, C., Beume, N. (eds.) PPSN 2008. LNCS, vol. 5199, pp. 889–898. Springer, Heidelberg (2008)
8. Said, M., Ahamed, A.: Hybrid periodic boundary condition for particle swarm optimization. IEEE Transactions on Antennas and Propagation 55(11), 3251–3256 (2007)
9. Tang, Y., Wang, Z.-D., Fang, J.-A.: Feedback Learning Particle Swarm Optimization. Applied Soft Computing 11, 4713–4725 (2011)

# An Analysis of Nonlinear Acceleration Coefficients Adjustment for PSO

Gang Wang and Zhikun Liu

School of Electronics and Information Engineering, Beihang University
37 Xueyuan Road, Haidian District, Beijing, 100191, China
gwang@buaa.edu.cn, liuzhikunlzk@126.com

**Abstract.** Linear acceleration coefficients adjustment had been widely used in particle swarm optimization (PSO). In this paper, a novel nonlinear strategy is developed, where the acceleration coefficients including both cognitive component and social component are adjusted nonlinearly to improve the optimization performance within a reasonable iteration times. Furthermore, the novel adjustment is deeply analyzed by experimental simulations based on four standard test functions. The results confirm the validity of the nonlinear parameter adjustment method in terms of the balance between convergence rate and optimization accuracy.

**Keywords:** particle swarm optimization, cognitive component, social component, nonlinear parameters adjustment.

## 1 Introduction

Particle swarm optimization is widely used in various optimization problems. Since PSO was proposed by Kennedy and Eberhart [1], a large body of research has focused on obtaining a better understanding of the acceleration coefficients choices, including cognitive component and social component. The cognitive component allows the particles to move toward their own best positions found so far represents the individual experience of each particle. The social component represents the collaborative effect of the particles, which pulls the particles toward the global best particle found so far.

In previous papers the cognitive component $c_1$ and social component $c_2$ always change linearly, either decreasing or increasing [2-5]. Compared with nonlinear inertia weight [6], another important component in PSO, the nonlinear cognitive component and social component are less focused by recent researchers. In particular, as in [7], the authors proposed nonlinear arccosine function strategy to adjust the cognitive and social component. The method in the latter stages set the ideal value of $c_1$ and $c_2$, so that the particles maintain a certain search speed, to avoid local convergence. However, the improvement is not obvious, mainly due to lack of detailed analysis of nonlinear curve slope.

Although the performance of some influential strategies is impressive [8-10], several questions remain unclear: (1) PSO with linearly changing parameters can

J. Lei et al. (Eds.): AICI 2012, LNAI 7530, pp. 698–705, 2012.

obtain superior results, yet there is not detailed analysis with nonlinear cognitive and social component. (2) Good convergence ability does not mean good optimization performance, but it is difficulty to analyze how to balance the convergence rate and optimization accuracy. The objective of the current research is not only to clarify these important issues, but also to explore novel parameter adjustment schemes. In particular, based on the time-varying acceleration coefficients PSO-TVAC [4], we explored the nonlinear cognitive and social component adjustment to improve the optimization performance within a reasonable iteration times.

The paper is organized as follows. Section II provides a review of PSO, followed by Section III discussing the issue of nonlinear parameters. Section IV then gives the experimental setup and detailed numerical results to support the theoretical analyses. Finally, Section V gives the concluding remarks.

## 2 Particle Swarm Optimization

Each particle of the swarm is characterized by its position $x$ and velocity $v$. Position of each particle represents a candidate solution to the problem under consideration, and each particle is capable of memorizing the personal best position $x_p$ and the global best position $x_g$. At each of algorithm iteration, every particle is attracted both by its personal best position and the global best position. The velocity and position update rules of standard PSO (STDPSO) are given by

$$v_{id}(t+1)=\omega\times v_{id}(t)+c_1\times r_{id}(t)\times\left(x_{pd}(t)-x_{id}(t)\right)+c_2\times R_{id}(t)\times\left(x_{gd}(t)-x_{id}(t)\right) \quad (1)$$

$$x_{id}(t+1)=x_{id}(t)+v_{id}(t+1) \quad (2)$$

where $t(t=1,2,\cdots,MAXiter)$, $i(i=1,2,\cdots,M)$, $d(d=1,2,\cdots,D)$ represent $t$-$th$ iteration, $i$-$th$ particle and $d$-$th$ dimension respectively. $M$ and $D$ are the total number; $MAXiter$ is the maximum number of allowable iterations. $\omega$, $c_1$ and $c_2$ are parameters of the algorithm, namely inertia weight, cognitive component and social component respectively. $r$ and $R$ are random positive numbers uniformly distributed in (0, 1).

The parameter adjustment scheme of time-decreasing inertia weight is given by

$$\begin{aligned}\omega&=(\omega_{\max}-\omega_{\min})\times\frac{(MAXiter-iter)}{MAXiter}+\omega_{\min}\\&=\omega_{\max}-(\omega_{\max}-\omega_{\min})\times(iter/MAXiter)\end{aligned} \quad (3)$$

where $\omega_{\max}$ and $\omega_{\min}$ (usually $\omega_{\max}$=0.9, $\omega_{\min}$=0.4) are the initial and final values of the inertia weight, respectively. $iter$ is the current iteration number.

The cognitive component $c_1$ and social component $c_2$ adjustment scheme of PSO-TVAC [4] is given by

$$\begin{aligned}c_1&=c_{\max}-(c_{\max}-c_{\min})\times(iter/MAXiter)\\c_2&=c_{\min}+(c_{\max}-c_{\min})\times(iter/MAXiter)\end{aligned} \quad (4)$$

where $c_{\max}$ and $c_{\min}$ are the maximum and minimum value of allowable $c_1$ and $c_2$ respectively. The iteration number $iter$ is noted as $x$ below.

# 3 Nonlinear Acceleration Coefficients Adjustment

The trajectory of cognitive component and social component given by PSO-TVAC is a straight line as shown in Figure 1 when they are linearly decreasing and increasing respectively. In this paper, we use nonlinear power function parameters to balance the particle's exploration-exploitation tradeoff.

The nonlinear strategy of cognitive component $c_1$ is as follows. We use high-order and low-order curves respectively instead of straight lines to indicate the change of cognitive component as formulas (5) and (6).

$$c_1 = c_{\max} - (c_{\max} - c_{\min}) \times (iter / MAXiter)^n \tag{5}$$

$$c_1 = c_{\max} - (c_{\max} - c_{\min}) \times (iter / MAXiter)^{1/n} \tag{6}$$

where $n = 1, 2, 3, \cdots$.

The corresponding curves of increasing social component $c_2$ can be formulated as (7) and (8).

$$c_2 = c_{\min} + (c_{\max} - c_{\min}) \times (iter / MAXiter)^n \tag{7}$$

$$c_2 = c_{\min} + (c_{\max} - c_{\min}) \times (iter / MAXiter)^{1/n} \tag{8}$$

where the value of $n = 1, 2, 3, \cdots$ determines the trajectory of social component. The linear and nonlinear parameter adjustment curves are shown in Figure 1.

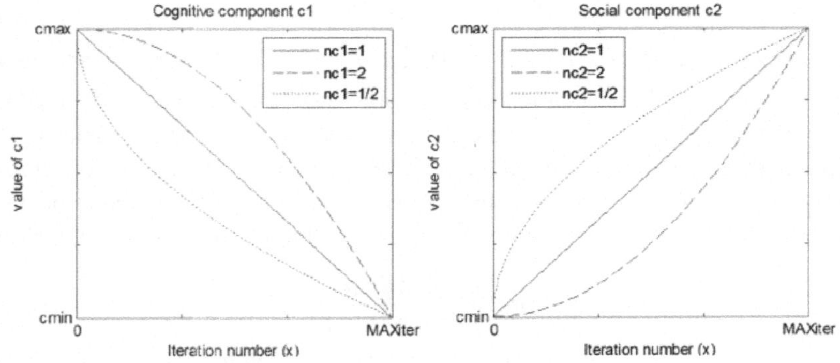


**Fig. 1.** Linear and Nonlinear $c_1$ and $c_2$

# 4 Experiments and Discussion

## 4.1 Experimental Setup

In order to see the effect of the nonlinear cognitive and social coefficients in PSO-TVAC, four benchmark problems are adopted as test functions. The details of these functions are summarized in Table 1, which includes the functions' names, the number of dimensions, the range of initial position, the optimal and the goal values.

**Table 1.** Benchmark Problems

| Name | Dim | Range | Optimal | Goal |
|---|---|---|---|---|
| Sphere | 30 | [-100,100] | 0 | 0.01 |
| Rosenbrock | 30 | [-30, 30] | 0 | 100 |
| Rastrigin | 30 | [-5.12,5.12] | 0 | 100 |
| Griewank | 30 | [-600,600] | 0 | 0.1 |

In our experiments, the parameters are set as the commonly employed parameter settings in the known literatures, which are $\omega_{max}$ =0.9, $\omega_{min}$ =0.4, $c_{max}$ =2.5, $c_{min}$ =0.5. So the nonlinear cognitive and social coefficients can be formulated as (9).

$$\begin{aligned} c_1 &= 2.5 - 2(x / MAXiter)^n \\ c_2 &= 0.5 + 2(x / MAXiter)^n \end{aligned} \tag{9}$$

Initially, simulations are carried out to observe the convergence rate and optimization accuracy of the new methods, nonlinear adjustment strategies of $c_1$ and $c_2$ respectively, in which the five numbers $n$=1/3, 1/2, 1, 2, and 3 mean five kinds of parameter adjustment curves. Second, we compare the optimal combination with both the original linear parameter strategy (PSO-TVAC) [4] and the arccosine function strategy (acos-PSO) [7].

The maximum iteration number is fixed to 1500 and population sizes were tested at M=40. A fully connected topology (the neighbors of each particle are the whole swarm) is used in all cases. Experimental test of the algorithm runs 100 times and in the end the iterative processes are given.

### 4.2 Results and Discussion

**Result 1: Nonlinear Adjustment Strategies of $c_1$ and $c_2$ Respectively**

The iterative processes over time of the four benchmarks are indicated in Figure 2 and Figure 3 for non-linear $c_1$ and $c_2$, respectively. From Figure 2, it is clear that the value of $n_{c1}$ is smaller, and the early stage of the optimization process for all the functions converges significantly faster. Further, the highest convergence rate is obtained at $n_{c1}$=1/3. In contrast, Figure 3 shows with the decrease of value of $n_{c2}$, the convergence rate is found to decline observably, and the convergence rate when it uses $n_{c2}$=3 parameter adjustment curve is almost as optimal as $n_{c2}$=2.

In Table 2, we summarize the results related to the Mean Best Value and the Standard Deviation of four benchmarks. The results show the change of $n_{c1}$ has extremely limited influence on optimization accuracy except Sphere function in general. For $n_{c2}$>1, except Rastrigin function, the optimization accuracy becomes too poor to meet the optimization goal. However, the method to use $n_{c2}$=1/3, especially $n_{c2}$=1/2 parameter adjustment curve has shown to have improved the optimization accuracy.

Based on the series of cases above, it can be concluded that on the one hand, a smaller $n_{c1}$ should be chosen to achieve a faster convergence rate at early stage of the optimization process, while it has extremely limited influence on optimization accuracy. On the other hand, a larger $n_{c2}$ is conducive to speed up the convergence rate, but it can easily result in local optimum prematurely. Meanwhile, a smaller $n_{c2}$ has opposite effect, is to improve the optimization accuracy and avoid local optimum prematurely, but result in low convergence rate.

**Table 2.** Mean Best Value and Standard Deviation of each benchmark for different combination of $n_{c1}$ and $n_{c2}$

| | | Mean Best Value | | | | Standard Deviation | | | |
|---|---|---|---|---|---|---|---|---|---|
| | | Sphere | Rosenbrock | Rastrigin | Griewenk | Sphere | Rosenbrock | Rastrigin | Griewenk |
| $n_{c2}$=1 | $n_{c1}$=1/3 | 2.71e-004 | 84.7874 | 52.7732 | 0.0313 | 0.0011 | 78.1809 | 11.97 | 0.0332 |
| | $n_{c1}$=1/2 | 1.06e-005 | 82.4708 | 49.4936 | 0.0249 | 3.05e-005 | 115.0539 | 13.4839 | 0.0264 |
| | $n_{c1}$=1 | 2.11e-009 | 55.8565 | 41.8878 | 0.0225 | 9.90e-009 | 47.7372 | 11.3579 | 0.0236 |
| | $n_{c1}$=2 | 1.01e-012 | 45.4374 | 38.077 | 0.0142 | 9.60e-012 | 43.5175 | 9.6494 | 0.0187 |
| | $n_{c1}$=3 | 2.85e-012 | 52.9643 | 34.0077 | 0.0141 | 8.93e-012 | 52.9226 | 9.5772 | 0.0161 |
| $n_{c1}$=1 | $n_{c2}$=1/3 | 3.99e-010 | 50.7481 | 47.1809 | 0.0144 | 1.26e-009 | 44.744 | 12.226 | 0.0158 |
| | $n_{c2}$=1/2 | 1.28e-010 | 47.6905 | 45.5591 | 0.0131 | 9.21e-010 | 45.2359 | 11.1579 | 0.0138 |
| | $n_{c2}$=1 | 2.11e-009 | 55.8565 | 41.8878 | 0.0225 | 9.90e-009 | 47.7372 | 11.3579 | 0.0236 |
| | $n_{c2}$=2 | 0.0356 | 189.0259 | 39.6711 | 0.0621 | 0.0626 | 274.2627 | 10.1028 | 0.0788 |
| | $n_{c2}$=3 | 0.3195 | 325.3113 | 39.7862 | 0.2081 | 0.6991 | 636.3718 | 10.024 | 0.197 |

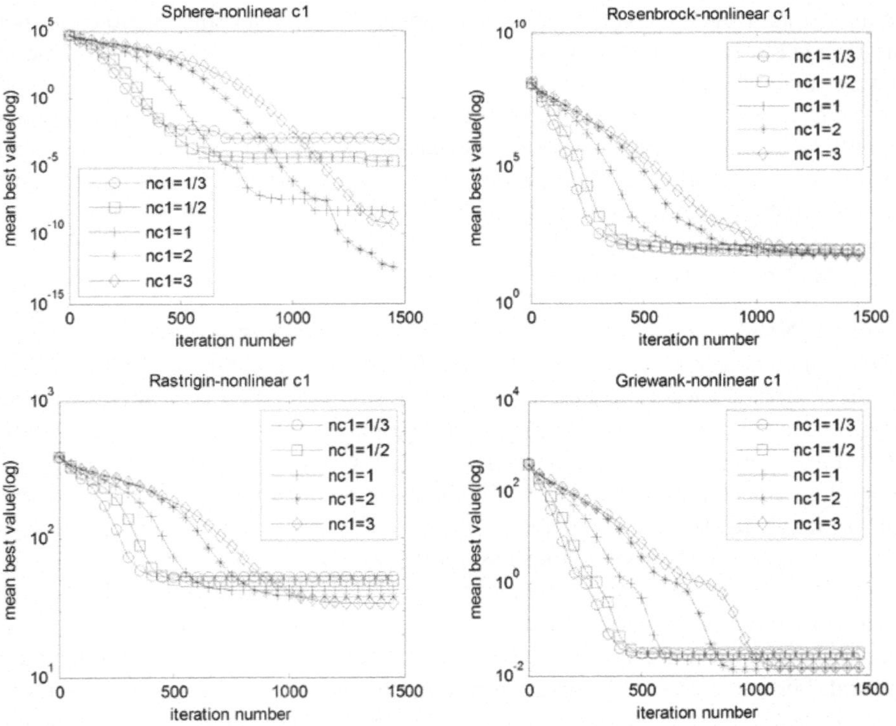


**Fig. 2.** Variation of the mean best value of each benchmark over time for $c_1$ respectively, in which the five numbers $n$=1/3, 1/2, 1, 2, and 3 mean five kinds of parameter adjustment curves

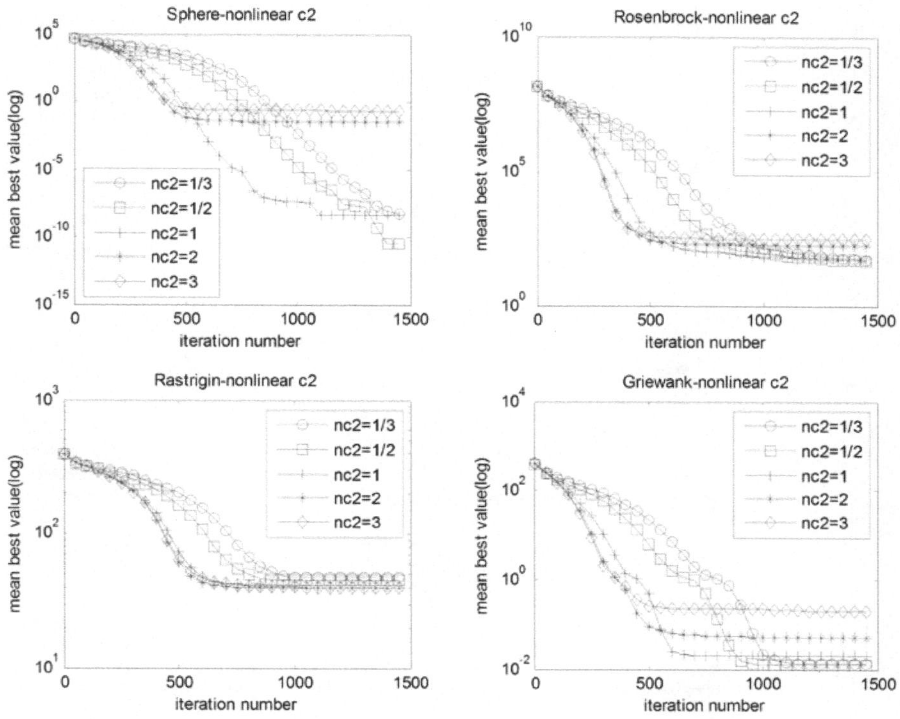


**Fig. 3.** Variation of the mean best value of each benchmark over time for $c_2$ respectively, in which the five numbers $n$=1/3, 1/2, 1, 2, and 3 mean five kinds of parameter adjustment curves

**Result 2: Comparison with PSO-TVAC and acos-PSO**

In order to balance the convergence rate and optimization accuracy during the optimization process to improve the performance of the algorithm, we are further concerned the combination of minimum of $n_{c1}$ ($n_{c1}$=1/3) and smaller $n_{c2}$ ($n_{c2}$=1/3, 1/2, 1), that is a novel parameter adjustment strategy for the PSO we proposed. Simulations are carried out to observe the convergence rate and the optimization accuracy of the new methods in comparison with both the original linear parameter strategy (PSO-TVAC) [4] and the arccosine function strategy (acos-PSO) [7].

**Table 3.** Mean Best Value and Standard Deviation of each benchmark for five parameter adjustment strategies

| | Mean Best Value | | | | Standard Deviation | | | |
|---|---|---|---|---|---|---|---|---|
| | Sphere | Rosen | Rastrigin | Griewenk | Sphere | Rosen | Rastrigin | Griewenk |
| $n_{c1}$=1, $n_{c2}$=1 | 2.11e-009 | 55.8565 | 41.8878 | 0.0225 | 9.90e-009 | 47.7372 | 11.3579 | 0.0236 |
| acos-PSO | 1.17e-008 | 59.1130 | 46.5839 | 0.0178 | 6.45e-008 | 61.2599 | 12.1276 | 0.0324 |
| $n_{c1}$=1/3, $n_{c2}$=1/3 | 2.01e-010 | 41.3538 | 51.8074 | 0.0175 | 1.43e-009 | 41.8942 | 12.4344 | 0.0186 |
| $n_{c1}$=1/3, $n_{c2}$=1/2 | 1.69e-010 | 44.8034 | 54.6928 | 0.0157 | 1.17e-009 | 39.1451 | 14.0953 | 0.0176 |
| $n_{c1}$=1/3, $n_{c2}$=1 | 2.71e-004 | 84.7874 | 52.7732 | 0.0313 | 0.0011 | 78.1809 | 11.9700 | 0.0332 |

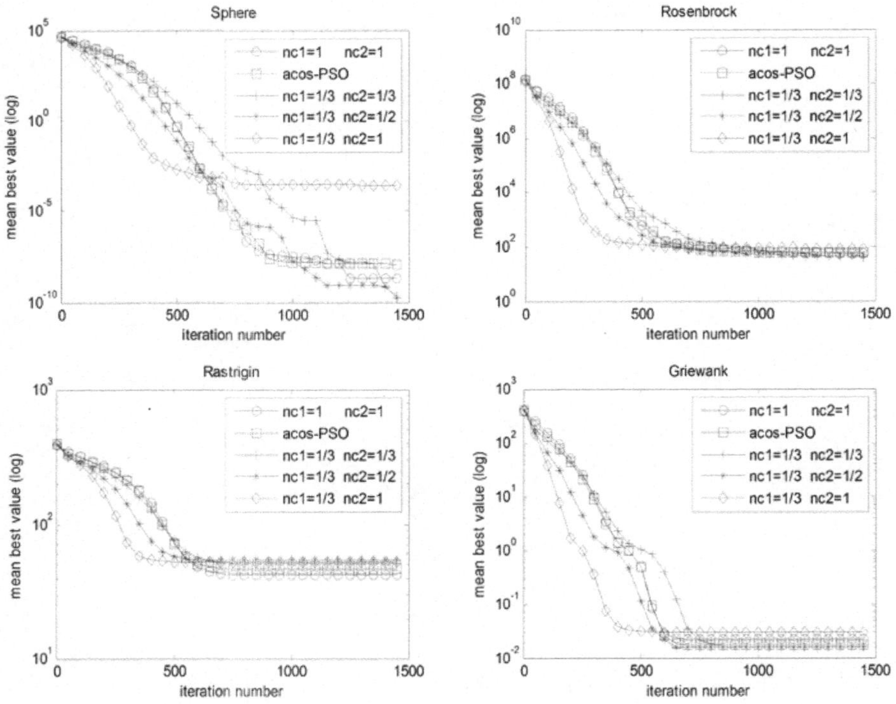


**Fig. 4.** Variation of the mean best value of each benchmark over time for nonlinear adjustment, PSO-TVAC and acos-PSO

Experimental results show that the nonlinear parameters can achieve good results even better than the PSO-TVAC and acos-PSO. From Table 3, it can be concluded that the optimization accuracy of all methods is almost the same expect Sphere function, and has an optimal solution to meet the optimization goal. For convergence rate, compared with PSO-TVAC and acos-PSO, the nonlinear parameter adjustment curve ($n_{c1}$=1/3 and $n_{c2}$=1) can do the best to improve the convergence ability to optimization goal without losing accuracy.

# 5 Conclusion

In this paper, the specific parameter is set to: $n_{c1}$=1/3 and $n_{c2}$=1, which is a combination of linear social component and nonlinear low-order power function cognitive component. During the early stages of the search, cognitive component decreases rapidly to strengthen the influence of social component, and so that particles are encouraged to wander through the entire search space and converge to the optima as far as possible. On the other hand, during the latter stages, cognitive component decreases slowly to keep particles a certain speed, so as to get rid of the interference of local optima.

This paper focused on cognitive component and social component adjustment to balance convergence rate and optimization accuracy as well as exploration and exploitation. We have developed a novel nonlinear adjustment strategy to improve the optimization performance within a reasonable iteration times.

## References


[1] Kennedy, J., Eberhart, R.: Particle swarm optimization. In: IEEE International Conference on Neural Networks, pp. 1942–1948. IEEE Press, Perth (1995)

[2] Shi, Y., Eberhart, R.: Empirical study of particle swarm optimization. In: Proceeding of Congress on Evolutionary Computation, pp. 1945–1950. IEEE Service Center, Piscataway (1999)

[3] Ozcan, E., Mohan, C.K.: Particle swarm optimization: surfing the waves. In: Proceeding of Congress on Evolutionary Computation, Piscataway, NJ (1999)

[4] Ratnaweera, A., Halgamuge, S.K., Watson, H.C.: Self-organizing hierarchical particle swarm optimizer with time-varying acceleration coefficients. IEEE Trans. on Evolutionary Computation 8(3), 240–255 (2004)

[5] Chang, X.-Y., Li, R.-J.: Experimental Analysis of Acceleration Coefficient in Particle Swarm Optimization Algorithm. Computer Engineering 36(4), 183–186 (2010)

[6] Zhao, X., Wang, G.: Nonlinear PSO—Convergence analysis and parameter adjustment schemes. In: BIC-TA, pp. 1111–1115 (2010)

[7] Chen, S., Cai, G., Guo, W., Chen, G.: An analysis of nonlinear acceleration coefficients adjustment schemes for PSO. Journal of Yangtze University (Nat. Sci. Edit.) Sci. & Eng. V 14(14) (2007)

[8] Hashemi, A.B., Meybodi, M.R.: A note on the learning automata based algorithms for adaptive parameter selection in PSO. Applied Soft Computing 11, 689–705 (2011)

[9] Zhan, Z.-H., Zhang, J., Li, Y., Shi, Y.-H.: Orthogonal Learning Particle Swarm Optimization. IEEE Trans. on Evolutionary Computation 15(6), 832–847 (2011)

[10] Jiang, M., Luo, Y., Yang, S.: Stochastic convergence analysis and parameter selection of the standard particle swarm optimization algorithm. Information Processing Letters 102, 8–16 (2007)

# Acoustic Sensor Network Node Self-localization Based on Adaptive Particle Swarm Optimization

Jinjie Yao[1], Yan Han[1,*], Liming Wang[1], Jinxiao Pan[1], Peirui Bai[2], and Jianhui Zhou[3]

[1] National Key Laboratory of Electronic Testing Technology,
North University of China, 030051, Taiyuan, China
[2] College of Information & Electrical Engineering,
Shandong University of Science and Technology, 266510, Qingdao, China
[3] Sichuan of Institute Aerospace Electronic Equipment
610100, Chengdu, China
zhong_bei@126.com

**Abstract.** It is quite important to obtain the sensor nodes location information in the underwater acoustic sensor networks localization. A method of acoustic sensor network node self-localization based on adaptive particle swarm optimization is proposed aiming at the stringent difficulties of the underwater acoustic sensor node localization and the shortage of standard particle swarm optimization (PSO) algorithm which is easily trapped into the local optimum. In the method, the global search ability and the local performance of the PSO algorithm are effectively improved by balancing the stochastic inertia weight. At the same time, the proposed method finds easy and elegant solutions to get rid of the local optimization by adopting the adaptive mutation strategy. The experimental results indicated that the new method can effectively solve the current problem in the underwater acoustic sensor node localization, and the pointing accuracy achieves 0.605m.

**Keywords:** Self-localization, Particle swarm optimization, Comprehensive learning, Adaptive mutation.

## 1 Introduction

With the development of the mobile communication, wireless localization has been applied to civil and military field. In order to improve the localization accuracy, it is important that enhance the performance of algorithm besides reinforcing the hardware equipment [1].

The sensor nodes self-localization is a very important research point in the wireless sensors network [2,3]. In order to obtain the location of the measured targets in the underwater acoustic sensor networks localization, the self-location of the sensor nodes must be specified primarily [4]. As the acoustic sensor nodes are distributed in the

---

* Corresponding author.

J. Lei et al. (Eds.): AICI 2012, LNAI 7530, pp. 706–713, 2012.

particular test area stochastically that their locations are unknown and randomly floating. Therefore, the self-localization of the sensor nodes, which can be realized through the information gained from the limited nodes, is the basis and primary point of the acoustic sensor network node self-location.

In this paper, a method based on adaptive particle swarm optimization is proposed to meet the strong demand of the acoustic sensor nodes self-localization. The experimental results show that: the proposed method has a very effective way to solve the current problem in the underwater sound sensor node localization, and the pointing accuracy achieves 0.605m.

The organization of the remaining content is as follows. In Section 2, we introduce the principle of acoustic sensor network node self-localization. And in Section 3 we propose the adaptive particle swarm optimization (APSO) algorithm based on stochastic inertia weight and adaptive mutation. Then, in Section 4 we present the experimental configuration and results, and finally end with some conclusions.

## 2 Principles

In the underwater acoustic sensor network positioning system, the sensor nodes are composed of buoys, wireless transceiver system, the system module of decode and control system module, embedded time synchronization circuits, acoustic transmitters and hydrophones. During the sensor nodes self-localization process, the master station transmits trigger single to every sensor nodes by the radio, thus every sensor node gets into the ready condition and takes turns to send out the underwater acoustic positioning signal at the fixed frequency and varied code; meanwhile, the master station sends out synchronism signal periodically to every sensor node in the way of radio broadcast. Due to the smaller range of the sensor nodes, that is, the master station is configured in the measurement and control boat which is in the hundreds meters away from the test proving ground, the transmission time between the master station and each sensor nodes is too short to be ignored. According to the received sound ranging signal delay and correspondingly signal processing algorithm, the self-localization of every sensor nodes can be realized finally [5].

Take the four floating underwater acoustic sensor network nodes as an example, assuming the sensor nodes $P_0$, $P_1$, $P_2$, $P_3$ distributes as a tetrahedron, as shown in Fig.1.

We establish a coordinate system where $P_0$ is the reference origin, X axis, Y axis and Z axis are also shown in Fig.1. The velocity between the floating sensor nodes of the underwater acoustic positioning single is $v$ which is known, and the transmission time is $t_{ij}$ $(i=0,1,2,3; j=0,1,2,3; i \neq j)$。

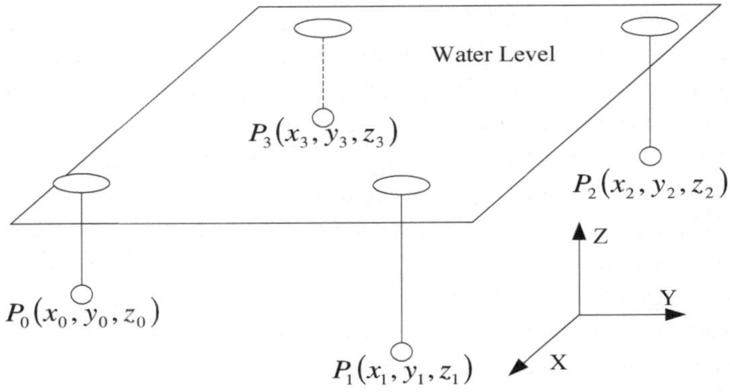


**Fig. 1.** The floating sensor nodes self-localization schematic diagram

In a location cycle, assuming that there is no position variation between the floating sensor nodes, and according to the hyperbolic intersection principle, the geometry position relations of the sensor nodes $P_0$, $P_1$, $P_2$, $P_3$ can be expressed as the following nonlinear equations:

$$\begin{cases} (x_1 - x_0)^2 + (y_1 - y_0)^2 + (z_1 - z_0)^2 = t_{10}^2 v^2 \\ (x_2 - x_0)^2 + (y_2 - y_0)^2 + (z_2 - z_0)^2 = t_{20}^2 v^2 \\ (x_3 - x_0)^2 + (y_3 - y_0)^2 + (z_3 - z_0)^2 = t_{30}^2 v^2 \\ (x_2 - x_1)^2 + (y_2 - y_1)^2 + (z_2 - z_1)^2 = t_{21}^2 v^2 \\ (x_3 - x_1)^2 + (y_3 - y_1)^2 + (z_3 - z_1)^2 = t_{31}^2 v^2 \\ (x_3 - x_2)^2 + (y_3 - y_2)^2 + (z_3 - z_2)^2 = t_{32}^2 v^2 \end{cases} \tag{1}$$

In (1), the constrained condition is that: $z_0$, $z_1$, $z_2$ and $z_3$ are all below zero, $x_2>x_0$, $x_3>x_0$, $y_2>y_0$, $y_3>y_0$, $z_0>z_1$, $z_0>z_2$, $z_0>z_3$, and $t_{ij}$ is the arrival time to be measured. The constrained formulas above shows that to get the self-localization of the sensor nodes is to get the solution of the nonlinear equations.

# 3 PSO and APSO Algorithm

## 3.1 PSO Algorithm

In $D$ dimension searching space, assuming that a swarm is composed of $m$ particles, the $i$ particle's position can be expressed as the following vector $x_i = [x_{i1}, x_{i2}, \cdots, x_{iD}]$, and the flight velocity is $v_i = [v_{i1}, v_{i2}, \cdots, v_{iD}]$, while the local optimum position is $p_i = [p_{i1}, p_{i2}, \cdots, p_{iD}]$, and the global optimum position searching by the whole particle swarm is $p_g = [p_{g1}, p_{g2}, \cdots, p_{gD}]$. The single particle update its own velocity and position according to the equations below [6]:

$$v_i(n+1) = \omega v_i(n) + c_1 r_1 (p_i - x_i(n)) + c_2 r_2 (p_g - x_i(n)) \tag{2}$$

$$x_i(n+1) = x_i(n) + v_i(n) \tag{3}$$

$$\omega = \omega_{\max} - t \times \frac{w_{\max} - \omega_{\min}}{T} \tag{4}$$

In (2) and (3), $i$=1, 2, …, m, and the non-negative constants $c_1$, $c_2$ are the learning factor; $\omega$ is the inertia weight, and the maximum and minimum inertia weight are $\omega_{\max}$ and $\omega_{\min}$ respectively; the current iteration number is $t$ and the total iteration number is $T$[7,8]. The particles do not stop searching for the individual extreme and global extreme in the solution space till the defined iteration number or the regulated error criterion is met. The main advantage of PSO is the quick convergence rate and easy to realize in the project, both of which standing out in the early period of this algorithm, while in the latter, the searching precision of PSO is gradually trending down and the algorithm is easy to get into the local optimum.

### 3.2 APSO Algorithm

In view of the unavoidable disadvantages of PSO, this paper proposes the adaptive particle swarm optimization algorithm (APSO) to modify it mainly in the two aspects of inertia weight updating and adaptive mutation.

**1) Inertia Weight Updating.** In the standard PSO algorithm, inertia weight $\omega$ is a kind of random numbers that are set to a random distribution, to some extent to overcome the flaw caused by the lack of linear inertia weight. If the early evolution were close to the optimal point, stochastic inertia weight $\omega$ may have a relatively small value to accelerate the convergence speed; else, it can still make the algorithm converge to the optimum. The calculation of stochastic inertia weight$\omega$can be expressed as:

$$\begin{cases} \omega = \mu + \sigma \cdot N(0,1) \\ \mu = \mu_{miin} + (\mu_{\max} - \mu_{\min}) \cdot rand(0,1) \end{cases} \tag{5}$$

Where $N(0,1)$ represents the random number subjecting to the standard normal distribution, $\sigma$ is variance of stochastic weight, $\mu$ is the average of that stochastic weight, $\mu_{min}$ is minimum value of average stochastic weight, $\mu_{max}$ is maximum value of average stochastic weight, $rand(0,1)$ stands for a random number between 0 and 1.

**2) Adaptive Mutation.** The adaptive mutation is put forward to let each particle toward new searching direction when the algorithm falling into local optimum [9], while the colony fitness variance determines the time of mutation, namely:

$$p_m = (p_{\max} - p_{\min})\left(\frac{\sigma^2}{N}\right) + (p_{\min} - p_{\max})\left(\frac{2\sigma^2}{N}\right) + p_{\max} \tag{6}$$

In Equation 6, $p_m$ is the mutation probability of colony global extreme value, $\sigma^2$ is the colony fitness variance, $p_{max}$ is maximum value of the mutation probability, $p_{min}$ is minimum mutation probability value. It indicates that the smaller the swarm fitness variance value is, that is, the greater the particles concentrate, the greater probability of swarm extreme value variation will be. Otherwise, when the swarm fitness variance value is greater, namely, the swarm maintains a higher diversity, the probability of swarm extreme value variation will be smaller[10]. The algorithm adaptively adjusts the mutation based on the location of particles in the swarm, and then gets rid of the local optimum.

### 3.3 Node Self-localization Algorithm Flow

In view of the unavoidable disadvantages of PSO, this paper proposes the adaptive particle swarm optimization algorithm (APSO) to modify it mainly in the two aspects, one is inertia weight updating and the other is adaptive mutation.

The main steps of the method based on adaptive particle swarm optimization algorithm in acoustic sensor network node self-localization are:

*Step* 1. According to the practical acoustic sensor network localization, initialize the position and velocity randomly in a variable values search range;

*Step* 2. Evaluate the fitness, store the current position of each particle in *pbest*, and the entire optimum adaptive value individual's position among *pbest* stored in the *gbest*;

$$fitness = \min \sum_{i=1}^{6} f_i^2 \tag{7}$$

where $f_i = (x_i - x_j)^2 + (y_i - y_j)^2 + (z_i - z_j)^2 - t_{ij}^2 v^2$, $i \neq j$.

*Step* 3. Calculate the random inertia weight according to equation 5;

*Step* 4. Update particle velocity and position according to equation 2 and 3;

*Step* 5. Calculate fitness variance $\sigma^2$ and mutation probability;

*Step* 6. Generate the random number $r \in [0,1]$, if $r<p_m$, variation by equation 7:

$$gbest = gbest \cdot (1 + 0.5\eta) \tag{8}$$

where $\eta$ is a random variable which obey the Gauss (0,1) distribution.

*Step* 7. For each particle, compare its fitness with the best experienced positions, if better, will be the best as the current position;

*Step* 8. Compare the value of all the *pbest* and *gbest*, update *gbest*;

*Step* 9. If the process reached termination condition or achieve maximum iterating time, returns the current global optimal individuals *gbest*, turn to *step* 7; otherwise, $k=k+1$, turn to *step* 3;

*Step* 10. Output global optimal individual *gbest*, which is the sensor network node location.

## 4 Experimental

In the underwater acoustic sensor network position, the experimental region is 200*200m 4m depth, the propagation velocity of the signal is 1500 m/s. The sensor nodes are set 0.5m, 2.5m, 1m, 1.5m below the water level, respectively. Triggered by the master station, the four sensor nodes send acoustic signal alternatively, which are received by the each assembly hydrophone of the sensor nodes. The signals collected by No.4 sensor nodes are shown in Fig.2.

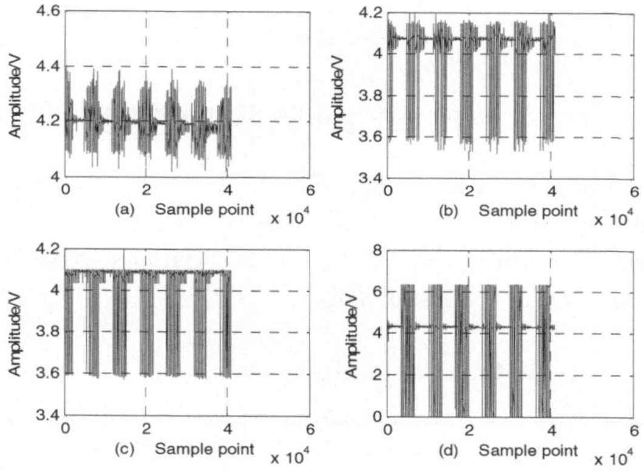


(a) Sensor node 1, (b) Sensor node 2, (c) Sensor node 3, (d) Sensor node 4.

**Fig. 2.** The underwater acoustic signal collected by sensor nodes

Measure the time difference of the acoustic signal acquainted by sensor, using particle swarm optimization algorithm and adaptive PSO algorithm to realize node self-localization. Choosing the following parameters in PSO algorithm: learning factor $c_1$ and $c_2$ is 2, the maximum weighting factor $\omega_{\max}$ is 0.9, the minimum weighting factor $\omega_{\min}$ is 0.4, the population size of the particle swarm is 20, 300 iterations, then the shows on Fig.3.

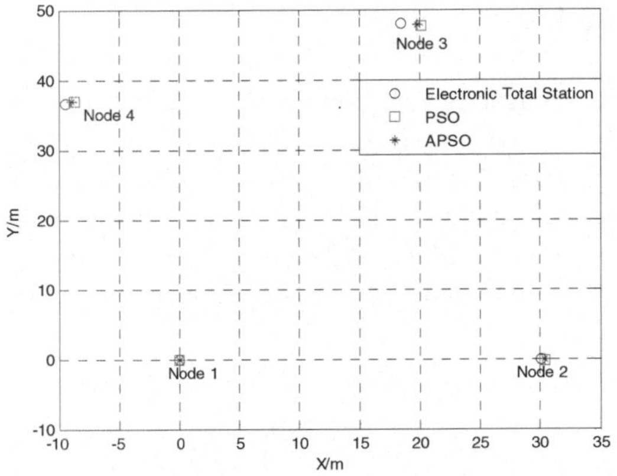


**Fig. 3.** Sensor node self-localization results

Using the constrained condition, Set the Node 1 as the reference node, the experimental results are obtained by PSO algorithm and adaptive PSO algorithm with total station measurements. Its mean error analysis based on total station measurements, and the results shows on Tab.1.

**Table 1.** Sensor localization error analysis

| | Node1 (m) | Node2 (m) | Node3 (m) | Node4 (m) | Mean (m) |
|---|---|---|---|---|---|
| PSO | 0.650 | 0.633 | 0.472 | 0.814 | 0.730 |
| APSO | 0.594 | 0.345 | 0.393 | 0.605 | 0.586 |

As can be seen from Table 1, the average positioning accuracy based on APSO sensor node localization is 0.605m, improving 0.1m comparing with the accuracy by PSO algorithm. Furthermore, the convergence rate of APSO is significantly faster than it of PSO, the convergence curve is shown in Fig.4.

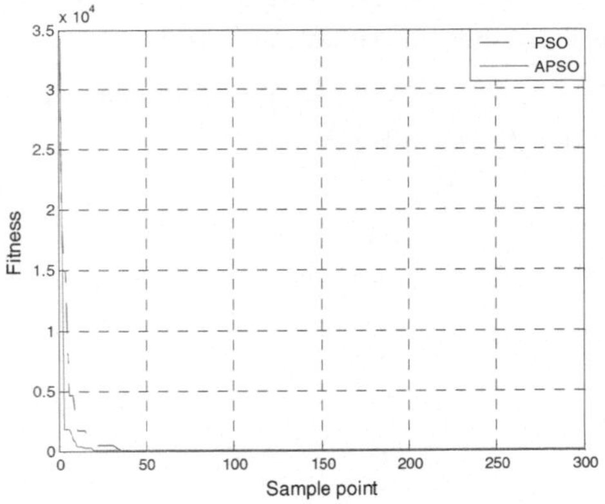


**Fig. 4.** Mean fitness curves with the sample point

## 5 Conclusions

In this paper, aiming at the disadvantages which are the difficulties in node localization of the acoustic sensors and the standard PSO is easy to fall into local optimum, a method of acoustic sensors network node self-localization based on adaptive particle swarm algorithm is proposed. Meanwhile, comparing the localization performance of the Total station measurement, PSO and APSO algorithm. It can be drawn from the analysis that APSO algorithm can effetely solve the problem of network node self-localization of the acoustic sensors, and overcome the shortcomings of PSO algorithm which are easy to fall into local optimum. This method has a certain value on solving the engineering problems, such as network node localization of wireless sensor, and mobile communication positioning.

## References

1. Chu, H.C., Jan, R.H.: A GPS-less, outdoor, self-positioning method for wireless sensor networks. Ad Hoc Networks 5, 547–557 (2007)
2. Mao, G.Q., Fidan, B., Anderson, B.D.O.: Wireless sensor network localization technique. Computer Networks 51, 2529–2553 (2007)
3. Bahi Jacques, M., Abdallah, M., Mostefaoui, A.: Localization and Coverage for high density sensor networks. Computer Communications 31, 770–781 (2008)
4. Vemula, M., Bugallo, M.F., Djuric, P.M.: Sensor self-localization with beacon position uncertainty. Signal Processing 89, 1144–1154 (2009)
5. Zhou, Z., Cui, J.H., Zhou, S.L.: Efficient localization for large-scale underwater sensor networks. Ad Hoc Networks 8, 267–279 (2010)
6. Cai, X.J., Cui, Z.H., Zeng, J.C., Tan, Y.: Dispersed particle swarm optimization. Information Processing Letters 105, 231–235 (2008)
7. Chatterjee, A., Siarry, P.: Nolinear inertia weight variation for dynamic adaptation in particle swarm optimization. Computers & Operations Research 33, 859–871 (2006)
8. Jiao, B., Lian, Z.G., Gu, X.S.: A dynamic inertia weight particle optimization algorithm. Chaos, Solutions and Fractals 37, 698–705 (2008)
9. Wu, Q.: Power load forecasts based on hybrid PSO with Gaussian and adaptive mutation and Wv-SVM. Expert Systems with Applications 37(1), 94–201 (2010)
10. Zheng, X.W., Liu, H.: A hybrid vertical mutation and self-adaptation based MOPSO. Computers and Mathematics with Applications 57, 2030–2038 (2009)

# Bimodal Biometrics Based on a Two-Stage Test Sample Representation

Yu Hou and Caikou Chen

Information Engineering College, Yangzhou University, Yangzhou 225127, China
houyu2005023106@126.com, cck.yzu@gmail.com

**Abstract.** Bimodal biometrics based on a two-stage test sample representation method for use with face recognition is presented in this paper. Until now a large amount of research has been proved that multi-biometrics can outperform single biometrics. The proposed method first let the test sample be linearly constructed from all the training samples each with a complex vector. By this step we find $k$ 'nearest neighbors' for the test sample. Then we re-expressed the test sample as a linear combination of the $k$ samples obtained above and classify the test sample into the class that makes the greatest contribution. The experimental results on CSIST and AR face image database demonstrate the efficiency and effectiveness of our method.

**Keywords:** biometrics, Face recognition, linear representation.

## 1 Introduction

It has been found that the use of multimodal biometric traits can produce better results than the use of a single biometric alone. This is mainly because multi biometric systems enable more information and rely on different pieces of evidence before taking a decision. Bimodal biometrics[3-4] is the simplest multi-biometric system. In this paper, we use the feature level fusion approach to bimodal biometrics and first denote the biometric trait samples including all the training samples and the test samples by complex vectors.

Our method first assumes that the test sample can be represented as a linear combination of all the training samples. We exploit the representation ability to choose the $k$ nearest neighbors for the test sample. Then we re-expressed the test sample as a linear combination of the determined $k$ training samples. After evaluating the contribution in representing the test sample of each class, the test sample is assigned to the class that makes the greatest contribution.

We note that linear representation is carried out in the original space and does not need any transforming. The first phase of our method identifies the $k$ nearest neighbors that have the $k$ greatest contributions and regards the other samples has no effect on the ultimate classification of the test sample. This can be viewed as a supervised sparse representation, what's more, in the second phase this treatment can increase the probability of the test sample is one class the $k$ training samples belong to. Our method not only provides a novel feature level fusion approach to bimodal biometrics, but also

J. Lei et al. (Eds.): AICI 2012, LNAI 7530, pp. 714–720, 2012.

expresses the test sample more accurately when the dimension of the observation data is larger than the number of all samples. Because all the representation coefficients are calculated by $\ell_2$-norm method[5]. This approach is superior to the state-of-the-art to face recognition in terms of efficiency, performance, and robustness. The experimental results on CSIST and AR database show that the proposed method can obtain a better classification performance than some tradition fusion approaches.

The rest of this paper is outlined as follows: In Section 2, we introduce the details of our method. Section 3 reports the experimental results and provides some analysis of them. Finally, we conclude our work in Section 4.

## 2 Our Two-Stage Method to Represent the Test Sample

Suppose there are $\boldsymbol{C}$ classes and $\boldsymbol{n}$ training samples in a high-dimensional image space $\mathbf{R}^{\mathbf{d}}$. Let $\mathbf{a}_i$ and $\mathbf{b}_i$ denote two biometric trait in the original image space. We arrange the given $\boldsymbol{n}$ training images of as columns of a single matrix. We use $\mathbf{m}_i=\mathbf{a}_i+i\mathbf{b}_i$ $(i=1,2\ldots n)$ to represent the $i$th training sample of the bimodal biometric trait. Let $\mathbf{y}$ stand for the test sample. $\mathbf{m}_i$ and $\mathbf{y}$ are all complex vectors.

### 2.1 The First Stage of the Two-Stage Test Sample Representation Method

We know, given sufficient training images, a test image $\mathbf{y}$ will approximately lie in the linear span of all training images:

$$\mathbf{y} = c_1\mathbf{m}_1 + c_2\mathbf{m}_2 + \cdots + c_n\mathbf{m}_n = \mathbf{Mc} \tag{1}$$

Where $c_i$ is a coefficient, $\mathbf{M}=[\mathbf{m}_1, \mathbf{m}_{2,\ldots}\mathbf{m}_n]^{\mathrm{T}}$ denote the entire training set. If $\mathbf{M}$ is a nonsingular square matrix, $\mathbf{c}$ can be estimated using $\mathbf{c}=\mathbf{M}^{-1}\mathbf{y}$; otherwise, we can obtain the solution by solving the normal equations: $\mathbf{M}^{\mathrm{T}}\mathbf{Mc}=\mathbf{M}^{\mathrm{T}}\mathbf{y}$. Thus, we have

$$\mathbf{c} = (\mathbf{M}^{\mathrm{T}}\mathbf{M})^{-1}\mathbf{M}^{\mathrm{T}}\mathbf{y} \ \text{ or } \ (\mathbf{M}^{\mathrm{T}}\mathbf{M} + \mu\mathbf{I})^{-1}\mathbf{M}^{\mathrm{T}}\mathbf{y} \tag{2}$$

where $\mathbf{I}$ is the identity matrix and $\mu$ is a small positive constant. In fact (2) is the solution of the following minimum $\ell_2$-norm problem:

$$\min_c \|\mathbf{c}\|_2, \text{ subject to } \mathbf{y} = \mathbf{Mc} \tag{3}$$

From (1), we can see that each training sample makes its own effort in expressing the test sample. $c_i\mathbf{m}_i$ denote the contribution of every training sample $\mathbf{m}_\mathbf{i}$ to reconstructing the test sample $\mathbf{y}$. The residual between the test sample $\mathbf{y}$ and the $i$th training sample is defined by $e_i(\mathbf{y})=||\mathbf{y}-c_i\mathbf{m}_i||^2$.

We assume that a smaller $e_i(\mathbf{y})$ means that the $i$th training sample has greater contribution on the reconstruction of the test sample $\mathbf{y}$. We exploit $e_i(\mathbf{y})$ to select $k$ training samples with the first $k$ greatest effect, simultaneously, record which class these samples belong to. Then the $k$ training samples compose a new set, $\tilde{\mathbf{M}} = [\tilde{\mathbf{m}}_1, \tilde{\mathbf{m}}_2 \cdots \tilde{\mathbf{m}}_k]$ and $L = \{l_1, l_2 \cdots l_d\}$ stand for the set of class labels of the $\boldsymbol{k}$ nearest neighbors.

### 2.2 The Second Stage of the Two-Stage Test Sample Representation Method

After obtaining a new set of training vectors $\tilde{\mathbf{M}}$, we re-express the test sample $\mathbf{y}$ as a linear combination of the determined $k$ nearest neighbors. i.e.

$$\mathbf{y} = b_1\tilde{\mathbf{M}}_1 + b_2\tilde{\mathbf{M}}_2 + \cdots b_k\tilde{\mathbf{M}}_k \tag{4}$$

Where $b_i (i = 1, 2 \cdots k)$ is the recovered representation coefficient. Since the neighbors might be from different classes, we use only the coefficients associated with same class and calculate the sum of the effect of the training samples from each class. For example, if all the $k$ neighbors from the $j$th $(j \in L)$ class are $\{\tilde{\mathbf{m}}_m, \tilde{\mathbf{m}}_n \cdots \tilde{\mathbf{m}}_t\}$, then the sum of the contribution in representing the test sample of the $j$th class will be $\mathbf{g}_j = b_m\tilde{\mathbf{m}}_m + b_n\tilde{\mathbf{m}}_n + \cdots b_t\tilde{\mathbf{m}}_t$. The deviation of $\mathbf{g}_j$ from $\mathbf{y}$ can be estimated by using

$$r_j(\mathbf{y}) = \left\| \mathbf{y} - \mathbf{g}_j \right\|^2 \ (j \in L) \tag{5}$$

Finally, the decision rule is: $r_l(\mathbf{y}) = \min_j r_j(\mathbf{y})$, $\mathbf{y}$ is assigned to Class $l$. We solve (4) by using $\mathbf{b} = (\tilde{\mathbf{M}}^{\mathrm{T}}\tilde{\mathbf{M}} + \mu\mathbf{I})^{-1}\tilde{\mathbf{M}}^{\mathrm{T}}\mathbf{y}$.

## 3 Experiments

In this section, we investigate the applications of our method in face recognition. The CSIST and AR face image database is used for the experiments.

### 3.1 Database Description

CSIST face database contains 100 subjects with 800 face images as a whole. Each subject simultaneously provides four near-infrared face images and four visible light face images. The face images were captured under varying illumination, facial details and expression. The size of every face images is $128 \times 128$. Fig. 1 shows the images of one subject in the CSIST database.

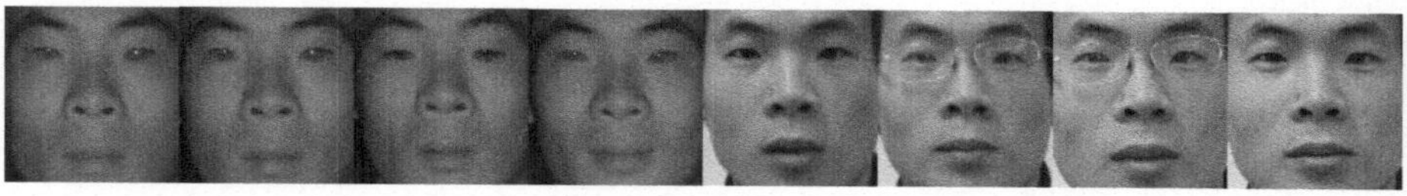

**Fig. 1.** The images of one subject in the CSIST database

AR database also simultaneously contains color images and gray images. The images with different facial expressions, lighting conditions and occlusions of 120 individuals were taken in two sessions and each section contains 13 color images. The size of every face images is 50×40.

Because the images are all RGB color images, we must transform them to gray images firstly. We use:

$$Gray = 0.299 \times R + 0.587 \times G + 0.114 \times B \tag{6}$$

where *Gray* is the pixel value of the obtained gray images. After converting the reading images into a one-dimensional vector **a**, we will normalize every vector using $\mathbf{a}_i' = \mathbf{a}_i / \|\mathbf{a}_i\|$.

### 3.2 Experimental Setting and Parameter Selection

In experiment, for CSIST database, we select the first and the third images per subject as training sample and the remaining samples are used for testing. For AR database, the first 7 samples of each class are used for training and the next 6 samples are used for testing. We estimate the coefficients by solving the normal equations.

We compare the performance of our new method with Principal Component Analysis (PCA) and Linear Discriminant Analysis (LDA). Take LDA as an example, there are two ways dealing with LDA. First one, we call LDA score level fusion. We applied LDA to gray images and color images, respectively. It will produce two features for a sample, then for a training sample and a test sample, we use the matching score level fusion scheme to calculate the sum of the distance between the two features. We classify the test sample into the class that the sum reaches its minimum value. Second one, we refer as complex LDA proposed in [1]. We also deal with PCA in the same way.

For PCA, we choose the number of principal components as 120 and use the nearest neighbor classifiers with Euclidean distance for final classification. For LDA, in order to prevent the within-class scatter matrix ($\mathbf{S}_w$) singular, we regularize $\mathbf{S}_w = \mathbf{S}_w + \mu\mathbf{I}$ and use PCA for dimension reduction firstly. Here we select the number of projection axes as 120 for feature extraction. Let $\mu = 0.01$ in all the experiments.

### 3.3 Experiment Result and Analysis on the CSIST Database

Table 1 shows the minimal recognition error rates of five different methods. Fig. 2 shows the classification error rates of our method and its global version.

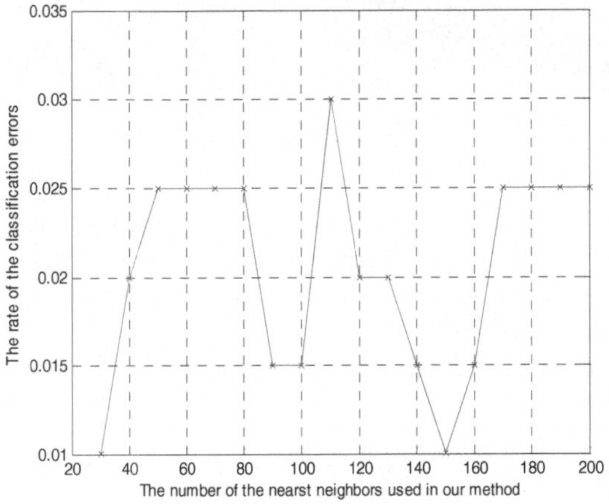


**Fig. 2.** The classification error rates of our method versus different number of neighbors

**Table 1.** Comparison on The Minimal Recognition Rates of Five Different Methods

| Methods | Minimal recognition rate(%) |
|---|---|
| PCA score level fusion | 8% |
| Complex PCA | 8% |
| LDA score level fusion | 1.5% |
| Complex LDA | 1% |
| Our method | 1% |

Fig. 2 clearly show that our method achieves a minimum error rate of 1% , which is 1.5% lower than the error rate of its global version. When we use a suitable number of nearest neighbors to express the test sample, it can achieve a very low error rate. From Table 1, we can see that our method outperforms PCA and LDA. Because it is carried out in the original space and no information is lost.

### 3.4 Experiment Result and Analysis on the AR Database

The maximal recognition rate of the five methods is given by table 2. Fig. 3 shows the recognition rates of our method versus different number of nearest neighbors used in second phase. Table 3 reports the time taken by different algorithms. Here we choose the number of the nearest neighbors as 50 and the number of optimal projection is 120 for PCA and LDA.

**Table 2.** Comparison on The Maximal Recognition Rates

| Methods | Maximal recognition rate(%) |
|---|---|
| PCA score level fusion | 67.08% |
| Complex PCA | 67.22% |
| LDA score level fusion | 67.08% |
| Complex LDA | 80.56% |
| Our method | 86.11% |

**Table 3.** Comparision on Recognition Speed

| Algorithms | Time(seconds) | Recognition rate(%) |
|---|---|---|
| PCA score level fusion | 219.8110 | 67.08% |
| Complex PCA | 112.6610 | 67.22% |
| LDA score level fusion | 143.3630 | 67.08% |
| Complex LDA | 76.0790 | 80.56% |
| Our method | 43.5510 | 82.78% |

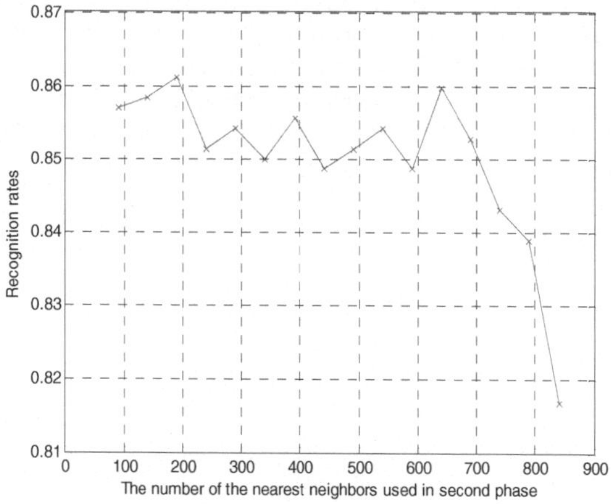


**Fig. 3.** The recognition rates of our method versus different number of neighbors

From fig. 3, we can see that the proposed method is always able to obtain a higher classification rates than its global version. Table 2 demonstrates the robustness and effectiveness of our method. Table 3 indicates that our method took much less time than complex LDA, LDA score level fusion, complex PCA and PCA score level fusion while the result is best.

# 4 Conclusions

Bimodal biometrics based on two-stage test sample representation method is proposed in this paper. It adopts two-phase strategy to evaluate the power of expressing the test sample of the bimodal training samples, then classify the test sample into the class that has the maximum power. A large number of experiments show a good performance of our method.

**Acknowledgments.** We wish to thank the National Science Foundation of China under Grant No. 60875004, the Natural Science Foundation of Jiangsu Province of China under Grant No.BK2009184, and the Natural Science Foundation of the Jiangsu Higher Education Institutions of China under Grant No.07KJB520133 for supporting this work.

## References

1. Yang, J., Yang, J.-Y., Frangi, A.F.: Combined Fisherfaces framework. Image Vision Computer 21(12), 1037–1044 (2003)
2. Xu, Y., Zhang, D., Yang, J., Yang, J.-Y.: A two-phase test sample sparse representation method for use with face recognition. 2011 IEEE Transactions on Circuits and Systems for Video Technology 21(9), 1255–1262 (2011)
3. Xu, Y., Zhong, A., Yang, J., Zhang, D.: Bimodal biometrics based on a representation and recognition approach. Optical Engineering 50(3), 037202-1–037202-7 (2011)
4. Xu, Y., Zhang, D., Yang, J.-Y.: A feature extraction method for use with bimodal biometrics. Pattern Recognition 43(3), 1106–1115 (2010)
5. Shi, Q., Eriksson, A., van den Hengel, A., Shen, C.: Is Face Recognition really a Compressive Sensing problem? In: Proc. of IEEE Conference on Computer Vision and Pattern Recognition, pp. 553–560 (2011)

# Sparse Discriminating Neighborhood Preserving Embedding

Li Guo, ZhongLong Zheng*, Jiong Jia, Huawen Liu, and Daohong Xiang

Department of Computer Science, Zhejiang Normal University,
321004 Zhejiang, China
lotus210902@126.com, zhonglong@zjnu.cn

**Abstract.** Dimensionality reduction (DR) methods have commonly been used as a principled way to process the high-dimensional data such as face images. In this paper, a novel linear DR method called discriminating neighborhood preserving embedding (DNPE), which incorporates between-class scatter matrix and within-class scatter matrix into neighborhood preserving embedding (NPE), is proposed. It has been shown that DNPE has stronger discriminating power than NPE does. Meanwhile, this paper also proposes sparse discriminating neighborhood preserving embedding (SDNPE) based on sparse representation theory, which directly generates the weight matrix without constructing adjacency graphs. Experimental results on Yale, ORL, AR and Extended YaleB face databases verify the efficacy of the proposed methods.

**Keywords:** Dimensionality reduction, Neighborhood Preserving Embedding, Sparse Representation, Maximum Margin Criterion.

## 1 Introduction

In many application domains, such as pattern recognition, information retrieval and text categorization, the data are usually provided in high-dimensional form. Dimensionality reduction is an effective approach to deal with such data, due to its potential to alleviate the so-called "curse of dimensionality"[1] and to improve the computational efficiency. Over the last century, various dimensionality reduction (DR) methods including linear DR methods and nonlinear DR methods have been developed.

Principal component analysis (PCA)[2; 3] and linear discriminant analysis (LDA)[4] are the commonly used linear DR methods. As an unsupervised technique, PCA aims to seek the linear combinations of the original data such that the derived principal components capture maximal variance. Different from PCA, LDA is a supervised learning algorithm which derives a projection basis that maximizes the between-class scatter and minimizes the within-class scatter. Both of them are simple and linear, thus they are extensively served as a feature extraction step for modeling, clustering, classification and visualization problems. However, PCA and LDA only

* Corresponding author.

J. Lei et al. (Eds.): AICI 2012, LNAI 7530, pp. 721–728, 2012.

see the global Euclidean structure, and they fail to discover the intrinsic data structures that are nonlinear.

In the past decade years, some nonlinear manifold learning algorithms have been proposed to discover the nonlinear structure of the manifold by investigating the local geometry of samples, such as Locally Linear Embedding (LLE)[5], Isometric Mapping (ISOMAP)[6], and Laplacian Eigenmaps (LE)[7]. However, these methods yield maps that can only be defined on the training samples and how to assess the maps on testing samples remains unclear.

As a linear version of LLE, neighborhood preserving embedding (NPE)[8; 9] is a recently proposed linear dimensionality reduction algorithm, which can be defined everywhere, rather than only on the training samples. Unlike PCA and LDA, which only preserve the global Euclidean structure, NPE aims at preserving the local neighborhood structure on the data manifold. However, it has an inherent limitation that it does not encode discriminant information, while discriminant information is important in recognition problem.

By incorporating between-class scatter and within-class scatter into the objective function of NPE, we propose a discriminating neighborhood preserving embedding (DNPE) algorithm, which not only considers the local neighborhood information but also emphasizes the discriminant information when applied to recognition tasks. Simultaneously, inspired by sparse representation theory, we derive the sparse version of DNPE (SDNPE) by adding $l_1$ regularization to DNPE. The SDNPE method constructs an adjacent weight matrix of the data points based on a sparse representation framework. The sparsity property is more effective than the dense representation in practically recognition applications with labeled information.

The rest of the paper is organized as follows: Section 2 reviews neighborhood preserving embedding (NPE), Maximum Margin Criterion (MMC)[10] and the related works. The discriminating neighborhood preserving embedding (DNPE) and sparse discriminating neighborhood preserving embedding (SDNPE) algorithms are introduced in Section 3 and Section 4, respectively. The experimental results are presented in Section 5. Finally, the conclusions are provided in Section 6.

## 2 Relative Works

The generic problem of linear dimensionality reduction can be stated in the following. Denote by $X=[x_1,x_2,\ldots,x_m]\in\mathbb{R}^n$ a set of points in high-dimensional space, we suppose to find a transformation matrix $A$ that maps these $m$ samples to a set of points $Y=[y_1,y_2,\ldots,y_m]\in\mathbb{R}^l (l<<n)$ in low-dimensional subspace, where $y_i=A^T x_i$ such that $y_i$ can "represent" $x_i$ in the sense of some criterion. The same notations mentioned below have the same meaning here.

### 2.1 Neighborhood Preserving Embedding (NPE)

As a recent promising method, NPE[8] which derives from LLE inherits the properties of LLE and then owns more merits. The objective function of NPE is defined as:

$$\min \sum_i (\mathrm{y}_i - \sum_j \mathrm{W}_{ij}\mathrm{y}_j)^2 \ , \tag{1}$$

where $\mathrm{y}_i = \mathrm{A}^T \mathrm{x}_i (i = 1,...,m)$, and $\mathrm{W}=(\mathrm{W}_{ij})$ is the weight matrix. In [8], an adjacency graph $G$ is first constructed with $m$ nodes. The *i-th* node corresponds to the data point $\mathrm{x}_i$. Then we compute the weight matrix $\mathrm{W}$. Each entry $\mathrm{W}_{ij}$ in $\mathrm{W}$ is the weight of the edge from node $i$ to node $j$ when node $j$ is among the k nearest neighbors of node $i$, and 0 if not. Minimizing the objective function is an attempt to ensure that if $\mathrm{x}_i$ can be linear represented by its nearest neighbors with linear coefficients $\mathrm{W}_{ij}$, then $\mathrm{y}_i$ can be as well. Therefore, the weights on the edges can be computed by minimizing the following cost function:

$$\begin{aligned} &\min \sum_i \| \mathrm{x}_i - \sum_j \mathrm{W}_{ij}\mathrm{x}_j \|^2 \\ &s.t. \ \sum_j \mathrm{W}_{ij} = 1, j = 1,2,...,m \end{aligned} \ . \tag{2}$$

Suppose $\mathrm{A}$ is a transformation matrix such that $\mathrm{Y}=\mathrm{A}^T\mathrm{X}$. Following some algebraic formulations, Eq. (1) can be reduced to

$$\begin{aligned} &\sum_i (\mathrm{y}_i - \sum_j \mathrm{W}_{ij}\mathrm{y}_j)^2 = \| \mathrm{Y}(I - \mathrm{W}) \|^2 \\ &= trace(\mathrm{Y}(I - \mathrm{W})(I - \mathrm{W})^T \mathrm{Y}^T) = trace(\mathrm{A}^T \mathrm{X} M \mathrm{X}^T \mathrm{A}) \end{aligned} \ , \tag{3}$$

where $M = (I - \mathrm{W})(I - \mathrm{W})^T$, $I$ represents an identity matrix. In order to eliminate an arbitrary scaling factor in the projection, a constraint is imposed as:

$$\mathrm{Y}^T\mathrm{Y}=I \Rightarrow \mathrm{A}^T\mathrm{X}\mathrm{X}^T\mathrm{A}=I \ . \tag{4}$$

Finally, the minimization problem reduces to be:

$$\begin{aligned} &\underset{\mathrm{A}}{\operatorname{argmin}} \ trace(\mathrm{A}^T \mathrm{X} M \mathrm{X}^T \mathrm{A}) \\ &\mathrm{A}^T\mathrm{X}\mathrm{X}^T\mathrm{A}=I \end{aligned} \ . \tag{5}$$

The transformation matrix that minimizes the objective function can be obtained by solving the following generalized eigenvector problem:

$$\mathrm{X} M \mathrm{X}^T a = \lambda \mathrm{X}\mathrm{X}^T a \ . \tag{6}$$

Obviously, matrices $\mathrm{X}M\mathrm{X}^{\mathrm{T}}$ and $\mathrm{X}\mathrm{X}^{\mathrm{T}}$ are symmetric and positive semi-definite. The vectors $a_i$ that minimize the objective function are given by minimum eigenvalues solutions to the above generalized eigenvector problem. Let the column vectors $a_0, a_1, ..., a_{l-1}$ be the solutions of Eq. (6), ordered according to their eigenvalues, $\lambda_0 \le \lambda_1 \le ... \le \lambda_{l-1}$. Thus, the embedding is $\mathrm{x}_i \to \mathrm{y}_i = \mathrm{A}^T \mathrm{x}_i$, $\mathrm{A} = (a_0, a_1, ..., a_{l-1})$ where $\mathrm{y}_i$ is a $l-$dimensional vector, and $\mathrm{A}$ is an $n \times l$ matrix.

### 2.2 Maximum Margin Criterion (MMC)

Recently, a feature extraction criterion, called maximum margin criterion (MMC)[10], is proposed, which can maximize the between-class distances after dimensionality reduction, while it does not depend on the nonsingularity of the within-class scatter matrix $S_w$. So it doesn't suffer from SSS (small sample size) problem. Denote $C=[C_1,C_2,...,C_c]$ as the whole data samples of $c$ classes. The objective function of Maximum Margin Criterion is defined as:

$$J=\frac{1}{2}\sum_{i=1}^{c}\sum_{j=1}^{c}p_i p_j(d(m_i,m_j)-s(C_i)-s(C_j)) \ , \tag{7}$$

where $p_i$ and $p_j$ are the prior probability of class $i$ and class $j$ respectively, $m_i$ and $m_j$ are the mean vectors of the class $C_i$ and $C_j$, respectively. $s(C_i)$, $s(C_j)$ are some measure of the scatter of the class $C_i$ and $C_j$, respectively.

Here, we use the overall variance $tr(S_i)$ to measure the scatter of the data.

$$s(C_i)=tr(S_i), \quad s(C_j)=tr(S_j) \ . \tag{8}$$

According to the equations defined above, Eq. (7) can be rewritten as:

$$J=\frac{1}{2}\sum_{i=1}^{c}\sum_{j=1}^{c}p_i p_j(d(m_i,m_j)-tr(S_i)-tr(S_j))=tr(S_b-S_w) \ . \tag{9}$$

So the goal of MMC is to maximize $tr(S_b-S_w)$, where the matrix $S_b$ is called between-class scatter matrix and $S_w$ is called within-class scatter matrix.

## 3 Discriminating Neighborhood Preserving Embedding

Due to the less discriminating power of NPE, we propose a novel linear DR method based on MMC and NPE, called Discriminating Neighborhood Preserving Embedding (DNPE). The objective function of DNPE is written as:

$$\begin{cases}\min \ trace(\mathrm{A}^T\mathrm{X}M\mathrm{X}^T\mathrm{A}) \\ \max \ trace(\mathrm{A}^T(S_b-S_w)\mathrm{A})\end{cases} \cdot \quad s.t. \ \mathrm{A}^T\mathrm{X}\mathrm{X}^T\mathrm{A}=I \tag{10}$$

Eq. (10) can be changed into the following constrained problem:

$$\min \ trace(\alpha\mathrm{A}^T\mathrm{X}M\mathrm{X}^T\mathrm{A}-(1-\alpha)\mathrm{A}^T(S_b-S_w)\mathrm{A}) \ , \quad s.t. \ \mathrm{A}^T\mathrm{X}\mathrm{X}^T\mathrm{A}=I \tag{11}$$

where $\alpha$ is a suitable constant, $0<\alpha\leq 1$. The parameter $\alpha$ can adjust the contributions of maximum margin criterion and the local structure according to different data sets. The transformation matrix A that minimizes the objective function (11) can be obtained by solving the generalized eigenvector problem:

$$(\alpha XMX^T - (1-\alpha)(S_b - S_w))a = \lambda XX^T a \ . \tag{12}$$

# 4 Sparse Discriminating Neighborhood Preserving Embedding

## 4.1 Sparse Representation

Sparse representation (SR) is initially proposed as an extension to traditional signal representations such as Fourier representation and wavelet representation. In the past few years, SR has been successfully applied to solve many practical problems in signal processing, statistics, and pattern recognition[11]. The essence of SR is that a sample can be linearly represented by a small number of all samples, mainly formed from samples belonging to the same class.

The description of SR is as follows. Given a data point(e.g. a face image with vector pattern ) $\mathrm{x} \in \mathbb{R}^n$ , and a matrix $\mathrm{X} = [\mathrm{x}_1, \mathrm{x}_2, \ldots, \mathrm{x}_m] \in \mathbb{R}^{n\times m}$ containing the elements of an overcomplete dictionary[12] in its columns, the goal of SR is to represent x using as few entries of X as possible. The optimization problem can be formally expressed as:

$$\begin{aligned} &\min_{\mathrm{s}} \ \|\mathrm{s}\|_1 \\ &s.t. \ \ \mathrm{x}=\mathrm{Xs} \end{aligned} \tag{13}$$

where $\mathrm{s} \in \mathbb{R}^m$ is the coefficient vector, and $\|\cdot\|_1$ denotes the $l_1$-norm. Here, $l_1$ replaces $l_0$. It can be shown that if the solution s is sparse enough, the solution of $l_0$-minimization problem is equal to the solution of $l_1$-minimization problem[13].

## 4.2 Sparse Discriminating Neighborhood Preserving Embedding

Motivated by sparse representation theory, a modified DNPE algorithm named sparse discriminating neighborhood preserving embedding (SDNPE) is proposed. The most difference between DNPE and SDNPE is the weight matrix. Unlike DNPE, which firstly constructs an adjacency graph and then computes the weight matrix W , SDNPE directly computes the weight matrix S by sparse coding.

Similar to NPE, we define the following objective function to seek the projections that best preserve the optimal weight vector $\mathrm{s}_i$ :

$$\min \sum_{i=1}^{m} \left\|\mathrm{y}_i - \mathrm{Ys}_i\right\|^2 \tag{14}$$

where $\mathrm{Y} = [\mathrm{y}_1, \mathrm{y}_2, \ldots, \mathrm{y}_m]$ .

Denote $e_i$ as an $m$-dimensional unit vector with the $i-th$ element 1, 0 otherwise. Then, Eq. (14) can be changed into:

$$\begin{aligned}\sum_{i=1}^{m}\|y_i - Ys_i\|^2 &= \sum_{i=1}^{m}\|Ye_i - Ys_i\|^2 = Y\left(\sum_{i=1}^{m}(e_i - s_i)(e_i - s_i)^T\right)Y^T \\ &= Y\sum_{i=1}^{m}(e_ie_i^T - e_is_i^T - s_ie_i^T + s_i^Ts_i)Y^T = trace(Y(I - S - S^T + S^TS)Y^T) \\ &= trace(A^TX(I - S - S^T + S^TS)X^TA)\end{aligned} \tag{15}$$

Thus, according to the DNPE and the constraint condition $A^TXX^TA = I$ , the objective function of SDNPE can be written as the following optimization problem:

$$\begin{aligned}&\min_{A}\ trace(\alpha - \alpha A^TX(S + S^T - S^TS)X^TA - (1-\alpha)A^T(S_b - S_w)A) \\ &\qquad s.t.\ A^TXX^TA = I\end{aligned} \tag{16}$$

where $\alpha$ is a suitable constant, $0 < \alpha \leq 1$. A is the transformation matrix. S is the weight matrix obtained by sparse representation. $S_b$ is the between-class scatter matrix and $S_w$ is the within-class scatter matrix.

For compact expression, the minimization problem can further be transformed to an equivalent maximization problem as follows:

$$\begin{aligned}&\max_{A}\ trace(\alpha A^TXS_\beta X^TA + (1-\alpha)A^T(S_b - S_w)A) \\ &\qquad s.t.\ A^TXX^TA = I\end{aligned} \tag{17}$$

where $S_\beta = S + S^T - S^TS$ . Thus, the transformation matrix that maximizes Eq. (17) can be obtained by solving the generalized eigenvector problem:

$$(\alpha XS_\beta X^T + (1-\alpha)(S_b - S_w))a = \lambda XX^Ta \tag{18}$$

## 5 Experimental Results

We evaluate the performance of DNPE and SDNPE on four representative facial image databases: Yale[11], ORL[14], AR[11] and Extended YaleB[11] face databases. Simultaneously, to verify the effectiveness of the proposed methods, we compare them with PCA, LDA, NPE and Sparsity Preserving Projections (SPP)[11]. These DR methods have been successfully applied to face recognition. In our experiment, the images are normalized and cropped to a size of 32×32.

For the four databases, we randomly select half of the images per class for training (i.e. 6, 5, 7 and 32 images per subject for Yale, ORL, AR and Extended YaleB databases, respectively), and the rest for testing. In the experiments, 20 training/testing pairs are randomly generated and the average recognition rates over these pairs are reported. And for simplicity, we use nearest neighborhood (NN) classifier for recognition.

The experimental results of these DR methods on the four face databases are shown in Fig. 1. Obviously, the recognition rates of SDNPE and DNPE are always better than those of other methods. On Yale, SDNPE and DNPE almost have the same best recognition rates. As discriminant version of NPE and SPP separately, DNPE and SDNPE both outperforms others. Note, LDA at most reduces to (c-1)-dimensions(c is the number of classes). Meanwhile we summarize the best average recognition rates of all competing methods on the four databases in Table 1. We can conclude that SDNPE and DNPE work stably across all the databases.

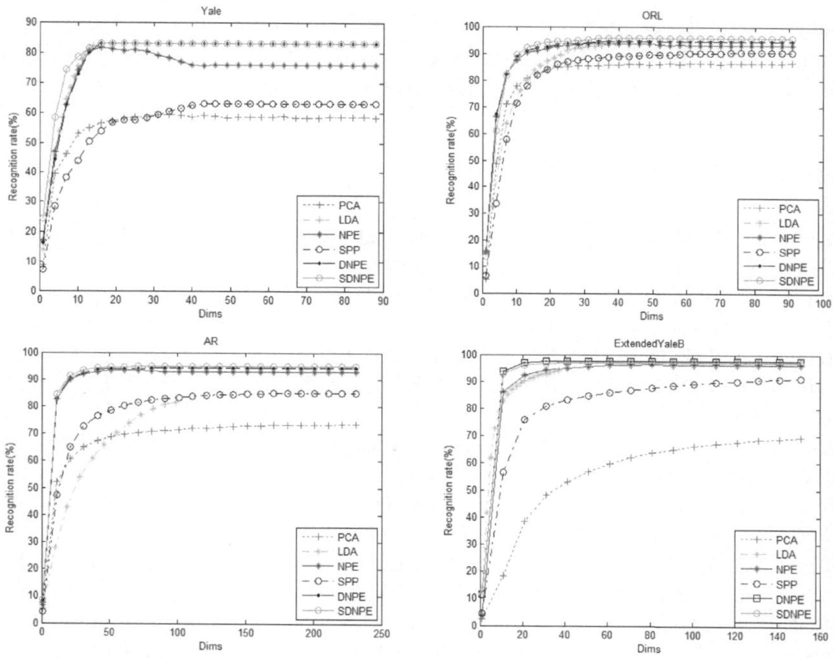


**Fig. 1.** Recognition rates of PCA, LDA, NPE, SPP, DNPE and SDNPE methods versus dimensions on the Yale, ORL, AR and Extended YaleB face databases

**Table 1.** The best recognition rates of 1NN classifier based on different LDR methods

| Databases | Methods | PCA | LDA | NPE | SPP | DNPE | SDNPE |
|---|---|---|---|---|---|---|---|
| Yale | Recognition rate (%) | 59.33 | 82.20 | 81.80 | 62.93 | 83.0 | 83.07 |
| | Dimensions | 31 | 14 | 16 | 43 | 16 | 16 |
| ORL | Recognition rate (%) | 86.58 | 93.88 | 93.88 | 90.60 | 94.63 | 95.75 |
| | Dimensions | 91 | 39 | 34 | 79 | 37 | 34 |
| AR | Recognition rate (%) | 73.56 | 81.51 | 93.39 | 84.82 | 94.20 | 94.83 |
| | Dimensions | 231 | 99 | 61 | 171 | 81 | 71 |
| Extended YaleB | Recognition rate (%) | 69.38 | 93.89 | 96.27 | 91.15 | 97.53 | 96.82 |
| | Dimensions | 151 | 37 | 81 | 151 | 41 | 41 |

## 6 Conclusion

In this paper, a new discriminating neighborhood preserving embedding algorithm is proposed. Compared with PCA and LDA, our DNPE can discover the intrinsic structure of the nonlinear data. Moreover, DNPE has better discriminating power than NPE. Then we modify the weight matrix of DNPE with sparse representation, and obtain a sparse discriminating neighborhood preserving embedding algorithm. Experimental results demonstrate the effectiveness of our algorithms.

**Acknowledgments.** The authors confirm that the research was supported by National Natural Science Foundation (No.61170109, No.61100119&No.11001247), and Zhejiang Natural Science Foundation (No.Y1090579&No.Y1100161), China. Opening Fund of Top Key Discipline of Computer Software and Theory in Zhejiang Provincial Colleges at Zhejiang Normal University.

## References

1. Jain, A.K., Duin, R.P.W., Mao, J.C.: Statistical Pattern Recognition: A Review. IEEE Trans. Pattern Anal. Mach. Intell. 22, 4–37 (2000)
2. Jolliffe, I.T.: Principal Component Analysis. Springer, New York (1986)
3. Turk, M., Pentland, A.: Eigenface for Recognition. J. Cog. Neuo. 3, 71–86 (1991)
4. Belhumeur, P.N., Hespanha, J.P., Kriegman, D.J.: Eigenfaces Vs. Fisherfaces: Recognition Using Class Specific Linear Projection. IEEE Trans. Pattern Anal. Mach. Intell. 19, 711–720 (1997)
5. Roweis, S.T., Saul, L.K.: Nonlinear Dimensionality Reduction by Locally Linear Embedding. Science 290, 2323–2326 (2000)
6. Tenenbaum, J.B., de Silva, V., Langford, J.C.: A Global Geometric Framework for Nonlinear Dimensionality Reduction. Science 290, 2319–2323 (2000)
7. Belkin, M., Niyogi, P.: Laplacian Eigenmaps for Dimensionality Reduction and Data Representaiton. Neural Comput. 15, 1373–1396 (2003)
8. He, X.F., Cai, D., Yan, S.C., Zhang, H.J.: Neighborhood Preserving Embedding. In: IEEE International Conference on Computer Vision (ICCV), Beijing, China (2005)
9. Pang, Y., Zhang, L., Liu, Z., Yu, N., Li, H.: Neighborhood Preserving Projections (NPP): A Novel Linear Dimension Reduction Method. In: Huang, D.-S., Zhang, X.-P., Huang, G.-B. (eds.) ICIC 2005, Part I. LNCS, vol. 3644, pp. 117–125. Springer, Heidelberg (2005)
10. Li, H.F., Jiang, T., Zhang, K.S.: Efficient and Robust Feature Extraction by Maximum Margin Criterion. IEEE Trans. Neural Networks 17, 157–165 (2006)
11. Qiao, L.S., Chen, S.C., Tan, X.Y.: Sparsity Preserving Projections with Applications to Face Recognition. Pattern Recognition 43, 331–341 (2010)
12. Wright, J., Yang, A.Y., Ganesh, A., Sastry, S.S., Ma, Y.: Robust Face Recognition Via Sparse Representation. IEEE Trans. Pattern Anal. Mach. Intell. 31 (2009)
13. Zou, H., Hastie, T.: Regularization and Variable Selection Via the Elastic Net. J. Roy. Stat. Soc. B 67, 301–320 (2005)
14. Speech, Vision and Robotics Group of the Cambridge University Engineering Department, `http://www.cl.cam.ac.uk/Research/DTG/attarchive:pub/data/att_faces.zip`

# Complex-Matrix-Based Horizontal and Vertical Discriminant Analysis for Feature Fusion

Xiuping Wang[1,2] and Caikou Chen[2]

[1] Information Engineering Department Jiangsu Animal Husbandry & Veterinary College, Taizhou 225300

[2] Information Engineering College Yangzhou University, Yangzhou 225012

**Abstract.** Considering the serial strategy generally used in feature fusion easily leads to curse of dimensionality and two-dimensional matrix for image representation outperforms one-dimensional vector, a novel strategy of parallel complex-matrix-based horizontal and vertical discriminant analysis is developed in this paper. It first respectively utilizes two different images of a subject as the real and imaginary part of a complex matrix, two-step discriminant analysis, namely horizontal LDA and vertical PCA, is then performed in the complex feature space. The experimental results demonstrate that the proposed method is more promising and effective.

**Keywords:** Feature fusion, Complex matrix, Two-step discriminant analysis.

## 1 Introduction

Recently, feature fusion has been widely developed in computer vision and pattern recognition, which utilizes the given multiple feature sets to generate new fused feature sets for effective classification. Feature extraction is the research priority of feature fusion. In general, it is used to construct a meaningful low-dimensional representation of high-dimensional data. Specifically, principal component analysis and linear discriminant analysis are two representative algorithms.

PCA [1] is originally devised to one-dimensional data, so when PCA is applied to images, the image matrices is first converted into one-dimensional vectors, which causes the covariance matrix is easy to inaccurately estimate. Recently, two-dimensional PCA (2DPCA) [2] is given more attention, which directly makes use of image matrices for feature extraction. Different from traditional PCA, 2DPCA ensures the covariance matrix more accurate. However, 2DPCA only focuses on the horizontal transformation of images, and ignores the discriminant information from vertical direction. In recent years, motivated by 2DPCA, 2DLDA [3] has been proposed. Although 2DLDA effectively uses class-label information and two-dimensional image matrices, it only does the horizontal transformation. Besides, the complex-matrix-based parallel scheme for feature fusion is better than original serial strategy [4]. Recently, Xu et al. have developed MCPCA [5] and IBLDA [6], which are based on the complex space. These two methods not only perform feature fusion well, but achieve good effects for feature extraction.

J. Lei et al. (Eds.): AICI 2012, LNAI 7530, pp. 729–734, 2012.

In this paper, a novel feature fusion strategy of parallel complex-matrix-based horizontal and vertical discriminant analysis is proposed. Firstly, two different image matrices from one subject are devoted to the real and imaginary part of a complex matrix and then it respectively constructs the between-class scatter and within-class scatter in the complex feature space, finally derives feature by applying 2DLDA for horizontal transformation and 2DPCA for vertical transformation on two-dimensional complex matrix. Based on this discriminant criterion, two different images from one subject are not only simply and intuitively integrated, but the computational cost is low, because the traditional high-dimensional vector is replaced with two-dimensional matrix. Experiments show that our proposed method is more efficient.

## 2 Complex-Matrix-Based Horizontal and Vertical Discriminant Analysis

### 2.1 Two-Dimensional Linear Discriminant Analysis (2DLDA)

Given a set of $M$ training samples that contains $C$ pattern classes, $\mathbf{x}_1, \mathbf{x}_2, \ldots, \mathbf{x}_M$ in $\mathrm{IR}^{m\times n}$, $\mathbf{x}_j (j = 1, \cdots, M)$ represents the *j*th image samples matrix, and the size is $m \times n$. $n_i$ is the number of training samples in class *i*. $\mathbf{x}_{ij}$ is the *j*th training sample in class *i*. The mean image of all training samples is denoted by $\mathbf{m}_0$ and $\mathbf{m}_i$ is the mean image of the training samples in class *i*, in particular, the size of each of the two mean images is $m \times n$.

Based on the given training samples, the image between-class scatter matrix $\mathbf{G}_\mathrm{b}$ and the image within-class scatter matrix $\mathbf{G}_\mathrm{w}$ are defined by Eq. (1) and Eq. (2). The generalized Fisher discriminant criterion is given by Eq. (3).

$$\mathbf{G}_\mathrm{b} = \frac{1}{M} \sum_{i=1}^{C} n_i (\mathbf{m}_i - \mathbf{m}_0)^\mathrm{T} (\mathbf{m}_i - \mathbf{m}_0) \tag{1}$$

$$\mathbf{G}_\mathrm{w} = \frac{1}{M} \sum_{i=1}^{C} \sum_{j=1}^{n_i} (\mathbf{x}_{ij} - \mathbf{m}_i)^\mathrm{T} (\mathbf{x}_{ij} - \mathbf{m}_i) \tag{2}$$

$$J(\mathbf{w}) = \frac{\mathbf{w}^\mathrm{T} \mathbf{G}_\mathrm{b} \mathbf{w}}{\mathbf{w}^\mathrm{T} \mathbf{G}_\mathrm{w} \mathbf{w}} \tag{3}$$

In contrast to one-dimensional conventional Linear Discriminant Analysis (LDA), two matrices $\mathbf{G}_\mathrm{b}$ and $\mathbf{G}_\mathrm{w}$ formed directly based on two-dimensional patterns are both $n \times n$ positive definite. Besides, the set of discriminant vectors $\mathbf{w}_i (i = 1, \ldots, q)$ can be obtained by maximizing the Rayleigh quotient function $J(\mathbf{w})$ so that the feature matrix $\mathbf{Y}$ extracted by 2DLDA is given by Eq. (4), where $\mathbf{X} = (\mathbf{x}_1, \mathbf{x}_2, \ldots, \mathbf{x}_M)$, $\mathbf{W} = (\mathbf{x}_1, \mathbf{w}_2, \ldots, \mathbf{w}_q)$.

$$\mathbf{Y} = \mathbf{X}\mathbf{W} \tag{4}$$

## 2.2 Complex-Matrix-Based Horizontal and Vertical Discriminant Analysis

Suppose that two matrices **A** and **B** are from two different images of one subject. The traditional strategy of feature fusion is serial, namely the fused feature **C** is constructed by serial connection between **A** and **B**. Although the performance of serial fusion is effective [4], the dimension of fused feature caused by serial strategy is often high, so that curse of dimensionality is likely to happen. In this paper, the complex matrix **C** is used for parallel feature fusion, that is $\mathbf{C} = \mathbf{A} + i\mathbf{B}$, so that the high dimension of serial feature fusion can be avoided.

Based on the complex matrix, the image between-class scatter matrix $\hat{\mathbf{G}}_{\mathrm{b}}$ and the image within-class scatter matrix $\hat{\mathbf{G}}_{\mathrm{w}}$ in complex space are constructed as follows:

$$\hat{\mathbf{G}}_{\mathrm{b}} = \frac{1}{M}\sum_{i=1}^{C} n_i (\mathbf{C}_i - \mathbf{C}_0)^{\mathrm{H}} (\mathbf{C}_i - \mathbf{C}_0) \tag{5}$$

$$\hat{\mathbf{G}}_{\mathrm{w}} = \frac{1}{M}\sum_{i=1}^{C}\sum_{j=1}^{n_i} (\mathbf{C}_j^{(i)} - \mathbf{C}_i)^{\mathrm{H}} (\mathbf{C}_j^{(i)} - \mathbf{C}_i) \tag{6}$$

where $\mathbf{C}_j^{(i)}$ is the jth training sample in class *i*. $\mathbf{C}_0$ and $\mathbf{C}_i$ are respectively the whole mean of all training samples and the mean image of the training samples in class *i*.

As we know from Section 2.1, after construction of $\hat{\mathbf{G}}_{\mathrm{b}}$ and $\hat{\mathbf{G}}_{\mathrm{w}}$, the feature matrix of LDA can be obtained by Eq. (7), where $\mathbf{C} = (\mathbf{C}_1, \mathbf{C}_2, \ldots, \mathbf{C}_M)$, $\mathbf{W} = (\mathbf{x}_1, \mathbf{w}_2, \ldots, \mathbf{w}_q)$. The procedure of feature extraction

$$\mathbf{Y} = \mathbf{CW} \tag{7}$$

can be viewed as the dimension reduction in horizontal direction. however, the correlation between image rows and the dimension reduction in vertical direction is ignored. So after conducting horizontal LDA in complex space, the vertical LDA then need to be performed.

Through horizontal LDA in complex space, the feature **Y** in horizontal direction can be gained by Eq. (7). And then the image covariance matrix **Z** in vertical direction is given by Eq. (8).

$$\mathbf{Z} = \frac{1}{M}\sum_{i=1}^{M} (\mathbf{Y}_i - \mathbf{Y}_0)(\mathbf{Y}_i - \mathbf{Y}_0)^{\mathrm{H}} \tag{8}$$

where $\mathbf{Y}_i$ is the horizontal feature of the *j*th training sample in class *i*, and $\mathbf{Y}_0$ is the whole mean of all horizontal features.

The novel discriminant criterion function is given by Eq. (9). The set of discriminant vectors **v***i* (*i* = 1, … , *t*) can be gained by obtaining the *t* eigenvectors of $\mathbf{Zv} = \lambda\mathbf{v}$ corresponding to the top *t* largest eigenvalues of Z.

$$J(\mathbf{v}) = \mathbf{v}^{\mathrm{T}}\mathbf{Zv} \tag{9}$$

Finally, we get the vertical features **U**, namely the final features for classification, by Eq. (10).

$$\mathbf{U} = \mathbf{V}^{\mathrm{T}}\mathbf{Y} \tag{10}$$

# 3 Experiments and Analysis

In this section, we perform experiments on ORL face database and another data set, including visible-light and near-infrared face images, and they will show the performance difference between the proposed method and other methods.

## 3.1 Experiment on ORL Face Database

The ORL face database is from Olivetti-Oracle Research Lab, which contains 40 persons, and each person has 10 images under different ways, or example, smiling and no smiling, eye opening and eye closed and so forth. The image resolution is 112×92.

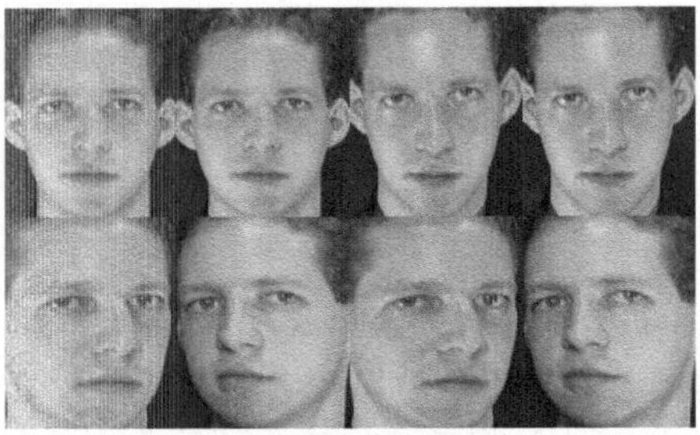

**Fig. 1.** The first four images of one person, the images of the first row are the real part, the images of the second row are the imaginary part

In this experiment, the first four images per class are used to construct complex training samples, specifically the first and third images per class are used as the real part and the second and fourth images are the imaginary part, which are shown in Fig. 1. The remaining images are utilized to test, specifically the seventh, eighth and ninth images per class are the real part, and the remaining are the imaginary part, which are shown in Fig. 2. Table 1 illustrates the classification performance of the proposed method and other methods under nearest neighbor classifier. And the number in the brackets is the corresponding dimension.

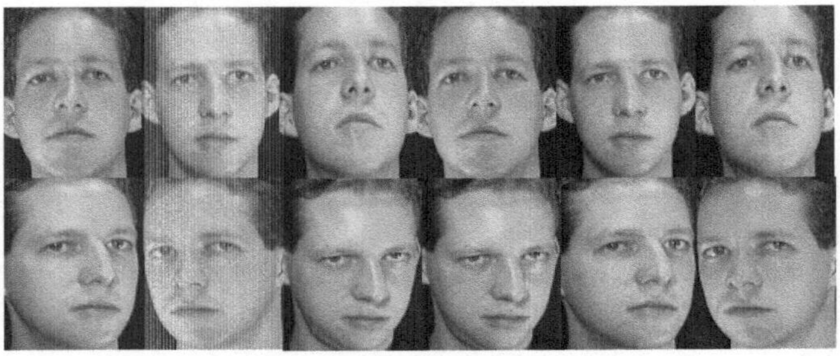

**Fig. 2.** The last six images of one person, the images of the first row are the real part, the images of the second row are the imaginary part

**Table 1.** The classification performance of the proposed method and other methods

| Methods | 1D-PCA | 1D-LDA | 2D-PCA | 2D-LDA | MCPCA | IBLDA | The proposed method |
|---|---|---|---|---|---|---|---|
| Recogntion Rate (%) | 87.5(40) | 90.42(35) | 91.67(6) | 94.17(5) | 91.25(6,8) | 27.5(1) | 95.17(7,3) |

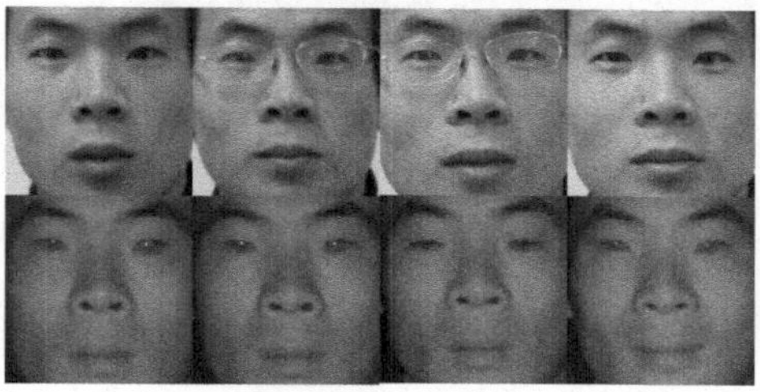

**Fig. 3.** The eight images of one person, the four images of the first row are visible-light, the four images of the second row are near-infrared

## 3.2 Experiment on the Face Database Including Visible-Light and Near-Infrared Face Images

In this experiment, a data set including near-infrared and visible-light face images are used to test the proposed method. There are 800 images of 100 persons in the data set, specifically each person has four near-infrared face images and visible-light face images, which are shown in Fig.3, and the image resolution is 128×128.

The first two visible-light and near-infrared images per class are used for the complex training samples, where the two visible-light images are used as the real part and the two near-infrared images are the imaginary part, which are displayed in Fig. 4. The remaining images are for test, specifically the last two visible-light images are the real part and the last two near-infrared images are the imaginary part, which are shown in Fig. 5. Table 2 shows the classification performance of the proposed method and other methods under nearest neighbor classifier, and the number in the brackets is the corresponding dimension.

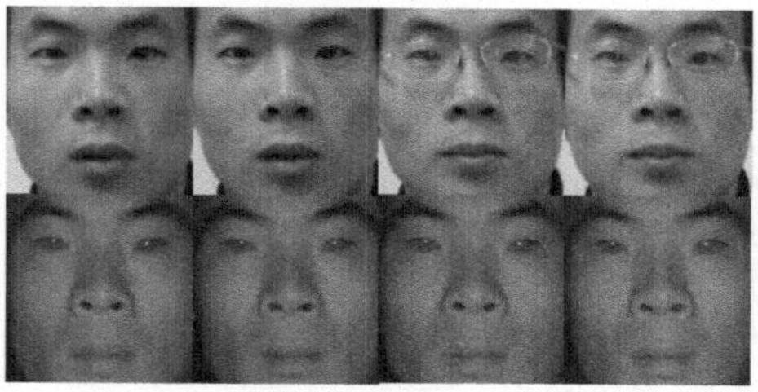

**Fig. 4.** The images of one person for training, the first row images are visible-light, the second row images are near-infrared

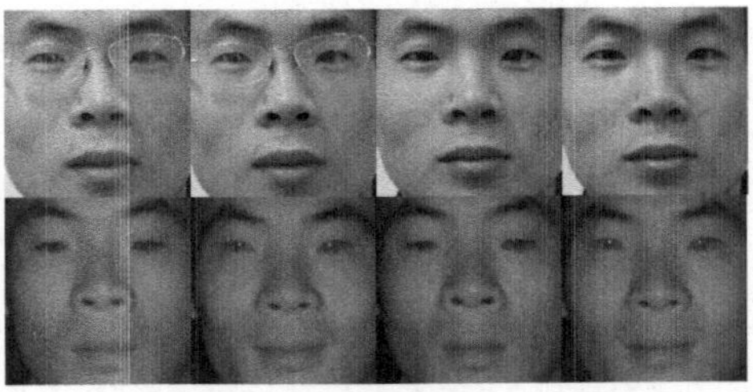

**Fig. 5.** The images of one person for test, the first row images are visible-light, the second row images are near-infrared

**Table 2.** The classification performance of the proposed method and other methods

| Methods | 1D-PCA | 1D-LDA | 2D-PCA | 2D-LDA | MCPCA | IBLDA | The proposed method |
|---|---|---|---|---|---|---|---|
| Recogntion Rate (%) | 77 (85) | 91.25(37) | 87.25(19) | 84 (13) | 95.75(20,19) | 63 (13) | 95.85(20,9) |

## 4 Conclusion

In this paper, a novel parallel feature fusion strategy of complex-matrix-based horizontal and vertical discriminant analysis is developed. It not only fuses the two different feature of one subject, but respectively perform 2DLDA in horizontal direction and 2DPCA in vertical direction. Experiments on two data sets demonstrate the effectiveness of the proposed method.

**Acknowledgment.** Supproted by the National Natural Science Foundation of China (No. 60875004), the Natural Science Foundation of Jiangsu Province of China (No. BK2009184), and the Natural Science Foundation of the Jiangsu Higher Education Institutions of China (No. 07KJB520133).

## References

[1] Turk, M., Pentland, A.: Eigenfaces for recognition. Journal of Cognitive Neuroscience 3(1), 71–86 (1991)

[2] Yang, J., Zhang, D., Frangi, A.F., Yang, J.Y.: Two dimensional PCA: a new approach to appearance-based face representation and recognition. IEEE Transactions on Pattern Analysis and Machine Intelligence 24(1), 131–137 (2004)

[3] Xiong, H., Swamy, M.N.S., Ahmad, M.O.: Two-dimensional FLD for face recognition. Pattern Recognition, 1121–1124 (2005)

[4] Yang, J., Yang, J.Y., Zhang, D., Lu, J.F.: Feature fusion: parallel strategy vs. serial strategy. Pattern Recognition 36(6), 1369–1381 (2003)

[5] Xu, Y., Zhang, D., Yang, J.-Y.: A feature extraction method for use with bimodal biometrics. Pattern Recognition 43(3), 1106–1115 (2010)

[6] Xu, Y., Zhang, D.: Represent and fuse bimodal biometric images at the feature level: complex-matrix-based fusion scheme. Optical Engineering 49(3) (2010)

# Segmentation of CAPTCHAs Based on Complex Networks

Kun Fang, Zhan Bu, and Zheng You Xia

College of Computer Science and Technology,
Nanjing University of Aeronautics and Astronautics, China
zhengyou_xia@nuaa.edu.cn

**Abstract.** CAPTCHA is a simple test that is designed to be easily generated by computers and easily recognized by humams, but difficult for computers to solve. It is now almost a standard security technology. The most widely deployed CAPTCHAs are text-based schemes, but to CAPTCHAs, segmenting the connected and distored characters is still an unsolving problem. In this paper, we proposed a Community Divided Model algorithm which based on complex networks to segment these CAPTCHAs. To evaluate the effectiveness of the proposed segmentation algorithm, we conducted several experiments on database which collected some CAPTCHAs from the Internet randomly. The results showed that the proposed algorithm is effective to segment two or more connected and distored characters.

**Keywords:** CAPTCHA, complex networks, segmentation, connected characters.

## 1 Introduction

CAPTCHA (Completely Automated Public Turing Test to Tell Computers and Humans Apart) is a program that generates and grades tests that are human solvable, but intend to be beyond the capabilities of current computer programs [1]. This technology is now almost a standard security mechanism that have been widely used in free email and forum accounts registration to against undesirable or malicious computer bot program registration and spam. The most widely deployed CAPTCHAs are text-based schemes, which typically require users to recongize connected and distorted characters [2]. It is well known that CAPTCHAs segmentation is much more diffcult than recongition since machine learning algorithms can efficiently solve the general recongition problem, but currently no effective general algorithm can be used to solve all CAPTCHAs segmentation [3]. If a scheme is vulnerable to be segmented, it is can be broken easily. Convolutional neural network had been widely used for recongising single character [4]. Chellapilla et al. [3] had studied that computers can recongnize single character even at high distortion and clutter settings with using a sequence of character transformations. A commonly accepted goal for CAPTCHAs design is that automated attacks should not be more than 0.01% successful but that the human success rate should be at least 90% [5].

J. Lei et al. (Eds.): AICI 2012, LNAI 7530, pp. 735–743, 2012.

Therefore, various CAPTCHAs segmentation mechanisms have been proposed. Mori and Malik [6] had broken the EZ-Gimpy and the Gimpy CAPTCHAs with sophisticated object recognition algorithms. Chellapilla et al.[7] used the machine learning algorithms to attack a number of early CAPTCHAs. Huang et al.[8] used a projection-based segmentation algorithm to break MSN and YAHOO CAPTCHAs. Bursztein et al. [9] detailedly describe of the text-based CAPTCHA Strengths and Weaknesses. Fig.1 shows an attack on a Microsoft CAPTCHA with using the vertical, color filling segmentation and thick arc removal algorithms [2].

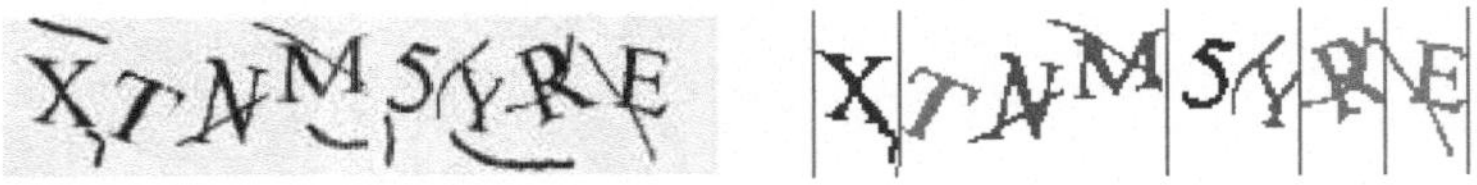

(a) (b)

**Fig. 1.** (a) Original image of Microsoft CAPTCHA. (b) Completely segmented image.

The CAPTCHAs made by Yan [2] and Huang [8] were mainly used straight and curved line as image clutter to confuse the defeating program. But when encounter CAPTCHAs with connected and distorted characters, these algorithms are not very effective. So, in this study, we propose a Community Divided Model segmentation algorithm which based on complex networks to segment these connected and distorted characters. And our experiment results indicate that this novel method is effective to segment a connected and distored characters.

The remainder of this paper are organized as follows: In section 2, we discuss our motivation and create the database. Section 3 presents our algorithm based on complex networks. Experiment results and analysis are discusseed in Section 4. We draw some conclusions and further studies in Section 5.

## 2 Motivation and Database Creation

In the last decade, complex networks[10,11] had become a new movement of research. Community structure is an important property of complex networks. Girvan and Newman [12] highlighted the property of community structure, in which network nodes are joined together in tight knit groups, between which there are only looser connections. They tested their method on computer-generated and real-world graphs and found the structure was high sensitivity and reliability. For CAPTCHAs, we find it has fixed length, that is to say we know the community number, and every single character tightly connected while touching characters are connected loosely. And the adjacent connected and distorted characters in most cases are touching at the middle of the core zone. The idea of this method is shown in Fig.2. In Fig.2(a), three characters consist of the CAPTCHAs and they connected each other. We can use the characteristic of these three characters and divide them into three community structures: red is 'q', green is 'o' and black is 's', the result is showed in Fig.2(b). In this paper, we modify the Girvan and Newman algorithm and propose a Community Divided Model which can segment a connected and distorted characters.

(a) (b)

**Fig. 2.** (a) A CAPTCHAs collected from internet. (b) Community structure of "qos". It use adjacency matrix for eight neighbors of each foreground pixels and the three communities were detected by the Girvan and Newman algorithm.

To determine whether the Community Divided Model can be used as an effiective method to segment the CAPTCHAs, we randomly collected hundreds of the CAPTCHAs from Authorize, 360buy, Tianya, Windows Live and Taobao [13,14,15,16,17,18], where the CAPTCHAs were used to register accounts. Fig.3 shows some examples. Our method does not consider the dot shape: "i" and "j", as Ahmad at el. [5] explained how to detect characters that contains a dot shape in detail.

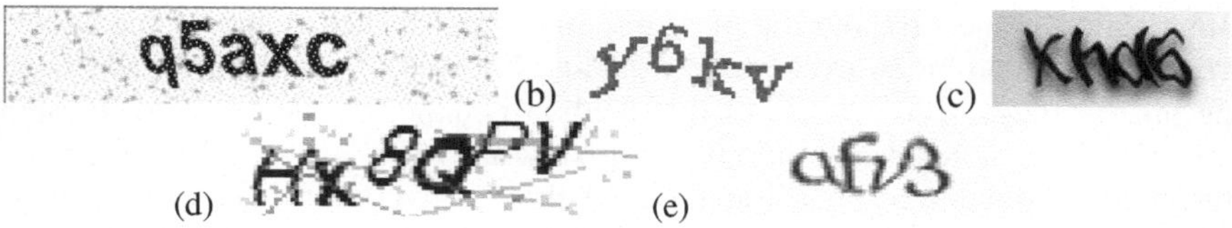

**Fig. 3.** Examples of Authorize(a), 360buy(b), Tianya(c), Windows Live(d) and Taobao(e)

According to our database, we discover these CAPTCHAs have following features: Some backgrounds are spotted lightly with different colors. Some characters are tilted, squeezed, bent, moved up and down, connected or distorted. But all of them contain digits and lowercase or uppercase letters. And the characters have fixed length.

# 3 Proposed Segmentation Algorithm

## 3.1 Community Divided Model

The connected graph $G(V,E)$ is composed of vertex set $V(G)$ and edge set $E(G)$. Vertex $v_i$ and $v_j$ are an elements of $V$ and $e_{ij}$ which is their edge weight is an element of $E$, all the initial values of $e_{ij}$ are 0. Let adjacency list $L[i]$ (i=1…n) is the element of $v_i$ neighbors. Queue $Q[k]$ is the element of k-th non-connected nodes group. $N$ is the community number. Since each CAPTCHAs has the fixed length, the algorithm we proposed for identify communities is stated as follows:

1. Calculate the adjacent list of all vertices: Use the eight neighbors of each foreground pixels on the thinned pattern to search the entire image. If $v_j$ is one of the eight neighbor nodes of $v_i$ (j≠i), then $v_j$ is the $v_i$ adjacent side, and put it into the $L[i]$.

2.Calculate the non-connected nodes group number: According to the adjacent list $L$, calculate all the connected nodes groups and add them into the $Q$, if the non-connected nodes group number euqals to $N$, the system breaks up.

3.Calculate the betweenness score for all edges: Use the DFS algorithm to calculate the shortest paths among the all vertices. If the edge $e_{ij}$ pass once, score-1 is added to it. Then calculate the highest betweenness score among them.

4.Remove the highest score edge: Remove the highest betweenness score $e_{ij}$ and update the edge set $E(G)$ .Meanwhile, delete the connected edge in the List $L[i]$.

5.Repeat from step 2 and recalculate betweennesses score for all edges affected by the removal until all the edges are removed and the system breaks up into $N$ non-connected nodes group.

### 3.2 Pre-processing

Binarization is the first and the most important step. Whether the segmentation step work well depend on bi-level image's quality. So, we convert the original image into the binary image. The process of this method is done via the standard thresholding method: those color value of all the pixels above a heuristically predetermined threshold is converted to black and those bellow converted to white [5]. Sometimes, the image has some noise points, we scan the entire image and if the unicom region's pixels are larger than a threshold, we regard it as noise and remove it. On the other hand, we care about the time we processing. In order to reduce the process time, we use Zhang's althorithm [19] to thin the images. Fig.4 shows the final images after the application of the standard thresholding method and thinning algorithm on Fig.3.

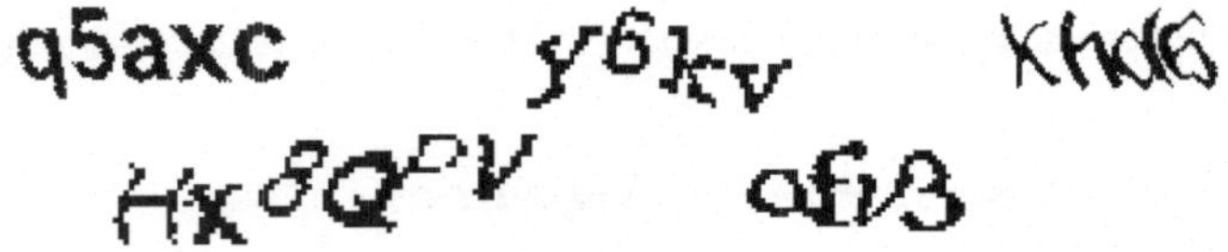


(a) Result images after the application of the standard thresholding method.

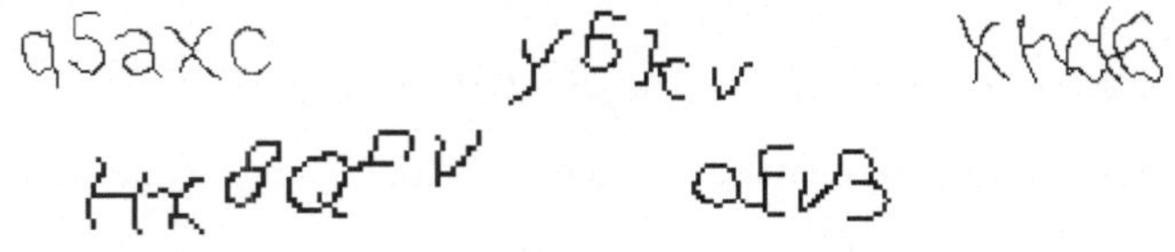


(b) Result images after the application of the thinning algorithm.

**Fig. 4.** Example of the Authorize, 360buy, Tianya, Windows Live, Taobao

### 3.3 Segmentation

We use the Community Divided Model that we proposed to segment the characters. If the characters are connected, we draw a red line to indicate it is the highest score that we should remove it, then we get the characters results of segmentation. Because most of Authorize and 360buy CAPTCHAs are non-connected and they have four or five communities, we get the results of segmentation directly. To, Tianya, Windows Live and Taobao, they have connected characters, we draw a red line to indicate the removing point. Fig.5 shows the removing point with a red line and Fig.6 shows the last results of segmentation. The original images are shown in Fig. 4(b).

**Fig. 5.** Removing point: red line indicate the highest score

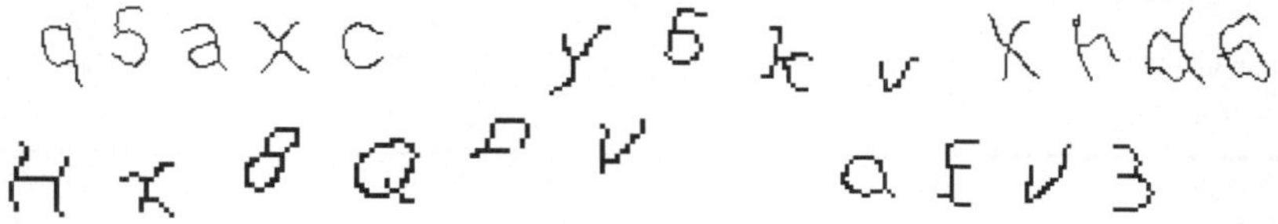

**Fig. 6.** Results of segmentation of the Authorize, 360buy, Tianya, Windows Live and Taobao

With the above results of segmentaition, we found this method has a great result of non-connected or connected characters segmentation even though these characters are distorted. We draw a red line to indicate the segmentation points, which are the highest score that should be removeed. And we can see that these experitmental results are basically consistent with the fact.

## 4 Experiment Results and Analysis

To evaluate the effectiveness of our techniques, we get more CAPTCHAs and designed three different difficulty experiments, they are presented as follows:

1) Select 100 CAPTCHAs randomly from 360buy and Authorize, respectively. Every CAPTCHAs from 360buy has four characters and Authorize has five characters, most of characters are non-touching expect few two characters touching.

2) Select 100 CAPTCHAs randomly from Tianya and Windows Live, respectively. Every CAPTCHAs from Tianya has four characters and Windows Live has six characters, most of characters are two characters touching.

3) Select 100 CAPTCHAs randomly from Taobao. Every CAPTCHAs has four characters,most of characters are more than two characters touching

In 2008, Huang et al. [8] proposed a Projection-based segmentation Alogorithm to break MSN and Yahoo CAPTCHAs, When the projection values in the sliding window were smaller than the threshold, the algorithm marked the position and erased these clutter items. Yan [2] proposed a low-cost attack on a CAPTCHA designed by Microsoft in 2008, when to segment the connected characters, they worked out the width of the object and then vertically divide the object into the same width. Here, we use the Huang's and Yan's algorithms to repeat the above three experimental steps to compare with the algorithm we proposed. If an algorithm can segment 60 from 100 images, the segmentation accuracy will be 60/100=0.6, or 60%. In this experiment, the number of two connected characters images is 153, three connected characters is 35 and four connected characters is 72. Table 1 displays results of the experiment by using Huang's ,Yan's and proposed algorithm. With the analysis of the results, we find the proposed algorithm is better than Huang's and Yan's algorithm.

**Table 1.** Results of segmentation different connected characters numbers by using Huang's, Yan's and proposed algorithm

| | *Huang's algorithm* | *Yan's algorithm* | *Proposed algorithm* |
|---|---|---|---|
| 2 characters | 39.22% | 60.13% | 66.01% |
| 3 characters | 22.86% | 22.86% | 31.43% |
| **≥4 characters** | 9.72% | 12.50% | 29.17% |

These algorithms are also applied to calculate segmentation rates of Authorize, 360buy,Tianya,Windows Live and Taobao systems. Table 2 displays the results of the calculation by using these algorithms. When to segment the Authorize system, the segmentation rate of the proposed algorithm is the same with Huang's, but 2% higher than Yan's. Similarly, when to segment the 360buy system, the segmentation rate of proposed algorithm is equal to Yan's and 5% higher than Huang's. And when to segment Tianya, Windows Live and Taobao systems, the proposed algorithm segmentation rates are higher than both of them. Therefore, the proposed algorithm is more effective than Huang's and Yan's when to segment two or more connected and distorted characters. Fig. 7 shows an example of the proposed algorithms which compared with Huang's and Yan's algorithm.

**Table 2.** Segmentation rates of the Huang's, Yan's and proposed algorithm

| | *Authorize* | *360buy* | *Tianya* | *Windows Live* | *Taobao* |
|---|---|---|---|---|---|
| Huang's algorithm | 98% | 90% | 58% | 40% | 15% |
| Yan's algorithm | 96% | 95% | 67% | 46% | 22% |
| Proposed algorithm | 98% | 95% | 71% | 55% | 33% |

In addition, the average run time is an another important factor when to segment a CAPTCHAs. The average run time of the proposed algorithm to segment Authorize, 360buy, Tianya, Windows Live and Taobao systems is 0.571s, 0.099s, 6.748s, 0.556s and 10.472s, respectively. The cause of these difference is that most of 360buy's and Authorize's characters are non-connected but most of Tianya and Windows Live are two characters connected and Taobao are more than two characters connected. Therefore, we can conclude that the more connected characters are, the more time-consuming is. The experiments are carried out on a PC with Core 2 processor, 2.29 GHz, 2GB RAM, with VC6.0, on Windows XP.

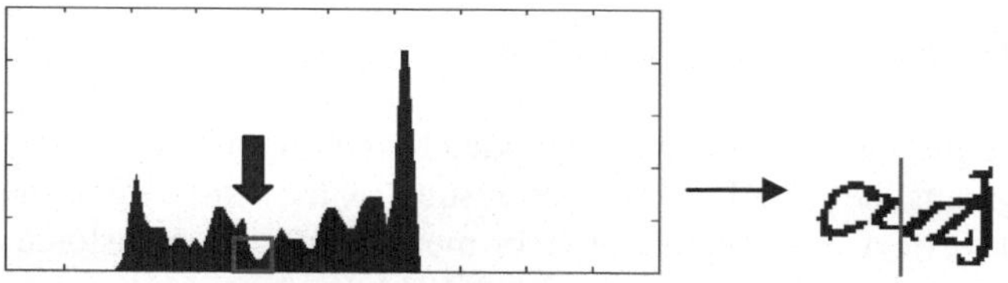

(b) Failure of segmentation from Huang's algorithm.

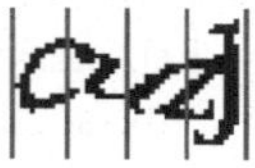

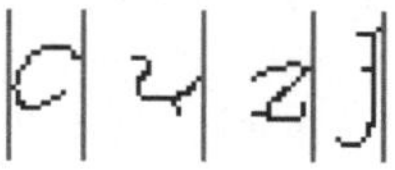

(c)Failure of segmentation from Yan's algorithm. (d)Successful segmentation of the proposed.

**Fig. 7.** An example of the proposed algorithm which compared with Huang's and Yan's

To evaluate the effectiveness of the proposed segmentation method, we collect more CAPTCHAs from different Internet websites. The results of the experiment, illustrated in Fig. 8, indicated that the proposed algorithm can also effectively segment these connected and distorted characters.

(a) Original images (b) Removing point (c) Results of segmentaion

**Fig. 8.** Some CAPTCHAs and their results of segmentation by using proposed algorithm

However, our method also fail to segment some images. In Fig.9 (a), the characters "5" and "c", "7" and "e" are overlapping and the red line marks the highest score but it fail to segment the images. In Fig. 9 (b), the characters "E" and "r", "M" and "c" also be failed to segment because character 'E' has much more pixels than 'r',and 'M' has much more pixels than 'c'. Therefore, our method is ineffective when the characters are overlapping or one character has much more pixels than another.

(a) (b)

**Fig. 9.** Failure results of segmentation. (a) Overlapping characters. (b) One character has much more pixels than another.

## 5 Conclusion and Further Work

In this paper, we propose a Community Divided Model algorithm based on complex networks to segment the CAPTCHAs. The results of the experiment show that the algorithm is more effective to segment two or more connected or distored characters. With the growing frequency of using the CAPTCHAs, our research can provide web designers some inspiration to improve the security of their website. To further work, we plan to study better method to segment overlapping characters and one character has much more pixels than another. Moreover, how to segment a text-based CAPTCHAs of Chinese is also our future work.

## References

1. von Ahn, L., Blum, M., Hopper, N.J., Langford, J.: CAPTCHA: Using Hard AI Problems for Security. In: Biham, E. (ed.) EUROCRYPT 2003. LNCS, vol. 2656, pp. 294–311. Springer, Heidelberg (2003)
2. Yan, J., El Ahmad, A.S.: A low-cost attack on a Microsoft CAPTCHA. In: 15th ACM Conference on Computer and Communications Security (2008)
3. Chellapilla, K., Larson, K., Simard, P., Czerwinski, M.: Computers beat humans at single character recognition in reading based Human Interaction Proofs (HIPs). In: 2nd Conference on Email and Anti-Spam (2005)
4. Simard, P.Y., Steinkraus, D., Platt, J.C.: Best Practice for Convolutional Neural Networks Applied to Visual Document Analysis. In: 7th International Conference on Document Analysis and Recognition, pp. 958–962. IEEE Computer Society, Los Alamitos (2003)
5. El Ahmad, A.S., Yan, J., Tayara, M.: The Robustness of Google CAPTCHAs. Bericht, Newcastle University (2011)
6. Mori, G., Malik, J.: Recognising objects in adversarial clutter: breaking a visual CAPTCHA. In: IEEE Conference on Computer Vision & Pattern Recognition (2003)
7. Chellapilla, K., Larson, K., Simard, P.Y., Czerwinski, M.: Building Segmentation Based Human-Friendly Human Interaction Proofs (HIPs). In: Baird, H.S., Lopresti, D.P. (eds.) HIP 2005. LNCS, vol. 3517, pp. 1–26. Springer, Heidelberg (2005)
8. Huang, S., Lee, Y., Bell, G., Ou, Z.: A projection-based segmentation algorithm for breaking MSN and YAHOO CAPTCHAs. In: Proceedings of the 2008 International Conference of Signal and Image Engineering, London, UK (2008)
9. Bursztein, E., Martin, M., Mitchell, C.: Text-based CAPTCHA Strengths and Weaknessses. In: 18th ACM Conference on Computer and Communications Security (2011)
10. Boccaletti, S., Latora, V., Moreno, Y., Chavez, M., Hwang, D.-U.: Complex networks: Structure and dynamics. Phys. Rep. 424(4-5), 175–308 (2006)
11. Freeman, L.C.: A set of measures of centrality based on betweenness. Sociometry 40, 35–41 (1977)
12. Girvan, M., Newman, M.E.J.: Community structure in social and biological networks. PNAS 99, 7821–7826 (2002)
13. Bursztein, E., Bethard, S., Fabry, C., Mitchell, J., Jurafsky, D.: How good are humans at solving CAPTCHAs? a large scale evaluation. In: 2010 IEEE Symposium on Security and Privacy (SP), pp. 399–413 (2010)
14. Authorize CAPTCHAs, `https://account.authorize.net/ui/themes/anet/Welcome/ForgottenLoginID.aspx` (accessed April 2012)

15. 360buy CAPTCHAs, https://passport.360buy.com/new/registpersonal.aspx (accessed April 2012)
16. Tianya CAPTCHAs, http://passport.tianya.cn/register (accessed March 2012)
17. Windows Live CAPTCHAs, https://signup.msn.cn/register (accessed April 2012)
18. Taobao CAPTCHAs, http://member1.taobao.com/member/new_register.jhtml (accessed April 2012)
19. Zhang, T.Y., Suen, C.Y.: A fast parallel algorithm for thinning digital patterns. Communications of the ACM 27(3), 236–239 (1984)

# A Robust Kernel-Based Fuzzy C-Means Algorithm by Incorporating Suppressed and Magnified Membership for MRI Image Segmentation

Hsu-Shen Tsai[1,*], Wen-Liang Hung[2], and Miin-Shen Yang[3]

[1] Department of Management Information System,
Takming University of Science and Technology, Taipei 11451, Taiwan
lv@takming.edu.tw
[2] Department of Applied Mathematics,
National Hsinchu University of Education, Hsin-Chu 30014, Taiwan
[3] Department of Applied Mathematics, Chung Yuan Christian University
Chung-Li 32023, Taiwan

**Abstract.** Bias-corrected fuzzy c-means (BCFCM) algorithm with spatial information has been proven effective for image segmentation. It still lacks enough robustness to noise and outliers. Some kernel versions of FCM with spatial constraints, such as KFCM_ $S_1$ , KFCM_ $S_2$ and GKFCM, were proposed to solve those drawbacks of BCFCM. However, the computational performances of these algorithms are still not good enough, especially for large data sets. In this paper, we adopt suppressed and magnified membership idea to speed the computation performance and propose a robust kernel-based fuzzy c-means algorithm (RKFCM). MRI image experiments illustrate that the proposed RKFCM is better than other algorithms in accuracy and computational efficiency. The RKFCM can exhibit the robustness to outlier, noise and weighting exponent m. Experimental results and comparisons indicate that the proposed RKFCM is a fast and robust clustering algorithm and suitable for MRI segmentation.

**Keywords:** Fuzzy c-means (FCM), Kernel-based FCM, Suppressed membership, Spatial bias correction, Image segmentation, Magic resonance image segmentation.

## 1 Introduction

Image segmentation plays a important role in many medical imaging applications [1]. In the recent years, fuzzy segmentation methods, especially the fuzzy c-means algorithms, have been widely used in the segmentation of brain medical images, because they can preserve more information from the original image than other segmentation methods. Since FCM did not consider the spatial information in the image space when applied in image segmentation, which make them very sensitive to noise and imaging artifacts. Recently, several researchers have considered to

---

* Corresponding author.

J. Lei et al. (Eds.): AICI 2012, LNAI 7530, pp. 744–754, 2012.

incorporate local spatial information into FCM algorithm to improve the performance of image segmentation, such as Tolias and Panas [2]; Pham [3]; Ahmed et al. [4]; Liew et al. [5]; Chen and Zhang [6]; Bandyopadhyay [7] and Yang and Tsai [8]. Ahmed et al. [4] first modified the FCM algorithm as a bias-corrected FCM (BCFCM) by regularizing the FCM objective function with a spatial neighborhood regularization term. Chen and Zhang [6] proposed kernel versions of FCM with spatial constraints, called KFCM_ $S_1$ and KFCM_ $S_2$, to improve the drawbacks of BCFCM. However, BCFCM, KFCM_ $S_1$ and KFCM_S $_2$ are heavily affected by their parameters. Afterward, Yang and Tsai [8] proposed a Gaussian kernel-based FCM (GKFCM) algorithm with a spatial bias correction to overcome their drawbacks. GKFCM can automatically learn the parameters by the prototype-driven learning scheme, and presents more efficiency and robustness. From our observations and experiments, GKFCM is very time consuming, especially for large data sets or a large image, and is sensitive to the weighting exponent m. For example, while Fuzzy clustering algorithm with spatial term is applied in image segmentation, a 256X256 image contains 65536 pixels, and compute the local spatial neighborhood term in each iteration step. That leads to heavily computational loads.

In this paper, we incorporate the suppressed and magnified membership ideas of Fan et al. [9] and Hung et al. [10] to kernel-based fuzzy c-means algorithm and propose the robust kernel-based fuzzy c-means algorithm (RKFCM) to improve the drawbacks of GKFCM. We present several examples and a real MRI case to assess the performance of RKFCM. The proposed RKFCM algorithm has fast convergence speed and robust properties: (1) robust to weighting exponent m;(2) robust to outlier and imaging artifacts. This paper is organized as follows. Section 2 is a brief survey of related work on FCM with spatial constraints. In section 3, A robust kernel-based Fuzzy c-means algorithm (RKFCM) is proposed. Experimental results and comparison are presented in Section 4. Finally, conclusions are made in section 5.

## 2 Related Work

### 2.1 Fuzzy Clustering Algorithms with a Spatial Correction

Ahmed et al. [4] considered a spatial neighborhood information to modify FCM with the following objective function $J_m^{BCFCM}$ :

$$J_m^{BCFCM}(\mu,a)=\sum_{i=1}^{c}\sum_{j=1}^{n}\mu_{ij}^m\left\|x_j-a_i\right\|^2+\frac{\alpha}{N_R}\sum_{i=1}^{c}\sum_{j=1}^{n}\mu_{ij}^m\sum_{x_r\in N_j}\left\|x_r-a_i\right\|^2 \tag{1}$$

where $N_j$ stands for the set of neighbors that exist in a window around $x_j$ and $N_R$ is the cardinality of $N_j$. Chen and Zhang [6] pointed out a shortcoming of the BCFCM and then replaced the Euclidean distance $\left\|x_j-a_i\right\|$ with a Gaussian kernel-induced distance $1-K(x_j,a_i)=1-\exp(-\left\|x_j-a_i\right\|^2/\sigma^2)$.

They gave the kernel version $JS_m^{KFCM\_S}$ of $JS_m^{BCFCM}$ as

$$JS_m^K(\mu,a)=\sum_{i=1}^{c}\sum_{j=1}^{n}\mu_{ij}^m(1-K(x_j,a_i))+\alpha\sum_{i=1}^{c}\sum_{j=1}^{n}\mu_{ij}^m(1-K(\bar{x}_j,a_i)) \tag{2}$$

## 2.2 Gaussian Kernel-Based FCM with a Spatial Bias Correction

The parameter $\alpha$ in BCFCM and KFCM_$S_1$ heavily affects the final clustering results. For estimating the parameter $\sigma$ and learning the parameter $\alpha$, Yang and Tsai [10] proposed a generalized type of BCFCM and KFCM_$S_1$ where the parameter $\alpha$ can be automatically estimated and learned from the data, called GKFCM. The modified objective function $JS_m^{GKFCM}$ with

$$JS_m^{GKFCM}=\sum_{i=1}^{c}\sum_{j=1}^{n}\mu_{ij}^m(1-K(x_j,a_i))+\sum_{i=1}^{c}\sum_{j=1}^{n}\eta_i\mu_{ij}^m(1-K(\bar{x}_j,a_i)) \tag{3}$$

where K(x, y)= exp(- $\|x-y\|^2/\sigma^2$) and $\sigma^2=\sum_{j=1}^{n}\left\|x_j-\bar{x}\right\|^2/n$.

The parameter $\eta_i$ is estimated as follows: $\eta_i=\dfrac{\min_{i'\neq i}(1-K(a_{i'},a_i))}{\max_k(1-K(a_k,\bar{x}))}$ (4)

## 2.3 Modify Membership Degree by Magnifying and Suppressing

To accelerate the convergence speed of FCM, there are many researches in the literature [9-13]. Among them, the suppressed and magnified membership idea is simple and effective.

The magnifying and suppressing membership degree are briefly below:[9-10].

If the data point $x_j$ has the largest membership in the $p$ th cluster of all c clusters, the value is noted as $u_{pj}$; $x_j$ has the second largest membership in the $s$ th cluster of all clusters, the value is noted as $u_{sj}$. After being modified and suppressed, the memberships of $x_j$ are modified as:

$$u_{pj}=1-\lambda\sum_{i\neq p}u_{ij}=1-\lambda+\lambda\, u_{pj} \tag{5}$$

$$u_{ij}=\lambda\, u_{ij}, i\neq p$$

where $\lambda = \exp(-\min_{i \neq t} \beta \|a_i - a_t\|^2)$, $\beta = \left( \frac{\sum_{j=1}^{n} \|x_j - \bar{x}\|}{n} \right)^{-1}$, $\bar{x} = \frac{\sum_{j=1}^{n} x_j}{n}$ (6)

## 3 The proposed Robust Kernel-Based Fuzzy c-Means Clustering Algorithm

Zhang and Chen [14] considered a kernel version of FCM by replacing the Euclidean distance $\|x_j - a_i\|$ with the kernel substitution as

$$\|\phi(x_j) - \phi(a_i)\|^2 = K(x_j, x_j) + K(a_i, a_i) - 2K(x_j, a_i) \quad (7)$$

where $\phi$ is a nonlinear map from the data space into the feature space with its corresponding kernel *K*, $K(x, y) = \phi(x)^T \phi(y)$ is an inner product kernel function. They specially assumed $K(x, x) = 1$, and then proposed the kernel-type objective function $JS_m^{KFCM}$ with $JS_m^{KFCM}(\mu, a) = \sum_{j=1}^{n}\sum_{i=1}^{c} (\mu_{ij})^m \|\phi(x_j) - \phi(a_i)\|^2$

$$= \sum_{j=1}^{n}\sum_{i=1}^{c} (\mu_{ij})^m (K(x_j, x_j) + K(a_i, a_i) - 2K(x_j, a_i))$$

$$= 2\sum_{j=1}^{n}\sum_{i=1}^{c} (\mu_{ij})^m (1 - K(x_j, a_i)) \quad (8)$$

If we take the Gaussian kernel with $K(x_j, a_i) = \exp(-\|x_j - a_i\|^2 / \sigma^2)$ then $K(x, x) = 1$, the update equations for the necessary conditions of minimizing $J_m^{\phi}(\mu, a)$ are as follows:

$$a_i = \sum_{j=1}^{n} \mu_{ij}^m K(x_j, a_i) x_j \Big/ \sum_{j=1}^{n} \mu_{ij}^m K(x_j, a_i), i = 1, 2, \ldots, c; j = 1, \ldots, n \quad (9)$$

$$\mu_{ij} = (1 - K(x_j, a_i))^{\frac{-1}{m-1}} \Big/ \sum_{k=1}^{c} (1 - K(x_j, a_k))^{\frac{-1}{m-1}}, i = 1, \ldots, c; j = 1, \ldots, n. \quad (10)$$

To speed the GKFCM, we adopt the the suppression idea of Fan et al. [9] and Hung et al. [10] to KFCM algorithm and propose the robust kernel-based Fuzzy c-means clustering algorithm (RKFCM) as follows:

**The Proposed RKFCM Algorithm**

Step 1: Fix $2 \le c \le n$ and fix any $\varepsilon > 0$.

Give c initials of cluster centers $a^{(0)} = (a_1^{(0)},...,a_c^{(0)})$ .and let s=1.

Step 2:Compute $\lambda$ with $a^{(s)}$ using (6)

Step 3: Compute the membership $\mu^{(s)}$ with $a^{(s-1)}$ using (10).

Step 4:Modify $\mu^{(s)}$ using (5)

Step 5: Update $a^{(s)}$ with $\mu^{(s)}$ using (9).

IF $\left\| a^{(s)} - a^{(s-1)} \right\| < \varepsilon$, STOP

ELSE $s$=$s$+1 and GOTO step 2.

## 4 Experimental Results and Comparisons

In this section, we perform experiments to compare the performances of these algorithms with MR image and a real MR image (Yang et al., [16]). All algorithms are implemented under the same initial values and stopping conditions.

**Example 1 (MR Images with 5% "Gaussian" Noise).** Figure 1 shows a comparison of segmentation results of these algorithms on a real T1-weighted MR image corrupted with 5 % "Gaussian" noise. The images are segmented into three clusters corresponding to background, gray matter (GM) and white matter (WM).

The segmentation results are shown in Figs. 1(b)~(j). We visually see that KFCM and GKFCM with $m$=5,10 are affected by the Gaussian noise such that the results are blurred. Figures 1(i) and 1(j) show that the proposed RKFCM with $m$=5,10 can achieve much stable segmentation results under Gaussian noise, compared to the KFCM and GKFCM algorithms. For quantitative measure, we use the comparison score defined in Masulli and Schenone [15]. The comparison score $S_{ik}$ was defined as

$$S_{ik} = \frac{|A_{ik} \cap A_{refk}|}{|A_{ik} \cup A_{refk}|}$$

where $A_{ik}$ represents the set of pixels belonging to the $k$th class found by the $i$th algorithm and $A_{refk}$ represents the set of pixels belonging to the $k$th class in the reference segmented image. The comparison scores of different segmentation classes as implementing KFCM , GKFCM and RKFCM algorithms for images in Fig. 1 are shown in Table 1. We see that RKFCM with a large $m$ value presents best comparison scores. We consider the number of iterations (NI) and the CPU time in second (s) for computational efficiency. The results are shown in Table 1 and Fig. 2. Overall, RKFCM is the best one in computational efficiency among these compared clustering algorithms. Our proposed RKFCM is much faster than KFCM and GKFCM. The RKFCM actually reduces the computational complexity, produces satisfactory results under noise and has robust property to weighting exponent $m$.

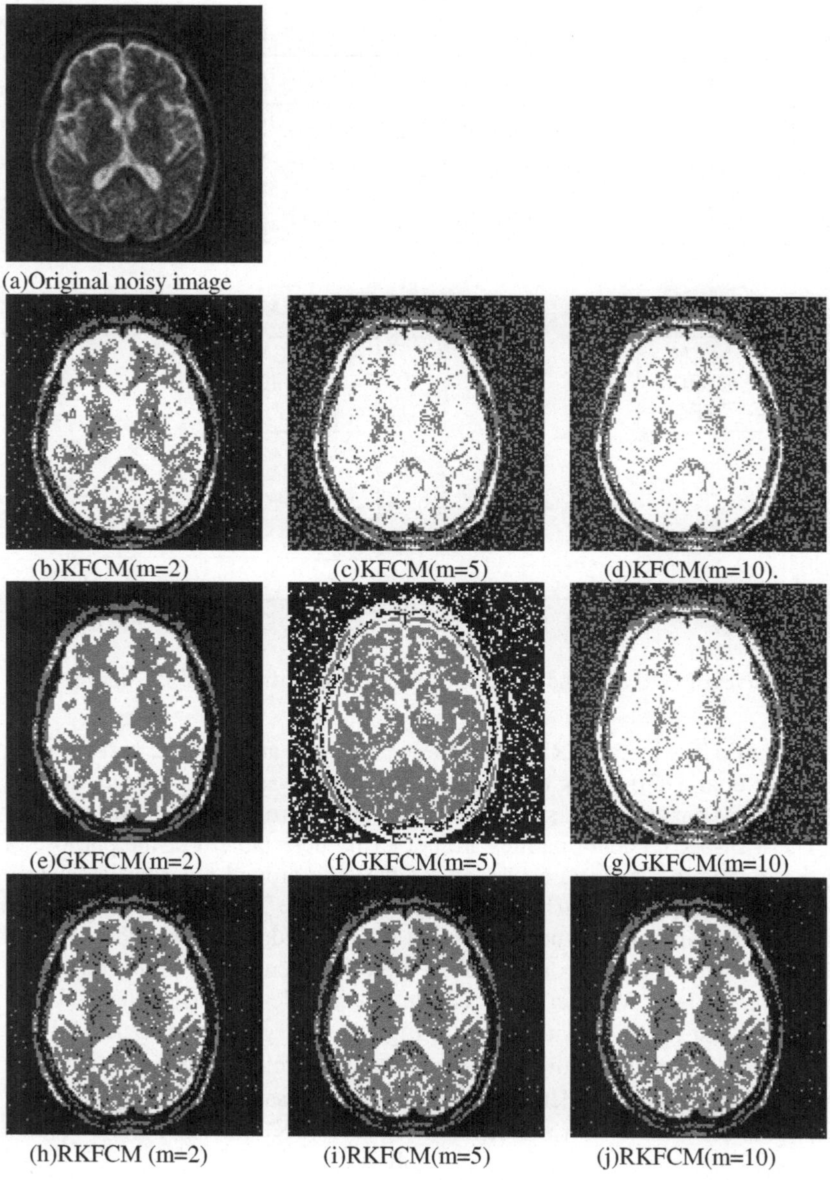


**Fig. 1.** Comparison of segmentation results on MR images with 5% Gaussian noises

**Table 1.** Comparison scores $S_{ik}$ of ith algorithm and kth class for the images in Fig. 1

| | m=2 | | | m=5 | | | m=10 | | |
|---|---|---|---|---|---|---|---|---|---|
| ith algorithm | kth class | | | kth class | | | kth class | | |
| | Class1 | Class2 | Class3 | Class1 | Class2 | Class3 | Class1 | Class2 | Class3 |
| KFCM | 0.9101 | 0.6072 | 0.6711 | 0.6858 | 0.1958 | 0.463 | 0.6373 | 0.1587 | 0.4464 |
| GKFCM | 0.9576 | 0.7601 | 0.7611 | 0.7846 | 0.475 | 0.1747 | 0.6803 | 0.5268 | 0.1861 |
| RKFCM | 0.9248 | 0.7254 | 0.7411 | 0.9248 | 0.7254 | 0.7411 | 0.9248 | 0.7254 | 0.7411 |

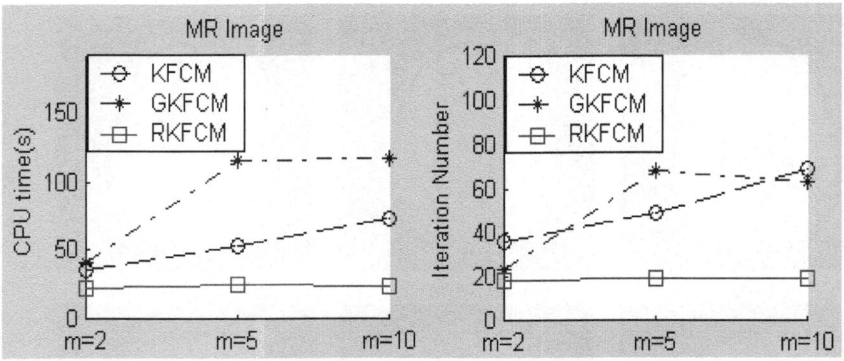


**Fig. 2.** Curves of CPU time and Iteration numbers for experiments on MR image

**Example 2.** Figure 3 is a real MRI data set was processed at 283×292 pixels (Yang et al.[16]; Hung et al. [17]). From this picture, one lesion was clearly seen in the MR images. However, some fuzzy shadows of lesions were suspected with tumor invasion. These suspected abnormalities are not easily detectable. For the purpose of detecting these abnormal tissues, a window of the area around chiasma is selected from the original MR image as shown in Fig. 3. Because AFCM can successfully differentiate the tumor from the normal tissues (Yang et al. [16]). We then applied AFCM ,GKFCM and RKFCM to the window selection picture and produced results as shown in Fig. 3(a)~(i). We can see occult lesions (circles) clearly enhanced with AFCM, GKFCM and RKFCM with $m$=2 in Fig. 3(a),(d),(g). However, AFCM and GKFCM with $m$=5,10 fail to show these occult lesions(circles), they yields incorrect segmentation. For evaluating detection of abnormal tissue with a quantitative measure, we use the comparison score, false negative (FN) and false positive (FP) [18]. The quantitative measures are defined as follows:

$$S=\frac{|A \cap A_{ref}|}{|A \cup A_{ref}|}$$

False Negative (FN)= $\frac{\left|A^C \cap A_{ref}\right|}{\left|A_{ref}\right|}$ , False Positive (FP)= $\frac{\left|A \cap A_{ref}^C\right|}{\left|A_{ref}^C\right|}$ ,

where $A$ represents the set of pixels belonging to the tumor tissue found by the $i$th algorithm, $A_{ref}$ represents the set of pixels belonging to the tumor tissue in the reference segmented image, $A_{ref} \cap A^C$ represents the set of pixels in $A_{ref}$ has not been detected to be tumor tissue by the ith algorithm, $A_{ref}^C \cap A$ represents the set of pixels in $A_{ref}^C$ has been detected to be tumor tissue by the ith algorithm, $A^C$ and $A_{ref}^C$ represent the complements of $A$ and $A_{ref}$, respectively. The quantitative measures S, FN, FP as implementing KFCM, AFCM and RKFCM algorithms for Fig. 3 are shown in Table 2. From Table 2, it clearly indicates that AFCM and GKFCM with $m$ =5, 10 have a poor performance, but RKFCM still has a very good performance. Furthermore, the performance of RKFCM is also robust to the weighting exponent $m$.

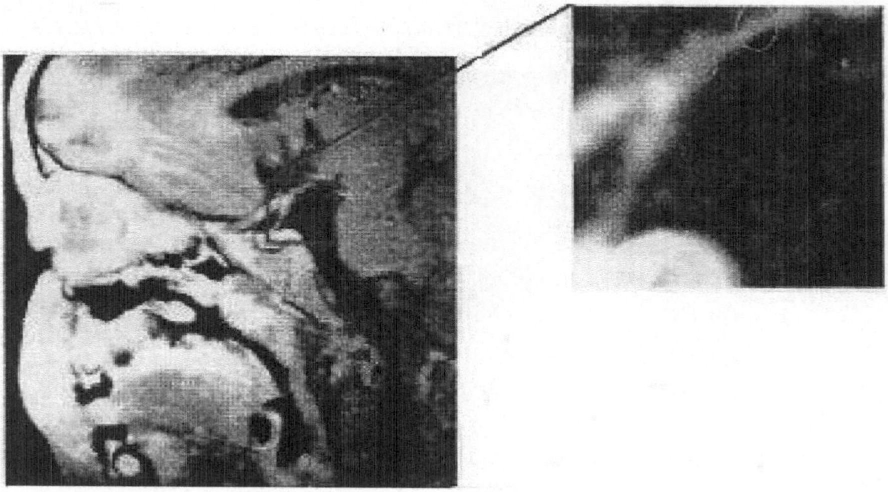

**Fig. 3.** Original MR image and its window selection

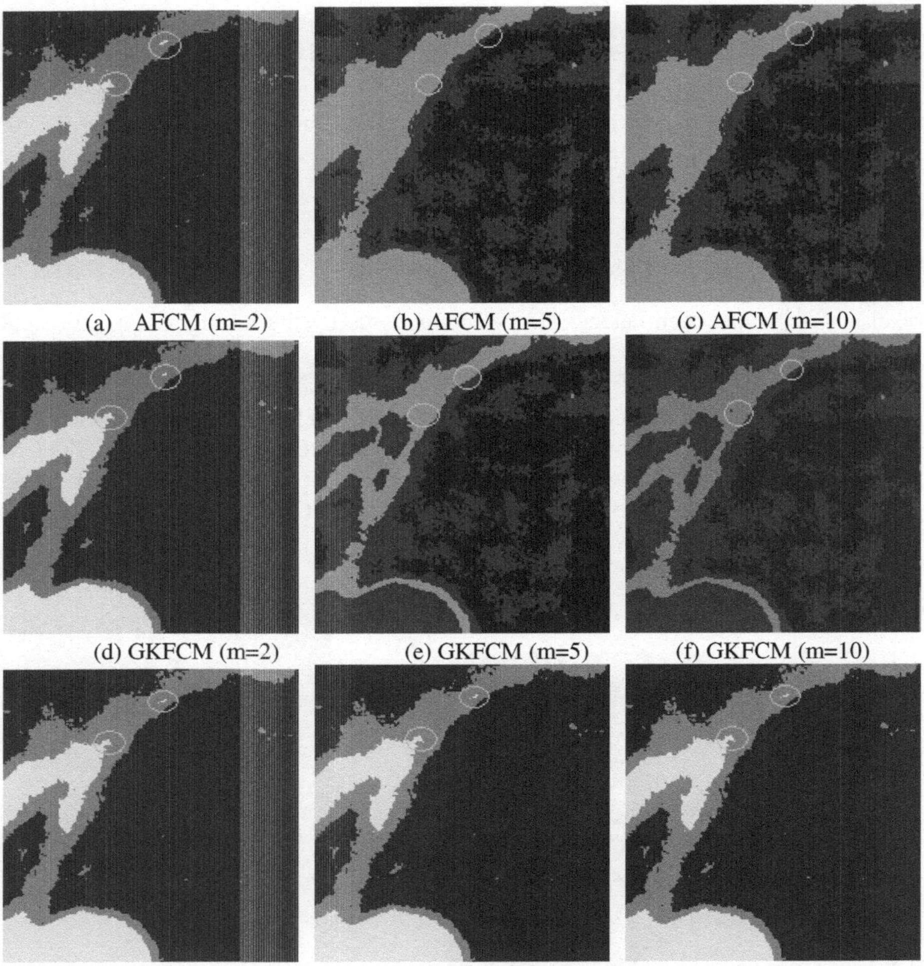

(a) AFCM (m=2) (b) AFCM (m=5) (c) AFCM (m=10)

(d) GKFCM (m=2) (e) GKFCM (m=5) (f) GKFCM (m=10)

(g) RKFCM (m=2) (h) RKFCM (m=5) (i) RKFCM (m=10)

**Table 2.** The values of S, FN and FP for Figs. 3(a)~(i)

| | m=2 | | | m=5 | | | m=10 | | |
|---|---|---|---|---|---|---|---|---|---|
| | **AFCM** | **GKFCM** | **RKFCM** | **AFCM** | **GKFCM** | **RKFCM** | **AFCM** | **GKFCM** | **RKFCM** |
| $\|A_{ref}\|$ | 135 | 135 | 135 | 135 | 135 | 135 | 135 | 135 | 135 |
| $\|A\|$ | 135 | 118 | 115 | 0 | 32 | 115 | 0 | 40 | 115 |
| $\|A \cap A_{ref}\|$ | 135 | 118 | 115 | 0 | 15 | 115 | 0 | 10 | 115 |
| $\|A \cup A_{ref}\|$ | 135 | 135 | 135 | 135 | 152 | 135 | 135 | 165 | 135 |
| $\|A^c \cap A_{ref}\|$ | 0 | 17 | 20 | 135 | 120 | 20 | 135 | 125 | 20 |
| $\|A_{ref}{}^c \cap A\|$ | 0 | 0 | 0 | 0 | 17 | 0 | 0 | 30 | 0 |
| **FN** | 0 | 0.1259 | 0.1481 | 1 | 0.8889 | 0.1481 | 1 | 0.9259 | 0.1481 |
| **FP** | 0 | 0 | 0 | 0 | 0.0044 | 0 | 0 | 0.0078 | 0 |
| **S** | 1 | *0.8741* | ***0.852*** | *0* | *0.0987* | ***0.852*** | *0* | *0.0606* | ***0.852*** |

# 5 Conclusion

In this paper, we adopt suppressed and magnified membership idea to speed up the computation performance and then proposed suppressed version KFCM clustering algorithm (RKFCM). From the simulation results, we find the advantages of RKFCM algorithm are more accurate and reduce computation time than KFCM and GKFCM. The RKFCM can exhibit the robustness to noise, image artifacts and weighting exponent m. Finally, the RKFCM algorithm is applied in the segmentation of a real MRI case. Results show that the RKFCM provides better detection of abnormal tissue than KFCM and GKFCM. Therefore, the proposed RKFCM is a good algorithm for real application.

# References


1. Pham, D.L., Xu, C.Y., Prince, J.L.: A survey of current methods in medical image segmentation. Annual Review of Biomedical Engineering 2, 315–337 (2000) (Technical report version, JHU/ECE 99-01, Johns Hopkins University)
2. Tolias, Y.A., Panas, S.M.: Image segmentation by a fuzzy clustering algorithm using adaptive spatially constrained membership functions. IEEE Trans. Systems Man Cybernet. A 28, 359–369 (1998)
3. Pham, D.L.: Spatial Models for Fuzzy Clustering. Computer Vision and Image Understanding 84, 285–297 (2001)
4. Ahmed, M.N., Yamany, S.M., Mohamed, N., Farag, A.A., Moriarty, T.: A modified fuzzy c-means algorithm for bias field estimation and segmentation of MRI data. IEEE Trans. Medical Imaging 21, 193–199 (2002)
5. Liew, A.W.C., Leung, S.H., Lau, W.H.: Segmentation of color lip images by spatial fuzzy clustering. IEEE Trans. Fuzzy Systems 11, 542–549 (2003)
6. Chen, S.C., Zhang, D.Q.: Robust image segmentation using FCM with spatial constrains based on new kernel-induced distance measure. IEEE Trans. Systems Man Cybernet.-Part B 34, 1907–1916 (2004)
7. Bandyopadhyay, S.: Satellite image classification using genetically guided fuzzy clustering with spatial information. International Journal of Remote Sensing 26, 579–593 (2005)
8. Yang, M.S., Tsai, H.S.: A Gaussian kernel-based fuzzy c-means algorithm with a spatial bias correction. Pattern Recognition Lett. 29, 1713–1725 (2008)
9. Fan, J.L., Zhen, W.Z., Xie, W.X.: Suppressed fuzzy c-means clustering algorithm. Pattern Recognition Lett. 24, 1607–1612 (2003)
10. Hung, W.L., Yang, M.S., Chen, D.H.: Parameter selection for suppressed fuzzy c-means with an application to MRI segmentation. Pattern Recognition Lett. 27, 424–438 (2006)
11. Pal, N.R., Bezdek, J.C.: Complexity reduction for "Large Image" processing. IEEE Trans. on Systems, Man, and Cybernetics-Part B 32, 598–611 (2002)
12. Eschrich, S., Ke, J., Hall, L.O., Goldgof, D.B.: Fast accurate fuzzy clustering through data reduction. IEEE Trans. Fuzzy Systems 11, 262–270 (2003)
13. Liao, L., Lin, T., Li, B.: MRI brain image segmentation and bias field correction based on fast spatially constrained kernel clustering approach. Pattern Recognition Lett. 29, 1580–1588 (2008)
14. Zhang, D.Q., Chen, S.C.: Clustering incomplete data using kernel-based fuzzy c-means algorithm. Neural Processing Letters 18, 155–162 (2003)

15. Masulli, F., Schenone, A.: A fuzzy clustering based segmentation system as support to diagnosis in medical imaging. Artif. Intell. Med. 16, 129–147 (1999)
16. Yang, M.S., Hu, Y., Lin, K.C.R., Lin, C.C.L.: Segmentation techniques for tissue differentiation in MRI of ophthalmology using fuzzy clustering algorithms. Magnetic Reson. Imag. 20, 173–179 (2002)
17. Hung, W.L., Chen, D.H., Yang, M.S.: Suppressed fuzzy-soft learning vector quantization for MRI segmentation. Artif. Intell. Med. 52, 33–43 (2011)
18. Fernandez-Garcia, N.L., Medina-Carnicer, R., Carmona-Poyato, A., Madrid-Cuevas, F.J., Prieto-Villegas, M.: Characterization of empirical discrepancy evaluation measures. Pattern Recognition Lett. 25, 35–47 (2004)

# Quantitative and Qualitative Coordination for Multi-robot Systems

Zhifeng Yao, Xuefeng Dai*, and Haimiao Ge

College of Computer and Control Engineering, Qiqihar University,
Qiqihar, Heilongjiang, 161006, China
daixuefeng203@163.com

**Abstract.** Coordination takes the role of integrating a set of individual robots into a whole multi-robot system to accompany tasks. During the past two decades, lots of achievements for the coordination of multi-robot systems were made. In this paper, these results were reviewed from two aspects. The first is from the point of view of that coordinated strategies were generated automatically by mathematical approaches, and the second is from the point of view that coordinated strategies were designed by control engineers. The approaches for generating and describing coordinated strategies are summarized, respectively. The potential future work was discussed especially for the case the goal of exploration is modeling an unknown indoor environment. It was pointed out that a few of coordinated behaviors can be realized by both of quantitative and qualitative approaches.

**Keywords:** Multi-robot Systems, Coordinated Algorithms, Coordinated Strategies, Coordination.

## 1 Introduction

Multi-robot systems (MRS) are superior to single robot systems in efficiency, robustness, and so on. There are lots of potential applications for MRS in civil life, agricultures, industry, and military. In order to make each robot in a team to operate cooperatively, coordinated algorithms or coordinated strategies take a key role. So, the development and research on coordination between robots has got more and more attention. The representative applications of MRS include multi-robot formation control, exploration and mapping, cooperative operation, etc.

According to the methods for developing coordinated strategies, the coordinated results includes market economy based [1-3], optimizing algorithm based [4-5], decision theoretic based [6], computing intelligence based [7-8] ones, and so on. They resolved the questions of task balancing [2], [4], exploration efficiency [6], adaptability [7-8], etc. For brevity, the results mentioned above are called as quantitative coordination in the following. On the other hand, there is a kind of results which maybe called as qualitative coordination. They dealt with the obstacle

* Corresponding author.

J. Lei et al. (Eds.): AICI 2012, LNAI 7530, pp. 755–761, 2012.

avoidance, deadlock resolution, faults handling for MRS in a logical level. Most of the results were based on the supervisory control of discrete event systems [9-18]. They included finite automata (FA) modeling based and Petri nets (PN) modeling based ones. FA and PN are used as mathematical models to describe coordinated strategies.

This paper is organized as follows. In sections 2 and 3, the state of the art of coordinated work for MRS was reviewed from the point of quantitative and qualitative coordination, respectively. Finally, future work in the fields was outlined, and our work is summarized in the last section.

## 2 Quantitative Coordination

One of the most distinct characteristics of coordinated results reviewed in this section is that the coordinated strategies were generated by a variety of mathematical tools, in an online or offline way.

The market economy based task assignment is a distributed approach through negotiation. It can realize global optimal assignment. However, the approach depends on communication seriously [1]. In order to implement balanced exploration, the $N$ tasks are grouped to $n$ clusters of tasks through $K$-means clustering algorithm [2]. Then, the problem was expressed as an equivalent multiple traveling salesman problem (TSP). A mathematical model for minimizing the distances traveled by all robots and balancing tasks between robots was established based on the sum of traveling costs and idle costs. The task allocation is conducted by auction mechanisms. Recently, a multi-agent coordination for disaster response with intra-path precedence constraints was put forward [3]. A time-extended schedules and allocations (more than one task is assigned to each robot) based on tiered auctions and two heuristic techniques, clustering and opportunistic path planning was presented.

Both supervised and unsupervised optimizing algorithms were used to realize task assignment for robots in a team. In exploration problems, it is not possible to know the optimal routes for the robots until the map of the workspace is revealed. In [4] the unknown workspace is partitioned into as many regions as robots by $K$-mean clustering. A coordinated algorithm which optimizes the on-line assignment of robots to targets, and keeps the robots working in separate areas was represented. The algorithm efficiently reduces the variance of average waiting time on those areas and fulfills balanced and sustained exploration for each teammate of MRS. On the other hand, the question of optimizing multi-robot behaviors was transferred into a linear programming question [5]. Observing meanwhile planning paths for two robots, the goal of optimized observing a moving target cooperatively in three dimensional environments is fulfilled in real time.

The decision theoretic based approach realized MRS task assignment by maximizing the utility and minimizing the cost. The work of [6] considered the problem of exploring an unknown environment with a team of heterogeneous robots. A technique that estimates the expected utility of a frontier cell based on the distance and visibility to cells that assigned to other robots was presented. The assignment is realized by an iterative calculation in a central fashion. However, the moving direction for each robot was adjusted based on the calculating results.

The coordinated strategies can be generated through the computing intelligence techniques [3], [7], [8]. A genetic algorithm (GA) was adopted as an alterative for generating time-extended coordination solution in city disaster response [3]. A two-layer multi-agent architecture was developed to implement multi-robot coordination [7]. Two type of machine learning, reinforcement learning (RL) and GA, were used to make decisions when the robots cooperatively transport an object to a goal location while avoiding obstacles. A modified RL algorithm called the sequential Q-learning algorithm was developed to deal with the issues of behavior conflict. Because of the off-line characteristics, the GA algorithm used to search for the optimal cooperation strategy in each step of object transportation. One of the two strategies was selected as winning one by an arbitrary mechanism. Later, the real time Asynchronous Situated Co-evolution algorithm (r-ASiCo) developed in [8] exploits natural open-ended evolution to generate emergent complex collective behaviors. In other words, the evolving is conducted through the real world instead of a simulated environment in order to obtain a self-organized collective solution in real time. The evolutionary process takes place in an asynchronous and decentralized way. The approaches for generating coordinated strategies are summarized in Table 1.

**Table 1.** Comparison of quantitative coordinats for MRS

| Applications | Mathematical tools | Coordinated Functions |
|---|---|---|
| Exploration [2] | *K*-means, auction, Multiple TSP | Balancing tasks, task allocation |
| Disaster response[3] | Auction, clustering | Time-extended schedules, allocations |
| Exploration [4] | *K*-means | Optimized assignment |
| Observation [5] | Linear programming | Optimized observing, planning |
| Exploration [6] | Decision theory | Task assignment, efficiency |
| Transportation [7] | RL, GA | Avoid obstacle, behavior conflict |
| Cleaning [8] | Open-ended evolution | Adaptability, self-organization, fault-tolerance |

## 3 Qualitative Coordination

In contrast with the quantitative approach, the coordinated strategies discussed in this section were designed by engineers in advance. The representative coordinated strategies developed by the supervisory control of discrete event systems can be categorized into centralized and decentralized ones. The former demands the central supervisor has high reliability and communication between robots has no delay. In addition, the architecture of the closed loop system is complicated. The state of the art about the later will be emphasized. The feature of the later is that each robot has its own supervisor. It determined the event that will be allowed to occur based on the states of the robot, environment and other robots. A variety of effective results based on decentralized supervisory have been appeared. The applications included operational coordination for two robots [9], multi-robot formation for land or aerial scenarios [10-11], shared mobile space robots coordination [12]. The questions resolved included obstacle avoidance[9], [12], deadlock resolution [12], [16], task assignment [15-18], etc. According to the mathematical models describing the coordinated behaviors, the results can be categorized into automata based and Petri nets based approaches.

The approach based on finite state automata models the behavior of each robot with an independent automaton. The model of MRS was constructed by the concurrent product of automata. The coordination was achieved by a group of supervisors resided on each robot [9, 11], or by a central supervisor [10].

In a tight-coupled task execution problem, each robot in the team was modeled as a fuzzy automaton [9]. The behaviors consist of two deliberative ones (*Route Follow*, *Go to Target*) and two reactive ones (*Avoid Obstacle*, *Wall Follow*). Each fuzzy state of the automaton represents the activation level of the corresponding behavior. A formal supervisory control architecture was proposed, and conjunctive and disjunctive rules were defined to fuse the decisions of each robot on their deliberative and reactive behaviors, respectively.

The supervisory control based on automata modeling was used to realized leader follower formation of mobile robots [10], or underwater vehicles [11]. A team of nonholonomic robots was considered in [10]. All of the mobile robots are required to navigate in an obstacle populated environment. The behaviors of leader robot and follower robots were modeled, respectively, for obstacle avoidance, navigation, and formation control. The supervisor accomplished the coordination between the continuous controllers so that the followers keep a predetermined geometric formation with the leader. Recently, a hybrid supervisory control approach of unmanned aerial vehicles for a two dimensional leader follower formation scenario was proposed [11]. A new method of abstraction which partitions the aerial space by a polar way and reduces the infinite space into a finite state automaton was introduced. Three supervisors for reaching and keeping the formation, and the collision avoidance were designed, respectively. The whole supervisor was achieved through the parallel composition of the above supervisors.

The applications of Petri nets based coordination include robot control software development [15], task plan and task assignment [13-14, 16-18], deadlock resolution [13, 18], etc.

A few of coordinated results utilized the supervisor based on place invariants (SBPI) for decentralized supervisory control of discrete event systems [19]. Three constraints which are mutual exclusion, task ordering, and synchronization were proposed for mission control of underwater vehicles [13]. The algorithm which checks if a system that is admissible in a centralized way is also admissible once distributed [19] was utilized under the constraints. In addition, the paper addressed the decentralization of the resulting discrete event system. It was discussed in [14] that using linear temporal logic (LTL) formulas as the specification language, the target behaviors can be stated in a more intuitive manner than using linear constraints [19]. In the approach, each robot behavior is modeled by a Petri net. To describe the change of shared state, two uncontrollable events and 4 transitions were added to the individual Petri nets. The coordinated rules between robots are defined by LTL. The supervisors are built by composing the *Buchi* automaton obtained for each LTL specification with the augmented individual PN model of the respective robot. However, only limited behavior was considered there.

A language Petri Net Plans (PNPs) which is based on PNs and allow for robot and multi-robot behavior design was presented in [15]. PNPs support a rich set of features that include sensing, interrupts and concurrency. The use of PNPs permit exploiting standard validation tools based on PNs to verify the consistency of the designed behaviors.

According the difference of onboard devices for each robot, the individual robots are categorized into skilled agents and helper agents [16]. The former find tasks and the latter which matched the task requirements assisted the corresponding former to finish the task. Two colored Petri net models were designed for the initiator and helper, respectively, based on the roles the agents could play. The robots can change their teammate selecting strategies dynamically to adapt to the environment. Nearest available, impatient, best available teammate, and best possible teammate strategies were designed.

The implemented coordinated functions, the mathematical models for describing coordinated behaviors, and the applications of MRS are summarized in Table 2.

**Table 2.** Comparison of qualitative coordinations for MRS

| Applications | Behavior model | Coordinated Functions |
|---|---|---|
| Transportation[9] | Fuzzy FA | Route follow, go to target, avoid obstacle, wall follow |
| Formation [10] | FA | Avoid obstacle, wall following, goal navigation |
| Formation [11] | FA | Reaching and keeping formation, inter-collision avoidance |
| Mission control [13] | PN | Avoid deadlock, mutual exclusion, ordering, synchronization |
| Soccer robot[14] | PN, LTL | Goal navigation, get ready, do not get ready |
| PNPs[15] | PN | The same as PN realized, joint commit |
| Disaster [16] | Colored PN | Adaptability, teammate selection |

In addition, the idea of [17] is similar to that of [6]. However, the information gain was instead of utility in [17]. It was computed based on the number of unknown cells in detecting regions that not overlapped with detected regions of other robots [17]. The robot bid for the right of exploring a region with others based on the information gain and the distance to the target. The waiting time for bidding is optimized through PN simulation software. For dynamic dispatching and routing of automated guided vehicles (AGV) systems, the static problem to determine an optimal firing sequence for PN is solved repeatedly [18]. Decomposing the PN describing the dynamic task into a few of subnets of tasks and robots, the optimal sequence of events is obtained. A penalty terms was added to the transitions that violate conditions to optimize the solutions obtained from the subnets to obtain a feasible resolution for the original PN.

## 4 Discussion and Conclusions

Summarizing, the potential future work includes the following. First, for the case the environment was separated physically in an exploration, after task assigned, e.g., by bidding approach, some of the robots in the MRS may have no task. The current approaches did not discuss the usage of idle these robots. The decentralized supervisory control was a promising tool to develop coordinated strategy which can deal with the issue.

Second, in most of the present work, occupancy grid map were used to represent the environment. However, there's a intrinsic shortage that the grid map consumes computer storage severely when the environment is large or the description of the environment should be more exact. How to develop coordinated algorithm suitable for adapting other kinds of maps should be investigated.

Finally, when the exploration is to establish an environment model, the trajectory for each robot should form a series of closure to optimize the local map relations to obtain a global map [20-21]. So, the task assignment for each member of a robot team needs taking into account the loop closure issue. In addition, the behavioral modeling approach based on partition the workspace in advance [11-12] can not be adopted for this case. For the sake of simplicity, we suggest that the exploring and coordinated behaviors for each robot are described by a few of normal or failure states of each robot. We are sure that the investigation on the ignored questions can promote cooperation between robots effectively.

From the contexts of the above two sections, we can draw the following conclusions. First of all, a few of issues related to the coordination of MSR can be solved by either quantitative approaches or qualitative approaches, for example, the behavior conflict discussed in [3] and [9], obstacle avoidance [7], [9], [10]. Second, dynamic task assignment problem can be resolved by static optimizing approaches through resolving the problem repeatedly. Finally, multi-agent framework was preferred as discussed in [7], [16], and [3].

**Acknowledgement.** This work was supported by the Education Department of the Heilongjiang Province Government, China, under Grant 12521601.

## References

1. Dias, M.B., Zlot, R., Kalra, N., Stentz, A.: Market-based Multirobot Coordination: A Survey and Analysis. Proceedings of the IEEE 94, 1257–1270 (2006)
2. Elango, M., Nachiappan, S., Tiwari, M.K.: Balancing Task Allocation in Multi-robot Systems Using K-means and Auction Based Mechanisms. Expert Systems with Applications 38, 6486–6491 (2011)
3. Jones, E.G., Dias, M.B., Stentz, A.: Time-extended Multi-robot Coordination for Domains with Intra-path Constraints. Autonomous Robots 30, 41–56 (2011)
4. Puig, D., Garcia, M.A., Wu, L.: A New Global Optimization Strategy for Coordinated Multi-robot Exploration: Development and Comparative Evaluation. Robotics and Autonomous Systems 59, 635–653 (2011)
5. Gu, F., He, Y.Q., Han, J.D., Wang, Y.C.: An Active Cooperative Observation Method for Multi-robots in Three Dimensional Environments. Acta Automatica Sinica 36, 1443–1453 (2010)
6. Burgard, W., Moors, M., Stachniss, C., Schneider, F.E.: Coordinated Multi-robot Exploration. IEEE Transactions on Robotics 21, 376–386 (2005)
7. Wang, Y., de Silva, C.W.: A machine-learning Approach to Multi-robot Coordination. Engineering Applications of Artificial Intelligence 21, 470–484 (2008)
8. Prieto, A., Becerra, J.A., Bellas, F., Duro, R.J.: Open-ended Evolution as a Means to Self-organize Heterogeneous Multi-robot Systems in Real Time. Robotics and Autonomous Systems 58, 1282–1291 (2010)

9. Jayasiri, A., Mann, G.K.I., Gosine, R.G.: Tightly-coupled Multi Robot Coordination Using Decentralized Supervisory Control of Fuzzy Discrete Event Systems. In: 2011 IEEE International Conference on Robotics and Automation, pp. 3358–3363. IEEE Press, New York (2011)
10. Gamage, G.W., Mann, G.K.I., Gosine, R.G.: Discrete event systems based formation control framework to coordinate multiple nonholonomic mobile robots. In: 2009 IEEE/RSJ International Conference on Intelligent Robots and Systems, pp. 4831–4836 (2009)
11. Karimoddini, A., Lin, H., Chen, B.M., Lee, T.H.: Hybrid Formation Control of the Unmanned Aerial Vehicles. Mechatronics 21, 886–898 (2011)
12. Roszkowska, E.: DES-Based Coordination of Space-Sharing Mobile Robots. In: Díaz, R.M., Pichler, F., Arencibia, A.Q. (eds.) EUROCAST 2007. LNCS, vol. 4739, pp. 1041–1048. Springer, Heidelberg (2007)
13. Palomeras, N., Ridao, P., Silvestre, C., El-fakdi, A.: Multiple Vehicles Mission Coordination Using Petri nets. In: 2010 IEEE International Conference on Robotics and Automation, pp. 3531–3536. IEEE Press, New York (2010)
14. Lacerda, B., Lima, P.U.: LTL-based Decentralized Supervisory Control of Multi-robot Tasks Modelled as Petri nets. In: 2011 IEEE/RSJ International Conference on Intelligent Robots and Systems, pp. 3081–3086. IEEE Press, New York (2011)
15. Ziparo, V.A., Iocchi, L., Lima, P.U., Nardi, D., Palamara, P.F.: Petri Net Plans: A Framework for Collaboration and Coordination in Multi-robot Systems. Autonomous Agents and Multi-Agent Systems 23, 344–383 (2011)
16. Ebadi, T., Purvis, M., Purvis, M.: A Framework for Facilitating Cooperation in Multi-Agent Systems. The Journal of Supercomputing 51, 393–417 (2010)
17. Sheng, W., Yang, Q.: Peer-to-peer Multi-robot Coordination Algorithms: Petri Net Based Analysis and Design. In: 2005 IEEE/ASME International Conference on Advanced Intelligent Mechatronics, pp. 1407–1412. IEEE Press, New York (2005)
18. Nishi, T., Tanaka, Y., Isoya, Y.: Petri Net decomposition for deadlock avoidance routing for bi-directional AGV systems. In: 2010 IEEE International Conference on Systems Man and Cybernetics, pp. 2453–2458. IEEE Press, New York (2010)
19. Iordache, M., Antsaklis, P.: Decentralized Supervision of Petri Nets. IEEE Transactions on Automatic Control 51, 376–381 (2006)
20. Bosse, M., Newman, P., Leonad, J., Teller, S.: Simultaneous Localization and Map Building in Large-scale Cyclic Environments Using the Atlas Framework. The International Journal of Robotics Research 23, 1113–1139 (2004)
21. Estrada, C., Neira, J., Tardos, J.D.: Hierarchical SLAM: Real-time Accurate Mapping of Large Environments. IEEE Transactions on Robotics 21, 588–596 (2005)
22. Balch, T.R., Arkin, R.C.: Behavior Based Formation Control for Multi-robot Teams. IEEE Transactions on Robotics and Automation 14, 926–939 (1998)

# Image Edge Detection Based on Relative Degree of Grey Incidence and Sobel Operator

Jing Sun

College of Information , Zhejiang Sci-Tech University, Hangzhou, China
jings531@163.com

**Abstract.** Edge points are characterized by sharp transitions in gray levels in adjacent pixels, and relative degree of grey incidence can just reflect the degree of variations. In this paper an image edge detection method integrating relative degree of grey incidence with Sobel operator is presented. Firstly, the comparison sequence is constructed by sequentially ranking a certain pixel and its eight neighborhood of the image, and the reference sequence is formed by taking two orientation operator of Sobel operator. Secondly, the quantitative level difference between reference sequence and behavior sequence is decreased using initialization operation. Then the pixel concerned can be judged as an edge point when there exists a higher relative degree of grey incidence, which means similar geometric shapes of two sequences. By comparing the experimental results, it is proved that the strategy proposed in this paper can detect more details many traditional methods can not find.

**Keywords:** relative degree of grey incidence, Sobel operator, edge detection.

## 1 Introduction

Edge means the abrupt intensity changes in local regions of an image, which is the key characteristic of an image. Edge detection can be thought of a key point in addressing many complicated problems, playing an important role in areas such as image segmentation, feature extracting, image recognition and so on. The most representative traditional edge detection methods nowadays include Sobel operator based on first order derivative, Canny operator, Prewitt operator and Roberts operator. These operators all utilize the weighted average mask of neighborhood intensity, that is, to complete edge detection by making convolution based on orientation derivative templates. It is well recognized that different types of operators have various degree of sensitivity to intensity changes of different directions, hence for edges with distinct properties, the detecting effects differ greatly. Moreover, the choice of threshold in a variety of algorithms will also influence the detecting effects directly. As a matter of fact there are multiple types of image embedded with diverse noises, however the mask used in conventional approaches is fixed, which can hardly satisfying various complicated circumstances of images. In an attempt to deal with this situation, many new mechanisms emerge constantly combining iterative adaptive RBF[1],bacterial foraging technique[2], interval type-2 fuzzy logic[3], cellular neural networks[4],

J. Lei et al. (Eds.): AICI 2012, LNAI 7530, pp. 762–768, 2012.

improved grey prediction model[5], mathematical morphology[6], lifting wavelet[7], gray system theory[8-9] etc. into edge detection, greatly improving the detecting accuracy.

When conducting edge detecting, for a given pixel there are two possibility whether it is an edge point, but we still can not eventually identify if it is an edge point, consequently, this is a typical issue of extension certain whereas intension uncertain. Starting from this point, the image can be viewed as a grey system, and we can exploit relative degree of grey incidence to complete edge detection, in a manner first taking the two direction mask of Sobel operator to form reference sequence in relative degree of grey incidence, then evaluating the larger one to identify the edge points. Thus, an image edge detection strategy based on relative degree of grey incidence and Sobel operator is put forward in this article.

## 2 Sobel Operator

Sobel operator is a kind of first-order derivative operator, which calculates the gradient magnitude of a pixel by using the neighborhood intensity values, the formulation is given by

$$S = \sqrt{d_x^2 + d_y^2} \tag{1}$$

Where $d_x$、$d_y$ can be implemented by convolution kernel

$$d_x = \begin{bmatrix} -1 & 0 & 1 \\ -2 & 0 & 2 \\ -1 & 0 & 1 \end{bmatrix} \tag{2}$$

$$d_y = \begin{bmatrix} 1 & 2 & 1 \\ 0 & 0 & 0 \\ -1 & -2 & -1 \end{bmatrix} \tag{3}$$

We know that one convolution kernel produces stronger response at vertical edges while another at horizontal edges, then we obtain edge points using the maximum value of the two convolution kernel.

## 3 Relative Degree of Grey Incidence

Basically, the idea behind grey relative incidence degree analysis is to calculate incidence degree according to the similarity degree of the variation tendency of date sequences curves (variation rate). The relative incidence degree method[10] is depicted as follows: Suppose reference sequence and comparison sequence to be $X_0=(x_0(1),x_0(2),\ldots,x_0(n)$ and $X_i=(x_i(1),x_i(2),\ldots,x_i(n))$, $i=1,2,\ldots,n$, respectively, then the initial image of $X_0$ and $X_i$ are:

$$X_0' = \frac{X_0}{x_0(1)} = \left(x_0'(1), x_0'(2), \ldots\ldots, x_0'(n)\right) = \left(1, \frac{x_0(2)}{x_0(1)}, \ldots\ldots, \frac{x_0(n)}{x_0(1)}\right) \tag{4}$$

$$X_i' = \frac{X_i}{x_i(1)} = \left(x_i'(1), x_i'(2), \ldots\ldots, x_i'(n)\right) = \left(1, \frac{x_i(2)}{x_i(1)}, \ldots\ldots, \frac{x_i(n)}{x_i(1)}\right) \tag{5}$$

after zero operation on initial point,

$$\begin{aligned} x_i'^{0} &= x_0' - x_0'(1) = \left(x_0'^{0}(1), x_0'^{0}(2), \ldots\ldots, x_0'^{0}(n)\right) \\ &= \left(0, x_0'(2) - x_0'(1), \ldots\ldots, x_0'(n) - x_0'(1)\right) \end{aligned} \tag{6}$$

then relative degree of grey incidence is

$$r_{0i} = \frac{1 + |s_0'| + |s_i'|}{1 + |s_0'| + |s_i'| + |s_i' - s_0'|} \tag{7}$$

where

$$|s_0'| = \left|\sum_{j=1}^{n-1} x_0'^{0}(j) + \frac{1}{2} x_0'^{0}(n)\right|$$

$$|s_i'| = \left|\sum_{j=1}^{n-1} x_i'^{0}(j) + \frac{1}{2} x_i'^{0}(n)\right|$$

The relative degree of grey incidence has the properties as follow: $0 < r_{0i} \leq 1$; $r_{0i}$ only relates to the variation rate of the sequence towards the initial point, while independent of the value of each observed data, in other words, the value of relative degree of grey incident will not change while multiplying by numbers; The variation rate of any two sequences are not independent of each other, that is , $r_{0i}$ is permanently nonzero; The more consistent the variation rate of the two sequences toward the initial point tend to be, the larger $r_{0i}$ is; $r_{0i}$ will change as any observed data in $X_0$ or $X_1$ alter; Similarly, $r_{0i}$ will change as the length of sequence $X_0$ and $X_i$ vary; $r_{00} = r_{ii} = 1$; $r_{0i} = r_{i0}$ 。

## 4 Image Edge Detection Based on Relative Degree of Grey Segmentation

Image edges are a series of points with step changes or roof changes in intensity. Every pixel should be judged as edge or non edge point when conducting edge detection. Edge points always exhibit abrupt intensity variation in neighbor pixels, while the relative degree of grey incidence indicates the variation degree exactly. Since an ideal edge point can be represented by the difference of the pixel value and its surrounding values, we take a pixel and its eight neighbors to form a comparison sequence $X_i$, then use Sobel operator to construct reference sequence. When the relative degree of grey incidence between comparison sequence and reference sequence is larger, which means the geometric shapes of the two sequences are similar, the pixel can be determined as an edge point; For the same reason, the pixel can be judged as an non edge point with smaller relative degree value. Thus , by means of relative degree of grey incidence, we perform edge detection for a $M \times N$ image by taking a certain pixel and its eight neighborhood to comprise the comparison sequence, consisting three principle steps:

(1) Expanding two $3 \times 3$ Sobel operators in a specific way to act as reference sequence $X_{01}=\{1,2,1,0,0,0,-1,-2,-1\}$; $X_{02}=\{-1,0,1,-2,0,2,-1,0,1\}$;

For each sequence, dividing sequence data by the corresponding maximum value in it respectively.

(2) For the gray-level image under detection, a point and its neighborhood can construct behavior sequence, also a $3 \times 3$ template is given as Fig 1 shows:

| | | |
|---|---|---|
| $\mathrm{Im}_{i-1,j-1}$ | $\mathrm{Im}_{i-1,j}$ | $\mathrm{Im}_{i-1,j+1}$ |
| $\mathrm{Im}_{i,j-1}$ | $\mathrm{Im}_{i,j}$ | $\mathrm{Im}_{i,j+1}$ |
| $\mathrm{Im}_{i+1,j-1}$ | $\mathrm{Im}_{i+1,j}$ | $\mathrm{Im}_{i+1,j+1}$ |

**Fig. 1.** Image 3x3 mask neighbor pixels

For image I with size of $m \times n$, the behavior sequence is denoted as:

$$I_{ij} = [\mathrm{Im}_{i-1,j-1}, \mathrm{Im}_{i-1,j}, \mathrm{Im}_{i-1,j+1}, \mathrm{Im}_{i,j-1}, \mathrm{Im}_{i,j}, \mathrm{Im}_{i,j+1}, \mathrm{Im}_{i+1,j-1}, \mathrm{Im}_{i+1,j}, \mathrm{Im}_{i+1,j+1}]$$

In an attempt to decrease the quantitive level variance in reference and behavior sequence, since the grey levels of an image is in the range 0~255, it seems necessary to carry out quantization operation on the behavior sequence. There are several approaches to achieve this goal such as initialization, averaging, maximum operation, median operation and so on, the suitable choice of one method over another resting on the impact on processed image made by each approach and detection result. In this paper, we adopt initialization method.

(3) Solving two relative incidence degree $r_{0i}^{(1)}$, $r_{0i}^{(2)}$ through computing the comparison sequence centered by every pixel with two reference sequence of Sobel operator respectively. The criteria of determining an edge point is: When $\max(r_{0i}^{(1)}, r_{0i}^{(2)})$ is greater than the preassigned threshold of incidence degree $r^{th}$, which means behavior and reference sequence have similar variation tendency, the pixel can be judged as edge point. Otherwise, it is not an edge point.

## 5 Experimental Result and Analysis

Assuming the original grey-scale image is I of size m×n, we add pixels around the image to avoid the limitation that the border pixels of the image can not be detected. The mechanism is depicted as follows: First, take the first and last row of the image and combine with the original image I to get a new image I1 of size (m+2) ×n, in which the pixels in first row and second row are same, and the pixels in the last row and the second row from bottom are equal. In a manner similar to the row, take the first and last column to combine with I1 to get new image I2 with size of (m+2) × (n+2), in which the pixels in first column and second column are identical, and the pixels in the last column and the second column from bottom are equal. Therefore, all the pixels in the image can be detected.

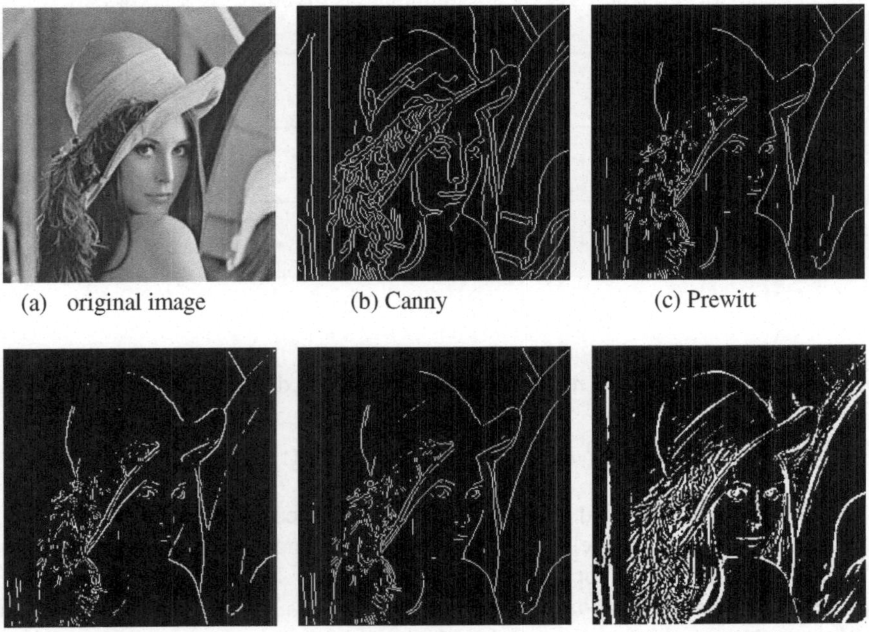

(a) original image (b) Canny (c) Prewitt

(d) Roberts (e) Soble (f) Our method

**Fig. 2.** Experimental results

Fig2(a) is the original image, Fig 2(b)(c)(d)(e) are detecting results based on traditional Canny , Prewitt, Roberts, Soble operators, Fig2(f) is experimental result combined relative degree of grey incidence and Sobel operator with threshold set to 0.5148. It is clear that the details of facial features, decorate plumage in the cap can be detected distinctly, achieving a good effect. Fig 3 are the results of applying the proposed method in this paper with threshold set to 0.5120、0.5130、0.5140、0.5150 respectively. It is evident that, as the threshold increases, more details information can be detected.

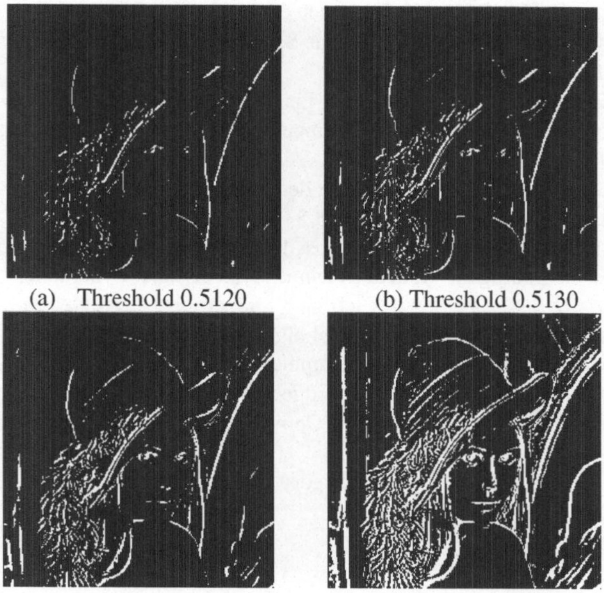

(a) Threshold 0.5120 (b) Threshold 0.5130

(c) Threshold 0.5140 (d) Threshold 0.5150

**Fig. 3.** Results under different threshold

## 6 Conclusions and Future Work

A novel image edge detection method is introduced in this paper, that is, an approach combining relative degree of grey incidence with Sobel operator, demonstrating the feasibility of applying grey theory on image segmentation once more, expanding the application of grey technique in image processing. The mechanism begins by sorting certain pixel and its eight neighbors to compose comparison sequence $X_i$, while taking Sobel operator to form reference sequence $X_0$, then using relative incidence degree to measure similarity degree of the geometric shape between comparison sequence and reference sequence, finally detecting whether the pixel is an edge point effectively. Moreover, when performing detection, we add pixels around the image to avoid the limitation that the border pixels can not be detected. Compared with traditional Canny , Prewitt, Roberts and Soble operators, the details detected are more clear using this method. However, one shortcoming is that the threshold must be determined artificially, so more research work on automatically determining threshold needs to be done in the future.

## References

1. Jung, J.-H., Gottlieb, S., Kim, S.O.: Iterative adaptive RBF methods for detection of edges in two-dimensional functions. Applied Numerical Mathematics 61(1), 77–91 (2011)
2. Verma, O.P., Hanmandlu, M., Kumar, P., et al.: A novel bacterial foraging technique for edge detection. Pattern Recognition Letters 32(8), 1187–1196 (2011)
3. Melin, P., Mendoza, O., Castillo, O.: An improved method for edge detection based on interval type-2 fuzzy logic. Expert Systems with Applications 37(12), 8527–8535 (2010)
4. Li, H., Liao, X., Li, C., et al.: Edge detection of noisy images based on cellular neural networks. Communications in Nonlinear Science and Numerical Simulation 16(9), 3746–3759 (2011)
5. Zhou, Z., Zheng, L., Xia, J., Yang, W., Lei, J.: Image Edge Detection Based on Improved Grey Prediction Model. Journal of Computational Information Systems 6(5), 1501–1507 (2010)
6. Sun, D., Cai, Y., Li, F., Wu, Y.: Edge detection based on mathematical morphology for color weld image. Welding in the World 53, 373–376 (2009)
7. Wang, X.: Image edge detection based on lifting wavelet. In: 2009 International Conference on Intelligent Human-Machine Systems and Cybernetics, vol. 1, pp. 25–27 (2009)
8. Ma, M., Fan, Y., Xie, S., et al.: A novel algorithm of image edge detection based on gray system theory. Journal of Image and Graphics: A Edition 8(10), 1136–1139 (2003)
9. Zhou, Z., Zhang, J., Lei, J., et al.: Edge detection based on soble operator and grey absolute correlation degree. Journal of Computational Information Systems 5(2), 967–974 (2009)
10. Liu, S., Dang, Y., Fang, Z.: Grey Theory and Applications, pp. 69–72. Science Press, Beijing (2004)

# Background Modeling from Surveillance Video Using Rank Minimization

Min Yang

College of Automation,
Nanjing University of Posts and Telecommunications,
Nanjing 210003
yangm@njupt.edu.cn

**Abstract.** Intelligent video surveillance systems can be applied to a wide range of potential applications. In this paper, we propose a new background modeling scheme that draws from the principles of low rank representation. We assume that the underlying background images are linearly correlated. Thus, the matrix composed of vectorized video frames can be approximated by a low-rank background matrix plus the sparse foreground components. Low rank representation can be exactly recovered via convex optimization that minimizes a combination of the nuclear norm and the $l_1$-norm, and this non-convex problem can be solved very efficiently in the inexact Augmented Lagrange Multiplier method. We tested our algorithm on real video, and our approach obtained good results, comparable to the Gaussian Mixture Model method.

**Keywords:** Background modeling, Nuclear norm minimization, $l_1$-norm minimization, Augmented Lagrange multipliers, Low-rank representation.

## 1 Introduction

Intelligent video surveillance systems can be applied to a wide range of potential applications, such as security and safety systems in important buildings, traffic surveillance in cities and highways, or supervision of suspicious behavior in supermarkets or in public means of transport.

Every video surveillance system starts its activity by detecting moving objects in the scene. In essence, the aim of this task is to separate pixels corresponding to the foreground motion objects from those corresponding to the stationary background ones. Different kinds of approaches can be found in the literature to model this problem, such as techniques based on optical flow, whose main disadvantage is that it requires very high computational time; frame difference, which is efficient but inaccurate and unreliable; or background subtraction, which models the background by comparison with the frames of the sequence[1]. Stauffer et al. in [2,3], present a well-known theoretical framework for the upgrade background, based on a mixing process of Gaussian distributions for each pixel, using an EM algorithm to update them. This statistical approach is more robust in scenes with many moving objects and lighting changes, and it is one of the most cited techniques in the literature.

J. Lei et al. (Eds.): AICI 2012, LNAI 7530, pp. 769–774, 2012.

Learning with sparsity has drawn a lot of attentions in recent machine learning and computer vision research[4], and it was recently shown that under broad conditions, the low-rank and the sparse components of a matrix can be exactly recovered via convex optimization that minimizes a combination of the nuclear norm and the $l_1$-norm[5]. The low-rank representation offers a blind separation of low-rank data and sparse noises.

In this paper, we propose a new background modeling scheme that draws from the principles of low rank representation. We assume that the underlying background images are linearly correlated. Thus, the matrix composed of vectorized video frames can be approximated by a low-rank matrix, and the moving objects can be detected as outliers in this low-rank representation. Formulating the problem as outlier detection allows us to get rid of many assumptions on the behavior of foreground. The low-rank representation of background makes it flexible to accommodate the global variations in the background. We pose background modeling as a rank minimization problem, where the goal is to decompose the data matrix as the sum of a low-rank background matrix plus a matrix of foreground, and this non-convex problem can be solved very efficiently in the inexact Augmented Lagrange Multiplier method.

## 2 Background Modeling Using Rank Minimization

### 2.1 Low Rank Minimization (LRM)

Given a noise corrupted data matrix D = A+E, where A is an unknown low-rank matrix and E represents the noise, the problem of finding a low rank approximation of D can be formulated as

$$\min_{A} \|D - A\|_F^2, s.t.rank(A) \leq r \tag{1}$$

The optimal solution to this (PCA) problem is given by $A = U_1 \Sigma_1 V_1^T$, where $U_1, \Sigma_1$ and $V_1$ are obtained from the top r singular values and singular vectors of the data matrix D. We may also write this as $A = UH_{\sigma_{r+1}}(\Sigma)V^T$, where $H_\varepsilon(x) = x1_{|x|>\varepsilon}$ is the hard thresholding operator.

When r is unknown, the problem of finding a low-rank approximation can be formulated as

$$\min_{A} rank(A) + \frac{\alpha}{2} \|D - A\|_F^2 \tag{2}$$

where $\alpha > 0$ is a parameter. Since this problem is in general NP hard, a common practice (see [6]) is to replace the rank of A by its nuclear norm $\|A\|_*$, i.e., the sum of its singular values, which leads to the following convex problem

$$\min_{A} \|A\|_* + \frac{\alpha}{2} \|D - A\|_F^2 \tag{3}$$

It is shown in [7] that the optimal solution to this problem is given by $A = US_\tau(\Sigma)V^T$, where $D = U\Sigma V^T$ is the SVD of D and $S_\tau(\Sigma)$ is the shrinkage-thresholding operator

$$S_\tau(\Sigma) = diag(\{\sigma_i - \tau\}_+), \tau \geq 0 \tag{4}$$

Where $t_+$ is the positive part of $t$, namely , $t_+ = \max(0,t)$. In words, this operator simply applies a soft-thresholding rule to the singular values of $\Sigma$, effectively shrinking these towards zero.

## 2.2 Principal Component Pursuit

While the above methods work well for data corrupted by Gaussian noise, they break down for data corrupted by gross errors. In [5] this issue is addressed by assuming that the outliers are sparse, i.e., only a small percentage of the entries of D are corrupted. Hence, the goal is to decompose the data matrix D as the sum of a low-rank matrix A and a sparse matrix E, i.e.,

$$\min_{A,E} rank(A) + \gamma\|E\|_0, s.t. D = A + E \tag{5}$$

where $\gamma > 0$ is a parameter. Since this problem is in general NP hard, a common practice is to replace the rank of A by its nuclear and the $\ell_0$ semi-norm by the $\ell_1$ norm. It is shown in [5] that, under broad conditions, the optimal solution to the problem in (5) is identical to that of the convex problem

$$\min_{A,E} \|A\|_* + \gamma\|E\|_1, s.t. D = A + E \tag{6}$$

While a closed form solution to this problem is not known, convex optimization techniques can be used to find the minimum. We refer the reader to [8] for a review of numerous approaches. One such approach is the Augmented Lagrange Multiplier (ALM) method, which minimizes

$$L(A,E,Y,\alpha) = \|A\|_* + \gamma\|E\|_1 + \langle Y, D - A - E\rangle + \frac{\alpha}{2}\|D - A - E\|_F^2 \tag{7}$$

The third term enforces the equality constraint via the matrix of Lagrange multipliers Y, while the fourth term (which is zero at the optimum) makes the cost function strictly convex and thus improves the convergence. The inexact ALM method then iterates the following steps till convergence. An inexact ALM (IALM) method, described in Algorithm 1.

**Algorithm 1**

**Input:** Observation matrix $D \in R^{m \times n}, \gamma$.

1: $Y_0 = D / J(D); E_0 = 0; \alpha_0 > 0; \rho > 1; k = 0.$

2: **while** not converged **do**

3: // Lines 4-5 solve $A_{k+1} = \arg\min_{A} L(A, E_k, Y_k, \alpha_k)$

4: $(U, S, V) = svd(D - E_k + \alpha_k^{-1} Y_k)$

5: $A_{k+1} = U S_{\alpha_k^{-1}}(S) V^T$

6: //Line 7 solves $E_{k+1} = \arg\min_{E} L(A_{k+1}, E, Y_k, \alpha_k)$

7: $E_{k+1} = S_{\gamma\alpha_k^{-1}}(D - A_{k+1} + \alpha_k^{-1} Y_k)$

8: $Y_{k+1} = Y_k + \alpha_k (D - A_{k+1} - E_{k+1}); \alpha_{k+1} = \rho a_k$.

9: k = k+1

10: **end while**

**Output:** $(A_k, E_k)$.

This ALM method is essentially an iterated thresholding algorithm, which alternates between thresholding the SVD of $D - E + Y / \alpha$ to get A and thresholding $D - E + Y / \alpha$ to get E. The update for Y is simply a gradient ascent step. Also, to guarantee the convergence of the algorithm, the parameter $\alpha$ is updated by choosing $\rho > 1$ so as to generate a sequence $\alpha_k$ that goes to infinity.

## 3 Experiments

In real applications, the foreground motion can be very complicated with nonrigid shape changes. Also, the background may be very complex, including illumination changes and varying textures such as waving trees and sea waves. Fig. 1(a) shows such a challenging example. The video includes an operating escalator, but it should be regarded as background for human tracking purpose.

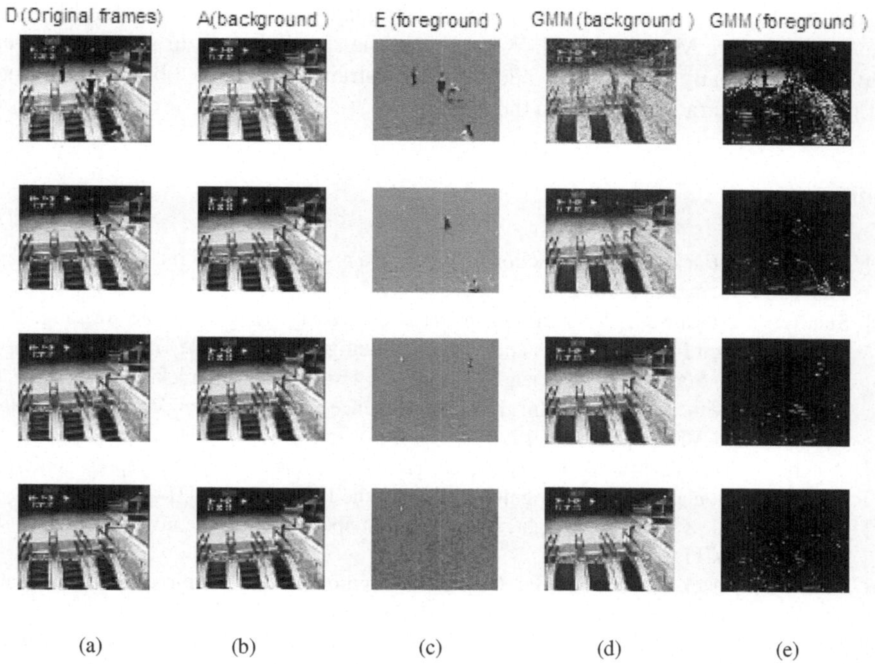


**Fig. 1.** Background modeling from video. four frames from a 200-frame escalator video sequence (a) Frames of original video *D*. (b)-(c) Low-rank background A and foreground sparse components E obtained by Low Rank Minimization, (d)-(e) competing approach based on Gaussian Mixture Model (GMM).

In this section we evaluate the effectiveness and the efficiency of our method based low-rank approximation. Test video data is a sequence of 200 grayscale frames taken in an airport. The frames have resolution 130×160; we stack each frame as a column of our matrix $D \in R^{20800\times 200}$. We decompose D into a low-rank term and a sparse term by solving the low rank minimization, our Matlab implementation requires 36 iterations. We compare our method to Mixture of Gaussian method from the computer vision literature.

Fig. 1(a) shows the $50^{th}$,$100^{th}$,$150^{th}$ ,and the last frame from the video; (b) and (c) show the corresponding columns of the low rank matrix and sparse matrix. Notice that correctly recovers the background, while correctly identifies the moving pedestrians. Figure 1(d) and (e) compares the result obtained by Mixture of Gaussian method. We can see that our method performs very well, our method yields a much more appealing result despite using less prior knowledge.

## 4 Conclusion

We propose a novel algorithm for background modeling. It solves the dynamic background challenges in the low rank minimization framework, and this problem can be solved in the inexact Augmented Lagrange Multiplier method. We tested our algorithm on real video, and our approach obtained good results, comparable to the Gaussian Mixture Model method. We note that the simplicity of our solution suggests that one can easily incorporate additional constraints so as to obtain much more efficient and accurate solutions in the future.

## References

[1] Piccardi, M.: Background subtraction techniques: a review. In: IEEE Int. Conf. on Systems, Man and Cybernetics (2004)
[2] Stauffer, C., Grimson, W.: Learning patterns of activity using real-time tracking. IEEE Transactions on Pattern Analysis and Machine Intelligence 22(8), 747–757 (2000)
[3] Grimson, W., Stauffer, C., Romano, R., Lee, L.: Using adaptive tracking to classify and monitor activities in a site. In: IEEE Conference on Computer Vision and Pattern Recognition, CVPR, pp. 22–29 (1998)
[4] Wright, J., Ma, Y., Mairal, J., Sapiro, G., Huang, T., Yan, S.: Sparse representation for computer vision and pattern recognition. Proc. of the IEEE 98(6), 1031–1044 (2010)
[5] Candes, E., Li, X., Ma, Y., Wright, J.: Robust principal component analysis. Journal of the ACM (May 2011)
[6] Recht, B., Fazel, M., Parrilo, P.: Guaranteed minimum-rank solutions of linear matrix equations via nuclear norm minimization. SIAM Review 52(3), 471–501 (2010)
[7] Cai, J.-F., Candès, E.J., Shen, Z.: A singular value thresholding algorithm for matrix completion. SIAM Journal of Optimization 20(4), 1956–1982 (2008)
[8] Lin, Z., Chen, M., Wu, L., Ma, Y.: The augmented Lagrange multiplier method for exact recovery of corrupted low-rank matrices (2009), http://arxiv.org/abs/1009.5055

# A Complete Set of Pseudo-Zernike Moment Invariants by Image Shape Description*

Lin Zheng[1], Sinan Zhao[2], Qian Liu[1], and Hongqing Zhu[1,**]

[1] Department of Electronics and Communications Engineering, East China University of Science and Technology, Shanghai 200237, China
[2] Department of Information and Control Engineering, Shanghai University of Electric Power, Shanghai 200090, China
hqzhu@ecust.edu.cn

**Abstract.** Orthogonal moments such as pseudo-Zernike moments have been successfully used in the field of image analysis. Conventionally, image function is mapped onto a set of orthogonal functions over the unit circle. If the origin of polar coordinate system is taken at the centroid, the rotation invariants will be easy to obtain. Based on pseudo-Zernike moments, this paper presents a new method to drive the complete rotation, scaling and translation (RST) invariants from the orthogonal projection transform (OPT). The efficiency and the robustness to different noises of the method for classification tasks are presented by comparing it with several existing methods.

**Keywords:** RST invariants, Shape description, Completeness, Pseudo-Zernike moments.

## 1 Introduction

In the past decades, invariants study on transformed image classification is widely applied in object recognition and scene matching. Ghorbel et al. verified the most important properties of invariant descriptors in [1]: (i) Invariance against some geometrical transformation such as translation, rotation and scaling: (ii) Stability to noise, to blur, to nonrigid and to small local deformation. (iii) Completeness. Since Hu [2] proposed the two-dimensional moment invariants of RST, researches in moment invariant theory have followed the above properties. Recently, a variety of literatures have reported moment functions and their invariants for pattern recognition tasks. For example, Chen et al. [3] presented a set of Zernike moment invariants for images under geometric transformation and blurring. Pseudo-Zernike moments were used by Zhang for constructing scale and rotation invariants [4]. Kan and Srinath [5] proposed an effective method of combining orthogonal Fourier-Mellin moments with centroid bounding circle scaling. Combined with blurred, Fourier-Mellin moment invariants were described by Bin et al. in [6]. Radial-Tchebichef moment invariants

---

* This work has been supported by National Natural Science Foundation of China under Grant No. 60975004.

** Corresponding author.

J. Lei et al. (Eds.): AICI 2012, LNAI 7530, pp. 775–783, 2012.

were proposed in the Cartesian coordinate by Xiao et al. [7], which were invariant to scaling and rotation. Gaussian-Hermite moments were used in rotation and translation invariants [8]. With the reference to the relevant literatures, all the above moment invariant theories have been successfully investigated in pattern recognition.

In [9], Crimmins derived a complete set of invariant descriptor using Fourier moments. Complete properties are very important in pattern recognition. A set of Fourier descriptors for two-dimensional shapes is defined which is complete in the sense that two objects have the same shape if and only if they have the same set of Fourier descriptors. Ghorbel et al. proposed a complete set of RST invariants from complex moments [10]. They discussed the robust and efficiency by using a complete and convergent set of feature invariants [1]. Lan [11] gave an approach named orthogonal projection transform for shape description and invariants based on Fourier-Mellin moments. Inspired by Lan et al. works, this paper proposed a set of invariants based on orthogonal projection transform with pseudo-Zernike moments. Unlike Lan's method, the invariants in currents study are complete. Based on the proposed complete set of invariants, the experiments show a good performance in image rotation, scaling and translation.

## 2 Mathematics Background

### 2.1 The Polar Coordinate System

*Im*(*x*, *y*) denotes the grey-level of the original image at the point (*x*, *y*). To facilitate the pseudo-Zernike polynomials calculation, the Cartesian coordinate system should be firstly converted into the polar coordinate system:

$$r = \sqrt{(x - x_0)^2 + (y - y_0)^2}, \quad \theta = \arctan(\frac{y - y_0}{x - x_0}) \tag{1}$$

where the point $(r, \theta)$ is determined by a distance from a fixed point $(x_0, y_0)$ and an angle from a fixed direction. Thus, we can map *Im*(*x*, *y*) into unit circle. The grey-level of *Im*(*x*, *y*) in the polar coordinate system is:

$$f(r, \theta) = Im(x, y) \tag{2}$$

### 2.2 Pseudo-Zernike Moments

The pseudo-Zernike polynomials of order $p$ and index $q$ are defined as:

$$V_{pq}(r, \theta) = R_{pq}(r)e^{jq\theta} \tag{3}$$

and the real-valued radial polynomials, $R_{pq}(r)$, is defined as

$$R_{pq}(r) = \sum_{k=0}^{p-|q|} (-1)^k \frac{(2p+1-k)!}{k!(p+|q|+1-k)!(p-|q|-k)!} r^{p-k} \tag{4}$$

If $q \geq 0$, it is obvious that $R_{p,-q}(r)=R_{p,q}(r)$. Just the nonnegative integer $q$ is considered. According to [4], letting $m = p - q$, the polynomials can be rewritten as:

$$R_{q+m,q}(r) = \sum_{k=0}^{m} c_{m,k}^{q} r^{q+k} \tag{5}$$

with

$$c_{m,k}^{q} = (-1)^{m-k} \frac{(2q+m+k+1)!}{k!(2q+1+k)!(m-k)!} \tag{6}$$

## 3 Orthogonal Projection Transforms

The OPT of image $f(r, \theta)$ is defined as

$$F_{q+m,q}(\theta) = \int_0^1 R_{q+m,q}(r) f(r,\theta) dr \tag{7}$$

where $R_{q+m,q}(r)$ is a set of radial orthogonal polynomials. In the definition of OPT, $\theta$ is an independent variable of function $F_{q+m,q}(\theta)$ with an interval in $[0, 360^0)$. If the function $F_{q+m,q}(\theta)$ is taken as the polar coordinate form and changed back into Cartesian coordinate, a closed contour can be derived.

Let $R = \max \sqrt{(x-x_0)^2 + (y-y_0)^2}$ be the maximum distance from point $(x_0, y_0)$ to any point of the image. Considering that $R_{q+m,q}(r)$ is the pseudo-Zernike radial polynomials with order $q + m$ and index $q$, then the discrete OPT can be obtained:

$$C_{q+m,q}(\theta) = \sum_{r=0}^{R} R_{q+m,q}(r/R) f(r/R,\theta) \tag{8}$$

In this equation, $f(r/R, \theta)$ and $R_{q+m,q}(r/R)$ are restricted in unit circle. Fig.1 shows an image and it's OPT curves.

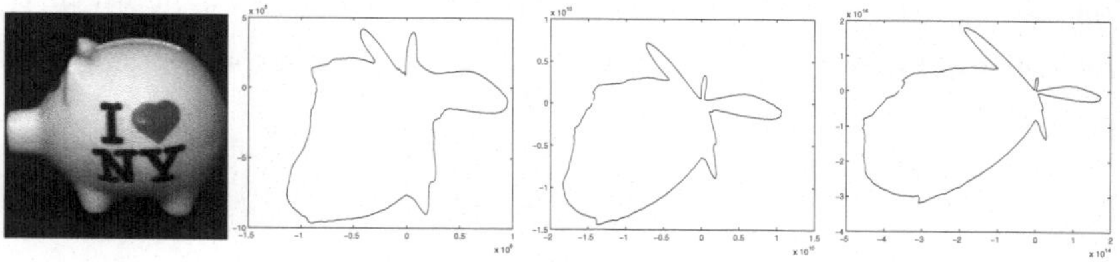

**Fig. 1.** Original object and OPT curves with $q$=0, $m$=1; $q$=1, $m$=2, and $q$=2, $m$=3, respectively

## 4 Invariant Descriptors

The transformation of an image $Im^f(x, y)$ into another image $Im^g(X, Y)$ can be represented as follows:

$$\begin{pmatrix} X \\ Y \end{pmatrix} = \begin{pmatrix} a_{11} & a_{12} \\ a_{21} & a_{22} \end{pmatrix} \begin{pmatrix} x \\ y \end{pmatrix} + \begin{pmatrix} x_t \\ y_t \end{pmatrix} \tag{9}$$

For the RST transformation, one has:

$$\begin{pmatrix} a_{11} & a_{12} \\ a_{21} & a_{22} \end{pmatrix} = \begin{pmatrix} \lambda\cos\beta & \lambda\sin\beta \\ -\lambda\sin\beta & \lambda\cos\beta \end{pmatrix} \tag{10}$$

Here, $Im^g(X, Y)$ has the same gray-level with $Im^f(x, y)$. $\lambda$, $\beta$ indicate the scaling factor, rotation angle, $x_t$ and $y_t$ indicate shifting along with the $x$ and $y$ axis, respectively.

If $(x_0, y_0)$ in Eq.(1) locates at the centroid, then translation invariants are available. Scaled and rotation images are discussed in the following part.

## 4.1 Scale Invariants

Let $g(r, \theta)$ be a scaled version of $f(r/\lambda, \theta)$, where $g$ contains the same shape information with $f$. The OPT of image function $g$ with pseudo-Zernike polynomials in Eq.(7) can be transformed as:

$$F^g_{q+m,q}(\theta) = \int_0^1 R_{q+m,q}(r)g(r,\theta)dr = \int_0^1 R_{q+m,q}(\lambda r)f(r,\theta)\lambda dr \tag{11}$$

**Table 1.** Classification results with different orders

| | $q$=1,$m$=2 | $q$=2,$m$=2 | $q$=2,$m$=3 | $q$=3,$m$=3 |
|---|---|---|---|---|
| CPZMs | 86% | 87% | 89% | 90% |

According to [4], the pseudo-Zernike polynomials $R_{q+m,q}(r)$ after scaling will be:

$$R_{q+m,q}(\lambda r) = \sum_{k=0}^{m} R_{q+k,q}(r) \sum_{l=k}^{m} \lambda^{q+l} c^q_{m,l} d^q_{l,k} \tag{12}$$

In Eq.(12), $d^q_{l,k}$ is an element of lower triangular matrix $D^q_m = (d^q_{i,j})$, which is the inverse matrix of lower triangular matrix $C^q_m = (c^q_{i,j})$. It can be expressed as:

$$d^q_{i,j} = \frac{(2q+2j+2)i!(2q+i+1)!}{(i-j)!(2q+i+j+2)!} \tag{13}$$

By substituting Eq.(12) into Eq.(11), the shape relationship between image function $g$ and $f$ is expressed as:

$$\begin{aligned} F^g_{q+m,q}(\theta) &= \int_0^1 R_{q+m,q}(r)g(r,\theta)dr = \int_0^1 R_{q+m,q}(\lambda t)f(t,\theta)\lambda dt \\ &= \lambda \int_0^1 \left(\sum_{k=0}^{m} R_{q+k,q}(t) \sum_{l=k}^{m} \lambda^{q+l} c^q_{m,l} d^q_{l,k}\right) f(t,\theta)dt = \sum_{k=0}^{m}\sum_{l=k}^{m} \lambda^{q+1+l} c^q_{m,l} d^q_{l,k} F^f_{q+k,q}(\theta) \end{aligned} \tag{14}$$

The OPT of scaled image function is a liner combination of original image function. After constructing the matrixes, the scale invariant can be obtained.

**Theorem 1.** For nonnegative integers $q$ and $m$, the following equation is a scale invariant.

$$I^{f}_{q+m,q}(\theta)=\sum_{k=0}^{m}\left(\sum_{l=k}^{m}\Gamma_{f}^{-(q+1+l)}c^{q}_{m,l}d^{q}_{l,k}\right)F^{f}_{q+k,q}(\theta) \tag{15}$$

with $\Gamma_f = F^{f}_{0+0,0}(\theta)$, $c^{q}_{m,k}$ and $d^{q}_{i,j}$ can be obtained from Eq.(6) and Eq.(13). The proof of Theorem 1 is given in Appendix A.

### 4.2 Rotation Invariants

Let $g(r, \theta) = f(r, \theta-\beta)$, the OPT of a rotation image function $g(r, \theta)$ can be written as:

$$F^{g}_{q+m,q}(\theta)=\int_0^1 R_{q+m,q}(r)g(r,\theta)dr=\int_0^1 R_{q+m,q}(r)f(r,\theta-\beta)dr=F^{f}_{q+m,q}(\theta-\beta) \tag{16}$$

Considering the Fourier transform of Eq.(16), we get:

$$\mathcal{F}[F^{f}_{q+m,q}(\theta-\beta)]=e^{-2\pi i\beta\xi}\mathcal{F}[F^{f}_{q+m,q}(\theta)] \tag{17}$$

Due to the modulus of $e^{-2\pi i\beta\xi}$ equals one, Eq.(17) has the same magnitude in both sides of the equal sign:

$$|\mathcal{F}[F^{f}_{q+m,q}(\theta-\beta)]|=|e^{-2\pi i\beta\xi}\mathcal{F}[F^{f}_{q+m,q}(\theta)]|=|\mathcal{F}[F^{f}_{q+m,q}(\theta)]| \tag{18}$$

**Theorem 2.** For non-negative integers $q$ and $m$, the pseudo-Zernike moment invariant set (PZM) for both scaling and rotation transformation is:

$$|\mathcal{F}[F^{f}_{q+m,q}(\theta)]|=|\mathcal{F}[\sum_{k=0}^{m}(\sum_{l=k}^{m}\Gamma_{f}^{-(q+1+l)}c^{q}_{m,l}d^{q}_{l,k})F^{f}_{q+k,q}(\theta)]| \tag{19}$$

The complete set of the proposed invariants of Eq.(19) can be obtained using method proposed by Ghosbel et al. in [10].

**Theorem 3.** For non-negative integers $q$ and $m$, the complete set of pseudo-Zernike moment invariants (CPZMs) for RST transformation is:

$$\left|\mathcal{F}[F^{f}_{q+m,q}(\theta)]\right|=|\mathcal{F}[\sum_{k=0}^{m}(\sum_{l=k}^{m}\Gamma_{f}^{(q+1+l)}c^{q}_{m,l}d^{q}_{l,k})I^{f}_{q+k,q}(\theta)]| \tag{20}$$

The proof is given in Appendix B.

# 5 Experimental Results

To verify the performance of the proposed moment invariants, in this section, several classification experiments are carried out. Fig.2 shows some test images which are chosen from the Columbia Coil-20 image data base [12].

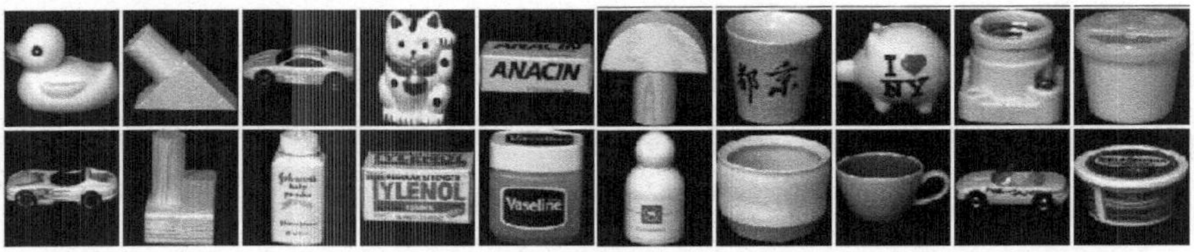


**Fig. 2.** Some of test images from the Columbia Coil-20 image data base

## 5.1 Invariants Calculation

First, convert image function into the polar coordinate form in terms of Eq.(1). $(x_0, y_0)$ denotes centroid. Thus, one can obtained polar coordinate form image $f(r, \theta)$. Then, the OPT $F_{q+m,q}(\theta)$ of image $f(r, \theta)$ and its curve can be computed by replacing Eq.(5) and Eq.(6) by Eq.(7). According to Eq.(15), one has a set of scale invariants $I^f_{q+m,q}(\theta)$ . Finally, the complete set of pseudo-Zernike moment invariants of a deformed image are calculated by Eq.(20).

**Table 2.** Classification results under salt-and-pepper noise with intensities from 0.01 to 0.04

| | Noise-free | 0.01 | 0.02 | 0.03 |
|---|---|---|---|---|
| CPZMs | 90% | 80.3% | 77.8% | 69.0% |
| OFMMs | 85% | 71.2% | 60.1% | 38.4% |
| ZMs | 100% | 81.0% | 66.0% | 51.0% |
| CMs | 96.8% | 85.5% | 81.5% | 80.1% |

**Table 3.** Classification results under Gaussian noise with variance from 0.005 to 0.015

| | Noise-free | 0.005 | 0.01 | 0.015 |
|---|---|---|---|---|
| CPZMs | 90% | 78.1% | 65.2% | 57.5% |
| OFMMs | 85% | 19.5% | 16.6% | 11.6% |
| ZMs | 100% | 76.3% | 49.8% | 43.5% |
| CMs | 96.8% | 60.0% | 53.9% | 45.1% |

## 5.2 Classification Results

In classification experiment, we choose objects from Columbia Coil-20 image data base. Let $x_t = 0$, $y_t = 0$. Set the scaling and rotation factors $\lambda \in \{0.8, 0.9, 1, 1.1, 1.2\}$, $\beta \in \{30^0, 60^0, 90^0, \ldots, 360^0\}$, respectively. Therefore each image derives 60 transformed images. Using 20 original images, 1200 images need to be verified. The classification measure used in this experiment is the Euclidean distance defined as:

$$dis(Im^f, Im^g) = \sqrt{\sum_{i=1}^{k} (Im^f(\theta_i) - Im^g(\theta_i))^2} \tag{21}$$

The classification accuracy $\eta$ is defined as:

$$\eta = \frac{Number\ of\ correctly\ classified\ images}{The\ total\ number\ of\ images\ used\ in\ the\ test} \times 100\% \tag{22}$$

The classification results in noise-free case are list in Table.1. From this table, one can see the proposed moment invariants have a good classification capability in noise free condition. The complete classification is also acceptable.

To test the robustness of complete invariants sets in noise case, two kinds of noises are added in each transformed image. Salt-and-pepper noise with intensities from 1% to 4% and Gaussian noise with mean $\mu = 0$, variance varying from 0.005 to 0.015 added in images before recognition. The CPZMs classification rate with orders $q = 3$ and $m$=3 is shown in Table 2 and Table 3, respectively. The results also compare with the other three methods [5, 10], where OFMMs, ZMs and CMs have 25, 9, 7 independent invariants, respectively. With increasing of noises, CPZMs show a better performance than the other three methods, especially in Gaussian noise case.

## 6 Conclusions

This paper proposed a complete set of invariants based on orthogonal projection transform with pseudo-Zernike moments. The simulation experiments demonstrate that the proposed invariant set has better discrimination capabilities over the existing ones. Further research should be focused on the selecting of the orders of CPZM.

## Appendix A

Before the proof of Theorem 1, three equations should be obtained. Firstly, the pseudo-Zernike polynomials $R_{q+m,q}(r)$ after scaling is written as:

$$R_{q+m,q}(\lambda r) = \sum_{k=0}^{m} R_{q+k,q}(r) \sum_{l=k}^{m} \lambda^{q+1} c_{m,l}^{q} d_{l,k}^{q} \tag{A1}$$

Secondly, if $m = 0$, the OPT $F_{q+m,q}(\theta)$ has the following form:

$$F_{q+0,q}^{g}(\theta) = \lambda^{q+1} F_{q+0,q}^{f}(\theta) \tag{A2}$$

Thirdly, the discrete OPT can be described as:

$$F_{q+m,q}^{g}(\theta) = \sum_{k=0}^{m} \sum_{l=k}^{m} \lambda^{q+l+1} c_{m,l}^{q} d_{l,k}^{q} F_{q+k,q}^{f}(\theta) \tag{A3}$$

A vector composed by translated invariants $(I^g_{q+i,q})^T$ can be written as:

$$[I^g_{q+0,q}, I^g_{q+1,q}, \ldots, I^g_{q+m,q}]^T = C^q_m \times diag(F^{g\ -(q+1)}_{0,0}, F^{g\ -(q+1+1)}_{0,0}, \ldots, F^{g\ -(q+m+1)}_{0,0}) D^q_m \times [F^g_{q+0,q}, F^g_{q+1,q}, \ldots, F^g_{q+m,q}]^T \tag{A4}$$

Substitution of Eqs.(A1)-(A3) into (A4), one has:

$$\begin{aligned}
&[I^g_{q+0,q}, I^g_{q+1,q}, \ldots, I^g_{q+m,q}]^T \\
&= C^q_m diag(F^{g\ -(q+1)}_{0,0}, \cdots, F^{g\ -(q+m+1)}_{0,0}) diag(\lambda^{-(q+1)}, \lambda^{-(q+1+1)}, \cdots, \lambda^{-(q+m+1)}) \\
&\times D^q_m C^q_m diag(\lambda^{q+1}, \lambda^{q+2}, \cdots, \lambda^{q+m+1}) D^q_m [F^g_{q+0,q}, F^g_{q+1,q}, \ldots, F^g_{q+m,q}]^T \\
&= C^q_m diag(F^{f\ \ -(q+1)}_{0+0,0}, \cdots, F^{f\ \ -(q+m+1)}_{0+0,0}) D^q_m [F^f_{q+0,q}, F^f_{q+1,q}, \ldots, F^f_{q+m,q}]^T \\
&= [I^f_{q+0,q}, I^f_{q+1,q}, \ldots, I^f_{q+m,q}]^T
\end{aligned} \tag{A5}$$

The proof has been completed.

## Appendix B

The invariants equation can be written in matrix form as:

$$[I^f_{q+0,q}, I^f_{q+1,q}, \ldots, I^f_{q+m,q}]^T = C^q_m \times diag(\Gamma^{-(q+1)}_f, \Gamma^{-(q+2)}_f, \cdots, \Gamma^{-(q+m+1)}_f) \times D^q_m [F^f_{q+0,q}, F^f_{q+1,q}, \ldots, F^f_{q+m,q}]^T \tag{B1}$$

Multiply $C^q_m \times diag(\Gamma^{q+1}_f, \Gamma^{q+2}_f, \cdots, \Gamma^{q+m}_f) D^q_m$ to both sides:

$$\begin{aligned}
&C^q_m \times diag(\Gamma^{q+1}_f, \Gamma^{q+2}_f, \cdots, \Gamma^{q+m+1}_f) \times D^q_m [I^f_{q+0,q}, I^f_{q+1,q}, \ldots, I^f_{q+m,q}]^T \\
&= [F^f_{q+0,q}, F^f_{q+1,q}, \ldots, F^f_{q+m,q}]^T
\end{aligned} \tag{B2}$$

Thus, after Fourier transform and modules operation, the complete format (20) is obtained. The proof has been completed.

## References


1. Derrode, S., Ghorbel, F.: Robust and Efficient Fourier-Mellin Transform Approximations for Invariant Grey-Level Image Description and Reconstruction. Comput. Vis. Image Understand. 83(1), 57–78 (2001)
2. Hu, M.K.: Visual Pattern Recognition by Moment Invariants. IRE Trans. Inf. Theory IT-8, 179–187 (1962)
3. Cheng, B.J., Shu, H.Z., Zhang, H., Luo, L.M., Coatrieux, J.L.: Combined Invariants to Similarity Transformation and to Blur Using Orthogonal Zernike Moments. IEEE Trans. Image Process. 20, 345–360 (2011)

4. Zhang, H., Dong, Z.F., Shu, H.Z.: Object Recognition by a Complete Set of Pseudo-Zernike Moment. In: ICASSP 2010, pp. 930–933 (2010)
5. Kan, C., Srinath, M.D.: Invariant Character Recognition with Zernike and Orthogonal Fourier-Mellin Moments. Pattern Recogn. 35, 143–154 (2002)
6. Bin, T.J., Lei, A., Cui, J.W., Kang, W.J.: Subpixel Edge Location Based on Orthogonal Fourier-Mellin Moments. Image and Vision Computing 26, 563–569 (2008)
7. Xiao, B., Ma, J.F., Cui, J.T.: Invariant Pattern Recognition using Radial Tchebichef Moments. In: CCPR, pp. 1–5 (2010)
8. Yang, B., Li, G.X., Zhang, H.L., Dai, M.: Rotation and Translation Invariants of Gaussian-Hermite Moments. Pattern Recognition Letters 32, 1283–1298 (2011)
9. Crimmins, T.R.: A Complete Set of Fourier Descriptors for Two-Dimensional Shape. IEEE Trans. Syst. Man Cybernet. 121, 848–855 (1982)
10. Ghorbel, F., Derrode, S., Mezhoud, R., Bannour, T., Dhahbi, S.: Image Reconstruction from a Complete Set of Similarity Invariants Extracted from Complex Moments. Pattern Recognition Letters 27, 1361–1369 (2006)
11. Lan, R.S., Yang, J.: Orthogonal Projection Transform with Application to Shape Description. In: ICIP 2010, pp. 281–284 (2010)
12. `http://www.cs.columbia.edu/CAVE/software/softlib/coil-20.php`

# Author Index